Omics-Nano-CRISPR Approaches for Oncology

Jen-Tsung Chen
https://orcid.org/0000-0002-3540-4449
National University of Kaohsiung, Taiwan

IGI Global
Scientific Publishing
Publishing Tomorrow's Research Today

VP of Publishing Ethics & Research Integrity	Melissa Wagner
Director of Acquisitions	Mikaela Felty
Director of Book Development	Jocelynn Hessler
Production Manager	Mike Brehm
Cover Design	Jose Rosado

Published in the United States of America by
IGI Global Scientific Publishing
701 East Chocolate Avenue
Hershey, PA, 17033, USA
Tel: 717-533-8845 | Fax: 717-533-7115
Website: https://www.igi-global.com E-mail: cust@igi-global.com

Library of Congress Cataloging-in-Publication Data

LCCN: 2026011178 (CIP Data Pending)
ISBN13: 9798337397153
Isbn13Softcover: 9798337397160
EISBN13: 9798337397177
British Cataloguing in Publication Data
A Cataloguing in Publication record for this book is available from the British Library.

Table of Contents

Detailed Table of Contents

barriers. Future directions i.e., digital twins, next-generation immunotherapies, and equitable global access underscore the transformative potential of personalized medicine in revolutionizing cancer management through precision, prediction, and participation.

Human Digital Twins (HDTs) are revolutionizing by offering dynamic, personalized virtual embodiments that integrate multi-omics, biosensor, and clinical information to simulate tumor biology and therapeutics. HDTs allow bidirectional data exchange in real time, with updates providing support for predictive, preventive, and personalized cancer therapy management. This chapter explores the complex architecture of HDTs, cyber-physical-human system functionality, and incorporation of artificial intelligence, machine learning, and multi-scale modelling of tumors in pursuit of precision oncology. It further explores their implications for drug discovery, optimization of radiation therapy and immunotherapy, and in silico clinical trials, and discusses sustainability metrics alongside ethical, legal, and societal implications. Incorporation of emerging fields like quantum computing, augmented reality, and Twin-as-a-Service platforms has enormous promise for sustainable healthcare through HDTs. By conjoining patient-level information, HDTs help advance towards superior outcomes in oncology.

Early cancer detection is essential to patients' survival, allowing for prompt treatment, and cutting down medical costs. Unfortunately, traditional diagnostic methods usually demonstrate low sensitivity, long processing time, and require advanced laboratory setup. Nevertheless, recent developments in the CRISPR technology have revolutionized the field of molecular diagnostics by introducing highly sensitive, sequence-specific, fast, and programmable platforms for detecting cancer-specific genetic and epigenetic biomarkers. Nano-biosensors that incorporate, quantum dots of gold nanoparticles graphene-related materials, Nanoparticles with magnetic properties, and plasmonic nanoparticles possess outstanding optoelectrical and fluorescence sensing abilities which ensure increased Both specificity and sensitivity. This chapter offers a thorough overview the recent progress in the integration of CRISPR-based diagnostics and nano-biosensors for early diagnosis of cancers.

Lung cancer remains a leading cause of cancer deaths worldwide, demanding innovative strategies for early detection, prognosis, and personalized therapy. This chapter highlights how artificial intelligence (AI) is reshaping precision medicine in lung cancer, beginning with the global burden and evolution of individualized care. It explores imaging-based AI approaches such as radiomics and deep learning for tumor detection and characterization, alongside AI-driven molecular and genomic profiling that interprets sequencing data and integrates multi-omics for biomarker discovery. Real world implementations including clinical decision support systems, AI-assisted drug discovery, and integration with electronic health records thereby demonstrate clinical impact. The chapter also addresses challenges of data quality, bias, interpretability, and ethics, while envisioning future directions such as digital twins and integrated AI-enabled oncology ecosystems, underscoring AI's role in advancing patient-centered lung cancer care.

Breast cancer is a major global health challenge due to its molecular complexity and diverse clinical outcomes. This chapter explores the integration of omics technologies, nanotechnology, CRISPR-based genome editing, and artificial intelligence (AI) in breast cancer research and management. Omics approaches enable molecular profiling, biomarker discovery, and tumor classification, while nanotechnology improves targeted drug delivery and diagnostic precision. CRISPR technologies support functional genomics and therapeutic target identification. AI enhances analysis of complex biological and clinical data, improving diagnosis, treatment planning, and personalized care. The convergence of these technologies offers promising opportunities to advance precision oncology and transform breast cancer diagnosis, therapy, and prognosis.

Colorectal cancer is a molecularly heterogeneous disease driven by complex genetic, epigenetic, metabolic, inflammatory, microbial, and microenvironmental alterations that influence diagnosis, progression, therapeutic response, and resistance. The recent developments in omics techniques such as genomics, transcriptomics, proteomics, metabolomics, epigenomics, and microbiomics give a complete overview of tumor biology and help in biomarker discovery, molecular classifications, patient stratification, and precision medicine. This chapter explores the integrated application of omics, nanotechnology, CRISPR, and artificial intelligence in advancing colorectal cancer diagnosis, targeted treatment, personalized medicine, and future precision oncology.

Prostate cancer remains a major global health challenge due to its molecular and phenotypic heterogeneity, disease progression, and therapeutic resistance. Advances in precision oncology integrating multi-omics technologies, artificial intelligence (AI), nanotechnology, and CRISPR-based genome editing are transforming prostate cancer diagnosis and treatment. Multi-omics approaches enable comprehensive molecular profiling for biomarker discovery, patient stratification, and therapeutic target identification, while AI integrates complex datasets to improve diagnosis, prognosis, and personalized treatment strategies. Nanotechnology enhances molecular imaging and targeted drug delivery, whereas CRISPR supports precise genome editing for functional validation and therapeutic development. This chapter examines the synergistic integration of these technologies and their potential to advance precision medicine and personalized prostate cancer management.

Skin cancer, including melanoma and non-melanoma skin cancer, remains a major global health challenge due to molecular heterogeneity, treatment resistance, and variable therapeutic responses. This chapter explores the integration of multi-omics technologies, nanotechnology, CRISPR-based genome editing, and artificial intelligence to advance precision dermato-oncology. Multi-omics approaches reveal driver mutations, dysregulated signalling pathways, and tumour immune microenvironment dynamics, while lipid- and polymer-based nanocarriers enable targeted topical, transdermal, and intratumoural drug delivery. CRISPR-Cas9 expands therapeutic opportunities through precise editing of oncogenic drivers and resistance pathways, although efficient delivery remains a key limitation. AI serves as the unifying computational framework by integrating omics data, optimizing nanocarrier design, prioritizing therapeutic targets, and supporting image-based diagnosis. Together, these technologies provide a promising framework for personalized, adaptive, and minimally invasive skin cancer management.

Artificial intelligence (AI) is transforming management of leukemia through integration of computation with precision medicine. The integration of machine learning, deep learning technologies has gained momentum in diagnostic, prognostic, and therapeutic applications. AI-enabled platforms boost the accuracy of subtyping, track minimal residual disease, and revise treatment through predictive modeling. Additional technologies like natural language processing, liquid biopsies, and federated learning boost the value of patient data while enhancing privacy and equity. In addition to diagnostic applications, AI accelerates drug development pipelines, informs adaptive clinical trials, and reinforces immunotherapy and gene editing. Prospective technologies such as quantum computing and synthetic biology emphasizes AI's transformational potential in the treatment of leukemia. This transdisciplinary integration reveals AI-enabled precision oncology, patient-specific plans to counter disease heterogeneity, therapeutic resistance, and socioeconomic disparities in management of leukemia.

Chapter 11

Vidyashree Kuballi, Sri Sathya Sai University for Human Excellence, India
Veda Paragonnavar, Sri Sathya Sai University for Human Excellence, India
Nagashri Nanjundeshwara, Sri Sathya Sai University for Human Excellence, India
Ramya Raghavan, Sri Sathya Sai University for Human Excellence, India

The convergence of omics technologies, nanofiber-based nanotechnology, and CRISPR-Cas systems is enabling precision diagnostics, targeted therapy, and real-time monitoring. These integrated platforms are advancing the management of benign tumors, tumor grading, metastasis, imaging, and theranostics, with demonstrated efficacy in both preclinical and clinical settings. Current review literature shows that electrospun nanofibers already support local chemotherapy, combinatorial therapy, cancer detection, cell capture, 3D tumor modeling, metastasis studies, and theranostics. At the same time, newer work extends this platform toward smarter, biomaterial-fabricated systems and CRISPR-enabled delivery concepts. Key challenges remain in optimizing delivery efficiency, minimizing off-target effects, and standardizing clinical translation. Future research will focus on expanding multiplexing capabilities, refining single-cell and spatial omics integration, and developing robust, patient-specific theranostic platforms to further advance the promise of precision oncology.

Chapter 12

S. Mayilazhagan, Department of Microbiology, Sacred Heart College, Tirupattur, India
S. Dayana, Department of Microbiology, Sacred Heart College, Tirupattur, India
T. Elavarasi, Department of Microbiology, Sacred Heart College, Tirupattur, India
M. Manigandan, Department of Microbiology, Sacred Heart College, Tirupattur, India
K. Gayathri, Department of Microbiology, Sacred Heart College, Tirupattur, India

Nanoparticle-mediated gene editing and immunotherapy have emerged as transformative approaches for the treatment of liver and prostate cancers, addressing limitations associated with conventional therapies such as chemotherapy, radiotherapy, and surgical interventions. The integration of nanotechnology with advanced gene-editing tools, particularly CRISPR-Cas systems, enables precise modification of cancer-associated genes while minimizing off-target effects. Nanoparticles serve as efficient delivery vehicles for gene-editing components, therapeutic nucleic acids, and immunomodulatory agents, enhancing cellular uptake, target specificity, and therapeutic efficacy. In liver cancer, nanoparticle-based platforms facilitate targeted disruption of oncogenic pathways, restoration of tumour suppressor genes, and modulation of the tumour microenvironment to suppress tumour progression. Similarly, in prostate cancer, nanoparticle-assisted gene editing enables the regulation of androgen receptor signalling and other molecular drivers responsible for tumour growth and metastasis.

Artificial Intelligence (AI) is now widely used not just in early drug discovery but also in drug manufacturing, quality control, and post market surveillance. This chapter looks at how AI is being used throughout the pharmaceutical product lifecycle, with a focus on new developments in manufacturing, better quality assurance, and real-time monitoring of drugs already on the market. The chapter starts by looking at how AI technologies like machine learning, deep learning, computer vision, and robotics are affecting pharmaceutical manufacturing. It covers topics such as predictive maintenance, real-time process monitoring, digital twins, and autonomous process control, which all help make manufacturing smarter as part of Industry 4.0. The chapter also covers the regulatory environment, concerns, and ethics of AI in the production of pharmaceuticals and their monitoring and provides a vision of how these technologies will ensure safety, effectiveness, and efficiency in the evolving AI-enabled pharmaceuticals sector.

Omics-Nano-CRISPR technologies are pioneering precision oncology by integrating multi-omics data, nanotechnology, and genome editing to improve cancer diagnosis and treatment. In addition to their therapeutic value, these technologies confront significant biosafety, bioethical and regulatory issues that must be addressed carefully, both scientifically and ethically within society. The future of Omics-Nano-CRISPR technologies in oncology is expected to be supported by ongoing advancements in explainable AI, enhanced bio-safety surveillance systems, and the development and harmonisation of international regulations and ethical governance structures. Finally, addressing bio-safety, bioethical and regulatory challenges will be crucial for the advancement of the safe, effective and socially responsible precision cancer medicine.

Preface

Cancer continues to represent one of the most formidable challenges in modern healthcare. Its remarkable molecular heterogeneity, complex tumor microenvironment, metastatic potential, and ability to develop resistance to therapy have exposed the limitations of conventional approaches based largely on generalized diagnostic and treatment strategies. The rapid emergence of precision oncology, however, is creating new opportunities to understand and manage cancer at an unprecedented level of resolution. In this evolving landscape, the convergence of multi-omics, nanotechnology, CRISPR-based genome editing, and artificial intelligence (AI) offers a particularly promising framework for transforming cancer diagnosis, prognosis, prevention, and treatment.

Omics-Nano-CRISPR Approaches for Oncology brings together contemporary perspectives on these rapidly developing fields and their convergence in cancer research and clinical translation. The volume is founded on the recognition that cancer is a multidimensional biological problem that requires equally multidimensional technological solutions. Genomics, transcriptomics, proteomics, metabolomics, epigenomics, and other omics disciplines provide increasingly comprehensive views of tumor biology; nanotechnology enables precise sensing, imaging, and delivery; CRISPR systems provide programmable tools for investigating and modifying disease-associated genes; and AI offers the computational capacity required to integrate, interpret, and translate the enormous datasets generated by these technologies.

A major theme of this book is the movement from conventional cancer management toward genuinely personalized medicine. The integration of AI with multiple omics layers can facilitate biomarker discovery, molecular classification, patient stratification, prediction of treatment response, and development of adaptive therapeutic strategies. Machine learning, deep learning, graph-based approaches, explainable AI, and federated learning are increasingly capable of addressing the scale and complexity of biomedical data while also raising important questions regarding transparency, privacy, and responsible use. These computational advances complement clinical developments such as liquid biopsies, next-generation sequencing, precision immunotherapy, and individualized treatment planning.

The volume also explores the emerging concept of human digital twins, which may provide dynamic virtual representations of individual patients by integrating clinical, molecular, and biosensor information. Such systems have the potential to support predictive, preventive, and personalized healthcare by enabling continuous updating of patient-specific models. Their prospective applications extend from drug discovery and treatment optimization to radiation therapy, immunotherapy, and in silico clinical trials. At the same time, digital twins introduce new considerations related to sustainability, data governance, cybersecurity, ethics, and the responsible management of highly personalized health information.

Early and accurate cancer detection represents another important focus. The combination of CRISPR-based molecular diagnostics with nano-biosensors may enable rapid and highly specific detection of cancer-associated genetic and epigenetic alterations. Advances involving gold nanoparticles, quantum dots, graphene-based materials, magnetic nanoparticles, and plasmonic systems demonstrate how nanoscale engineering can enhance the sensitivity and specificity of molecular detection. These approaches may ultimately contribute to diagnostic platforms that are faster, more precise, and potentially more accessible than many conventional methodologies.

The book then considers the application of integrated technologies to individual cancer types. Lung, breast, colorectal, prostate, and skin cancers are examined through the complementary perspectives of molecular profiling, AI-assisted diagnosis, targeted nanodelivery, genome editing, therapeutic resistance, and personalized treatment. These disease-focused discussions demonstrate that although cancers differ substantially in their biological characteristics, the underlying principles of precision oncology remain broadly applicable. Omics technologies can identify actionable molecular alterations; AI can integrate heterogeneous datasets; nanotechnology can improve the delivery and localization of therapeutic agents; and CRISPR can facilitate functional validation and targeted manipulation of cancer-associated pathways.

The scope of the volume extends beyond solid tumors to leukemia and other difficult-to-treat malignancies. AI-assisted approaches can improve disease classification, prognosis, minimal residual disease monitoring, treatment selection, and drug development. For rare cancers, where limited patient numbers and fragmented datasets frequently impede conventional research, emerging computational approaches such as transfer learning, few-shot learning, graph neural networks, federated learning, and generative methods may provide new opportunities for knowledge generation and therapeutic development.

An additional dimension of the book concerns nanofibers and nanoparticles as platforms for oncology theranostics and gene-based interventions. Nanofiber systems have applications in localized chemotherapy, combinatorial treatment, cancer detection, cell capture, three-dimensional tumor modeling, metastasis research, and real-time monitoring. Nanoparticle-mediated gene editing and immunotherapy similarly offer opportunities to improve the delivery of CRISPR components, therapeutic nucleic acids, and immunomodulatory agents. These technologies are particularly relevant to cancers in which targeted delivery and modulation of oncogenic signaling or the tumor microenvironment remain significant therapeutic challenges.

Importantly, technological innovation in oncology does not end with discovery and clinical intervention. The pharmaceutical lifecycle itself is being transformed by AI, including manufacturing, quality control, process monitoring, predictive maintenance, digital twins, robotics, and post-market surveillance. The incorporation of intelligent systems throughout pharmaceutical development and production may improve efficiency, consistency, quality, and safety while supporting the broader transition toward data-driven healthcare and Industry 4.0.

The transformative potential of Omics–Nano–CRISPR approaches must, however, be accompanied by careful consideration of biosafety, bioethics, regulation, and social responsibility. Genome editing, advanced nanomaterials, AI-driven decision-making, and highly integrated health-data systems each introduce distinctive risks and governance challenges. Issues involving off-target effects, data privacy, algorithmic bias, transparency, informed consent, accountability, equitable access, and international regulatory harmonization must be addressed alongside scientific and technological progress. Responsible innovation requires that advances in precision oncology remain firmly connected to patient autonomy, safety, fairness, and global health equity.

The organization of this volume reflects these interconnected themes. The chapters progress from the foundations of AI-enabled precision medicine and personalized healthcare to digital twins, advanced diagnostics, cancer-specific applications, nanotechnology-enabled theranostics, pharmaceutical applications, and the ethical and regulatory considerations surrounding these technologies.

CHAPTER OVERVIEW

1. The opening chapter examines the integration of AI with multiple omics datasets to support personalized cancer medicine, with emphasis on biomarker discovery, molecular classification, patient stratification, drug-response prediction, adaptive treatment, explainable AI, graph-based methods, and federated learning.
2. The second chapter provides a broad overview of personalized cancer healthcare, discussing genomics, biomarkers, multi-omics integration, next-generation sequencing, liquid biopsies, AI, targeted therapies, immunotherapy, cell-based interventions, predictive modeling, patient engagement, and future developments in precision oncology.
3. The third chapter explores human digital twins as dynamic, patient-specific models integrating multi-omics, biosensor, and clinical information. It considers their applications in tumor modeling, drug discovery, radiation and immunotherapy optimization, in silico trials, sustainability, and emerging technologies.
4. The fourth chapter addresses the integration of CRISPR-based diagnostics with nano-biosensors for early cancer detection, highlighting programmable molecular recognition and advanced nanoscale sensing materials capable of improving diagnostic sensitivity and specificity.
5. The fifth chapter focuses on AI-enabled precision medicine for lung cancer, including radiomics, deep learning, molecular and genomic profiling, biomarker discovery, clinical decision support, AI-assisted drug discovery, electronic health records, and future digital oncology ecosystems.
6. The sixth chapter applies the integrated technological framework to breast cancer, addressing molecular profiling, biomarker discovery, tumor classification, targeted drug delivery, functional genomics, diagnosis, treatment planning, and personalized patient care.
7. The seventh chapter considers colorectal cancer as a molecularly heterogeneous disease and explores the combined contribution of genomics, transcriptomics, proteomics, metabolomics, epigenomics, microbiomics, nanotechnology, CRISPR, and AI to diagnosis, treatment, and precision medicine.
8. The eighth chapter examines precision approaches to prostate cancer, focusing on multi-omics profiling, AI-assisted diagnosis and prognosis, biomarker discovery, therapeutic target identification, molecular imaging, targeted drug delivery, and CRISPR-based functional and therapeutic strategies.
9. The ninth chapter explores integrated approaches to skin cancer, including melanoma and non-melanoma disease, with emphasis on multi-omics characterization, nanocarrier-based delivery, CRISPR-mediated targeting of oncogenic and resistance pathways, and AI-assisted diagnosis and treatment optimization.
10. The tenth chapter discusses AI-enhanced precision medicine in leukemia, covering disease subtyping, diagnosis, prognosis, minimal residual disease monitoring, predictive treatment strategies, liquid biopsies, drug development, adaptive clinical trials, immunotherapy, and gene-editing approaches.
11. The eleventh chapter examines nanofibers as multifunctional platforms in oncology theranostics. It considers their roles in localized and combination therapy, cancer detection, cell capture, three-dimensional tumor modeling, metastasis studies, imaging, and emerging integration with omics and CRISPR technologies.
12. The twelfth chapter focuses on nanoparticle-mediated gene editing and immunotherapy for liver and prostate cancers. It discusses nanoparticles as delivery systems for gene-editing components, therapeutic nucleic acids, and immunomodulatory agents, as well as strategies for modifying oncogenic pathways and the tumor microenvironment.

13. The thirteenth chapter expands the discussion to the pharmaceutical product lifecycle, examining AI applications in drug manufacturing, quality control, predictive maintenance, real-time process monitoring, digital twins, autonomous production, regulatory compliance, and post-market surveillance.
14. The fourteenth and concluding chapter addresses biosafety, bioethical, and regulatory challenges associated with Omics–Nano–CRISPR oncology, emphasizing responsible innovation, explainable AI, safety monitoring, ethical governance, international regulatory harmonization, patient protection, and equitable access.

CONCLUSION

Taken as a whole, the contributions provide a multidisciplinary perspective on how emerging technologies can be brought together to address the biological and clinical complexity of cancer. The volume moves beyond viewing omics, nanotechnology, CRISPR, and AI as independent technological domains and instead considers their potential as interconnected components of an integrated precision-oncology ecosystem.

The future of cancer care is likely to be increasingly data-driven, molecularly informed, technologically enabled, and individualized. Earlier detection, more accurate diagnosis, improved prediction of therapeutic response, targeted delivery, programmable treatment strategies, continuous monitoring, and adaptive clinical decision-making are among the possibilities offered by the convergence of these fields. Yet the ultimate value of these innovations will depend on rigorous scientific validation, effective clinical translation, affordability, responsible governance, and equitable access.

It is hoped that *Omics-Nano-CRISPR Approaches for Oncology* will serve as a valuable reference for researchers, clinicians, pharmaceutical scientists, bioengineers, computational scientists, biotechnology professionals, policymakers, and students. By bringing together perspectives from molecular biology, artificial intelligence, nanotechnology, genome editing, diagnostics, therapeutics, pharmaceutical science, and ethics, this volume aims to encourage interdisciplinary collaboration and contribute to the development of safer, more precise, more effective, and more patient-centered approaches to cancer prevention and care.

Jen-Tsung Chen
National University of Kaohsiung, Taiwan

Chapter 1

AI-Driven Multiple Omics for Personalized Cancer Medicine

Manoj Mani
http://orcid.org/0009-0008-6601-0339
Bharathiar University, India

Yogasri Balamurugan
Bharathiar University, India

Prakash Dass Sundaram
Bharathiar University, India

Shakeera Banu Mohammed Jaffer
Bharathiar University, India

Nalini Devi Dharumaiyan
Bharathidasan University, India

Gavaskar Shanmugam
Bharathiar University, India

Antony Prabhu Jeyabal Philomenathan
http://orcid.org/0009-0001-6500-9538
Bharathiar University, India

Vijaya Anand Arumugam
http://orcid.org/0000-0001-7485-1586
Bharathiar University, India

ABSTRACT

Integrating artificial intelligence (AI) with multi-omics revolutionize personalized medicine by offering an integrated view of tumor biology and precision treatments. The integration of multi-omics using AI-based paradigms identifies biomarkers, optimizes molecular subtypes, and enhances treatment stratification. Machine learning and deep learning algorithms overcome high-dimensional data issues, including early detection, stratification of patients, drug response, and adaptive treatment strategies. New paradigms such as graph neural networks, explainable AI (XAI), and federated learning add clarity of explanation, and integration over diverse datasets. Translation of multi-omics-guided AI includes liquid biopsies, optimization of immunotherapy, and dynamic treatment regimens with AI. Challenges of data heterogeneity, explainability, and ethics notwithstanding, such integration forms the basis of next-generation care, wherein therapy is personalized to individual molecular signatures, leading to reduction of adverse effects.

DOI: 10.4018/979-8-3373-9715-3.ch001

1. INTRODUCTION: CONVERGENCE OF AI AND MULTI-OMICS IN ONCOLOGY

Multi-omics integration guided by artificial intelligence (AI) is transforming the horizon of precision oncology by providing an integrated view of cancer biology and stimulating personalized treatment approaches. Integration of the genomics, proteomics, and other omics domains by virtue of AI-guided analytical approaches strengthens the identification of new biomarkers as well as individualized treatment regimes, enhancing diagnostic specificity as well as treatment efficiency. In diffuse large B-cell lymphoma (DLBCL), for example, employment of AI coupled with multi-omics as well as cancer's spatial multi-omics technologies has realized unprecedented views of molecular complexities as well as spatial heterogeneities, determinants that have paramount implications towards optimized diagnostics as well as customized treatment programs (Schürch *et al.,* 2024; Shao *et al.,* 2024; Wei *et al.,* 2023a; Viduedo *et al.,* 2024). Machine learning (ML) as well as deep learning (DL) approaches play a cardinal role in high-dimensional omics data set analysis, enhancing cancer classification, cancer risk estimation, as well as clinical decision-making. Further still, novel AI platforms specifically engineered towards providing personalized counterfactual treatment advice offer glimpses towards integrating multi-omics data, providing probabilistic treatment advice with calibrated confidence, addressing issues such as treatment assignment bias, as well as the complexity involved within high-dimensional data.

Despite such technological advancements, single-omics approaches continue to reveal critical limitations that highlight the need for integrative models. Classical bulk sequencing, which examines bulk populations of cells, obfuscates intratumoral heterogeneity and neglects essential elements of clonal evolution and cancer development (Massimino *et al.,* 2023). Moreover, reliance on a linear correlation between gene expression and protein functionality overlooks the importance of post-translational modifications and higher-order regulatory mechanisms that ultimately dictate cellular phenotypes (Moreno-Sánchez *et al.,* 2016). Though single-cell genomics has increased resolution in deconstructing complexity within tumors, such approaches remain compromised by problems of sample processing, as well as sensitivity and specificity issues, coupled with inherent issues related to cross-platform data merging (Wills & Mead, 2015). Moreover, such methods often lack spatial resolution, an element crucial for deciphering multicellular interactions within the tumor microenvironment (De Vries *et al.,* 2020). Recent advancements in single-cell multi-omics technologies have begun to compensate for such shortcomings by combining multiple layers of molecules, thereby providing a deeper insight into tumor heterogeneity and the dynamics of immunological response. Nevertheless, the absence of spatial context continues to present a major challenge that underscores the need for approaches that combine single-cell resolution with space-resolved data to fully capture the complexity of tumor ecosystems (Liang *et al.,* 2024).

Within this context, state-of-the-art AI-centered approaches to holistic patient stratification have become central to advancing personalized oncology. By leveraging large, multi-modal datasets with sophisticated computational approaches, AI facilitates the grouping of heterogeneous malignancies by clinically relevant subgroups that are essential for personalized treatment as well as the identification of new drugs (Gardner *et al.,* 2020). AI-powered platforms not only suppot diagnostic sensitivity and treatment plans but also reduce false positive rates and end delays with treatment delays, thereby improving clinical outcomes (Hassane, 2024). Particular examples of ML algorithms have shown extraordinary potential in predicting cancer risk, monitoring treatment response, and identifying predictive biomarkers for immunotherapies (Damane *et al.,* 2023). In gynecological oncology, AI-assisted risk stratification of clinical data, imaging modalities, and molecular characterization demonstrates its translational potential despite significant barriers that prevent widespread use of such technologies within realistic clinical settings (Jiang *et al.*, 2023). The potential of AI to integrate and digest high-dimensional omics and clinical data enables the development of treatments that tailor therapies to each individual's molecular as well as clinical profile, leading to optimal effectiveness while reducing toxicities (Akter, 2024).

2. MULTI-OMICS LANDSCAPE IN CANCER

2.1 Genomics

Single-nucleotide variants (SNVs) are one of the most basic categories of genomic mutations underlying cancer initiation and progression, which include inherited predispositions and acquired somatic mutations. These variants are generally categorized as germline and somatic, wherein germline SNVs predispose individuals to hereditary cancer susceptibility while somatic SNVs are responsible for the accumulation of oncogenic mutations across tumorigenesis (Matsuno *et al.*, 2022). Genome-wide association analyses have recognized several cancer-related SNVs, including those involved in lung carcinoma, whereby coding sequence mutations as well as expression-modulating variants have been found to relate to tumor initiation as well as prognosis (Tian & Liu, 2024). Functionally, SNVs of essential genes may disrupt tumor suppressor pathways, of which an example is seen with coding and non-coding SNVs of the *SOCS3* gene that reduce its regulatory activity of the JAK/STAT signaling pathway and consequently reinforce oncogenic signaling (Fatemeh *et al.*, 2024). Importantly, from a functional standpoint, is the exposure that even synonymous SNVs, previously regarded as functionally silent, indeed regulate the stability of mRNA as well as translation efficiency. Selected examples include SNVs of *TP53* as well as *BCL2L12* that demonstrate their overestimated contribution towards oncogenesis (Soussi *et al.*, 2017).

The precise identification of SNVs presents notable technical difficulties, particularly in scenarios where matched normal samples are not available. This challenge has driven the innovation of sophisticated computational methodologies, such as the Linkage Disequilibrium-Based Method for the Detection of Somatic Single-Nucleotide Variants (LDSSNV), which enhances the sensitivity and specificity associated with the detection of somatic SNVs in genomes (Lan *et al.*, 2023). In addition to SNVs, the interaction between small-scale variants and larger structural variants (SVs) provides a more comprehensive perspective on genomic instability. SVs-which encompass chromosomal rearrangements, copy number variations, inversions, and translocations the potential to disrupt gene structure, alter gene expression regulation, and produce novel oncogenic fusion proteins. Such alterations are particularly significant in pediatric tumors, where approximately half of the cases reveal intricate SVs within driver genes, frequently associated with negative clinical outcomes (van Belzen *et al.*, 2024). As aging progresses and tumor evolution unfolds, the accumulation of somatic SVs aggravates genomic instability and contributes to intratumoral heterogeneity, thereby presenting additional challenges to therapeutic strategies (Hamdan & Ewing, 2022).

Copy number variants (CNVs), defined by amplification/deletion of discrete segments of genomes, are a defining feature of cancer genomes. CNVs drive tumor evolution by allowing genetically distinct subclones with selective growth advantages to expand. Such genetic mutations could be the outcome of several factors, such as replication stress, abnormalities at the time of mitosis, as well as chromosome instability, leading to aneuploidy and disruption of cellular homeostasis (Steele *et al.*, 2022). The copy number variation patterns have also been of use for differentiating between various types of tumors, showing association with clinical outcomes, and treatment resistance prediction. Pathogenic CNVs, especially those involving oncogenes and tumor suppressor genes, are increasingly seen as prognostic indicators and potential treatment targets (Becchi *et al.,* 2023).

Advancements in whole-genome sequencing (WGS) and whole-exome sequencing (WES) have transformed the exploration of mutational landscapes in cancer. These technologies have revolutionized the study of mutational profiles of cancer by enabling the identification of somatic alterations that include base substitutions, double-base changes, and small insertions/deletions, which collectively create separate mutational signatures that are due to carcinogen exposure as well as compromised processes of DNA repair (Alexandrov *et al.,* 2020). Applying such technologies to the study of non-small cell lung cancer (NSCLC), WES has contributed substantially towards the identification of mutational profiles that are predictive of immune checkpoint inhibitor response as well as response to cytotoxic chemotherapy (Peroz *et al.,* 2024). The expansion of the COSMIC database of cancer mutants to include SVs has substantially improved the cataloging of mutational signatures, thereby providing deeper insights into tumor etiology and informing precision oncology initiatives (Everall *et al.,* 2023). Similarly, with regard to studying breast cancer, the use of WGS and WES has recognized germline mutations that are relevant to cancer susceptibility, established risk stratification models, as well as discovered rare actionable variants that each support customization of diagnostic as well as therapeutic approaches (Ganatra *et al.,* 2024).

AI has become central to the interpretation of genomic variants, particularly at the levels of annotation, prioritization, and prediction of pathogenicity (**Figure 1**). Programs like DeepVariant and Combined Annotation-Dependent Depletion (CADD) use ML strategies to predict the functional consequence of genomic variants at high sensitivities. Enhancements of CADD by training protein language models in addition to regulatory element protein annotations have increased its proficiency at predicting coding as well as non-coding areas (Schubach *et al.,* 2024). The varCADD platform expands such proficiency by adding rich population variant data such that representative and dependable deleteriousness scores are obtained (Nazaretyan *et al.,* 2024). In NSCLC analysis, CADD demonstrates high sensitivity as well as negative predictive value for clinically optimized variants, but its specificity is restricted such that comprehensive interpretation within clinical scenarios is essential (Kim *et al.,* 2024). AI-centered variant interpretation strategies adhering to American College of Medical Genetics and Genomics/ Association for Molecular Pathology (ACMG/AMP) guidelines have progressively outperformed heuristic strategies, particularly towards variants of unknown/suspicious significance (Nicora *et al.,* 2022). Programs like CancerVar that combine DL with established clinical criteria have allowed scalable as well as consistent somatic mutation classification, such that clinical genomics workflow efficiency has improved (Li *et al.,* 2022).

While DeepVariant mainly aims at improving variant calling accuracy, CADD can predict the effect of both coding and noncoding variants on their functional impacts. With the same concept, varCADD takes advantage of incorporating population variation to improve predictions of pathogenicity. On the other hand, CancerVar is an effective approach for clinical interpretation because it combines ACMG/ AMP guidelines with the process of clinical decision-making. Despite that these techniques deal with different steps of the genomic analysis process, their performances depend on the quality of the reference datasets and validation experiments.

There have been significant developments in AI-driven genome data analysis, a few limitations still hold back its adoption in clinical applications. First, variant interpretation is hard because the accuracy of such algorithms is reliant on the type, representation, and diversity of training datasets that may be biased towards certain groups. Additionally, the majority of predictions of pathogenicity using AI technologies will need functional and clinical validations to be able to impact therapy. With the increased amount of genome variants, differentiating clinically relevant alterations from those of unknown significance has become a problem.

Figure 1. AI tools and their application in cancer

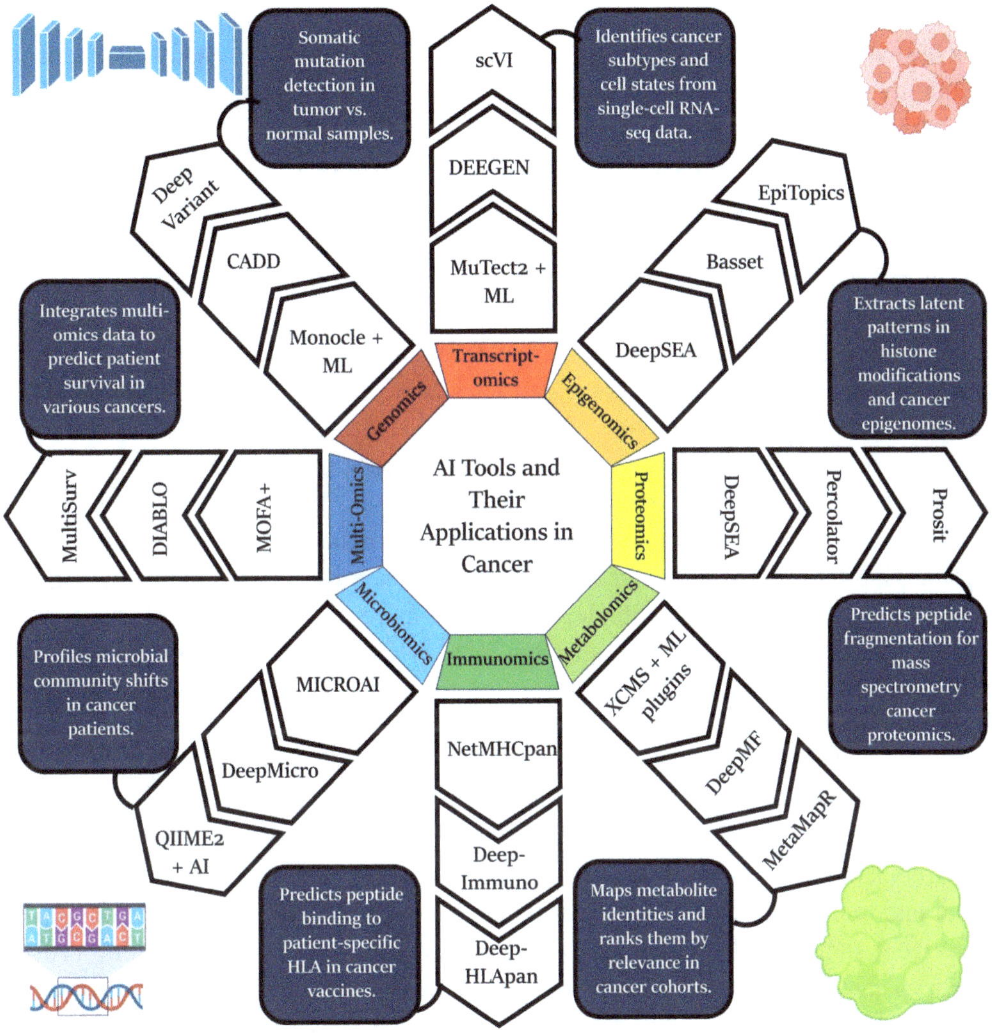

2.2 Transcriptomics

Single-cell RNA sequencing (scRNA-seq) has become an indispensable tool in today's oncology by facilitating an investigation of tumor heterogeneity, a key aspect of cancer biology responsible for variation in disease progression, treatment response, as well as outcome. By offering transcriptomic description at the single-cell resolution, scRNA-seq sketches the molecular signatures of discrete cellular subpopulations of tumors, thus offering unparalleled insight into the architecture of malignancies and their related microenvironments (Huang *et al.*, 2023). Methodological innovations such as single-nucleus full-length RNA sequencing (snHH-seq) have also increased scRNA-seq's analytical capabilities by allowing the identification of full-length transcripts, somatic mutations, as well as splice variant events of varying tumor types. These breakthroughs have revealed previously unknown malignant subclusters and contributed towards the development of an enhanced insight regarding their involvement in tumor evolution (Chen *et al.,* 2024).

Beyond profiling cellular heterogeneity, scRNA-seq has also been invaluable for dissecting the dynamic interactions of malignant cells with their immediate stromal, immune, and endothelial compartments. Such high-fidelity mapping reveals mechanisms of immune evasion, stromal remodeling, and metastatic potential while explaining the molecular basis of therapeutic resistance. Moreover, the potential for reconstructing lineage trajectories and clonal associations provides key insights into the evolution of the tumor and cellular plasticity, which are key determinants of disease recurrence and therapy failure (Cao *et al.,* 2022; Zhang *et al.,* 2021). In an effort to contend with the high-dimensionality of scRNA-seq data, AI approaches have become pivotal. Algorithms such as CellRank that combine RNA velocity with Markov state modeling yield probabilistic predictions of cellular fates such that differentiation hierarchies could be robustly charted in both normal development and malignancy (Lange *et al.,* 2022). Of particular relevance is oncology, whereby cell fates are frequently aberrant and difficult to trace. Similarly, unsupervised learning approaches that employ vector quantization and mutual information maximization have substantially improved cancer subtype prediction from transcriptomics data such that subtypes of direct clinical utility for prognosis and choice of therapy could be discerned (Chen *et al.,* 2022). Complementary therewith, graph abstraction paradigms have allowed the development of topology-preserving maps of cells that combine discrete clustering and continuous lineage inference such that scRNA-seq datasets of high complexity could be more readily interpreted (Wolf *et al.,* 2019).

The use of DL approaches towards transcriptomics has significantly fortified the identification of fusion genes that are pivotal oncogenic drivers of cancer and clinically translational biomarkers. Fusion transcripts, normally the result of chromosomal rearrangements, are nowadays identified with greater sensitivity and specificity by the use of deep neural network strategies, with existing literature finding thousands of fusion events across an assortment of cancer types (Hafstað *et al.,* 2023). Despite the persistent existence of error rates pertaining to long-read sequencing technologies, the strategies are increasingly facilitating thorough characterization of full-length fusion transcripts, thus yielding a richer insight into transcriptomic abnormalities (Masuda *et al.,* 2024). Multimodal models such as the dGMU framework showcase the potential of deep multimodal learning by integrating diverse omics layers towards characterizing tumor cell populations as well as disease condition prediction (Oni, 2019).

The broad applicability of DL across cancer transcriptomics goes beyond sheer sequencing data. In imaging-based oncology, convolutional neural networks (CNNs) like MobileNet and ResNet50 have been developed specifically for enhancing the identification of malignancies from radiological images, particularly for lung cancer, in which they have contributed considerably towards strengthening early detection capabilities (Shiva & Kalluri, 2024). Moreover, deep neural network-based information fusion systems are increasingly integrating transcriptomic, genome-wide genomic data, clinical data, and histopathological data for evaluating cancer mortality risk, identifying high-risk molecular signatures, and informing clinical decisions (Yang *et al.,* 2023a). Overall, these advancements highlight the increasing relevance of transcriptomic profiling and AI-based analytics towards advancing evidence-based precision oncology, from explaining basic principles of tumor biology to generating clinically relevant insights.

While CellRank is designed for the task of trajectory inference by taking into consideration RNA velocity, graph abstraction techniques focus on maintaining the cellular topology in the process of lineage reconstruction. In contrast, vector quantization techniques enhance the classification of subtypes in unsupervised learning processes. While multimodal approaches like dGMU can integrate the different types of omics data, CNN-based frameworks are mainly applied in transcriptomic analysis derived from imaging data.

Despite offering unparalleled insights into heterogeneity within cells, there are still some analysis and technical difficulties that need addressing in transcriptomic profiling. Some of the existing problems include batch effects, variability in sequencing, sparsity of data, and variation between different analytical pathways. Furthermore, the number of transcripts does not have to correspond to protein levels and biological activity, hence complicating biological understanding of transcriptomics conducted on its own. Combining transcriptomic information with other omic layers and clinical data will be crucial in increasing biological meaning and utility of the results obtained.

2.3 Epigenomics

The epigenomic landscape contributes a crucial function in selecting the choice of the transcriptional programs that characterize cancer initiation, progression, and treatment resistance. Two of the most thoroughly examined epigenetic changes are DNA methylation as well as enhancer remodeling, with newer advancements of AI-augmented computational platforms offering precious information regarding their functional outcome. Analytic platforms such as ELMER (Enhancer Linking by Methylation/Expression Relations) combine DNA methylation data with transcriptome data for the identification of enhancers plus the deduction of their regulatory networks. Scrutiny of over 2,000 profiles of tumors, The Cancer Genome Atlas by ELMER has shed light on transcription factor–mediated networks such as *GATA3/FOXA1* of breast cancer cells, as well as *SOX17/FOXA2* endometrial cancer cells that play prognostic as well as therapeutically relevant functions (Yao *et al.,* 2015). Complementarily, the MethylationToActivity (M2A) DL platform precisely predicts promoter activity from DNA methylation signatures from diverse types of cancer, offering an increased view of gene-level regulatory effects (Williams *et al.,* 2021). High-resolution methylation maps have shown widespread promoter hypermethylation of enhancers associated with tumor suppressor genes, suggesting that epigenetic gene silence is a fundamental mechanism of neoplastic transformation (Vidal *et al.,* 2017). Recently, enhancer-guided core transcriptional regulatory circuits have been recognized as crucial components of oncogenic programming, influenced by genomic mutations or epigenetic modifications, presenting prospective indicators for prognosis, therapy, and therapies (Xu *et al.,* 2023).

Along with DNA methylation, chromatin accessibility, and 3D genome structure are also established as oncogenic determinants of gene expression. Techniques such as ATAC-seq and Hi-C have spawned high-definition maps of regulatory landscapes and chromatin topology of cancer types. For example, in breast cancer, resistance towards doxorubicin has been attributed to 3D chromatin reorganization involving A/B compartment switching, TAD size reduction, and chromatin loop remodeling that is commensurate with differential gene expression of resistance proteins such as *FA2H* and *FOXA1* (Wang *et al.,* 2022). Similarly, in pancreatic cancer, 3D genome structures such as enhancer-promoter loops and higher-order chromatin domains dictate subtype-by-subtype specificity of gene expression. The *HNF1B* transcription factor has been identified as a catalyst for subtype switching through direct alteration of chromatin structures, exemplifying the dynamic interplay between chromatin architecture and gene expression. These interactions underpin malignant transformation by modifying genomic architecture spatially (Kato *et al.,* 2022). Systematic description of chromatin accessibility across tumor types by

ATAC-seq has also been defined as regulatory signatures that not only characterize tumor subtypes by their complexity but also identify functional enhancer-promoter interactions that often implicate oncogenic hub genes such as *MYC* and *SRC* (Corces *et al.,* 2018). SVs also highlight the oncogenic potential of genome alterations, as examples like the interchromosomal insertion of *MYC* towards breast cancer progression illustrate the reliance of malignant transformation upon genome reorganization by space (Rossini *et al.,* 2024). Next-gen technologies such as SCA-seq development also now allow exploration of high-order epigenomic interactions at previously unimaginable spatial resolution, such that our understanding of chromatin accessibility and high-range genome communication of tumors is expanding tremendously (Xie *et al.,* 2024).

The development of AI and DL has greatly increased the ability to map, predict, and functionally annotate cancer-related epigenomes. DeepSEA and Basset models use convolutional neural network architectures to predict the functional effects of noncoding variants and regulatory elements, while newer models extend these capabilities toward context-dependent uses within oncology. DeepRegFinder, for instance, integrates convolutional and recurrent neural networks to automatically classify enhancers and promoters, selectively separating active versus poised states with high proficiency across diverse cancer cell types (Ramakrishnan *et al.,* 2024). Similarly, DNAResDualNet (DARDN), an algorithm based upon a convolutional neural network architecture, predicts cancer-specialized CTCF binding sites, thus illuminating oncogenic transcriptional programs within cancers such as T-cell acute lymphoblastic leukemia and breast cancer (Cho *et al.,* 2024). Other models like PARM, trained upon massively parallel reporter assays, discern transcription factor binding motifs and predict promoter activity while detecting higher-order motif interactions essential for oncogenic control (Barbadilla-Martinez *et al.,* 2024). DECRES, by utilizing ENCODE data, identifies cis-regulatory elements within diverse tumor types and a range of tissues, providing an overarching map of enhancer and promoter sites within the cancer epigenome (Li *et al.,* 2018). These approaches illustrate how integrative epigenomic profiling has transformed our understanding of regulatory programs within cancer.

On the other hand, ELMER and M2A target DNA methylation-based regulation, but DeepSEA and Basset concentrate on non-coding variant prediction of functional impacts. The latest approaches like DeepRegFinder and DARDN combine these methods with regulatory elements detection and chromatin binding through deep learning structures. Despite all these tools' efforts to improve epigenomic annotation, differences in input data, biological hypothesis, and validation strategies impede comparison of the research findings.

AI-augmented epigenomic studies have significantly enhanced our knowledge of cancer regulatory processes, although some methodological challenges still persist. The nature of epigenetic changes is very dynamic and highly context-specific, which makes their interpretation for various types of cancers and stages of the disease very difficult. Moreover, there is a lack of experimental proof of the biological importance of most predictions generated through computer algorithms. There should be an increased standardization of epigenomic data and a better connection between epigenomics and transcriptomics/proteomics to make research more reliable and clinically applicable.

2.4 Proteomics and Phosphoproteomics

Mass spectrometry (MS)–based proteomics coupled with AI has become the hub of progress in cancer research by facilitating comprehensive characterization of protein landscapes as well as their regulatory modifications. The high dimensionality of proteomic data demands computational strength for unraveling high-dimensional spectral data into biologically interpretable patterns. AI, including ML and DL, is progressively incorporated across proteomics workflows, facilitating enhanced biomarker identification, tumor classification, and treatment stratification (Zeng *et al.,* 2024). Tools like MSpectraAI illustrate the translational applicability of such strategies by utilizing deep neural networks for proteomic spectral classification across diverse tumor types at high specificity and sensitivity, exemplifying their potential in multi-cancer diagnostics (Wang *et al.,* 2020). Moreover, the use of ML in tumor immunopeptidomics has refined the identification of neoantigens by combining features like HLA-binding affinity as well as peptide retention time, consequently optimizing pipes for clinical immunotherapy (Wei *et al.,* 2024). Higher-level algorithms combining dimensionality reduction strategies such as principal component analysis (PCA) with one-dimensional neural networks like ResNet and SENet have strengthen cancer diagnostics even further, with remarkable success in ovarian cancer detection (Ma *et al.,* 2024).

One of the key strengths of proteomics lies in its characterisation of post-translational modifications (PTMs), which serve as dynamic regulators of protein activity as well as signalling pathways of cancer. The use of AI-based computation approaches has transformed the study of PTMs, specifically around the domains of methylation and phosphorylation. For instance, ML models for the prediction of SET8-mediated methylation have recorded an up-to-19-fold improvement of correctness over conventional approaches, uncovering crucial alterations of substrate networks of breast cancer (Biggar *et al.,* 2024). Likewise, CancerOmicsNet, which applies deep graph learning, predicts the efficacy of multitargeted kinase inhibitors and demonstrates optimized specificity towards antiproliferative effects for a range of cancer cell types (Singha *et al.,* 2022). Furthermore, complementary tools such as PTMapper offer high-definition mapping of kinase–substrate interactomes that are crucial towards deciphering phosphorylation-dependent signalling tracks as well as the identification of potential drug targets (Narushima *et al.,* 2016). Data-informed approaches continue towards correcting shortcomings of kinase–substrate annotation, consequently enabling systematic prioritisation of interactions imperative towards oncogenic signalling cascades of cancer cells (Petursson & Petsalaki, 2022).

The phosphoproteomic profile of cancer reveals deregulation of pivotal signaling pathways such as *Hedgehog*, *phosphoinositide 3-kinase* (PI3K), *Notch*, *VEGF*, *WNT/β-catenin*, and *MAPK* that organize cell growth, survival, metabolism, and differentiation. Abnormal regulation of these cascades drives hallmark processes like epithelial–mesenchymal transition, cancer stem cell maintenance, angiogenesis, and metastatic dissemination (Papatsirou *et al.,* 2021; Rohilla *et al.,* 2024). Here, notably, noncoding regulators like circular RNAs (circRNAs) have been reported as regulators of such pathways by fine-tuning transcriptional crosstalk as well as post-transcriptional regulation, thus widening the regulatory axes of oncogenic networks. In order to capture the native complexity of these multi-layered signaling systems, computational modeling becomes an essential tool. By combining phosphoproteomic datasets with network-based algorithms, such models create pathway wiring diagrams, make predictions of molecular interactions, and uncover emergent vulnerabilities of tumor signaling (Garland, 2017). Such paradigms not only advance mechanistic insight but also aid strategy for designing targeted therapies as well as personalized treatment approaches by uncovering predictive biomarkers of response to therapy.

MSpectraAI is aimed at spectral classification and thus diagnosis of cancer diseases; CancerOmicsNet applies graph learning for drug response prediction, while PTMapper concentrates on kinase-substrate mapping and not on disease diagnosis. These three methods illustrate the development of AI applications in proteomics; however, implementation of these methods in the practice is still lacking. Although there have been rapid advancements in AI-assisted proteomics, applying such proteomic discoveries clinically is difficult. Mass spectrometry data are known to have significant technical variability, and variations in sample preparation, machine, and computational tools could affect the reproducibility of the results across different laboratories. Moreover, whole-phosphoproteomic profiling continues to be cost-intensive and technically difficult, which limits its application clinically. Further research will have to consider the development of standardized methods, validation of AI-derived biomarkers in multi-center patient cohorts, and integration of proteomic results with other omics.

2.5 Metabolomics

Metabolomics provides a functional summary of cellular states and has become pivotal in oncological studies. However, the complexity of metabolomic data demands advanced computational approaches for accurate interpretation. Integration of AI and ML paradigms within metabolomics workflows has progressively enabled advancements in metabolite identification, reconstruction of pathways, and biomarker discovery (**Figure 2**). When coupled with AI algorithms, software such as XCMS and MetaboAnalyst has shown improved data analysis, diagnostic specificity, and interpretability on a variety of cancer cases (Xu *et al.,* 2024). Automated machine learning (AutoML) integrated with explainable AI (XAI), for example, has shown superior predictive capabilities over conventional ML for distinguishing cancer patients from controls, such as manifested in renal cell carcinoma and ovarian cancer (Bifarin & Fernández, 2024). Moreover, these approaches provide mechanistic insights by identifying disturbed metabolic pathways, such as aminoacyl-tRNA biosynthesis and glycine metabolism, that occur in colorectal cancer (CRC) (Yang *et al.,* 2024a). In NSCLC cases, AI-augmented metabolomics has identified cancer-specific metabolic signatures that relate to radiation response therapeutics by establishing its potential use in developing diagnostic as well as predictive biomarkers (Murcia-Mejía *et al.,* 2024).

Figure 2. AI applications in metabolomics

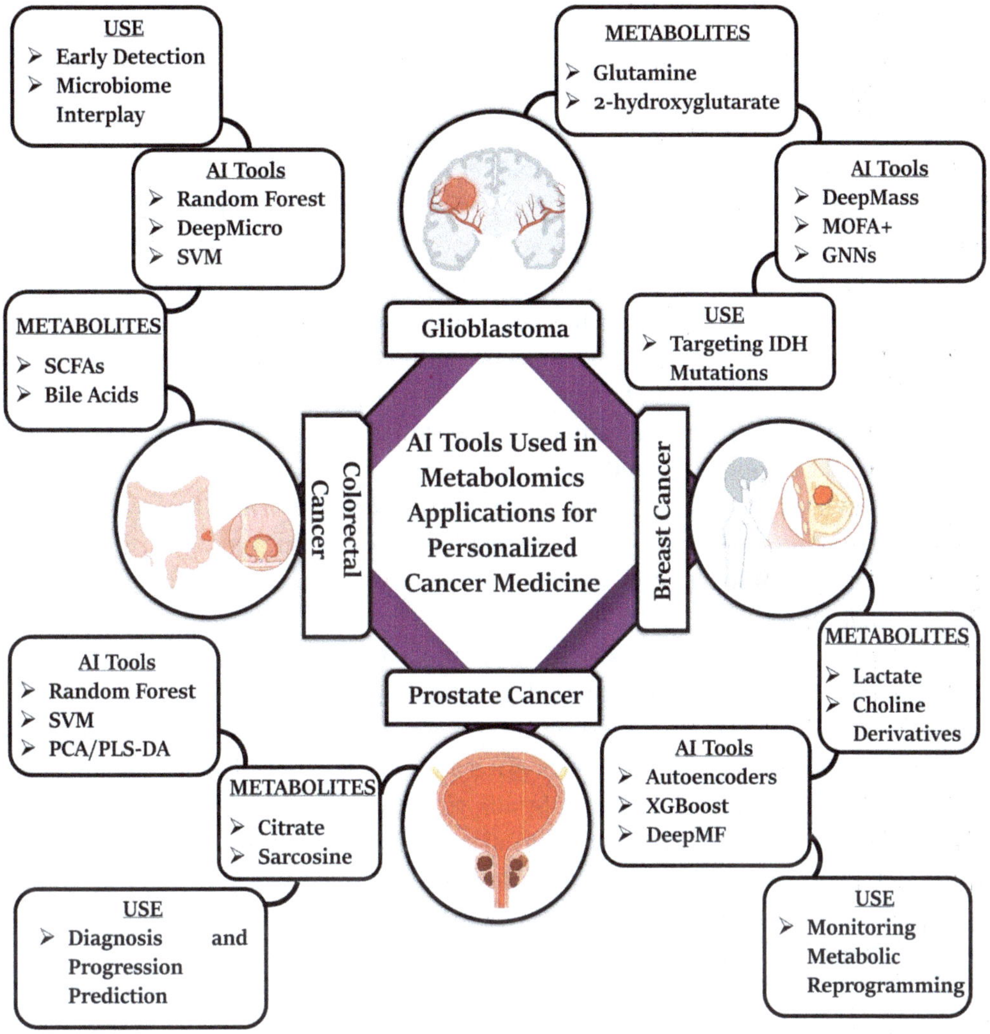

Metabolic reprogramming is a signature of cancer metabolism that supports uncontrolled proliferation, stress survival, and metastatic development. AI-based approaches have moved past the characterization of this phenomenon by correlating transcriptomic and metabolic signatures. TFmeta, an ML platform, discerned TFs that control glycolysis in NSCLC by merging TF-binding signatures with RNA-seq datasets (Zhang *et al.,* 2018). Supplemental DL approaches have broadened metabolomic usage to histopathological imaging such that CNNs can detect adenocarcinoma of the pancreas by glycolytic activity with clinically meaningful specificity (Tsou & Wu, 2020). Genome-scale metabolic models (GEMs) also offer predictive platforms by which mutation-related metabolic pathways that could serve as biomarkers and potential drugs could be discovered (Lee *et al.,* 2022). In triple-negative breast cancer (TNBC), an intelligent augmentation platform allowed the prediction of genes that were connected with metabolic reprogramming by integrating multi-omics layers (Jenna *et al.,* 2022). Correspondingly, in clear cell renal cell carcinoma (ccRCC), an ML–based metabolism-related prognostic signature (MRPS) indicated strong prognosis and predictive potential for immunological checkpoint treatment response (Liu *et al.,* 2025).

Flux balance analysis (FBA), in conjunction with AI optimization, has substantially augmented the analytical capabilities of metabolomics by offering a comprehensive perspective on metabolic fluxes within tumors. In the context of gynecologic cancers, FBA has successfully identified metabolic biomarkers, including succinate dehydrogenase and ornithine transaminase, which serve as indicators of response to immune checkpoint inhibitors (Idumah *et al.,* 2025). Regarding breast cancer, FBA has disclosed metabolic flux variations that are specific to subtypes, particularly in glycolysis and glutamate metabolism, thereby facilitating patient stratification and risk assessment (Trilla-Fuertes *et al.,* 2019). The incorporation of ML with FBA increases the interpretability of data by selecting refined flux variables from intricate networks, thus enhancing the discovery of biomarkers. Notable applications in renal cell carcinoma have underscored FBA's potential to forecast crucial metabolic genes, identifying vulnerabilities that may be targeted for therapeutic benefit (Gatto *et al.,* 2015). Furthermore, comparative analyses of NADPH production mechanisms across various cancer cell lines employing FBA have unveiled cancer-specific metabolic dependencies, providing valuable insights into redox regulation and metabolic weaknesses (Ahmad, 2018). Despite the persisting challenges associated with data integration and scalability, the synthesis of AI, FBA, and metabolomics establishes a formidable framework for elucidating the metabolic intricacies of tumors, stratifying patients, and informing personalized therapeutic strategies (Sahu *et al.,* 2021).

AutoML methods enhance metabolite classification by optimizing predictive models, while TFmeta unites transcription factors with metabolism pathways. Genome-scale metabolic modeling allows for performing simulation of metabolic processes on the systems level, while Flux Balance Analysis predicts metabolic fluxes in cells. Though metabolomics gives a functional view of cancer biology, metabolite levels are heavily influenced by external and physiological factors, which can make it difficult to understand the biological significance. The variations in metabolomic tools, metabolite identification, and pathway databases also play an important role. Many metabolomic biomarkers derived from AI need to be validated independently before their use for clinical purposes. In the future, emphasis needs to be put on standardized metabolomics procedures and multi-omics.

2.6 Immunomics and Microbiomics

Immunomics' emergence, which entails the thorough investigation of immune-related molecules as well as pathways, has greatly increased our insight into tumor-immune interactions as well as their treatment prospects. Herein, AI has become an imperative catalyst, specifically with reference to predicting neoantigens, an essential prerogative towards deriving personalized immunotherapies of cancer. AI-supportd platforms such as NetMHCpan, as well as DeepHLApan, have transformed tumor-specific antigen identification by investigating large-scale multi-omics datasets at tremendous specificity (Bulashevska *et al.,* 2024). These platforms take advantage of high-level ML algorithms towards deciphering voluminous interactions of tumor-derived peptides with human leukocyte antigens (HLAs), concentrating on both binding affinity as well as immunogenicity, the two essential determinants of effective neoantigens. DeepHLApan uses DL methodologies to incorporate binding and immunogenicity models within a comprehensive predictive framework that surpasses current neoantigen detection techniques (Wu *et al.,* 2019). Integration significantly elevates the specificity of clinically applicable neoantigens selected for cancer immunotherapy development. Beyond antigen discovery per se, AI provides an essential contribution towards treatment optimization by anticipating patient-specific immune reactions towards offsetting therapy-related toxicities, thus deriving more rationalized as well as effective immunotherapy strategies (Olawade *et al.,* 2025).

Though the immunomic aspect of cancer science is strengthened by AI-based inventions, the microbiomics realm adds another dimension of complexity and potential therapy. The human microbiome is central to the modulation of tumor immunity, impacting responses to immune checkpoint inhibitors and other immunotherapies. Incorporation of AI in microbiome profiling assists the discernment of microbial signatures that are amenable to cancer development/progression as well as response to treatment. Correlating microbial community structure with that of the host's immune profile, AI-based models identify microbiota-immune cross-talk mechanisms that hold new promise as biomarkers as well as adjuvant therapies. Notwithstanding, challenges continue with regard to clinical use of immunomic and microbiomic understandings. Difficulties related to data privacy, algorithmic explainability, as well as scalability of predictive models across diverse populations continue to need resolution if the potential of AI across precision oncology is to be fully actualized (Olawade *et al.,* 2025).

NetMHCpan is designed to predict peptide-HLA binding affinity, but DeepHLApan is designed to predict immunogenicity as well, hence providing better prioritization of neoantigens. While DeepHLApan is more informative overall, both algorithms require good quality immunopeptidomics datasets and population validation. The applications of immunomics and microbiomics in AI provide great possibilities for individual immunotherapies; nonetheless, there are some serious obstacles that must be addressed. There is significant variation in immune reactions and microbial communities between individuals and between regions, which reduces the ability of prediction models to generalize. Causal connections between microbial communities and the results of treatments are not entirely known. There are many ethical issues associated with patient confidentiality, data exchange, and algorithmic bias.

3. ADVANCED AI METHODOLOGIES FOR MULTI-OMICS INTEGRATION

3.1 Representation Learning Across Modalities

In the personal cancer medicine paradigm, representation learning has emerged as a pivotal paradigm for integrating and interpreting omics datasets of varying types. Embeddings, latent space modeling of autoencoders and variational autoencoders (VAEs), as well as contrastive learning, have demonstrated considerable promise towards improving the accuracy of cancer subtyping. In particular, VAEs support the uncovering of biologically informative latent representations from high-dimensional datasets covering DNA methylation and metabolomics, thereby facilitating subtype identification as well as providing mechanistic insights towards tumorigenesis (Wang & Wang, 2019). For example, the Deeply Integrating Latent Consistent Representations (DILCR) system utilizes VAEs to denoise omics data while at the same time learning consistent latent embeddings; these embeddings are then fused subsequently by an attention-based fusion network, enhancing cancer subtype classification accuracy (Cai & Wang, 2024). Complementarily, the MoCLIM system utilizes contrastive learning towards harmonizing multimodal representations, yielding higher subtype clustering within lower-dimensional latent spaces (Yang *et al.,* 2023b). Together, these approaches highlight the capability of VAEs towards modeling nonlinear, intricate relationships between molecular features, such that they advance precision oncology by enabling fine-level stratification of cancer subtypes whilst facilitating development of targeted treatment strategies

(Wei & Mahmood, 2020). Furthermore, the transferable nature of embeddings obtained from VAEs highlights their robustness across datasets, as such, it also provides credence towards their capacity to generate clinically relevant insights that generalize across datasets (Gomari *et al.,* 2022).

Building from these first principles, self-supervised learning (SSL) has been an auspicious strategy for tackling the challenges of high-dimensionality as well as the sparsity of annotations that characterize cancer omics data. The SubOmiEmbed framework is a typical example of this strategy by employing VAEs with feature subsetting that requires the model to reconstruct data from sparse subsets. This setup supports learning of low-dimensional but biologically relevant latent spaces and thereby mitigates overfitting while improving the unification of diverse modalities like DNA methylation and gene expression. Such embeddings have been successfully used in cancer type classification, demonstrating the potential of SSL for learning robust and interpretable features from sparse multi-omics data. Similarly, the Self-omics framework offers an integrated SSL-based pretraining framework that combines contrastive alignment with data recovery from degenerate samples, and embeddings that outperform state-of-the-art models are obtained on the TCGA pan-cancer dataset for classification tasks (Hashim *et al.,* 2022).

Important, the framework is engineered in a way, it accepts missing omic layers as well as diverse datasets while extending its use for zero-shot rare cancer classification, representing an important step towards clinically deployable generalizability. Representation learning has shown great potential in the integration of heterogeneous multi-omics data sets, but the latent representations have been hard to interpret from a biological perspective. Most deep learning techniques need significant computation power and a large number of annotated data sets, which may be hard to acquire in practice.

3.2 Graph Neural Networks (GNNs) and Network Biology

The integration of multi-omics data with advanced computational approaches, including heterogeneous graph neural networks (GNNs), is increasingly framed as an essential component of cancer subtype classification and biomarker identification. Compared with traditional linear models, GNNs take advantage of both topological structures and feature-level relationships, thus being specifically well-equipped to manage high-dimensional and multimodal cancer data. The multimodal GNN model framed by Li and Nabavi provides an example of such an approach by constructing heterogeneous multi-layer graphs that integrate both inter-omics and intra-omic interactions, leading to improved effectiveness of cancer subtype classification across comprehensive datasets such as TCGA (Li & Nabavi, 2024). Likewise, the MOCSC model created by Ma and Guan employs a late integrating strategy that combines a series of sparse denoising autoencoders with a view correlation exploration network that tackles the inherent noise and complexity of the multi-omics data while enhancing cancer subtype identification (Ma & Guan, 2022). Building upon these advancements, Islam and Hasan introduced a patient-oriented network approach that synthesizes subnetworks by random walk with restart and k-means clustering for subtype identification. This model retains the heterogeneity of individuals while combining an assortment of omic signals (Islam & Hasan, 2024). In parallel, Wei et al. developed a hierarchical multi-kernel learning model that fine-tunes each omics layer's kernel parameters before resorting to clustering for enhanced subtype classification interpretability as well as explorations of underlying potential biomarkers (Wei *et al.,* 2023b). Altogether, such studies identify the flexibility of graph-based learning frameworks towards overcoming the complex nonlinear relationships inherent in cancer multi-omics.

In addition to subtype classification, network biology also contributes towards unraveling the molecular mechanisms responsible for carcinogenesis as well as treatment response. Protein–protein interaction

(PPI) networks also present powerful platforms for the identification of crucial nodes as well as modules responsible for oncogenic processes. An example is the EPI-GBRWR approach that utilizes graph-based random walk algorithms for enhancing key protein prioritization in several cancers, like breast, lung, colorectal, as well as ovarian cancers, thereby yielding actionable information applicable towards precision oncology (Rout *et al.,* 2024). Correlatively, signaling networks present active illustrations of aberrated pathways frequently altered by genetic, genomic, as well as epigenetic mutations found within tumors. Recent developments like genome microarrays as well as proteomic profiling have allowed extensive mapping of those networks responsible for the identification of cancer-specifying signaling signatures of diagnostic as well as prognostic potential (Zhang *et al.,* 2009). Complexity of signaling networks defined by high crosstalk as well as modular structure necessitates high-level computational approaches like factor analysis that could dissect them into functional modules able to predict treatment success.

Targeting deregulated PPIs of these biological networks has high translational potential because they control key cellular processes like proliferation, differentiation, and apoptosis. Structural characterization of protein domains and interfaces has recognized multiple druggable sites that offer potential therapeutic opportunities (Dar *et al.,* 2019). In addition, the use of reverse-phase protein microarrays has allowed high-resolution mapping of signaling pathway activity, generating patient-distinct circuit diagrams that guide personalized treatments (Pierobon *et al.,* 2012). These approaches highlight GNN-guided computational modeling's interaction with network biology, demonstrating how integrated platforms are able to bridge the complexities of multi-omics data with systems-level insight, thereby propelling the next generation of precision diagnostics and personalized therapy plans.

While graph neural networks perform well when capturing complicated biological interactions, their accuracy is highly dependent upon the design of the graph structure, network architecture, and data quality. The differences between the biological network databases may lead to inaccurate predictions in different studies. Standardizing the graph building procedures and increasing computational speed will be vital for the widespread clinical application.

3.3 Federated and Transfer Learning for Distributed Omics

The development of federated and transfer learning frameworks has become essential for advancing cancer research in contexts where rare tumor subtypes or underrepresented datasets limit predictive accuracy. Domain adaptation techniques are particularly valuable for addressing these challenges, as they enable models to leverage information from well-annotated cancer types and transfer this knowledge to domains with scarce data. For example, the DAPM-CDR model effectively harnesses knowledge from cancers with abundant datasets to increased drug response prediction in rare cancers, achieving superior performance in scenarios traditionally limited by data scarcity (Sun *et al.,* 2024). Likewise, Generative Adversarial Networks (GANs) have been deployed to augment training cohorts, thereby enhancing the generalizability and robustness of DL frameworks. In thyroid cancer histopathology, GAN-based augmentation refined classification accuracy for minority class images, a common limitation in rare tumor settings (Dee *et al.,* 2024). Beyond augmentation, optimal transport methods provide an advanced strategy for mitigating batch effects across heterogeneous datasets, ensuring reliable classification of rare cancer phenotypes by reconciling distributional discrepancies in histopathology data. Importantly, knowledge transfer between related cancers, such as rectum-to-cervical cancer adaptation, further illustrates the potential of cross-domain learning to address sparse data environments (Zeng *et al.,* 2023).

In conjunction with domain adaptation, few-shot learning has surfaced as a significant approach for distributed omics, providing strong generalization abilities when faced with a limited number of training examples. The Gaussian Distributed Prototypical Network (GDPN) illustrates this capability by successfully identifying genetic mutations within cancer research literature, even amidst severe data scarcity (Cao *et al.,* 2023). Likewise, the implementation of prototypical networks in lung cancer classification has achieved an accuracy rate of 98% using minimal training samples, underscoring their potential for medical imaging applications where data imbalance poses an ongoing challenge. Furthermore, studies in lymphatic metastasis and prostate cancer detection demonstrate that few-shot learning may outperform traditional DL methods in terms of both predictive accuracy and adaptability to variations in domain, particularly in situations characterized by severely limited data availability (Yamagishi *et al.,* 2025). Federated and transfer learning approaches present an attractive alternative for collaboration in terms of privacy protection; still, the differences in data quality, organizational policies, and computational capabilities influence the performance of these methods. It will be crucial to solve the issues related to efficient communication, security of the models, and regulations.

3.4 Causal Inference and XAI

Causal inference and XAI have become two of the most important methods for improving the interpretability, transparency, and clinical validity of AI-based models of personalized oncology. Through the provision of mechanistic rationales of model predictions, these strategies surmount the often-described "black box" problem of many DL models, thereby building confidence and enabling their incorporation within decision-making pathways at the bedside. In breast cancer, explainable AI (XAI) strategies such as feature importance analysis and decision tree modeling have been employed towards developing interpretable AI platforms able to discern optimal treatment approaches of high specificity. These strategies permit clinician correlation of predictive outcomes with definable patient attributes, thereby building faith in AI-provided therapy guidelines (Lokare *et al.,* 2024). In high-grade serous ovarian cancer, biomarker identification by XAI usage has guided gene identification of potential therapeutics such as *IL27RA*. This demonstrates an XAI potential not only towards enhancing model explainability but also towards creating biologically reasonable hypotheses that could direct experimental validation as well as development of precision therapeutics (Chakraborty *et al.,* 2024). Likewise, when incorporated with multimodal real-world datasets- consisting of genomic data, imaging data, and clinical data-XAI platforms have enabled identification of crucial prognostic markers across a variety of cancer types, thereby improving personalized assessment of cancer-related risk as well as fine-tuning treatment approaches (Keyl *et al.,* 2025).

For NSCLC, XAI methods, such as Gradient-weighted Class Activation Mapping (Grad-CAM), have been used to integrate imaging and genetic data. These approaches shed light on the CT image features related to their respective genetic mutations, thus providing insight into the functional association between radiographic phenotypes of the disease and its underlying molecular drivers, which in turn increase the interpretable outcome of DL models and permit more accurate treatment interventions (Shariaty *et al.,* 2024). Beyond advocating model explainability, XAI also promotes higher accuracy and dependability of cancer diagnostics that widens access to high-quality predictive tools at the grassroots level, a feature that is particularly essential for early cancer detection and personalized treatment with restricted resources

(Olumuyiwa *et al.,* 2024). Despite the advances in explainable AI in terms of model predictability, the interpretability alone cannot guarantee the clinical effectiveness of the prediction model. Some of the most common interpretation techniques give post hoc explanations that do not represent biological processes.

4. TRANSLATIONAL APPLICATIONS IN PERSONALIZED CANCER MEDICINE

4.1 Early Detection and Liquid Biopsy Integration

Translational application of AI in personalized cancer medicine offers a wide range of applications in various steps (**Figure 3**). Liquid biopsies, particularly by analyzing circulating tumor DNA (ctDNA) and exosomal RNA, have become valuable tools of personalized cancer medicine that offer a minimally invasive approach to early diagnosis, disease monitoring, and therapeutic guidance. By separating tumor-derived molecular components from biological fluids like blood, liquid biopsies allow for real-time assessment of tumor dynamics, treatment response, and disease progression, thus offering a complement to conventional tissue-based diagnostic strategies (Bamodu *et al.,* 2023). The clinical utility of ctDNA and exosomal RNA has been greatly improved by advancements in molecular technologies like next-generation sequencing (NGS) and digital PCR, which provide high specificity and sensitivity for detecting actionable mutations and for preserving tumor heterogeneity (Turabi *et al.,* 2024). Such technological advancements warrant early cancer detection, aid in minimal residual disease (MRD) monitoring, and offer a dynamic assessment of treatment efficacy, thus advocating for precision-guided intervention. Importantly, the use of diverse analytes-e.g., ctDNA, exosomal RNA, and circulating proteins-through a multi-omics approach provides an integrated vision of tumor biology, thus enhancing treatment outcome predictions and potential resistance mechanisms (Coupland *et al.,* 2024). Despite their transformative potential, clinical utilization of liquid biopsies meets multiple challenges that include methodological standardization needs, assay quality controls, limitations by costs, and integration within prevailing clinical practices. Addressing such barriers holds promise towards unlocking the potential of liquid biopsy technologies for precision oncology. Nevertheless, by offering a dynamic, non-invasive, and longitudinal tracking paradigm, liquid biopsies are on the eve of revolutionizing cancer care by affording customization of treatments that best suit individual patient molecular signatures (Ho *et al.,* 2024).

Figure 3. Translational application of AI in personalized cancer medicine

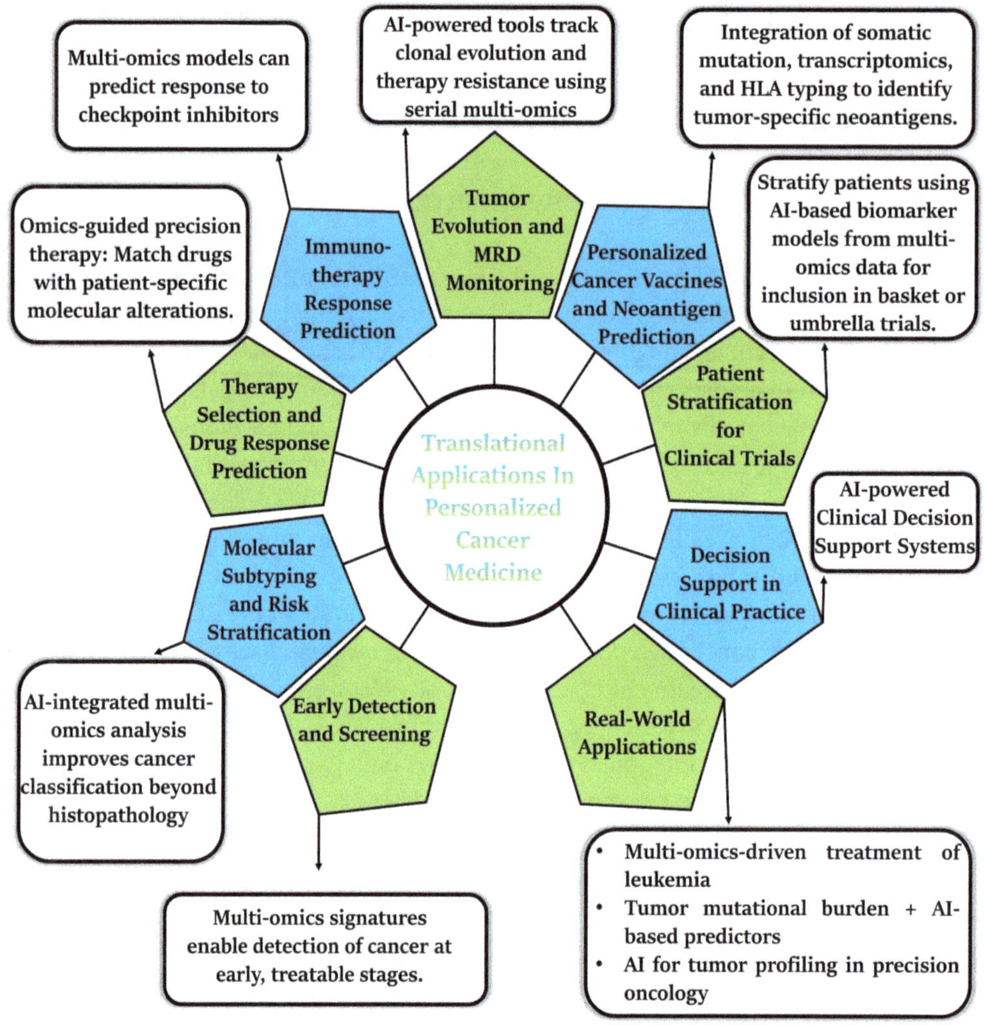

4.2 Patient Stratification and Molecular Subtyping

Integrating AI with longitudinal multi-omics has greatly strengthened tumor evolution modeling and patient stratification in precision oncology. Through modeling temporal dynamics of genomic, epigenomic, and transcriptomic landscapes, AI supports a dynamic view of clonal evolution, tumor development, and response to therapy, thus maximizing the design of personalized treatment regimens. In metastatic colorectal cancer (mCRC), Consensus Molecular Subtypes (CMS) use is a great example of the clinical application of molecular stratification. CMS subtyping unifies tumor heterogeneity at genetic, epigenetic, and transcriptomic levels to enable oncologists to tailor systemic chemotherapy, targeted therapy, and combination regimens in pre-identified patient subgroups (Rebersek, 2020). This molecularly rationalized paradigm is a shift from the conventional "one gene, one drug" dogma to a "multi-gene, multi-drug" approach to meet the sophisticated genomic landscape, convergent pathway dependencies, and dynamic clonal evolution patterns in CRC (Dienstmann *et al.,* 2017). AI-based models also increase personalized cancer therapy by using multi-omics data to produce counterfactual treatment predictions. For example, Schürch et al. developed a platform providing probabilistic treatment suggestions calibrated in confidence, thus facilitating clinical decision-making through the combination of patient-specific molecular signatures and possible therapeutic effects (Schürch *et al.,* 2024). Support for such an approach comes from clonal evolution models like those developed by Tao *et al.,* which offer predictive cancer evolution estimates after analyzing mutational phenotypes and their temporal evolution. These models have the potential to refine prognostic predictability beyond typical clinical and genomic characteristics, and guide treatment priority and surveillance schedules (Tao *et al.,* 2019).

4.3 Predictive Modelling for Therapy Response

A combination of AI with multi-omics is transforming personalized medicine in cancer by allowing accurate prediction of treatment response and identification of drug repurposing strategies. ML models have proven unparalleled potential in chemotherapy and target therapy response prediction in mCRC. With the support of radiomics and molecular biomarker signatures, it has been possible to have an accurate prediction of responders and non-responders, thus enhancing patient stratification and optimization of therapeutic effectiveness (Russo *et al.,* 2022). Integrating multi-omics, including genomics, transcriptomics, proteomics, and metabolomics, also enhances predictive modeling by offering a complete molecular snapshot of tumors. The crossroad of multi-omics and AI enables the discovery of new biomarkers and actionable targets, which are crucial in the development of customized therapeutic regimens. In ICIs, high-resolution multi-omics maximizes treatment personalization and patient selection, enhancing immunotherapy efficacy and translation to diverse patient populations (Limeta *et al.,* 2023). DL has progressed predictive modeling to the discovery of synergistic drug pairs. Software such as AuDNNsynergy combines multi-omics data-e.g., gene expression profiles, copy number alteration, and mutational signatures-to score potent drug combinations. It addresses the widespread cancer treatment resistance issue by discovering synergistic interventions that achieve a balance between high therapeutic efficacy and low toxicity (Zhang *et al.,* 2020). Additionally, in breast cancer, integration of multi-omics data with clinical factors on DL platforms has increased therapeutic and diagnostic accuracy and decision-making. The models permits the prediction of therapeutic response and likely recurrence risk, in spite of the complexity and heterogeneity of tumor biology (Varpe *et al.,* 2024).

4.4 Personalized Treatment Design and Clinical Trials

The application of AI has substantially optimized the design of individualized treatment schedules and clinical trials. AI-enabled methods make it possible to process genomic, proteomic, and clinical big data, enabling patterns to be identified to inform personalized therapeutic regimens of a patient's molecular and genetic make-up (Akter, 2024). Central to the approach is biomarker discovery that not only predicts therapy response but also mechanisms of resistance, thereby enhancing treatment effectiveness and patient outcome. N-of-1 trials are a shining example of AI application in precision oncology, where each patient acts as his/her own control. In this case, the "liquid dynamic medicine" idea uses genomic profiling and liquid biopsy-based biomarkers to dynamically adjust treatment based on tumor evolution and drug response (Silvestris *et al.,* 2017). The dynamic approach supports real-time modification of therapeutic strategy, thereby optimizing clinical effectiveness and reducing unwanted effects. Dynamic Treatment Regimes (DTRs) push the capability of AI further through the use of reinforcement learning (RL) algorithms for iterative improvement of treatment alternatives. With the incorporation of patient-specific information, such as genomic, clinical, and longitudinal treatment response, DTRs allow for the development of adaptive treatment regimens that evolve based on individual patient requirements, increasing efficacy and safety (Yazzourh *et al.,* 2025). The incorporation of AI into Clinical Decision Support Systems (CDSS) also suppots simpler practical application of such insights into clinical operations. AI-driven CDSS systems integrate multi-omics, drug, and clinical data to suggest the best individualized treatment plans, facilitating oncologists in clinical decision-making and enhancing therapeutic outcomes (Molefi *et al.,* 2023). In spite of the potential of advancing translational medicine through this technology, the adoption of AI-based multi-omics in oncological clinical practice faces several barriers that include regulatory, technological, and economic issues. The integration of such tools into everyday practice needs standardization of workflow processes, validation at multiple centers, interoperability with electronic medical records, and regulation.

5. CASE STUDIES

AI solutions to pan-cancer analysis through TCGA multi-omics datasets have significantly strengthened cancer diagnosis, prognosis, and classification precision. Platforms like HTML utilize multi-omics self-adaptive dynamic learning to analyze patient-specific data and recognize individualized abnormalities for personalized and accurate cancer prognostics (Lu *et al.,* 2023). Likewise, DeepProg utilizes deep-learning and machine-learning ensembles for the prediction of patient survival subtypes in 32 cancer cohorts and the revelation of genomic signatures of poor prognosis (Poirion *et al.,* 2020). The OmiVAE model integrates VAEs and classifiers to derive low-dimensional features from high-dimensional omics data with high accuracy for classifying 33 types of tumors and proving the usefulness of multi-omics integration in pan-cancer classification (Zhang *et al.,* 2019). Other programs, such as TCGAplot, also enable integrative analysis, visualization, and permit differential gene expression studies and correlation studies with tumor mutational burden and microsatellite instability (Liao & Wang, 2023). DL methods have also been utilized to detect cancer-type gene expression patterns, which will serve as possible biomarkers for future targeted therapy (Divate *et al.,* 2021).

DL models have also maximized therapy selection in TNBC using multi-modal data fusion. The Self-Normalizing Balanced (SNB) model can predict patients' chemotherapy eligibility and is more effective than conventional clinical guidelines (Yang *et al.,* 2024b). Secondary learning frameworks like ARML and AdaLoss improve pCR prediction by reweighting secondary tasks, thus allowing appropriate treatment response (Moon & Zhang, 2024). DL combined with radiomics and MRI data makes it easier to predict and assess treatment response and prognosis without invasive procedures. It also makes it possible to predict patient-specific neoadjuvant treatment outcomes prior to intervention using MRI-based digital twins (Stowers *et al.,* 2023; Cheng *et al.,* 2025). DL multi-omics models have better prediction of TNBC subtype, exemplifying the merit of integrated intensive molecular analysis in the more discriminatory clinical outcome optimization (Yang *et al.,* 2024b).

For Glioblastoma (GBM), AI-driven multi-omics models have refined the invasive subtype classification by incorporating transcriptomic, methylomic, and proteomic information. Transcriptome profiling detects invasion-associated gene sets in bad-prognosis subgroups and cell cycle-associated genes in good-prognosis subtypes (Park *et al.*, 2019). ML networks have also made discoveries of the master regulators of GBM subtypes such as *PKCδ* and *DNA-PK*, which have been validated by complementing omics analysis (Migliozzi *et al.*, 2023). DeepAutoGlioma is constructed using autoencoders to combine transcriptome and methylome data and project them into a lower-dimensional space for high-accuracy subtype prediction (Munquad & Das, 2023). Multimodal models that combine CNNs, radiomics, and semantic features optimize subtype and prognosis prediction even more, predicting key mutations and methylation patterns, with transcription factors *STAT3* and *SRF* associated with invasiveness and poor prognosis (Park *et al.*, 2019).

AI-powered omics analysis also contributes to CAR-T therapy personalization by deconstructing the tumor immune microenvironment (TIME). The combination of genomics, proteomics, and AI allows the discovery of novel biomarkers and tailored treatment regimens, overcoming TIME heterogeneity that usually thwarts CAR-T efficacy, especially against solid tumors (Zhou *et al.*, 2024b). ML models deliver individualized counterfactual treatment recommendations, optimizing patient selection while avoiding side effects (Schürch *et al.*, 2024; Limeta *et al.*, 2023). AI algorithms regulated by multi-omics are able to predict ideal treatment regimens with continuous treatment optimization through decision-support systems and feedback loops (Vegesna, 2024). Through solving data integration and treatment assignment bias problems, these AI-driven methods allow clinicians to customize CAR-T therapy based on patient molecular profiles to maximize efficacy and advance precision oncology (Zhou *et al.*, 2024b).

6. CHALLENGES, LIMITATIONS, AND ETHICAL CONSIDERATIONS

The integration of multi-omics with AI is marred by a number of key challenges that impact model performance and clinical utility. The initial of them is heterogeneity and high dimensionality of the omics datasets, which makes it difficult to extract biologically relevant information along with defining causal relationships between genotypes and phenotypes. This is the lack of annotated data that acts as a barrier to the capacity of ML algorithms to generalize across various biological contexts, confining their predictive potential (Wu & Xie, 2025). Also, missing or incomplete data sets, in most cases caused by the limitations of experimental design or budgetary issues, is a refined deterrent because several AI methods require input data in complete (Flores *et al.*, 2023). Although DL models can uncover intricate, non-linear interactions within multi-omics data, their interpretability continues to be limited, especially when forecasting information into patient treatment (Gonesh *et al.*, 2023). Practical and ethical considerations like handling sensitive health data and introducing streamlined longitudinal sampling add to challenges in multi-omics integration (Mohr *et al.*, 2024). Notwithstanding these challenges, AI-based methods have shown promise in improving disease prediction efficiency and pinpointing major genetic loci contributing to cancer, propelling precision oncology forward (Xin *et al.*, 2024).

Explainability and reproducibility of AI-based cancer models must be ensured for their application in a clinical setting. XAI techniques such as Local Interpretable Model-agnostic Explanations (LIME), Shapley Additive Explanations (SHAP), and Grad-CAM offer human-interpretable explanations of model output. SHAP, among others, is specifically noteworthy for being model-agnostic, thus it can explain predictions for a wide range of algorithms, including ensemble tree-based algorithms commonly used in oncology (Ghasemi *et al.*, 2024). In digital pathology, the shift from traditional microscopy to whole-slide image analysis by AI requires interpretable models to support diagnostic confidence and pathologists' trust (Faa *et al.*, 2024). Combining with domain knowledge and visualization methods supports clinical interpretability of AI predictions (Damaševičius, 2024). Reproducibility is also of similar concern, solved using standardized measures and validation procedures. Tree ensemble models like boosted forests are used in research employing electronic health records for treatment failure prediction due to their consistent rule generation and ensuring interpretability across heterogeneous data (Farooq & Shahid, 2023).

AI-driven multi-omics cancer research also generates ethical and privacy concerns that need to be addressed diligently. The "black box" quality of the majority of AI models creates transparency issues, potentially impacting clinical trust and accountability. Biased AI algorithms, internal to the system, can reinforce health inequities and underserve minority groups (Shreve *et al.,* 2022). Additionally, dealing with patient data of a sensitive nature poses privacy threats, most notably in training the models, data exchange, and publication. Confidentiality and integrity of data must be ensured to preserve patient trust (Zhou *et al.,* 2024a). These ethical frameworks should also maintain patient dignity, autonomy, safety, equity, and inclusivity, so that applications of AI in oncology do not perpetuate entrenched inequalities or undermine patient rights (Kelkar *et al.,* 2024). More specifically, for malignant gliomas, the incorporation of high-throughput multi-omics data and AI is technically as well as ethically demanding, such as coping with intratumor heterogeneity, creating stable biomarkers, and protecting against ethical abuse of patient-derived information (Tambi *et al.,* 2024).

7. FUTURE DIRECTIONS

Converging with radiomics, digital pathology, and multi-modal AI are revolutionizing precision oncology by facilitating the extraction and analysis of high-dimensional data from heterogeneous inputs, such as imaging, pathology, and molecular data. This facilitates the discovery of prognostic and predictive biomarkers and, hence, personalized treatment guidance and better clinical outcomes (Li, 2025). AI's application scope is broad and includes immunogenomics, radiomics, and pathomics, wherein AI analyzes high-scale genomic and multi-omic data to discover novel biomarkers, optimized immunotherapy regimens, and efficacy of therapeutic interventions (Chang *et al.,* 2025). Examples include models like BioGPT and AlphaFold represent the potential of AI to analyze large biomedical datasets and make sense of them. These models support the discovery of driver cancer mutations and pathways, enabling targeted therapy development and progress toward personalized cancer modeling. In addition, the convergence of AI with quantum computing and synthetic biology will be poised to expand next-generation cancer treatment. Quantum AI, taking advantage of increased computational capability, will accelerate genomic screening, drug design, and optimization of patient-specific therapy approaches, revolutionizing diagnostics and immunotherapy design. Multi-modal fusion of datasets, including pathological images with genomic and clinical data, is required to advance the understanding of tumor heterogeneity and improve survival prediction models. Algorithms such as PONET, for example, make use of multi-modal learning to predict more accurately and gain mechanistic insights into the disease biology. Integration enables characterization of tumor subtypes, detection of high-risk patients, and personalization of treatment protocols. Additionally, digital pathology, when enhanced by multi-modal AI strategies, presents an integrated comprehension of the intricate nature of the cancer entity, contributing to diagnostic accuracy, therapy planning, and prognosis evaluation by combining clinical, radiological, and molecular data synergistically (Waqas *et al.,* 2024).

One of the most exciting advances in the field of precision medicine includes the development of biomedical foundation AI models that learn general biological representation through large multimodal datasets. In contrast to task-specific AI models, such an approach can be used in various downstream tasks, including biomarker discovery, treatment response predictions, and patient stratifications. Future research should be concentrated on building an integration of foundation models with multimodal omics data, digital pathology, radiomics, and longitudinal clinical data.

8. CONCLUSION

The intersection of AI and multi-omics is a revolution in precision oncology, evolving from descriptive cancer biology to actionable personalized treatment. The convergence of genomics, transcriptomics, proteomics, metabolomics, and immunomics through sophisticated computational models allows novel biomarkers to be identified, patient stratification at high resolution, and prediction of treatment response in real-time. ML-based solutions like DL, VAEs, GNNs, and XAI frameworks allow thorough exploration of high-dimensional data, refine interpretability, and ease clinical translation. Emphatically, their applications in integrating liquid biopsies, developing immunotherapies, drug repurposing, and adaptive clinical trials promise their revolutionary application in oncology. Although heterogeneity of data, ethical dilemmas, and bias in algorithms are formidable challenges, current advancements in federated learning, causal inference, and multimodal representation learning promise to surmount these challenges. Future possibilities of the integration of quantum AI, digital pathology, and multi-modal imaging will continue to refine personalized cancer treatment. Finally, AI-governed multi-omics enables oncology to be patient-focused, in which the therapy is optimized for efficacy, safety, and long-term survival.

DATA AVAILABILITY

The authors confirm that the data supporting the findings of this study are available within the book/chapter.

ACKNOWLEDGEMENT

All the authors are thankful to their respective Universities and Institutions for their valuable supports.

REFERENCES

Ahmad, Z. (2018). Flux balance analysis identifies distinct NADPH production strategies across NCI 60 cancer cell lines. *bioRxiv*, 332262.

Akter, S. (2024). AI-driven precision medicine: Transforming personalized cancer treatment. *JAPMI*, *2*(1), 10–21. DOI: 10.60087/vol2iisue1.p21

Alexandrov, L. B., Kim, J., Haradhvala, N. J., Huang, M. N., Tian Ng, A. W., Wu, Y., & Stratton, M. R. (2020). The repertoire of mutational signatures in human cancer. *Nature*, *578*(7793), 94–101. DOI: 10.1038/s41586-020-1943-3

Bamodu, O. A., Chung, C. C., & Pisanic, T. R.II. (2023). Harnessing liquid biopsies: Exosomes and ctDNA as minimally invasive biomarkers for precision cancer medicine. *The Journal of Liquid Biopsy*, *2*, 100126. DOI: 10.1016/j.jlb.2023.100126

Barbadilla-Martínez, L., Klaassen, N., Franceschini-Santos, V. H., Breda, J., Hernandez-Quiles, M., van Lieshout, T., & van Steensel, B. (2024). The regulatory grammar of human promoters uncovered by MPRA-trained deep learning. bioRxiv, 2024-07.

Becchi, T., Beltrame, L., Mannarino, L., Calura, E., Marchini, S., & Romualdi, C. (2023). A pan-cancer landscape of pathogenic somatic copy number variations. *Journal of Biomedical Informatics*, *147*, 104529. DOI: 10.1016/j.jbi.2023.104529

Bifarin, O. O., & Fernández, F. M. (2024). Automated machine learning and explainable AI (AutoML-XAI) for metabolomics: Improving cancer diagnostics. *Journal of the American Society for Mass Spectrometry*, *35*(6), 1089–1100. DOI: 10.1021/jasms.3c00403

Biggar, K., Ridgeway, N., Chopra, A., Lukinovic, V., Feldman, M., Charih, F., & Green, J. (2024). Machine learning-based exploration of enzyme-substrate networks: SET8-mediated methyllysine and its changing impact within cancer proteomes. https://doi.org/DOI: 10.21203/rs.3.rs-3771179/v1

Bulashevska, A., Nacsa, Z., Lang, F., Braun, M., Machyna, M., Diken, M., & König, R. (2024). Artificial intelligence and neoantigens: Paving the path for precision cancer immunotherapy. *Frontiers in Immunology*, *15*, 1394003. DOI: 10.3389/fimmu.2024.1394003

Cai, Y., & Wang, S. (2024). Deeply integrating latent consistent representations in high-noise multi-omics data for cancer subtyping. *Briefings in Bioinformatics*, *25*(2), bbae061. DOI: 10.1093/bib/bbae061

Cao, J., Peek, N., Renehan, A., & Ananiadou, S. (2023). Gaussian distributed prototypical network for few-shot genomic variant detection. In *The 22nd Workshop on Biomedical Natural Language Processing and BioNLP Shared Tasks* (pp. 26-36). DOI: 10.18653/v1/2023.bionlp-1.2

Cao, Y. H., Ding, J., Tang, Q. H., Zhang, J., Huang, Z. Y., Tang, X. M., & Fu, D. (2022). Deciphering cell–cell interactions and communication in the tumor microenvironment and unraveling intratumoral genetic heterogeneity via single-cell genomic sequencing. *Bioengineered, 13*(7-12), 14974–14986. DOI: 10.1080/21655979.2023.2185434

Chakraborty, D., Gutierrez-Chakraborty, E., Rodriguez-Aguayo, C., Başağaoğlu, H., Lopez-Berestein, G., & Amero, P. (2024). Discovering genetic biomarkers for targeted cancer therapeutics with eXplainable Artificial Intelligence. *Cancer Communications (London, England), 44*(5), 584–588. DOI: 10.1002/cac2.12530

Chang, L., Liu, J., Zhu, J., Guo, S., Wang, Y., Zhou, Z., & Wei, X. (2025). Advancing precision medicine: The transformative role of artificial intelligence in immunogenomics, radiomics, and pathomics for biomarker discovery and immunotherapy optimization. *Cancer Biology & Medicine, 22*(1), 33–47. DOI: 10.20892/j.issn.2095-3941.2024.0376

Chen, H., Fang, X., Shao, J., Zhang, Q., Xu, L., Chen, J., & Guo, G. (2024). Pan-Cancer Single-Nucleus Total RNA Sequencing Using snHH-Seq. *Advanced Science (Weinheim, Baden-Württemberg, Germany), 11*(5), 2304755. DOI: 10.1002/advs.202304755

Chen, Z., Zhu, L., Yang, Z., & Matsubara, T. (2022). Automated cancer subtyping via vector quantization mutual information maximization. In *Joint European Conference on Machine Learning and Knowledge Discovery in Databases* (pp. 88-103). Cham: Springer International Publishing.

Cheng, C., Wang, Y., Zhao, J., Wu, D., Li, H., & Zhao, H. (2025). Deep learning and radiomics in triple-negative breast cancer: Predicting long-term prognosis and clinical outcomes. *Journal of Multidisciplinary Healthcare, 18*, 319–327. DOI: 10.2147/JMDH.S509004

Cho, H. J., Wang, Z., Cong, Y., Bekiranov, S., Zhang, A., & Zang, C. (2024). DARDN: A Deep-learning approach for CTCF binding sequence classification and oncogenic regulatory feature discovery. *Genes, 15*(2), 144. DOI: 10.3390/genes15020144

Corces, M. R., Granja, J. M., Shams, S., Louie, B. H., Seoane, J. A., Zhou, W., & Chang, H. Y. (2018). The chromatin accessibility landscape of primary human cancers. *Science, 362*(6413), eaav1898.

Coupland, S. E., Sonntag, S. R., Heimann, H., & Grisanti, S. (2024). The concept of liquid biopsy in the treatment of malignant eye tumors. *Die Ophthalmologie, 121*(12), 946–953. DOI: 10.1007/s00347-024-02132-3

Damane, B. P., Mkhize-Kwitshana, Z. L., Kgokolo, M. C., Luvhengo, T., & Dlamini, Z. (2023). Applying artificial intelligence prediction tools for advancing precision oncology in immunotherapy: future perspectives in personalized care. In *Artificial intelligence and precision oncology: bridging cancer research and clinical decision support* (pp. 239–258). Springer Nature Switzerland. DOI: 10.1007/978-3-031-21506-3_12

Damaševičius, R. (2024). Explainable artificial intelligence methods for breast cancer recognition. *Innov Discov, 1*(3), 25. DOI: 10.53964/id.2024025

Dar, K. B., Bhat, A. H., Amin, S., Anjum, S., Reshi, B. A., Zargar, M. A., & Ganie, S. A. (2019). Exploring proteomic drug targets, therapeutic strategies and protein-protein interactions in cancer: Mechanistic view. *Current Cancer Drug Targets*, *19*(6), 430–448. DOI: 10.2174/1568009618666180803104631

De Vries, N. L., Mahfouz, A., Koning, F., & De Miranda, N. F. (2020). Unraveling the complexity of the cancer microenvironment with multidimensional genomic and cytometric technologies. *Frontiers in Oncology*, *10*, 1254. DOI: 10.3389/fonc.2020.01254

Dee, W., Alaaeldin Ibrahim, R., & Marouli, E. (2024). Histopathological domain adaptation with generative adversarial networks: Bridging the domain gap between thyroid cancer histopathology datasets. *PLoS One*, *19*(12), e0310417. DOI: 10.1371/journal.pone.0310417

Dienstmann, R., Vermeulen, L., Guinney, J., Kopetz, S., Tejpar, S., & Tabernero, J. (2017). Consensus molecular subtypes and the evolution of precision medicine in colorectal cancer. *Nature Reviews. Cancer*, *17*(2), 79–92. DOI: 10.1038/nrc.2016.126

Divate, M., Tyagi, A., Richard, D. J., Prasad, P. A., Gowda, H., & Nagaraj, S. H. (2021). Deep learning-based pan-cancer classification model reveals cancer-specific gene expression signatures. bioRxiv, 2021-03.

Everall, A., Tapinos, A., Hawari, A., Cornish, A., Sud, A., Chubb, D., & Wedge, D. C. (2023). Comprehensive repertoire of the chromosomal alteration and mutational signatures across 16 cancer types from 10,983 cancer patients. MedRxiv, 2023-06. DOI: 10.1101/2023.06.07.23290970

Faa, G., Fraschini, M., & Barberini, L. (2024). Reproducibility and explainability in digital pathology: The need to make black-box artificial intelligence systems more transparent. *Journal of Public Health Research*, *13*(4), 22799036241284898. DOI: 10.1177/22799036241284898

Farooq, M., & Shahid, M. U. (2023). Explainable AI and machine learning algorithms to predict treatment failures for patients with cancer. *Journal of Clinical Oncology : Official Journal of the American Society of Clinical Oncology*, *41*(16), e13577–e13577. DOI: 10.1200/JCO.2023.41.16_suppl.e13577

Fatemeh, S., Mahboobeh, Z., Khadijeh, A., Amirhossein, M. K., & Pegah, M. (2024). An in-silico study to determine susceptibility to cancer by evaluating the coding and non-coding non-synonymous single nucleotide variants in the SOCS3 gene. *Journal of Biomolecular Structure & Dynamics*, *42*(16), 8281–8292. DOI: 10.1080/07391102.2023.2256408

Flores, J. E., Claborne, D. M., Weller, Z. D., Webb-Robertson, B. J. M., Waters, K. M., & Bramer, L. M. (2023). Missing data in multi-omics integration: Recent advances through artificial intelligence. *Frontiers in Artificial Intelligence*, *6*, 1098308. DOI: 10.3389/frai.2023.1098308

Ganatra, H., Tan, J. K., Simmons, A., Bigogno, C. M., Khurana, V., Ghose, A., & Ayodele, O. (2024). Applying whole-genome and whole-exome sequencing in breast cancer: A review of the landscape. *Breast Cancer (Tokyo, Japan)*, *31*(6), 999–1009. DOI: 10.1007/s12282-024-01628-9

Gardner, S., Das, S., & Taylor, K. (2020). AI enabled precision medicine: patient stratification, drug repurposing and combination therapies. In *Artificial intelligence in oncology drug discovery and development*. IntechOpen. DOI: 10.5772/intechopen.92594

Garland, J. (2017). Unravelling the complexity of signalling networks in cancer: A review of the increasing role for computational modelling. *Critical Reviews in Oncology/Hematology*, *117*, 73–113. DOI: 10.1016/j.critrevonc.2017.06.004

Gatto, F., Miess, H., Schulze, A., & Nielsen, J. (2015). Flux balance analysis predicts essential genes in clear cell renal cell carcinoma metabolism. *Scientific Reports*, *5*(1), 10738. DOI: 10.1038/srep10738

Ghasemi, A., Hashtarkhani, S., Schwartz, D. L., & Shaban-Nejad, A. (2024). Explainable artificial intelligence in breast cancer detection and risk prediction: A systematic scoping review. *Cancer Innovation*, *3*(5), e136. DOI: 10.1002/cai2.136

Gomari, D. P., Schweickart, A., Cerchietti, L., Paietta, E., Fernandez, H., Al-Amin, H., & Krumsiek, J. (2022). Variational autoencoders learn transferrable representations of metabolomics data. *Communications Biology*, *5*(1), 645. DOI: 10.1038/s42003-022-03579-3

Gonesh, C., Saha, M. B., Khan, M. J., Saha, H., & Avinash, D. K. (2023). Integrative analysis of multi-omics data with deep learning: Challenges and opportunities in bioinformatics. *J. Propuls. Technol*, *44*, 1384–1389.

Hafstað, V., Häkkinen, J., Larsson, M., Staaf, J., Vallon-Christersson, J., & Persson, H. (2023). Improved detection of clinically relevant fusion transcripts in cancer by machine learning classification. *BMC Genomics*, *24*(1), 783. DOI: 10.1186/s12864-023-09889-y

Hamdan, A., & Ewing, A. (2022). Unravelling the tumour genome: The evolutionary and clinical impacts of structural variants in tumourigenesis. *The Journal of Pathology*, *257*(4), 479–493. DOI: 10.1002/path.5901

Hashim, S., Ali, M., Nandakumar, K., & Yaqub, M. (2022). SubOmiEmbed: Self-supervised representation learning of multi-omics data for cancer type classification. In *2022 10th International Conference on Bioinformatics and Computational Biology (ICBCB)* (pp. 66-72). IEEE.

Hassane, M. (2024). Artificial intelligence-driven precision medicine in cancer treatment. *Science*, *1*, 100041.

Ho, H. Y., Chung, K. S., Kan, C. M., & Wong, S. C. (2024). Liquid biopsy in the clinical management of cancers. *International Journal of Molecular Sciences*, *25*(16), 8594. DOI: 10.3390/ijms25168594

Huang, D., Ma, N., Li, X., Gou, Y., Duan, Y., Liu, B., & Zhang, X. (2023). Advances in single-cell RNA sequencing and its applications in cancer research. *Journal of Hematology & Oncology*, *16*(1), 98. DOI: 10.1186/s13045-023-01494-6

Idumah, G., Li, L., Yehia, L., Mahdi, H., & Ni, Y. (2025). Metabolic signatures of immune checkpoint inhibitor response in gynecologic cancers: insights from flux balance analysis. bioRxiv, 2025-01.

Islam, S., & Hasan, M. N. (2024). Personalized graph feature-based multi-omics data integration for cancer subtype identification. *arXiv preprint arXiv:2408.08832*.

Jenna, S., Boucher, B., Dudragne, L., & Diallo, A. B. (2022). Augmented intelligence to define drug targets associated with triple negative breast cancer (TNBC) metabolic reprogramming. *Cancer Research*, *82*(12, Supplement), 6375–6375. DOI: 10.1158/1538-7445.AM2022-6375

Jiang, Y., Wang, C., & Zhou, S. (2023). Artificial intelligence-based risk stratification, accurate diagnosis and treatment prediction in gynecologic oncology. [). Academic Press.]. *Seminars in Cancer Biology, 96*, 82–99. DOI: 10.1016/j.semcancer.2023.09.005

Kato, H., Iwadate, D., Yamamoto, K., Fujiwara, H., Ijichi, H., Fujishiro, M., & Tateishi, K. (2022). Hi-C analysis in patient-derived organoids reveals the impact of three-dimensional chromatin structures on pancreatic cancer transcriptional subtypes. *Cancer Research, 82*(22), 47–48. DOI: 10.1158/1538-7445.PANCA22-A069

Kelkar, A. H., Hantel, A., Koranteng, E., Cutler, C. S., Hammer, M. J., & Abel, G. A. (2024). Digital health to patient-facing artificial intelligence: Ethical implications and threats to dignity for patients with cancer. *JCO Oncology Practice, 20*(3), 314–317. DOI: 10.1200/OP.23.00412

Keyl, J., Keyl, P., Montavon, G., Hosch, R., Brehmer, A., Mochmann, L., & Kleesiek, J. (2025). Decoding pan-cancer treatment outcomes using multimodal real-world data and explainable artificial intelligence. *Nature Cancer, 6*(2), 307–322. DOI: 10.1038/s43018-024-00891-1

Kim, E., Duarte, S. E., Yu, E., Hong, I., Lee, G., & Chae, Y. K. (2024). Evaluation of Artificial Intelligence (AI)-based in silico tools for variant classification in clinically actionable NSCLC variants. medRxiv, 2024-04.

Lan, J., Chen, W., Yin, G., Haque, A. A., Xie, K., Yu, Q., & Yuan, X. (2023). LDSSNV: A linkage disequilibrium-based method for the detection of somatic single-nucleotide variants. *IEEE/ACM Transactions on Computational Biology and Bioinformatics, 20*(5), 3020–3032. DOI: 10.1109/TCBB.2023.3291134

Lange, M., Bergen, V., Klein, M., Setty, M., Reuter, B., Bakhti, M., & Theis, F. J. (2022). CellRank for directed single-cell fate mapping. *Nature Methods, 19*(2), 159–170. DOI: 10.1038/s41592-021-01346-6

Lee, S. M., Lee, G., Lee, S., Song, H., Yoon, S. S., Yun, H., & Kim, H. U. (2022). A computational model of cellular metabolism with mutation data predicts metabolites associated with somatic mutations in cancers. *Cancer Research, 82*(12, Supplement), 2733–2733. DOI: 10.1158/1538-7445.AM2022-2733

Li, B., & Nabavi, S. (2024). A multimodal graph neural network framework for cancer molecular subtype classification. *BMC Bioinformatics, 25*(1), 27. DOI: 10.1186/s12859-023-05622-4

Li, Q., Ren, Z., Cao, K., Li, M. M., Wang, K., & Zhou, Y. (2022). CancerVar: An artificial intelligence–empowered platform for clinical interpretation of somatic mutations in cancer. *Science advances, 8*(18), eabj1624.

Li, R. (2025). Abstract IA05: Toward multi-modal foundation AI for precision oncology. *Clinical Cancer Research, 31*(2, Supplement), IA05–IA05. DOI: 10.1158/1557-3265.TARGETEDTHERAP-IA05

Li, Y., Shi, W., & Wasserman, W. W. (2018). Genome-wide prediction of cis-regulatory regions using supervised deep learning methods. *BMC Bioinformatics, 19*(1), 1–14. DOI: 10.1186/s12859-018-2187-1

Liang, A., Kong, Y., Chen, Z., Qiu, Y., Wu, Y., Zhu, X., & Li, Z. (2024). Advancements and applications of single-cell multi-omics techniques in cancer research: Unveiling heterogeneity and paving the way for precision therapeutics. *Biochemistry and Biophysics Reports, 37*, 101589. DOI: 10.1016/j.bbrep.2023.101589

Liao, C., & Wang, X. (2023). TCGAplot: An R package for integrative pan-cancer analysis and visualization of TCGA multi-omics data. *BMC Bioinformatics*, *24*(1), 483. DOI: 10.1186/s12859-023-05615-3

Limeta, A., Gatto, F., Herrgård, M. J., Ji, B., & Nielsen, J. (2023). Leveraging high-resolution omics data for predicting responses and adverse events to immune checkpoint inhibitors. *Computational and Structural Biotechnology Journal*, *21*, 3912–3919. DOI: 10.1016/j.csbj.2023.07.032

Liu, Y., Yan, Z., Liu, C., Yang, R., Zheng, Q., Jian, J., & Liu, X. (2025). Integrated RNA sequencing analysis and machine learning identifies a metabolism-related prognostic signature in clear cell renal cell carcinoma. *Scientific Reports*, *15*(1), 1691. DOI: 10.1038/s41598-025-85618-7

Lokare, R. R., Wadmare, J., Patil, S., Wadmare, G., & Patil, D. (2024). Transparent precision: Explainable AI empowered breast cancer recommendations for personalized treatment. *IAES Int J Artif Intell*, *13*(3), 2694–2694. DOI: 10.11591/ijai.v13.i3.pp2694-2702

Lu, Y., Peng, R., Dong, L., Xia, K., Wu, R., Xu, S., & Wang, J. (2023). Multiomics dynamic learning enables personalized diagnosis and prognosis for pancancer and cancer subtypes. *Briefings in Bioinformatics*, *24*(6), bbad378. DOI: 10.1093/bib/bbad378

Ma, L., Gao, W., Hu, X., Zhou, D., Wang, C., Yu, J., & Tang, K. (2024). An improved cancer diagnosis algorithm for protein mass spectrometry based on PCA and a one-dimensional neural network combining ResNet and SENet. *The Analyst*, *149*(23), 5675–5683. DOI: 10.1039/D4AN00784K

Ma, Y., & Guan, J. (2022). MOCSC: a multi-omics data based framework for cancer subtype classification. In *2022 IEEE international conference on bioinformatics and biomedicine (BIBM)* (pp. 2853-2859). IEEE. DOI: 10.1109/BIBM55620.2022.9995564

Massimino, M., Martorana, F., Stella, S., Vitale, S. R., Tomarchio, C., Manzella, L., & Vigneri, P. (2023). Single-cell analysis in the omics era: Technologies and applications in cancer. *Genes*, *14*(7), 1330. DOI: 10.3390/genes14071330

Masuda, K., Sota, Y., & Matsuda, H. (2024). Gene fusion detection in long-read transcriptome datasets from multiple cancer cell lines. *Frontiers in Bioscience (Landmark Edition)*, *29*(12), 413. DOI: 10.31083/j.fbl2912413

Matsuno, Y., Kusumoto-Matsuo, R., Manaka, Y., Asai, H., & Yoshioka, K. I. (2022). Echoed induction of nucleotide variants and chromosomal structural variants in cancer cells. *Scientific Reports*, *12*(1), 20964. DOI: 10.1038/s41598-022-25479-6

Migliozzi, S., Oh, Y. T., Hasanain, M., Garofano, L., D'angelo, F., Najac, R. D., & Iavarone, A. (2023). Integrative multi-omics networks identify PKCδ and DNA-PK as master kinases of glioblastoma subtypes and guide targeted cancer therapy. *Nature Cancer*, *4*(2), 181–202. DOI: 10.1038/s43018-022-00510-x

Mohr, A. E., Ortega-Santos, C. P., Whisner, C. M., Klein-Seetharaman, J., & Jasbi, P. (2024). Navigating challenges and opportunities in multi-omics integration for personalized healthcare. *Biomedicines*, *12*(7), 1496. DOI: 10.3390/biomedicines12071496

Moon, C. I., & Zhang, B. (2024). Optimizing neoadjuvant treatment response prediction for triple-negative breast cancer using clinical trial data and deep auxiliary learning. bioRxiv, 2024-11.

Moreno-Sánchez, R., Saavedra, E., Gallardo-Pérez, J. C., Rumjanek, F. D., & Rodríguez-Enríquez, S. (2016). Understanding the cancer cell phenotype beyond the limitations of current omics analyses. *The FEBS Journal, 283*(1), 54–73. DOI: 10.1111/febs.13535

Munquad, S., & Das, A. B. (2023). DeepAutoGlioma: A deep learning autoencoder-based multi-omics data integration and classification tools for glioma subtyping. *BioData Mining, 16*(1), 32. DOI: 10.1186/s13040-023-00349-7

Murcia-Mejía, M., Canela-Capdevila, M., García-Pablo, R., Jiménez-Franco, A., Jiménez-Aguilar, J. M., Badía, J., & Joven, J. (2024). Combining metabolomics and machine learning to identify diagnostic and prognostic biomarkers in patients with non-small cell lung cancer pre-and post-radiation therapy. *Biomolecules, 14*(8), 898. DOI: 10.3390/biom14080898

Narushima, Y., Kozuka-Hata, H., Tsumoto, K., Inoue, J. I., & Oyama, M. (2016). Quantitative phosphoproteomics-based molecular network description for high-resolution kinase-substrate interactome analysis. *Bioinformatics (Oxford, England), 32*(14), 2083–2088. DOI: 10.1093/bioinformatics/btw164

Nazaretyan, L., Rentzsch, P., & Kircher, M. (2024). varCADD: large sets of standing genetic variation enable genome-wide pathogenicity prediction. bioRxiv, 2024-09.

Nicora, G., Zucca, S., Limongelli, I., Bellazzi, R., & Magni, P. (2022). A machine learning approach based on ACMG/AMP guidelines for genomic variant classification and prioritization. *Scientific Reports, 12*(1), 2517. DOI: 10.1038/s41598-022-06547-3

Olawade, D. B., Clement David-Olawade, A., Adereni, T., Egbon, E., Teke, J., & Boussios, S. (2025). Integrating ai into cancer immunotherapy-a narrative review of current applications and future directions. *Diseases (Basel, Switzerland), 13*(1), 24. DOI: 10.3390/diseases13010024

Olumuyiwa, B. I., Han, T. A., & Shamszaman, Z. U. (2024). Enhancing cancer diagnosis with explainable & trustworthy deep learning models. *arXiv preprint arXiv:2412.17527.*

Oni, O. (2019). *Multi-Platform Genomic Data Fusion with Integrative Deep Learning* (Doctoral dissertation).

Papatsirou, M., Artemaki, P. I., Karousi, P., Scorilas, A., & Kontos, C. K. (2021). Circular RNAs: Emerging regulators of the major signaling pathways involved in cancer progression. *Cancers (Basel), 13*(11), 2744. DOI: 10.3390/cancers13112744

Park, J., Shim, J. K., Yoon, S. J., Kim, S. H., Chang, J. H., & Kang, S. G. (2019). Transcriptome profiling-based identification of prognostic subtypes and multi-omics signatures of glioblastoma. *Scientific Reports, 9*(1), 10555. DOI: 10.1038/s41598-019-47066-y

Peroz, M., Mananet, H., Roussot, N., Kaderbhai, C. G., Derangère, V., Truntzer, C., & Ghiringhelli, F. (2024). Clinical interest in exome-based analysis of somatic mutational signatures for non-small cell lung cancer. *Cancers (Basel), 16*(17), 3115. DOI: 10.3390/cancers16173115

Petursson, B., & Petsalaki, E. (2022). Data-driven extraction of human kinase-substrate relationships from omics datasets.

Pierobon, M., Wulfkuhle, J., Liotta, L. A., & Petricoin, E. F.III. (2012). Development and clinical implementation of reverse phase protein microarrays for protein network activation mapping: personalized cancer therapy. In *Systems Biology in Cancer Research and Drug Discovery* (pp. 309–323). Springer Netherlands. DOI: 10.1007/978-94-007-4819-4_13

Poirion, O. B., Chaudhary, K., Huang, S., & Garmire, L. X. (2020). Multi-omics-based pan-cancer prognosis prediction using an ensemble of deep-learning and machine-learning models. *medRxiv*, 19010082.

Ramakrishnan, A., Wangensteen, G., Kim, S., Nestler, E. J., & Shen, L. (2024). DeepRegFinder: Deep learning-based regulatory elements finder. *Bioinformatics Advances*, *4*(1), vbae007. DOI: 10.1093/bioadv/vbae007

Rebersek, A. P. M. (2020). Consensus molecular subtypes (CMS) in metastatic colorectal cancer-personalized medicine decision. *Radiology and Oncology*, *54*(3), 272–277. DOI: 10.2478/raon-2020-0031

Rohilla, J., Mishra, R., Varshney, S., Mazumder, R., Mazumder, A., Chauhan, A., & Tyagi, P. K. (2024). An insight into signalling pathways in cancer: Hedgehog, PI3K, and notch pathways and therapeutic perspectives. *Letters in Drug Design & Discovery*, *21*(18), 4148–4160. DOI: 10.2174/0115701808353919241126110520

Rossini, R., Oshaghi, M., Nekrasov, M., Bellanger, A., Domaschenz, R., Dijkwel, Y., & Paulsen, J. (2024). Loss of multi-level 3D genome organization during breast cancer progression. bioRxiv, 2023-11.

Rout, T., Mohapatra, A., Kar, M., & Muduly, D. K. (2024). Essential cancer protein identification using graph-based random walk with restart. *Computer Methods in Biomechanics and Biomedical Engineering*, •••, 1–14.

Russo, V., Lallo, E., Munnia, A., Spedicato, M., Messerini, L., D'Aurizio, R., & Peluso, M. (2022). Artificial intelligence predictive models of response to cytotoxic chemotherapy alone or combined to targeted therapy for metastatic colorectal cancer patients: A systematic review and meta-analysis. *Cancers (Basel)*, *14*(16), 4012. DOI: 10.3390/cancers14164012

Sahu, A., Blätke, M. A., Szymański, J. J., & Töpfer, N. (2021). Advances in flux balance analysis by integrating machine learning and mechanism-based models. *Computational and Structural Biotechnology Journal*, *19*, 4626–4640. DOI: 10.1016/j.csbj.2021.08.004

Schubach, M., Maass, T., Nazaretyan, L., Röner, S., & Kircher, M. (2024). CADD v1. 7: Using protein language models, regulatory CNNs and other nucleotide-level scores to improve genome-wide variant predictions. *Nucleic Acids Research*, *52*(D1), D1143–D1154. DOI: 10.1093/nar/gkad989

Schürch, M., Boos, L., Heinzelmann-Schwarz, V., Gut, G., Krauthammer, M., Wicki, A., & the Tumor Profiler Consortium. (2024). Towards AI-based precision oncology: a machine learning framework for personalized counterfactual treatment suggestions based on multi-omics data. *arXiv preprint arXiv:2402.12190.*

Shao, Y., Lv, X., Ying, S., & Guo, Q. (2024). Artificial intelligence-driven precision medicine: Multi-omics and spatial multi-omics approaches in Diffuse Large B-Cell Lymphoma (DLBCL). *Frontiers in Bioscience (Landmark Edition)*, *29*(12), 404. DOI: 10.31083/j.fbl2912404

Shariaty, F., Pavlov, V. A., Fedyashina, S. V., & Serebrennikov, N. A. (2024). Integrating deep learning and explainable AI for non-invasive prediction of EGFR and KRAS mutations in NSCLC: a novel radiogenomic approach. In *2024 V International Conference on Neural Networks and Neurotechnologies (NeuroNT)* (pp. 32-35). IEEE. DOI: 10.1109/NeuroNT62606.2024.10585441

Shiva, S., & Kalluri, H. K. (2024). Lung Cancer Detection Using Fusion-Based Deep Learning Techniques. In *2024 IEEE Students Conference on Engineering and Systems (SCES)* (pp. 1-5). IEEE. DOI: 10.1109/SCES61914.2024.10652562

Shreve, J. T., Khanani, S. A., & Haddad, T. C. (2022). Artificial intelligence in oncology: Current capabilities, future opportunities, and ethical considerations. In *American Society of Clinical Oncology Educational book. American Society of Clinical Oncology. Annual Meeting* (Vol. 42, pp. 1-10).

Silvestris, N., Ciliberto, G., De Paoli, P., Apolone, G., Lavitrano, M. L., Pierotti, M. A., & Stanta, G. (2017). Liquid dynamic medicine and N-of-1 clinical trials: A change of perspective in oncology research. *Journal of Experimental & Clinical Cancer Research : CR*, *36*(1), 128. DOI: 10.1186/s13046-017-0598-x

Singha, M., Pu, L., Stanfield, B. A., Uche, I. K., Rider, P. J., Kousoulas, K. G., & Brylinski, M. (2022). Artificial intelligence to guide precision anticancer therapy with multitargeted kinase inhibitors. *BMC Cancer*, *22*(1), 1211. DOI: 10.1186/s12885-022-10293-0

Soussi, T., Taschner, P. E., & Samuels, Y. (2017). Synonymous somatic variants in human cancer are not infamous: A plea for full disclosure in databases and publications. *Human Mutation*, *38*(4), 339–342. DOI: 10.1002/humu.23163

Steele, C. D., Abbasi, A., Islam, S. A., Bowes, A. L., Khandekar, A., Haase, K., & Pillay, N. (2022). Signatures of copy number alterations in human cancer. *Nature*, *606*(7916), 984–991. DOI: 10.1038/s41586-022-04738-6

Stowers, C. E., Wu, C., Kumar, S., Lima, E. A., Zu, X., Yam, C., & Yankeelov, T. E. (2023). Developing MRI-based digital-twins via mathematical modeling and deep learning to predict the response of triple-negative breast cancer to neoadjuvant therapy. *Cancer Research*, *83*(7, Supplement), 845–845. DOI: 10.1158/1538-7445.AM2023-845

Sun, Y., Qiao, G., Gao, B., & Li, Y. (2024). DAPM-CDR: A domain adaptation prompting model for drug response prediction. *Future Generation Computer Systems*, *160*, 316–324. DOI: 10.1016/j.future.2024.06.009

Tambi, R., Zehra, B., Vijayakumar, A., Satsangi, D., Uddin, M., & Berdiev, B. K. (2024). Artificial intelligence and omics in malignant gliomas. *Physiological Genomics*, *56*(12), 876–895. DOI: 10.1152/physiolgenomics.00011.2024

Tao, Y., Rajaraman, A., Cui, X., Cui, Z., Eaton, J., Kim, H., & Schwartz, R. (2019). Improving personalized prediction of cancer prognoses with clonal evolution models. *bioRxiv*, 761510.

Tian, X., & Liu, Z. (2024). Single nucleotide variants in lung cancer. *Chinese Medical Journal Pulmonary and Critical Care Medicine*, *2*(2), 88–94. DOI: 10.1016/j.pccm.2024.04.004

Trilla-Fuertes, L., Gámez-Pozo, A., Díaz-Almirón, M., Prado-Vázquez, G., Zapater-Moros, A., López-Vacas, R., & Fresno Vara, J. A. (2019). Computational metabolism modeling predicts risk of distant relapse-free survival in breast cancer patients. *Future Oncology*, *15*(30), 3483–3490. DOI: 10.2217/fon-2018-0698

Tsou, P., & Wu, C. J. (2020). Seeing glycolysis on PDAC: Applying deep learning convolutional neural network model. *Cancer Research*, *80*(16, Supplement), 2107–2107. DOI: 10.1158/1538-7445.AM2020-2107

Turabi, K., Klute, K., & Radhakrishnan, P. (2024). Decoding the dynamics of circulating tumor DNA in liquid biopsies. *Cancers (Basel)*, *16*(13), 2432. DOI: 10.3390/cancers16132432

van Belzen, I. A., van Tuil, M., Badloe, S., Janse, A., Verwiel, E. T., Santoso, M., & Hehir-Kwa, J. Y. (2024). Complex structural variation is prevalent and highly pathogenic in pediatric solid tumors. *Cell Genomics*, *4*(11), 100675. DOI: 10.1016/j.xgen.2024.100675

Varpe, A., Bhujbal, R., & Kumar, R. (2024). Enhancing predictive power in personalized breast cancer treatment through the integration of multi-omics data and clinical information with deep learning models. In *2024 MIT Art, Design and Technology School of Computing International Conference (MITADTSoCiCon)* (pp. 1-7). IEEE. DOI: 10.1109/MITADTSoCiCon60330.2024.10575890

Vegesna, V. V. (2024). Medicine: A framework for tailored cancer treatment. *International journal of innovative research in advanced engineering, 6*(11), 747-752.

Vidal, E., Sayols, S., Moran, S., Guillaumet-Adkins, A., Schroeder, M. P., Royo, R., & Esteller, M. (2017). A DNA methylation map of human cancer at single base-pair resolution. *Oncogene*, *36*(40), 5648–5657. DOI: 10.1038/onc.2017.176

Viduedo, P. H. Z., Candeia, V. C. U., Martins, V. B. L., Destefani, A. C., & Destefani, V. C. (2024). Harnessing the power of AI and machine learning for next-generation sequencing data analysis: A comprehensive review of applications, challenges, and future directions in precision oncology. *Revista Ibero-Americana de Humanidades. Ciência & Educação (Bauru)*, *10*(8), 2898–2904.

Wang, S., Zhu, H., Zhou, H., Cheng, J., & Yang, H. (2020). MSpectraAI: A powerful platform for deciphering proteome profiling of multi-tumor mass spectrometry data by using deep neural networks. *BMC Bioinformatics*, *21*(1), 1–15. DOI: 10.1186/s12859-020-03783-0

Wang, X., Yan, J., Ye, Z., Zhang, Z., Wang, S., Hao, S., & Wei, G. (2022). Reorganization of 3D chromatin architecture in doxorubicin-resistant breast cancer cells. *Frontiers in Cell and Developmental Biology*, *10*, 974750. DOI: 10.3389/fcell.2022.974750

Wang, Z., & Wang, Y. (2019). Extracting a biologically latent space of lung cancer epigenetics with variational autoencoders. *BMC Bioinformatics*, *20*(S18, Suppl 18), 568. DOI: 10.1186/s12859-019-3130-9

Waqas, A., Naveed, J., Shahnawaz, W., Asghar, S., Bui, M. M., & Rasool, G. (2024). Digital pathology and multimodal learning on oncology data. *BJR|. Artificial Intelligence*, *1*(1), ubae014.

Wei, F., Kouro, T., Nakamura, Y., Ueda, H., Iiizumi, S., Hasegawa, K., & Sasada, T. (2024). Enhancing Mass spectrometry-based tumor immunopeptide identification: Machine learning filter leveraging HLA binding affinity, aliphatic index and retention time deviation. *Computational and Structural Biotechnology Journal*, *23*, 859–869. DOI: 10.1016/j.csbj.2024.01.023

Wei, L., Niraula, D., Gates, E. D., Fu, J., Luo, Y., Nyflot, M. J., & Cui, S. (2023a). Artificial intelligence (AI) and machine learning (ML) in precision oncology: A review on enhancing discoverability through multiomics integration. *The British Journal of Radiology*, *96*(1150), 20230211. DOI: 10.1259/bjr.20230211

Wei, R., & Mahmood, A. (2020). Recent advances in variational autoencoders with representation learning for biomedical informatics: A survey. *IEEE Access : Practical Innovations, Open Solutions*, *9*, 4939–4956. DOI: 10.1109/ACCESS.2020.3048309

Wei, Y., Li, L., Zhao, X., Yang, H., Sa, J., Cao, H., & Cui, Y. (2023b). Cancer subtyping with heterogeneous multi-omics data via hierarchical multi-kernel learning. *Briefings in Bioinformatics*, *24*(1), bbac488. DOI: 10.1093/bib/bbac488

Williams, J., Xu, B., Putnam, D., Thrasher, A., Li, C., Yang, J., & Chen, X. (2021). MethylationToActivity: A deep-learning framework that reveals promoter activity landscapes from DNA methylomes in individual tumors. *Genome Biology*, *22*(1), 1–27. DOI: 10.1186/s13059-020-02220-y

Wills, Q. F., & Mead, A. J. (2015). Application of single-cell genomics in cancer: Promise and challenges. *Human Molecular Genetics*, *24*(R1), R74–R84. DOI: 10.1093/hmg/ddv235

Wolf, F. A., Hamey, F. K., Plass, M., Solana, J., Dahlin, J. S., Göttgens, B., & Theis, F. J. (2019). PAGA: Graph abstraction reconciles clustering with trajectory inference through a topology preserving map of single cells. *Genome Biology*, *20*(1), 1–9. DOI: 10.1186/s13059-019-1663-x

Wu, J., Wang, W., Zhang, J., Zhou, B., Zhao, W., Su, Z., & Chen, S. (2019). DeepHLApan: A deep learning approach for neoantigen prediction considering both HLA-peptide binding and immunogenicity. *Frontiers in Immunology*, *10*, 2559. DOI: 10.3389/fimmu.2019.02559

Wu, Y., & Xie, L. (2025). AI-driven multi-omics integration for multi-scale predictive modeling of genotype-environment-phenotype relationships. *Computational and Structural Biotechnology Journal*, *27*, 265–277. DOI: 10.1016/j.csbj.2024.12.030

Xie, Y., Ruan, F., Li, Y., Luo, M., Zhang, C., Chen, Z., & Tang, C. (2024). Spatial chromatin accessibility sequencing resolves high-order spatial interactions of epigenomic markers. *eLife*, *12*, RP87868. DOI: 10.7554/eLife.87868

Xin, L., Huang, C., Li, H., Huang, S., Feng, Y., Kong, Z., & Tang, H. (2024). Artificial intelligence for central dogma-centric multi-omics: challenges and breakthroughs. *arXiv preprint arXiv:2412.12668.*

Xu, J., Jiang, T., Guo, J., Pan, T., & Li, Y. (2023). Landscape and perturbation of enhancer-driven core transcription regulatory circuits in cancer. *Molecular Therapy. Nucleic Acids*, *32*, 872–874. DOI: 10.1016/j.omtn.2023.05.011

Xu, Y., Cao, L., Chen, Y., Zhang, Z., Liu, W., Li, H., & Xu, W. (2024). Integrating machine learning in metabolomics: A path to enhanced diagnostics and data interpretation. *Small Methods*, *8*(12), 2400305. DOI: 10.1002/smtd.202400305

Yamagishi, Y., Baba, Y., Suzuki, J., Okada, Y., Kanao, K., & Oyama, M. (2025). Few-shot learning for prostate cancer detection on MRI: Comparative analysis with radiologists' performance. *Journal of Imaging Informatics in Medicine*, *39*(2), 1–13. DOI: 10.1007/s10278-025-01581-9

Yang, C. H., Moi, S. H., Chuang, L. Y., & Lin, Y. D. (2023a). An information fusion system-driven deep neural networks with application to cancer mortality risk estimate. *IEEE Transactions on Neural Networks and Learning Systems*.

Yang, R., Tsigelny, I. F., Kesari, S., & Kouznetsova, V. L. (2024a). Colorectal cancer detection via metabolites and machine learning. *Current Issues in Molecular Biology*, *46*(5), 4133–4146. DOI: 10.3390/cimb46050254

Yang, X., Vladmirovich, R. I., Georgievna, P. M., Sergeevna, A. Y., He, M., Zeng, Z., & Sergeevich, K. T. (2024b). Personalized chemotherapy selection for patients with triple-negative breast cancer using deep learning. *Frontiers in Medicine*, *11*, 1418800. DOI: 10.3389/fmed.2024.1418800

Yang, Z., Chen, Z., Matsubara, Y., & Sakurai, Y. (2023b). MoCLIM: Towards accurate cancer subtyping via multi-omics contrastive learning with omics-inference modeling. In *Proceedings of the 32nd ACM International Conference on Information and Knowledge Management* (pp. 2895-2905).

Yao, L., Shen, H., Laird, P. W., Farnham, P. J., & Berman, B. P. (2015). Inferring regulatory element landscapes and transcription factor networks from cancer methylomes. *Genome Biology*, *16*(1), 1–21. DOI: 10.1186/s13059-015-0668-3

Yazzourh, S., Savy, N., Saint-Pierre, P., & Kosorok, M. R. (2025). Medical Knowledge Integration Into Reinforcement Learning Algorithms for Dynamic Treatment Regimes. *International Statistical Review = Revue Internationale de Statistique*.

Zeng, J., Han, Z., Peng, X., Xiao, J., Wang, P., & Wang, Y. (2023). Polymerized feature-based domain adaptation for cervical cancer dose map prediction. In *2023 IEEE 20th International Symposium on Biomedical Imaging (ISBI)* (pp. 1-4). IEEE. DOI: 10.1109/ISBI53787.2023.10230491

Zeng, W. F., Mann, M., & Strauss, M. T. (2024). Artificial intelligence in mass spectrometry-based proteomics. In *Artificial Intelligence in Clinical Practice* (pp. 389–394). Academic Press. DOI: 10.1016/B978-0-443-15688-5.00010-3

Zhang, D. Y., Ye, F., Gao, L., Liu, X., Zhao, X., Che, Y., & Lee, P. (2009). Proteomics, pathway array and signaling network-based medicine in cancer. *Cell Division*, *4*(1), 20. DOI: 10.1186/1747-1028-4-20

Zhang, T., Zhang, L., Payne, P. R., & Li, F. (2020). Synergistic drug combination prediction by integrating multiomics data in deep learning models. In *Translational bioinformatics for therapeutic development* (pp. 223–238). Springer US.

Zhang, X., Zhang, J., Sun, K., Yang, X., Dai, C., & Guo, Y. (2019, November). Integrated multi-omics analysis using variational autoencoders: application to pan-cancer classification. In *2019 IEEE International Conference on Bioinformatics and Biomedicine (BIBM)* (pp. 765-769). IEEE. DOI: 10.1109/BIBM47256.2019.8983228

Zhang, Y., Wang, D., Peng, M., Tang, L., Ouyang, J., Xiong, F., & Xiong, W. (2021). Single-cell RNA sequencing in cancer research. *Journal of Experimental & Clinical Cancer Research : CR*, *40*(1), 1–17. DOI: 10.1186/s13046-021-01874-1

Zhang, Y., Zhang, X., Lane, A. N., Fan, T. W., & Liu, J. (2018, August). TFmeta: a machine learning approach to uncover transcription factors governing metabolic reprogramming. In *Proceedings of the 2018 ACM International Conference on Bioinformatics, Computational Biology, and Health Informatics* (pp. 351-359). DOI: 10.1145/3233547.3233580

Zhou, J., Huang, C., & Gao, X. (2024a). Patient privacy in AI-driven omics methods. *Trends in Genetics : TIG*, *40*(5), 383–386. DOI: 10.1016/j.tig.2024.03.004

Zhou, Z., Wang, J., Wang, J., Yang, S., Wang, R., Zhang, G., & Lu, Q. (2024b). Deciphering the tumor immune microenvironment from a multidimensional omics perspective: Insight into next-generation CAR-T cell immunotherapy and beyond. *Molecular Cancer*, *23*(1), 131. DOI: 10.1186/s12943-024-02047-2

APPENDIX

Abbreviations

AI - Artificial Intelligence
XAI - Explainable AI
DLBCL - Diffuse Large B-Cell Lymphoma
SNVs - Single-Nucleotide Variants
SVs - Structural Variants
LDSSNV - Linkage Disequilibrium-Based Method for the Detection of Somatic Single-Nucleotide Variants
CNVs - Copy Number Variants
WGS - Whole-Genome Sequencing
WES - Whole-Exome Sequencing
NSCLC - non-small cell lung cancer
CADD - Combined Annotation-Dependent Depletion
ACMG/AMP - American College of Medical Genetics and Genomics/Association for Molecular Pathology
scRNA-seq - Single-cell RNA sequencing
snHH-seq - Single-Nucleus Full-Length RNA Sequencing
ELMER - Enhancer Linking by Methylation/Expression Relations
M2A - MethylationToActivity
MS - Mass Spectrometry
DARDN - DNAResDualNet
ML - Machine Learning
DL - Deep Learning
PCA - Principal Component Analysis
PTMs - Post-Translational Modifications
PI3K - Phosphoinositide 3-Kinase
circRNAs - Circular RNAs
AutoML - Automated Machine Learning
CNNs - Convolutional Neural Networks
GEMs - Genome-Scale Metabolic Models
TNBC - Triple-Negative Breast Cancer
ccRCC - Clear Cell Renal Cell Carcinoma
MRPS - Metabolism-Related Prognostic Signature
FBA - Flux Balance Analysis
HLAs - Human Leukocyte Antigens
VAEs - Variational Autoencoders
DILCR - Deeply Integrating Latent Consistent Representations
SSL - Self-Supervised Learning
GNNs - Graph Neural Networks
PPI - Protein–Protein Interaction

GANs - Generative Adversarial Networks
GDPN - Gaussian Distributed Prototypical Network
Grad-CAM - Gradient-weighted Class Activation Mapping
ctDNA - Circulating Tumor DNA
NGS - Next-Generation Sequencing
MRD - Minimal Residual Disease
CRC - Colorectal Cancer
CMS - Consensus Molecular Subtypes
mCRC - Metastatic Colorectal Cancer
DTRs - Dynamic Treatment Regimes
RL - reinforcement learning
CDSS - Clinical Decision Support Systems
SNB - Self-Normalizing Balanced
GBM - Glioblastoma
TIME - Tumor Immune Microenvironment
LIME - Local Interpretable Model-agnostic Explanations
SHAP - Shapley Additive Explanations

Chapter 2
Personalized Healthcare for Cancers:
The Latest Overview

Kazi Asraf Ali
http://orcid.org/0000-0002-1753-7504
Maulana Abul Kalam Azad University of Technology, India

Amlan Bishal
Bharat Technology, Uluberia, India

Sabyasachi Choudhuri
http://orcid.org/0000-0002-7736-8490
Maulana Abul Kalam Azad University of Technology, India

Sourav Dutta
http://orcid.org/0009-0003-3632-8622
Maulana Abul Kalam Azad University of Technology, India

Tanushree Das
http://orcid.org/0009-0008-4084-7106
Eminent College of Pharmaceutical Technology, India

Sanjit Kumar Roy
Maulana Abul Kalam Azad University of Technology, India

ABSTRACT

Cancer treatment has shifted from uniform protocols to precision oncology, where personalized medicine drives patient-centered care. This chapter reviews the foundations of precision oncology, emphasizing genomics, biomarkers, and multi-omics integration in guiding diagnosis and therapy selection. It highlights the role of next-generation sequencing, liquid biopsies, artificial intelligence, and big data analytics in advancing cancer care. Personalized strategies across major cancer types are explored, including targeted therapies, immunotherapy, and cell-based interventions. Clinical implementation, patient engagement, and predictive modeling are addressed, alongside challenges such as ethical, economic, and translational barriers. Future directions i.e., digital twins, next-generation immunotherapies, and equitable global access underscore the transformative potential of personalized medicine in revolutionizing cancer management through precision, prediction, and participation.

DOI: 10.4018/979-8-3373-9715-3.ch002

1. INTRODUCTION

Precision oncology, or individualized medicine in oncology, has enjoyed remarkable growth by 2025, transforming the paradigm of cancer treatment whereby the diagnostics, risk assessment, and therapy of an individual patient and his or her idiosyncratic features are customized based on the unique biological composition of the tumor of that patient (Rituraj et al., 2025). This approach takes advantage of a deep understanding of genetic mutations, molecular profiles, and environmental conditions and is no longer guided by the one-size-fits-all therapies but increases the effectiveness of the treatment and the tolerance of the treatment (Hamdy et al., 2024). Precision oncology phenomenon can be attributed to the application of landmark targeted therapies like imatinib to treat chronic myeloid leukemia (Ikhtiar et al., 2025), and EGFR inhibitors of lung adenocarcinoma that generated the idea that molecular drivers could be adopted to create effective less toxic therapy (Deshmukh et al., 2025). The use of modern technologies as illustrated in Figure 1, such as Next-Generation Sequencing (NGS), enables conducting a comprehensive molecular profiling of tumors of different types, at their early stages, and in the progressive disease, and discovering the actionable genetic alterations to inform treatment (Ghoreyshi et al., 2025). Predictive biomarkers such as tumor mutational burden or PD-L1 expression are also selective immune checkpoint inhibitors that have enhanced individualized treatment, which leads to long-term responses in patients in subsets (Jin et al., 2024).

Among the most significant issues that personalized oncology addresses is the issue of tumor heterogeneity, biological diversity of the cancers in patients and within a particular tumor. It is this heterogeneity that normally leads to resistance to treatment and poor consistency of outcomes, which decrease the effectiveness of standardized treatment (Maleki et al., 2024). Individual modalities strive to get rid of this by matching patients with treatments tailored to the genetic and molecular context of their tumors and minimize unnecessary side effects and preclude non-responding therapy. This method also fosters the incorporation of the patients by formulating the treatment plans, which are founded on the biology of the patients (Singh et al., 2024).

Accuracy oncology is currently the new standard in various malignancies, including breast cancer, lung cancer, and melanoma, and biomarker-based regimens have shown to lead to improved survival and life quality compared to traditional chemotherapy as of practice (Fujiwara et al., 2024). The advancement of artificial intelligence (AI) into clinical processes provokes the fast processing of intricate genomic data that can be applied to find new targets and choose patients in the best way during treatment. To give an example, there are now AI apps that can process extensive amounts of cancer genomic data to determine homologous recombination deficiency with greater accuracy and precision, to guide treatment, such as the use of PARP inhibitors (Chong et al., 2025).

Nevertheless, in spite of the advancement, it is costly, lack of sophisticated molecular testing in some locations, and complicated tumors that lack actionable mutations are some of the obstacles that precision oncology has to struggle with (Chhabra, 2024). Another ethical issue that complicates the adoption of genetic information is the privacy of the genetic information and the disparity in the access to healthcare. Moreover, resistance to treatment and tumor alterations require the incessant re-evaluation of treatment plans, which in many cases requires the sequential monitoring of the cancer through the use of liquid biopsies (Masucci, Karlsson, Blomqvist, & Ernberg, 2024). The multi-omics combinatory, adaptive trial design, and AI-based predictive models are also a subject of focus in the contemporary literature to tackle the existing issues. These inventions can change the personalized cancer care into a more effective, available, and customized care (Ali, Choudhuri, Bishal et al, 2025). The concept of personalized healthcare has already redefined the paradigm of treating cancers by 2025 with the hope of better patient outcomes as a result of exquisitely-tuned and biology-guided treatment (Molla & Bitew, 2024).

Figure 1. Integration of Next-Generation Medical Imaging Analytics (NGMIA) with precision medicine [Adapted from (Roisman et al., 2023) under License CC-By-4.0]

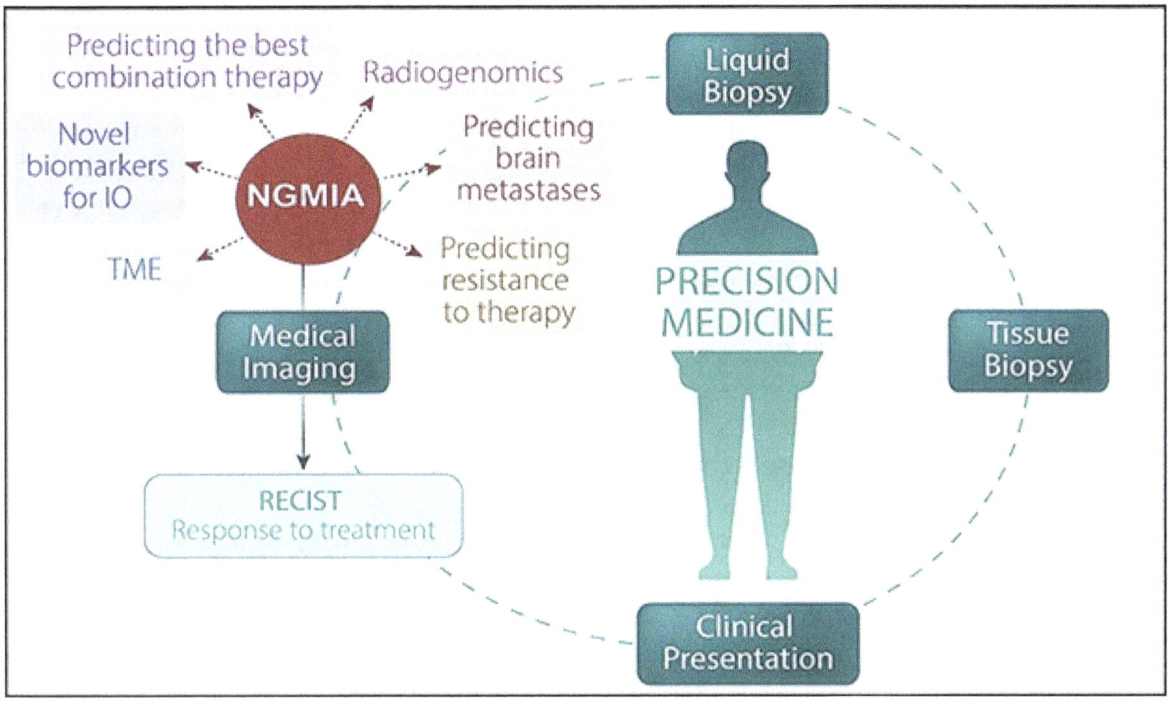

2. FOUNDATIONS OF PERSONALIZED CANCER CARE

2.1 Genetic and Genomic Insights

However, the development of genomics has resulted in widespread profiling of tumors in a comprehensive manner that involves a study of both germline and somatic mutations, that has significantly influenced the contemporary cancer care (Jamalinia & Weiskirchen, 2025). Nowadays, next-generation sequencing (NGS) technologies can be used to perform high-throughput sequencing of whole genomes, exomes, transcriptomes, and targeted panels of genes to provide detailed maps of genetic alterations, such as single nucleotide variants, gene fusions, insertions, deletions, and copy number changes (Satam et al., 2023). NGS enables clinicians to detect driver mutations, anticipate resistance, as well as category cancers and this is the basis of precision oncology (Isaic et al., 2025). Single-cell sequencing and bioinformatics/artificial intelligence tools can help find clonal heterogeneity and the evolution history of the tumor, whereas the analysis of intricate genomic information can be improved by single-cell sequencing, thereby speeding up the process of finding clinically relevant mutations (Chen, Zhou, Li et al, 2023).

Non-invasive detection of circulating tumor DNA can also be used by these technologies with the support of liquid biopsies, which can give real-time information on the development of the disease and the effectiveness of treatment. Altogether, the combination of NGS, single-cell analysis, and AI-powered interpretation facilitated a more accurate risk identification, therapeutic choice, and cancer recurrence early detection, which leads to personalized treatment approaches in cancer therapy (Pandey & Yadav, 2025; Velpula & Buddolla, 2025).

2.2 Biomarkers in Cancer Diagnosis and Management

The biomarkers in cancer play a central role in the development of diagnosis, prognosis, and targeted therapy based on a broad range of molecular markers, which include mutations (e.g., BRCA1/2, ALK fusions), protein expression patterns, circulating tumor DNA (ctDNA), and even circulating tumor cells (CTCs) (Qoronfleh & Al-Dewik, 2025). The established examples of biomarkers, including prostate-specific antigen (PSA), carcinoembryonic antigen (CEA), and alpha-fetoprotein (AFP) continue to play a role in screening and monitoring, whereas the new panels have been designed to incorporate multi-omics and bioinformatics tools to increase sensitivity and specificity of various types of cancer (Zhou et al., 2024). As a case in point, estrogen, progesterone and HER2 receptor status are directly used to subtype breast cancers, with direct effect on the prognosis and the targeted therapy. Liquid biopsies are a paradigm shift, as they allow the detection of ctDNA, CTCs, and exosomes in blood and other fluids in noninvasive manner, which, in addition to early diagnosis, may help to monitor diseases in real time and detect low-level residual disease after treatment (Schiavone et al., 2025). Identification of certain mutations in ctDNA may be used to personalize treatment (e.g. EGFR T790M in lung cancer), predict drug resistance, and detect recurrence long before it may manifest itself (Bartolomucci et al., 2025). Moreover, there are the additions of epigenetic changes and microRNA signatures that add more prognostic and predictive power particularly when dealing with hard-to-biopsy tumors. Combination of these biomarkers into multidimensional risk models using machine learning and high-throughput technologies has continued to facilitate development of accurate, patient-specific management strategies and better patient outcomes in the field of oncology (Kamińska et al., 2019).

2.3 Multi-Omics Integration (Genomics, Proteomics, Metabolomics)

Integration of multi-omics in cancer care is a revolutionary step to converting other pieces of data such as genomics, transcriptomics, proteomics, metabolomics and other molecular layers to build advanced profiles of tumors to further refine the diagnosis process and personalize the treatment (Menyhárt & Győrffy, 2021). Genomics determines mutant and chromosome changes that cause cancer, whereas transcriptomics can determine dynamic changes in gene expression (Ronemus et al., 2025); proteomics and metabolomics explain the role of proteins and metabolic restructuring that are vital in tumor development and resistance to treatment (Ji et al., 2025). The combination of the above layers makes the integration process very powerful in enabling clinicians and researchers to identify convergent pathways and functional tumor subtypes that would not be revealed by single-omics methods. In the example, the combination of genomic and metabolomic results will enable the identification of metabolic patterns, specific to an aggressive tumor phenotype, which enhances biomarker sensitivity to prognosis and early detection (Dong et al., 2025). Spatial proteomics integration and epigenomics identification are additional tools in the discovery of new immune evasion mechanisms and drug response, which promotes the development of new therapeutic targets. Multi-omics datasets, which are rapidly growing in size, are now becoming the focus of artificial intelligence and machine learning, thereby facilitating dynamic stratification of patients, prediction of outcomes, and the discovery of actionable molecular features (Hai et al., 2024). These integrative methods are prepping the way to pathway-based and no single-gene-based clinical decision-making, and better patient outcomes and highly personalized regimens of cancer patients. Integration Multi-omics integration in cancer medicine is the process of integrating genomics (DNA sequence and mutations), transcriptomics (RNA expression), proteomics (protein abundance and activity), and metabolomics (metabolic pathways and small molecules) and is also increasingly incorporating epigenomics and radiomics (Chakraborty et al., 2024). This is a multifaceted strategy that uses the latest computational infrastructure and AI to merge dissimilar molecular layers that enable researchers to describe tumor heterogeneity (Ali et al., 2024), plot disease progression, and identify mechanisms of therapeutic resistance (Vosoughi et al., 2025). As an illustration, the combination of genomic and metabolomic data can provide distinct metabolic signatures and driver mutations characterizing aggressive cancer subtypes and enable more reasonable risk evaluation (Cararo Lopes et al., 2023). The multi-omics information is also essential in defining the immunological escape mechanisms (Chen et al., 2025) and relevant biomarkers which makes it possible to monitor real-time and make interventions based on the predictive modeling (Liu et al., 2025). Still troublesome like aligning different datasets and ensuring reproducibility but standardized data formats, better machine learning analytics, and group work are quickly emerging new applications in clinical settings (Večurkovská et al., 2025). With the maturity of these integrative tools, multi-omics profiling is set to be a clinical bedside tool in the early diagnosis, prediction, and individualized treatment of cancer.

3. TECHNOLOGICAL ADVANCEMENTS DRIVING PERSONALIZATION

The achievement of personalized cancer medicine depends on the ongoing creation of advanced technologies that have the ability to examine the multifaceted, heterogeneous molecular environment of each specific tumour and transform the information into a clinical decision-making action plan.

3.1 Next-Generation Sequencing and Liquid Biopsies

Individualized oncology needs to be molecularly profiled precisely, which can be highly achieved by Next-Generation Sequencing (NGS) tools, such as whole-exome sequencing (WES) and RNA sequencing (RNA-seq) (Khoury et al., 2025). NGS forms the basis upon which one can identify a set of genomic, transcriptomic, and proteomic abnormalities that make the cancer of one patient different than the other (Bode & Dong, 2018).

As an adjunct to tissue-based NGS, liquid biopsies (LB) are currently a tool of vital importance, albeit of minimal invasiveness, in the monitoring of tumour genetics and progression (Palacín-Aliana, García-Romero, Asensi-Puig, Carrión-Navarro, González-Rumayor, & Ayuso-Sacido, 2021). Liquid biopsies usually examine circulating tumour cells (CTCs) and circulating tumour DNA (ctDNA) extracted out of a variety of biofluids, most of which are blood plasma (Snow et al., 2019). These assays have significant clinical benefits compared to standard invasive biopsies: they are noninvasive, can give longitudinal temporal information on the dynamics of tumour heterogeneity in response to therapy, and can be used to provide rapid molecular data (42). The ability is especially essential in very heterogeneous tumours whereby standard biopsy might fail to capture the full complexity of the heterogeneity (Palacín-Aliana, García-Romero, Asensi-Puig, Carrión-Navarro, González-Rumayor, & Ayuso-Sacido, 2021).

A major technological problem is the ability to detect genomic changes existing at very low allele fractions especially in early-stage disease. Droplet Digital PCR (ddPCR) has been found to effectively address this problem because it is a high-specificity technology that has high sensitivity. In, e.g., breast cancer at the three-yearly stage, ddPCR of the circulating tumour DNA following surgery can indicate tumour recurrence up to 11 months prior to the traditional clinical techniques. This also applies to metastatic breast cancer (mBC) patients, in which large volumes of the ctDNA are shed into the bloodstream, which can be of prognostic and Therapy guidance use at all stages of the disease (Palacín-Aliana, García-Romero, Asensi-Puig, Carrión-Navarro, González-Rumayor, & Ayuso-Sacido, 2021).

Whereas the technological potential of sensitive detection is real, most liquid biopsy tests are still in the active development phase and have no available clinical evidence to be approved by the U.S. Food and Drug Administration (FDA) to use as companion diagnostics or prognostic markers (Snow et al., 2019). The regulatory constraint implies that the existing implementation barrier is not only the technological capability of detection, but the highly demanding need throughout a high clinical utility validation in large-scale research.

Moreover, NGS data requires special equipment due to the complexity of its interpretation, particularly when incidental results, potentially including possible germline pathogenic variants (PGPVs), are obtained. Molecular Tumour Boards (MTBs) are part of an institution involved in critical triage and the assessment of such reports (Rives et al., 2024). A review of MTB has a clinical value, with the detection of PGPVs among 12% of patients receiving tumour NGS. Despite the high-efficiency review, the research shows that the rates of the follow-up genetic counselling are low because physicians who treat patients under-refer patients, which means that NGS implementation will need organizational restructuring to provide patients with optimal management (Rives et al., 2024).

3.2 Artificial Intelligence (AI) and Machine Learning (ML) Applications

AI and ML play a crucial role in converting complicated genomic and clinical data into clinical prediction algorithms which are the key elements of precision medicine (Wang et al., 2022). As predictive analytics, ML algorithms develop a highly complex model of the interaction between patient biomarkers and therapeutic responses, used in clinical decision-making. A particularly important use is to incorporate ML in the Response Adaptive Randomization (RAR) designs in clinical trials. With advanced modelling such as nine varying algorithms such as parametric (Logistic, LASSO, Ridge regression), nonparametric (Gradient Boosting Machine, Random Forest, Support Vector Machine, Naive Bayes, k-Nearest Neighbours, and Neural Networks) ones, the researchers are able to predict responses of individual patients and reassign treatments as needed (Wang et al., 2022).

Simulation studies reveal that RAR designs based on ML yield better clinical outcomes than the traditional Equal Randomization designs. In particular, the ensemble technique, the combination of several ML methods, achieved better results than all the individual algorithms. This group methodology had an estimated 5 percent greater response rate and allocated more than 20 percent of the patients to their individualized optimum treatment. The excellent ability of the ensemble model proves that non-linear complexity in modelling techniques is required to prove the complete biological complexity of cancer response. This solution does not only improve the efficacy of the therapeutic procedure, but also ethically more acceptable clinical trials, especially in patients of late stage who rely on trial participation as a source of novel treatment (Wang et al., 2022).

Besides clinical trial optimization, AI can be used in single-cell analysis to enable accurate oncology precision in the stratification of different tumour models. ML clustering and feature selection have been useful in the discovery of biomarkers like PD-L1, TIM-3, and TIGIT, which are predictors of positive responses to Therapy (Rafiq et al., 2025). AI is also useful to interpret single-cell metabolic profiling and to find common metabolic weaknesses across species, e.g. KRAS-mutant tumours.

3.3 Role of Big Data and Bioinformatics

Cancer is complex and requires a genomic, epigenetic, and proteomic perspective of aberrations. The complexity of oncogenesis and progression cannot be fully unraveled by single-platform analysis and multi-omics analysis is crucial to cancer studies. The science of big data leads to the creation of new statistical and mathematical processes to combine the high-dimensional data, with somatic mutation, Copy Number Variation (CNV), gene expression, methylation, and proteomic data (Heo et al., 2021).

These data layers are then combined to find coherent and conserved molecular features across patients using bioinformatics frameworks, e.g. multi-view clustering (Heo et al., 2021). The major objective is to categorize cancers into new subtypes that define genetic and phenotypic heterogeneity. The recent literature points to the use of deep learning models on multi-omics data to perform vital functions, such as cancer classification, driver genes, survival analysis, and drug response prediction (Sartori et al., 2025).

Multi-omics integration can be seen to be powerful in large-scale proteogenomic datasets such as the work on lung adenocarcinoma (LUAD) by the Clinical Proteomic Tumour Analysis Consortium (CPTAC). The combination of genomics, transcriptomics, and high-resolution mass spectrometry-based proteomics helped scientists to identify complicated biological processes, such as the KEAP1/NFE2L2 network that is a distinctive biomarker in LUAD. This capacity to predict the complex interaction in different pathways can help precision medicine (Ali, Choudhuri, Chakraborty et al, 2025) go beyond mere identification of targets to predictive drug failure. The integrative approaches to the understanding of a non-responsive or resistant tumour will allow achieving higher therapeutic efficacy and the possibility to prevent the mechanisms of drug resistance in advance (Heo et al., 2021). This is an all-embracing, data-oriented framework which is an indication of a paradigm shift to systems biology in oncology.

4. PERSONALIZED APPROACHES IN MAJOR CANCER TYPES

4.1 Lung Cancer

Molecular analysis is very crucial in the clinical management of Non-Small Cell Lung Cancer (NSCLC) because it has a high mutational burden. Therefore, the NGS should be introduced into routine genomic biomarker testing in all patients with advanced NSCLC as suggested by the key clinical guidelines. 1 NGS is needed to detect alterations in the targetable proteins, such as EGFR mutations (including Exon 19 deletions or L858R point mutations), ALK rearrangements, ROS1 fusions, BRAF V600E, MET Exon

14 deletions, KRAS G12C, and HER2 Exon 20 insertions. The result of the presence of these biomarkers defines the choice of targeted therapies (Ramos et al., 2025).

The BATTLE trial was the first to confirm this strategy as the first prospective, adaptively randomized trial in heavily pretreated NSCLC which necessitated the use of real-time tumour profiling to inform individual selection of treatment choice (Kim et al., 2011). Personalized treatment has more recently been applied outside metastatic disease. Clinical trials in the phase 3 stage have shown that neoadjuvant immunochemotherapy, including the use of nivolumab and chemotherapy, would significantly increase the event-free survival of patients with resectable stage IIA to IIIB NSCLC with the use of chemotherapy alone (50). This confirmation of molecular-based treatment in the earlier phases solidifies NSCLC as one of the successful models of personalized oncology application.

4.2 Breast Cancer

Liquid biopsies are also of great importance in augmenting personalized approaches in breast cancer. The measurement of circulating cell-free DNA (cfDNA) offers valuable prognostic data and informs treatment at all stages of the disease and particularly in the metastatic breast cancer (mBC) where high levels of the latter are observed (Palacín-Aliana, García-Romero, Asensi-Puig, Carrión-Navarro, González-Rumayor, & Ayuso-Sacido, 2021). Detection of ctDNA in the early stages of BC after surgery can be done by using high-sensitivity technologies, including ddPCR. This ability of the cfDNA testing to serve as a prognostic surrogate with a 11-month higher predictability of tumour recurrence as compared to conventional clinical tests changes its purpose into a powerful predictive surrogate. This early diagnosis potential opens the possibility of very individualized interventions that are perhaps preemptive based on molecular relapse overtaking clinical disease (Bishal et al., 2025).

4.3 Colorectal Cancer

Colorectal cancer (CRC) has been noted to be highly heterogeneous at the molecular level and lethal. Genomic profiling has radically transformed the management of CRC and further research is needed to determine new prognostic and predictive markers. Precision strategies are currently targeting more sophisticated methods of analysis that goes beyond bulk tumour traits. Tumour Single-cell analysis has been critical in the identification of new subtypes of cells that are important in tumour progression and can be used as a therapeutic target. Moreover, transcriptome and proteome spatial analysis offers important information on the cellular interactions that are pathogenic in the tumour microenvironment. This shift to spatial and cellular profiling at the high-resolution level is critical to eliminate the heterogeneity of tumours and optimize personalised treatment regimens in the case of CRC (Ahluwalia et al., 2024).

4.4 Hematological Malignancies

The field of hematological malignancies is developing rapidly using precision medicine, with NGS, liquid biopsy, and more complex minimal residual disease (MRD) monitoring (Khoury et al., 2025). These illnesses usually have distinct, hegemonic molecular motivators, which can be precisely interfered with. There are a number of common molecular changes that lead Therapy in Acute Myeloid Leukemia (AML). The targeted inhibitors used to treat FLT3 mutations (mostly FLT3-ITD) include gilteritinib and midostaurin. Ivosidenib and enasidenib effectively target IDH1 and IDH2 mutations respectively, which result in epigenetic dysregulation. Newer treatment options involve Menin inhibitors of NPM1-mutated AML. Risk stratification and identification of clonal evolution, which is critical in the process of adaptation of therapy, require comprehensive genomic profiling (Khoury et al., 2025).

An example of precision oncology is Chronic Myeloid Leukemia (CML), which is characterized by the BCR-ABL1 fusion gene. Treatment is with Tyrosine Kinase Inhibitors (TKIs) and the choice of one agent (e.g., imatinib, nilotinib, dasatinib) is often specific to the comorbidities of the patient and also whether or not they carry the BCR-ABL1 resistance mutations. Individualized therapies in Multiple Myeloma involve immunotherapies which target the B-cell maturation antigen (BCMA) (Khoury et al., 2025).

4.5 Rare and Paediatric Cancers

Paediatric cancers also present a special problem, since all the particular diagnoses qualify as rare or orphan diseases, restricting the number of patients that can be used in conventional clinical trials. Moreover, in paediatric tumours, the number of targetable mutations is much smaller, and often targets are hard to drug, like non-kinase fusion oncoproteins. There are also regulatory and cost obstacles that have brought about setbacks where the potential and effective agents are left off-label even after encouraging academic testing (Laetsch et al., 2021).

In order to address these constraints, there has been a strong adoption of biomarker-based histology-agnostic trial designs in the field of paediatric oncology. The most effective case study is the phase 1 clinical trial of the Trk inhibitor larotrectinib in malignancies of TRK fusion (Forrest et al., 2018). A response rate of 91% in patients with tumours that had the fusion was achieved by this Therapy regardless of the site of origin of the tumour. This strategy makes paediatric oncology the best case in favor of pan-cancer molecular targeting. Precision genomics has been proven in early implementation programs that have shown up to 70% of patients treated with recommended personalized Therapy have a clinical response or stabilized disease (Marshall et al., 2023). Innovative trials and the combination of genomic information requires the collaboration of the entire world in order to combine the geographically scattered patient data which are essential to create the strong cohorts required to be used in the clinical validation and regulatory approval (Langenberg et al., 2021).

5. TARGETED AND TAILORED THERAPIES

Precision cancer medicine focuses on the fundamental principles of personalized medicines that target specific, unique molecular alterations that cause cancer progression and dissemination (Gambardella et al., 2020). Although monoclonal antibodies are selective to cell-surface receptors, which is why they are the best option compared to conventional chemotherapeutic agents due to their specificity and a reduced toxicity profile, small-molecule inhibitors disrupt altered intracellular signaling pathways in tumor cells (Zhong et al., 2021). The immunotherapeutic approaches, like immune checkpoint inhibitors, activate and modify the host immune system to identify and eliminate cancerous cells. In parallel, alternative forms of cell-based therapy, like chimeric antigen receptor T-cell (CAR-T) Therapy, entail genetic alterations of the immune cells of a patient in order to be able to detect and destroy tumors precisely. As an additional way of enhancing clinical efficacy, combination and adaptive treatment strategies have been formulated to respond to tumor heterogeneity and resistance mechanisms. Collectively, these developments can be seen as the precision medicine paradigm whereby the treatment regimens are customized based on unique molecular and biological characteristics of the cancer of a particular patient.

5.1 Targeted Small-Molecule Inhibitors

One of the most novel methods of precision cancer therapy is the use of small-molecule inhibitors (Raghavan et al., 2024; Zhong et al., 2021). These compounds target to specifically interfere with molecular processes that are necessary to any tumor formation and survival, and are usually targeted to enzymes, receptors, or signaling intermediates. Kinase inhibitors which act on EGFR, BCR-ABL, and ALK are examples of this type. They are good in therapy since their small size enables them to enter the cell and be absorbed by mouth. Medicinal chemistry has been advanced to be more selective, decrease the off-target toxicity, and increase bioavailability. Nevertheless, the issues remain, which necessitate the further research of drugs, and the creation of the next-generation inhibitors, including intertumoral heterogeneity and the appearance of resistance mutations (EGFR T790M, BCR-ABL T315I).

EGFR T790M mutations that lead to resistance were discovered by Kobayashi et al. (2005), and inhibitors that target the mutation are third-generation inhibitors. In a study carried out by Chapman et al. (2011), vemurafenib enhanced the survival of the BRAF V600E-mutant melanoma, and Davies et al. (2002) found widespread BRAF V600E mutations in melanoma and made it the target of small-molecule inhibition.

5.2 Monoclonal Antibodies and Immunotherapy

Because of their particular binding to extracellular targets, monoclonal antibodies (mAbs) are an essential part of targeted Therapy. It can be used to kill cells with antibodies, inhibit growth factor receptors directly (e.g., trastuzumab against HER2), and deliver toxic drugs via antibody-drug conjugates (ADCs). Immunotherapy targets tumor-specific antigens and leverages the patient's immune system to fight cancer (Kumar et al., 2021). Immune checkpoint inhibitors, such as anti-PD-1/PD-L1 and anti-CTLA-4, have revolutionized the treatment of lung cancer, melanoma, and other cancers, often producing long-lasting effects. However, the range of response rates underscores the importance of biomarker-driven patient selection, including MSI-H status, tumor mutational burden, and PD-L1 expression. To provide a more straightforward overview of these therapeutic agents, the key monoclonal antibodies and immune checkpoint inhibitors are listed in Table 1.

Table 1. Different monoclonal antibodies and Immune checkpoint inhibitors with their applications

Antibody	Target	Application (Cancer Type)	Key Observations	Reference
Trastuzumab	HER2 receptor	HER2-positive breast cancer	Improves both overall survival and progression-free survival in patients whose tumours overexpress HER2	(Yang et al., 2017)
Rituximab	CD20 on B-cells	Non-Hodgkin lymphoma	Eliminates malignant B-cells and, when added to chemotherapy, improves response rates	(Pérez-Callejo et al., 2015)
Pembrolizumab	PD-1 immune checkpoint	Melanoma; non-small cell lung cancer (NSCLC); several other solid tumours	Boosts anti-tumour immunity and delivers long-lasting responses in a subset of patients	(Garon et al., 2015)
Nivolumab	PD-1 immune checkpoint	Melanoma, NSCLC, renal cell carcinoma	Shows durable responses and better survival versus older therapies; side effects are manageable.	(Topalian et al., 2019)
Ipilimumab	CTLA-4 immune checkpoint	Advanced melanoma	First immune-checkpoint antibody showing a survival benefit; carries risk of immune-related adverse events.	(Chang et al., 2020)
Bevacizumab	VEGF ligand (angiogenesis inhibition)	Colorectal cancer, lung cancer, renal cell carcinoma	Slows tumour blood-vessel growth and improves progression-free survival when used with other treatments	(Chitoran et al., 2025)
Daratumumab	CD38 on plasma cells	Multiple myeloma	Targets a key antigen on myeloma cells, improving outcomes in both newly diagnosed and relapsed settings	(Lokhorst et al., 2015)
Sacituzumab govitecan	TROP-2 (antibody-drug conjugate)	Triple-negative breast cancer	Delivers a chemotherapy payload via the antibody, improving response rates in a difficult-to-treat group	(Schreiber et al., 2021)
Isatuximab	CD38 on multiple myeloma cells	Relapsed/refractory multiple myeloma	Demonstrates anti-tumour activity by several immune mechanisms and direct apoptosis of tumour cells	(Frampton, 2021)
Zanidatamab	HER2 (bispecific antibody)	HER2-positive biliary tract cancer (and other HER2-driven tumours)	A newer dual-target approach to HER2, expanding treatment beyond breast cancer	(Harding et al., 2023)

5.3 CAR-T and Cell-Based Therapies

The potential of cell-based immunotherapy is demonstrated by Chimeric Antigen Receptor T-cell (CAR-T) therapy. CAR-T therapy redirects immune effector activity toward cancer cells by genetically altering a patient's T cells to generate synthetic receptors targeting tumor-specific antigens (e.g., CD19 in B-cell malignancies). Approved CAR-T therapies have shown promising results in treating blood cancers. However, issues include neurotoxicity, limited efficacy in solid tumors, cytokine release syndrome (CRS), and high production costs, which highlight areas that need innovation. Targeting two antigens simultaneously, employing "off-the-shelf" CAR-Ts from other individuals, and other cell-based therapies (such as TIL therapy and CAR-NK cells) are new strategies that aim to increase the accessibility of treatments (Wang et al., 2024).

The revolutionary impact of CAR-T cell therapies in the treatment of hematologic malignancies was well established by early clinical studies. The therapeutic effectiveness of CD19-targeted CAR-T cells in patients with refractory chronic lymphocytic leukemia was demonstrated in the groundbreaking work of Kalos et al. (2011), which established proof of concept for this strategy and produced sustained complete remissions. CAR-T therapy was then established as a significant breakthrough in the treatment of juvenile leukemia when Maude et al. (2014) reported remission rates of 71% in pediatric and young adult patients with acute lymphoblastic leukemia treated with tisagenlecleucel (CTL019) (Maude et al., 2014). Following the results of Neelapu et al. (2017), who demonstrated high response rates and durable remissions with axicabtagene ciloleucel in refractory large B-cell lymphoma, the field made further strides. Eventually, it helped the FDA approve the first CAR-T product for lymphoma.

5.4 Combination and Adaptive Therapies

Due to compensatory molecular mechanisms, clonal development within the tumor microenvironment, or signaling pathway reactivation, acquired resistance often develops during monotherapy. By simultaneously blocking multiple molecular targets within a single route or across interconnected signaling networks, combination treatment strategies have been developed to address this challenge and reduce the risk of adaptive therapeutic escape. Examples with clinical validation include BRAF/MEK targeting in melanoma, the use of trastuzumab and pertuzumab for dual HER2 inhibition, and the combination of VEGF inhibition and immune checkpoint blockade (Welti et al., 2022). Furthermore, adaptive therapy strategies adjust dosage and treatment combinations over time to delay the emergence of resistance while reducing cumulative toxicity, supported by mathematical modeling and real-time illness monitoring. This approach demonstrates how precision oncology has progressed from static treatment plans to patient-specific, dynamically responsive therapy paradigms (Welti et al., 2022). Furthermore, Table 2 highlights different combination therapy approaches and outlines their respective drawbacks.

Table 2. Different combinational therapies along with their limitations

Strategy	What is Targeted	Cancer Type	Key Findings	Main Challenges	Reference
Targeted + Targeted	BRAF inhibitor (dabrafenib) + MEK inhibitor (trametinib)	Melanoma with BRAF V600E mutation	Shows better survival and delays resistance compared to using a BRAF inhibitor alone	Resistance eventually appears; side effects like fever and skin issues	(Greco et al., 2019)
Targeted + Immunotherapy	EGFR inhibitors + PD-1/PD-L1 immune checkpoint blockade	NSCLC	Early research shows they work well together by reducing immune escape	Can lead to serious immune-related side effects	(Yi et al., 2022)
Immunotherapy + Immunotherapy	Nivolumab (anti-PD-1) + Ipilimumab (anti-CTLA-4)	Advanced melanoma, renal cell carcinoma	Markedly boosts response and survival compared to single-agent Therapy	High frequency of immune-related toxic reactions	(Larkin et al., 2019)

continued on following page

Table 2. Continued

Strategy	What is Targeted	Cancer Type	Key Findings	Main Challenges	Reference
Targeted + Chemotherapy	PARP inhibitor (olaparib) + platinum chemotherapy	BRCA-mutated ovarian and breast cancers	Enhances DNA damage response, leading to better treatment outcomes	Overlapping blood-related toxicities	(McMullen et al., 2020)
Adaptive Therapy (dose modulation)	Intermittent BRAF inhibitor use	Metastatic melanoma	Adjusting doses allows sensitive cancer cells to persist, slowing resistance.	Needs close monitoring; not widely adopted clinically yet	(Kavran et al., 2022)
Oncolytic Virus + Immunotherapy	T-VEC (oncolytic Virus) + Pembrolizumab	Melanoma	Strengthens anti-tumor immune response and increases long-lasting benefits	Limited to injectable lesions; accessibility issues	(Larocca et al., 2020)
Cell Therapy + Immuneodulation	CAR-T cells + PD-1 blockade	B-cell cancers; some solid tumors	Early evidence shows better T-cell persistence and tumor killing	Still experimental; risk of cytokine release syndrome	(Rafii et al., 2025)
Radiotherapy + Immunotherapy	Radiotherapy + PD-1/ PD-L1 blockade	NSCLC; head and neck cancers	Radiation can boost immune activation, improving immunotherapy response	Best timing and dose combinations are still being optimized	(Procureur et al., 2021)
Targeted + Hormone Therapy	CDK4/6 inhibitors + endocrine therapy	HR-positive, HER2-negative breast cancer	Significantly extends time without disease progression	Neutropenia and eventual resistance may occur	(Lloyd et al., 2022)

6. CLINICAL APPLICATIONS AND PATIENT-CENTERED CARE

The main aim of the practical application of precision oncology is to choose the right treatment of the right patient at the right time. Whereas predictive models utilize clinical and genetic data to guide the choice of treatment and predict resistance, molecular diagnostics and patient stratification plan who will most probably respond to specific medication. Outside the clinic, digital health technology including wearables and smartphone platforms provides continuous monitoring of a patient health and medication compliance with it. As a way of making sure that the treatment options are based on the preferences of the patient and the scientific evidence, this approach focuses on the participation of the patient and collaborative decision-making (Dinh-Le et al., 2019; Khan et al., 2017). Precision oncology improves outcomes without losing focus on patient values and quality of life due to the presence of sophisticated procedures and personalized care.

6.1 Stratifying Patients for Treatment Selection

Precision oncology cannot be effective without proper patient stratification. The molecular diagnostics of actionable genomic changes (e.g. driver mutations, gene fusions, and epigenetic modifications) is applied with molecular diagnostics, e.g. next-generation sequencing (NGS), fluorescence in situ hybridization (FISH), as well as liquid biopsies. Stratification is boosted by proteomic, metabolomic, and immunological phenotyping and genomic data. With these multiomic insights and clinical concerns, one can be able to make sure that those who are likely to achieve the greatest benefits of treatment get the same treatment hence exposure to hazardous or unproductive regimens is minimized. Real world therapy choices are gradually being guided by the use of molecular tumor profiling technologies and tumor boards.

Molecular stratification has been incorporated into precision oncology as a necessary step towards being able to match treatments to particular oncogenic drivers. The relevance of biomarker-directed treatment was defined by Lynch et al. (2004), who formulated innovative evidence that therapeutic value of gefitinib was restricted to those patients who had non-small cell lung cancer and EGFR mutations. Inhibitors that bind BRAF were subsequently created by discovering activating BRAF mutations as actionable targets as described by Davies et al. (2002). Soda et al. (2007) achieved a breakthrough in the area, discovering the EML4-ALK fusion, which has allowed introducing ALK-targeted treatment of a particular group of lung malignancies. Moreover, FLT3 inhibitors have to be added into clinical trials in populations of patients with biomarkers, since Nakao et al. (1996) demonstrated that a mutation in FLT3 in acute myeloid leukemia was a result of a poor prognosis.

6.2 Predictive Models for Treatment Response

Predictive models are based on computational and machine learning models to predict the results of treatment on the basis of the genomic, clinical, and demographic characteristics. Clinicians can use these models to improve the sequencing of therapies, determine resistance, and quantify response time. As an example, the early response assessment can be improved with the help of prediction systems that combine both movements of imaging information and circulating tumor DNA (ctDNA). Integration of artificial intelligence into electronic health records also supports real-time decision making whereby oncologists are able to make the best decisions as they learn the journey of the patient. Table 3 is a summary of the key types of predictive models currently used in oncology, their clinical applications, and limitations

Table 3. Predictive Models in Oncology and Their Applications

Model Type	Example Models	What They Are Used For	Main Drawbacks	Reference
Genomic-based models	• GDL (Genome Deep Learning) • TCGA (The Cancer Genome Atlas) • MammaPrint • CP-GEP (Melanoma Gene Expression Profile) • VeriStrat	• Predict how a tumor will behave • Understand how well a treatment may work • Estimate the risk of cancer coming back	• Need patient tissue samples • May overlook factors beyond genetics • Performance depends on available datasets	(Bao et al., 2022; Pang et al., 2024; Sun et al., 2019; Zhang et al., 2025)
Imaging-based models (Radiomics)	• Deep Radiomics • Foundation Model for Cancer Imaging Biomarkers • Radiomics for Tumor Prognosis • DeepSurv	• Predict disease progression • Monitor how patients respond to treatment • Help guide personalized Therapy	• Requires high-quality medical images • Different imaging methods can affect results • Needs strong validation before clinical use	(Bao et al., 2022; Pai et al., 2024; Pang et al., 2024)
Machine learning / AI models	• IBM Watson for Oncology • Deep Survival Models • AI-driven breast cancer recurrence tools • Continuous Individualized Risk Index (CIRI) • DeepSurv	• Predict individual drug responses • Forecast patient outcomes • Continuously assess patient risk	• Requires huge datasets • Can introduce hidden biases • Often difficult to interpret	(Pang et al., 2024; Yousefi et al., 2017)
Clinical biomarker models	• HER2 testing for breast cancer • PSA testing for prostate cancer • ThroLy score • VeriStrat	• Track treatment effects • Predict chances of relapse • Help assess risk of blood-related complications	• Sometimes lacks accuracy • May be influenced by unrelated biological factors	(Chung et al., 2023; Tarighati et al., 2022)
Integrated digital models	• Digital Oncology Platform (DOP) • Digital Twins in Oncology • Predictive models for treatment side effects • AI tools for drug-related adverse effects	• Real-time tracking of patient outcomes • Helps doctors make tailored decisions • Predicts treatment-related risks	• Hard to adopt in hospitals • Data security concerns • Challenging to integrate with current systems	(Pang et al., 2024; Shen et al., 2025)
Preclinical models	• Patient-Derived Xenografts (PDX) • Genetically Engineered Mouse Models (GEMMs) • Organoids	• Study how cancers grow and spread • Evaluate potential drug effectiveness • Help guide personalized therapies	• Limited to certain cancers • Ethical concerns with animals • Time-consuming and costly	(Garcia et al., 2020; Honkala et al., 2022)
QSAR models (Quantitative Structure–Activity Relationship)	• QSAR-based drug screening tools • ADMET predictors	• Identify promising drug candidates early • Predict how drugs will act in the body	• Needs large chemical databases • May overlook complex biological processes	(Achary, 2020; Shah et al., 2024)

continued on following page

Table 3. Continued

Model Type	Example Models	What They Are Used For	Main Drawbacks	Reference
Radiogenomic models	• GARD (Genomic Adjusted Radiation Dose) • RSI (Radiation Sensitivity Index)	• Tailors radiation therapy dose to the patient • Predicts tumor sensitivity to radiation	• Needs genomic input • Not suitable for all cancer types	(Juang et al., 2025; Mondal et al., 2023)
Microbiome-based models	• DeepGeni •Microbiome-guided prediction tools	• Evaluate cancer risk •Predict how patients may respond to Therapy	•Accuracy depends on sample quality •Microbiome varies widely between individuals	(Teixeira et al., 2024)

6.3 Role of Digital Health and Wearables

Digital health technology that are slowly being adopted in the treatment of cancer include wearable sensors, mobile applications, and remote monitoring systems. Wearables help to notify in advance about treatment-related toxicities or disease progression due to the constant monitoring of physiological parameters, including heart rate, activity level, and sleeping patterns (Garg et al., 2018; Liao et al., 2019). Through these tools proactive therapeutic intervention is made possible and allows the patient to receive immediate feedback (Liao et al., 2019). The digital health platforms promote the continuity and access to care by enabling remote consultations, symptom monitoring, and adherence monitoring. Wearable data is enhanced with clinical and molecular data to become a more precise medicine ecosystem.

6.4 Patient Engagement and Shared Decision-Making

Another aspect of precision cancer medicine, there is matching treatment to patients values, preferences and quality-of-life goals, on top of molecular targeting. Shared decision-making focuses on transparent communication to be able to make patients aware of their treatment options, possible risks, benefits, and uncertainties. Some of the tools that facilitate this process are decision aids, patient portals and structured communication frameworks (Irizarry et al., 2015; Smallwood et al., 2017). Engaging the patient actively will make the treatment plans physiologically correct and human-centred, as well as enhance adherence and establish trust. Finally, patient viewpoints will turn precision oncology into a comprehensive approach incorporating both science and humane care.

7. CHALLENGES AND BARRIERS

Though the idea of personalized cancer medicine is revolutionary and has a tremendous potential, it meets severe, many-sided challenges in the Ethical, Economic, and Translational domains. Such barriers should be handled in a systematic manner to maximize the advantages of precision oncology and provide equal access to personalized treatment to a variety of patients.

7.1 Data Privacy, Ethics, and Equity in Personalized Care

Precision oncology depends on genomic and clinical data that is of high quality and large volumes of data. The privacy, consent, and ownership concerns are compromised with the help of some of the most sensitive personal data (germline variants, ancestry signals, incidental findings) that can be obtained by the same datasets. Patients will also be reluctant to give data because of discrimination (insurance, employment) and commercialization of data and biospecimens. These complications increase the burden on sharing information across the board and postpone the progression of reference sets that can be utilized to render precision instruments precise throughout the population (Lajmi et al., 2024).

The ethical issues related to personalized cancer medicine are centred on informed consent especially in light of the complicated nature of the genomic information. It is not easy to make sure that the patients are actually aware of what they are signing, especially when there is a wide range of patients with different levels of literacy and awareness of their health (O'Doherty et al., 2021). Most recent cases such as the one in the Stanford Medical center demonstrate that the consent process may fail. Ethical issues can be caused by germline variants detected in tumour sequencing. Policies, standardized consent forms and better genetic counselling are needed by institutions (Tripura et al., 2025).

Most reference databases and clinical trials are primarily based on individuals of European descent, which is damaging to use biomarkers to other underrepresented groups (Lajmi et al., 2024). The result is that there is an unequal access to biomarker-driven medications, particularly to those patients in the African, East, and South Asian regions who are usually less likely to be eligible to receive targeted medicine. It is significant to the science as well as equity to enhance the genomic resources and develop region-specific cancer atlases. The existing failings in the healthcare disparity is aggravated by the fact that major genomic databases are not diverse enough and this restricts the efficacy of carrier screening and denies many patients access to crucial genomic data regarding their treatment of cancer (Green et al., 2024).

7.2 Cost and Accessibility of Personalized Medicine

Personalized cancer medicine is not performed widely because it requires a high amount of money, more than 20,000 dollars per patient. Genomic sequencing cost is considered as a huge challenge to many healthcare providers (Adeniji et al., 2021). Increased out-of-pocket costs result in medication adherence problems among the uninsured people. Low- and middle-income nations have a particular acute problem with this issue, as out-of-pocket expenses occupy most healthcare expenditures (Morsi et al., 2025).

Some patients can find it difficult to access care due to the geographic disparities in testing infrastructure. Five countries (South Africa, Kenya, Morocco, Nigeria, and Egypt) in Africa have the majority of next-generation sequencers, which restricts the choice of underserved regions. Delay in getting molecular test results also exists with most of them taking over 9 weeks to do this which can deteriorate the patient's conditions. The insurance coverage is complex, and there are few safeguards against the act of Genetic Information Non-discrimination. Also, adoption of precision medicine presents reimbursement issues to healthcare providers particularly in cases of rare cancers in which evidence of cost-effectiveness is needed (Moffitt et al., 2024).

7.3 Translating Research to Clinical Practice

Precision medicine has major obstacles to its application in normal healthcare. The issues that medical people are challenged with include time constraints, training deficit, and the accuracy issue. New drugs are being developed at a high pace, and it is not easy to keep up. Studies indicate that a significant number of patients fail to get their treatments and some of the concerns raised include the samples of biopsy, failure to order the test as well as a mistake in the treatment procedure despite positive information received. Individualized cancer therapy demands a shift in medical training and continuous education. Healthcare providers have a gap in knowledge about precision medicine (Sadik et al., 2022).

It requires effective cooperation between different specialists, which is hard to achieve due to the current structure of the medical practice. Genomic technology is a challenge because of the intricacy of the interpretation and testing standards (Chambers et al., 2016). Effective collaboration among various specialists is needed, but the current medical practice structure makes this difficult. Advancements in genomic technology create challenges due to the complexity of interpretation and testing standards (Shaaban & Ji, 2024).

The regulatory systems are failing to match with the rapid development of the precision medicine technologies. The rules of the gene therapies and tailored treatment in different countries vary, and this makes the regulatory environment confusing. The purpose of the revision of the GDPR by the EU is to ensure that it is more effective in safeguarding the data of patients and enhance innovation. The resemblance in the type of cancer however complicates standard clinical trials. There is a need to implement new regulations and quality checks of adaptive trials and real-world evidences (Masucci, Karlsson, Blomqvist, & Ernberg, 2024).

Precision medicine implementation in healthcare needs new infrastructure and new workflows. Healthcare systems can contribute to learning by leveraging evidence in clinical encounters to promote better care and require organizations with quality improvement as its main objective. Specialist shortages can be resolved by task-sharing in cancer care and training primary care providers. Nonetheless, it is difficult to translate molecular tumour boards and interdisciplinary teams to different environments (Bamodu & Chung, 2024).

Clinical benefit and analytic validity require evidence on the part of payers and regulators. The exact design used is sometimes inappropriate and therefore, other designs like basket, umbrella and platform trials and real-world evidence are vital in approval and reimbursement (Jazieh et al., 2024). Evidence expectations can be balanced to aid in adoption. Precision medicine needs good validation studies to ensure quality assurance; however, cancer is diverse, and the conventional trials do not suffice. The novel control strategies and surveillance mechanisms are required in adaptive trial design and generation of real-world evidence (Masucci, Karlsson, Blomqvist, & Ernberg, 2024).

8. FUTURE DIRECTIONS

Combining digital models, immune engineering and policies, precision oncology is rapidly gaining ground in the quest to provide affordable personalized care. Digital twins, new immunotherapies and universal access strategies are the three areas.

8.1 Integrating Digital Twins and Virtual Patients

Digital twins (DTs) are new technologies in the medical field of cancer that develop virtual replicas of a given patient. They simulate disease progression and effects of medication using different types of data. These models contribute to the prediction of successful treatment and strategy changes, which is possible in the case of personalized and proactive patient care (Giansanti & Morelli, 2025). The mathematical concepts of digital twins in oncology are the modelling of health trajectory with the help of the differential equations, the multi-omics data integration with the help of the Bayesian networks, the disease progression modelling with the help of Markov chains and the treatment optimization with the help of the reinforcement learning (Sun et al., 2022). The more recent developments have been applying machine learning with the intention of enhancing predictive capabilities such as recurrent neural networks and transformers to predict future health events using time-series clinical data (Vallée, 2025). This diversity of approaches and outcomes between studies highlights the need to have guidelines in the development and certification of digital twins.

There are also very limited studies which have been clinically validated with most showing partial validity and this speaks of the long strides which are required to get out of the experimental models to clinical use (Ştefănigă et al., 2024). The second step in the development of precision medicine is the merging of digital twins and artificial intelligence with big language models. The current progress of digital twins with the help of LLM points at the possibility of developing a more sophisticated model of a patient that could provide personalized, compassionate, and accurate performance in clinical management. It is possible to apply federated learning methods to allow a large number of healthcare providers to develop AI models without sharing personal identifiable information about a patient, which will allow them to collaborate at scale and protect privacy (Lammert et al., 2025). Digital twins are likely to go beyond treating individual patients to include public health and resource management. The further steps are possible to involve real-time monitoring and sophisticated data integration (Sadée et al., 2025).

8.2 Next-Generation Immunotherapies

Immune checkpoint inhibitors (ICIs) have revolutionized the treatment of cancer since its introduction, and over 150 immunotherapy approvals have been given by the FDA. Nevertheless, since the year 2011 the area has been evolving rapidly to more personalized and complex cancer treatments. The FDA approved the first TIL-based Therapy (lifileucel) against advanced melanoma and the first TCR-based Therapy (afamitresgene autoleucel) against severe melanoma. The first IL-15 agonist (nogapendekin alfa) received a license on the treatment of cancer of the bladder, and solid tumors are treated (Meng et al., 2024). Checkpoint inhibitors are no longer limited to the traditional PD-1/PD-L1 and CTLA-4 targets, but these targets alone represent 81% of all checks to date. The next-generation techniques have many different targets. Synthetic immunomodulators which trigger innate immune sensors, cytokine-based strategies, oncolytic viruses, cancer vaccines, adoptive cell therapies, bispecific antibodies and layers of immune regulation, including co-inhibitory and co-stimulatory checkpoint modulators (Kyriakidis et al., 2025). Chimeric antigen receptor (CAR) T-cell therapy is among the most significant personalized immunotherapies developed. The technology is at the fifth generation whereby each generation addresses the inadequacies of the other. The development of CAR-T therapy into solid tumors is another remarkable achievement because this has been a major impediment in the past because of the immunosuppressive tumor microenvironment (Bui et al., 2024). Fifth-generation CARs use drug-dependent switches to regulate the fourth-generation CARs which are dual-targeted and logic-gated designs designed to enhance specificity and minimize off-target effects. Mayo Clinic scientists have designed MC9999, a PD-L1 target but at the same time targets both tumour cells and immunosuppressed cells in the tumour microenvironment. Another paradigm shifts in precision medicine of using the own cells of the patient to fight the cancer is this approach which has been successful in a variety of solid tumour models, where the off-the-shelf allogeneic CAR-T therapies are readily accessible instead of needing weeks to make a single patient treatment. Such a solution would bring much greater accessibility to the system and lower the cost of specialized cells (Chen, Abila, & Mostafa Kamel, 2023). Bispecific T-cell engagers (BiTEs) are an immunotherapy method that is quite a recent, fast-growing modality that transfers the T cell to tumor cells without genomic modification. The FDA has granted the utilization of clinical proof-of-concept to counteract ICI resistance, which is based on the distinctive immunotherapeutic actions offered by tarlatamab, a bispecific DLL3 × CD3 T-cell enganger of small cell lung cancer (Zhao et al., 2025). The mechanism of action of these agents is that they jointly bind with tumour-associated antigens, and T-cell receptors and they effectively bridge the immune cells to the cancer cells to facilitate cancer destruction by the immune cells. In addition to traditional monoclonal approaches, the therapeutic arsenal is also receiving the addition of bispecific antibodies which bind specific combinations of immune checkpoints, tumour antigens, and immune effector molecules (Kyriakidis et al., 2025).

8.3 Towards Universal Access to Personalized Oncology

The inequity in access to precision oncology is pronounced on a global level and may contemplate additional issues with the existing healthcare issues. Not everyone can afford about two-thirds of the total precision medicines, non-Europeans, non-North Americans, non-Australians, and non-Japanese (Kyriakidis et al., 2025). In low- and middle-income countries out-of-pocket healthcare spending takes over fifty percent of the healthcare spending, in contrast to considerably less in rich countries. Limited access is caused by financial barriers, insufficient infrastructures, disparities in education, and rules. The population in five African countries is 71 percent, and it can take up to over 9 weeks to have delays in genomic testing. This may cause the patient to receive regular care until the result of the test (Johnson et al., 2021). The technology and distributed care networks are aimed at democratization of precision medicine by focusing on the new methods of accessing cancer treatment.

The network established by Karkinos in India is to provide care outside the major cities where most of the resources are centralized. With AI and digital integration, the accessibility can be increased, the under-served areas can be served, and remote consultations can be held with specialists (Ramachandran et al., 2022). New manufacturing approaches are being created to ensure personalized treatments have become more affordable and accessible. Serum Institute of India and Univercells Technologies are collaborating to reduce costs by 90 percent by local and small-scale manufacturing. This may help in cutting down on the time it takes to get the cancer treatments to the affected patients by up to 15 years to only three. Quality is guaranteed through automated systems where new therapies are accessed to patients at a faster rate (Lamichhane & Agrawal, 2023). The universal access to personalized oncology has a number of priorities. To decrease health inequities, first, there would be a high and continued investment in cancer management in low- and middle-income countries. Second, policymakers, business leaders, healthcare workers, and patient advocates should collaborate with each other to identify long-term solutions. Also, more training activities should be conducted to train locals on precision medicine. Lastly, to deal with equity in precision medicine, special attention should be paid to eliminating healthcare barriers and the introduction of new technologies in practice (Masucci, Karlsson, Blomqvist, & Ernberg, 2024). The advancement of the science of precision oncology should lead to fair access to such applications in both revenue settings. Equity in the workforce capacity, cost of drugs and diagnostic infrastructure, especially in the low- and middle-income countries (LMICs) can be improved through a tiered diagnostic approach. Regional genomic hubs and tele-MTBs can leverage on the experience and minimizing the per-sample cost of triage and then implement high-throughput sequencing based diagnosis materials with low costs. At the same time, an urgent necessity exists to invest in workforce training in genetic counselling, bioinformatics and molecular pathology that would result in equity by design (Al Saber et al., 2021).

9. CONCLUSION

Individualized care is a radical change in the cancer treatment process, which presents therapies based on the genetic, molecular, and clinical characteristics of each patient. With the combination of genomics, biomarkers, and use of high-level computational programs, clinicians will gain enhanced understanding of disease behavior, maximize treatment efficacy, and minimize toxicity. The continued interplay between next-generation sequencing, artificial intelligence and real-time patient monitoring is currently transforming oncology into a patient-centered discipline that is both data-driven and patient-centered. Nevertheless, the issues of data privacy, high-cost, and inaccessibility have to be overcome in order to make sure that these innovations promote the good of all populations in the world in an equitable manner. In the future, the combination of digital twins, adaptive therapies, and universal precision oncology systems has tremendous potential to transform cancer care in a new dimension- as an active versus reactive treatment, proactive versus predictive therapy, participatory versus passive therapy.

REFERENCES

Achary, P. G. R. (2020, September 1). Applications of Quantitative Structure-Activity Relationships (QSAR) based Virtual Screening in Drug Design: A Review. *Mini Reviews in Medicinal Chemistry*, *20*(14), 1375–1388. DOI: 10.2174/1389557520666200429102334 PMID: 32348219

Adeniji, A. A., Dulal, S., & Martin, M. G. (2021, June). Personalized Medicine in Oncology in the Developing World: Barriers and Concepts to Improve *Status Quo. World Journal of Oncology*, *12*(2–3), 50–60. DOI: 10.14740/wjon1345 PMID: 34046099

Ahluwalia, P., Ballur, K., Leeman, T., Vashisht, A., Singh, H., Omar, N., Mondal, A. K., Vaibhav, K., Baban, B., & Kolhe, R. (2024, January 23). Incorporating Novel Technologies in Precision Oncology for Colorectal Cancer: Advancing Personalized Medicine. *Cancers (Basel)*, *16*(3), 480. DOI: 10.3390/cancers16030480 PMID: 38339232

Al Saber, M., Biswas, P., Dey, D., Kaium, M. A., Islam, M. A., Tripty, M. I. A., Rahman, M. D. H., Rahaman, T. I., Biswas, M. Y., Paul, P., Rahman, M. A., Hasan, M. N., & Kim, B. (2021, December 23). A Comprehensive Review of Recent Advancements in Cancer Immunotherapy and Generation of CAR T Cell by CRISPR-Cas9. *Processes (Basel, Switzerland)*, *10*(1), 16. DOI: 10.3390/pr10010016

Ali, K. A., Choudhuri, S., Bishal, A., Das, T., Nandy, S., & Hossain, C. M. (2025). Approaches of AI and Advanced Computational Biology for Combating HIV Infections. In *AI and Precision Medicine in Infectious Disease Management* (1st ed., pp. 149–166). CRC Press. DOI: 10.1201/9781003615699-8

Ali, K. A., Choudhuri, S., Chakraborty, R., Pradhan, A., Kotal, A., & Shaharyar, M. A. A. I. (2025). Technology-Powered Precision Medicine and Smart Healthcare. In *AI and Precision Medicine in Infectious Disease Management* (1st ed., pp. 51–70). CRC Press. DOI: 10.1201/9781003615699-3

Ali, K. A., Mohin, S., Mondal, P., Goswami, S., Ghosh, S., & Choudhuri, S. (2024, March 29). Influence of artificial intelligence in modern pharmaceutical formulation and drug development*Future Journal of Pharmaceutical Sciences*, *10*(1), 53. DOI: 10.1186/s43094-024-00625-1

Bamodu, O. A., & Chung, C. C. (2024, August). Cancer Care Disparities: Overcoming Barriers to Cancer Control in Low- and Middle-Income Countries. *JCO Global Oncology*, (10), e2300439. DOI: 10.1200/GO.23.00439 PMID: 39173080

Bao, D., Liu, Z., Geng, Y., Li, L., Xu, H., Zhang, Y., Hu, L., Zhao, X., Zhao, Y., & Luo, D. (2022, December). Baseline MRI-based radiomics model assisted predicting disease progression in nasopharyngeal carcinoma patients with complete response after treatment. *Cancer Imaging; the Official Publication of the International Cancer Imaging Society*, *22*(1), 10. DOI: 10.1186/s40644-022-00448-4 PMID: 35090572

Bartolomucci, A., Nobrega, M., Ferrier, T., Dickinson, K., Kaorey, N., Nadeau, A., Castillo, A., & Burnier, J. V. (2025, March 24). Circulating tumor DNA to monitor treatment response in solid tumors and advance precision oncology. *NPJ Precision Oncology*, *9*(1), 84. DOI: 10.1038/s41698-025-00876-y PMID: 40122951

Bishal, A., Paul, P., Bera, S., Patra, D., Bandyopadhyay, B., Debnath, B., & Ali, K. A. (2025, August 5). Targeted Drug Delivery for Breast Cancer using Functionalized Liposomes: Preparation Methods, Challenges, and Clinical Translation. *AAPS PharmSciTech*, *26*(7), 207. DOI: 10.1208/s12249-025-03206-4 PMID: 40764411

Bode, A. M., & Dong, Z. (2018). Recent advances in precision oncology research. *NPJ Precision Oncology*, *2*(1), 11. DOI: 10.1038/s41698-018-0055-0 PMID: 30202789

Bui, T. A., Mei, H., Sang, R., Ortega, D. G., & Deng, W. (2024, August). Advancements and challenges in developing in vivo CAR T cell therapies for cancer treatment. *EBioMedicine*, *106*, 105266. DOI: 10.1016/j.ebiom.2024.105266 PMID: 39094262

Cararo Lopes, E., Sawant, A., Moore, D., Ke, H., Shi, F., Laddha, S., Chen, Y., Sharma, A., Naumann, J., Guo, J. Y., Gomez, M., Ibrahim, M., Smith, T. L., Riedlinger, G. M., Lattime, E. C., Trooskin, S., Ganesan, S., Su, X., Pasqualini, R., & White, E. (2023, June). Integrated metabolic and genetic analysis reveals distinct features of human differentiated thyroid cancer. *Clinical and Translational Medicine*, *13*(6), e1298. DOI: 10.1002/ctm2.1298 PMID: 37317665

Chakraborty, S., Sharma, G., Karmakar, S., & Banerjee, S. (2024). Multi-OMICS approaches in cancer biology: New era in cancer therapy. *Biochim Biophys Acta BBA - Mol Basis Dis.*, *1870*(5), 167120.

Chambers, D. A., Feero, W. G., & Khoury, M. J. (2016, May 10). Convergence of Implementation Science, Precision Medicine, and the Learning Health Care System: A New Model for Biomedical Research. *Journal of the American Medical Association*, *315*(18), 1941. DOI: 10.1001/jama.2016.3867 PMID: 27163980

Chang, C. Y., Park, H., Malone, D. C., Wang, C. Y., Wilson, D. L., Yeh, Y. M., Van Boemmel-Wegmann, S., & Lo-Ciganic, W.-H. (2020, March 25). Immune Checkpoint Inhibitors and Immune-Related Adverse Events in Patients With Advanced Melanoma: A Systematic Review and Network Meta-analysis. *JAMA Network Open*, *3*(3), e201611. DOI: 10.1001/jamanetworkopen.2020.1611 PMID: 32211869

Chapman, P. B., Hauschild, A., Robert, C., Haanen, J. B., Ascierto, P., Larkin, J., Dummer, R., Garbe, C., Testori, A., Maio, M., Hogg, D., Lorigan, P., Lebbe, C., Jouary, T., Schadendorf, D., Ribas, A., O'Day, S. J., Sosman, J. A., Kirkwood, J. M., & McArthur, G. A. (2011, June 30). Improved Survival with Vemurafenib in Melanoma with BRAF V600E Mutation. *The New England Journal of Medicine*, *364*(26), 2507–2516. DOI: 10.1056/NEJMoa1103782 PMID: 21639808

Chen, S., Zhou, Z., Li, Y., Du, Y., & Chen, G. (2023, October 27). Application of single-cell sequencing to the research of tumor microenvironment. *Frontiers in Immunology*, *14*, 1285540. DOI: 10.3389/fimmu.2023.1285540 PMID: 37965341

Chen, Y. J., Abila, B., & Mostafa Kamel, Y. (2023, January 21). CAR-T: What Is Next? *Cancers (Basel)*, *15*(3), 663. DOI: 10.3390/cancers15030663 PMID: 36765623

Chhabra, R. (2024, October 23). Molecular and modular intricacies of precision oncology. *Frontiers in Immunology*, *15*, 1476494. DOI: 10.3389/fimmu.2024.1476494 PMID: 39507541

Chitoran, E., Rotaru, V., Stefan, D. C., Gullo, G., & Simion, L. (2025, March 27). Blocking Tumoral Angiogenesis VEGF/VEGFR Pathway: Bevacizumab—20 Years of Therapeutic Success and Controversy. *Cancers (Basel), 17*(7), 1126. DOI: 10.3390/cancers17071126 PMID: 40227654

Chong, P. L., Vaigeshwari, V., Mohammed Reyasudin, B. K., Noor Hidayah, B. R. A., Tatchanaamoorti, P., Yeow, J. A., & Kong, F. Y. (2025, December 31). Integrating artificial intelligence in healthcare: Applications, challenges, and future directions. *Future Science OA, 11*(1), 2527505. DOI: 10.1080/20565623.2025.2527505 PMID: 40616302

Chung, C., Yeung, V. T. Y., & Wong, K. C. W. (2023, September). Prognostic and predictive biomarkers with therapeutic targets in breast cancer: A 2022 update on current developments, evidence, and recommendations. *Journal of Oncology Pharmacy Practice : Official Publication of the International Society of Oncology Pharmacy Practitioners, 29*(6), 1343–1360. DOI: 10.1177/10781552221119797 PMID: 35971313

Davies, H., Bignell, G. R., Cox, C., Stephens, P., Edkins, S., Clegg, S., Teague, J., Woffendin, H., Garnett, M. J., Bottomley, W., Davis, N., Dicks, E., Ewing, R., Floyd, Y., Gray, K., Hall, S., Hawes, R., Hughes, J., Kosmidou, V., & Futreal, P. A. (2002, June). Mutations of the BRAF gene in human cancer. *Nature, 417*(6892), 949–954. DOI: 10.1038/nature00766 PMID: 12068308

Deshmukh, V. G., Sapkal, S. B., Gadekar, S. S., & Deshmukh, V. (2025, September). EGFR inhibitors across generations: Progress, challenges, and future directions. *Journal of Molecular Structure, 1339*, 142326. DOI: 10.1016/j.molstruc.2025.142326

Dinh-Le, C., Chuang, R., Chokshi, S., & Mann, D. (2019, September 11). Wearable Health Technology and Electronic Health Record Integration: Scoping Review and Future Directions. *JMIR mHealth and uHealth, 7*(9), e12861. DOI: 10.2196/12861 PMID: 31512582

Dong, M., Wang, L., Hu, N., Rao, Y., Wang, Z., & Zhang, Y. (2025, August 29). Integration of multi-omics approaches in exploring intra-tumoral heterogeneity. *Cancer Cell International, 25*(1), 317. DOI: 10.1186/s12935-025-03944-2 PMID: 40883767

Forrest, S. J., Geoerger, B., & Janeway, K. A. (2018, February). Precision medicine in pediatric oncology. *Current Opinion in Pediatrics, 30*(1), 17–24. DOI: 10.1097/MOP.0000000000000570 PMID: 29189430

Frampton, J. E. (2021, September). Isatuximab: A Review of Its Use in Multiple Myeloma. *Targeted Oncology, 16*(5), 675–686. DOI: 10.1007/s11523-021-00827-0 PMID: 34351561

Fujiwara, Y., Kato, S., & Kurzrock, R. (2024, April). Evolution of Precision Oncology, Personalized Medicine, and Molecular Tumor Boards. *Surgical Oncology Clinics of North America, 33*(2), 197–216. DOI: 10.1016/j.soc.2023.12.004 PMID: 38401905

Gambardella, V., Tarazona, N., Cejalvo, J. M., Lombardi, P., Huerta, M., Roselló, S., Fleitas, T., Roda, D., & Cervantes, A. (2020, April 19). Personalized Medicine: Recent Progress in Cancer Therapy. *Cancers (Basel), 12*(4), 1009. DOI: 10.3390/cancers12041009 PMID: 32325878

Garcia, P. L., Miller, A. L., & Yoon, K. J. (2020, May 22). Patient-Derived Xenograft Models of Pancreatic Cancer: Overview and Comparison with Other Types of Models. *Cancers (Basel), 12*(5), 1327. DOI: 10.3390/cancers12051327 PMID: 32456018

Garg, S., Williams, N. L., Ip, A., & Dicker, A. P. (2018, December). Clinical Integration of Digital Solutions in Health Care: An Overview of the Current Landscape of Digital Technologies in Cancer Care. *JCO Clinical Cancer Informatics*, (2), 1–9. DOI: 10.1200/CCI.17.00159 PMID: 30652580

Garon, E. B., Rizvi, N. A., Hui, R., Leighl, N., Balmanoukian, A. S., Eder, J. P., Patnaik, A., Aggarwal, C., Gubens, M., Horn, L., Carcereny, E., Ahn, M.-J., Felip, E., Lee, J.-S., Hellmann, M. D., Hamid, O., Goldman, J. W., Soria, J.-C., Dolled-Filhart, M., & Gandhi, L. (2015, May 21). Pembrolizumab for the Treatment of Non–Small-Cell Lung Cancer. *The New England Journal of Medicine*, *372*(21), 2018–2028. DOI: 10.1056/NEJMoa1501824 PMID: 25891174

Ghoreyshi, N., Heidari, R., Farhadi, A., Chamanara, M., Farahani, N., Vahidi, M., & Behroozi, J. (2025, April 20). Next-generation sequencing in cancer diagnosis and treatment: Clinical applications and future directions. *Discover. Oncology*, *16*(1), 578. DOI: 10.1007/s12672-025-01816-9 PMID: 40253661

Giansanti, D., & Morelli, S. (2025, May 20). Exploring the Potential of Digital Twins in Cancer Treatment: A Narrative Review of Reviews. *Journal of Clinical Medicine*, *14*(10), 3574. DOI: 10.3390/jcm14103574 PMID: 40429568

Greco, A., Safi, D., Swami, U., Ginader, T., Milhem, M., & Zakharia, Y. (2019, December 5). Efficacy and Adverse Events in Metastatic Melanoma Patients Treated with Combination BRAF Plus MEK Inhibitors Versus BRAF Inhibitors: A Systematic Review. *Cancers (Basel)*, *11*(12), 1950. DOI: 10.3390/cancers11121950 PMID: 31817473

Green, S., Prainsack, B., & Sabatello, M. (2024, January). The Roots of (in)Equity in Precision Medicine: Gaps in the Discourse. *Personalized Medicine*, *21*(1), 5–9. DOI: 10.2217/pme-2023-0097 PMID: 38088178

Hai, L., Jiang, Z., Zhang, H., & Sun, Y. (2024, December 23). From multi-omics to predictive biomarker: AI in tumor microenvironment. *Frontiers in Immunology*, *15*, 1514977. DOI: 10.3389/fimmu.2024.1514977 PMID: 39763649

Hamdy, N. M., Basalious, E. B., El-Sisi, M. G., Nasr, M., Kabel, A. M., Nossier, E. S., & Abadi, A. H. (2024, November). Advancements in current one-size-fits-all therapies compared to future treatment innovations for better improved chemotherapeutic outcomes: A step-toward personalized medicine. *Current Medical Research and Opinion*, *40*(11), 1943–1961. DOI: 10.1080/03007995.2024.2416985 PMID: 39412377

Harding, J. J., Fan, J., Oh, D. Y., Choi, H. J., Kim, J. W., Chang, H. M., Bao, L., Sun, H.-C., Macarulla, T., Xie, F., Metges, J.-P., Ying, J., Bridgewater, J., Lee, M.-A., Tejani, M. A., Chen, E. Y., Kim, D. U., Wasan, H., Ducreux, M., & Zhao, H. (2023, July). Zanidatamab for HER2-amplified, unresectable, locally advanced or metastatic biliary tract cancer (HERIZON-BTC-01): A multicentre, single-arm, phase 2b study. *The Lancet. Oncology*, *24*(7), 772–782. DOI: 10.1016/S1470-2045(23)00242-5 PMID: 37276871

Heo, Y. J., Hwa, C., Lee, G. H., Park, J. M., & An, J. Y. (2021, July 31). Integrative Multi-Omics Approaches in Cancer Research: From Biological Networks to Clinical Subtypes. *Molecules and Cells*, *44*(7), 433–443. DOI: 10.14348/molcells.2021.0042 PMID: 34238766

Honkala, A., Malhotra, S. V., Kummar, S., & Junttila, M. R. (2022, February). Harnessing the predictive power of preclinical models for oncology drug development. *Nature Reviews. Drug Discovery, 21*(2), 99–114. DOI: 10.1038/s41573-021-00301-6 PMID: 34702990

Ikhtiar, F., Jamal, A., & Bokhari, S. M. S. M. (2025, September 16). Deciphering and targeting oncogenic pathways through integrated approaches and amino acid metabolism in hematologic malignancies. *Discover. Oncology, 16*(1), 1687. DOI: 10.1007/s12672-025-03535-7 PMID: 40956459

Irizarry, T., DeVito Dabbs, A., & Curran, C. R. (2015, June 23). Patient Portals and Patient Engagement: A State of the Science Review. *Journal of Medical Internet Research, 17*(6), e148. DOI: 10.2196/jmir.4255 PMID: 26104044

Isaic, A., Motofelea, N., Hoinoiu, T., Motofelea, A. C., Leancu, I. C., Stan, E., Gheorghe, S. R., Dutu, A. G., & Crintea, A. (2025, September 23). Next-Generation Sequencing: A Review of Its Transformative Impact on Cancer Diagnosis, Treatment, and Resistance Management. *Diagnostics (Basel), 15*(19), 2425. DOI: 10.3390/diagnostics15192425 PMID: 41095644

Jamalinia, M., & Weiskirchen, R. (2025, April). Advances in personalized medicine: Translating genomic insights into targeted therapies for cancer treatment. *Annals of Translational Medicine, 13*(2), 18–18. DOI: 10.21037/atm-25-34 PMID: 40438512

Jazieh, A. R., El Rouby, N., Guinigundo, A., Huelsman, K. M., Curran, E., Khan, R., Grund, J., Calvo, A. R., Claes, J. J., Overton, S. C., Hellard, S., Vasiliadis, L., Liu, M., & Blaxall, B. C. (2024, August 1). A Systematic Approach to Optimize the Implementation of Precision Oncology in Clinical Practice: A Meeting Proceeding. *Journal of Immunotherapy and Precision Oncology, 7*(3), 210–216. DOI: 10.36401/JIPO-23-41 PMID: 39219991

Ji, Y., Li, R., Tang, G., Xiao, Y., Ye, R., Shi, J., Geng, C., Ran, R., Zhu, C., Wang, W., Chen, C., & Yang, Q. (2025, July 14). Integrative proteomics and metabolomics analyses reveals the regulation of autophagy and ferroptosis by RAB10 through Slc37a2/mTOR pathway in breast cancer. *Cellular and Molecular Life Sciences : CMLS, 82*(1), 275. DOI: 10.1007/s00018-025-05788-5 PMID: 40658144

Jin, M., Fang, J., Peng, J., Wang, X., Xing, P., Jia, K., Hu, J., Wang, D., Ding, Y., Wang, X., Li, W., & Chen, Z. (2024, November 29). PD-1/PD-L1 immune checkpoint blockade in breast cancer: Research insights and sensitization strategies. *Molecular Cancer, 23*(1), 266. DOI: 10.1186/s12943-024-02176-8 PMID: 39614285

Johnson, K. B., Wei, W., Weeraratne, D., Frisse, M. E., Misulis, K., Rhee, K., Zhao, J., & Snowdon, J. L. (2021, January). Precision Medicine, AI, and the Future of Personalized Health Care. *Clinical and Translational Science, 14*(1), 86–93. DOI: 10.1111/cts.12884 PMID: 32961010

Juang, Y. R., Ang, L., & Seow, W. J. (2025, March 4). Predictive performance of risk prediction models for lung cancer incidence in Western and Asian countries: A systematic review and meta-analysis. *Scientific Reports, 15*(1), 4259. DOI: 10.1038/s41598-024-83875-6 PMID: 40038330

Kamińska, K., Nalejska, E., Kubiak, M., Wojtysiak, J., Żołna, Ł., Kowalewski, J., & Lewandowska, M. A. (2019, February). Prognostic and Predictive Epigenetic Biomarkers in Oncology. *Molecular Diagnosis & Therapy, 23*(1), 83–95. DOI: 10.1007/s40291-018-0371-7 PMID: 30523565

Kavran, A. J., Stuart, S. A., Hayashi, K. R., Basken, J. M., Brandhuber, B. J., & Ahn, N. G. (2022, March 22). Intermittent treatment of $BRAF^{V600E}$ melanoma cells delays resistance by adaptive resensitization to drug rechallenge. *Proceedings of the National Academy of Sciences of the United States of America, 119*(12), e2113535119. DOI: 10.1073/pnas.2113535119 PMID: 35290123

Khan, N., Marvel, F. A., Wang, J., & Martin, S. S. (2017, August). Digital Health Technologies to Promote Lifestyle Change and Adherence. *Current Treatment Options in Cardiovascular Medicine, 19*(8), 60. DOI: 10.1007/s11936-017-0560-4 PMID: 28647844

Khoury, R., Raffoul, C., Khater, C., & Hanna, C. (2025, July 7). Precision Medicine in Hematologic Malignancies: Evolving Concepts and Clinical Applications. *Biomedicines, 13*(7), 1654. DOI: 10.3390/biomedicines13071654 PMID: 40722725

Kim, E. S., Herbst, R. S., Wistuba, I. I., Lee, J. J., Blumenschein, G. R.Jr, Tsao, A., Stewart, D. J., Hicks, M. E., Erasmus, J.Jr, Gupta, S., Alden, C. M., Liu, S., Tang, X., Khuri, F. R., Tran, H. T., Johnson, B. E., Heymach, J. V., Mao, L., Fossella, F., & Hong, W. K. (2011, June 1). The BATTLE Trial: Personalizing Therapy for Lung Cancer. *Cancer Discovery, 1*(1), 44–53. DOI: 10.1158/2159-8274.CD-10-0010 PMID: 22586319

Kobayashi, S., Boggon, T. J., Dayaram, T., Jänne, P. A., Kocher, O., Meyerson, M., Johnson, B. E., Eck, M. J., Tenen, D. G., & Halmos, B. (2005, February 24). *EGFR* Mutation and Resistance of Non–Small-Cell Lung Cancer to Gefitinib. *The New England Journal of Medicine, 352*(8), 786–792. DOI: 10.1056/NEJMoa044238 PMID: 15728811

Kumar, A. R., Devan, A. R., Nair, B., Vinod, B. S., & Nath, L. R. (2021, December). Harnessing the immune system against cancer: Current immunotherapy approaches and therapeutic targets. *Molecular Biology Reports, 48*(12), 8075–8095. DOI: 10.1007/s11033-021-06752-9 PMID: 34671902

Kyriakidis, N. C., Echeverría, C. E., Bautista, J., Rivera-Orellana, S., Ramos-Medina, M. J., Salazar-Santoliva, C., Izquierdo-Condoy, J. S., Ortiz-Prado, E., Guerrero, S., & López-Cortés, A. (2025, August 28). Reprogramming cancer immunity with next-generation combination therapies. *Frontiers in Cell and Developmental Biology, 13*, 1652047. DOI: 10.3389/fcell.2025.1652047 PMID: 40950404

Laetsch, T. W., DuBois, S. G., Bender, J. G., Macy, M. E., & Moreno, L. (2021, March). Opportunities and Challenges in Drug Development for Pediatric Cancers. *Cancer Discovery, 11*(3), 545–559. DOI: 10.1158/2159-8290.CD-20-0779 PMID: 33277309

Lajmi, N., Alves-Vasconcelos, S., Tsiachristas, A., Haworth, A., Woods, K., Crichton, C., Noble, T., Salih, H., Várnai, K. A., Branford-White, H., Orrell, L., Osman, A., Bradley, K. M., Bonney, L., McGowan, D. R., Davies, J., Prime, M. S., & Hassan, A. B. (2024). Challenges and solutions to system-wide use of precision oncology as the standard of care paradigm. *Cambridge Prisms. Precision Medicine, 2*, e4. DOI: 10.1017/pcm.2024.1 PMID: 38699518

Lamichhane, P., & Agrawal, A. (2023, April). Precision medicine and implications in medical education. *Annals of Medicine and Surgery (London), 85*(4), 1342–1345. DOI: 10.1097/MS9.0000000000000298 PMID: 37113952

Lammert, J., Pfarr, N., Kuligin, L., Mathes, S., Dreyer, T., Modersohn, L., Metzger, P., Ferber, D., Kather, J. N., Truhn, D., Adams, L. C., Bressem, K. K., Lange, S., Schwamborn, K., Boeker, M., Kiechle, M., Schatz, U. A., Bronger, H., & Tschochohei, M. (2025, July 9). Large language models-enabled digital twins for precision medicine in rare gynecological tumors. *NPJ Digital Medicine*, *8*(1), 420. DOI: 10.1038/s41746-025-01810-z PMID: 40634659

Langenberg, K. P. S., Looze, E. J., & Molenaar, J. J. (2021, August 27). The Landscape of Pediatric Precision Oncology: Program Design, Actionable Alterations, and Clinical Trial Development. *Cancers (Basel)*, *13*(17), 4324. DOI: 10.3390/cancers13174324 PMID: 34503139

Larkin, J., Chiarion-Sileni, V., Gonzalez, R., Grob, J. J., Rutkowski, P., Lao, C. D., Cowey, C. L., Schadendorf, D., Wagstaff, J., Dummer, R., Ferrucci, P. F., Smylie, M., Hogg, D., Hill, A., Márquez-Rodas, I., Haanen, J., Guidoboni, M., Maio, M., Schöffski, P., & Wolchok, J. D. (2019, October 17). Five-Year Survival with Combined Nivolumab and Ipilimumab in Advanced Melanoma. *The New England Journal of Medicine*, *381*(16), 1535–1546. DOI: 10.1056/NEJMoa1910836 PMID: 31562797

Larocca, C. A., LeBoeuf, N. R., Silk, A. W., & Kaufman, H. L. (2020, December). An Update on the Role of Talimogene Laherparepvec (T-VEC) in the Treatment of Melanoma: Best Practices and Future Directions. *American Journal of Clinical Dermatology*, *21*(6), 821–832. DOI: 10.1007/s40257-020-00554-8 PMID: 32767272

Liao, Y., Thompson, C., Peterson, S., Mandrola, J., & Beg, M. S. (2019, May). The Future of Wearable Technologies and Remote Monitoring in Health Care. *American Society of Clinical Oncology Educational Book*, (39), 115–121. DOI: 10.1200/EDBK_238919 PMID: 31099626

Liu, J., Yang, L., Liu, D., Wu, Q., Yu, Y., Huang, X., Li, J., & Liu, S. (2025, September). The role of multi-omics in biomarker discovery, diagnosis, prognosis, and therapeutic monitoring of tissue repair and regeneration processes. *Journal of Orthopaedic Translation*, *54*, 131–151. DOI: 10.1016/j.jot.2025.07.006 PMID: 40822515

Lloyd, M. R., Spring, L. M., Bardia, A., & Wander, S. A. (2022, March 1). Mechanisms of Resistance to CDK4/6 Blockade in Advanced Hormone Receptor–positive, HER2-negative Breast Cancer and Emerging Therapeutic Opportunities. *Clinical Cancer Research*, *28*(5), 821–830. DOI: 10.1158/1078-0432.CCR-21-2947 PMID: 34725098

Lokhorst, H. M., Plesner, T., Laubach, J. P., Nahi, H., Gimsing, P., Hansson, M., Minnema, M. C., Lassen, U., Krejcik, J., Palumbo, A., van de Donk, N. W. C. J., Ahmadi, T., Khan, I., Uhlar, C. M., Wang, J., Sasser, A. K., Losic, N., Lisby, S., Basse, L., & Richardson, P. G. (2015, September 24). Targeting CD38 with Daratumumab Monotherapy in Multiple Myeloma. *The New England Journal of Medicine*, *373*(13), 1207–1219. DOI: 10.1056/NEJMoa1506348 PMID: 26308596

Maleki, E. H., Bahrami, A. R., & Matin, M. M. (2024, January). Cancer cell cycle heterogeneity as a critical determinant of therapeutic resistance. *Genes & Diseases*, *11*(1), 189–204. DOI: 10.1016/j.gendis.2022.11.025 PMID: 37588236

Marshall, M., Ivanovich, J., Schmitt, M., Helvie, A., Langsford, L., Casterline, J., & Ferguson, M. (2023, October 30). Pediatric precision oncology: "better three hours too soon than a minute too late.". *Frontiers in Oncology*, *13*, 1279953. DOI: 10.3389/fonc.2023.1279953 PMID: 38023209

Masucci, M., Karlsson, C., Blomqvist, L., & Ernberg, I. (2024, May 24). Bridging the Divide: A Review on the Implementation of Personalized Cancer Medicine. *Journal of Personalized Medicine*, *14*(6), 561. DOI: 10.3390/jpm14060561 PMID: 38929782

Maude, S. L., Frey, N., Shaw, P. A., Aplenc, R., Barrett, D. M., Bunin, N. J., Chew, A., Gonzalez, V. E., Zheng, Z., Lacey, S. F., Mahnke, Y. D., Melenhorst, J. J., Rheingold, S. R., Shen, A., Teachey, D. T., Levine, B. L., June, C. H., Porter, D. L., & Grupp, S. A. (2014, October 16). Chimeric Antigen Receptor T Cells for Sustained Remissions in Leukemia. *The New England Journal of Medicine*, *371*(16), 1507–1517. DOI: 10.1056/NEJMoa1407222 PMID: 25317870

McMullen, M., Karakasis, K., Madariaga, A., & Oza, A. M. (2020, June 17). Overcoming Platinum and PARP-Inhibitor Resistance in Ovarian Cancer. *Cancers (Basel)*, *12*(6), 1607. DOI: 10.3390/cancers12061607 PMID: 32560564

Meng, L., Wu, H., Wu, J., Ding, P., He, J., Sang, M., & Liu, L. (2024, January 4). Mechanisms of immune checkpoint inhibitors: Insights into the regulation of circular RNAS involved in cancer hallmarks. *Cell Death & Disease*, *15*(1), 3. DOI: 10.1038/s41419-023-06389-5 PMID: 38177102

Menyhárt, O., & Győrffy, B. (2021). Multi-omics approaches in cancer research with applications in tumor subtyping, prognosis, and diagnosis. *Computational and Structural Biotechnology Journal*, *19*, 949–960. DOI: 10.1016/j.csbj.2021.01.009 PMID: 33613862

Moffitt, L. R., Karimnia, N., Wilson, A. L., Stephens, A. N., Ho, G. Y., & Bilandzic, M. (2024, December 18). Challenges in Implementing Comprehensive Precision Medicine Screening for Ovarian Cancer. *Current Oncology (Toronto, Ont.)*, *31*(12), 8023–8038. DOI: 10.3390/curroncol31120592 PMID: 39727715

Molla, G., & Bitew, M. (2024, November 30). Revolutionizing Personalized Medicine: Synergy with Multi-Omics Data Generation, Main Hurdles, and Future Perspectives. *Biomedicines*, *12*(12), 2750. DOI: 10.3390/biomedicines12122750 PMID: 39767657

Mondal, D., Pareek, V., & Barthwal, M. (2023, January). Personalized medicine in radiation oncology and radiation sensitivity index: Pathbreaking genomic way to define the role of radiation in cancer management. *Journal of Cancer Research and Therapeutics*, *19*(Suppl 2), S508–S512. DOI: 10.4103/jcrt.jcrt_508_23 PMID: 38384012

O'Doherty, K. C., Shabani, M., Dove, E. S., Bentzen, H. B., Borry, P., Burgess, M. M., Chalmers, D., De Vries, J., Eckstein, L., Fullerton, S. M., Juengst, E., Kato, K., Kaye, J., Knoppers, B. M., Koenig, B. A., Manson, S. M., McGrail, K. M., McGuire, A. L., Meslin, E. M., & Burke, W. (2021, January). Toward better governance of human genomic data. *Nature Genetics*, *53*(1), 2–8. DOI: 10.1038/s41588-020-00742-6 PMID: 33414545

Pai, S., Bontempi, D., Hadzic, I., Prudente, V., Sokač, M., Chaunzwa, T. L., Bernatz, S., Hosny, A., Mak, R. H., Birkbak, N. J., & Aerts, H. J. W. L. (2024, March 15). Foundation model for cancer imaging biomarkers. *Nature Machine Intelligence*, *6*(3), 354–367. DOI: 10.1038/s42256-024-00807-9 PMID: 38523679

Palacín-Aliana, I., García-Romero, N., Asensi-Puig, A., Carrión-Navarro, J., González-Rumayor, V., & Ayuso-Sacido, Á. (2021, July 28). Clinical Utility of Liquid Biopsy-Based Actionable Mutations Detected via ddPCR. *Biomedicines, 9*(8), 906. DOI: 10.3390/biomedicines9080906 PMID: 34440110

Pandey, S., & Yadav, P. (2025, January). Liquid biopsy in cancer management: Integrating diagnostics and clinical applications. *Practical Laboratory Medicine, 43*, e00446. DOI: 10.1016/j.plabm.2024.e00446 PMID: 39839814

Pang, W., Chen, M., & Qin, Y. (2024, May 9). Prediction of anticancer drug sensitivity using an interpretable model guided by deep learning. *BMC Bioinformatics, 25*(1), 182. DOI: 10.1186/s12859-024-05669-x PMID: 38724920

Pérez-Callejo, D., González-Rincón, J., Sánchez, A., Provencio, M., & Sánchez-Beato, M. (2015, September). Action and resistance of monoclonal CD20 antibodies therapy in B-cell Non-Hodgkin Lymphomas. *Cancer Treatment Reviews, 41*(8), 680–689. DOI: 10.1016/j.ctrv.2015.05.007 PMID: 26045227

Morsi, , M.H., Elawfi, , B., Alsaad, , S.A., Nazar, . A., Mostafa, , H.A., & Awwad, , S.A. (2025, August). Unveiling the Disparities in the Field of Precision Medicine: A Perspective. *Health Science Reports, 8*(8), e71102. DOI: 10.1002/hsr2.71102 PMID: 40735262

Procureur, A., Simonaggio, A., Bibault, J. E., Oudard, S., & Vano, Y. A. (2021, February 8). Enhance the Immune Checkpoint Inhibitors Efficacy with Radiotherapy Induced Immunogenic Cell Death: A Comprehensive Review and Latest Developments. *Cancers (Basel), 13*(4), 678. DOI: 10.3390/cancers13040678 PMID: 33567530

Qoronfleh, M. W., & Al-Dewik, N. (2025, September 12). Cancer Biomarkers: Reflection on Recent Progress, Emerging Innovations, and the Clinical Horizon. *Cancers (Basel), 17*(18), 2981. DOI: 10.3390/cancers17182981 PMID: 41008828

Rafii, S., Mukherji, D., Komaranchath, A. S., Khalil, C., Iqbal, F., Abdelwahab, S. I., Abyad, A., Abuhelwa, A. Y., Gandikota, L., & Al-Shamsi, H. O. (2025, September 3). Advancing CAR T-Cell Therapy in Solid Tumors: Current Landscape and Future Directions. *Cancers (Basel), 17*(17), 2898. DOI: 10.3390/cancers17172898 PMID: 40940995

Rafiq, Z., Bashir, T., Lu, W., Puebla-Osorio, N., & Integrating, A. I. (2025, October 10). Machine Learning, and Animal Models for Precision Oncology: Bridging Preclinical and Clinical Gaps. *ACS Pharmacology & Translational Science, 8*(10), 3677–3683. DOI: 10.1021/acsptsci.5c00543 PMID: 41098564

Ramachandran, V., Pradhan, A., Kumar, A., Sarvepalli, B., Rao, S., Oswal, K., Kommu, R., Sharma, M., Pathak, S., Kunnambath, R., Kuriakose, M. A., Rengaswamy, S., Alajlani, M., & Arvanitis, T. N. (2022, September 1). A Distributed Cancer Care Model with a Technology-Driven Hub-and-Spoke and further Spoke Hierarchy: Findings from a Pilot Implementation Programme in Kerala, India. *Asian Pacific Journal of Cancer Prevention : APJCP, 23*(9), 3133–3139. DOI: 10.31557/APJCP.2022.23.9.3133 PMID: 36172676

Ramos, R., Moura, C. S., Costa, M., Lamas, N. J., Correia, R., Garcez, D., Pereira, J. M., Lindahl, T., Sousa, C., & Vale, N. (2025, February 21). Lung Cancer Therapy: The Role of Personalized Medicine. *Cancers (Basel), 17*(5), 725. DOI: 10.3390/cancers17050725 PMID: 40075573

Rituraj, , Pal, R. S., Wahlang, J., Pal, Y., Chaitanya, M. V. N. L., & Saxena, S. (2025, June 9). Rituraj, Pal RS, Wahlang J, Pal Y, Chaitanya M, Saxena S. Precision oncology: Transforming cancer care through personalized medicine. *Medical Oncology*, *42*(7), 246. DOI: 10.1007/s12032-025-02817-y PMID: 40488843

Rives, T. A., Collard, J., Li, N., Yan, D., Dietrich, C. S., Miller, R. W., Ueland, F. R., Pickarski, J., & Kolesar, J. M. (2024, September). Clinical Utility of Molecular Tumor Board Review for Identification of Possible Germline Pathogenic Variants on Tumor Next-Generation Sequencing Reports. *JCO Precision Oncology*, (8), e2400301. DOI: 10.1200/PO.24.00301 PMID: 39259913

Roisman, L. C., Kian, W., Anoze, A., Fuchs, V., Spector, M., Steiner, R., Kassel, L., Rechnitzer, G., Fried, I., Peled, N., & Bogot, N. R. (2023, November 21). Radiological artificial intelligence - predicting personalized immunotherapy outcomes in lung cancer. *NPJ Precision Oncology*, *7*(1), 125. DOI: 10.1038/s41698-023-00473-x PMID: 37990050

Ronemus, M., Bradford, D., Laster, Z., & Li, S. (2025, February 27). Exploring genome-transcriptome correlations in cancer. *Biochemical Society Transactions*, *53*(01), 113–120. DOI: 10.1042/BST20240108 PMID: 39910794

Sadée, C., Testa, S., Barba, T., Hartmann, K., Schuessler, M., Thieme, A., Church, G. M., Okoye, I., Hernandez-Boussard, T., Hood, L., Shmulevich, I., Kuhl, E., & Gevaert, O. (2025, July). Medical digital twins: Enabling precision medicine and medical artificial intelligence. *The Lancet. Digital Health*, *7*(7), 100864. DOI: 10.1016/j.landig.2025.02.004 PMID: 40518342

Sadik, H., Pritchard, D., Keeling, D. M., Policht, F., Riccelli, P., Stone, G., Finkel, K., Schreier, J., & Munksted, S. (2022, October). Impact of Clinical Practice Gaps on the Implementation of Personalized Medicine in Advanced Non–Small-Cell Lung Cancer. *JCO Precision Oncology*, (6), e2200246. DOI: 10.1200/PO.22.00246 PMID: 36315914

Sartori, F., Codicè, F., Caranzano, I., Rollo, C., Birolo, G., Fariselli, P., & Pancotti, C. (2025, May 28). A Comprehensive Review of Deep Learning Applications with Multi-Omics Data in Cancer Research. *Genes*, *16*(6), 648. DOI: 10.3390/genes16060648 PMID: 40565540

Satam, H., Joshi, K., Mangrolia, U., Waghoo, S., Zaidi, G., Rawool, S., Thakare, R. P., Banday, S., Mishra, A. K., Das, G., & Malonia, S. K. (2023, July 13). Next-Generation Sequencing Technology: Current Trends and Advancements. *Biology (Basel)*, *12*(7), 997. DOI: 10.3390/biology12070997 PMID: 37508427

Schiavone, M. L., Scarpitta, R., Ravera, F., Bleve, S., Reduzzi, C., Di Cocco, F., Dameri, M., Zoppoli, G., Naccarato, A. G., Nuzzo, P. V., Cristofanilli, M., Fanelli, G. N., & Scatena, C. (2025, September). Liquid biopsy in breast cancer: Redefining precision medicine. *The Journal of Liquid Biopsy*, *9*, 100312. DOI: 10.1016/j.jlb.2025.100312 PMID: 40740670

Schreiber, A. R., Andress, M., & Diamond, J. R. (2021, December 2). Tackling metastatic triple-negative breast cancer with sacituzumab govitecan. *Expert Review of Anticancer Therapy*, *21*(12), 1303–1311. DOI: 10.1080/14737140.2021.1993065 PMID: 34651524

Shaaban, S., & Ji, Y. (2024). Challenges of Clinical Pharmacogenomics Implementation in the Era of Precision Medicine. *Medical Research Archives*, *12*(11). DOI: 10.18103/mra.v12i11.5955

Shah, M., Patel, M., Shah, M., Patel, M., & Prajapati, M. (2024, October). Computational transformation in drug discovery: A comprehensive study on molecular docking and quantitative structure activity relationship (QSAR). *Intell Pharm.*, *2*(5), 589–595. DOI: 10.1016/j.ipha.2024.03.001

Shen, S., Qi, W., Liu, X., Zeng, J., Li, S., Zhu, X., Dong, C., Wang, B., Shi, Y., Yao, J., Wang, B., Jing, L., Cao, S., & Liang, G. (2025, March 19). From virtual to reality: Innovative practices of digital twins in tumor therapy. *Journal of Translational Medicine*, *23*(1), 348. DOI: 10.1186/s12967-025-06371-z PMID: 40108714

Singh, D., Dhiman, V. K., Pandey, M., Dhiman, V. K., Sharma, A., Pandey, H., Verma, S. K., & Pandey, R. (2024). Personalized medicine: An alternative for cancer treatment. *Cancer Treatment and Research Communications*, *42*, 100860. DOI: 10.1016/j.ctarc.2024.100860 PMID: 39827574

Smallwood, A. J., Schapira, M. M., Fedders, M., & Neuner, J. M. (2017, February). A pilot randomized controlled trial of a decision aid with tailored fracture risk tool delivered via a patient portal. *Osteoporosis International : A Journal Established as Result of Cooperation between the European Foundation for Osteoporosis and the National Osteoporosis Foundation of the USA*, *28*(2), 567–576. DOI: 10.1007/s00198-016-3767-4 PMID: 27647529

Snow, A., Chen, D., & Lang, J. E. (2019, November). The current status of the clinical utility of liquid biopsies in cancer. *Expert Review of Molecular Diagnostics*, *19*(11), 1031–1041. DOI: 10.1080/14737159.2019.1664290 PMID: 31482746

tefănigă, S. A., Cordo , A. A., Ivascu, T., Feier, C. V. I., Muntean, C., Stupinean, C. V., Călinici, T., Alua , M., & Bolboacă, S. D. (2024, November 13). Advancing Precision Oncology with Digital and Virtual Twins: A Scoping Review. *Cancers (Basel)*, *16*(22), 3817. DOI: 10.3390/cancers16223817 PMID: 39594772

Sun, T., He, X., Song, X., Shu, L., & Li, Z. (2022, July 14). The Digital Twin in Medicine: A Key to the Future of Healthcare? *Frontiers in Medicine*, *9*, 907066. DOI: 10.3389/fmed.2022.907066 PMID: 35911407

Sun, Y., Zhu, S., Ma, K., Liu, W., Yue, Y., Hu, G., Lu, H., & Chen, W. (2019, November 21). Identification of 12 cancer types through genome deep learning. *Scientific Reports*, *9*(1), 17256. DOI: 10.1038/s41598-019-53989-3 PMID: 31754222

Tarighati, E., Keivan, H., & Mahani, H. (2022). A review of prognostic and predictive biomarkers in breast cancer. *Clin Exp Med.* https://link.springer.com/10.1007/s10238-021-00781-1

Teixeira, M., Silva, F., Ferreira, R. M., Pereira, T., Figueiredo, C., & Oliveira, H. P. (2024, May 30). A review of machine learning methods for cancer characterization from microbiome data. *NPJ Precision Oncology*, *8*(1), 123. DOI: 10.1038/s41698-024-00617-7 PMID: 38816569

Topalian, S. L., Hodi, F. S., Brahmer, J. R., Gettinger, S. N., Smith, D. C., McDermott, D. F., Powderly, J. D., Sosman, J. A., Atkins, M. B., Leming, P. D., Spigel, D. R., Antonia, S. J., Drilon, A., Wolchok, J. D., Carvajal, R. D., McHenry, M. B., Hosein, F., Harbison, C. T., Grosso, J. F., & Sznol, M. (2019, October 1). Five-Year Survival and Correlates Among Patients With Advanced Melanoma, Renal Cell Carcinoma, or Non–Small Cell Lung Cancer Treated With Nivolumab. *JAMA Oncology*, *5*(10), 1411. DOI: 10.1001/jamaoncol.2019.2187 PMID: 31343665

Tripura, U., Nagrale, N. V., Singh, O. G., Dey, A., & Venkatesh, J. (2025). Ethical, Social, and Legal Issues Related to Precision Medicine in India: Challenges and Solutions. *Ann Afr Med.* https://journals.lww.com/10.4103/aam.aam_13_25

Vallée, A. (2025, August 5). Digital Twins for Personalized Medicine Require Epidemiological Data and Mathematical Modeling: Viewpoint. *Journal of Medical Internet Research*, *27*, e72411. DOI: 10.2196/72411 PMID: 40762974

Večurkovská, I., Roškovičová, V., & Kaťuchová, J. (2025, September 5). Challenges in the Clinical Application of Machine Learning for Pancreatic Cancer. *Bratislavské Lekárske Listy*, *126*(10), 2437–2450. DOI: 10.1007/s44411-025-00312-4

Velpula, T., & Buddolla, V. (2025, June). Enhancing detection and monitoring of circulating tumor cells: Integrative approaches in liquid biopsy advances. *The Journal of Liquid Biopsy*, *8*, 100297. DOI: 10.1016/j.jlb.2025.100297 PMID: 40453103

Vosoughi, P., Naghib, S. M., & Takdehghan, G. (2025, December). Machine learning in cancer prognostic and diagnostic biomarkers: A promising approach for early cancer detection. *Sensors and Actuators Reports*, *10*, 100385. DOI: 10.1016/j.snr.2025.100385

Wang, Y., Carter, B. Z., Li, Z., & Huang, X. (2022, January 7). Application of machine learning methods in clinical trials for precision medicine. *JAMIA Open*, *5*(1), ooab107. DOI: 10.1093/jamiaopen/ooab107 PMID: 35178503

Wang, Z., Zhou, L., & Wu, X. (2024). Influencing factors and solution strategies of chimeric antigen receptor T-cell therapy (CAR–T) cell immunotherapy. *Oncology Research*, *32*(9), 1479–1516. DOI: 10.32604/or.2024.048564 PMID: 39220130

Welti, M., Dimitriou, F., Gutzmer, R., & Dummer, R. (2022, November 8). Triple Combination of Immune Checkpoint Inhibitors and BRAF/MEK Inhibitors in BRAFV600 Melanoma: Current Status and Future Perspectives. *Cancers (Basel)*, *14*(22), 5489. DOI: 10.3390/cancers14225489 PMID: 36428582

Yang, H. Y., Ma, D., Liu, Y. R., Hu, X., Zhang, J., Wang, Z. H., Di, G.-H., Hu, X.-C., & Shao, Z.-M. (2017, April 25). Impact of hormone receptor status and distant recurrence-free interval on survival benefits from trastuzumab in HER2-positive metastatic breast cancer. *Scientific Reports*, *7*(1), 1134. DOI: 10.1038/s41598-017-00663-1 PMID: 28442763

Yi, M., Zheng, X., Niu, M., Zhu, S., Ge, H., & Wu, K. (2022, January 21). Combination strategies with PD-1/PD-L1 blockade: Current advances and future directions. *Molecular Cancer*, *21*(1), 28. DOI: 10.1186/s12943-021-01489-2 PMID: 35062949

Yousefi, S., Amrollahi, F., Amgad, M., Dong, C., Lewis, J. E., Song, C., Gutman, D. A., Halani, S. H., Velazquez Vega, J. E., Brat, D. J., & Cooper, L. A. D. (2017, September 15). Predicting clinical outcomes from large scale cancer genomic profiles with deep survival models. *Scientific Reports*, *7*(1), 11707. DOI: 10.1038/s41598-017-11817-6 PMID: 28916782

Zhang, Z., Xiao, Y., Liu, J., Xiao, F., Zeng, J., Zhu, H., Tu, W., & Guo, H. (2025, June 19). Machine learning-based MRI radiomics predict IL18 expression and overall survival of low-grade glioma patients. *NPJ Precision Oncology*, *9*(1), 196. DOI: 10.1038/s41698-025-00966-x PMID: 40533462

Zhao, S., Zhao, H., Yang, W., & Zhang, L. (2025, August). The next generation of immunotherapies for lung cancers. *Nature Reviews. Clinical Oncology*, *22*(8), 592–616. DOI: 10.1038/s41571-025-01035-9 PMID: 40528044

Zhong, L., Li, Y., Xiong, L., Wang, W., Wu, M., Yuan, T., Yang, W., Tian, C., Miao, Z., Wang, T., & Yang, S. (2021, May 31). Small molecules in targeted cancer therapy: Advances, challenges, and future perspectives. *Signal Transduction and Targeted Therapy*, *6*(1), 201. DOI: 10.1038/s41392-021-00572-w PMID: 34054126

Zhou, Y., Tao, L., Qiu, J., Xu, J., Yang, X., Zhang, Y., Tian, X., Guan, X., Cen, X., & Zhao, Y. (2024, May 20). Tumor biomarkers for diagnosis, prognosis and targeted therapy. *Signal Transduction and Targeted Therapy*, *9*(1), 132. DOI: 10.1038/s41392-024-01823-2 PMID: 38763973

Chapter 3
Human Digital Twin for Sustainable Cancer Healthcare

Manoj Mani
http://orcid.org/0009-0008-6601-0339
Bharathiar University, India

Antony Prabhu Jeyabal Philomenathan
http://orcid.org/0009-0001-6500-9538
Bharathiar University, India

Preethy Ganesa Moorthy
Bharathiar University, India

Velayuthaprabhu Shanmugam
Bharathiar University, India

Barkavi Chandrasekaran
Bharathiar University, India

Gavaskar Shanmugam
Bharathiar University, India

Sharmila Nagaraj
Bharathiar University, India

Vijaya Anand Arumugam
http://orcid.org/0000-0001-7485-1586
Bharathiar University, India

ABSTRACT

Human Digital Twins (HDTs) are revolutionizing by offering dynamic, personalized virtual embodiments that integrate multi-omics, biosensor, and clinical information to simulate tumor biology and therapeutics. HDTs allow bidirectional data exchange in real time, with updates providing support for predictive, preventive, and personalized cancer therapy management. This chapter explores the complex architecture of HDTs, cyber-physical-human system functionality, and incorporation of artificial intelligence, machine learning, and multi-scale modelling of tumors in pursuit of precision oncology. It further explores their implications for drug discovery, optimization of radiation therapy and immunotherapy, and in silico clinical trials, and discusses sustainability metrics alongside ethical, legal, and societal implications. Incorporation of emerging fields like quantum computing, augmented reality, and Twin-as-a-Service platforms has enormous promise for sustainable healthcare through HDTs. By conjoining patient-level information, HDTs help advance towards superior outcomes in oncology.

DOI: 10.4018/979-8-3373-9715-3.ch003

1. INTRODUCTION

The concept of the digital twin (DT) has progressed significantly, and its application has brought tremendous transformations in the field of healthcare, especially in oncology. Initially imagined to be static computational models, digital twins have transformed into active systems with a true representation of individual patient portraits by integrating medical data, physiological inputs, and environmental factors. This has led to personalized and accurate interventions with enhanced disease understanding and optimized efficacy to facilitate care planning. In oncology, DTs mimic intricate interactions of genetic and environmental factors and gain deeper disease mechanism insights to allow personalized therapeutic plans. Digital twins are virtual copies of patients and mimic intricate biological processes to allow clinicians to anticipate outcomes and customize care based on patient-specific attributes (Papachristou *et al.,* 2024). Digital twins have been further employed to aid patient-specific clinical decisions and improve research efforts beyond theoretical modelling. By modelling patient reactions at a molecular biological level, DTs provide predictive analyses to help determine optimal treatment paths based on the most appropriate treatments and identify adverse reactions proactively at population scales (Jensen & Deng, 2023). Moreover, tailored digital twins like Medical Necessity Twin and Care Navigator Twin have become integrated with healthcare infrastructure to aid clinical operations, track patient progress, and customize care interventions in real-time. These models rely on a patient's past data and adapt continuously to evolving health states (Pandey *et al.,* 2024).

A key feature distinguishing the Human Digital Twin (HDT) from earlier models is its ability to maintain active, two-directional interaction between the physical patient and their DT. Unlike static models based upon sporadic updates, HDTs constantly exchange data in real-time and thus revise patient records as fresh physiological signals, clinical findings, and environmental variables are recorded. This interactivity supports the incorporation of large historical data sets with artificial intelligence (AI) programs to create comprehensive predictions about disease advancement, drug response, and possible complications. By allowing active adjustments to therapy plans, HDTs facilitate personalized medicine and support individualized care plans. In addition to supporting healthcare delivery by analyzing workflows, resource assignments, and patient interaction, and thus enhancing clinical efficiency and patient safety, DTs further present transformative possibilities by addressing broader concerns beyond clinical outcomes alone. The approach to cancer care with DTs involves all clinical, economic, and ecological elements involved with healthcare and is known as the triple bottom line. At a clinical level, HDTs allow therapy plans to be clearly personalized to the individual patient by modelling therapeutic outcomes and interactions among genetic predispositions and environmental demands. This approach allows safer and more precise intervention by reducing unwarranted treatments and optimizing patient benefit. These modelling exercises enhance care delivery by highlighting patient-specific factors involved with disease advancement or drug resistance (Kemkar *et al.,* 2024).

The incorporation of AI and machine learning (ML) to high-dimensional treatments further enhances patient modelling by providing direct perspectives of disease progression, therapeutic outcomes, and patient reactions in real time. Such technology advancement enables healthcare practitioners to refine therapeutic plans based on new data developed to ensure plans of care align with changing health statuses among patients. Such technology fosters a data-driven, flexible approach accommodating precision medicine and furthering patient safety and compliance. Economically, HDTs create avenues for cancer therapy expenditure reduction through process streamlining and enhanced paradigms for decision-making (Basu, 2024). By modelling different options for therapy delivery, healthcare experts can evaluate the effectiveness and cost prospects of different therapy pathways prior to execution. This planning reduces excessive dependency upon trial-and-error techniques, and it reduces resources while lessening financial burden upon health facilities and patients equally. These process efficiencies enhance cost-effectiveness and care access by providing clinical practitioners with tools accommodating optimal allocation of resources (Pandey *et al.,* 2024).

Besides clinical and economic benefits, HDTs contribute significantly to promoting sustainable healthcare practices by reducing reliance upon tangible resources and minimizing wastage generation. For instance, with respect to nuclear medicine, theranostic DTs (TDTs) ensure precise administration of radiopharmaceuticals and thus minimize the need for excess materials and adverse environmental effects. This concerted approach enhances precise therapy delivery while supporting greener practices by lessening excessive exposure to radioactive agents. The application of DTs in this context is aligned with global efforts to reduce environmental footprint with respect to medical procedures and promote sustainable healthcare delivery. By harnessing advanced data modelling, real-time tracking with AI-driven insights, HDTs are revolutionizing cancer care delivery. By incorporating biological complexities with computational intelligence, they permit personalized, cost-effective, and environment-conscious therapy plans. As research unfolds further, it is predicted to see HDTs will take an increasingly prominent position in oncology with respect to proposing original solutions to challenges posed by healthcare today while opening doors to a more accurate and sustainable approach to patient care delivery (Brosch-Lenz *et al.,* 2023).

The current chapter relies on a narrative literature review of peer-reviewed scientific publications available from 2015 to 2025. The relevant literature was found by searching such databases as PubMed, Scopus, Web of Science, IEEE Xplore, and Google Scholar by means of the following keywords: Human Digital Twin, Digital Twin, Cancer, Precision Oncology, Artificial Intelligence, Machine Learning, Multi-omics, Sustainable Healthcare, and Digital Health. The original research articles, systematic reviews, methodological papers, and international reports related to oncology and digital twin technologies were included. The main focus was on recent studies of high quality while also considering seminal works that outlined the basic concepts of digital twin technology, precision medicine, and artificial intelligence in health care.

While there is much progress made in Human Digital Twin technology, the current body of research literature is somewhat scattered, with a vast majority of papers being concentrated around some particular aspects, namely computational modeling, artificial intelligence, medical imaging, and digital health infrastructure. There is currently a lack of review papers that comprehensively discuss the technological architecture, multi-scale tumor modeling, data standardization, prediction, sustainability, ethical regulation, and clinical application of Human Digital Twins in oncology. Moreover, insufficient effort is devoted to the analysis of the convergence of the mentioned cutting-edge technologies, such as federated learning, explainable AI, digital biomarkers, quantum computing, and Twin-as-a-Service technologies. To fill this gap, this chapter offers an integrated view on Human Digital Twin ecosystems in oncology by critically reviewing current evidence from biological, computational, clinical, and ethical perspectives. This work will not only summarize current trends in HDT research but will also discuss challenges, implementation barriers, sustainability issues, and future research directions in this field.

2. FOUNDATIONS OF HDTS IN CANCER CARE

Cyber-physical-human systems (CPHS) greatly enhance oncology through the efficient integration of patients', healthcare practitioners', and digital system roles to create an integrated healthcare system for enhanced cancer care. By combining cyber-physical components consisting of sensors and computational platforms with human interaction, CPHS creates an integrated environment whereby data-driven information reinforces clinical activities. With AI, CPHS ensures data capture is done instantly while allowing for predictive data analysis and decision aid tools to empower healthcare practitioners to make efficient care decisions (Matthies *et al.,* 2023). For oncology applications, CPHS addresses longstanding challenges like fragmented patient records, outdated communications plans, and limited interoperability through the adoption of wireless sensor networks and cloud information technology. This facilitates constant monitoring to aid early detection of disease progression and foster data-driven adaptivity through predictive models (Padmavathi *et al.,* 2024).

The fast adoption of digital health technology, spurred by the COVID-19 pandemic, has further accelerated the incorporation of CPHS into cancer care. Remote monitoring technology, consisting of wearable sensors and telemedicine platforms, has become a key element in cancer patient management, providing continuous and non-invasive health data capture with reduced hospital visits (Georgescu *et al.,* 2024). The technology facilitates patients' and healthcare providers' constant interaction and monitors patients' health information in real time, and can facilitate timely intervention irrespective of geographic boundaries. Moreover, discoveries in translational oncology have brought with them digital biomarkers and virtual studies consisting of simulated treatment situations and evaluation of patient-specific reactions, and thus have informed personalized medicine platforms. The application of DTs and virtual patient models to simulate diseases is one outstanding example of how the incorporation of CPHS can advance cancer research and clinical decision-making. Digital health tools now incorporate beyond telemedicine to wellness technology, early cancer screenings, and digital therapeutics, all helping to prevent cancer and improve its diagnosis and management (Cohen & Schwamm, 2024). Incorporation of all these technologies demands strong digital infrastructures, regulatory environment, and clinical environments to mitigate challenges concerned with cost barriers, technology oversaturation, and data security, and ultimately bring enhanced patient outcomes and transformed cancer care model (Georgescu *et al.,* 2024).

Interplay of data, modelling, insight, and action forms the fundamental basis of HDTs and hence advances precision medicine by virtue of a holistic and integrative approach. Data is the first pillar in the whole architecture, in that real-time pathophysiological signal streams are obtained from wearable sensors as well as varied monitoring devices. The information is complemented by patient-specific data obtained from multiomics analysis, including genomic, proteomic, and metabolomic sequencing, to collectively encompass the heterogeneity of cancer and its complicated biological circuits (Basu, 2024). Integration of diversified data streams ensures that the DT adequately mirrors the individualized disease signature of every patient, to allow for modelling of individualized cancer dynamics. Advanced computational models are the next important pillar of HDT frameworks, wherein mechanistic models simulate the biological processes, ML algorithms detect patterns from the high-dimensional data sets, and stochastic models account for variability and uncertainty. These models are successively refined through sensitivity analysis and uncertainty quantification methods to ensure that the predictions are tightly asymptotic to actual patient outcomes (Shahriyari *et al.,* 2022). Simulating tumor proliferation, interactions of the immune compartment, and mechanisms of drug resistance, these models give clinicians the capability to predict disease progression and tailor therapeutic intervention suitably.

Insights derived from these models are validated with proven clinical guidelines, such as those from the National Comprehensive Cancer Network (NCCN), to generate dynamic and individualized care pathways for patients. These pathways allow for actionable advice concerning treatment sequencing, dose changes, and monitoring procedures, thus matching customized simulations to evidence-based guidelines (Pandey *et al.,* 2024). The final element-action-is obtained through healthcare decision technology's provision of real-time output to healthcare practitioners, permitting adaptive treatment plans that can respond to changes in the patient's clinical state. The combination of these four elements-data, model, insight, and action-creates a closed-loop system that enhances the prediction of cancer progression, supports early onset of adverse reactions, and improves treatment methodologies, thus driving precision oncology and enhancing clinical results (Stahlberg *et al.,* 2022). Interdisciplinary collaborations between experts in AI, computational biology, and systems medicine underlie the development and deployment of healthcare decision technologies for oncological practice. These collaborations leverage advances in technology and subject matter expertise to build patient-specific models from various data sources. Technical advancements in data analytics, interoperability platforms, and cloud-based architecture are required to synthesize various patient data streams into integrated digital representations (Thirupathi *et al.,* 2024). Computational biology plays a pivotal role in modelling biological structures, allowing for simulations that predict disease pathways, therapeutic outcomes, and patient trajectories, while AI algorithms enhance the precision and scalability of these models (Kausar *et al.,* 2024).

Integration of generative AI with HDTs is revolutionizing the world of drug discovery by allowing for the exploration of vast chemical realms, predicting molecular interactions, and augmenting the characteristics of drugs, thereby accelerating pharmaceutical research and development (Mariam *et al.,* 2024). Moreover, the advancement of on-body sensor technology and multimodal AI is propelling the field of human body DTs, which allows for pervasive health monitoring, prevention of diseases, and personalized treatment plans (Tang *et al.,* 2024). Cooperative endeavors among computer scientists, biologists, and medical practitioners are fostering HDT application in clinical practice, while at the same time addressing ethical concerns such as data confidentiality, algorithmic fairness, and informed consent (Kausar *et al.,* 2024). Distinctive conceptual and practical distinctions of HDTs from other computational models, including simulation models and digital avatars, only serve to highlight the underlying primacy of HDTs in the realm of oncology. HDTs are accurate digital surrogates of individuals or their anatomical systems, directly connected to real-time streams of patient data through sensor networks and communication interfaces. Two-way interconnectivity enables HDTs to run AI and ML algorithms that constantly evaluate incoming data and provide actionable intelligence capable of influencing treatment decisions (Basu, 2024). Unlike simulation models, which often operate in isolation and rely upon static data sets, HDTs integrate live data to augment predictions and respond to changes in the patient state in real-time.

Digital avatars, which are frequently utilized in visualization and entertainment domains, generally serve as static representations that do not possess real-time interactive features. Conversely, HDTs are characterized by their ability to continuously refresh based on physiological signals, thereby ensuring that simulations accurately reflect the patient's shifting health conditions. This capacity for dynamism is essential in the realm of precision medicine, where treatment protocols must be adaptable to subtle fluctuations in disease advancement (Park *et al.,* 2023). Furthermore, HDTs utilize cognitive mimetic strategies, which emulate human perception, reasoning, and decision-making processes to enhance the simulation of patient reactions, a characteristic that is largely absent in traditional simulation frameworks and digital avatars (Saariluoma *et al.,* 2020). This cognitive modelling significantly improves the twins' capacity to decode intricate biological signals, thereby enabling more precise evaluations and informed clinical decision-making. The capability of HDTs to merge real-time physiological information with sophisticated computational models and cognitive paradigms distinguishes them as revolutionary instruments in the field of healthcare and beyond. Their implementation within oncology reinforces precision medicine, mitigates trial-and-error treatment approaches, and promotes patient safety through ongoing monitoring and responsive feedback mechanisms. As scholarly research advances and interdisciplinary partnerships proliferate, HDTs are likely to transform clinicians' methodologies in cancer treatment, providing novel avenues for personalized, efficient, and ethically sound healthcare delivery (el Ata, 2024).

3. ARCHITECTURE AND INFRASTRUCTURE OF CANCER-SPECIFIC HDTS

3.1 Layered Architecture

Optimally layered architecture design is at the heart of developing and fielding HDTs for cancer-specific use cases, as it facilitates the untethered combination of physical devices, computational models, and protocols of communication to provide bespoke healthcare solutions (Figure 1). It is the physical layer at the foundation of this architecture that assimilates patient-centric data acquisition through a network of clinical monitoring devices, wearable devices, and biosensors. These technologies, integrated into the backdrop of the Internet of Medical Things (IoMT), continuously capture real-time physiological health signs and biomarkers that are vital for early cancer identification and disease tracking. It is the DT-IoMT model that valorizes the importance of assimilating smart medical devices and sensors in facilitating continuous data acquisition, thus supporting dynamic adaptive healthcare environments (Kabir & Ray, 2024). The physical layer's efficacy is also augmented through the use of advanced AI-based wearable electronics and nanoscale devices that track biomarkers for cancer growth and treatment efficacy. Loaded with ML algorithms, devices interpret large datasets in real time, giving clinicians actionable information that enhances diagnostic precision and optimally improves outcomes for patients (Adeniyi *et al.,* 2024). The real-time output of continuously streamed patient-specific data is pivotal to the construction of the high-fidelity DT, as it ensures simulations and predictive models bear similarities to the physiological state of the patient at that moment. By assimilating such streams of data into a structured architecture, HDTs can simulate disease progression, anticipate treatment outcomes, and inform clinical intervention with heretofore unprecedented precision.

Figure 1. Layered Architecture of Cancer-Specific HDTs

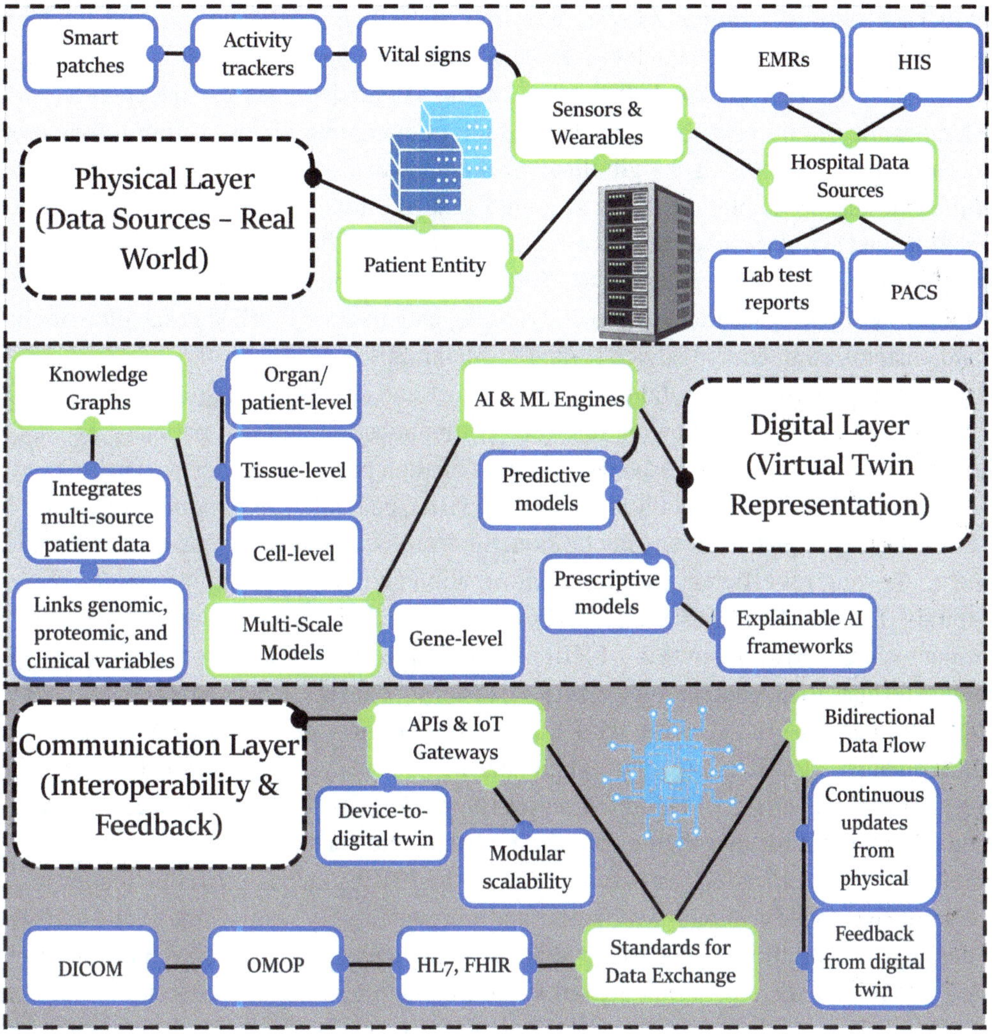

Digital architecture's computational nucleus is where disparate sets of data are fused, processed, and interpreted into meaningful health insights. An integrated reference information model facilitates such fusion by permitting the aggregation, normalizing, and visualization of information such that patient-specific data are fused in an optimal fashion. Multisource data, including real-time sensor inputs, genomic profiles, and stored clinical histories, are referenced against standard clinical guidelines such as those of the NCCN, allowing for the generation of customizable Cancer Care Paths (Pandey *et al.*, 2024). Also, the use of multiphysics, multiscale computational models, for instance, that account for tumor heterogeneity and patient-specific responses, enhances HDT's potential to model cancer development and adapt therapeutic interventions in such a manner that precision oncology is facilitated (Kemkar *et al.*, 2024).

To replicate tumor progression and treatment dynamics realistically, the digital layer enlists hybrid modelling platforms that assimilate biological complexity at scales. For instance, Hybrid Discrete Continuum (HDC) and Evolutionary Hybrid Cellular Automata (EHCA) model tumor growth by representing cellular behavior and extracellular matrix interactions, gaining insights into invading activities and morphological transformations during disease onset (Soos, 2012). Likewise, Multi-Level Parameterized Automata (MLPA) allows for tumor behavior to be modeled with immense precision by combining macroscopic electronic health records and microscopic genome data, permitting customized treatment plans (Gu & Chen, 2024). Hidden Markov Inducing a Dynamic Hierarchical Bayesian Network (HMDHBN) model enhances the digital layer by assimilating radiomic features for predicting tumor progression over space and time, augmenting longitudinal tracking of disease progression (Amiri & Mahjoub, 2019). Multiscale cell-based platforms that account for intracellular regulations, extracellular chemical communication, and cellular movement give detailed simulations that reproduce experimental findings and augment clinical decision-making (Jiang, 2007). Together, these modelling techniques strengthen the digital layer by allowing for the creation of dynamic, patient-specific HDTs that allow for predictive analysis and customized treatment advice (Wang & Maini, 2017).

The communication layer is pivotal to assuring secure and seamless data transmission between the physical and digital layers, thus allowing real-time communication between patient sensors, computational platforms, and clinical teams. Advanced communication protocols, underwritten by standards like HL7 and Fast Healthcare Interoperability Resources (FHIR), guarantee interoperability among disparate systems, ensuring regular data flow and integration. Such protocols are indispensable for chronic disease management, allowing for efficient sharing of information and underwriting infrastructure needed for smart hospitals (Kabir & Ray, 2024). The ONCO-FAIR (Findable, Accessible, Interoperable, and Reusable) project is illustrative of how HL7 FHIR can be adapted to oncology-specific needs by adapting data exchange formats to chemotherapy workflows, fostering data reuse, and adhering to the FAIR principles (Guilbert *et al.,* 2024). Likewise, the RECONNECT project in Ireland has shown the successful implementation of HL7 FHIR to integrate complicated healthcare systems, guaranteeing interoperability and bolstering the management of chronic diseases (Ngo *et al.,* 2024). These projects testify to the role of communication standards in assuring data-driven healthcare services.

Evermore, the communication tier is reinforced by the FHIR practice guideline that provides a single, universal data standardization, governance, and interoperability framework to allow HDT scaling to every healthcare institution. By adopting standard resources and secure communication protocols, healthcare systems can share data unhindered while maintaining patient confidentiality and ensuring regulatory compliance. As the communication tier assumes higher sophistication, it creates the foundation for interconnected healthcare environments wherein patient information is aggregated, analyzed, and responded to in real-time. Standardizing the physical, digital, and communication layers of this layered architecture enables the build-up of high-fidelity HDTs that can simulate, monitor, and optimize cancer care pathways. An amalgamation of real-time biosensor data, advanced computational simulations, and interoperable communication networks ensures that insights obtained by patients are continually updated and shaped to clinical guidelines. Standardizing data acquisition, computational analysis, and information exchange, HDTs provide a single frame of reference that propels precision medicine, improves outcomes for patients, and enables sustainable cancer care solutions. However, concomitant to innovation in technology is the consideration of ethical, security, and costs that are at the core of responsible and equitable implementation of HDT into clinical practice (Tang *et al.,* 2024).

3.2 Digital Thread and Twin Graph Networks

Digital threads and twin graph networks are innovative solutions to maintaining the continuity and integrity of patient data throughout the cancer care continuum-from initial diagnosis to treatment and survivorship. Underpinning that solution is the application of HDTs, which are holistic, virtual representations of individual patients. DTs combine vast amounts of personalized data, such as genetic profiles, biometric measures, and environmental exposures, to generate a multi-dimensional model that can simulate and foretell patient-specific reactions to various therapeutic interventions (Brosch-Lenz *et al.,* 2023). Through their provision of a living, holistic picture of the state of health of a patient, DTs extend the capability of clinicians to provide precision medicine, individually tailoring treatment regimens to the unique biological and lifestyle environment of each patient. The application of DTs in oncology care takes precision medicine to another level through the possibility of generating bespoke treatment policies and bidirectional therapeutic response tracking. It allows for adaptive treatment changes, improves clinical results, and enhances patient safety (tefănigă *et al.,* 2024). Significantly, theranostic digital twins (TDTs) have become an important asset in nuclear medicine, with diagnostic value to allow for the personalization of therapies by radiopharmaceutical. Using patient-specific information, TDTs optimize doses and delivery regimens, thus achieving maximal efficacy of systemic therapies and elevating survival rates for cancer patients. Real-time feeding of streams of integrated data into twin graphs ensures that clinicians access the latest and most relevant information from patients, permitting informed decision-making at every point of care (Hull *et al.,* 2023).

This continuous data integration is the foundation for the build-up of a resilient cancer data ecosystem, in which insights from the journey of each patient inform wider clinical practice and underpin the refinement of algorithms for therapies. Such data ecosystems drive efficiency, support care coordination, and lower healthcare costs by learning from trends and patterns within cohorts of patients (Alfano *et al.,* 2021). Nevertheless, the bringing together of disparate sources of heterogeneous data is challenging in terms of interoperability, scalability, and data stewardship. Securing the confidentiality and integrity of patient information is of critical importance, as virtual replicas, under the heading of DTs, depend upon the collation and analysis of sensitive health information from disparate domains (tefănigă *et al.,* 2024). Overcoming these challenges necessitates not only innovations in technology but also powerful ethical underpinnings to influence data stewardship and safeguard patients' rights. For secure, scalable, and resilient linking of disparate sets of data within a twin graph infrastructure, various methodologies for proposing and validating it in recent literature encompass hierarchical graph-based algorithms, including the Pyramid framework, that integrate heterogeneous data effectively by extracting relationships between disparate databases. The approach utilizes top-down matching and bottom-up encoding techniques to boost elements of schema transformation and entity matching by advanced computational methods of transformer models and contrastive learning for unsupervised feature synthesis (Jiang *et al.,* 2024). These advances allow for alignment of data structures as they boost the precision and efficacy of record linking across disparate sets of data.

Aside from algorithmic improvements, distributed computing platforms like Hadoop and Spark allow the computational infrastructure to be deployed to accommodate large-scale data integration. Though such systems allow for scalability and fault tolerance, they require extensive knowledge and resources to deploy, acting as impediments to mainstream use in clinical settings (Rozony *et al.,* 2024). Moreover, techniques like the three-party secure multi-party computation (MPC) framework have become important methodologies for the protection of sensitive health data during record linkage procedures. Enabling data to be safely integrated in a non-disclosing fashion to prevent exposure of patient identity, MPC frameworks enjoy dramatic increases in processing speeds and efficiency over comparable methods and become especially useful in healthcare application domains (Mağara *et al.,* 2024).

Resilient graph-based record linkage platforms that take advantage of parallel processing over distributed networks have been cornerstones to handling large datasets while supporting reliable and accurate data fusion. These platforms enhance real-time analytics by coping with the volume and variability of healthcare data to allow for seamless patient tracking and predictive modelling (Jupin & Shi, 2015). Challenges, nevertheless, can be seen in reconciling data from disparate sources, real-time fusion, and cross-domain interoperability. The future research and innovation are favorable to achieving said challenges, as future technologies must strike a balance between computational efficiency and the ethical implications that accompany data fusion (Rozony *et al.,* 2024). Under the union of digital thread and twin graph networks, persistence and contextual richness of the patient data can be preserved and utilized throughout the continuum of cancer care. Through supporting secure, scalable, and intelligent fusion of disparate datasets, those platforms lay the groundwork for a data-centric oncology environment that improves not only individual care but also adaptive learning and population health management. While many of the hurdles may be overwhelming, the continued progress of graph-based algorithms, distributed processing, and privacy-preserving methodologies predicts a future of integrated, patient-centric care.

3.3 Secure Real-Time Bidirectional Data Flow

Guaranteeing secure, real-time, bidirectional exchange of data is paramount to the successful operation of HDTs in oncology, wherein sensitive patient data needs to be constantly shared across various institutions without breaching confidentiality or data integrity. To tackle these issues, various mechanisms have been advanced, including blockchain technology, encryption-based communication channels, and federated learning (FL) protocols. These technologies collectively constitute a secure and reliable environment for sharing data while adhering to regulatory and ethical guidelines. FL has become an important approach to allow for collective model training of AI models by healthcare institutions without necessitating sensitive data exchange. The decentralized learning methodology enables various parties to train models collectively while maintaining patient data locally in servers, hence maintaining confidentiality. In oncology, wherein patient information is quite sensitive and governed by stringent protection policies, FL provides a scalable solution to amalgamate insights from disparate sources without jeopardizing data guardianship. Advanced versions of FL, including homomorphically encrypted federated learning (HEFL) and somewhat-homomorphically-encrypted federated learning (SHEFL), extend data protection by ensuring that only encrypted model parameters are exchanged during model training. These protocols guard against adversarial assaults, including model inversion and membership inference, by safeguarding sensitive health information even when model update institutes exchange model updates (Truhn *et al.,* 2024).

Blockchain technology enables FL through the provision of a tamper-proof and decentralized system for managing data exchanges. Through the use of secure and transparent ledgers, blockchain guarantees that data transactions are traceable and tamper-proof, hence stifling malicious activities such as poisoning of data or unauthorized tampering of data. An example of blockchain's application within FL is the FedTwin paradigm that employs consensus algorithms specifically tailored for use within DT networks. It applies differential methodologies of privacy and generative adversarial networks for the protection of local model parameters and elimination of spurious data, hence securing the validity and reliability of the global model (Qu *et al.,* 2022). Besides security, reduction of latency and maximization of data streaming's reliability are critical to enabling real-time clinical decision-making under HDT architectures. Smart streaming architectures ensure a dramatic reduction of data transmission and decoding times to enable fast processing of medical signals and clinical images. Optimized streaming architectures ensure maximal throughput for image segmentation and classification tasks to enable timely diagnosis with reduced operating costs. Such enhancements find application in oncology, where rapid interpretation of medical imaging can inform critical treatment decisions (Kulkarni *et al.,* 2023).

Real-time analytics is also needed in genomics, for which large volumes of sequencing data need to be processed through high-throughput processing. Prodduturi's research, cited by Ramakrishnan et al. highlights how stream processing and distributed computing platforms enable genomic data to be processed efficiently, speeding up diagnostic interpretation to inform preventive care plans. By making genomic analysis faster and more accurate, such systems enable clinicians to provide timely personalizations (Ramakrishnan *et al.,* 2025). Edge AI contributes complementarily by doing local processing of data at devices, thus lessening the computational burden of cloud infrastructure and cutting down latency. Desani and Reddy discuss how local data processing allows for real-time feedback for emergency intervention and chronic disease care. By carrying out analytics on-device, Edge AI safeguards patient confidentiality by not moving sensitive information to third-party server sites, as it also allows for quicker clinical actions (Desani & Reddy, 2020). In addition, research by Hirani and Zhao documents methods of reducing data-transfer delay. Having accelerated buffer-draining techniques allows for the quick identification and processing of incoming data streams, so that healthcare systems may instantly respond to patients' needs. Such latency-reduction measures are critical in oncology high-stakes domains, for which treatment refinement may be necessitated by abrupt changes in patients' status (Hirani & Zhao, 2021).

4. MULTI-SCALE MODELLING OF TUMOR BIOLOGY

Formation of hierarchical models that realistically represent tumor biology at various scales-from molecular and genetic to organ-system scales-requires integration of heterogeneous biological datasets with advanced computational methods (Figure 2). Such hierarchical models represent a fundamental necessity for de-mixing complex interactions intrinsic to biological systems, manifesting with feedback loops interconnecting higher-level physiological mechanisms with lower-level cellular-molecular dynamics (Nikolov *et al.,* 2019). Integrating data from heterogeneous sources, these models formulate executable schematics that can realistically simulate complex biological phenomena across different spatial and temporal scales, thereby providing a deeper insight into cancer pathogenesis and therapy responses (Fink & Herren, 1996). In silico cancer models take recourse to high-throughput datasets to generate diagnostic/prognostic molecular signatures, identify aberrant pathways, and reveal intrinsic mechanisms via both mechanistic simulations and statistical approaches (Edelman *et al.,* 2010). Mathematical modelling plays a central role here, given that it deals with biological complexities beyond several tiers-spanning DNA mutation events and signal transduction pathways to cell proliferation as well as tumor growth dynamics (Chowbina *et al.,* 2011). This multi-scale modelling framework integrates processes unfolding across different orders of magnitude within both spatial as well as temporal scales, which is a prerequisite for design as well as evaluation of targeted treatment strategies within clinical oncology. These models facilitate not only testing of a hypothesis related to biology but also offer predictive abilities; however, roadblocks persist in translational application to clinical scenarios of results emanating from pre-clinical simulations (Yankeelov *et al.,* 2016).

Figure 2. Multi-Scale Tumor Biology Modelling

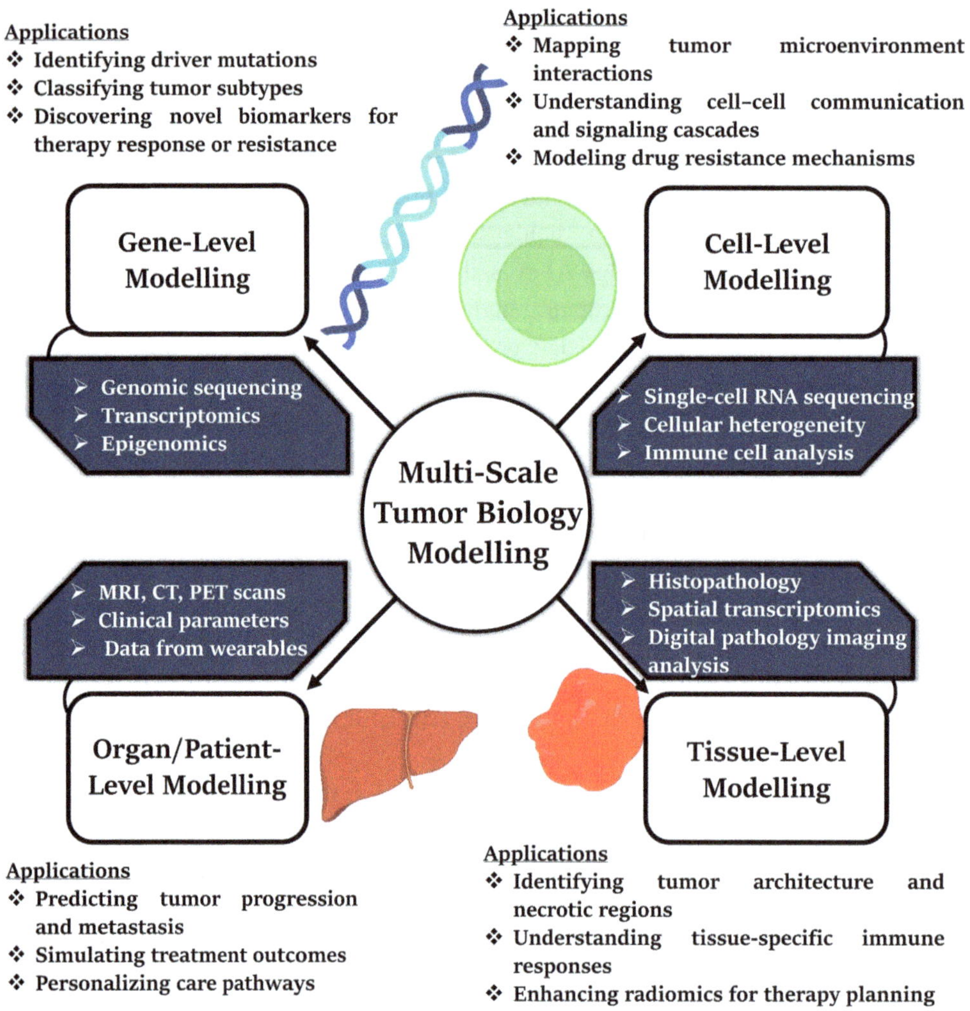

Advanced development of HDTs, dynamic predictive models tailored for individualized therapy, can be significantly enhanced with the integration of multi-omics data, spatial transcriptomics, digital pathology, radiomics, and multi-modal imaging modalities. Spatial multi-omics protocols that combine genomic, transcriptomic, proteomic, metabolomic, and epigenomic data yield a comprehensive view of cellular interplays and spatial heterogeneity found within tumors. These combined protocols represent a prerequisite to the construction of rich spatial atlases that deepen our knowledge regarding disease pathogenesis and support the design of treatment strategies (Liu *et al.,* 2024a). For example, the StereoMM is a tool that employs graph fusion models to combine histological images with spatial transcriptomic data, such that it enhances identification of tumor heterogeneity and spatial domains to boost DT representation accuracy (Luo *et al.,* 2025). Advanced computational algorithms such as the dual-path graph attention auto-encoder allow the integration of single-cell with space-mapped datasets, such that they can facilitate vigorous examination of spatial heterogeneity and raise tumor model fidelity (Lv *et al.,* 2024). Similar to this, the Multi-Omics Imaging Integration Toolset (MIIT) offers a collection of tools that facilitate the alignment of spatial transcriptomics and molecular imaging data, including mass spectrometry imaging, in order to capture biological variation in heterogeneous tumor complexes into DT systems (Wess *et al.,* 2024). These integration methods not only let us simulate tumor activity at scales beyond other dimensions, but they also serve as a foundation for global precision oncology projects where dynamic and rich models help tailor healthcare approaches (Li *et al.,* 2025).

A comprehensive understanding of tumor biology remains insufficient without the modelling of the tumor microenvironment (TME) and the immune landscape, both of which are pivotal in cancer progression, metastasis, and drug resistance. The TME comprises stromal cells, immune cells, cytokines, and extracellular matrix components, which together form a supportive niche conducive to tumor proliferation and immune evasion. Interactions within the TME promote mechanisms such as immunosuppression and therapy resistance, which complicate treatment outcomes (Liu *et al.,* 2024b). The immune landscape, characterized by the diversity and spatial arrangement of immune cells, further affects tumor behavior through processes like immunoediting, which imposes selective pressures that drive the evolution of immune-resistant cancer cells (Pérez-Romero *et al.,* 2020). The dynamic interactions between tumor cells and the immune microenvironment serve as a crucial determinant of cancer evolution. Immune checkpoints, cytokine signaling pathways, and cell-cell communication networks regulate the immune response, influencing the tumor's capacity to withstand therapeutic interventions (Popel, 2020). Furthermore, self-organizing behaviors within the TME, where tumor cells and adjacent stromal and immune constituents interact adaptively, establish a feedback loop that enhances tumor resilience and complicates treatment protocols. Addressing these adaptive interactions is essential for surmounting drug resistance and enhancing therapeutic efficacy (Asadi *et al.,* 2024).

Studying cancer evolution helps reveal much about tumor diversity and plasticity. Immune selection pressures and therapeutic interventions can lead to clonal evolution, yielding resistant subpopulations that subvert conventional treatment strategies. An awareness of this dynamic interplay enables novel strategies to be developed that anticipate and counteract tumor plasticity by exploiting vulnerabilities within these evolutionary programs (McCoach & Bivona, 2019). Through multi-scale modelling, spatial analysis, and immune kinetics, researchers can generate integrated tumor models that not only emulate disease progression but also help design adaptable, targeted therapies to confine tumor growth and counteract resistance.

5. DATA ACQUISITION, INTEGRATION, AND HARMONIZATION

Implementation of durable ETL (Extract, Transform, Load) pipelines in the sphere of cancer healthcare involves meticulous consideration with respect to gathering, integration, and harmonizing structured and unstructured data emanating from disparate data sources such as Electronic Medical Records (EMRs), Hospital Information Systems (HIS), Picture Archiving and Communication Systems (PACS), as well as instruments related to the IoMT. Thanks to its variety, volume, and volatility, complex oncology data necessitate the deployment of optimal practices to ensure data quality, interoperability, and reliability across clinical workflows. An elementary strategy is to utilize advanced analytical paradigms such as

Big Data architectures, AI, and ML algorithms to evaluate, standardize, and purify preponderant healthcare data before integration. Such paradigms play a critical role in discerning discrepancies, removing duplicates, and maintaining the semantic integrity of datasets to ensure interoperability among disparate systems (Manias *et al.,* 2023). AI-facilitated data validation pipeline deployment plays a critical role in verifying that structured data (e.g., laboratory data and imaging data) and unstructured data (e.g., clinical narrations and pathology reports) align with standardized terminologies of medicine to boost subsequent analytics with increased accuracy.

Given the increasingly dynamic landscape of healthcare data sources and regulatory frameworks, it is critical to adopt dynamic ETL approaches that enable scalable and adaptable pipelines. These utilize schema discovery algorithms coupled with rule-based transformations that share the ability to support automatic adaptation to new data sources with a consequent decrease in extensive manual reconfigurations. These adaptive solutions enable both increased operational efficiency as well as assurance with respect to regulatory mandates surrounding data integrity, data privacy, as well as data security (Mishra *et al.,* 2025). Through machine-readable rules integrated into the ETL workflow, organizations can support flexible pipelines that scale alongside progress within healthcare. To effectively bridge the automation with complex clinical data intricacies, hybrid ETL approaches-such as Dynamic ETL (D-ETL)-present a workable solution by synchronizing automated data extraction routines with leeway for manual overrides where complex transformations present a challenge. Through such approaches, transparency and traceability within data management increase to enable collaboration between data engineers as well as clinicians to work together efficiently with datasets requiring extensive coding or contextual interpretation (Ong *et al.,* 2017).

The Structural Data Capture (SDC) framework greatly enhances data harmonization by preserving the semantic, contextual, and structural content of oncology data throughout its exchange between systems. Aligning datasets with standardized medical terminologies such as Systematized Nomenclature of Medicine – Clinical Terms (SNOMED CT), SDC ensures that clinical data retains its definition across multiple platforms as a means to ensure interoperability and avoid misinterpretation (Goel *et al.,* 2021). Such standardization is especially important within oncology, where precise data representation is central to successful diagnosis, treatment planning, as well as outcome measurement. The establishment of complete clinical data warehouses is yet another core component of effective data gathering as well as harmonization. Precision medicine is a good example where data aggregation across a large scale can lead to systematic analysis of patient charts, imaging examinations, as well as genomic information to inform individualized treatment strategies. Clinicians can be able to discern trends, discover new biomarkers, as well as optimize treatment paths such that they lead to better care for the patient as well as durable long-term outcomes (Foran *et al.,* 2017).

Implementation of international standards like HL7, FHIR, Observational Medical Outcomes Partnership (OMOP), and Digital Imaging and Communications in Medicine (DICOM) supports the harmonization effort by providing structured frameworks for interoperability and semantic modelling. The extension of the OMOP Common Data Model (CDM), i.e., the Medical Imaging CDM (MI-CDM), plays a pivotal role in imaging data unification. By providing a means to connect DICOM sources and to monitor image feature origin, MI-CDM contributes to building standardized, computable imaging biomarkers and to devising phenotype definitions based on radiological data (Park *et al.,* 2024). LOINC coding with the Human Phenotype Ontology (HPO) aligned with laboratory data through FHIR standards is an example of a successful semantic integration strategy. It has been verified with real-world EMR data to be an effective means to discover biomarkers for diseases like asthma and has the potential to be applied to oncological research to reveal novel diagnostic markers (Zhang *et al.,* 2019). FAIRness for FHIR also stresses the importance of remaining compliant with the FAIR principles in health data management. In their FHIR4FAIR Implementation Guide, structured guidelines for standardizing and managing health datasets are established to ensure interoperability and reusability across both clinical and research environments (Martínez-García *et al.,* 2022). It promotes effective information exchange about health data with simultaneous safeguarding of patient anonymity and support for multi-institution collaborative research endeavors.

6. AI-DRIVEN PREDICTIVE AND PRESCRIPTIVE ANALYTICS

The areas of deep learning and ML have brought unprecedented advancement in cancer detection, radiogenomics, and survival prognosis for HDTs, thus enhancing oncological practices' accuracy and individualization. Radiomics-a computational framework to extract clinically relevant information from radiological images-has seen significant improvement with these AI-driven methods to allow for more advanced feature extraction and robust analytical procedures. This has led to enhanced outcomes for cancer diagnosis and therapy by providing superior imaging biomarkers to support clinicians' decision-making procedures (Vial *et al.,* 2018). The development of cancer DTs-virtual representations at a patient-specific scale to mimic tumorous dynamics-has been further enhanced by combining sophisticated mathematical modelling with ML proxies. These combined models efficiently predict tissue-level responses by addressing tumor heterogeneity and refining therapeutic protocols, and hence improving oncological intervention (Kemkar *et al.,* 2024).

The revolutionary effect of AI goes beyond diagnostic ability and has greatly contributed to cancer research through improvement in the ability to predict survival and improve therapy effectiveness for various cancer subtypes. Various methods, consisting of Support Vector Machines (SVM), Naïve Bayes classifiers, and Convolutional Neural Networks (CNN), have been widely employed for purposes ranging from classification of tumors to prognostic modelling. There is evidence to demonstrate that these models enhance diagnostic accuracy, improve workflow-based classification processes, and achieve firmer outcomes-based patient prediction, and hence ultimately improve the patient management and care processes (Zadeh Shirazi *et al.,* 2024). Moreover, AI incorporation with medical imaging and genomic analyses has resulted in decreased operational cost and increased access, significantly benefiting disadvantaged areas mainly because it makes possible enhanced diagnostic methods with reduced compromise on accuracy (Aftab *et al.,* 2025). The interpretive capacity of deep learning to handle complex genomic data sets has revolutionized cancer prognosis and makes it possible to perform accurate analyses for survival and contribute to the discovery of new biomarkers and hence create fresh directions for oncology research (Lee, 2023).

Reinforcement learning (RL), classified as a segment of AI, has revealed significant potential in modelling therapeutic responses and refining treatment protocols in the field of oncology. RL methodologies adjust treatment approaches dynamically, utilizing patient-specific information to facilitate personalized interventions that enhance clinical results. In the context of cancer therapy, RL frameworks have been demonstrated to improve overall survival rates by roughly 15%, prolong progression-free survival by 20%, and decrease severe adverse effects by 25%, illustrating the capability for more patient-focused care approaches (Dixit, 2025). Specifically in ovarian cancer, RL has been effectively utilized to model treatment pathways and discover innovative therapeutic regimens, highlighting its capacity to customize interventions and enhance survival rates (Murphy *et al.,* 2021). Applications within radiotherapy, exemplified by the MOAC-RT model, illustrate how RL refines treatment strategies by adjusting dosages and schedules in real-time based on tumor dynamics, achieving improved therapeutic outcomes with fewer treatment sessions and reduced cumulative doses (Khosravi & Roayaei, 2024). Hybrid models that merge mechanistic insights with data-driven learning significantly advance treatment planning by integrating historical patient data, genomic information, and continuous clinical monitoring. These models proficiently forecast therapeutic outcomes while considering obstacles such as data sparsity and computational complexity. Through the amalgamation of simulation-based reasoning and adaptive learning, hybrid models establish robust treatment pathways that are both accurate and ethically sound. Furthermore, they promote equity and alleviate biases that may emerge from diverse datasets, thereby enhancing clinical trust and decision-making processes (Ekundayo, 2024).

The integration of causal inference and explainable artificial intelligence (XAI) in HDTs is critical to achieving transparency and reliability necessary for clinical deployment. While AI systems show superior performance through pattern recognition, techniques for causal inference elucidate the basic cause-and-effect mechanisms embedded in the data and hence provide actionable knowledge for therapeutic interventions. This dual approach-employing predictive data analytics and causal reasoning-increases the efficiency of disease control and personalized therapy methods. The models of XAI and interpretable ML are critical to rendering complicated AI models interpretable to healthcare experts, thus facilitating their incorporation as normative clinical practices. Techniques like Shapley values and Local Interpretable Model-Agnostic Explanations (LIME) provide interpretable findings to doctors, transforming dark box models to credible decision-support tools (Oh, 2024).

The deployment of XAI is broader than healthcare and has implications for engineering applications like predictive maintenance, whereby AI models predict key parameters like the remaining useful life (RUL) of elements. By increasing interpretability and prediction accuracy, XAI frameworks aid proactive maintenance planning and enhance system reliability and performance (Kobayashi & Alam, 2024). Similarly, XAI aids in interpreting oncology diagnostic and survival prediction models and enables informed incorporation by clinicians of AI-driven recommendations into planning for treatments. By ensuring AI predictions are sound and actionable, causal inference and explainable tools improve the implementation of HDTs for precision medicine and allow physicians to tap AI's full benefits while ensuring accountability and ethical rigor. By integrating strategic deployment of ML, deep learning, reinforcement learning, causal inference, XAI, predictive and prescriptive analytics with HDTs are transforming cancer care. These tools allow for precise diagnosis and personalized planning for treatments and monitoring patients' health in real time while ensuring transparency and reliability, and an ethical foundation for AI-driven recommendations.

7. PERSONALIZATION OF CANCER THERAPY THROUGH HDTS

Virtual twin simulations, generally referred to as DTs, are increasingly recognized as novel tools in oncology to improve drug selection, select optimal drug dosages, and improve adaptive radiotherapy protocols. Based on patient-specific computational models, DTs bring a fundamentally personalized approach to therapeutic planning, allowing practitioners to tailor interventional measures to unique physiological and pathological characteristics. In the application to radiotherapy, DTs allow optimal radiation doses to be simulated and predicted, permit individualization of treatment schedules, and permit therapeutic response measurements to occur in near-real time. These capabilities improve therapeutic effectiveness while minimizing adverse side effects and improving patient outcomes and general quality of life (Sumini *et al.,* 2024). In pharmacology applications, HDTs bring data-driven drug discovery and development by integrating patient physiology and disease dynamics with corresponding virtual models. This approach allows for more accurate target selection, efficient drug delivery, and improvement of precision therapy through simulated representations of biological systems in humans. By incorporating intricate interactions at various biological scales, HDTs allow drug development campaigns to occur not only more quickly but also to a greater degree tailored to the unique requirements at the individual-patient level (Venkatesh *et al.,* 2024). Especially with reference to application to radiopharmaceutical therapy applications, TDTs apply physiologically based pharmacokinetic modelling to simulate drug behavior and therapeutic effect so as to permit virtual trial recommendations for optimal therapeutic regimens tailored to unique patient profiles (Abdollahi & Rahmim, 2022).

The implementation of in silico clinical trials via patient-specific DTs illustrates the significant potential of computational models within precision medicine. These virtual trials enhance pharmacological treatment options by replicating large patient populations and various therapeutic contexts, consequently diminishing dependence on conventional clinical testing and expediting drug development processes. Through the continuous refinement of treatment protocols via iterative modelling, in silico trials enhance the predictive precision concerning patient responses while safeguarding safety and efficacy (Sinisi *et al.,* 2020). Models such as the Multi-level Parameterized Automata (MLPA) framework amalgamate macroscopic patient data with microscopic molecular information, facilitating simulations of cancer progression and treatment reactions across different scales. This methodology offers a comprehensive platform for optimizing personalized cancer therapies and informing clinical decisions (Gu & Chen, 2024). The predictive abilities of HDTs extend into the domain of immunotherapy and combination treatments, where the personalization of therapy poses particular difficulties due to inherent biological complexities. Utilizing case-based reasoning and twin libraries enables DTs to emulate individual patient responses to various therapeutic strategies, thereby directing the formulation of personalized treatment regimens. For example, in the context of triple-negative breast cancer (TNBC), DT models have been utilized to forecast patient-specific reactions to neoadjuvant therapies. These models have shown a capacity to enhance pathological complete response rates by fine-tuning treatment schedules without increasing toxicity, thereby facilitating safer and more effective treatment modalities (Wu *et al.,* 2024b).

The integration of host-disease interactions with the planning of immunotherapies is of utmost significance, considering immune-mediated diseases show significant heterogeneity in patient outcomes owing to the intricate interplay present at the level of the TME. By integrating multimodal predictors to capture antitumor immune reactions with all their subtleties, DTs allow optimal combinations of immunotherapy to identify the best therapeutic strategies, at once addressing different suppressive pathways (Roelofsen & Thommen, 2022). Immune DTs, reflecting the fluctuating and multi-faceted nature of immune interactions, further enhance prediction by modelling immune cells', cytokines', and signal networks' actions to ultimately determine therapy outcomes. These models hold significant promise for furthering personalized immunotherapy strategies by acknowledging immune reaction heterogeneity amongst varied populations of patients. Despite DTs' enormous potential to revolutionize cancer therapy, however, some significant hurdles remain to their clinical implementation. The seamless incorporation of data from a range of sources-various and varied data from genomically encoded information to imaging experiments-forms an indispensable precursor to aggregating strong and reliable models reflecting biological complexity adequately. Moreover, designing computational models with simulable interactions present at all levels within the immune system and other biological networks is a task of considerable difficulty. Verification of models to ensure validity and overcoming data sparsity and biases are critical areas needing concerted effort to facilitate DT inventions' translation to mainstream clinical utility (Niarakis *et al.,* 2024).

8. HDTS IN MONITORING, PROGNOSTICS, AND VIRTUAL TRIALS

Digital biomarkers (DBs) represent a paradigm change in disease monitoring, prognostic assessments, and therapeutic management by employing advanced digital health tools consisting of wearables, mobile phones, and medical sensors. These tools facilitate constant and non-invasive tracking of physiological, behavioral, and environmental variables and hence permit comprehensive assessment of a person's state of health in real time. By integrating data sources from these tools, healthcare experts and scientists are able to gain critical information about disease advancement, relapse risk, and therapy-related toxicities, and hence allow timely and personalized healthcare administration (Smokovski *et al.,* 2024). For individuals with substance use disorder, with a spotlight on opioid use disorder, digital biomarkers have proven significant utility for detecting and tracking physiological changes associated with opioid consumption, withdrawal syndromes, and situations of overdose. By continuously monitoring vital signs, circadian patterns of mobility and motion, and biochemical signals, DBs aid in controlled administration of opioids and implementation of harm reduction strategies to minimize risks for addiction and death (Rigatti *et al.,* 2023). Online patient monitoring tools like RADAR-based are prime expressions of employing digital biomarkers towards long-term tracking of health for a range of state situations for diseases. By collecting data from true situations over extended periods, these tools aid clinical decision-making processes and help towards personalizing management strategies (Rashid *et al.,* 2023).

Aside from clinical monitoring, digital biomarkers are increasingly being integrated with preclinical drug discovery and testing frameworks. The continuous, non-obtrusive monitoring of experimental subjects in their native environments allows scientists to assess drug safety and efficacy with greater ecological validity. The approach reduces the need to perform invasive measures and repetitive clinical visits while offering a more accurate representation of physiological reactions to drug prospects (Baran *et al.,* 2022). With further breakthroughs in data science and computerized intelligence, the notion of digital biomarkers is expanding and emphasizing its prospects to revolutionize early diagnosis and personalized drug therapy techniques. By supporting bedside health assessments in real time, these tools are poised to improve patient outcomes further, reduce healthcare costs, and align resource allocation more effectively (Au *et al.,* 2022). HDTs improve digital biomarker functionality with sophisticated, computational models that capture human physiology, disease trajectories, and pharmacological intervention responses. In silico clinical trials using HDTs offer data-driven paradigms to simulate drug effects and biological interactions and thereby enhance drug discovery and development workflows' effectiveness and accuracy. In silico clinical trials allow for precision medicine and personalized clinical decision-making by modelling patient-specific variations in disease mechanisms and therapeutic outcomes, and have significant applications for public health surveillance and interventional approaches (Venkatesh *et al.,* 2024).

Unlike traditional clinical trials, in silico trials conducted via HDTs hold significant advantages when it comes to cost-effectiveness, scalability, and safety. By generating models of simulated patient populations, these trials allow drug prospects and medical equipment to be tested with minimal need for exhaustive in vivo studies and thus relieve concerns associated with animal research while further reducing trial phases and financial burdens (Badano, 2021; Nicolo *et al.,* 2023). Moreover, adoption of virtual trials can facilitate quicker commercialization of new therapeutic agents and secure broader and sooner access to key therapies to wider populations while, at the same time, enhancing medical equipment safety profiles by allowing adverse reactions to be identified at an early stage. Despite these promising breakthroughs, the adoption of HDTs and in silico trials via current regulatory schemes poses challenges owing to the nascent stage of governance frameworks and overseeing agencies. The absence of standardized guidelines and regulatory clarity has prevented large-scale adoption by emphasizing initiatives geared towards the establishment of protocols favoring innovation and ensuring data integrity while fostering healthy competition at an industry level. Overcoming these hurdles requires collaborative efforts among regulatory agencies, industry players, and research scholars to develop protocols striking a balance between patient safety and technological advancement (Venkatesh *et al.,* 2024; Roesler *et al.,* 2024).

9. SUSTAINABILITY METRICS AND DECISION INTELLIGENCE

HDTs represent a major breakthrough in healthcare by offering a system to reduce incidents of overdiagnosis, overtreatment, and unnecessary diagnostic procedures while maintaining or improving clinical outcomes. By creating digital models reflecting an individual's biological, behavioral, and environmental variables, HDTs allow for extremely personalized healthcare measures. The approach encourages precise risk profiling and personalized management measures, especially for non-communicable diseases (NCDs), in which standardized methods are prone to producing unnecessary measures (el Ata, 2024). By integrating sprawling datasets ranging from EMRs to genomics data, HDTs can simulate patient-specific disease trajectories and thus enhance early discovery and preventive care measures. In managing cardiovascular diseases, for instance, digital constructs generated by HDTs assist healthcare practitioners in identifying individuals at high risk and thus allow for strategic measures and reduction of unnecessary investigation and procedures (Dziopa *et al.,* 2025). The application of HDTs in personalized care has brought measurable improvement to clinical outcomes, with studies pointing to a success level of up to 80% in reducing unwanted measures among various diseases (Shen *et al.,* 2024). By tailoring health management measures to individual risk portfolios, HDTs empower healthcare practitioners to adopt precise care measures and thus mitigate nonessential diagnostic appraisals and invasive treatments. Additionally, combining advanced ML technology with comprehensive longitudinal health data sets enables risk assessments and reduction for digital overdiagnosis-a growing menace with growing dependency by the healthcare industry on superior diagnostic tools. This is critical to address since unchecked overdiagnosis is not only expensive but further exposes patients to preventable risks perpetrated by unnecessary explorations and treatments (Capurro *et al.,* 2022).

Given the increasing computational demands of AI-driven healthcare solutions, ensuring energy efficiency and minimizing carbon footprints within HDT pipelines has become an essential aspect of sustainable healthcare delivery. The exponential growth in AI's energy consumption-surging by approximately 300,000-fold since 2012-highlights the need for integrative strategies that optimize energy use throughout the AI lifecycle, from algorithm development to deployment (Hidalgo *et al.,* 2023). A holistic, full-stack approach is recommended, wherein energy efficiency is embedded at multiple levels, including model configuration, hardware optimization, and software-hardware co-design, to balance computational performance with sustainability objectives. Reinforcement learning and predictive analytics have been applied to enhance energy-aware scheduling and workload allocation in high-performance computing environments, enabling smarter resource utilization while maintaining system responsiveness. By optimizing computational tasks based on energy demand forecasts, these AI-driven techniques can dynamically adapt workloads to achieve maximum efficiency without compromising performance (Jalasri *et al.,* 2025). Moreover, the development of standardized methodologies for calculating carbon emissions and assessing energy footprints is critical for establishing transparent sustainability metrics. Future research efforts should aim to refine emission calculation protocols and develop energy-efficient AI architectures that can be systematically evaluated and benchmarked across healthcare applications (Ariyanti & Suryanegara, 2024).

Beyond promoting algorithmic efficiency, sustainable computing environments demand a comprehensive approach entailing operational and embodied emissions of carbon. Green initiatives aimed at reducing the carbon footprint left by AI technologies need to cover elements like data center management, hardware creation, and lifecycle tuning. Adopting smart energy management schemes integrated with AI-driven analytics has significant potential for supporting simultaneous energy reduction while maintaining high computational performance levels (Wu *et al.,* 2024a). These schemes can efficiently schedule process activities, allocate loads to diverse servers, and adjust cooling requirements and hence limit excessive energy consumption. Addressing AI-driven healthcare solutions' environmental footprint also demands setting up platforms supporting responsible innovation and the dissemination of sustainable technology to all populations. By integrating energy optimization methods with clinical decision-support schemes and by supporting inter-disciplinary research among data scientists, engineers, and healthcare researchers and practitioners, the AI community can aid equitable access to environment-friendly healthcare technology. Integrating precision health models like HDTs with smart energy schemes opens avenues towards a path with enhanced healthcare environments with reduced carbon footprints and hence ultimately supports global initiatives to reduce greenhouse emissions and promote sustainable development (Hidalgo *et al.,* 2023).

10. ETHICAL, LEGAL, AND SOCIAL IMPLICATIONS

The establishment of trust, agency, and accountability when applying ethical HDTs necessitates integrating comprehensive frameworks and strategies to manage the complexities of human-centric design, data security, and governance. As part of this process, a Human Factors Engineering (HFE) approach is crucial and emphasizes humanity-centric methods with trust and usability at their foundation when running DT systems. By managing data privacy issues, security issues, and user experience issues, HFE frameworks promote clarity and functionality when it comes to employing HDTs and ensure confidence among stakeholders and broader application (Pietrantoni *et al.,* 2024). On this basis, a Trust Framework for DTs-initially conceived with industrial applications-offers useful recommendations to transfer to healthcare facilities. The recommendations range from explaining operational DT mechanics to promoting intellectual property protection and employing strict IT security protocols-crucial components to ensure user and stakeholder trust (Trauer *et al.,* 2022). Beyond this basis, other novel ideas like Semantic Twins and Self-Sovereign Identities (SSIs) assist with establishing trustworthiness with high-risk applications by allowing individuals to control data while securing its integrity. Though SSIs satisfy minimum trust requirements, actual implementation must be supplemented with broader governance constructs by virtue of architectural blueprints and ecosystems with governance models to provide complete oversight and enforceable ethics adherence (Kääriäinen & Kortesniemi, 2024).

A vibrant protocol for continuous co-evolution of DTs with their corresponding physical entities is critical to maintaining trust over the long term. This protocol ensures high-fidelity DTs always correlate with their counterparts in the real world with respect to configuration and behavior, and thus prevent data drift and model bias that can undermine clinical decision-making. Its assurance promotes confidence in the DT's ability to provide actionable and accurate insights, thus supporting ongoing care and patient safety (Bersani *et al.,* 2022). Parallel to trust-building measures, privacy-respecting technologies are critical to aligning patient rights with data-driven breakthroughs in healthcare technology. De-identification techniques to remove personally identifiable information from data sets are critical to protecting patient confidentiality while allowing data sharing to support AI and ML applications. Compliance with data privacy legislation like the Health Insurance Portability and Accountability Act (HIPAA) and the General Data Protection Regulation (GDPR) ensures patient privacy is protected while allowing advances to occur in healthcare analytics (Singhal, 2024). FL, a distributed approach to AI model training, reinforces further privacy by allowing data sets to remain local while supporting group learning. By only moving model parameters and not raw data, FL protects data and maintains patient confidentiality during the training process (Yapar, 2024).

The application of FL in healthcare lets organizations benefit from large and diverse datasets to promote medical research, improve patient outcomes, and reduce costs while preserving privacy. Secure multi-party calculation techniques ensure data protection during learning processes, and model updates are performed securely and in accordance with privacy protocols (Kondaveeti *et al.,* 2024). Nevertheless, there are challenges like data heterogeneity and maintaining model performance, areas for further research and optimization. Supporting technologies like differential privacy and encryption methods enhance data protection to allow secure storage, transfer, and analysis of health information while preserving its integrity for AI-related use cases (Dhade & Shirke, 2024). Moreover, to deal with AI incorporation in healthcare from an ethical standpoint, there is a need to create complete governance models with mandates to mitigate algorithmic bias and health disparities while ensuring digital sovereignty and informed consent. XAI protocols and dynamic consent instruments are imperative to ensure transparency and patient autonomy by explaining AI-based decisions and personal data usage processes and flows. These measures improve accountability and confidence and allow patients to make informed decisions about joining AI-enabled healthcare environments.

Algorithmic bias, potentially introduced at any point during AI construction-during data acquisition, preprocessing, and modelling-poses an overriding ethical challenge. Techniques such as decentralized learning, separation of confounding variables, and explainable models are essential to alleviating bias and supporting fairness in AI deployment (Chen *et al.,* 2021; Nazer *et al.,* 2023). Moreover, encryption and differential privacy measures secure patient data during AI training, achieving a balance between preserving privacy and gaining advantage through analytics (Williamson & Prybutok, 2024). Active monitoring of algorithms is a second overriding approach to governance and is one by which healthcare systems can identify and rectify biases susceptible to exacerbating health disparities and those tied to variables such as race, socioeconomic environment, or population group membership. Regular auditing and monitoring processes ensure AI systems meet ethics and regulations and therefore increase their reliability and universal coverage (Sendak *et al.,* 2023). Combined governance processes, consolidated through work involving collaborations with AI engineers, healthcare practitioners, policymakers, and patient constituencies, are vital to ensuring AI-enhanced healthcare solutions are legally sufficient and ethically supportable. By creating frameworks supporting equitable access and benefit consent, and accountability, stakeholders can stimulate innovation while ensuring AI deployment supports broader objectives for health equity and patient autonomy (Williamson & Prybutok, 2024).

11. CASE STUDIES AND REAL-WORLD APPLICATIONS

HDTs have proved to be successfully applied to personalize therapeutic regimens in various cancers, ranging from malignancies like breast cancer to lung and prostate cancers, using advanced computational models to allow for extremely personalized regimens of therapy administration. For TNBC, specifically, HDTs have been utilized to predict personalized patient reactions to protocols of neoadjuvant therapy administration. By modelling various regimens of therapy administration, DTs greatly improve prediction performance with respect to pathological complete response, with an observed 21.62% improvement upon optimization of administration protocols for Adriamycin/Cytoxan and Taxol (Wu *et al.,* 2024b). These digital models account for biological variables unique to patients to personalize regimens to administer therapy to achieve maximal therapeutic benefit while reducing unnecessary medical procedures. Similarly, for lung and prostate cancers, TDTs have been theorized to serve as a new modality for individualizing radiopharmaceutical therapies. TDTs incorporate physiologically based pharmacokinetic modelling to simulate patient-specific reactions to radioactive tracers and therapeutic agents separately and uniquely to individual patients, and thus allow therapists to modulate therapy parameters like dose and radioactivity to timing. By precise tuning to the variables involved, TDTs enhance therapy specificity, and hence outcomes improve with respect to both progression-free and overall survival for patients (Abdollahi & Rahmim, 2022; Brosch-Lenz *et al.,* 2023).

The MLPA framework offers a prime exemplification of the collective capabilities of DTs in oncology by integrating information gleaned on the basis of EMRs and genomic studies to accurately simulate therapeutic outcomes and tumour growth. The methodology permits investigators to study individualized regimens of therapy matched to unique clinical and genetic attributes of the patient and has prototype potential to revolutionize clinical outcomes by modulating therapy accordingly based on unique tumour dynamics (Gu & Chen, 2024). Further breakthroughs at tissue-level modelling protocols and agent-based modelling platforms have permitted investigators to integrate advanced cancer system models with DT technology and hence enhance tissue-specific reaction anticipation and fine-tune oncology therapy based on accurate biological interactions (Kemkar *et al.,* 2024). Global initiatives like the NIH Human Tumor Atlas, EU DigiTwins, and Project CancerID present key templates and evidence bases to expedite large-scale implementation of Hybrid DTs to clinical sites. These initiatives reflect mandates for integrating state-of-the-art computational infrastructures, advanced modelling paradigms, and assorted sets of observational data to generate predictive models propelling personalized medicine. The NIH Human Tumor Atlas, for instance, emphasizes profound phenotyping and adaptive modelling paradigms to dynamically plot therapeutic reactions and customize therapy at unprecedented scales (Stahlberg *et al.,* 2022). Similarly, the EU DigiTwins programme interprets the paradigm-shifting potential of HDTs to facilitate predictive, preventive, and personalized care by exploiting data flows in near-real time and simulation paradigms to facilitate therapeutic decision-making (Li *et al.,* 2025).

The referenced projects identify major challenges needing resolutions to achieve widespread implementation of HDT, involving challenges related to data interoperability, regulatory adherence, ethical governance, and designing user-accessible interfaces amenable to being successfully utilized by healthcare practitioners and patients equally. Suitable deployment of HDTs demands strict data standardization and full-scale cybersecurity to ensure patient information while permitting seamless interfacing with healthcare infrastructures. Moreover, modelling complicated multi-omics interactions through these tools significantly enhances an organization's capacity to enhance treatment regimens, enhance organizational workflow process flows, and enhance comprehensive patient management plans (Cellina *et al.,* 2023). In spite of these possibilities, significant challenges persist and include ensuring data protection and integrity, and redressing disparities with respect to healthcare infrastructure and access. These challenges necessitate collaborative efforts by various stakeholders, involving healthcare service providers, policymakers, researchers, and technology innovators to develop standardized guidelines and facilitate equitable access to solutions grounded in HDT. Knowledge acquired through global initiatives like the NIH Human Tumor Atlas, EU DigiTwins, and Project CancerID about their successes and failures can inform stakeholders about overcoming challenges involved with implementing HDT and catalyzing wiser, streamlined, and efficient healthcare solutions globally (Li *et al.,* 2025; el Ata, 2024).

12. CONVERGENCE OF EMERGING TECHNOLOGIES

The alignment of emerging technologies-such as metaverse platforms, quantum computing, neuromorphic engineering, and augmented/virtual reality (AR/VR)-holds the ability to significantly augment the application of HDTs for oncology practice (Figure 3.). The combination of these advanced technologies yields a healthcare experience that is more immersive, precise, and efficient, and thus optimizes patient outcomes while streamlining clinical workflows. The metaverse platforms like Dr. Meta illustrate this combination by providing multidisciplinary digital spaces to facilitate cancer care through virtual multidisciplinary conferences, learning centers, and VR healing theaters. These spaces overcome space and logistical challenges and create enhanced interaction between healthcare practitioners and patients while enhancing continuous learning and therapy interventions (Kim *et al.,* 2023). Oncology DTs, as elaborated upon previously, represent advanced computational models to simulate cancer pathways for individual patients. These models allow clinicians to make personalized decisions regarding therapy, to probe various treatments, and to undertake virtual clinical trials. The application of quantum computing and neuromorphic engineering to HDT ecosystems holds a particular opportunity to aid these capabilities. The capacity for quantum computing to process complex data sets at exponential speeds, coupled with neuromorphic architecture to emulate neural networks, lends itself to vast-scale biological data analysis at a raw and continuous level-covering genotypic and phenotypic variables and environmental factors. This computational synergy lends itself to increased accurate simulation and predictive modelling for oncology and thus boosts therapy and diagnostic specificity (Jensen & Deng, 2023).

Figure 3. Convergence of Emerging Technologies for HDTs

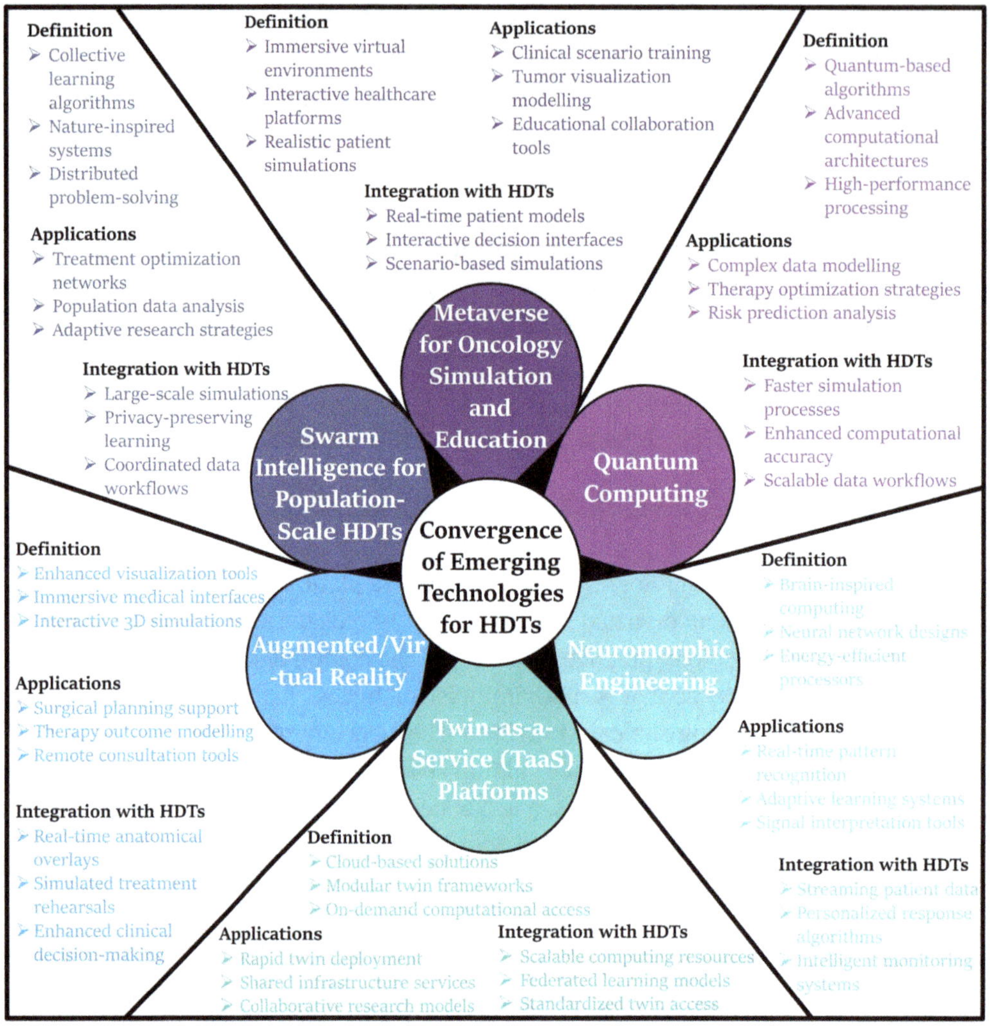

Augmented reality and virtual reality technology greatly reinforce the DT framework through immersive environments that enhance interaction and data visualization. For patients, these environments can allow for therapeutic interventions through interactive interfaces for symptom tracking and adherence to therapy protocols. For healthcare professionals, AR/VR aids enhanced diagnostics and superior understanding of tumour advancement, refining intervention strategies. ML techniques coupled with AR/VR platforms have shown their mettle in literature regarding breast cancer studies, whereby generation of DTs in real time permitted accurate diagnostic assessments and optimisation of therapy in simulated environments (Moztarzadeh *et al.,* 2023). Despite these prospects, several challenges persist to be adequately addressed to fully leverage the promise offered by these converging sciences to oncology research and practice. Data privacy concerns, regulation, and equitable access to advanced computing resources are major hurdles necessitating system-level solutions. Without effective governance frameworks and standardised ontology, adoption of these sciences can deepen health inequities already present through differences in healthcare delivery (Turab & Jamil, 2023). Solutions to these challenges are necessary to fully leverage metaverse-driven, quantum-powered, and AI-facilitated oncology solutions while protecting patient safety and confidence.

The incorporation of emerging digital technologies in the HDT environment has the capacity to transform clinical work in oncology. Advanced computational environments are used to streamline workstreams, enhance decision-making processes, and allow personalized paths to therapy based upon real-time patient information. This personalized approach enhances therapeutic outcomes while reducing unnecessary procedures and diminishing healthcare spend, and creates a culture of efficient and patient-specific care (Pandey *et al.,* 2024). At the same time, techniques like Twin-as-a-Service (TaaS) and swarm intelligence create scalable paths to growing HDTs from point applications to full-scale population health solutions. TaaS paradigms facilitate the creation and deployment of DTs by offering reusable templates, standardized data sets, and kits for streamlining management and oversight of infrastructure. This approach enables healthcare experts to focus on clinical applications at a clinical level while data integrity and security, and system scale are managed by a foundational framework (Talasila *et al.,* 2023). In healthcare environments, TaaS makes possible simulated situations in near-real time by correlating data flows from heterogeneous sensors and medical equipment through to enhanced diagnosis and personalized therapy plans (Tang *et al.,* 2024).

The incorporation of swarm intelligence, defined by decentralized and self-organizing data processing systems, significantly improves the scalability of health decision technologies. This methodology is particularly beneficial in the realm of personalized preventive medicine, where predictive models leverage biological, behavioral, and environmental data at the individual level to effectively diagnose and manage NCDs with customized interventions. Algorithms rooted in swarm intelligence facilitate efficient data aggregation and decision-making among large populations, thus fostering more responsive and adaptable healthcare solutions (el Ata, 2024). Although the amalgamation of Technology as a Service (TaaS) and swarm intelligence provides promising avenues for the extensive deployment of HDTs, it is imperative to systematically tackle challenges related to data security, interoperability, and quality assurance. It is essential to ensure that healthcare datasets are standardized and safeguarded against unauthorized access to uphold trust and precision within clinical applications. Furthermore, collaborative efforts across institutions are required to develop scalable infrastructures capable of accommodating various healthcare settings without undermining ethical principles or patient privacy (Cellina *et al.,* 2023). The convergence of metaverse platforms, quantum computing, neuromorphic engineering, augmented AR/VR technologies, and scalable frameworks such as TaaS and swarm intelligence signifies a transformative evolution in oncology care. By facilitating immersive, data-driven simulations and personalized interventions at both individual and population scales, these integrated solutions possess the potential to redefine cancer treatment and healthcare delivery-provided that governance, security, and equitable access challenges are addressed through cooperative, multidisciplinary initiatives (Park *et al.,* 2023).

13. FUTURE DIRECTIONS AND RESEARCH FRONTIERS

HDTs represent a significant advancement in the fields of anticipatory medicine and preventive oncology by allowing for extremely individualized healthcare measures based upon predictive analytics and simulation modelling. These digital doubles can emulate complex disease processes and provide complete information regarding the dynamic interaction between genes, environment, and physiologies involved in the onset of NCDs and various cancers. By offering sophisticated simulations of disease courses, HDTs improve early diagnostics and risk stratification and provide tailored therapeutic approaches and thus change traditional clinical pathways to predictive and preventive pathways (Subasi & Subasi, 2024). In oncology, the combination of medical imaging information with mechanism-based computational modelling through HDTs provides an unprecedented level of accuracy in therapy planning. By simulating tumor outgrowth dynamics, therapeutic effect, and biological interaction with patients individually, HDTs provide researchers and attending physicians with effective tools to formulate individually tailored therapy regimens, taking into account inter-individual variability. This approach enhances personalization of therapy while at the same time improving prognostic assessments with subsequent improvement in patient care and reduction in adverse outcomes (Checcucci *et al.,* 2025).

Virtual patient replica construction enables researchers to study interactions between physical and digital entities, supporting early disease detection and health outcomes improvement. By running comprehensive simulation schemes, HDTs can determine the effect of many therapeutic agents and quantitatively appraise desired side effects and therapeutic effectiveness to inform optimized therapy protocols. These simulation schemes are essential to building global health models capable of overcoming geographical boundaries and promoting equal access to personalized medicine and democratizing healthcare solutions for diverse populations. Despite data complexity and integration challenges and ethics-related concerns, there is a revolutionary opportunity for health DT applications to transform the healthcare paradigm towards predictive, preventive, and personalized models (Wu *et al.,* 2022). To ensure seamless implementation and scale application of HDTs to clinical and research environments, validation metrics standardization and interoperability frameworks are necessary. The definition of universal DT technology standards is critical to allowing seamless flows of technical information and data across multi-system environments and enhancing collaborative activities among institutions. International institutions like ISO, IEC, ITU, and IEEE are actively involved in devising frameworks to create standardized data formats and interface and validation techniques, protocols, and thus create foundations for global DT implementation interoperability (Wang *et al.,* 2024; Piroumian, 2021).

The integration of blockchain technology with DTs provides an additional degree of security and standardization, especially in those systems with high data complexity, like the Industrial Internet of Water Things (IIoWT). The distributed ledger feature of blockchain provides secure data transfer, while protocols such as Message Queuing Telemetry Transport (MQTT) ensure real-time interoperability among different data sources and thus help to achieve seamless system interlocking across different domains (Mohammed *et al.,* 2024). The installation of an aggregated and hierarchical DT architecture further supports interoperability by permitting co-simulation techniques and distributed computational procedures to achieve multiscale dataset synchronization. As a combined process, these necessitate a standardized DT platform to allow for seamless interlocking as well as risk reduction and efficiency improvement in healthcare and other areas (David *et al.,* 2024; Klar & Angelakis, 2023). The effective execution of HDT-driven medicine for oncology demands a very skilled workforce with clinical and technological expertise. As a result, there is an imperative to establish a complete policy framework and education initiative (Pandey *et al.,* 2024).

However, significant technical and ethical challenges persist, involving intricacies in data amalgamation, scalability limitations, and challenges regarding patient privacy and algorithm transparency. Overcoming these challenges demands inter-disciplinary cooperation involving clinicians, data scientists, regulatory agencies, and ethicists to develop certified protocols, ensure data interoperate, and foster DT technology trust. Systematic training programs like those offered by the St. Jude Global Academy are an excellent illustration of how global education initiatives can augment workforce capacity through clinical education, online modules, and distance education tools like Cure4Kids, which have successfully trained thousands of healthcare practitioners all over the world (Moreira *et al.,* 2025). Policy initiatives also have to foster innovation while at the same time remedying present infrastructural and knowledge deficiencies. As pointed out by Patel et al. DT technology incorporation in oncology lags compared to other therapy areas owing to an insufficient clarion call for technical issues and insufficient incentives to stimulate research and application. Infrastructure standardization and provision of regulatory advice are vital to elicit innovation stimulus, adoption facilitation, and translate technological breakthroughs to equitable and efficient healthcare solutions (Patel *et al.,* 2023).

14. CONCLUSION

The emergence of HDTs represents a paradigm-shifting breakthrough in cancer care, moving oncology from reactive therapy to predictive, preventive, and extremely personalized medicine. By continuously blending biosensor data, multi-omics, and sophisticated computational models, HDTs facilitate real-time tracking, simulation of tumor kinetics, and optimization of therapy plans. This application decreases overdiagnosis, reduces needless interventions, and increases therapeutic effectiveness while aligning with ecological and economic sustainability paradigms. The application of HDTs to radiotherapy planning, immunotherapy, drug discovery, and virtual clinical trials provides evidence for their capacity to rationalise healthcare delivery and expedite translational research. To realise their full benefits, however, critical challenges like data amalgamation, interoperability, privacy, and algorithmic bias need to be overcome. Ethical paradigms, governance models, and standard validation metrics need to follow technological breakthroughs to ensure trustworthiness, accountability, and equitable access. As global initiatives show, HDTs hold the capacity to transform oncology by driving innovation, efficiency, and sustainability, and eventually transform healthcare to a more responsive and patient-centric paradigm.

DATA AVAILABILITY

The authors confirm that the data supporting the findings of this study are available within the book/chapter.

ACKNOWLEDGEMENT

All the authors are thankful to their respective Universities and Institutions for their valuable supports.

REFERENCES

Abdollahi, H., & Rahmim, A. (2022). DTs for personalized healthcare: Application to radiopharmaceutical therapies. *Frontiers in Biomedical Technologies*, *10*(1), 1–5.

Adeniyi, M., Ayoola, V. B., Samuel, T. E., & Awosan, W. (2024). Artificial intelligence-driven wearable electronics and smart nanodevices for continuous cancer monitoring and enhanced diagnostic accuracy. [IJSRMT]. *Int J Sci Res Modern Technol*, *3*(11), 3–18. DOI: 10.38124/ijsrmt.v3i11.106

Aftab, M., Mehmood, F., Zhang, C., Nadeem, A., Dong, Z., Jiang, Y., & Liu, K. (2025). AI in oncology: Transforming cancer detection through machine learning and deep learning applications. *arXiv preprint arXiv:2501.15489*.

Alfano, C. M., Suarez-Almazor, M., Rodriguez, M. A., Palos, G. R., & Gilmore, K. R. (2021). Innovating the personalization of stratified survivorship care pathways: Using a cancer data ecosystem to improve care access, outcomes, efficiency, and costs. *Journal of the National Cancer Institute. Monographs*, *2021*(57), 3–9. DOI: 10.1093/jncimonographs/lgaa011 PMID: 34478510

Amiri, S., & Mahjoub, M. A. (2019). HMDHBN: Hidden Markov inducing a dynamic hierarchical Bayesian network for tumor growth prediction. In *International Conference on Computer Analysis of Images and Patterns* (pp. 3-14). Cham: Springer International Publishing. DOI: 10.1007/978-3-030-29888-3_1

Ariyanti, S., & Suryanegara, M. (2024). Current research themes and future research needs on making ai's energy consumption efficient: a review. In *2024 4th International Conference on Electronic and Electrical Engineering and Intelligent System (ICE3IS)* (pp. 99-104). IEEE. DOI: 10.1109/ICE3IS62977.2024.10775966

Asadi, M., Zafari, V., Sadeghi-Mohammadi, S., Shanehbandi, D., Mert, U., Soleimani, Z., & Zarredar, H. (2024). The role of tumor microenvironment and self-organization in cancer progression: Key insights for therapeutic development. *BioImpacts*, *15*, 30713. DOI: 10.34172/bi.30713 PMID: 40256216

Au, R., Kolachalama, V. B., & Paschalidis, I. C. (2022). Redefining and validating digital biomarkers as fluid, dynamic multi-dimensional digital signal patterns. *Frontiers in Digital Health*, *3*, 751629. DOI: 10.3389/fdgth.2021.751629 PMID: 35146485

Badano, A. (2021). In silico imaging clinical trials: Cheaper, faster, better, safer, and more scalable. *Trials*, *22*(1), 64. DOI: 10.1186/s13063-020-05002-w PMID: 33468186

Baran, S. W., Bratcher, N., Dennis, J., Gaburro, S., Karlsson, E. M., Maguire, S., & LaFollette, M. R. (2022). Emerging role of translational digital biomarkers within home cage monitoring technologies in preclinical drug discovery and development. *Frontiers in Behavioral Neuroscience*, *15*, 758274. DOI: 10.3389/fnbeh.2021.758274 PMID: 35242017

Basu, A. (2024). Human digital twins and machine learning applications in precision medicine and surgery: Current state and future directions. *Artificial Intelligence in Medicine and Surgery-An Exploration of Current Trends, Potential Opportunities, and Evolving Threats-Volume 2.*

Bersani, M. M., Braghin, C., Cortellessa, V., Gargantini, A., Grassi, V., Presti, F. L., & Scandurra, P. (2022). Towards trust-preserving continuous co-evolution of digital twins. In *2022 IEEE 19th International Conference on Software Architecture Companion (ICSA-C)* (pp. 96-99). IEEE. DOI: 10.1109/ICSA-C54293.2022.00024

Brosch-Lenz, J., Uribe, C., Rahmim, A., & Saboury, B. (2023). Theranostic digital twins: An indispensable prerequisite for personalized cancer care. *Journal of Nuclear Medicine*, *64*(3), 501. DOI: 10.2967/jnumed.122.264929 PMID: 36863782

Capurro, D., Coghlan, S., & Pires, D. E. (2022). Preventing digital overdiagnosis. *Journal of the American Medical Association*, *327*(6), 525–526. DOI: 10.1001/jama.2021.22969 PMID: 35061004

Cellina, M., Cè, M., Alì, M., Irmici, G., Ibba, S., Caloro, E., & Papa, S. (2023). Digital twins: The new frontier for personalized medicine? *Applied Sciences (Basel, Switzerland)*, *13*(13), 7940. DOI: 10.3390/app13137940

Checcucci, E., Oing, C., Amparore, D., Porpiglia, F., & Rescigno, P. (2025). Digital twins in urological oncology: Precise treatment planning via complex modelling. *European Urology Oncology*, *8*(1), 1–4. DOI: 10.1016/j.euo.2024.10.005 PMID: 39443222

Chen, R. J., Chen, T. Y., Lipkova, J., Wang, J. J., Williamson, D. F., Lu, M. Y., & Mahmood, F. (2021). Algorithm fairness in AI for medicine and healthcare. *arXiv preprint arXiv:2110.00603.*

Chowbina, S., Janes, K. A., Peirce, S. M., & Papin, J. A. (2011). Mathematical and computational models in cancer. In *Targeted Therapies: Mechanisms of Resistance* (pp. 113–126). Humana Press. DOI: 10.1007/978-1-60761-478-4_6

Cohen, A. B., & Schwamm, L. H. (2024). Digital health for oncological care. *The Cancer Journal (Sudbury, Mass.)*, *30*(1), 34–39. DOI: 10.1097/PPO.0000000000000693 PMID: 38265925

David, I., Shao, G., Gomes, C., Tilbury, D., & Zarkout, B. (2024). Interoperability of digital twins: Challenges, success factors, and future research directions. In *International Symposium on Leveraging Applications of Formal Methods* (pp. 27-46). Cham: Springer Nature Switzerland.

Desani, N. R., & Reddy, R. R. K. (2020). Edge AI for real-time health monitoring using streaming data. *international Journal of Science and Research*, *10*(1), 1669-1674.

Dhade, P., & Shirke, P. (2024). Federated learning for healthcare: A comprehensive review. *Engineering Proceedings*, *59*(1), 230.

Dixit, R. R. (2025). Applying reinforcement learning to optimize cancer treatment protocols in machine learning frameworks. In *Artificial Intelligence Revolutionizing Cancer Care* (pp. 20–35). CRC Press.

Dziopa, K., Lekadir, K., van der Harst, P., & Asselbergs, F. W. (2025). Digital twins: Reimagining the future of cardiovascular risk prediction and personalised care. *Hellenic Journal of Cardiology : HJC = Hellēnikē Kardiologikē Epitheōrēsē*, *81*, 4–8. DOI: 10.1016/j.hjc.2024.06.001 PMID: 38852883

Edelman, L. B., Eddy, J. A., & Price, N. D. (2010). In silico models of cancer. *Wiley Interdisciplinary Reviews. Systems Biology and Medicine*, *2*(4), 438–459. DOI: 10.1002/wsbm.75 PMID: 20836040

Ekundayo, F. (2024). Reinforcement learning in treatment pathway optimization: A case study in oncology. *International Journal of Science and Research Archive*, *13*(02), 2187–2205. DOI: 10.30574/ijsra.2024.13.2.2450

el Ata, N. A. (2024). Perspective chapter: making the shift to personalized preventive medicine with human digital twins. In *Personalized Medicine-New Perspectives*. IntechOpen. DOI: 10.5772/intechopen.1003639

Fink, P. K., & Herren, L. T. (1996). Modelling disease processes for drug development: bridging the gap between quantitative and heuristic models. In *Proceedings of the 28th conference on Winter simulation* (pp. 1183-1190). DOI: 10.1145/256562.256930

Foran, D. J., Chen, W., Chu, H., Sadimin, E., Loh, D., Riedlinger, G., & DiPaola, R. S. (2017). Roadmap to a comprehensive clinical data warehouse for precision medicine applications in oncology. *Cancer Informatics*, *16*, 1176935117694349. DOI: 10.1177/1176935117694349 PMID: 28469389

Georgescu, I., Dricu, A., Artene, S. A., Vrăjitoru, N. R., Barcan, E., Tache, D. E., & Popescu, S. O. (2024). Digital-focused approaches in cancer patients' management in the post-covid era: Challenges and solutions. *Applied Sciences (Basel, Switzerland)*, *14*(18), 8097. DOI: 10.3390/app14188097

Goel, A. K., Campbell, W. S., & Moldwin, R. (2021). Structured data capture for oncology. *JCO Clinical Cancer Informatics*, *5*(5), 194–201. DOI: 10.1200/CCI.20.00103 PMID: 33591796

Gu, J., & Chen, J. Y. (2024). MLPA: A multi-scale digital twin framework for personalized cancer simulation and treatment optimization. In *Proceedings of the 15th ACM International Conference on Bioinformatics, Computational Biology and Health Informatics* (pp. 1-1). DOI: 10.1145/3698587.3701425

Guilbert, B., Rieu, T., Delamarre, D., Laurent, F., Serre, F., Leroy, S., & Cuggia, M. (2024). ONCO-FAIR project: Improving data interoperability in oncology chemotherapy treatments for data reuse. *Studies in Health Technology and Informatics*, *316*, 1373–1377. DOI: 10.3233/SHTI240667 PMID: 39176636

Hidalgo, I. Fenández-de_Vega, F., Ceberio, J., Garnica, O., Velasco, J. M., Cortés, J. C., & Díaz, J. (2023). Sustainable artificial intelligence systems: an energy efficiency approach. Authorea Preprints.

Hirani, A., & Zhao, X. B. (2021). *U.S. Patent No. 10,916,252*. Washington, DC: U.S. Patent and Trademark Office.

Hull, R., Chauke-Malinga, N., Gaudji, G. R., Blenman, K. R., & Dlamini, Z. (2023). The role of digital twinning, the next generation of EMR/EHR in healthcare in a society 5.0: collecting patient data from birth to the grave. In *Society 5.0 and Next Generation Healthcare: Patient-Focused and Technology-Assisted Precision Therapies* (pp. 179–200). Springer Nature Switzerland. DOI: 10.1007/978-3-031-36461-7_8

Jalasri, M., Panchal, S. M., Mahalingam, K., Venkatasubramanian, R., Hemalatha, R., & Boopathi, S. (2025). AI-powered smart energy management for optimizing energy efficiency in high-performance computing systems. In *Future of Digital Technology and AI in Social Sectors* (pp. 329–366). IGI Global.

Jensen, J., & Deng, J. (2023). Digital twins for radiation oncology. In *Companion Proceedings of the ACM Web Conference 2023* (pp. 989-993). DOI: 10.1145/3543873.3587688

Jiang, S., Lan, Y., Wang, W., & Guo, Z. (2024). Pyramid: A heterogeneous data integration algorithm based on hierarchical graph. In *ICASSP 2024-2024 IEEE International Conference on Acoustics, Speech and Signal Processing (ICASSP)* (pp. 6220-6224). IEEE. DOI: 10.1109/ICASSP48485.2024.10447879

Jiang, Y. (2007). A multiscale, cell-based framework for modelling cancer development. In *International Conference on Computational Science* (pp. 770-777). Berlin, Heidelberg: Springer Berlin Heidelberg.

Jupin, J., & Shi, J. Y. (2015). A proposition for resilient graph-based record linkage using parallel processing on distributed networks. In *2015 Resilience Week (RWS)* (pp. 1-3). IEEE. DOI: 10.1109/RWEEK.2015.7287442

Kääriäinen, T. P., & Kortesniemi, Y. (2024). Building trustworthy twin-based systems with self-sovereign identities. *IEEE Access : Practical Innovations, Open Solutions*, *12*, 182101–182123. DOI: 10.1109/ACCESS.2024.3507924

Kabir, M. R., & Ray, S. (2024). DT-IoMT: A digital twin reference model for secure internet of medical things. In *2024 IEEE Computer Society Annual Symposium on VLSI (ISVLSI)* (pp. 433-438). IEEE. DOI: 10.1109/ISVLSI61997.2024.00084

Kausar, A., Kausar, A., & Das, S. (2024). AI in bioinformatics and computational biology. In *Future of AI in Biomedicine and Biotechnology* (pp. 278–290). IGI Global Scientific Publishing. DOI: 10.4018/979-8-3693-3629-8.ch014

Kemkar, S., Tao, M., Ghosh, A., Stamatakos, G., Graf, N., Poorey, K., & Radhakrishnan, R. (2024). Towards verifiable cancer digital twins: Tissue level modelling protocol for precision medicine. *Frontiers in Physiology*, *15*, 1473125. DOI: 10.3389/fphys.2024.1473125 PMID: 39507514

Khosravi, A., & Roayaei, M. (2024). MOAC-RT: Personalized multi-objective radiotherapy using actor-critic deep reinforcement learning. PREPRINT (Version 1) available at Research Square [https://doi.org/10.21203/rs.3.rs-5637868/v1] DOI: 10.21203/rs.3.rs-5637868/v1

Kim, S., Jung, T., Sohn, D. K., & Chang, Y. J. (2023). Enhancing cancer care services with the multi-domain metaverse cancer care digital platform. In *International XR Conference* (pp. 254-264). Cham: Springer Nature Switzerland.

Klar, R., & Angelakis, V. (2023). Standardized and interoperable digital twins. In *2023 IEEE Conference on Standards for Communications and Networking (CSCN)* (pp. 382-382). IEEE. DOI: 10.1109/CSCN60443.2023.10453189

Kobayashi, K., & Alam, S. B. (2024). Explainable, interpretable, and trustworthy AI for an intelligent digital twin: A case study on remaining useful life. *Engineering Applications of Artificial Intelligence*, *129*, 107620. DOI: 10.1016/j.engappai.2023.107620

Kondaveeti, H. K., Simhadri, C. G., Mangapathi, S., & Vatsavayi, V. K. (2024). Federated learning for privacy preservation in healthcare. In *Federated Learning and Privacy-Preserving in Healthcare AI* (pp. 121–136). IGI Global Scientific Publishing. DOI: 10.4018/979-8-3693-1874-4.ch009

Kulkarni, P., Garin, S., Kanhere, A. U., Siegel, E., Yi, P. H., & Parekh, V. S. (2023). High-throughput AI inference for medical image classification and segmentation using intelligent streaming. *Clinical Orthopaedics and Related Research*. arXiv2305.15617v2

Lee, M. (2023). Deep learning techniques with genomic data in cancer prognosis: A comprehensive review of the 2021–2023 literature. *Biology (Basel)*, *12*(7), 893. DOI: 10.3390/biology12070893 PMID: 37508326

Li, X., Loscalzo, J., Mahmud, A. F., Aly, D. M., Rzhetsky, A., Zitnik, M., & Benson, M. (2025). Digital twins as global learning health and disease models for preventive and personalized medicine. *Genome Medicine*, *17*(1), 11. DOI: 10.1186/s13073-025-01435-7 PMID: 39920778

Liu, X., Peng, T., Xu, M., Lin, S., Hu, B., Chu, T., & Wu, P. (2024a). Spatial multi-omics: Deciphering technological landscape of integration of multi-omics and its applications. *Journal of Hematology & Oncology*, *17*(1), 72. DOI: 10.1186/s13045-024-01596-9 PMID: 39182134

Liu, Y., Liang, J., Zhang, Y., & Guo, Q. (2024b). Drug resistance and tumor immune microenvironment: An overview of current understandings. *International Journal of Oncology*, *65*(4), 96. DOI: 10.3892/ijo.2024.5684 PMID: 39219258

Luo, B., Teng, F., Tang, G., Cen, W., Liu, X., Chen, J., & Zhang, J. (2025). StereoMM: A graph fusion model for integrating spatial transcriptomic data and pathological images. *Briefings in Bioinformatics*, *26*(3), bbaf210. DOI: 10.1093/bib/bbaf210 PMID: 40407386

Lv, T., Zhang, Y., Liu, J., Kang, Q., & Liu, L. (2024). Multi-omics integration for both single-cell and spatially resolved data based on dual-path graph attention auto-encoder. *Briefings in Bioinformatics*, *25*(5), bbae450. DOI: 10.1093/bib/bbae450 PMID: 39293805

Mağara, Ş. S., Dietrich, N., Ünal, A. B., & Akgün, M. (2024). Accelerating privacy-preserving medical record linkage: A three-party MPC approach. *arXiv preprint arXiv:2410.21605.*

Manias, G., Kouremenou, E., Alzúaz, A. A., Kranas, P., Melillo, F., & Kyriazis, D. (2023). An optimized pipeline for the processing of healthcare data towards the creation of holistic health records. In *2023 International Conference on Applied Mathematics & Computer Science (ICAMCS)* (pp. 50-56). IEEE. DOI: 10.1109/ICAMCS59110.2023.00015

Mariam, Z., Niazi, S. K., & Magoola, M. (2024). Unlocking the future of drug development: Generative AI, digital twins, and beyond. *BioMedInformatics*, *4*(2), 1441–1456. DOI: 10.3390/biomedinformatics4020079

Martínez-García, A., Cangioli, G., Chronaki, C., Löbe, M., Beyan, O., Juehne, A., & Parra-Calderón, C. L. (2022). FAIRness for FHIR: towards making health datasets FAIR using HL7 FHIR. In *MEDINFO 2021: One World, One Health–Global Partnership for Digital Innovation* (pp. 22-26). IOS Press. DOI: 10.3233/SHTI220024

Matthies, D. J., Gabrecht, M., & Hellbrück, H. (2023). Cyber-physical & human systems (cphs)–a review and outlook. *Proceedings of Mensch und Computer*, *2023*, 364–369. DOI: 10.1145/3603555.3608527

McCoach, C. E., & Bivona, T. G. (2019). Engineering multidimensional evolutionary forces to combat cancer. *Cancer Discovery*, *9*(5), 587–604. DOI: 10.1158/2159-8290.CD-18-1196 PMID: 30992280

Mishra, A. D., Shrivastava, A. K., & Bhushan, M. (2025). Disease diagnosis and control by reinforcement learning techniques: A systematic literature review. *Discover Computing*, *28*(1), 129. DOI: 10.1007/s10791-025-09539-9

Mohammed, M. A., Lakhan, A., Abdulkareem, K. H., Abd Ghani, M. K., Marhoon, H. A., Kadry, S., & Zapirain, B. G. (2024). Industrial Internet of Water Things architecture for data standarization based on blockchain and digital twin technology. *Journal of Advanced Research*, *66*, 1–14. DOI: 10.1016/j.jare.2023.10.005 PMID: 37839503

Moreira, D. C., Hashmi, S. K., Andujar, A., Berg, F., Conger, K., Fox Irwin, L., & Rodríguez-Galindo, C. (2025). The St. Jude Global Academy: A multifaceted education program to expand pediatric oncology workforce capacity. *Pediatric Blood & Cancer*, *72*(3), e31499. DOI: 10.1002/pbc.31499 PMID: 39757452

Moztarzadeh, O., Jamshidi, M., Sargolzaei, S., Jamshidi, A., Baghalipour, N., Malekzadeh Moghani, M., & Hauer, L. (2023). Metaverse and healthcare: Machine learning-enabled digital twins of cancer. *Bioengineering (Basel, Switzerland)*, *10*(4), 455. DOI: 10.3390/bioengineering10040455 PMID: 37106642

Murphy, B., Nasir-Moin, M., von Oiste, G., Chen, V., Riina, H. A., Kondziolka, D., & Oermann, E. K. (2021). Patient level simulation and reinforcement learning to discover novel strategies for treating ovarian cancer. *arXiv preprint arXiv:2110.11872*.

Nazer, L. H., Zatarah, R., Waldrip, S., Ke, J. X. C., Moukheiber, M., Khanna, A. K., & Mathur, P. (2023). Bias in artificial intelligence algorithms and recommendations for mitigation. *PLOS Digital Health*, *2*(6), e0000278. DOI: 10.1371/journal.pdig.0000278 PMID: 37347721

Ngo, V. M., Sood, G., Donohue, F., Kearney, P., Buckley, C., & Roantree, M. (2024). Using hl7-fhir as an integration platform for chronic disease services management and planning in the irish healthcare sector. In *The Joint Conference of ISEH ICEPH & ISEG on Environment and Health.* (pp. 324) DOI: 10.20944/preprints202408.0681.v1

Niarakis, A., Laubenbacher, R., An, G., Ilan, Y., Fisher, J., Flobak, Å., & Glazier, J. A. (2024). Immune digital twins for complex human pathologies: Applications, limitations, and challenges. *NPJ Systems Biology and Applications*, *10*(1), 141. DOI: 10.1038/s41540-024-00450-5 PMID: 39616158

Nicolo, C., Sips, F., Vaghi, C., Baretta, A., Carbone, V., Emili, L., & Bursi, R. (2023). Accelerating digitalization in healthcare with the insilicotrials cloud-based platform: Four use cases. *Annals of Biomedical Engineering*, *51*(1), 125–136. DOI: 10.1007/s10439-022-03052-6 PMID: 36074307

Nikolov, S., Dimitrov, A., & Vera, J. (2019). Hierarchical levels of biological systems and their integration as a principal cause for tumour occurrence. *Nonlinear Dynamics Psychology and Life Sciences*, *23*(3), 315. PMID: 31173701

Oh, T. R. (2024). Integrating predictive modelling and causal inference for advancing medical science. *Childhood Kidney Diseases*, *28*(3), 93–98. DOI: 10.3339/ckd.24.018

Ong, T. C., Kahn, M. G., Kwan, B. M., Yamashita, T., Brandt, E., Hosokawa, P., & Schilling, L. M. (2017). Dynamic-ETL: A hybrid approach for health data extraction, transformation and loading. *BMC Medical Informatics and Decision Making*, *17*(1), 134. DOI: 10.1186/s12911-017-0532-3 PMID: 28903729

Padmavathi, U., Harshitha, R. S., & Jayashre, K. (2024). Integrating cyber-physical systems for enhanced efficiency in healthcare solutions. In *Technologies for Sustainable Healthcare Development* (pp. 300–324). IGI Global.

Pandey, H., Amod, A., Jaggi, K., Garg, R., Jain, A., & Tantia, V. (2024). Digital twin ecosystem for oncology clinical operations. *arXiv preprint arXiv:2409.17650.*

Papachristou, K., Katsakiori, P. F., Papadimitroulas, P., Strigari, L., & Kagadis, G. C. (2024). Digital twins' advancements and applications in healthcare, towards precision medicine. *Journal of Personalized Medicine*, *14*(11), 1101. DOI: 10.3390/jpm14111101 PMID: 39590593

Park, J., Park, E., Lee, D., Kang, S., Lee, T., Lee, S., & Lee, U. (2023). (2023). Human digital twins for pervasive healthcare: A scoping review. *Interactive Journal of Medical Research.* Advance online publication. DOI: 10.2196/46012

Park, W. Y., Jeon, K., Schmidt, T. S., Kondylakis, H., Alkasab, T., Dewey, B. E., & Nagy, P. (2024). Development of medical imaging data standardization for imaging-based observational research: OMOP common data model extension. *Journal of Imaging Informatics in Medicine*, *37*(2), 899–908. DOI: 10.1007/s10278-024-00982-6 PMID: 38315345

Patel, S., Goldsack, J. C., Cordovano, G., Downing, A., Fields, K. K., Geoghegan, C., & Robinson, E. J. (2023). Advancing digital health innovation in oncology: Priorities for high-value digital transformation in cancer care. *Journal of Medical Internet Research*, *25*, e43404. DOI: 10.2196/43404 PMID: 36598811

Pérez-Romero, K., Rodríguez, R. M., Amedei, A., Barceló-Coblijn, G., & Lopez, D. H. (2020). Immune landscape in tumor microenvironment: Implications for biomarker development and immunotherapy. *International Journal of Molecular Sciences*, *21*(15), 5521. DOI: 10.3390/ijms21155521 PMID: 32752264

Pietrantoni, L., San Román-Niaves, M., & De Angelis, M. (2024). Human Trust and Digital Twins in a Human Factors and Ergonomic Framework. In *2024 IEEE 29th International Conference on Emerging Technologies and Factory Automation (ETFA)* (pp. 1-9). IEEE. DOI: 10.1109/ETFA61755.2024.10711109

Piroumian, V. (2021). Digital twins: Universal interoperability for the digital age. *Computer*, *54*(1), 61–69. DOI: 10.1109/MC.2020.3032148

Popel, A. S. (2020). Immunoactivating the tumor microenvironment enhances immunotherapy as predicted by integrative computational model. *Proceedings of the National Academy of Sciences of the United States of America*, *117*(9), 4447–4449. DOI: 10.1073/pnas.2001050117 PMID: 32102915

Qu, Y., Gao, L., Xiang, Y., Shen, S., & Yu, S. (2022). FedTwin: Blockchain-enabled adaptive asynchronous federated learning for digital twin networks. *IEEE Network*, *36*(6), 183–190. DOI: 10.1109/MNET.105.2100620

Ramakrishnan, R., Washington, A., Suveena, S., Rani, J. R., & Oommen, O. V. (2025). From DNA to Big Data: NGS technologies and their applications. *Artificial Intelligence (AI) in Cell and Genetic Engineering*, 459-482.

Rashid, Z., Folarin, A. A., Ranjan, Y., Conde, P., Sankesara, H., Zhang, Y., & Dobson, R. J. (2023). disease insight through digital biomarkers developed by remotely collected wearables and smartphone data. *arXiv preprint arXiv:2308.02043.*

Rigatti, M., Chapman, B., Chai, P. R., Smelson, D., Babu, K., & Carreiro, S. (2023). Digital biomarker applications across the spectrum of opioid use disorder. *Cogent Mental Health*, *2*(1), 2240375. DOI: 10.1080/28324765.2023.2240375 PMID: 37546179

Roelofsen, L. M., Kaptein, P., & Thommen, D. S. (2022). Multimodal predictors for precision immunotherapy. *Immuno-oncology Technology*, *14*, 100071. DOI: 10.1016/j.iotech.2022.100071 PMID: 35755892

Roesler, D., Czypionka, T., Eisenberg, S., Kraus, M., Reiss, M., & Zech, C. (2024). Impact of in-silico technologies in the development of high-risk medical devices on healthcare. *European Journal of Public Health*, *34*(Supplement_3), ckae144–1210. DOI: 10.1093/eurpub/ckae144.1210

Rozony, F. Z., Aktar, M. N. A., Ashrafuzzaman, M., & Islam, A. (2024). A systematic review of big data integration challenges and solutions for heterogeneous data sources. *Academic Journal on Business Administration. Innovation & Sustainability*, *4*(04), 1–18.

Saariluoma, P., Cañas, J., & Karvonen, A. (2020). Human digital twins and cognitive mimetic. In *International conference on human interaction and emerging technologies* (pp. 97-102). Cham: Springer International Publishing.

Sendak, M., Balu, S., & Hernandez, A. F. (2023). Proactive algorithm monitoring to ensure health equity. *JAMA Network Open*, *6*(12), e2345022–e2345022. DOI: 10.1001/jamanetworkopen.2023.45022 PMID: 38100115

Shahriyari, L., Abdel-Rahman, M., Asadpoure, A., Cebulla, C., Chang, Y. H., Hao, W., & Zervantonakis, I. (2022). *My Virtual Cancer* (No. DOE-UMASSAMHERST-21630). Univ. of Massachusetts, Amherst, MA (United States).

Shen, M. D., Chen, S. B., & Ding, X. D. (2024). The effectiveness of digital twins in promoting precision health across the entire population: A systematic review. *NPJ Digital Medicine*, *7*(1), 145. DOI: 10.1038/s41746-024-01146-0 PMID: 38831093

Singhal, S. (2024). Data privacy, compliance, and security including AI ML: healthcare. In *Practical Applications of Data Processing, Algorithms, and Modelling* (pp. 111-126). IGI Global Scientific Publishing. DOI: 10.4018/979-8-3693-2909-2.ch009

Sinisi, S., Alimguzhin, V., Mancini, T., Tronci, E., Mari, F., & Leeners, B. (2020). Optimal personalised treatment computation through in silico clinical trials on patient digital twins. *Fundamenta Informaticae*, *174*(3-4), 283–310. DOI: 10.3233/FI-2020-1943

Smokovski, I., Steinle, N., Behnke, A., Bhaskar, S. M., Grech, G., Richter, K., & Golubnitschaja, O. (2024). Digital biomarkers: 3PM approach revolutionizing chronic disease management-EPMA 2024 position. *The EPMA Journal, 15*(2), 149–162. DOI: 10.1007/s13167-024-00364-6 PMID: 38841615

Soos, B. (2012). *Multi-scale Models of Tumor Growth and Invasion* (Doctoral dissertation, University of Waterloo).

Stahlberg, E. A., Abdel-Rahman, M., Aguilar, B., Asadpoure, A., Beckman, R. A., Borkon, L. L., & Zervantonakis, I. (2022). Exploring approaches for predictive cancer patient digital twins: Opportunities for collaboration and innovation. *Frontiers in Digital Health, 4*, 1007784. DOI: 10.3389/fdgth.2022.1007784 PMID: 36274654

tefănigă, S. A., Cordo , A. A., Ivascu, T., Feier, C. V. I., Muntean, C., Stupinean, C. V., & Bolboacă, S. D. (2024). Advancing precision oncology with digital and virtual twins: A scoping review. *Cancers (Basel), 16*(22), 3817. DOI: 10.3390/cancers16223817 PMID: 39594772

Subasi, A., & Subasi, M. E. (2024). Digital twins in healthcare and biomedicine. In *Artificial Intelligence, Big Data, Blockchain and 5G for the Digital Transformation of the Healthcare Industry* (pp. 365-401). Academic Press. DOI: 10.1016/B978-0-443-21598-8.00011-7

Sumini, M., Teodori, F., & Isolan, L. (2024). Digital twins in dosimetry and radiotherapy, a survey and some applications. *Radiation Physics and Chemistry, 218*, 111649. DOI: 10.1016/j.radphyschem.2024.111649

Talasila, P., Gomes, C., Mikkelsen, P. H., Arboleda, S. G., Kamburjan, E., & Larsen, P. G. (2023). Digital twin as a service (DTaaS): a platform for digital twin developers and users. In *2023 IEEE Smart World Congress (SWC)* (pp. 1-8). IEEE. DOI: 10.1109/SWC57546.2023.10448890

Tang, C., Yi, W., Occhipinti, E., Dai, Y., Gao, S., & Occhipinti, L. G. (2024). A roadmap for the development of human body digital twins. *Nature Reviews Electrical Engineering, 1*(3), 199–207. DOI: 10.1038/s44287-024-00025-w

Thirupathi, L. SrihaReddy, E., & Deepika, J. U. (2024). Challenges and innovations in the creation of digital twins in healthcare. In *Exploring the Advancements and Future Directions of Digital Twins in Healthcare 6.0* (pp. 69-85). IGI Global.

Trauer, J., Schweigert-Recksiek, S., Schenk, T., Baudisch, T., Mörtl, M., & Zimmermann, M. (2022). A digital twin trust framework for industrial application. *Proceedings of the Design Society, 2*, 293–302. DOI: 10.1017/pds.2022.31

Truhn, D., Arasteh, S. T., Saldanha, O. L., Müller-Franzes, G., Khader, F., Quirke, P., & Kather, J. N. (2024). Encrypted federated learning for secure decentralized collaboration in cancer image analysis. *Medical Image Analysis, 92*, 103059. DOI: 10.1016/j.media.2023.103059 PMID: 38104402

Turab, M., & Jamil, S. (2023). A comprehensive survey of digital twins in healthcare in the era of metaverse. *BioMedInformatics, 3*(3), 563–584. DOI: 10.3390/biomedinformatics3030039

Venkatesh, K. P., Brito, G., & Kamel Boulos, M. N. (2024). Health digital twins in life science and health care innovation. *Annual Review of Pharmacology and Toxicology, 64*(1), 159–170. DOI: 10.1146/annurev-pharmtox-022123-022046 PMID: 37562495

Vial, A., Stirling, D., Field, M., Ros, M., Ritz, C., Carolan, M., & Miller, A. A. (2018). The role of deep learning and radiomic feature extraction in cancer-specific predictive modelling: A review. *Translational Cancer Research, 7*(3), 803–816. DOI: 10.21037/tcr.2018.05.02

Wang, K., Wang, Y., Li, Y., Fan, X., Xiao, S., & Hu, L. (2024). A review of the technology standards for enabling digital twin [version 2; peer review: 2 approved]. *Digital Twin, 1*(1), 4. DOI: 10.12688/digitaltwin.17549.2

Wang, Z., & Maini, P. K. (2017). Editorial special section on multiscale cancer modelling. *IEEE Transactions on Bio-Medical Engineering, 64*(3), 501–503. DOI: 10.1109/TBME.2017.2655439 PMID: 28824199

Wess, M., Andersen, M. K., Midtbust, E., Guillem, J. C. C., Viset, T., Størkersen, Ø., & Tessem, M. B. (2024). Spatial integration of multi-omics data using the novel multi-omics imaging integration toolset. bioRxiv, 2024-06.

Williamson, S. M., & Prybutok, V. (2024). Balancing privacy and progress: A review of privacy challenges, systemic oversight, and patient perceptions in AI-driven healthcare. *Applied Sciences (Basel, Switzerland), 14*(2), 675. DOI: 10.3390/app14020675

Wu, C., Lima, E., Stowers, C., Xu, Z., Yam, C., Son, J. B., & Yankeelov, T. (2024b). Abstract PO5-03-06: Optimizing therapeutic regimens via digital twins to improve triple negative breast cancer response to neoadjuvant therapy. *Cancer Research, 84*(9, Supplement), PO5–PO03. DOI: 10.1158/1538-7445.SABCS23-PO5-03-06

Wu, C., Lorenzo, G., Hormuth, D. A., Lima, E. A., Slavkova, K. P., DiCarlo, J. C., & Yankeelov, T. E. (2022). Integrating mechanism-based modelling with biomedical imaging to build practical digital twins for clinical oncology. *Biophysics Reviews, 3*(2), 021304. DOI: 10.1063/5.0086789 PMID: 35602761

Wu, C. J., Acun, B., Raghavendra, R., & Hazelwood, K. (2024a). Beyond efficiency: Scaling AI sustainably. *IEEE Micro, 44*(5), 37–46. DOI: 10.1109/MM.2024.3409275

Yankeelov, T. E., An, G., Saut, O., Luebeck, E. G., Popel, A. S., Ribba, B., & Genin, G. M. (2016). Multi-scale modelling in clinical oncology: Opportunities and barriers to success. *Annals of Biomedical Engineering, 44*(9), 2626–2641. DOI: 10.1007/s10439-016-1691-6 PMID: 27384942

Yapar, O. (2024). Federated learning for national healthcare systems: Balancing privacy and innovation. *World Journal of Advanced Engineering Technology and Sciences, 13*(1), 10–30574.

Zadeh Shirazi, A., Tofighi, M., Gharavi, A., & Gomez, G. A. (2024). The application of artificial intelligence to cancer research: A comprehensive guide. *Technology in Cancer Research & Treatment, 23*, 15330338241250324. DOI: 10.1177/15330338241250324 PMID: 38775067

Zhang, X. A., Yates, A., Vasilevsky, N., Gourdine, J. P., Callahan, T. J., Carmody, L. C., & Robinson, P. N. (2019). Semantic integration of clinical laboratory tests from electronic health records for deep phenotyping and biomarker discovery. *NPJ Digital Medicine, 2*(1), 32. DOI: 10.1038/s41746-019-0110-4 PMID: 31119199

APPENDIX

Abbreviations

AI - Artificial Intelligence
AR/VR - Augmented/Virtual Reality
CDM - Common Data Model
CNN - Convolutional Neural Networks
CPHS - Cyber-Physical-Human Systems
DBs - Digital Biomarkers
DICOM - Digital Imaging and Communications in Medicine
DT - Digital Twin
EHCA - Evolutionary Hybrid Cellular Automata
EMRs - Electronic Medical Records
ETL - Extract, Transform, Load
FAIR - Findable, Accessible, Interoperable, and Reusable
FHIR - Fast Healthcare Interoperability Resources
FL - Federated Learning
GDPR - General Data Protection Regulation
HDC - Hybrid Discrete Continuum
HDTs - Human Digital Twins
HEFL - Homomorphically Encrypted Federated Learning
HFE - Human Factors Engineering
HIPAA - Health Insurance Portability and Accountability Act
HIS - Hospital Information Systems
HMDHBN - Hidden Markov Inducing a Dynamic Hierarchical Bayesian Network
IIoWT - Industrial Internet of Water Things
IoMT - Internet of Medical Things
LIME - Local Interpretable Model-Agnostic Explanations
MI-CDM - Medical Imaging CDM
MIIT - Multi-Omics Imaging Integration Toolset
ML - Machine Learning
MLPA - Multi-Level Parameterized Automata
MLPA - Multi-level Parameterized Automata
MPC - Multi-Party Computation
MQTT - Message Queuing Telemetry Transport
NCCN - National Comprehensive Cancer Network
NCDs - non-communicable diseases
OMOP - Observational Medical Outcomes Partnership
PACS - Picture Archiving and Communication Systems
SDC - Structural Data Capture
SHEFL - Somewhat-Homomorphically-Encrypted Federated Learning
SNOMED CT - Systematized Nomenclature of Medicine – Clinical Terms

SVM - Support Vector Machines
TaaS - Technology as a Service
TDTs - Theranostic Digital Twins
TME - Tumor Microenvironment
TNBC - Triple-Negative Breast Cancer
XAI - Explainable Artificial Intelligence

Chapter 4

CRISPR Diagnostics Coupled With Nano-biosensors for Early Cancer Detection

T. Elavarasi
http://orcid.org/0009-0003-0792-4484
Department of Microbiology, Sacred Heart College, Tirupattur, India

K. Kesavardhini
http://orcid.org/0009-0005-8235-042X
Department of Microbiology, Sacred Heart College, Tirupattur, India

S. Dayana
http://orcid.org/0009-0008-7135-5606
Department of Microbiology, Sacred Heart College, Tirupattur, India

A. Mubarak Ali
http://orcid.org/0009-0009-6367-7864
Department of Microbiology, Sacred Heart College, Tirupattur, India

B. Lokeshwari
http://orcid.org/0009-0005-6612-5134
Department of Microbiology, Sacred Heart College, Tirupattur, India

P. Saranraj
http://orcid.org/0000-0002-5052-6532
Department of Microbiology, Sacred Heart College, Tirupattur, India

ABSTRACT

Early cancer detection is essential to patients' survival, allowing for prompt treatment, and cutting down medical costs. Unfortunately, traditional diagnostic methods usually demonstrate low sensitivity, long processing time, and require advanced laboratory setup. Nevertheless, recent developments in the CRISPR technology have revolutionized the field of molecular diagnostics by introducing highly sensitive, sequence-specific, fast, and programmable platforms for detecting cancer-specific genetic and epigenetic biomarkers. Nano-biosensors that incorporate, quantum dots of gold nanoparticles graphene-related materials, Nanoparticles with magnetic properties, and plasmonic nanoparticles possess outstanding optoelectrical and fluorescence sensing abilities which ensure increased Both specificity and sensitivity. This chapter offers a thorough overview the recent progress in the integration of CRISPR-based diagnostics and nano-biosensors for early diagnosis of cancers.

DOI: 10.4018/979-8-3373-9715-3.ch004

1. INTRODUCTION

Cancer is multifaceted and extremely complicated disease that involves features of uncontrolled growth, resistance to programmed cell death, continued angiogenesis, abnormal intracellular signaling, invasion of tissues, and metastasis. All these features contribute to the tumour's growth, cancer progression, and drug resistance (Dessale et al., 2022). Due to development of changes at the genetic and epigenetic levels, Tumour cells' capacity to spread and infiltrate tissues to other sites of the body via blood and lymph vessels increases, thus posing greater difficulties for treatment. One of the leading causes of mortality is cancer. on a global scale as its frequency keeps increasing due to a number of reasons including inherited genetic mutations, cancer-causing substances, radiation, smoking, excessive alcohol use, and unhealthy living style (Ferlay et al., 2021). In addition to that, certain pathogens have also been identified as carcinogens. For example, Helicobacter pylori has been shown to be connected to the development of stomach cancer, whereas One of the main causes is thought to be HPV behind cervical cancer. In terms of genetics, genetic abnormalities Tumour suppressors in DNA repair genes, and proto-oncogenes are of critical importance in carcinogenesis (Sun et al., 2023).

One of the most efficient methods in decreasing cancer-induced deaths and improving the effectiveness of treatment is early diagnosis. Nevertheless, traditional methods of tumor detection, such as tissue biopsies, X-ray imaging, MRI, ultrasonography, CT scanning, PET, and standard blood tests, can demonstrate insufficient sensitivity in detecting early-stage tumors, micrometastases, and minimal residual diseases (Wang et al., 2024). Furthermore, these approaches can be invasive, costly, time-consuming, and require specialized equipment, which makes them available in few facilities. Their efficacy can be significantly reduced due to high rates of false positive and false negative diagnoses, which may delay timely treatment or cause overdiagnosis. As a result, liquid biopsy has emerged as a potentially useful method for both treatment monitoring and cancer detection (Tavakoli et al., 2022). By identifying biomarkers released by tumor cells, such as circulating tumor DNA (ctDNA), circulating tumor RNA (ctRNA), microRNAs (miRNAs), proteins, extracellular vesicles, and circulating tumor cells (CTCs), liquid biopsy enables non-invasive and continuous monitoring of tumor development. However, the employment of highly sensitive and selective biosensors is necessary because biological samples contain incredibly tiny quantities of biomarkers (Gong et al., 2021).

The CRISPR-associated (Cas) clustered regularly interspaced short palindromic repeats system has been game-changing in terms of molecular analysis because of its programmable nucleic acid recognition ability, specificity, and unparalleled analytical sensitivity (Dubey et al., 2022). Having been identified as an adaptive immune response against bacteriophage infection in bacteria and archaea, CRISPR/Cas system was later harnessed as a genome-editing tool for performing specific genetic modifications. However, besides being used as genome editors, Cas nucleases including Cas12 and Cas13 have been discovered to have collateral nuclease activity which becomes activated (Li & Hu, 2022). After the target DNA or RNA is identified, thus enabling the cleavage of reporter molecules. This unique feature of Cas systems has made it possible to create biosensors based on CRISPR which are able to find even minuscule levels of nucleic acid biomarkers. In comparison with PCR-based assays, CRISPR-based diagnostics are rapid, specific, programmable, easy to use, and compatible with point-of-care testing (POCT) (Kaminski et al., 2021).

Despite such benefits, CRISPR/Cas-based biosensors continue to struggle due to several problems including low concentration levels of biomarkers, poor signal amplification, inadequate signal-to-noise ratios, and low levels of analytical stability (Aman et al., 2020). The using nanotechnology in overcoming

These problems have shown themselves to be a useful instrument in improving the efficiency of these biosensors. Nanoparticles, like plasmonic nanoparticles, magnetic nanoparticles, quantum dots, carbon nanomaterials, and metal-organic frameworks have distinct optical, electrical, magnetic, and catalytic properties that help in improving signal transduction, target enrichment, and detection sensitivity (van Dongen et al., 2020). Through the mixture of the molecular recognition capability of CRISPR/Cas technology with the signal enhancement ability of nanomaterials, highly efficient nano-biosensors have been designed that show immense promise in detecting cancer biomarkers quickly, sensitively, and accurately (Bondy-Denomy, 2018).

2. NANOTECHNOLOGY FOR BIOSENSORS

The discipline of nanotechnology refers to the scientific study of designing, manipulating, and using materials on the scale of 1-100 nm (Bao et al., 2021). This discipline is highly multidisciplinary in nature and includes the utilization of principles from chemistry, physics, biology, materials science, and engineering to design novel nanostructures with unprecedented levels of control over composition and architecture (Cheng et al., 2022). As materials of nanoscale have unique physical and chemical characteristics unlike any seen in bulk form, the ability to exploit these differences for the benefit of mankind has enabled researchers to develop functional materials that have found their use in the field of medicine, electronics, environment, and biotechnology. In the medical sciences, nanotechnology holds immense promise and has been used to great success for diagnosing diseases, delivering drugs to targeted sites, molecular imaging, and many more applications (Afzal et al., 2022).

Nanotechnology has greatly influenced The layout and capabilities of biomarkers through increased sensitivity, signal enhancement, and miniaturization (Ramesh et al., 2023). Nanomaterials have a major role in designing biosensors that are efficient and consume fewer raw materials and energy for the production of eco-friendly technologies. The large high electrical conductivity, surface-to-volume ratio, and catalytic effect, and optics of nanomaterials improve the interaction between biomolecules and signal generation (Bhardwaj et al., 2024). Therefore, nanomaterials have become very essential elements in bioanalytical applications, especially in molecular diagnostics, biomedical imaging, therapeutics, and controlled delivery of drugs. The application of nanotechnology to biosensor design has resulted in the emergence of fast, sensitive, and portable diagnostic systems which can detect biomarkers of diseases even at lower concentrations (Liu et al., 2025).

2.1 Nano-biosensors for Cancer Detection

Nano-biosensors are high-end instruments used to detect, quantify, and analyze biological analytes at molecular and cellular level using engineered nanomaterials and biological recognition units (Hassan, 2022). In these biosensors, various types of nanostructures have been used which include carbon nanowires, graphenequantum dots, metallic nanoparticles, carbon nanotubes, and metal oxide nanoparticles, magnetic nanoparticles, and thin films of nanoparticles which are known for enhancing signal production and amplification (Hemdan et al., 2025). Because of excellent analytical capabilities, nano-biosensors are gaining significance in the early detection of cancer by recognizing various biomarkers in the body like circulating tumor cells (CTCs) and circulating tumor DNA (ctDNA), miRNAs, proteins, exosomes, and other vesicles outside of cells with great accuracy and sensitivity. Their ability to detect the markers in a non-invasive way through liquid biopsy is an advantage over traditional invasive tissue biopsy (Liu et al., 2023).

Liquid biopsy-based nano-biosensors have gained much significance in the field of precision oncology since they enable doctors to follow tumor evolution and effectiveness of treatments through minimally invasive biological samples, including urine and blood, and saliva (Duan et al., 2026). Developments in nanomaterials Recent years have seen resulted in important advancements in the performance of biosensors in terms of sensitivity, lowering of detection limits, and capability of detecting multiple biomarkers at one time. Such multiplexed detection not only saves the time but also costs required in diagnosing diseases. Moreover, nano-biosensors have been extensively studied for POCT, personalized medicine, and disease monitoring purposes (Pulumati et al., 2023).

3. BIOSENSOR

A biosensor can be defined as an analytical instrument used for identifying, and Measuring biological molecules in complex biological samples via conversion of certain biological reactions into physical or chemical signals (Chen et al., 2023). Biosensors offer quick, precise, highly sensitive, and real-time measurements of target analytes, hence forming an integral component in diagnostics, environmental, food safety testing, and drug discovery applications. Nanobiosensors further improve the performance of biosensors through their capability of working on a molecular scale in order to identify very small levels disease biomarkers that very high accuracy (Banerjee et al., 2021). In essence, a biosensor comprises four key elements including target analyte, aspect of biological recognition (bioreceptor), transducer as well as signal processing/detection. The interaction of the analyte and bioreceptors causes certain physicochemical modifications that are then transformed by transducers into electrical, optical, electrochemical, piezoelectric, or thermal signals for subsequent processing of analytical data (Gulati et al., 2025).

Most of the developments of biosensors have been due to advancements in integration of the transducer and biological recognition element (Khazaei et al., 2023). The electrochemical signals that result from enzyme activities taking place on the transducer's surface are detected by the first generation of biosensors, also known as mediator-free amperometric biosensors. The second generation biosensors include the use of electron transfer mediators, co-reactants, auxiliary enzymes, as well as nanomaterials for effective electron transfer (John et al., 2022). While electron transfer mediators greatly increase the efficiency of sensors, they might also raise problems such as toxicity, instability and lack of biocompatibility. The third generation biosensors solve these challenges by creating direct electrical contact between the electrode and the immobilized enzyme without using diffusible electron transfer mediators. In this case, the enzyme and the mediator are immobilized on the electrode, resulting into effective electron transfer, increased analytical stability, fast responses, good reproducibility, as well as improved performance (Naresh & Lee, 2021).

3.1 Features of Biosensors

3.1.1 Selectivity and Sensitivity

Selectivity and sensitivity are key parameters for designing an effective biosensor. The biological element must be able to differentiate the specific target analyte even when surrounded by many other interfering compounds in very complicated biological samples (Huang et al., 2021). At the same time, the biosensor must be sufficiently sensitive in order to recognize extremely small concentrations of biomarkers, sometimes in the nanogram, picogram, or femtogram per milliliter level, and thus provide an opportunity for early diagnosis of diseases (Du et al., 2025). The recent advancements in biosensors based on nanomaterials have been able to greatly enhance these parameters. For instance, a g-C_3N_4/chitosan-based biosensor exhibited remarkable selectivity in relation to flutamide, while preserving the high sensitivity in the presence of interfering agents like glucose, lactose, and urea (Park et al., 2018).

3.1.2 Linearity and Limit of Detection (LOD)

An excellent biosensor must display a strong linear relationship between the quantity of the substance being measured and the response signal produced by the biosensor over a broad variety of analyte concentrations to ensure the accuracy of measurement. The limit of detection (LOD) is another crucial feature for the biosensor, since LOD indicates The lowest detectable concentration of an analyte against noise (Nasiri & Abbasian, 2025).

3.1.3 Stability

Stability of analysis is another essential feature that indicates the ability of biosensors to be reliable and applicable for a long time. Due to the fact that bioreceptors, including nucleic acids, enzymes, and antibodies are prone to decomposition when exposed to different environmental factors, it is necessary to keep the biosensors' ability to work properly while being stored and used many times (Ramesh et al., 2023). A high level of stability guarantees reliable work regardless of temperature, humidity, and other factors. For example, Nasiri et al. (2025) designed a biosensor that was highly stable; it had only about 3% difference of signals obtained in repeated experiments (Nasiri & Abbasian, 2025).

3.2 Architectural Design of Biosensors

The fabrication of advanced biosensors depends significantly on the choice of advanced materials and structural design of biosensors. The conductive polymers (CPs), transition metal oxides, two-dimensional nanomaterials, and hybrid nanocomposites constitute the core materials that can be used to design high-performing biosensor devices (Kurul et al., 2025). CPs can be fabricated through chemical, electrochemical, and biological approaches. These have been widely used for enzyme immobilization in electrochemical biosensors. Among all the CPs, GOx biosensors based on conductive polymers like polypyrrole and polythiophene have shown superior catalytic performance, effective electron transfer, and enhanced sensitivity towards glucose detection (Hsu & Wu, 2022). Despite the effectiveness of traditional analytical methods like ELISA in target molecule detection, they involve costly instrumentation, complex methodologies, and laboratory testing. Thus, affinity-based biosensors are preferred choices owing to their ease, quickness, and specificity (Huang et al., 2021).

The affinity Biosensors are capable of broadly categorized in four groupings based on their mode of recognition: antibody biosensors, DNA aptamer- biosensors, biosensors based on molecularly imprinted polymers (MIPs), involving the use of natural or artificial receptor molecules (Naresh & Lee, 2021). Recently, Nanomaterials in two dimensions, like MXenes and transition metal oxides, have become widely studied because of their exceptional electrical conductivity, large surface area, and mechanical flexibility, and superior biocompatibility for fabrication of biosensors (Tai et al., 2022). MXenes are characterized by graphene-like structure, improved hydrophilicity, and surface functionalization. Consequently, they can be efficiently used for biomolecule immobilization and electron transfer. All these physicochemical properties make MXenes a highly promising group of nanomaterials for fabrication of sensitive and multifunctional biosensors (Horgan et al., 2020).

4. NANOMATERIAL-BASED MODALITIES FOR CANCER DETECTION

4.1 Nanoparticle-based Biosensors

Biosensors based on nanoparticles have made significant contributions towards the diagnosis of cancer because of the heightened sensitivity, and accuracy in identifying biomarkers (Vaughan, 2024). The physical characteristics, such biocompatibility and a high surface area to volume ratio, tunability of optical properties, and enhanced catalytic ability have made the nanoparticles an essential part of the next generation biosensing systems. The combination of gold nanoparticles and aptamers, which is specific to cancer cells, has helped discriminate leukemia cells from Burkitt's lymphoma cells, leading to more accurate diagnosis of cancer and identifying it at its earliest stage (Timilsina & Li, 2024). In a similar way, Nano-Interfaced Microfluidic Exosome (nano-IMEX) biosensor utilizes the integration of nanomaterials in microfluidics to detect exosomes that cause ovarian cancer. Functionalization of gold nanoparticles with the cancer markers causes optical properties of nanoparticles, thus allowing rapid and precise detection of cancer via plasmonic signal amplification. More specifically, the recent developments in plasmonic biosensors using phase singularity with two-dimensional nanomaterials have allowed the ultrasonic detection of cancer biomarkers (Díaz-García et al., 2024).

Recent advancements in nanoparticle-based biosensing systems that incorporate the concept of theranostics have significantly improved the therapeutic uses of nanotechnology by combining the identification and management of cancer in one device (Dkhar et al., 2024). For example, Raman spectroscopy combined with photodynamic therapy provides simultaneous real-time detection and treatment of tumors, thus providing opportunities for personalized treatment (Achi et al., 2024). Nanoparticles used as graphene brain-computer interfaces (BCIs), which are now being tested clinically, have also proven to be very promising in neurosurgical oncology since they help to precisely differentiate between cancerous tissues and healthy brain cells during surgery. Moreover, paper-in-polymer-pond (PiPP) biosensor is a relatively cheap, mobile and fast device capable of diagnosing biomarkers of colorectal and prostate cancer within an hour (Sadeghi et al., 2023).

4.2 Quantum Dot-based Biosensors

Semiconductor nanocrystals known as quantum dots (QDs) have extraordinary optical and electronic properties, making them one of the most interesting materials for the advancement of innovative nanodevices for the diagnosis and treatment of cancer (Lu et al., 2023). The extremely high wide excitation and fluorescence spectrum, narrow emission, photostability, and adjustability of optical properties of QDs make them perfect optical probes for the observation of tumor development both in vivo and in vitro. Using the help of recent advances in the field, multifunctional quantum dot theranostic systems were developed, which integrate sensitive disease diagnosis and targeted drug delivery into one nanosystem

(Yasamineh et al., 2024). Moreover, quantum dots allow detecting various tumor biomarkers, including circulating tumor cells (CTCs), circulating tumor DNA (ctDNA), and nucleic acid biomarkers, thus increasing the effectiveness of liquid biopsy diagnostics. If quantum dots are conjugated with drugs, it becomes possible to deliver anticancer drugs selectively to tumor sites with minimal toxicity to normal tissues (Wang et al., 2024).

Quantum dot-based biosensors have proven to be effective tools for diagnosing and treating highly aggressive cancers such as TNBC, where an early diagnosis is very problematic from a clinical perspective (Montalvan et al., 2022). QD-based biosensor chips that can be miniaturized allow for the continuous monitoring of cancer markers in real-time, hence promoting prompt clinical decisions and individualized treatments. The development of environmentally friendly cadmium-free QDs has increased their medical application by offering biocompatible QDs without compromising their optical properties (Hsu & Wu, 2022). Besides cancer imaging and drug delivery, QDs have been used in hydrogels biosensing system for detecting cancer markers in body fluids such as blood, urine, and saliva. They are highly biocompatible for use in clinical diagnostics and can be functionalized with therapeutic antibodies, peptides, and other targeting ligands to achieve high selectivity in recognizing cancer cells (Wang et al., 2024). As a result, QDs have been transformed into multifunctional nanoplatforms that combine molecular imaging, biomarker detection, drug delivery, and therapy monitoring, hence playing a key role in precision oncology (Dezhakam et al., 2024).

4.3 Paramagnetic Nanoparticle-based Biosensors

Biosensors using paramagnetic nanoparticles have been found to be highly efficient devices that increase sensitivity, accuracy, and clinical usefulness in the diagnosis and treatment of cancers (Kuntamung et al., 2021). The special property of the nanoparticles allows fast target separation, biomarker enrichment, and sensitive signal production without any complex sample preparation process. A good example of such biosensor technology is the magnetic particle spectroscopy, which can detect cancer biomarkers based on the magnetic relaxation response of the paramagnetic nanoparticles (Mohan et al., 2025). The advantage of this technique in detecting cancer biomarkers is that it does not require any sample wash procedure, thus making the process simple and sensitive (Li et al., 2023).

Another notable progress includes the employment of superparamagnetic iron oxide nanoparticles (SPIONs) conjugated with polyethylenimine (PEI) that act as efficient carriers for directly administering chemotherapy medications inside the cancerous cells (Farzin et al., 2022). Apart from being used therapeutically, SPION-based biosensors have exhibited remarkable efficiency in diagnosing cancers at the point of care because of extremely more sensitivity, biocompatibility, fast analytical process. Magnetic devices Magseed and Magtrace, for example, have shown promise in surgical oncology. through assisting the precise localization of tumors and mapping of sentinel lymph nodes during surgery (Vijayakumar, 2023). These technologies aid the decision-making process of surgeons to accurately differentiate between cancerous and normal tissues. Moreover, the biosensors developed using magnetic nanoparticles have been able to detect RNA molecules contained within exosomes, such as cancer-specific miRNAs and mRNAs, which are highly promising biomarkers in detecting different cancers in their early stages (Zhao et al., 2022).

4.4 Nano-shell-based Biosensors

Nano-shells refer to nanostructures made up of a thin metal encircling a dielectric (insulating) core shell, with gold being the most used metal shell (Smith et al., 2020). Just like quantum dots, the optical behavior of nano-shells is tuned by altering the size of the dielectric core and the thickness of the metal shell. The tunability feature makes it possible for nano-shells to support localized surface plasmon resonance (LSPR) and therefore make them suitable for optical biosensing purposes (Sun & Chen, 2024). The latest improvements in the analytical capabilities of nano-shells have been achieved via the formation of hybrid nanostructures utilizing covalent-organic frameworks (COFs), metal-organic frameworks (MOFs), and gold nanoparticles. The hybrid structures allow for better surface area, signal amplification, and biomolecular recognition, allowing for ultrasensitive detection of cancer biomarkers (Pan et al., 2022).

Biosensors based on nano-shells have shown tremendous ability to detect the clinically relevant cancer biomarkers, specifically breast cancer biomarker CA15-3, with impressive sensitivity and specificity even in trace amounts; this helps in early cancer diagnosis (Yin et al., 2022). Additionally, multiplexed nano-shell biosensors facilitate parallel and label-free identification of many cancer biomarkers in a single experiment, significantly improving the diagnostic procedure's effectiveness. Functionalization of nano-shells with biomolecules helps in electrochemical amplification of signals, leading to efficient, real-time, and high-throughput cancer diagnostics (Zhuang et al., 2022). Additionally, the application of electroactive nanoparticles in electrochemical biosensors has helped increase the sensitivity of these biosensors, which makes them ability to identify traces of biomarkers in very small biological samples (Zhang et al., 2021).

4.5 Gold Nanoparticle-based Biosensors

AuNP-based biosensors have been identified as one of the most studied nanotechnological tools for the diagnosis of cancer Because of their distinct optical, electrical, catalytic, and biocompatibility properties. AuNPs significantly increase the sensing signals obtained due to biomolecular reactions, thereby increasing the sensitivity of the biosensor and providing the ability to diagnose cancerous biomarkers at an early stage (Yuan et al., 2020). AuNPs are made more specific to target tumor biomarkers with the functionalization of antibodies, peptides, aptamers, and other ligands. Additionally, AuNPs can be used for non-invasive diagnosis of cancer using liquid biopsy of the biomarkers available in biological fluids like blood and saliva. A number of novel AuNP biosensing techniques can diagnose multiple biomarkers in a single analysis to increase diagnostic accuracy and decrease the time and cost involved in analysis (Cheng et al., 2021). The fast response rate, portable nature, and POCT suitability of AuNP biosensors make them a very promising tool for decentralized cancer diagnostics. Recently, the integration of artificial intelligence (AI) technology into AuNP biosensors is gaining popularity for automated data analysis and improved diagnostic precision (Huang et al., 2020).

AuNP and metal oxide biosensor development for cancer screening purposes in saliva provides yet another option for non-invasive techniques. Saliva sampling is easy, quick, and non-painful, therefore suitable for regular cancer screening and monitoring (Bogers et al., 2021). Moreover, besides being an important tool for diagnostics, gold nanoparticles can be used as very efficient nanocarriers that deliver drugs directly to tumors, which enhances treatment effectiveness and minimizes the risk of general toxicity of treatment (Fu et al., 2021). Finally, gold Nanoparticles have been effective in photothermal treatment, a technique where the nanoparticles turn the absorbed light into localized heat that destroys the tumor cells without damaging adjacent healthy tissues. Thus, AuNP biosensors can serve as multidimensional nanoplatforms, capable of integrating several aspects of diagnostics and therapy in one complex structure (Gupta et al., 2023).

4.6 Nanowire-based Biosensors

Nanowire biosensors have become very efficient analysis tools for detecting cancers because of the unique electrical conductivity of the nanowires and the high rate of signal transduction. The silicon nanowire field-effect transistor (SiNW-FET) biosensor is capable of detecting even the smallest amounts of tumor-derived exosomes, thus being useful for early detection of cancers (Liu et al., 2025). It has also been found that vertically aligned silicon nanowires that have been modified on the surface show great capacity in isolation of circulating tumor cells from the blood. This allows for continuous monitoring of cancer progression without performing standard tissue biopsies (Hassan, 2022). Recent advances have further widened the scope of clinical use of nanowire biosensors owing to the integration of nanowire biosensor devices that are based on hydrogel, which possess high biocompatibility, environmental stability, and biomarker detection capabilities (Hemdan et al., 2025). The introduction of plasmonic sensing systems along with two-dimensional (2D) nanomaterials has led to the improvement of sensitivity in detecting cancer biomarkers down to sub-attomolar concentration ranges.

Recent advances in sustainable nanotechnology have introduced environmentally benign nanomaterials that combine excellent biosensing performance with reduced ecological impact (Liu et al., 2023). Green synthesis approaches employing biological resources minimize hazardous chemicals while producing highly biocompatible nanomaterials suitable for CRISPR-based diagnostic platforms. Despite challenges related to large-scale production and reproducibility, sustainable nanomaterials represent a promising direction for developing safer and environmentally responsible cancer diagnostic technologies (Duan et al., 2026). Figure 1 illustrates how different nanomaterials, such as quantum dots, gold nanoparticles, magnetic nanoparticles, carbon-based nanostructures, and nanosensors, are used to detect cancer through imaging, biomarker sensing, and targeted diagnostic approaches. These nanomaterial-based modalities improve early cancer detection by enhancing sensitivity, specificity, signal amplification, and real-time monitoring of tumor-related molecular changes.

Figure 1. Nanomaterial-Based Modalities for Cancer Detection

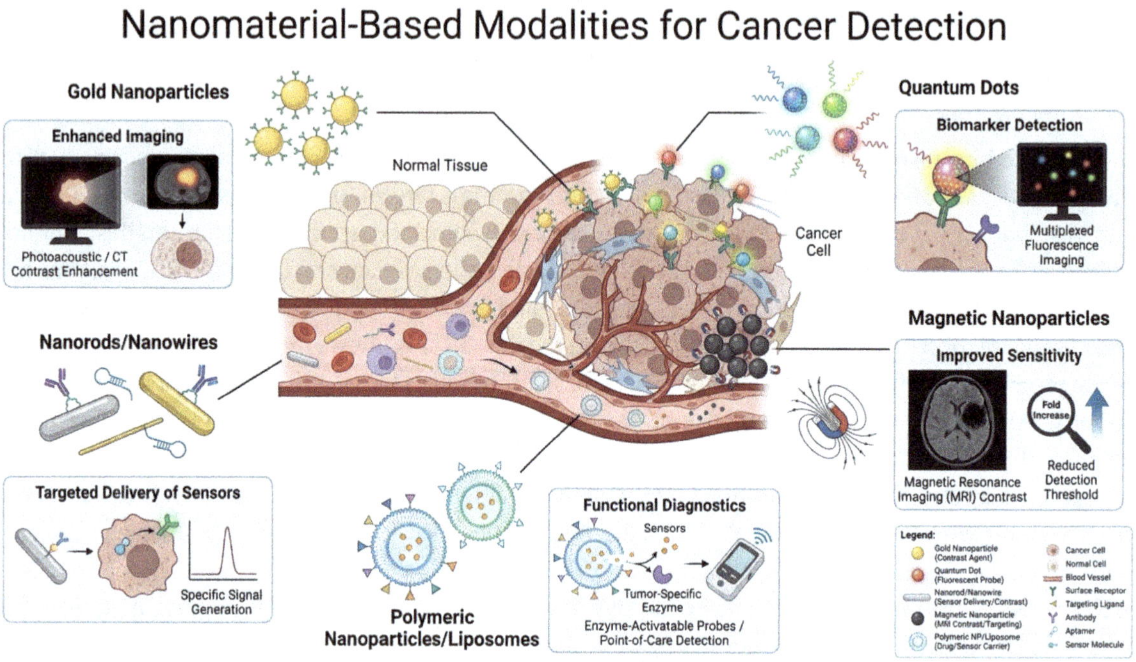

5. TYPES OF NANO-BIOSENSORS AND THEIR MECHANISMS OF CANCER DETECTION

5.1 Electrochemical Nano-biosensors

The electrochemical nano-biosensors are considered as one of the most popular types of biosensors used for cancer detection due to their great sensitivity, quickness of operation, very low detection limit, and great selectivity. The electrochemical biosensor measures any chemical reaction between an analyte and the component of biological recognition in the form of electrical signal such as current, voltage, impedance, or potential (Pulumati et al., 2023). The application of nanomaterials such as gold nanoparticles (AuNPs), carbon nanotubes (CNTs), and graphene, platinum nanoparticles, and other nanoscale conductive materials greatly improves the sensitivity of biosensors because to their large surface area, strong electrical conductivity, and catalytic property, and good ability of transferring electrons (Chen et al., 2023).

Electrochemical nano-biosensor development generally entails the modification of conductive electrodes through the deposition of nanoparticles to boost their active surface area (Banerjee et al., 2021). Bioreceptors such as antibodies, enzymes, and nucleic acids are then immobilized on the nano-modified electrode to specifically capture cancer biomarkers. Electrochemical biosensors can use potentiometric, amperometric, voltammetric, chronopotentiometric, impedimetric, or field effect transistor-based methods depending on the principle of operation (Dessale et al., 2022). The working electrode (WE), reference electrode (RE), and counter electrode (CE) make up this three-electrode configuration. The three electrodes are controlled by a potentiostat with signal processing software (Ferlay et al., 2021).

When the biological sample is added, the cancer biomarkers react specifically with the recognition elements, which are immobilized on the electrode surface, leading to modifications in electrochemical characteristics as voltage, current, impedance, or electron transfer rate (Sun et al., 2023). The extent of electron transfer between the bio-receptor and the electrode determines the biosensor performance. The role of nanomaterials in improving the electron transfer between the bio-receptors and the electrode is critical in enhancing the biosensor performance (Wang et al., 2024). Nano materials enhance electrocatalysis, accelerate electron transfer rates, amplify detection signals, increase adsorption of biomolecules, and provide direct enzyme immobilization on electrode surfaces. As a result, nano electrochemical biosensors have high sensitivity and rapid response for detection of clinically important cancer biomarkers such as AFP, CEA, and PSA, even at very low concentration levels (Tavakoli et al., 2022).

5.2 Optical Nano-biosensors

The optical nano- biosensor makes advantage of the optical characteristics of nanomaterials to detect the biological recognition event via an optical signal, which may include absorbance and fluorescence., chemiluminescence, colorimetry, and surface plasmon resonance (SPR) (Gong et al., 2021). The optical response generated after binding of the target analyte to the biological recognition component is then measured and quantified in order to identify the level of the biomarker. Analytical selectivity is attained through selective biological recognition agents like antibodies, aptamers, enzymes, and nucleic acid probes that interact with the target cancer biomarker (Dubey et al., 2022).

Quantum dots, gold nanoparticles, carbon nanotubes, graphene derivatives, and plasmonic nanostructures are among some of the nanomaterials that enhance optical biosensors' performance and sensitivity due to high fluorescence intensities, localized surface plasmon resonances, and amplification of light (Ying et al., 2022). The optical nano-biosensors have been widely used in detecting a number of clinically important biomarkers of cancer, such as PSA, CEA, and HER2. Moreover, the biosensors allow for a complete non-invasive detection of biomarkers in biological fluids, including blood, urine, and saliva, which means no need for invasive tissue biopsies (Kaminski et al., 2021). The optical nano-biosensors allow for real-time monitoring of biomarker expression while disease is developing and being treated. Multiplex detection of different biomarkers is possible due to the specific optical properties of nanomaterials, offering an opportunity to obtain comprehensive molecular information from one analysis. Besides, optical biosensors provide great stability and reproducibility of results, as well as the durability of these biosensors, which makes them extremely effective for diagnostics in clinics (Aman et al., 2020).

5.3 Magnetic Nano-biosensors

Magnetic nanobiosensors consist of magnetic nanoparticles and an external magnetic field, enabling precise measurement and quantitation of the biological samples. Magnetic biosensors have their own strength because of being more effective in conditions where optical or electrochemical techniques fail due to opaque samples, presence of interfering signals, and complicated bio-matrices (van Dongen et al., 2020). The sensing principle is based on the changes in the magnetic property that take place after the reaction of magnetic nanoparticles with the biological targets. The lack of the magnetic signal in biological samples makes magnetic sensors more sensitive (Bondy-Denomy, 2018).

SPIONS especially those filled of magnetite (Fe_3O_4), are commonly employed nanomagnetic materials due to their superior biocompatibility, chemical stability, and magnetic qualities. SPIONs are coated with very specific biological recognition materials capable of binding specifically with cancer biomarkers (Bao et al., 2021). Upon binding to the cancer biomarker, there is an observable difference in magnetic relaxation property, magnetic susceptibility, or magnetoresistance. The change is observed through highly advanced techniques including magnetic relaxation spectroscopy (MRS), giant magnetoresistance (GMR), magnetic particle imaging (MPI), and nuclear magnetic resonance (NMR) (Cheng et al., 2022). Magnetic nanobiosensors have shown great potential in detecting exosomes, circulating tumor cells (CTCs), and other cancer-related proteins has high analytical sensitivity. In contrast to the optical biosensors, the magnetic Biosensors are unaffected by optical interference, which makes them highly efficient even in highly turbid biological samples (Afzal et al., 2022).

5.4 Mechanical Nano-biosensors

The mechanical nano-biosensor operates on the principle of sensing biological events based on physical phenomena like variations in mass, mechanical stress, elasticity, stiffness, or resonant frequency caused by biomolecular recognition (Ramesh et al., 2023). While other conventional methods of biosensing often need labels, mechanical biosensors make use of the conversion of the above physical phenomena into electrical/mechanical signals for Direct and label-free detection (Gulati et al., 2025). Many types of mechanical transducers utilized in biosensing include microcantilevers, Surface acoustic wave (SAW) sensors with quartz crystal microbalance (QCM) (Khazaei et al., 2023). Microcantilever-based biosensors comprise extremely small cantilevers made up of silicon, silicon nitride, or polymers, with their surfaces modified using aptamers, or probes of nucleic acid type specific to target cancer biomarkers. The adsorption of biomarkers leads to the change in surface stress or mass load, and consequently, cantilever deflection or resonance frequency is changed. In the same manner (John et al., 2022). SAW-based biosensors identify how variations in mass and viscoelastic characteristics affect the way sound waves travel through piezoelectric substrates., or acoustic impedance of the surface as a result of biomolecular interactions (Naresh & Lee, 2021).

5.5 Piezoelectric Nano-biosensors

Nano-biosensors based on piezoelectric materials work on the principle of piezoelectricity, which refers to the ability of certain materials to produce electricity when subjected to mechanical pressure. Nano-biosensors work based on the idea of converting biochemical recognition processes into measurable electrical signals by detecting mass change in materials with piezoelectric properties (Huang et al., 2021). The piezoelectric materials that are frequently employed for the fabrication of biosensors include quartz, lead zirconate titanate (PZT), polyvinylidene fluoride (PVDF), zinc oxide, aluminum nitride, and aluminum phosphate. Unlike other electrodes, the surface of piezoelectric materials is decorated with highly specific biorecognition molecules like antibodies, antigens, aptamers, and nucleic acid probes (Du et al., 2025).

After introduction of the biological sample, the target cancer biomarkers selectively attach to the immobilized receptors, resulting in an high in the weight of the surface. The results in a proportionate decrease in the resonant frequency of the piezoelectric crystal, which is the measure of the concentration of the biomarker (Park et al., 2018). In the process of mass attachment, mechanical pressure also produces an electric signal as a result of the piezoelectric effect. This makes it possible for piezoelectric nano-biosensors to detect even minute levels of changes in mass. Hence, they are capable of detecting low concentrations of cancer biomarkers (Nasiri & Abbasian, 2025).

6. CRISPR/CAS SYSTEMS FOR BIOSENSING APPLICATIONS

The CRISPR/Cas platform has gained popularity as one of the best molecular diagnostics owing to its more programmability, target specificityas well as signal amplification properties. Based on molecular characteristics, there are various CRISPR systems. Among them, Type II (Cas9), Type V (Cas12), and Type VI (Cas13) have been applied widely for the purpose of biosensing (Ramesh et al., 2023). Whereas Cas9 functions by sequence-specific cleavage (cis-cleavage) of double-stranded DNA, Cas12 and Cas13 show the property of collateral (trans) cleavage, which means that the process of activation upon the recognition of target results in non-specific cleavage of nearby single-stranded RNA (ssRNA) or single-stranded DNA (ssDNA), respectively. This property has made Cas12 and Cas13 useful for more sensitive identification of nucleic acid using the process of signal amplification through fluorescent reporter molecules (Rabiee, 2025).

Cas13 proteins, which are comprised of approximately 900 to 1300 amino acids, are responsible for recognition and cleavage of single-stranded RNA as well as simultaneous collateral cleavage of adjacent ssRNA reporters. SHERLOCK (Specific High-sensitivity enzymatic Reporter UNlocking) is an example of a technology which utilizes Cas13 enzyme for attomolar (10^{-18} M) sensitive detection of viruses, bacteria, cancer-associated mutations and SNPs. In the same way, Cas12 proteins recognize both double- and single-stranded DNA targets and form the basis of the DNA Endonuclease-Targeted CRISPR Trans Reporter (DETECTR) platform (Hsu & Wu, 2022). After complementary annealing of the CRISPR RNA (crRNA) with target DNA, Cas12 makes sequence-specific cleavage of the complex, followed by uncontrolled cleavage of ssDNA probes attached utilizing quencher reporters and fluorophore, producing a fluorescence signal proportionate to the desired DNA levels (Huang et al., 2021). Despite the fact that SHERLOCK, DETECTR, and HOLMESv2 have outstanding Most CRISPR/Cas-based sensing technologies need nucleic acid amplification for analytical sensitivity and specificity. prior to the detection due to inability of recognition of very small amounts of biomarkers (Naresh & Lee, 2021).

7. PLASMONIC NANOMATERIALS FOR BIOSENSING

The unique properties of plasmonic nanomaterials make them essential building blocks of modern biosensors because of their exceptional sensitivity to biomolecular interactions and distinct optical characteristics. At the interface between dielectric materials and plasmonic metals like Au and Ag, light interacts with the oscillations of free electrons in metallic nanoparticles to produce surface plasmon resonance (SPR) (Tai et al., 2022). Such interactions generate highly localized electromagnetic fields, which increase the light absorption and dispersion by plasmonic nanoparticles significantly. Thus, plasmonic nanomaterials are more sensitive to the variations of the dielectric environment. In addition, These nanoparticles' size can be used to adjust their optical characteristics., morphology, chemical composition, and surface properties. During biosensing, interactions with biomolecules cause variations in the optical properties of plasmonic nanomaterials like resonance wavelength, absorbance, and scattering intensity (Horgan et al., 2020).

Plasmon-mediated processes contributing to the signal amplification include fluorescence quenching, wavelength shift caused by fluorescence resonance energy transfer (FRET), fluorescence enhancement, and enhancement of localized electromagnetic fields. Nanostructures having sharp edges or corners create zones of highly concentrated electromagnetic enhancement referred to as "hot spots," forming the foundation of surface-enhanced Raman scattering (SERS) technology (Vaughan, 2024). The use of such hot spots causes an incredible amplification of Raman signals, providing a very sensitive method for the identification of molecular vibrational fingerprints, thereby enabling the detection ofbiomarkers even at very small levels. As such, plasmonic nanomaterials has received a lot of attention in recent years in developing detection methods for viruses, cancer biomarkers, and pregnancy disorders among others. In fact, the combination of plasmonic nanomaterials and CRISPR/Cas biosensors has resulted in a highly sensitive platform for next-generation biosensing (Timilsina & Li, 2024).

8. CRISPR/CAS-BASED SERS AND LSPR NANO-BIOSENSORS FOR DETECTING DISEASE BIOMARKERS (SUMMARY)

CRISPR/Cas convergence with localized surface plasmon resonance (LSPR) and surface enhanced Raman scattering (SERS)nanotechnology has revolutionized disease biomarker detection through high speed, sensitivity, and specificity (Díaz-García et al., 2024). The combination of CRISPR/Cas12a based SERS biosensor with gold nanostructures, magnetic nanoparticles and Raman probes makes it possible to perform detection of viral DNA, bacteria and cancer biomarkers in attomolar concentration without any amplification step (Dkhar et al., 2024). Similarly, LSPR-based CRISPR biosensors use plasmonic gold nanomaterials that produce optical resonance shift for fast and label-free nucleic acid and protein biomarker detection, including the PSA protein marker. Examples of such highly sensitive and rapid diagnostic platforms include SHERLOCK based SERS system, RPA-Cas12a-mPAD, HCR-powered LSPR sensor, and RT-RPA/CRISPR-Cas12a colorimetric assay (Achi et al., 2024). These diagnostic platforms offer highly sensitive detection, wide range and great specificity for applications such as cancer diagnosis, virus detection (SARS-CoV-2, HIV-1), food safety and biomarker monitoring. Thus, convergence of CRISPR/Cas with SERS and LSPR technologies is extremely promising for development of highly sensitive, rapid and multiplexed point-of-care diagnostic platforms (Sadeghi et al., 2023).

9. CRISPR-CAS-BASED NANOSENSORS FOR CANCER DIAGNOSIS

9.1 Electrochemical Signal

Electrochemical nanosensors that are based on CRISPR-Cas have emerged as a highly efficient method for detecting cancer because to their extreme sensitivity, speed, cheapness, and capacity to identify in real-time. Most of the electrochemical biosensors lack efficiency when it comes to sensing nucleic acid in very low concentrations (Lu et al., 2023). The use of CRISPR-Cas systems especially Cas12a and Cas13a has tremendously improved the ability of electrochemical biosensing regarding its specificity and accuracy. Through this method, The intended nucleic acid will get identified by the biosensor and activate the CRISPR system, leading to the cutting of reporter molecules and generation of electrical signals (Yasamineh et al., 2024).

Gold nanoparticles, magnetic beads, titanium carbide nanosheets, and conductive polymers are some of the nanomaterials whose emergence has helped improve electrochemical sensing through improved electron transfer, signal amplification, and increased sensor stability (Wang et al., 2024). The new PEC (photoelectrochemical) sensors have been found to be quite efficient due to their high sensitivity, which has made them detect even traces of cancer biomarkers thanks to their extremely low detection limits. There are initiatives to develop the E-CRISPR testing devices that will be portable and may even involve use of smartphones (Montalvan et al., 2022).

9.2 Electrochemiluminescence Signal

Electrochemiluminescence (ECL) represents a highly advanced method of diagnostics which combines electrochemistry and luminescence To enhance the sensitivity of detecting biomarkers (Hsu & Wu, 2022). Unlike the conventional fluorescent methods, the electrochemiluminescence does not require an external light source, which means that there is less background noise. In conjunction with the CRISPR-Cas systems, ECL provides exceptional sensitivity in detecting biomarkers of cancer, namely microRNAs and DNA mutations. CRISPR serves as a target recognition system, triggering signal amplification and subsequent luminescence (Wang et al., 2024).

Further advancements in ECL biosensor sensitivity and efficiency have been made through use of nanomaterials, like gold nanoclusters, carbon dots, perovskite nanocrystals, and graphitic carbon nitride nanosheets into the system. These nanomaterials have some different optical and electrochemical properties which make them brighter and more efficient at sensing (Dezhakam et al., 2024). Very sophisticated signal amplification processes such as DNA walkers and cyclic signal amplification process have made it possible to detect as low as femtomolar level. This high sensitivity makes it easy to identify the cancer biomarker at an starting stage. It is predicted that CRISPR-ECL biosensor will lead significant role in portable cancer diagnosis kits in future (Kuntamung et al., 2021).

9.3 Integration with Imaging Modalities (MRI, CT, PET)

The combination of nanosensors with imaging techniques like MRI, CT, and PET has resulted in significant advancements in the diagnosis and management of cancers. By functionalizing the nanosensors with targeting biomolecules, they will bind specifically to the cancerous cells and enhance the imaging process, making it easily identifiable (Mohan et al., 2025). Nanosensors have the capability to give a detailed account of cancer development processes without any invasion into the body. Combining molecular imaging with sensing gives an instant picture of cancer progression (Li et al., 2023).

Nanoparticles of many varieties, including but not limited to Carbon nanotubes, gold nanoparticles, magnetic nanoparticles, and superparamagnetic iron oxide nanoparticles (SPIONs) have all been successfully employed as imaging contrast agents. Moreover, apart from enhancing image quality, nanoparticles can also be used for the delivery of drugs and thus creating multimodal theranostics devices (Farzin et al., 2022). Indeed, using magnetic nanoparticles to enhance the visibility of MR images and deliver antitumor drugs into the right place at the same time is one example of that. Another example is represented by gold nanoparticles, which can penetrate biological barriers and thus perform imaging and treatment.

Combining CRISPR technology and MR imaging, CT and PET imaging is quite promising in oncology (Vijayakumar, 2023). Figure 2 illustrates how CRISPR-Cas systems integrated with nanosensors detect cancer-related biomarkers such as DNA mutations, RNA, microRNAs, and proteins with high sensitivity. These nanosensors enable rapid, specific, and early cancer diagnosis by converting molecular recognition signals into measurable optical, electrochemical, or fluorescent outputs.

Figure 2. **CRISPR-Cas-Based Nanosensors for Cancer Diagnosis**

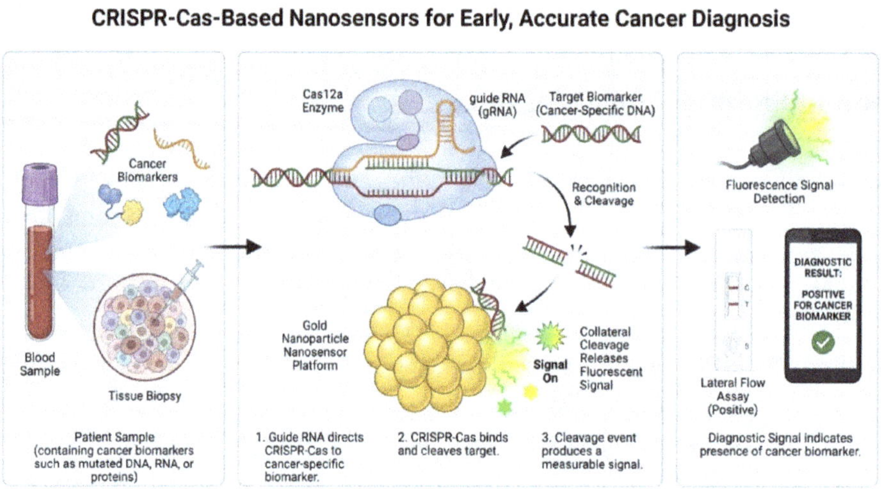

10. APPLICATIONS OF NANO-BIOSENSORS IN VARIOUS CANCER THERAPIES

The field of oncology has been greatly transformed by the emergence of nano-biosensors, which provide an innovative way of diagnosing, managing, and treating various forms of cancer through the precise and accurate identification of associated biomarkers (Zhao et al., 2022). With the use of advanced nanotechnology in conjunction with biosensors, it has become possible to develop a variety of diagnostic tools that integrate nanoparticles, nanowires, carbon quantum dots, nanotubes, and other types of engineered materials that can detect CTCs, ctDNA, nucleic acids, proteins, exosomes, and VOCs in biological fluids like blood, saliva, urine, and even breath (Smith et al., 2020). This allows for an improved chance of identifying cancer at an early stage by being able to detect very small quantities of biomarkers associated with the disease. In addition to their role in the diagnosis of cancer, nano-biosensors can also serve as potential therapeutic methods by acting as intelligent carriers of chemotherapeutic drugs that target specific tumor sites.

Moreover, the use of nano-biosensors in combination with more sophisticated techniques of medical imaging like MRI, CT, PET, and various other multimodal imaging techniques had resulted in a remarkable improv in the accuracy of tumor localization, disease staging, monitoring treatment, and assessing the therapeutic response (Sun & Chen, 2024). Due to their capabilities in the form of being capable of performing early diagnosis, real-time disease monitoring, targeted therapy, and personalized treatment strategy from a single platform, nano-biosensors have become an essential part of precision oncology. Various applications of these sensors have been explored in many different kinds of cancers, especially in cases of prostate, breast, lung, and brain cancer (Pan et al., 2022). Table 1 summarizes the major applications of nano-biosensors in cancer diagnosis and therapy, including early tumor detection, biomarker identification, drug delivery monitoring, treatment response evaluation, and image-guided therapy. It highlights how nano-biosensors improve cancer management by offering high sensitivity, rapid detection, targeted action, and real-time monitoring of therapeutic outcomes.

Table 1. Applications of Nano-biosensors in Various Cancer Therapies and Diagnosis

Cancer Type	Nano-biosensor Application	Target Biomarker	Diagnostic/Therapeutic Role	Advantages	References
Breast Cancer	Gold nanoparticle biosensors	HER2, CA15-3	Early diagnosis and monitoring	High sensitivity and rapid detection	(Yin et al., 2022)
Lung Cancer	Graphene-based biosensors	EGFR, CEA	Cancer screening and prognosis	Real-time analysis	(Zhuang et al., 2022)
Prostate Cancer	Electrochemical nano-biosensors	PSA	Early-stage diagnosis	Low detection limit	(Zhang et al., 2021)
Brain Cancer	Magnetic nanoparticle biosensors	GFAP, EGFRvIII	Tumor identification	Non-invasive detection	(Yuan et al., 2020)
Colorectal Cancer	Quantum dot biosensors	CEA, KRAS	Biomarker detection	High specificity	(Cheng et al., 2021)
Liver Cancer	Carbon nanotube biosensors	AFP	Disease monitoring	Fast response time	(Huang et al., 2020)
Leukemia	DNA nano-biosensors	BCR-ABL	Genetic mutation analysis	Accurate molecular detection	(Bogers et al., 2021)
Ovarian Cancer	Fluorescent nano-biosensors	CA125	Early diagnosis	Enhanced signal amplification	(Fu et al., 2021)
Skin Cancer	Plasmonic nano-biosensors	Melanoma markers	Cancer detection	Portable and sensitive	(Aman et al., 2020)
Theranostic Applications	Multifunctional nano-biosensors	Multiple biomarkers	Combined diagnosis and therapy	Personalized treatment	(Bondy-Denomy, 2018)

11. CHALLENGES FACING CRISPR-BASED POINT-OF-CARE (POC) DIAGNOSTICS

Even though CRISPR-enabled detection devices represent very effective technologies, there are still some technical, logistical, regulatory, and economic constraints associated with their implementation in the clinic (Li & Hu, 2022). One of the most serious concerns is related to the necessity of performing nucleic acid extraction and purification, which often requires advanced laboratory equipment and trained personnel, and therefore cannot be performed easily. Development of simplified protocols based on lysis reactions, one-pot reaction chemistries, microfluidic technologies, and fully automated devices may help to solve this problem and improve the accessibility of CRISPR-enabled detection systems (Dubey et al., 2022). In addition, it is necessary to consider the issue of CRISPR reagents storage and transport, since a lot of CRISPR components have to be kept in continuous cold chain in order to maintain the activity of enzymes. Development of lyophilized CRISPR reagents would solve this problem and simplify the process of reagents transport to remote medical facilities (Gong et al., 2021).

CRISPR-based POC diagnostic devices can also be impacted by the effect of various environmental conditions, like temperature and humidity, and operator variability, making it necessary to conduct comprehensive testing in different settings to achieve consistent performance (Tavakoli et al., 2022). Moreover, the high price of specialized reagents and nanomaterials, as well as the process of manufacturing itself, continues to be an obstacle for large-scale commercialization. As a result, there is a necessity to come up with scalable and affordable production methods without sacrificing the diagnostic processes' sensitivity and specificity (Wang et al., 2024). The integration of CRISPR with artificial intelligence (AI) diagnostics is a very promising method to address many challenges of using these tools. Machine learning algorithms allow quick data analysis, detection of target genetic sequences, improvement of gRNA design, reduction of off-target effects, and optimization of CRISPR-based tests (Sun et al., 2023).

Furthermore, AI-aided automation is helpful in performing high-throughput sample analysis, instant data analysis, and effective management of large diagnostic data sets with reduced human involvement, thereby increasing the efficiency of workflow in large-scale clinical and epidemiologic studies (Dessale et al., 2022). Predictive analytics enabled by the AI also aid in disease surveillance by predicting outbreaks in advance, tracking infectious disease transmission, and providing fast responses to public health. Apart from being used in clinical diagnosis, the CRISPR-based biosensor technologies are increasingly being used for applications like food safety, environmental monitoring, agriculture pathogens' detection, and biodefense (Ferlay et al., 2021). The emerging technologies like CRISPR-based electrochemical biosensor technologies provide quick, highly sensitive, and portable platforms that do not need any equipment, which have the capability to revolutionize the point-of-care diagnostic technology. However, the broad clinical acceptance of these technologies would require the thorough consideration of issues like regulatory approval, quality assurance, data privacy, ethics, and biosafety (Huang et al., 2021). Incorporating environmentally sustainable nanomaterials and eco-friendly fabrication technologies into CRISPR-based diagnostic platforms will improve scalability, minimize hazardous waste generation, and support the long-term sustainability of precision diagnostics (Pan et al., 2022).

12. FUTURE PERSPECTIVES

The tremendous progress in nanotechnology has greatly promoted the development of the miniaturized diagnostic tools, thus creating immense possibilities for the advancement of precision medicine, target drug delivery, and personalized cancer treatment (Nasiri & Abbasian, 2025). Due to the distinct physical characteristics of nanomaterials, like a high surface-to-volume ratio, tuneable surface chemistry, and quantum confinement effects, highly sensitive biosensing devices that are able to identify even very small concentrations of disease biomarkers have been developed. These advancements have led to the invention of point-of-care (POC) diagnostic devices for tuberculosis, malaria, and various types of cancer. Moreover, the combination of nanofluidics and microfluidics in the lab-on-a-chip (LOC) system has appeared to be an efficient approach enables the quick identification of infections and cancers (Ramesh et al., 2023). The LOC platform is characterized by precise handling of nanoliter volume samples, simultaneous performing of several bioassays, low amounts of reagents needed, automation of sample preparation, quick analysis, minimal production of hazardous waste, and high accuracy of biomarker detection. Apart from that, there is more and more interest in integrating the nano-biosensors with smartphones, wearable devices, wireless communication, and Internet of Medical Things (IoMT) system (Kurul et al., 2025).

Incorporation of nanosensors in microfluidics has shown tremendous promise in bioanalysis at point-of-care, especially in regions where advanced laboratory equipment is not available. Due to the small size, inexpensive manufacturing process, quick detection capability, and reduced amounts of reagents used, microfluidics offer an ideal solution for diagnostics in clinical practice (Dezhakam et al., 2024). Microfluidics have been extensively used in the detection of single cancer cells, liquid biopsies, drug screening, and studies on angiogenesis. At the same time, the emergence of the use of AI in conjunction with nanotechnology revolutionizes precision medicine through personalized treatment modalities derived from thorough molecular analysis of patients (Kuntamung et al., 2021). Sophisticated nano-biosensors provide biomarker information that can be analyzed by means of AI-powered analysis and algorithms that enable accurate patient stratification, predicting drug efficacy, optimal design of nanomaterials, and individualized therapy regimes. Despite the ongoing difficulties associated with heterogeneous nature of cancer tumors and biological diversity of patients, which makes diagnostics and therapy quite challenging, advanced AI-based techniques such as pattern recognition, predictive modeling, and classification algorithms can be used to solve this problem (Mohan et al., 2025). The further combination of CRISPR technology, nanotechnology, microfluidics, artificial intelligence, and digital health technologies will result in an intelligent, portable, and completely automated diagnostic system for rapid and highly accurate cancer diagnostics (Li et al., 2023). Future developments should emphasize environmentally sustainable biosensor design by combining biodegradable nanomaterials, renewable biopolymers, solvent-free synthesis techniques, and low-energy manufacturing processes. Integrating these eco-friendly approaches with CRISPR diagnostics will facilitate the development of highly sensitive, scalable, and environmentally responsible diagnostic platforms for precision oncology (Pan et al., 2022).

13. CONCLUSION

The nanotechnology and CRISPR/Cas based molecular diagnostic technologies is a major revolution towards early cancer diagnosis and personalized cancer treatments. Traditional diagnostic methods like biopsy and imaging technologies usually suffer from low sensitivity in identifying small tumors and micrometastasis, which may delay diagnosis and treatment. Though the use of liquid biopsy is a small invasive technique In contrast to tissue biopsies, the low concentration of biomarkers of cancer present in biological fluids makes the need for a highly sensitive analytical platform necessary. CRISPR/Cas technology has become a promising and programmable technology for diagnostics due to high specificity and versatility, but it faces some challenges with respect to sensitivity, signal amplification, and molecular stability. However, the addition of nanomaterials overcomes all such problems of CRISPR/Cas technology as it amplifies fluorescence, colorimetric, electrochemical, and plasmonic signals, and reduces the detection limits up to attomolar concentrations. Thus, CRISPR-based nano-biosensors are very effective tools in the detection of various cancer biomarkers of prostate, breast, lung, pancreas, brain cancers, and many others, with an excellent sensitivity and specificity. However, there are a few obstacles that need to be cleared for implementing CRISPR-based nano-biosensors on a large scale.

Innovations that include the use of lyophilized reagents, one-pot reaction assays, automation in sample preparation, artificial intelligence for data interpretation, and improved microfluidics technologies have increasingly overcome the above challenges and made CRISPR-based diagnostic tools more relevant. For the future, the combination of CRISPR/Cas technology with nano-biosensors, microfluidics, wearables, artificial intelligence, and smart health care systems is poised to revolutionize cancer diagnosis from being an intermittent clinical process to becoming a continuous, real-time and personalized process. It is anticipated that continued innovations in the field will speed up the adoption of these advanced diagnostic platforms in clinical practice, thus enhancing Early cancer identification and therapy outcomes. The integration of sustainable nanomaterials with CRISPR-based biosensors represents an important future direction, offering environmentally responsible diagnostic platforms that combine high sensitivity, excellent biocompatibility, and scalable manufacturing.

REFERENCES

Achi, F., Attar, A. M., & Ait, L. A. (2024). Electrochemical nanobiosensors for the detection of cancer biomarkers in real samples: Trends and challenges. *Trends in Analytical Chemistry*, *170*, 117423. DOI: 10.1016/j.trac.2023.117423

Afzal, O., Altamimi, A. S. A., Nadeem, M. S., Alzarea, S. I., Almalki, W. H., Tariq, A., Mubeen, B., Murtaza, B. N., Iftikhar, S., Riaz, N., & Kazmi, I. (2022). Nanoparticles in drug delivery: From history to therapeutic applications. *Nanomaterials (Basel, Switzerland)*, *12*(24), 4494. DOI: 10.3390/nano12244494

Aman, R., Mahas, A., & Mahfouz, M. (2020). Nucleic acid detection using CRISPR/Cas biosensing technologies. *ACS Synthetic Biology*, *9*(6), 1226–1233. DOI: 10.1021/acssynbio.9b00507

Banerjee, A., Maity, S., & Mastrangelo, C. H. (2021). Nanostructures for biosensing, with a brief overview on cancer detection, IoT, and the role of machine learning in smart biosensors. *Sensors (Basel)*, *21*(4), 1253. DOI: 10.3390/s21041253

Bao, M., Chen, Q., Xu, Z., Jensen, E. C., Liu, C., Waitkus, J. T., Yuan, X., He, Q., Qin, P., & Du, K. (2021). Challenges and opportunities for clustered regularly interspaced short palindromic repeats based molecular biosensing. *ACS Sensors*, *6*(7), 2497–2522. DOI: 10.1021/acssensors.1c00530

Bhardwaj, H., Archana, N., Noumani, A., Himanshu, J. K., Chakravorty, S., & Solanki, P. R. (2024). Recent advancement in the detection of potential cancer biomarkers using the nanomaterial integrated electrochemical sensing technique: A detailed review. *Materials Advances*, *5*(2), 475–503. DOI: 10.1039/D3MA00621B

Bogers, J. F., Berghuis, N. F., Busker, R. W., van Booma, A., Paauw, A., & van Leeuwen, H. C. (2021). Bright fluorescent nucleic acid detection with CRISPR-Cas12a and poly(thymine) templated copper nanoparticles. *Biology Methods and Protocols*, *6*(1), 20–26. DOI: 10.1093/biomethods/bpaa020

Bondy-Denomy, J. (2018). Protein inhibitors of CRISPR-Cas9. *ACS Chemical Biology*, *13*(2), 417–423. DOI: 10.1021/acschembio.7b00831

Chen, Y., Ma, B., Zuo, Y., Chen, G., Hao, Q., Zhao, C., Chen, Y., & Zhang, X. (2023). Versatile sweat bioanalysis on demand with hydrogel-programmed wearables. *Biosensors & Bioelectronics*, *235*, 115412. DOI: 10.1016/j.bios.2023.115412

Cheng, M., Xiong, E., Tian, T., Zhu, D., Ju, H. Q., & Zhou, X. (2021). A CRISPR-driven colorimetric code platform for highly accurate telomerase activity assay. *Biosensors & Bioelectronics*, *172*, 112749. DOI: 10.1016/j.bios.2020.112749

Cheng, M., Xiong, E., Tian, T., Zhu, D., Ju, H. Q., & Zhou, X. (2021). A CRISPR-driven colorimetric code platform for highly accurate telomerase activity assay. *Biosensors & Bioelectronics*, *172*, 112749. DOI: 10.1016/j.bios.2020.112749

Dessale, M., Mengistu, G., & Mengist, H. M. (2022). Nanotechnology: A promising approach for cancer diagnosis, therapeutics and theragnosis. *International Journal of Nanomedicine*, *17*, 3735–3749. DOI: 10.2147/IJN.S378074

Dezhakam, E., Vayghan, R. F., Dehghani, S., Kafili-Hajlari, T., Naseri, A., Dadashpour, M., & Hasanzadeh, M. (2024). Highly efficient electrochemical biosensing platform in breast cancer detection based on MOF-COF@Au core-shell like nanostructure. *Scientific Reports, 14*(1), 1–13. DOI: 10.1038/s41598-024-78836-y

Díaz-García, D., Díaz-Sánchez, M., Álvarez-Conde, J., & Gómez-Ruiz, S. (2024). Emergence of quantum dots as innovative tools for early diagnosis and advanced treatment of breast cancer. *ChemMedChem, 19*(16), e202400172. DOI: 10.1002/cmdc.202400172

Dkhar, D. S., Kumari, R., Patel, V., Srivastava, A., Prasad, R., Srivastava, R., & Chandra, P. (2024). Versatile approaches of quantum dots in biosensing and imaging. *Wiley Interdisciplinary Reviews. Nanomedicine and Nanobiotechnology, 16*(5), e1998. DOI: 10.1002/wnan.1998

Du, B., Tian, X., Chen, Z., Ge, Y., Chen, C., Gao, H., Li, S., & Zhang, Y. (2025). Ultrasensitive optoelectronic biosensor arrays based on twisted bilayer graphene superlattice. *National Science Review, 12*(10), nwaf357. Advance online publication. DOI: 10.1093/nsr/nwaf357

Duan, L., Zhang, S., Yan, Q., Chen, Y., Su, Y., & Hu, X. (2026). Comparative efficacy of e-health interventions for quality of life in patients with cancer: A network meta-analysis of randomized controlled trials. *Worldviews on Evidence-Based Nursing, 23*(1), e70103. Advance online publication. DOI: 10.1111/wvn.70103

Dubey, A. K., Gupta, V. K., Kujawska, M., Orive, G., Kim, N.-Y., Li, C.-Z., Mishra, Y. M., & Kaushik, A. (2022). Exploring nano-enabled CRISPR-Cas-powered strategies for efficient diagnostics and treatment of infectious diseases. *Journal of Nanostructure in Chemistry, 12*(5), 833–864. DOI: 10.1007/s40097-022-00472-7

Farzin, M. A., Abdoos, H., & Saber, R. (2022). AuNP-based biosensors for the diagnosis of pathogenic human coronaviruses: COVID-19 pandemic developments. *Analytical and Bioanalytical Chemistry, 414*(24), 7069–7084. DOI: 10.1007/s00216-022-04193-2

Ferlay, J., Colombet, M., Soerjomataram, I., Parkin, D. M., Piñeros, M., Znaor, A., & Bray, F. (2021). Cancer statistics for the year 2020: An overview. *International Journal of Cancer, 149*(4), 778–789. DOI: 10.1002/ijc.33588

Fu, X., Shi, Y., Peng, F., Zhou, M., Yin, Y., Tan, Y., Chen, M., Yin, X., Ke, G., & Zhang, X. B. (2021). Exploring the trans-cleavage activity of CRISPR/Cas12a on gold nanoparticles for stable and sensitive biosensing. *Analytical Chemistry, 93*(11), 4967–4974. DOI: 10.1021/acs.analchem.1c00027

Gong, S., Zhang, S., Lu, F., Pan, W., Li, N., & Tang, B. (2021). CRISPR/Cas-Based In Vitro Diagnostic Platforms for Cancer Biomarker Detection. *Analytical Chemistry, 93*(35), 11899–11909. DOI: 10.1021/acs.analchem.1c02533

Gulati, S., Yadav, R., Kumari, V., Nair, S., Gupta, C., & Aishwari, M. (2025). Nanosensors in healthcare: Transforming real-time monitoring and disease management with cutting-edge nanotechnology. *RSC Pharmaceutics, 2*(5), 1003–1018. Advance online publication. DOI: 10.1039/D5PM00125K

Gupta, D., Boora, A., Thakur, A., & Gupta, T. K. (2023). Green and Sustainable Synthesis of Nanomaterials: Recent Advancements and Limitations. *Environmental Research*, *231*, 116316. DOI: 10.1016/j.envres.2023.116316

Hassan, R. Y. A. (2022). Advances in electrochemical nano-biosensors for biomedical and environmental applications: From current work to future perspectives. *Sensors (Basel)*, *22*(19), 7539. DOI: 10.3390/s22197539

Hemdan, M., Abuelhaded, K., Shaker, A. A. S., Ashour, M. M., Abdelaziz, M. M., Dahab, M. I., Nassar, Y. A., Sarguos, A. M. M., Zakaria, P. S., Fahmy, H. A., & El-Sherbiny, I. M. (2025). Recent advances in nano-enhanced biosensors: Innovations in design, applications in healthcare, environmental monitoring, and food safety, and emerging research challenges. *Sensing and Bio-Sensing Research*, *48*, 100783. DOI: 10.1016/j.sbsr.2025.100783

Horgan, C. C., Bergholt, M. S., Nagelkerke, A., Thin, M. Z., Pence, I. J., Kauscher, U., & Stevens, M. M. (2020). Integrated photodynamic Raman theranostics for cancer diagnosis, treatment, and post-treatment molecular monitoring. *Theranostics*, *11*(4), 2006–2019. DOI: 10.7150/thno.53031

Hsu, C. C., & Wu, Y. (2022). Recent advances in nanotechnology-enabled biosensors for detection of exosomes as new cancer liquid biopsy. *Experimental Biology and Medicine (Maywood, N.J.)*, *247*(23), 2152–2172. DOI: 10.1177/15353702221110813

Huang, D., Qian, J., Shi, Z., Zhao, J., Fang, M., & Xu, Z. (2020). CRISPR-Cas12a-assisted multicolor biosensor for semiquantitative point-of-use testing of the Nopaline synthase terminator in genetically modified crops by unaided eyes. *ACS Synthetic Biology*, *9*(11), 3114–3123. DOI: 10.1021/acssynbio.0c00365

Huang, X., Zhu, Y., & Kianfar, E. (2021). Nano biosensors: Properties, applications and electrochemical techniques. *Journal of Materials Research and Technology*, *12*, 1649–1672. DOI: 10.1016/j.jmrt.2021.03.048

John, S. A., Chattree, A., Ramteke, P. W., Shanthy, P., Nguyen, T. A., & Rajendran, S. (2022). Nanosensors for plant health monitoring. *Nanosensors for Smart Agriculture*, *1*, 449–461. DOI: 10.1016/B978-0-12-824554-5.00012-4

Kaminski, M. M., Abudayyeh, O. O., Gootenberg, J. S., Zhang, F., & Collins, J. J. (2021). CRISPR-based diagnostics. *Nature Biomedical Engineering*, *5*(7), 643–656. DOI: 10.1038/s41551-021-00760-7

Khazaei, M., Hosseini, M. S., Haghighi, A. M., & Misaghi, M. (2023). Nanosensors and their applications in early diagnosis of cancer. *Sensing and Bio-Sensing Research*, *41*, 100569. DOI: 10.1016/j.sbsr.2023.100569

Kuntamung, K., Jakmunee, J., & Ounnunkad, K. (2021). A label-free multiplex electrochemical biosensor for the detection of three breast cancer biomarker proteins employing dye/metal ion-loaded and antibody-conjugated polyethyleneimine-gold nanoparticles. *Journal of Materials Chemistry. B*, *9*(33), 6576–6585. DOI: 10.1039/D1TB00940K

Kurul, F., Turkmen, H., Cetin, A. E., & Topkaya, S. N. (2025). Nanomedicine: How nanomaterials are transforming drug delivery, bio-imaging, and diagnosis. *Next Nanotechnology*, *7*, 100129. DOI: 10.1016/j.nxnano.2024.100129

Li, C. H., Chan, M. H., Chang, Y. C., & Hsiao, M. (2023). Gold nanoparticles as a biosensor for cancer biomarker determination. *Molecules (Basel, Switzerland), 28*(1), 364. DOI: 10.3390/molecules28010364

Li, C.-Z., & Hu, T. Y. (2022). Nanotechnology Powered CRISPR-Cas Systems for Point of Care Diagnosis and Therapeutic. *AAAS Research, 9810237*, 2022/9810237. Advance online publication. DOI: 10.34133/2022/9810237

Liu, X., Yu, L., Xiao, A., Sun, W., Wang, H., Wang, X., Feng, S., & Zhang, Y. (2025). Analytical methods in studying cell force sensing: Principles, current technologies and perspectives. *Regenerative Biomaterials, 12*, rbaf007. Advance online publication. DOI: 10.1093/rb/rbaf007

Liu, Z., Zhang, S., Wu, Z., Mu, E., Wei, H., Liu, Y., Li, Z., & Chen, Y. (2023). High-performance integrated chip-level thermoelectric device for power generation and microflow detection. *Nano Energy, 114*, 108611. DOI: 10.1016/j.nanoen.2023.108611

Lu, W., Wang, Y., & Cao, X. (2023). Editorial: Advanced nanobiosensors for non-invasive disease diagnosis. *Frontiers in Bioengineering and Biotechnology, 11*, 1208679. DOI: 10.3389/fbioe.2023.1208679

Mohan, B., Modi, K., Singh, G., Paul, A., Garazade, I. M., Pombeiro, A. J. L., & Kumar, S. (2025). Understanding the electrochemical MOF sensors in detecting cancer with special emphasis on breast carcinoma biomarkers. *Topics in Current Chemistry (Cham), 383*(1), 1–48. DOI: 10.1007/s41061-025-00493-0

. Montalvan, J., Healy, L., Javaid, H., Marin, I., Menegaz, B., Hsu, C., & Kuerer, H. (2022). Abstract P3-20-06: magseed/magtrace-guided surgery in breast cancer—an institutional perspective. *Cancer Research, 82*(4_Supplement), P3-20-06. https://doi.org/DOI: 10.1158/1538-7445.SABCS21-P3-20-06

Naresh, V., & Lee, N. (2021). A review on biosensors and recent development of nanostructured materials-enabled biosensors. *Sensors (Basel), 21*(4), 1109. DOI: 10.3390/s21041109

Nasiri, H., & Abbasian, K. (2025). Surface plasmon resonance detection of anti-cancer drug flutamide by graphitic carbon nitride/chitosan nanocomposite. *Scientific Reports, 15*(1), 1–9. DOI: 10.1038/s41598-025-86665-w

Pan, R., Li, Y., Chen, Y., Li, G., Zhang, Y., Wang, X., & Li, J. (2022). Ultrasensitive CRISPR/Cas12a-driven SERS biosensor for on-site nucleic acid detection and its application to milk authenticity testing. *Journal of Agricultural and Food Chemistry, 70*(14), 4484–4491. DOI: 10.1021/acs.jafc.1c08262

Park, M., Kang, B. H., & Jeong, K. H. (2018). Paper-based biochip assays and recent developments: A review. *Biochip Journal, 12*(1), 1–10. DOI: 10.1007/s13206-017-2101-3

Pulumati, A., Pulumati, A., Dwarakanath, B. S., Verma, A., & Papineni, R. V. L. (2023). Technological advancements in cancer diagnostics: Improvements and limitations. *Cancer Reports, 6*(2), e1764. DOI: 10.1002/cnr2.1764

Rabiee, N. (2025). Programmable Biointerfaces and Adaptive Functionality in Next-Generation Green Nanomaterials. *ACS Synthetic Biology, 14*(10), 3862–3879. DOI: 10.1021/acssynbio.5c00620

Ramesh, M., Janani, R., Deepa, C., & Rajeshkumar, L. (2023). Nanotechnology-enabled biosensors: A review of fundamentals, design principles, materials, and applications. *Biosensors (Basel), 13*(1), 40. DOI: 10.3390/bios13010040

Sadeghi, M., Sadeghi, S., Naghib, S. M., & Garshasbi, H. R. (2023). A comprehensive review on electrochemical nano biosensors for precise detection of blood-based oncomarkers in breast cancer. *Biosensors (Basel), 13*(4), 481. DOI: 10.3390/bios13040481

Smith, R., Geary, S. M., & Salem, A. K. (2020). Silicon nanowires and their impact on cancer detection and monitoring. *ACS Applied Nano Materials, 3*(9), 8522–8536. DOI: 10.1021/acsanm.0c01572

Sun, S., & Chen, J. (2024). Recent advances in hydrogel-based biosensors for cancer detection. *ACS Applied Materials & Interfaces, 16*(36), 46988–47002. DOI: 10.1021/acsami.4c02317

Sun, X., Zhang, M., Xie, L., Ren, Q., Chang, J., & Jiang, W. (2023). Material-enhanced biosensors for cancer biomarkers detection. *Microchemical Journal : Devoted to the Application of Microtechniques in All Branches of Science, 194*, 109298. DOI: 10.1016/j.microc.2023.109298

Tai, J., Fan, S., Ding, S., & Ren, L. (2022). Gold nanoparticles based optical biosensors for cancer biomarker proteins: A review of the current practices. *Frontiers in Bioengineering and Biotechnology, 10*, 877193. DOI: 10.3389/fbioe.2022.877193

Tavakoli, H., Mohammadi, S., Li, X., Fu, G., & Li, X. (2022). Microfluidic platforms integrated with nano-sensors for point-of-care bioanalysis. *Trends in Analytical Chemistry, 157*, 116806. DOI: 10.1016/j.trac.2022.116806

Timilsina, S. S., & Li, X. J. (2024). A paper-in-polymer-pond (PiPP) hybrid microfluidic microplate for multiplexed ultrasensitive detection of cancer biomarkers. *Lab on a Chip, 24*(21), 4962–4973. DOI: 10.1039/D4LC00485J

van Dongen, J. E., Berendsen, J. T. W., Steenbergen, R. D. M., Wolthuis, R. M. F., Eijkel, J. C. T., & Segerink, L. I. (2020). Point-of-care CRISPR/Cas nucleic acid detection: Recent advances, challenges and opportunities. *Biosensors & Bioelectronics, 166*, 112445. DOI: 10.1016/j.bios.2020.112445

Vaughan, O. (2024). Building brain–computer interfaces with graphene. *Nature Electronics, 7*(12), 1061–1062. DOI: 10.1038/s41928-024-01308-8

Vijayakumar, S. (2023). Cancer targeted contrast studies and photothermal therapy using engineered gold nanoparticles. *Applied Physics. A, Materials Science & Processing, 129*(6), 1–9. DOI: 10.1007/s00339-023-06736-5

Wang, B., Hu, S., Teng, Y., Chen, J., Wang, H., Xu, Y., Wang, K., Li, Y., & Zhang, J. (2024). Current advance of nanotechnology in diagnosis and treatment for malignant tumors. *Signal Transduction and Targeted Therapy, 9*(1), 1–65. DOI: 10.1038/s41392-024-01889-y

Wang, M., Jin, L., Hang-Mei Leung, P., Wang-Ngai Chow, F., Zhao, X., Chen, H., Pan, W., Liu, H., & Li, S. (2024). Advancements in magnetic nanoparticle-based biosensors for point-of-care testing. *Frontiers in Bioengineering and Biotechnology, 25*(12), 1393789. DOI: 10.3389/fbioe.2024.1393789

Yasamineh, S., Nikben, N., Hamed Ahmed, M., Abdul Kareem, R., Kadhim Al Aridhy, A., & Hosseini Hooshiar, M. (2024). Increasing the sensitivity and accuracy of detecting exosomes as biomarkers for cancer monitoring using optical nanobiosensors. *Cancer Cell International*, *24*(1), 1–44. DOI: 10.1186/s12935-024-03379-1

Yin, B., Wang, J., Zhang, Y., Liu, C., Wang, Y., & Li, Q. (2022). A CRISPR-Cas12a integrated SERS nanoplatform with chimeric DNA/RNA hairpin guide for ultrasensitive nucleic acid detection. *Theranostics*, *12*(13), 5914–5930. DOI: 10.7150/thno.75816

Ying, S., Guan, Z., Ofoegbu, P. C., Clubb, P., Rico, C., He, F., & Hong, J. (2022). Green Synthesis of Nanoparticles: Current Developments and Limitations. *Environmental Technology & Innovation*, *26*, 102336. DOI: 10.1016/j.eti.2022.102336

Yuan, C., Tian, T., Sun, J., Hu, M., Wang, X., Xiong, E., Cheng, M., Bao, Y., Lin, W., Jiang, J., Yang, C., Chen, Q., Zhang, H., Wang, H., Wang, X., Deng, X., Liao, X., Liu, Y., Wang, Z., & Zhou, X. (2020). Universal and naked-eye gene detection platform based on the clustered regularly interspaced short palindromic repeats/Cas12a/13a system. *Analytical Chemistry*, *92*(5), 4029–4037. DOI: 10.1021/acs.analchem.9b05597

Zhang, W. S., Pan, J., Li, F., Zhu, M., Xu, M., Zhu, H., Yu, Y., & Li, G. (2021). Reverse transcription recombinase polymerase amplification coupled with CRISPR-Cas12a for facile and highly sensitive colorimetric SARS-CoV-2 detection. *Analytical Chemistry*, *93*(8), 4126–4133. DOI: 10.1021/acs.analchem.1c00013

Zhao, W., Hu, J., Liu, J., Li, X., Sun, S., Luan, X., & Zhang, Y. (2022). Si nanowire Bio-FET for electrical and label-free detection of cancer cell derived exosomes. *Microsystems & Nanoengineering*, *8*(1), 1–12. DOI: 10.1038/s41378-022-00387-x

Zhuang, J., Zhao, Y., Yang, M., Li, J., Chen, Y., & Liu, H. (2022). SERS-based CRISPR/Cas assay on microfluidic paper analytical devices for supersensitive detection of pathogenic bacteria in foods. *Biosensors & Bioelectronics*, *207*, 114167. DOI: 10.1016/j.bios.2022.114167

Chapter 5
AI-Enhanced Precision Medicine for Lung Cancer:
Future of AI in Oncology

Kazi Asraf Ali
http://orcid.org/0000-0002-1753-7504
Maulana Abul Kalam Azad University of Technology West Bengal, India

Sabyasachi Choudhuri
http://orcid.org/0000-0002-7736-8490
Maulana Abul Kalam Azad University of Technology West Bengal, India

Naureen Afrose
http://orcid.org/0000-0001-8334-9053
Division of Pharmaceutics, Dr. B.C. Roy College of Pharmacy and Allied Health Sciences, Durgapur, India

Rideb Chakraborty
http://orcid.org/0009-0007-3931-8349
Division of Pharmaceutics, Dr. B.C. Roy College of Pharmacy and Allied Health Sciences, Durgapur, India

Chanchal Koley
http://orcid.org/0009-0001-8453-1364
Bengal College of Pharmaceutical Technology, India

ABSTRACT

Lung cancer remains a leading cause of cancer deaths worldwide, demanding innovative strategies for early detection, prognosis, and personalized therapy. This chapter highlights how artificial intelligence (AI) is reshaping precision medicine in lung cancer, beginning with the global burden and evolution of individualized care. It explores imaging-based AI approaches such as radiomics and deep learning for tumor detection and characterization, alongside AI-driven molecular and genomic profiling that interprets sequencing data and integrates multi-omics for biomarker discovery. Real world implementations including clinical decision support systems, AI-assisted drug discovery, and integration with electronic

DOI: 10.4018/979-8-3373-9715-3.ch005

health records thereby demonstrate clinical impact. The chapter also addresses challenges of data quality, bias, interpretability, and ethics, while envisioning future directions such as digital twins and integrated AI-enabled oncology ecosystems, underscoring AI's role in advancing patient-centered lung cancer care.

1. INTRODUCTION

Lung cancer remains the preeminent global oncological challenge, consistently ranking as the leading cause of cancer-related mortality in both men and women internationally (Barta et al., 2019). The burden of disease is significant, with global data from 2020 estimating 2.2 million new cases and 1.8 million lung cancer-related deaths, constituting approximately 18.0% of all cancer deaths worldwide (Li et al., 2023).

Despite advancements, the historical five-year survival rate following diagnosis has remained low, recorded at only 20% in the United States between 2010 and 2016 (Jani et al., 2023). This low survival rate is critically linked to the high proportion of diagnoses occurring at advanced stages. While recent modelled trends show a promising increase in the five-year relative survival rate, reaching 34.90% by 2021, the persistent survival gap underscores the urgent necessity for innovations in early detection and personalized therapeutic strategies.

The epidemiology of lung cancer displays considerable geographical and demographic heterogeneity. As an example, in 2020, the age-standardized incidence rate (ASIR) differed radically among countries with 36.8 per 100,000 in Denmark and 5.9 per 100,000 in Mexico. Moreover ASIR, as well as age-standardized mortality rate (ASMR), is approximately two times higher in men compared to women. Although some developed nations, such as the United States, have observed a downward trend in ASIR, particularly among men, other regions, such as China, have shown upward trends in age-specific incidence rates among individuals aged 50 to 59. This variability highlights that current screening and risk assessment protocols are neither universally effective nor adequately distributed, pointing to substantial deficiencies, particularly in developing countries (Li et al., 2023).

Etiologically, while active smoking remains the most significant risk factor, growing evidence implicates long-term exposures to environmental compounds, including radon, asbestos, polycyclic aromatic hydrocarbons (PAHs), and fine particulate matter (PM2.5), alongside specific occupational exposures (e.g., arsenic, nickel, and chromium compounds) ((Smolarz et al., 2025). The high mortality rates stemming from delayed diagnosis, coupled with the heterogeneity of disease presentation globally, establish a clear mandate to leverage advanced computational tools to enhance screening efficiency and standardize detection accuracy, thereby optimizing early intervention, as shown in Table 1.

Table 1. Global Lung Cancer Burden and Survival Metrics

Metric	Value	Source Context	Source
New Cases	2.2 million	Global estimate	(Li et al., 2023)
Cancer-Related Deaths	1.8 million	Accounts for total cancer deaths of 18%	(Li et al., 2023)
5-Year Relative Survival Rate	20%	Low survival due to advanced-stage diagnosis	(Jani et al., 2023)
5-Year Relative Survival Rate	34.90%	Reflects recent improvements	(Li et al., 2023)

1.1 Evolution of Precision Medicine

The management of non-small cell lung cancer (NSCLC) has shifted dramatically over the past two decades, transitioning from broad cytotoxic regimens to a highly personalized approach defined by precision oncology (Fujiwara et al., 2024). Precision oncology dictates that treatment decisions should be tailored to each patient's distinct molecular and tumour mutation profiles. This evolution has fundamentally redefined clinical trial design, moving from tumour type-centric models to histology-agnostic, gene-directed approaches that utilize innovative adaptive designs centered on comprehensive biomarker profiling (Tsimberidou et al., 2020).

The cornerstone of this revolution was the discovery of "targetable" mutations. Early successes, such as the use of imatinib for BCR-ABL-positive chronic myeloid leukaemia, set the stage for breakthroughs in lung cancer, notably the successful deployment of erlotinib for patients harbouring epidermal growth factor receptor (EGFR)-mutant lung adenocarcinoma (Fujiwara et al., 2024). Tyrosine Kinase Inhibitors (TKIs) have been applied in advanced NSCLC through the detection of EGFR mutations and anaplastic lymphoma kinase (ALK) rearrangements. These targeted agents have offered significant gains in progression-free survival and survival in general as compared to the traditional chemotherapy (Bennati et al., 2016). ALK rearrangement, while less frequent (found in approximately 5% of NSCLC patients), also benefits significantly from specific ALK-TKIs as optimal first-line therapy (Nguyen et al., 2025).

Effective implementation of precision medicine requires complete tumour profiling to be performed early in the disease course (Fujiwara et al., 2024). This is typically achieved through modern molecular diagnostics, which highly recommend Next-Generation Sequencing (NGS) using multigene panels (Tsimberidou et al., 2020). The logistical complexity of tissue acquisition and processing, along with the inherent long turnaround times for molecular testing, creates significant challenges for accelerating time-to-treatment initiation. Consequently, high-quality precision medicine requires specialized expertise delivered by a well-coordinated multidisciplinary team (MDT) that employs standardized tissue processing protocols and reflex testing strategies (Fujiwara et al., 2024).

1.2 Role of AI in Oncology

Artificial Intelligence (AI), particularly via deep learning (DL), provides the necessary computational infrastructure to manage and interpret the expansive, complex datasets—often termed "big data"—generated by high-throughput genomic, proteomic, and imaging techniques (Chang et al., 2025). These datasets offer rich insights into cancer biology, but their high complexity and subtle patterns often render their analysis time-consuming and difficult for humans. AI addresses this by processing vast datasets with enhanced efficiency and objectivity (Huhulea et al., 2025).

AI applications span the entire cancer care continuum, encompassing prevention, screening, detection, diagnosis, prognosis, treatment optimization, and clinical trial design (Chang et al., 2025). A key strength of AI is its capacity for multimodal integration (Ali et al., 2024), allowing it to objectively incorporate and synthesize complementary information from diverse sources, including radiology, pathology, and genomics, thereby supporting personalized treatment decisions. Indicatively, scholarly studies have managed to integrate the radiomics, pathomics, and genomic measurements to forecast essential biomarkers of immunotherapy, including programmed death-ligand 1 (PD-L1) expression and tumour mutation burden (TMB) in patients with lung cancer (Huhulea et al., 2025). This capability is critical for assessing and predicting immune response.

In the evolving clinical landscape, AI serves as a central orchestrator, managing complex targeted therapies, liquid biopsies, and advanced genomic technologies (Penault-Llorca & Socinski, 2025). The goal of AI integration is not merely to improve accuracy but to standardize and enhance the objectivity of diagnostic processes, mitigating human fatigue and interpretation variability (Huhulea et al., 2025).

2. AI IN LUNG CANCER DIAGNOSIS

2.1 Imaging-Based AI Approaches (CT, PET, MRI)

Imaging modalities like Magnetic Resonance Imaging (MRI), Computed Tomography (CT), Positron Emission Tomography (PET), and form the initial basis of lung cancer diagnosis and staging. Intensive learning through Convolutional Neural Networks (CNNs) is also used extensively in Computer Aided Detection (CADe) and Computer Aided Diagnosis (CADx) machines to examine pulmonary nodules on CT scans (Cheo et al., 2025).

The use of AI in Low-Dose CT (LDCT) screening regimens has the potential to enhance human perception and overcome the issue of high-volume screening, specifically, the laborious system of category, and a considerable variability of readers (Schreuder et al., 2021). The nodule detectors have demonstrated high levels of efficacy exposing sensitivities of 79% (95% CI: 72 to 86%) when applied to independent datasets and 85% (95% CI: 83 to 88%) when applied to all datasets. (Cheo et al., 2025; T.-W. Wang et al., 2025).

While AI models often exhibit a wide range of specificity (64% to 95.93%), potentially exceeding the range observed in human radiologists (61.67% to 84%) (Cheo et al., 2025), comparative studies show nuanced performance. In some instances, expert human readers achieved significantly higher accuracy than AI-based CAD systems, particularly for smaller lesions measuring 5, 8, 10, or 12 mm (Peters et al., 2021). This suggests that the current clinical utility of AI is primarily as an augmented tool (CADe) to reduce the workload and minimize diagnostic misses, mainly benefiting less-experienced radiologists operating under suboptimal imaging conditions (S. Chen et al., 2025). Continued efforts are focused on optimizing the generalizability of these models through standardized evaluation protocols and diverse datasets (T.-W. Wang et al., 2025).

In addition to the CT, AI plays an important role in the improvement of functional imaging, including ^{18}F-fluorodeoxyglucose PET/CT (^{18}F-FDG PET/CT). This mode is crucial in the description of indeterminate nodules of the lungs. The introduction of AI allows quantifying and analyzing semi-quantitative PET measures, such as standardized uptake value (SUVmax), metabolic tumour volume (MTV), and total lesion glycolysis (TLG) (Kapsoritakis et al., 2025). High values of these metrics are significantly associated with malignancy and tumour aggressiveness. AI models aid in processing these parameters, often combining them with dual-time-point imaging (DTPI) data to enhance the specificity and accuracy of characterizing the tumour's biological activity, thereby providing a more biologically informed assessment than traditional morphological imaging (Rodriguez et al., 2025). AI systems are also being investigated to improve staging and prediction of treatment response using other modalities, including MRI (Logullo et al., 2023).

2.2 Radiomics and Deep Learning in Early Detection

Radiomics is a technique of high-throughput extraction of quantitative features of standard medical images, which results in radiomic signatures, which encode the complex tumour properties that are usually not visible to the human eye. These features, derived from CT or PET images, are pivotal in characterizing tumours, aiding in diagnosis, and evaluating treatment response and patient prognosis (W. Zhang et al., 2023).

Radiomic signatures identified using machine learning techniques, such as LASSO regression, have demonstrated significant prognostic value for metrics including Disease-Free Survival (DFS) in early-stage NSCLC. Furthermore, combining radiomics features with clinical factors (e.g., age, histology, and tumour stage) using probability-weighted models has shown potential to improve prognostic accuracy and facilitate personalized treatment decisions, despite modest initial accuracy values (Tang et al., 2023).

The Sybil deep learning model (Table 2) exemplifies this advancement, leveraging volumetric LDCT data to predict future lung cancer risk over several years. Sybil achieved high accuracy in predicting 1-year lung cancer risk, with Area Under the Receiver-Operator Curve (AUC) values of 0.92 on the held-out NLST validation set, 0.86 on the Massachusetts General Hospital (MGH) set, and 0.94 on the Chang Gung Memorial Hospital (CGMH) cohort in Taiwan.1 Over a 6-year window, the model demonstrated strong performance, with Concordance Index (C-index) values ranging from 0.75 to 0.81 across these diverse cohorts (Mikhael et al., 2023).

Table 2. Performance of Deep Learning Models in Lung Cancer Risk Prediction (Sybil Model)

Validation Cohort	Prediction Target	AUC (1 Year Risk)	C-index (6-Year Risk)
NLST (Held-out)	Lung Cancer Risk	0.92 (95% CI, 0.88–0.95)	0.75 (95% CI, 0.72–0.78)
MGH	Lung Cancer Risk	0.86 (95% CI, 0.82–0.90)	0.81 (95% CI, 0.77–0.85)
CGMH (Taiwan)	Lung Cancer Risk	0.94 (95% CI, 0.91–1.00)	0.80 (95% CI, 0.75–0.86)

2.3 Integration of Pathology and Histopathological AI Models

Digital pathology, which involves scanning Hematoxylin and Eosin (H&E)-stained tissue sections into Whole Slide Images (WSIs), creates a high-dimensional data source suitable for AI analysis. Deep learning models, predominantly Convolutional Neural Networks (CNNs), are successfully applying computer vision techniques to these histopathological images for accurate diagnosis, subtyping, prognosis prediction, and characterization of molecular status ((Davri et al., 2023). Furthermore, DL models trained on WSI data have demonstrated the ability to predict patient response to specific treatments, such as chemoradiotherapy (CRT) in NSCLC ((Pan et al., 2023).

A breakthrough is the capacity of these DL models to predict oncogenic driver alterations—such as EGFR mutations, ALK rearrangements, ROS, and KRAS—directly from standard H&E WSIs (Parra-Medina et al., 2025). This ability transforms the morphological diagnosis into an initial molecular screen, offering a noninvasive, high-throughput approach that addresses the logistical bottlenecks and extended turnaround times associated with traditional molecular testing (Park et al., 2025). The mechanism relies on the detection of specific, quantifiable pathomic signatures that correlate with the genetic alterations, even if these features are not easily discernible by human pathologists.

The results of systematic reviews and meta-analyses confirm the excellent predictive functionality of DL models of major drivers of NSCLC. The ALK variations showed a high level of performance with a sensitivity of 84% (95% CI: 62 to 95%) and specificity of 85% (95% CI: 55 to 96%). It was also found that EGFR mutation prediction was also highly effective, with a sensitivity of 80% (95% CI: 72 to 86%) and a specificity of 77% (95% CI: 69 to 83) ((Parra-Medina et al., 2025). Predictability of these biomarkers on H&E slides has been demonstrated in a wide range of multi-scanner cohorts around the world, which supports the applicability of this method (Table 3).

Table 3. Deep Learning Model Performance in Molecular Prediction from H&E Whole Slide Images (NSCLC)

Molecular Alteration	DL Model Architecture	Sensitivity	Specificity	Source
ALK Rearrangement	Primarily CNNs	84% (CI: 62–95%)	85% (CI: 55–96%)	(Parra-Medina et al., 2025)
EGFR Mutation	Primarily CNNs	80% (CI: 72–86%)	77% (CI: 69–83%)	
TP53 Mutation	Primarily CNNs	70%	70%	

3. AI IN MOLECULAR AND GENOMIC PROFILING

The development of high-throughput sequencing technologies has generated large, multidimensional genomic data sets on somatic mutations, copy-number variations, gene fusions, and transcriptomic signatures, the manual interrogation of which is an impeding bottleneck to clinical translation. AI, and more specifically deep learning architectures, is better placed to break down this complexity by identifying higher-order, finer-grain patterns and interactions within genomic data that classical statistical methods typically cannot detect (or learn). An illustration is that convolutional neural networks (CNNs) and recurrent neural networks (RNNs) can be trained to take raw sequencing strictly reads or variant-call format (VCF) files (Ahmad et al., 2025), resulting in more precisely identifying driver mutations and, at the same time, determining the functional implications of rare or novel variants of unknown importance. In addition to its discrete changes, AI can perform integrative analyses that combine genomic data with transcriptomic, epigenomic, and proteomic profiles to define more robust molecular subtypes of lung adenocarcinoma or squamous-cell carcinoma (Gandhewar et al., 2025). The solution goes beyond the constraints of single-gene biomarkers like EGFR or ALK and is moving towards a systems-level understanding of oncogenic drivers and resistance mechanisms. One of the most fascinating uses is predictive biomarker discovery, where models can predict mutational status, microsatellite instability (MSI), or tumour-mutational burden (TMB) directly based on conventional histopathology whole-slide images- a paradigm referred to as "pathomic analysis". This approach has the potential to increase access to genomic profiling in resource-constrained environments (Chakraborty et al., 2024).

Further, graph neural networks (GNNs) have the flexibility to capture the complexity of biological pathways and protein-protein interaction networks, and to clarify how co-mutations and epistatic interactions form to determine therapeutic vulnerability or acquired resistance. This understanding is essential for predicting the evolution of tumours under the selective pressure of targeted therapies, which can then be used to pre-design combination regimens (Bishal et al., 2025). Transforming entire genomic landscapes into actionable clinical information puts AI-based profiling on the path from a passive diagnostic tool to a proactive prognostic and predictive engine. It enables a dynamic, longitudinal view of a patient with disease and the ever-refining of treatment interventions in line with the tumour's changing genomic structure, thereby providing the basis for a truly adaptive and personalized therapeutic paradigm in modern lung cancer (Afrose et al., 2025; Chakraborty et al., 2025).

3.1. Next-Generation Sequencing Data Interpretation

The unprocessed results of next-generation sequencing systems (Chakraborty et al., 2024), which contain billions of short reads of DNA and RNA, make a daunting analytical problem. Alignment to the reference genome, variant calling, and annotation are computationally intensive processes that necessitate sophisticated computational pipelines. The traditional methods of bioinformatics are fundamental, but often face constraints when challenged by inherent noise, the complex genomic environment (e.g., paralogous genes), low tumour purity, and the daunting search space of possible mutations. This pivot point is changing with the implementation of convolutional neural networks (CNNs) along with recurrent neural networks (RNNs) to identify subtle, complex patterns in sequencing data that traditional statistical tools cannot reveal. The models have been trained on large, carefully curated genomic datasets, making them outperform current accuracy in distinguishing between genuine somatic mutations and sequencing artefacts and germline polymorphisms, even at the tiny allele frequency scale - a feature that is paramount when liquid biopsies are being analyzed, where circulating tumour DNA is sparse((Chakraborty et al., 2025).

In addition to variant calling, AI-based interpretation systems integrate variants with multidimensional knowledge graphs, such as predicting protein structure, gene-gene interactions, enriching pathways, and integrating real-world clinical evidence, to enable systems-level exploration beyond a simplistic oncogenic driver hot-spot report (Afrose et al., 2025). As an example, an AI model can be used to contextualise a new, non-canonical EGFR mutation, not only by identifying its presence but also predicting its effects on protein conformation, location within a functional domain, and similarity to known resistance mechanisms, which subsequently predicts its functional relevance and therapeutic implications (Ali et al., 2025). In addition, NLP algorithms automatically search the continuously growing pool of oncology literature and clinical trial databases to correlate complex mutational signatures, including those caused by APOBEC enzymes or homologous recombination deficiency, with specific therapeutic or prognostic phenotypes, which would itself be unsustainable through human curation. The final result of this AI-mediated synthesis is the creation of ranked, evidence-based recommendations that not only prioritize FDA-approved targeted therapies and outline eligibility to participate in specific clinical trials based on strict molecular guidelines, but also predict acquired resistance to particular agents. The AI-enhanced NGS data interpretation, as a result, transforms into an adaptive, learning system that automatically improves its output over time, thereby providing clinicians with a highly contextualized, patient-specific genomic narrative and making the broad potential of precision oncology real at the point of care (Athanasopoulou et al., 2025).

3.2. Predicting Driver Mutations and Biomarkers

Convolutional Neural Networks (CNNs) are becoming widely used in modern studies to not only analyze genomic sequences, but more enticingly, to process histopathological whole-slide images (WSIs), thus allowing the prediction of mutational statuses of key oncogenic drivers such as EGFR, KRAS, and ALK fusions, to be predicted simply by the routine use of H&E-stained histopathological specimens. This will remove assumptions about the logistical and financial costs of next-generation sequencing and effectively support a form of virtual genotyping that can identify patients likely to harbor actionable genomic alterations. Further, these AI systems combine multimodal data, such as radiological CT scans, transcriptomic profiles, and a range of clinical variables, to reveal latent, nonlinear associations that traditional statistical approaches can obscure. As a result, they will be able to predict the expression states of programmed death-ligand 1 (PD-L1), tumor mutational burden (TMB), and the presence of novel biomarkers, thereby effectively stratifying patients for immunotherapeutic and targeted therapeutic strategies (Budczies et al., 2024; N. Zhang et al., 2024).

In addition to identification, state-of-the-art algorithms can predict the functional implications of rare or novel variants of uncertain importance, thereby inferring their oncogenic potential and guiding clinical interpretation. By combining these heterogeneous data streams, AI systems can build a holistic digital representation of the tumor, enable the identification of new biomarker patterns and provide a dynamic, predictive model that is invaluable for personalizing therapies to the distinctive molecular structure of the individual malignancy. By so doing, this methodology will not be confined to a mere gene-by-gene paradigm but will move toward a genuinely holistic, systems-level understanding of tumorigenesis.

3.3. AI-Driven Multi-Omics Integration

The main challenge lies in the sheer scale, inherent heterogeneity, and complex interdependence of the distinct layers of omics such as genomics, transcriptomics, epigenomics, proteomics, and metabolomics that are not susceptible to traditional statistical dogmas, as shown in Figure 1. AI has a unique ability to address this complexity through unsupervised and semi-supervised deep learning architectures, such as autoencoders and graph neural networks. These models perform nonlinear dimensionality reduction, revealing latent representations and shared structures that capture the synergistic and often compensatory interactions among molecular strata (Wu & Xie, 2025). For example, an AI framework can combine somatic mutation data with DNA methylation and gene expression network data to define discrete molecular subtypes of lung adenocarcinoma that predict therapeutic response and the composition of the immune microenvironment. These subtypes cannot be identified when omic data are analyzed independently.

Furthermore, such integrated models can identify new biomarkers that are not based solely on single genomic changes, but rather on dynamic multispectral ones —such as a composite score that integrates a specific mutation, its resulting transcriptional consequences, and related metabolic changes (Camps et al., 2025). Such a systems-level view enables the isolation of patients into significantly refined cohorts, thereby improving the prediction of the predictive sensitivity to specific therapies and immunotherapies, as well as the discovery of new synthetically lethal therapeutic targets. Finally, AI-based multi-omics integration goes beyond data fusion, catalyzing discovery, untangling the deep heterogeneity of lung cancer, and transforming the complex mosaic of molecular information into actionable clinical outputs, which, in turn, will enable the implementation of truly personalized treatment approaches.

Figure 1. Multiomics integration showing features extracted from a patient. Reproduced from (Sartori et al., 2025) under Creative Commons Attribution (CC BY 4.0).

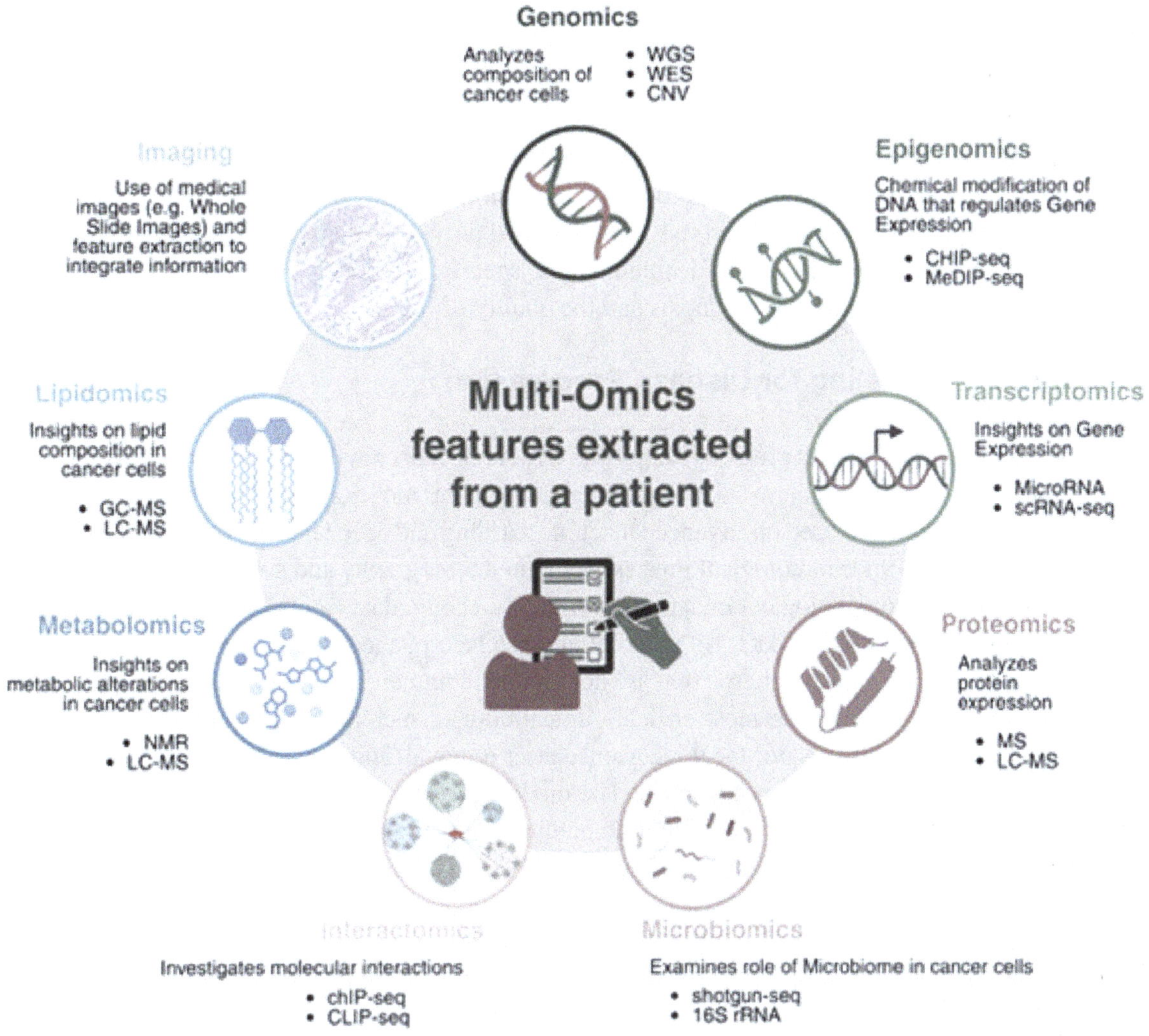

4. AI FOR PROGNOSIS AND RISK STRATIFICATION

In the area of prognosis and risk stratification, AI is essentially rebalancing our ability to predict the clinical progression of patients with lung carcinoma, beyond the limitations of the traditional TNM staging and histopathological subtyping. Through the process of integrating and running complex, nonlinear analyses on multiscale datasets, such as high-resolution radiological modalities like computed tomography (CT) as well as positron emission tomography (PET), digitized whole-slide histopathology images (WSIs), and genomic profiles, AI models, especially those with deep-learning architectures, reveal latent prognostic signals that are otherwise invisible to the human viewer (Sharma et al., 2024). For example, convolutional neural networks (CNNs) can identify subtle patterns in CTs or WSIs to generate so-called radiomics and pathomics signatures that quantitatively predict the likelihood of disease recurrence, the potential for metastasis, and overall survival with excellent accuracy. These models can be used to stratify early-stage disease patients into separate risk groups—those who can benefit from more aggressive adjuvant therapy and those for whom surveillance is sufficient —and thus personalize postoperative care (Yu et al., 2024). Also, the integrative systems that combine clinical variables, mutational state, and a treatment history can dynamically forecast each case and deliver an ever-renewed risk analysis generated by AI. This ability is important for streamlining surveillance and proactively addressing potential complications. This paradigmatic shift in prognostic assessment of a set of patients with no opportunity to change it into a dynamic, personalized risk trajectory will provide clinicians with granular prognostic models based on data, allowing them to decide on more specific and timely therapeutic actions, which will eventually benefit the survival prognosis and the quality of life.

4.1. Predictive Modeling for Disease Progression

The development of powerful predictive algorithms for lung cancer in the field of AI-enhanced precision medicine represents a paradigm shift in patient management, moving from a reactive to a proactive approach. These models are based on advanced machine learning and deep learning structures, trained on multimodal data that include radiological images (computed tomography and positron emission tomography), whole-slide histopathological images, genomic sequencing data (mutational burden and fusion events), and longitudinal clinical data. By combining these heterogeneous data streams, AI algorithms can identify nonlinear, complex patterns that are not visible to human analysis. In turn, they help make precise prognostications about the most critical clinical course, such as the probability of metastatic spread to specific organs, the risk of local recurrence after removal, and the rate of disease progression in patients with inoperable or advanced cancer. The models use convolutional neural networks to identify subtle prognostic features in serial CT scans, enabling predictions of tumor growth kinetics and the possible development of treatment-resistant subclones. Image-based features are further enhanced using generative adversarial networks, whereas transformer-based architectures combine genomic and clinical data into a single representation. Ensemble learning methods maximize the benefit of all modalities and produce an overall risk score that is superior to conventional clinicopathologic staging systems. These predictive insights are ground-breaking clinically ((Terranova & Venkatakrishnan, 2024; Young et al., 2024). They enable oncologists to select and sort patients based on their personal risk level to allow high-risk patients, including the administration of adjuvant chemotherapy or targeted radiation, to be prespecified, and low-risk patients to avoid needless treatment-related morbidity. This individual risk stratification forms the basis of a genuinely personalized surveillance program and of the prompt initiation of subsequent lines of treatment, with the ultimate goal of augmenting survival rates and resource utilization by matching the intensity of therapeutic effort with the anticipated aggressiveness of the disease(Karthik et al., 2025).

4.2. Survival Analysis Using Machine Learning

Modern biomedical analytics regularly face challenges with high-dimensional data, complex interactions among variables, and the nonlinearity inherent in multi-omics data and extensive clinical phenotypic descriptions. Survival forests, support vector machines trained on censored data, gradient-boosting ensembles, including XGBoost, and other machine-learning algorithms adeptly avoid these challenges and uncover subtle patterns in prognostics. More specifically, deep learning paradigms, including survival networks and temporal neural networks, can learn latent representations of unstructured inputs, namely whole-slide histopathological images and radiomic features derived from computed tomography scans (X. Li et al., 2025). These models produce highly individualized risk scores that provide a more dynamic prediction of a patient's survival probability over time, rather than a single flat estimate, allowing for finer stratification of patients beyond broad TNM staging. Such methods can be used to provide more personalized clinical treatment, whether by escalating treatment for those at high risk or by avoiding overtreatment for those deemed at low risk. Moreover, it is possible to explain the main contributors to the personal risk the model predicts (a genetic mutation, a radiomic signature, or a clinical comorbidity) using techniques of model interpretability, such as SHAP values, thereby building trust and translating biological knowledge into action. Subsequently, machine-learned survival analysis will mark a breakthrough, bringing population-based methods into actual, individualized prognostic predictions—a vital decision-supporting foundation of precision medicine in lung cancer (Wiegrebe et al., 2024).

4.3. Identification of High-Risk Patient Subgroups

A key use of AI in prognostic stratification for lung cancer lies in the intelligent recognition of high-risk patient subsets an exercise that outstrips the fine-tuning provided by routine clinicopathological staging. By applying unsupervised learning algorithms (e.g., hierarchical clustering and deep autoencoders) to multidimensional datasets combining genomic, transcriptomic, proteomic, and digitized histopathology data, AI can break down the deep heterogeneity of lung cancer into discrete molecular and phenotypic subunits. These empirical subgroups often shed light on aggressive tumor subtypes that share common biological characteristics, including specific co-mutation status, dysregulated immune-evasion pathways, and specific spatial arrangements of tumor-infiltrating lymphocytes, patterns that cannot be seen with the naked eye (Nielsen et al., 2024). Nomadic with on-top longitudinal real-world supervised models further combine latent clinical variables, treatment history, and serial imaging to predict rapid disease progression or early recurrence at high time resolution. Such risk stratification provides the granularity to pre-emptively allocate more intensive surveillance protocols and to consider adjuvant or neoadjuvant therapies for patients with a high predicted risk of micro metastatic disease, even where traditional indicators are equivocal. As a result, the clinical paradigm has shifted from a reactive stance (as illustrated by the conventional delineation of high-risk cohorts) to a proactive one (based on AI-enhanced delineation) (Nielsen et al., 2024). This move will ensure that the most potent and potentially deadly therapeutic arsenals are reserved for patients with the highest biological need, thereby directly addressing the twin problems of over- and under-treatment in modern oncology.

5. AI IN TREATMENT PLANNING AND THERAPEUTIC OPTIMIZATION

The implementation of AI in treatment planning represents a shift from standardized guidelines to highly personalized treatment plans. AI systems offer clinicians unparalleled decision-support tools by synthesizing and analyzing multimodal patient data to support their decision-making (Xu et al., 2024; Xuyan et al., 2025). Such systems may simulate the potential efficacy and toxicity of different treatment options, producing a dynamic risk-benefit profile tailored to the individual's unique disease biology and overall health, as shown in Figure 2.

This type of computation opens the door to abandoning one-size-fits-all regimens and to optimizing therapeutic sequences and combinations that humans alone cannot determine, thus preparing the way for personalized intervention plans.

Figure 2. Diagram outlining the automated treatment planning workflow. Reproduced from (Zhong et al., 2021) under Creative Commons Attribution (CC BY 4.0).

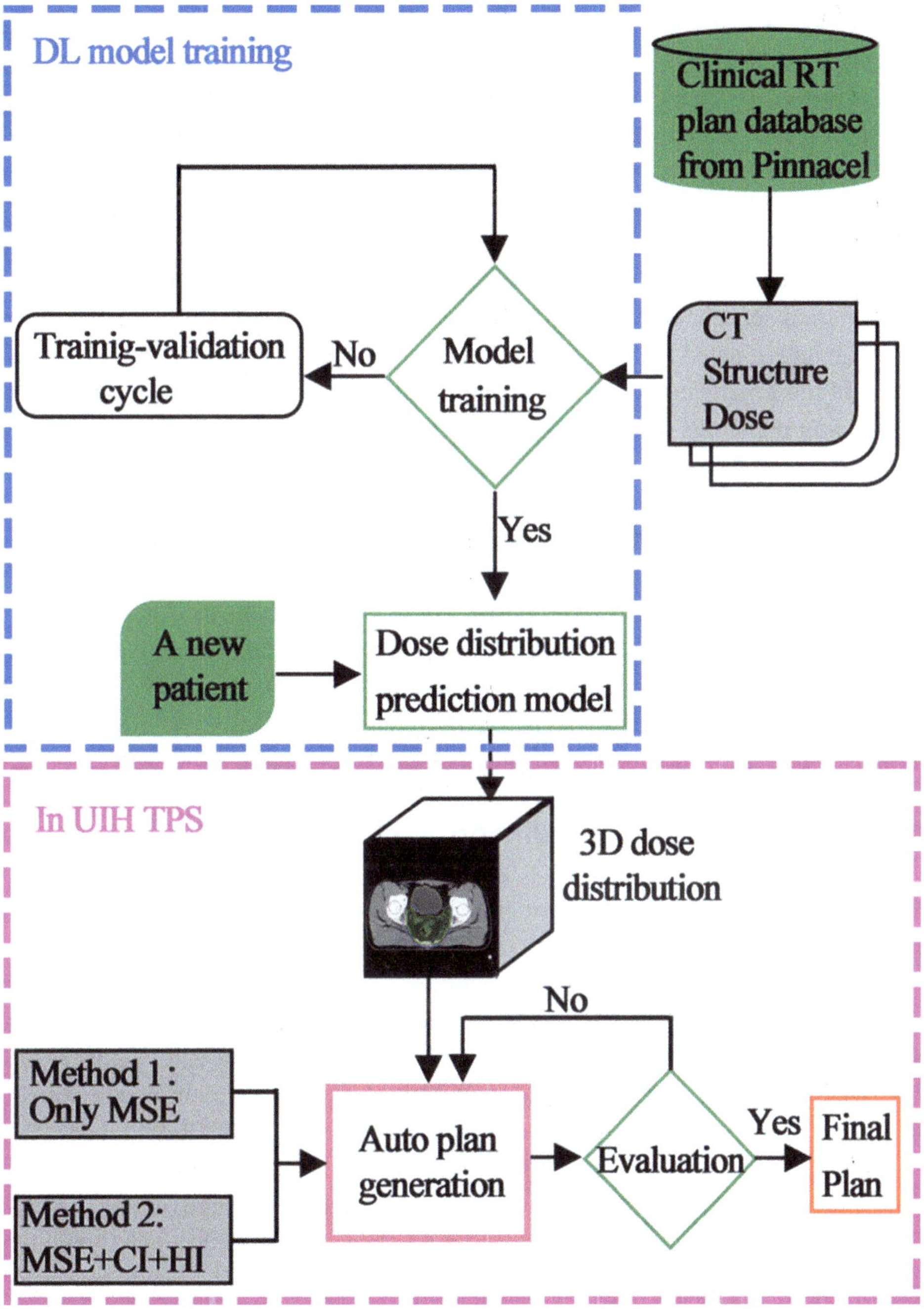

5.1 AI-Guided Targeted Therapies

The introduction of targeted therapies has fundamentally changed the paradigm of treating lung cancer, replacing nonspecific cytotoxic chemotherapies with targeted drugs that inhibit molecular signaling pathways that enable tumor growth (De Jong et al., 2023; La'ah & Chiou, 2024). The enormous potential of such agents, however, is set against a significant clinical challenge: the astounding complexity and heterogeneity of the genomic changes present in lung cancer. It is in this complexity that AI becomes a disruptive force, facilitating a degree of accuracy in pairing patients with ideal targeted interventions that was once unachievable. AI has much more than a mere pattern recognition role; it is an advanced computational co-pilot that synthesises, processes, and forecasts based on multidimensional information to inform therapeutic decisions on a scale and depth never before achievable (Bertolaccini et al., 2024). Recent next-generation sequencing (NGS)plays a crucial role in identifying actionable mutations, the initial step in deploying targeted therapies (Hassan et al., 2023; H. Wang et al., 2025). As NGS creates an in-depth map of the genetic code of a tumor, the resulting data is enormous and replete with minor differences, including numerous "variants of unknown significance (VUS) (Kwong et al., 2022; Ying et al., 2021). These are genetic mutations, the clinical impact of which on cancer progression and drug response remains completely unknown. Conventional procedures fail to categorize such VUS, and clinicians tend to be ambiguous in their findings and decisions. This task, however, is unique to deep learning algorithms. They may be trained on large genomic databases that associate specific patterns of mutation with patient outcomes and drug responses. These models can be used to predict the functional impact of a mutation by examining its sequence, structural context, and evolutionary conservation, and distinguishing between a driver mutation, which is important to the survival of the cancer, and a benign passenger mutation. This ability to read the functional language of the genome is critical, making ambiguous VUS clinically actionable and preventing any potential therapeutic opportunity from slipping by.

5.2 Immunotherapy Response Prediction

The ability to predict which lung cancer patients can benefit from immunotherapy is one of the most impactful challenges and opportunities in contemporary oncology, and AI is already proving beneficial in this field. The response to immune checkpoint inhibitors is also determined by a complex interplay of factors, including tumor mutational burden, patterns of PD-L1 expression, the tumor microenvironment, and host immune status (Chi et al., 2021; Dall'Olio et al., 2022). Digital pathology images, also referred to as radiomics and pathomics, can be incorporated into AI models and combined with genomic and clinical data to form a comprehensive response signature (L. Chang et al., 2025; Lu et al., 2021; Song et al., 2023). Such models can detect spatial interactions between immune cells and tumor cells in biopsy samples or extract subtle biomarkers from CT scans that predict a receptive microenvironment. By correctly identifying who will respond and who will not, AI can reduce unnecessary toxicity in patients who will not respond and can make the use of this potent drug class as efficient as possible.

5.3 AI in Radiation Therapy Planning

Planning radiation therapy is a complicated, time-consuming task that AI is radically improving. Many key steps are automated and optimized by AIs, starting with the accurate segmentation of tumors and sensitive organs-at-risk on CT or MRI scans, a process called contouring, as shown in Figure 3.

Deep learning algorithms achieve this with the same efficiency and precision as human professionals, ensuring the tumor receives a lethal dose of radiation while minimizing exposure to normal tissues (Shahid & Imran, 2025; Sharma et al., 2022). In addition to contouring, AI has found applications in dose optimization, where it can compute the most efficient arrangement of radiation beams and intensity distributions to produce a highly conformal dose distribution. Potential toxicities (including radiation pneumonitis) can also be predicted using the planned dose and patient-specific anatomy, and pre-emptive changes to the plan can be made to enhance safety and quality of life (Liu & Imai, 2025; Parhi et al., 2025; Rankin et al., 2025).

Figure 3. Diagram showing the workflow for AI-automated palliative dosimetric plan development. Reproduced from (Jones et al., 2024) under Creative Commons Attribution (CC BY 4.0).

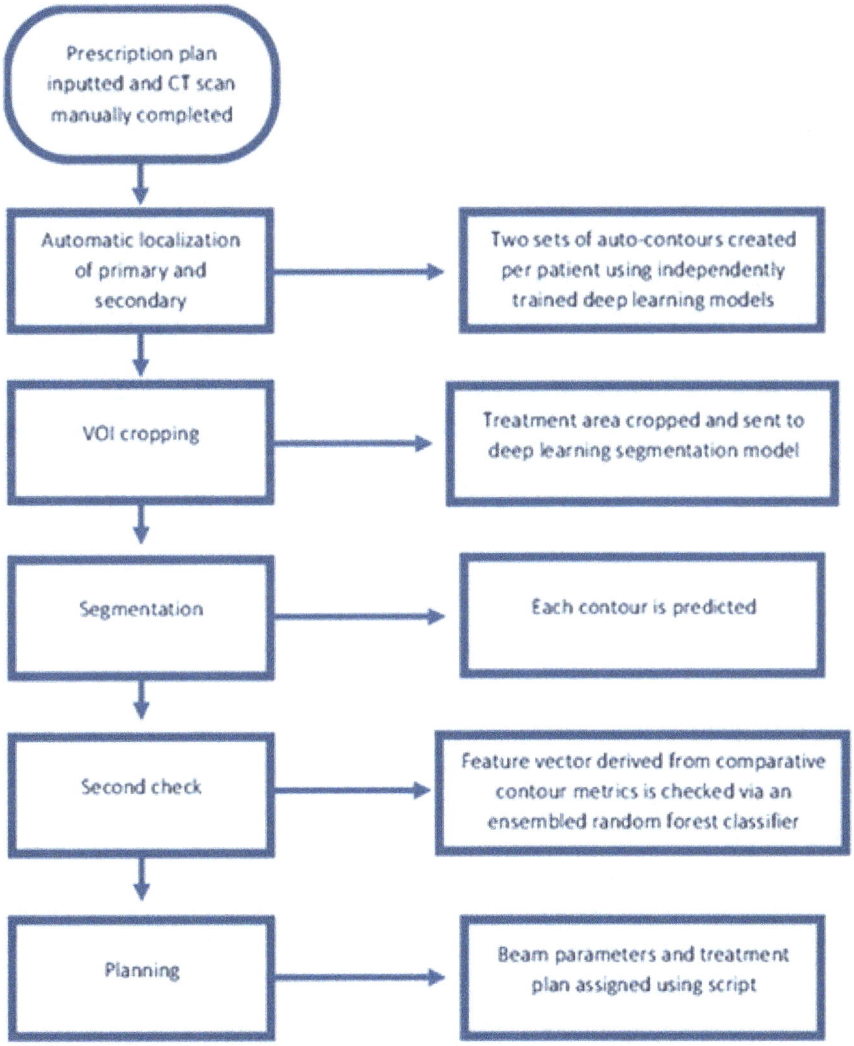

5.4 Adaptive and Personalized Treatment Algorithms

The highest level of AI in treatment is the creation of dynamic, adaptive, personalized treatment algorithms. These AI-driven systems, unlike static plans, allow adaptive treatment, in which the plan changes based on tumor behavior and the patient's condition. By analyzing real-time data streams from regular follow-up scans, liquid biopsies, and even wearable devices, AI models can help spot early indicators of progress or treatment response (Dubey & Sikarwar, 2025; Tiwari et al., 2025). It can be used to make timely adjustments to therapy, e.g., to discontinue treatment at the first sign of resistance or to de-escalate treatment upon achieving a deep response. These closed-loop systems capture the fundamental idea of precision medicine, which involves building a continuously learning, patient-specific therapy course that dynamically adjusts to overcome the changing defenses of the cancer and achieve the best long-term outcomes.

6. REAL-WORLD APPLICATIONS AND CLINICAL DECISION SUPPORT

The shift of AI into a practical resource in the clinical workflow fulfills its promise in precision medicine. This integration centers on implementing AI systems at the point of care as a powerful ally rather than an independent agent, supporting oncologists by expanding their knowledge base and facilitating seamless integration of the latest evidence and multifaceted data analyses to meet patients' daily needs. These practical applications transcend theoretical models to provide decision support, reduce clinicians' cognitive load, reduce diagnostic and treatment delays, and ultimately equalize excellence in lung cancer care across multiple healthcare institutions and geographic regions (Orstad et al., 2023). The vision is to build a synergistic collaboration between human intuition and machine intelligence, so each patient reaps the benefits of a data-driven, thoroughly analyzed approach to treatment.

6.1 Clinical Decision Support Systems (CDSS) in Lung Cancer

AI-operated CDSS are transforming oncologists' workflows, serving as a second opinion that is both integrated and intelligent. They are information systems integrated into the clinical setting and programmed to handle individual patient data on the fly against a significant, continuously refreshed knowledge source containing clinical guidelines, published oncology literature, drug databases, and institutional protocols. In a patient with lung cancer who has recently been diagnosed, an AI-based CDSS can automatically process structured and unstructured data, including pathology reports (Dasgupta et al., 2024; Elhaddad & Hamam, 2024; Ogut, 2025), genomic sequencing, and radiology scans, to generate a prioritized list of NCCN guideline-compliant treatment alternatives. In addition, it can alert not only to the possibility of rare yet actionable mutations that a busy clinician may miss, but also to relevant clinical trials the patient is eligible to participate in, and even to warn the care team about possible drug-drug interactions or contraindications based on the patient's comorbidities and drug list. This does not supplant but enhances the oncologist's judgment, ensuring that complex, multi-factorial decisions are made on a multifaceted, evidence-based basis and resulting in practice variation and increased adherence to the highest standards of care.

6.2 AI-Powered Clinical Trials and Drug Discovery

Clinical trial designs and drug discovery are known to be incredibly slow, expensive, and inefficient. They are undergoing a transformation driven by AI, developing a new era of speed and precision for both. In drug discovery, AI algorithms can analyze large chemical libraries and biological samples to predict novel drug candidates targeting a specific pathway in lung cancer, significantly reducing time spent in the initial discovery phase. More directly, AI is changing the way patients are recruited to clinical trials, which can be a significant bottleneck. Rather than manual screening, natural language processing (NLP) algorithms can sift through electronic health records to automatically identify patients who meet the complex, specific inclusion and exclusion criteria of a trial (Sim et al., 2023), accelerating enrollment and speeding trial completion. In addition to matching, AI can be used to create more innovative, more flexible trial protocols. Such trials, often called baskets or umbrellas, can dynamically assign patients to new treatment arms based on revised molecular and clinical profiles, thereby making the trial process more efficient and more likely to generate meaningful results about which therapies are effective in specific patient subgroups.

6.3 Integration With Electronic Health Records

The electronic health record (EHR) contains abundant patient data, much of which is stored in unstructured formats such as physician notes, radiology reports, and discharge summaries (Tayefi et al., 2021). Commonly used in natural language processing, the profound synergy of AI is the key to unlocking this latent data and transforming the EHR into a dynamic, intelligent tool rather than a passive digital utility cabinet. The accuracy of NLP algorithms in reading and interpreting clinical notes is high, enabling them to isolate key information such as family history of cancer, performance status, symptom progression, and response-to-treatment data points, which, although often important in decision-making, are otherwise hard to compile and analyze. This naturalistic connection can enable the persistent execution of AI models in the background, forming a dynamic digital twin of the patient that supports each new clinical interaction. This will facilitate proactive monitoring of patients, including those most susceptible to complications or disease progression, early detection, and provide an overarching, real-time data source for the CDSS. Using AI to turn the EHR into a structured, queryable, and analytically ready resource has become the critical data backbone that enables all other advanced real-world applications and individual treatment algorithms (Sousa et al., 2024).

7. CHALLENGES AND LIMITATIONS

7.1 Data Quality, Bias, and Interpretability

AI models require large-scale, high-quality datasets to produce reliable predictions; however, because sequencing methods, imaging modalities, and patient demographics vary, cancer data exhibit substantial heterogeneity (Chukwunweike et al., 2024). Model generalization across institutions and patient populations is hampered by discrepancies resulting from variations in sample collection, preprocessing techniques, and annotation standards(Okenwa et al., 2024). Furthermore, the majority of publicly accessible cancer datasets, such as The Cancer Genome Atlas (TCGA), are skewed towards particular geographic and racial groups, limiting the potential for AI models to be applied across a variety of demographics (Ahmed, 2022). Standardization initiatives are being investigated, including harmonized multi-omics datasets and federated learning methodologies, to improve data quality in oncology AI research (Biswas & Chakrabarti, 2020).

7.2 Ethical and Legal Considerations

Patient privacy, informed consent, and algorithmic fairness are among the ethical and legal issues brought up by the use of AI in cancer. AI models may reinforce biases in conventional cancer treatment methods trained on historical clinical data, leading to different therapy recommendations for underrepresented groups (Ahmed, 2020). Furthermore, mistakes in AI predictions may lead to less-than-ideal or even dangerous treatment decisions. Hence, medical decision-making using black-box AI systems may increase the clinician's responsibility (Chakravarty et al., 2021).

Guidelines for AI-driven medical technology have been published by regulatory agencies, including the European Medicines Agency (EMA) and the US Food and Drug Administration (FDA), which stress the importance of open validation procedures and continuous post-market monitoring. Since clinical trials must demonstrate consistent safety, efficacy, and real-world applicability before being used in clinical settings, obtaining regulatory approval for AI-based precision oncology technologies remains difficult. Although AI-based oncology research has demonstrated consistent safety and efficacy in applications, obtaining regulatory approvals remains difficult. Cooperation among oncologists, AI developers, and regulatory bodies is required for efficient and effective

AI-based approaches, as the absence of standardized AI validation procedures can make it challenging to approve AI-assisted drugs. Regulatory bodies require evidence on how AI can identify specific diseases and the associated drugs and medications(Kumar, 2025).

Researchers are attempting to overcome these issues by:

- Creating frameworks for explainable AI (XAI) to increase the transparency of medication predictions powered by AI.
- Standardising validation pathways by incorporating AI models into regulatory workflows.
- Promoting cooperative AI research projects across regulatory agencies, business, and academia to guarantee adherence to medication development protocols.

To ensure that AI-based therapeutic advances remain safe and can be effectively implemented in clinical practice, these technological and regulatory obstacles must be removed (Ahn et al., 2023).

7.3 Barriers to Clinical Adoption

The lack of transparency and interpretability is a significant drawback of AI in oncology, making it difficult to win over clinicians. The black-box nature of deep learning models, especially CNNs and transformer networks, makes it challenging for oncologists to comprehend the reasoning of the recommendations generated by AI(Ahluwalia et al., 2024). This lack of explainability hampers the use of AI in clinical processes, as doctors need interpretable frameworks for decision-making to validate therapy recommendations and inform patients of findings (Kann et al., 2021).

Researchers are investigating explainable AI (XAI) strategies to improve model transparency and address this problem. These strategies include rule-based hybrid models, attention mechanisms, and feature attribution techniques. To improve interpretability for oncologists, LIME (Local Interpretable Model-Agnostic Explanations) and SHAP (Shapley Additive Explanations) provide visualizations to prioritize specific genomic or clinical features of AI models. Furthermore, hybrid neuro-symbolic AI models provide a structured and comprehensible approach to decision-making in cancer therapy by combining deep learning with explicit rule-based reasoning (Kann et al., 2021).

8. FUTURE PERSPECTIVES

8.1 Emerging Trends in AI for Precision Oncology

In recent years, emerging AI has demonstrated significant potential for addressing lung cancer. The comprehensive definition of AI is expansive, encompassing applications or technologies that can learn and identify patterns and characteristics from extensive datasets, thereby mimicking cognitive functions akin to human thought (Y. Wang et al., 2025). The utility of AI in clinical decision-making for lung cancer is increasing across a growing array of clinical and experimental studies, including lung cancer screening, aiding diagnosis, predicting outcomes, and evaluating treatment efficacy and prognosis. This will entice both novices and experienced individuals to explore study subjects in this domain, necessitating a comprehensive delineation of the current research landscape, development methodologies, and prospective research focal points (Chouaïd et al., 2023)). This AI-driven method offers the benefit of repeated and noninvasive determination of mutation status. This methodology was endorsed by a PET/CT-based deep learning model for patients with NSCLC, utilizing radiomic features to differentiate EGFR-mutant variants from wild-type, achieving an AUC of 0.81 (Z. Chen et al., 2021). Consequently, precise forecasting of treatment response is essential to prevent superfluous toxicity and surgical delays, achieving an overall accuracy of 88%. Besides forecasting patient responses to treatments, AI now provides methods to adjust drug dosages for single or combination therapy tailored to individual patients, using patient-specific data gathered over time (Z. Chen et al., 2021). Deep learning is an emerging AI technique in cancer research that is advancing rapidly. The expansion of high-quality medical data and the advancement of algorithms position deep learning approaches as highly promising for enhancing the accuracy and efficacy of cancer diagnosis and treatment.

Furthermore, the FDA's favorable stance on AI medical devices enhances the potential for the practical implementation of deep learning in oncology. Future research should concentrate on enhancing the reliability and interpretability of deep learning systems to facilitate their clinical adoption (Z. Chen et al., 2021). Deep learning methodologies have been utilized across multiple facets of cancer research, including the exploration of biological mechanisms, the development of anticancer therapies, and the conduct of randomized controlled trials (RCTs). Research has employed deep learning to elucidate the fundamental causes of cancer by examining correlations between genotypes and phenotypes, yielding numerous documented accomplishments. A recent study using deep learning techniques revealed that F-box/WD repeat-containing protein 7 (Fbw7) is involved in the oxidative metabolism of cancer cells, based on gene expression profiles from the Cancer Genome Atlas dataset. Genomics also found genetic changes with possible therapeutic implications that were not detected by standard molecular tumor boards across many cancer types. The identification of these genetic variants not only delineates pertinent biological pathways but also proposes targets for pharmacological development. Machine learning techniques have been used to expedite the preliminary identification of potential anticancer compounds (Z. Chen et al., 2021).

8.2 Digital Twins and Virtual Patient Models

This application encompasses datasets for training, preprocessing techniques, algorithms for predictive model generation and model development acceleration, and pretrained models that leverage expertise from prior iterations. Machine learning, encompassing robust techniques such as deep learning, convolutional neural networks (CNNs), and decision trees, belongs to the core of AI. These sophisticated AI algorithms construct models that improve medical image and data analysis, facilitating the efficient examination of multidimensional datasets, augmenting image analysis and interpretation, and delivering decision support systems that assist researchers and clinicians in navigating complex lung cancer data to yield valuable insights and recommendations for precision medicine (Y. Wang et al., 2025). Patients with distinct inflammatory characteristics can be identified. Two cohorts of patients with resected NSCLCs, one retrospective and the other prospective, established and validated three inflammatory phenotypes based on the spatial distributions of CD8þ T cells: inflamed, altered, and desert. The inflammatory phenotype in retrospective cohort patients was an independent predictor of extended disease-free survival and time to recurrence. The results offer significant insights for the more tailored management of resected NSCLCs, achievable solely using AI methodologies. Time constraints and inter observer variability render manual classification of a tumor's immunological profile unfeasible. Standard biostatistical techniques also struggle to establish robust associations between inflammatory phenotypes and numerous genomic alterations.

However, clinicians need to assume responsibility for AI methodologies. AI tools, including natural language processing, deep learning, machine learning, and reinforcement learning, can analyze large datasets and enhance classification algorithms in real time as learning progresses (Chouaïd et al., 2023). Ultimate imaging and prognostic systems have great survival benefits. Radiomic-clinical 18F-FDG PET/CT model enables a more accurate prediction of tumour immune status in non-small cell lung carcinoma than traditional clinical and radiomic models. The hypertension status and the size of lung nodules are considered, which provides a valuable prognosis of NSCLC with a significant net reclassification improvement (NRI) and integrated discrimination enhancement (IDI). The article critically examined the CD8 expression, which is a phenomenon to evaluate the tumour immune microenvironment qualities (TIME) as explained by Tong *et al.* The work confirms that a hybrid radiomic-clinical PET/CT model is capable of predicting tumour immune state in NSCLC more accurately than known clinical models as well as individual radiomic models (El Sabrouty & Elouadi, 2024).

The combination of imaging data and clinical data described herein makes an integrative approach more complete to provide a comprehensive evaluation of the immunological microenvironment of the tumor. Overlays of radiomic phenotypes and clinical data enhances the predictive ability of the immunological phenotype of non-small cell lung cancer tumours. The increased predictive capacity of the model is necessary to the personalization of immunotherapeutic regimens as it allows to stratify the patients that are likely to respond positively to the immune checkpoint inhibitors (ICIs).

These results therefore highlight the fact that modern imaging modalities when used in combination with clinical parameters provide oncologists with a powerful tool that will enable them gain a better understanding of the tumour immunological milieu. This is critical to changing therapeutic approaches and improving the clinical outcome of such patients with non-small cell lung cancer. With the help of the EfficientNetV2 architecture, the Efficient Survival Benefit Prediction System (ESBP) has proved to significantly increase the diagnostic accuracy in the group of practitioners. It has been consistently used to foretell the survival advantages of EGFR-TKI and ICI treatments against an advanced-stage NSCLC. Two-year experience clinicians indicated a rise in the accuracy of diagnosis by 47.91 per cent to 66.32 per cent and the five-year experience clinicians recorded an increase of 53.12 per cent to 61.41 per cent. An artificial-intelligence-generated score incorporated into the primary CT scan during the management of incidentally detected tumours may have avoided subsequent surveillance in half of benign pulmonary nodules according to the guidelines (6 of 12), and lessened the costs and logistical implications of both patient and health care provider (El Sabrouty & Elouadi, 2024).

8.3 Towards fully Integrated AI-Enabled Precision Medicine

Lung cancer is a ubiquitous worldwide health problem which imposes significant impacts to both the individuals and the society. The imbalance in the occurrence and demise of non-small cell lung cancer (NSCLC) highlights the imperative to implement combined initiatives that include preventative measures, early detection guidelines, the best treatment choices, and effective support systems to care-givers of the affected individuals. Recent empirical results also underscore the need to adopt a multidisciplinary approach, which is comprehensive, to face this malignancy. The preventive measures, especially the smoking cessation programs and the population health education programs can be considered as the invaluable levers to improving the prevalence of lung cancer. Initial diagnostic methods, namely low-dose computer tomography (CT) screening, have proved to be effective in improving the survival rates as they allow the detection of tumours at earlier stages that are more manageable. The area of therapy of NSCLC is actively developing, and it is driven by the current research and collaboration between countries and has the common goal, which is to improve the outcome of patients. Targeted therapies and immunotherapies have shown potential to extend life and increase the quality of life. At the same time, comprehensive support networks that involve psychological, social and palliative services are invaluable in meeting the complex needs of patients and their families.

Artificial intelligence (AI) has become a revolution in the fight against lung cancer in the modern era. The use of AI is found in different aspects of care, such as real-time monitoring, treatment planning, predictive modelling, and accurate tumour detection. When subjected to complex algorithms on large genomic and imaging datasets, AI is able to identify subtle patterns that can be missed by the human eye, enhancing diagnostic accuracy. Moreover, the possibility to predict and analyze data is especially relevant to AI; by combining the heterogenous data streams, one can predict the progression of the disease and the effectiveness of treatment, and discuss really personalized treatment regimen. This informal, personalized data-based method promises more effective treatments accompanied by a subsequent decrease in side effects.

The AI also supports the evidence-based decision making whereby the information of various sources can be combined to suggest the best therapeutic processes. Real-time monitoring, powered by AI, means that the patients will have an opportunity to adjust their treatment plans on time and, therefore, achieve the maximum efficacy and the minimum of complications. However, despite the promising future of AI, there are many challenges that still exist that exclude its smooth adoption into a traditional healthcare system. The need to earn the trust of both patients and clinicians requires the concerns associated with the data privacy, the algorithmic transparency, and the necessity to conduct clinical validation on the highest level. Additionally, for AI solutions to be implemented successfully, healthcare professionals must be trained in their practical use. Prospects for NSCLC care are promising, marked by ongoing innovation and the potential to reduce the disease's global impact significantly. With its promise of more individualized care, more effective medicines, and improved patient outcomes, AI is at the vanguard of this change. We are getting closer to a time when lung cancer is easier to treat and may even be cured, as we overcome the obstacles to integrating AI(El Sabrouty & Elouadi, 2024).

CONCLUSION

AI is reshaping the landscape of lung cancer care by bridging the gap between molecular biology, clinical data, and therapeutic decision-making. Through advanced analytics and deep learning, AI enhances diagnostic accuracy, enables comprehensive molecular characterization, and supports predictive modeling for prognosis and therapy response. AI-powered systems are increasingly being integrated into clinical workflows, driving personalized treatment strategies that optimize outcomes while minimizing toxicity. However, challenges such as data heterogeneity, model interpretability, and ethical governance remain significant hurdles to large-scale clinical adoption. Looking ahead, the convergence of AI with precision medicine supported by innovations such as digital twins and real-time decision support heralds a new era of intelligent, adaptive oncology. By transforming complex biomedical data into actionable insights, AI is not merely augmenting human expertise but redefining the future of lung cancer management.

REFERENCES

Afrose, N., Chakraborty, R., Bhowmick, P., & Bhowmick, M. (2025). Advancing drug delivery systems through biomedical engineering: Innovations and future directions. In Innovations in Biomedical Engineering [Internet]. (pp. 1–32). Elsevier., Available from https://linkinghub.elsevier.com/retrieve/pii/B9780443301469000010, Retrieved October 28, 2025, from. DOI: 10.1016/B978-0-443-30146-9.00001-0

Afrose, N., Chakraborty, R., & Rajendran, K. (2025, August). Plasma-polymerized chitosan/agarose/chitosan-agarose composite coatings for antibacterial and antibiofilm wound dressings: A review. [Internet]. *International Journal of Biological Macromolecules, 320 (Pt 1)*, 145873. Retrieved October 27, 2025, from https://linkinghub.elsevier.com/retrieve/pii/S0141813025064281. DOI: 10.1016/j.ijbiomac.2025.145873 PMID: 40645242

Ahluwalia, P., Ballur, K., Leeman, T., Vashisht, A., Singh, H., Omar, N., Mondal, A. K., Vaibhav, K., Baban, B., & Kolhe, R. (2024, January 23). Incorporating Novel Technologies in Precision Oncology for Colorectal Cancer: Advancing Personalized Medicine. [Internet]. *Cancers (Basel)*, *16*(3), 480. Retrieved November 2, 2025, from https://www.mdpi.com/2072-6694/16/3/480. DOI: 10.3390/cancers16030480 PMID: 38339232

Ahmad, S., Bashar, A., Khanna, K., Bano, N., & Raza, K. (2025). Deep learning in variant detection and annotation. In Deep Learning in Genetics and Genomics [Internet]. (pp. 383–396). Elsevier., Available from https://linkinghub.elsevier.com/retrieve/pii/B9780443275746000102, Retrieved October 27, 2025, from. DOI: 10.1016/B978-0-443-27574-6.00010-2

Ahmed, Z. (2020, December). Practicing precision medicine with intelligently integrative clinical and multi-omics data analysis. [Internet]. *Human Genomics*, *14*(1), 35. Retrieved November 2, 2025, from https://humgenomics.biomedcentral.com/articles/10.1186/s40246-020-00287-z. DOI: 10.1186/s40246-020-00287-z PMID: 33008459

Ahmed, Z. (2022, April 15). Multi-omics strategies for personalized and predictive medicine: Past, current, and future translational opportunities. [Internet]. *Emerging Topics in Life Sciences*, *6*(2), 215–225. Retrieved November 2, 2025, from https://portlandpress.com/emergtoplifesci/article/6/2/215/230971/Multi-omics-strategies-for-personalized-and. DOI: 10.1042/ETLS20210244 PMID: 35234253

Ahn, J. S., Shin, S., Yang, S. A., Park, E. K., Kim, K. H., Cho, S. I., Ock, C.-Y., & Kim, S. (2023). Artificial Intelligence in Breast Cancer Diagnosis and Personalized Medicine. [Internet]. *Journal of Breast Cancer*, *26*(5), 405. Retrieved November 2, 2025, from https://ejbc.kr/DOIx.php?id=10.4048/jbc.2023.26.e45. DOI: 10.4048/jbc.2023.26.e45 PMID: 37926067

Ali, K. A., Choudhuri, S., Chakraborty, R., Pradhan, A., Kotal, A., & Shaharyar, M. A. A. I. (2025). Technology-Powered Precision Medicine and Smart Healthcare. In AI and Precision Medicine in Infectious Disease Management [Internet]. 1st ed., pp. 51–70). CRC Press., Available from https://www.taylorfrancis.com/books/9781003615699/chapters/10.1201/9781003615699-3, Retrieved October 29, 2025, from. DOI: 10.1201/9781003615699-3

Ali, K. A., Mohin, S., Mondal, P., Goswami, S., Ghosh, S., & Choudhuri, S. Influence of artificial intelligence in modern pharmaceutical formulation and drug development. Futur J Pharm Sci [Internet]. 2024 Mar 29 [cited 2025 Feb 20];10(1):53. Available from: https://fjps.springeropen.com/articles/10.1186/s43094-024-00625-1

Athanasopoulou, K., Michalopoulou, V. I., Scorilas, A., & Adamopoulos, P. G. Integrating Artificial Intelligence in Next-Generation Sequencing: Advances, Challenges, and Future Directions. CIMB [Internet]. 2025 June 19 [cited 2025 Oct 28];47(6):470. Available from: https://www.mdpi.com/1467-3045/47/6/470

Barta, J. A., Powell, C. A., & Wisnivesky, J. P. (2019, January 22). Global Epidemiology of Lung Cancer. *Annals of Global Health*, *85*(1), 8. DOI: 10.5334/aogh.2419 PMID: 30741509

Bennati, C., Paglialunga, L., Ricciuti, B., Metro, G., Marcomigni, L., Gili, A., & Crinò, L. (2016, June). Targeting EGFR and ALK in NSCLC: Current evidence and future perspective. *Lung Cancer Management*, *5*(2), 79–90. DOI: 10.2217/lmt-2016-0005 PMID: 30643552

Bertolaccini, L., Casiraghi, M., Uslenghi, C., Maiorca, S., & Spaggiari, L. (2024, October). Recent advances in lung cancer research: Unravelling the future of treatment. [Internet]. *Updates in Surgery*, *76*(6), 2129–2140. Retrieved November 2, 2025, from https://link.springer.com/10.1007/s13304-024-01841-3. DOI: 10.1007/s13304-024-01841-3 PMID: 38581618

Bishal, A., Paul, P., Bera, S., Patra, D., Bandyopadhyay, B., Debnath, B., & Ali, K. A. (2025, August 5). Targeted Drug Delivery for Breast Cancer using Functionalized Liposomes: Preparation Methods, Challenges, and Clinical Translation. [Internet]. *AAPS PharmSciTech*, *26*(7), 207. Retrieved August 19, 2025, from https://link.springer.com/10.1208/s12249-025-03206-4. DOI: 10.1208/s12249-025-03206-4 PMID: 40764411

Biswas, N., & Chakrabarti, S. (2020, October 14). Artificial Intelligence (AI)-Based Systems Biology Approaches in Multi-Omics Data Analysis of Cancer. [Internet]. *Frontiers in Oncology*, *10*, 588221. Retrieved November 2, 2025, from https://www.frontiersin.org/article/10.3389/fonc.2020.588221/full. DOI: 10.3389/fonc.2020.588221 PMID: 33154949

Budczies, J., Kazdal, D., Menzel, M., Beck, S., Kluck, K., Altbürger, C., Schwab, C., Allgäuer, M., Ahadova, A., Kloor, M., Schirmacher, P., Peters, S., Krämer, A., Christopoulos, P., & Stenzinger, A. (2024, October). Tumour mutational burden: Clinical utility, challenges and emerging improvements. [Internet]. *Nature Reviews. Clinical Oncology*, *21*(10), 725–742. Retrieved November 1, 2025, from https://www.nature.com/articles/s41571-024-00932-9. DOI: 10.1038/s41571-024-00932-9 PMID: 39192001

Camps, J., Jimenez-Franco, A., García-Pablo, R., Joven, J., & Arenas, M. (2025). Artificial intelligence-driven integration of multi-omics and radiomics: A new hope for precision cancer diagnosis and prognosis. Biochimica et biophysica acta (BBA)-Molecular basis of disease, 1871(6), 167841.

Chakraborty, R., Afrose, N., Bhowmick, P., & Bhowmick, M. (2025). Biomedical sensors in wearable health technologies. In Innovations in Biomedical Engineering [Internet]. (pp. 159–184). Elsevier., Available from https://linkinghub.elsevier.com/retrieve/pii/B9780443301469000058, Retrieved October 28, 2025, from. DOI: 10.1016/B978-0-443-30146-9.00005-8

Chakraborty, R., Jana, B., Koner, S., Kanta Maiti, S., Afrose, N., Bhowmick, P., ... Exploring the transcriptomic landscape of rheumatoid arthritis using next-generation RNA sequencing. ITPS [Internet]. 2024 Nov 22 [cited 2025 Oct 27];8(1):14. Available from: https://accscience.com/journal/ITPS/8/1/10.36922/itps.4449

Chakraborty, R., Kannan, S. P., Afrose, N., & Narayanasamy, D. (2025, September). Tumor microenvironment expressed enzymes (Matrix metalloproteinases, cathepsins, urokinase-type plasminogen activator) triggered polymersomes for liquid biopsy and cancer diagnostics: A review. [Internet]. *International Journal of Biological Macromolecules, 321(2)*, 146375. Retrieved October 27, 2025, from https://linkinghub.elsevier.com/retrieve/pii/S0141813025069326. DOI: 10.1016/j.ijbiomac.2025.146375 PMID: 40738429

Chakravarty, K., Antontsev, V., Bundey, Y., & Varshney, J. (2021, June). Driving success in personalized medicine through AI-enabled computational modeling. [Internet]. *Drug Discovery Today*, *26*(6), 1459–1465. Retrieved November 2, 2025, from https://linkinghub.elsevier.com/retrieve/pii/S1359644621000738. DOI: 10.1016/j.drudis.2021.02.007 PMID: 33609781

Chang, L., Liu, J., Zhu, J., Guo, S., Wang, Y., Zhou, Z., & Wei, X. (2025, January 2). Advancing precision medicine: The transformative role of artificial intelligence in immunogenomics, radiomics, and pathomics for biomarker discovery and immunotherapy optimization. [Internet]. *Cancer Biology & Medicine*, 1–15. Retrieved November 2, 2025, from https://www.cancerbiomed.org/content/early/2025/01/02/j.issn.2095-3941.2024.0376. DOI: 10.20892/j.issn.2095-3941.2024.0376 PMID: 39749734

Chang, T. G., Park, S., Schäffer, A. A., Jiang, P., & Ruppin, E. (2025, March). Hallmarks of artificial intelligence contributions to precision oncology. *Nature Cancer*, *6*(3), 417–431. DOI: 10.1038/s43018-025-00917-2 PMID: 40055572

Chen, S., Gao, L., Tan, M., Zhang, K., & Lv, F. (2025, June 26). Performance Evaluation of Four Deep Learning-Based CAD Systems and Manual Reading for Pulmonary Nodules Detection, Volume Measurement, and Lung-RADS Classification Under Varying Radiation Doses and Reconstruction Methods. *Diagnostics (Basel)*, *15*(13), 1623. DOI: 10.3390/diagnostics15131623 PMID: 40647622

Chen, Z., Lin, L., Wu, C., Li, C., Xu, R., & Sun, Y. (2021, November). Artificial intelligence for assisting cancer diagnosis and treatment in the era of precision medicine. [Internet]. *Cancer Communications*, *41*(11), 1100–1115. Retrieved November 2, 2025, from https://onlinelibrary.wiley.com/doi/10.1002/cac2.12215. DOI: 10.1002/cac2.12215 PMID: 34613667

Cheo, H. M., Ong, C. Y. G., & Ting, Y. (2025, June 24). A Systematic Review of AI Performance in Lung Cancer Detection on CT Thorax. *Healthcare (Basel)*, *13*(13), 1510. DOI: 10.3390/healthcare13131510 PMID: 40648536

Chi, A., He, X., Hou, L., Nguyen, N. P., Zhu, G., Cameron, R. B., & Lee, J. M. (2021, June 11). Classification of Non-Small Cell Lung Cancer's Tumor Immune Micro-Environment and Strategies to Augment Its Response to Immune Checkpoint Blockade. [Internet]. *Cancers (Basel)*, *13*(12), 2924. Retrieved November 2, 2025, from https://www.mdpi.com/2072-6694/13/12/2924. DOI: 10.3390/cancers13122924 PMID: 34208113

Chouaïd, C., Gendarme, S., & Auliac, J. B. (2023, July). Artificial intelligence to finally enable precision medicine for the management of resected non-small-cell lung cancer. [Internet]. *Annals of Oncology : Official Journal of the European Society for Medical Oncology*, *34*(7), 565–566. Retrieved November 2, 2025, from https://linkinghub.elsevier.com/retrieve/pii/S0923753423006750. DOI: 10.1016/j.annonc.2023.05.001 PMID: 37182802

Chukwunweike, J. N., Ayodele, P. A., & Atata, B. B. (2024, August). HARNESSING MACHINE LEARNING FOR CYBERSECURITY: HOW CONVOLUTIONAL NEURAL NETWORKS ARE REVOLUTIONIZING THREAT DETECTION AND DATA PRIVACY. *Int J Res Publ Rev [Internet]*, *5*(8), 3608–3617. Retrieved November 2, 2025, from https://ijrpr.com/uploads/V5ISSUE8/IJRPR32530.pdf. DOI: 10.55248/gengpi.5.0824.2402

Dall'Olio, F. G., Marabelle, A., Caramella, C., Garcia, C., Aldea, M., Chaput, N., Robert, C., & Besse, B. (2022, February). Tumour burden and efficacy of immune-checkpoint inhibitors. [Internet]. *Nature Reviews. Clinical Oncology*, *19*(2), 75–90. Retrieved November 2, 2025, from https://www.nature.com/articles/s41571-021-00564-3. DOI: 10.1038/s41571-021-00564-3 PMID: 34642484

Dasgupta, S., Gayen, S., Chakraborty, T., Afrose, N., Pal, R., Mahata, S., Nasare, V., & Roy, S. (2024, March 27). Potential role of immune cell therapy in gynecological cancer and future promises: A comprehensive review. [Internet]. *Medical Oncology*, *41*(5), 98. Retrieved November 2, 2025, from https://link.springer.com/10.1007/s12032-024-02337-1. DOI: 10.1007/s12032-024-02337-1 PMID: 38536512

Davri, A., Birbas, E., Kanavos, T., Ntritsos, G., Giannakeas, N., Tzallas, A. T., & Batistatou, A. (2023, August 5). Deep Learning for Lung Cancer Diagnosis, Prognosis and Prediction Using Histological and Cytological Images: A Systematic Review. *Cancers (Basel)*, *15*(15), 3981. DOI: 10.3390/cancers15153981 PMID: 37568797

de Jong, D., Das, J. P., Ma, H., Pailey Valiplackal, J., Prendergast, C., Roa, T., Braumuller, B., Deng, A., Dercle, L., Yeh, R., Salvatore, M. M., & Capaccione, K. M. (2023, May 21). Novel Targets, Novel Treatments: The Changing Landscape of Non-Small Cell Lung Cancer. [Internet]. *Cancers (Basel)*, *15*(10), 2855. Retrieved November 2, 2025, from https://www.mdpi.com/2072-6694/15/10/2855. DOI: 10.3390/cancers15102855 PMID: 37345192

Dubey, S., & Sikarwar, S. S. (2025). Applications of AI in Cancer Detection — A Review of the Specific Ways in which AI Is Being Used to Detect and Diagnose Various Types of Cancer. In R. Singh, A. Gehlot, N. Rathour, & S. V. Akram (Eds.), AI in Disease Detection [Internet]. 1st ed., pp. 147–166). Wiley., Available from https://onlinelibrary.wiley.com/doi/10.1002/9781394278695.ch7, Retrieved November 2, 2025, from. DOI: 10.1002/9781394278695.ch7

El Sabrouty, R., & Elouadi, A. (2024, August 2). Harnessing AI for Precision Oncology: Transformative Advances in Non-Small Cell Lung Cancer Treatment. [Internet]. *Statistics, Optimization & Information Computing*, *13*(1), 347–367. Retrieved November 2, 2025, from http://www.iapress.org/index.php/soic/article/view/2078. DOI: 10.19139/soic-2310-5070-2078

Elhaddad, M., & Hamam, S. (2024, April 6). AI-Driven Clinical Decision Support Systems: An Ongoing Pursuit of Potential. [Internet]. *Cureus*, *16*(4), e57728. Retrieved November 2, 2025, from https://www.cureus.com/articles/244463-ai-driven-clinical-decision-support-systems-an-ongoing-pursuit-of-potential. DOI: 10.7759/cureus.57728 PMID: 38711724

Fujiwara, Y., Kato, S., & Kurzrock, R. (2024, April). Evolution of Precision Oncology, Personalized Medicine, and Molecular Tumor Boards. *Surgical Oncology Clinics of North America*, *33*(2), 197–216. DOI: 10.1016/j.soc.2023.12.004 PMID: 38401905

Gandhewar, N., Pimpalkar, A., Jadhav, A., Shelke, N., & Jain, R. (2025). Leveraging Deep Learning for Genomics Analysis: Advances and Applications. In S. Choudhary, S. Kumar, S. Gowroju, M. Gulhane, & R. Sri Lakshmi (Eds.), Genomics at the Nexus of AI, Computer Vision, and Machine Learning [Internet]. 1st ed., pp. 191–225). Wiley., Available from https://onlinelibrary.wiley.com/doi/10.1002/9781394268832.ch9, Retrieved October 27, 2025, from. DOI: 10.1002/9781394268832.ch9

Hassan, S., Bahar, R., Johan, M. F., Mohamed Hashim, E. K., Abdullah, W. Z., Esa, E., Abdul Hamid, F. S., & Zulkafli, Z. (2023, January 19). Next-Generation Sequencing (NGS) and Third-Generation Sequencing (TGS) for the Diagnosis of Thalassemia. [Internet]. *Diagnostics (Basel)*, *13*(3), 373. Retrieved November 2, 2025, from https://www.mdpi.com/2075-4418/13/3/373. DOI: 10.3390/diagnostics13030373 PMID: 36766477

Huhulea, E. N., Huang, L., Eng, S., Sumawi, B., Huang, A., Aifuwa, E., Hirani, R., Tiwari, R. K., & Etienne, M. (2025, April 13). Artificial Intelligence Advancements in Oncology: A Review of Current Trends and Future Directions. *Biomedicines*, *13*(4), 951. DOI: 10.3390/biomedicines13040951 PMID: 40299653

Jani, C. T., Singh, H., Abdallah, N., Mouchati, C., Arora, S., Kareff, S., Salciccioli, J., Thomson, C. C., & Velcheti, V. (2023, September). Trends in Lung Cancer Incidence and Mortality (1990-2019) in the United States: A Comprehensive Analysis of Gender and State-Level Disparities. *JCO Global Oncology*, (9), e2300255. DOI: 10.1200/GO.23.00255 PMID: 38127772

Jones, S., Thompson, K., Porter, B., Shepherd, M., Sapkaroski, D., Grimshaw, A., & Hargrave, C. (2024, June). Automation and artificial intelligence in radiation therapy treatment planning. [Internet]. *Journal of Medical Radiation Sciences*, *71*(2), 290–298. Retrieved November 2, 2025, from https://onlinelibrary.wiley.com/doi/10.1002/jmrs.729. DOI: 10.1002/jmrs.729 PMID: 37794690

Kann, B. H., Hosny, A., & Aerts, H. J. W. L. (2021, July). Artificial intelligence for clinical oncology. [Internet]. *Cancer Cell*, *39*(7), 916–927. Retrieved November 2, 2025, from https://linkinghub.elsevier.com/retrieve/pii/S1535610821002105. DOI: 10.1016/j.ccell.2021.04.002 PMID: 33930310

Kapsoritakis, N., Tsitoura, F., Stathaki, M., Bourogianni, O., Georgoulias, P., Barmparis, G. D., Bertsias, A., Tsironis, G. P., & Koukouraki, S. (2025, October 17). Clinical Impact of FDG PET/CT in Pulmonary Nodule Characterization: Current Perspectives on Dual-Time-Point Imaging and Semi-Quantitative Imaging Metrics. *Cancers (Basel)*, *17*(20), 3353. DOI: 10.3390/cancers17203353 PMID: 41154408

Karthik, A., Patthi, S., Maheswari, B. U., Devi, J. S., Pradhan, A. K., Balaji, G., Nallasivan, G., & Rajaram, A. (2025, February). Advancing idiopathic pulmonary fibrosis prognosis through integrated CNN-LSTM predictive modeling and uncertainty quantification. [Internet]. *Biomedical Signal Processing and Control*, *100*, 106811. Retrieved November 1, 2025, from https://linkinghub.elsevier.com/retrieve/pii/S1746809424008693. DOI: 10.1016/j.bspc.2024.106811

Kumar, A. AI Driven Precision Oncology: Predictive Biomarker Discovery and Personalized Treatment Optimization Using Genomic Data. 2025 Mar 17 [cited 2025 May 7]; Available from: https://zenodo.org/doi/10.5281/zenodo.15037946

Kwong, A., Ho, C. Y. S., Shin, V. Y., Au, C. H., Chan, T. L., & Ma, E. S. K. (2022, December). How does re-classification of variants of unknown significance (VUS) impact the management of patients at risk for hereditary breast cancer? [Internet]. *BMC Medical Genomics*, *15*(1), 122. Retrieved November 2, 2025, from https://bmcmedgenomics.biomedcentral.com/articles/10.1186/s12920-022-01270-4. DOI: 10.1186/s12920-022-01270-4 PMID: 35641994

La'ah, A. S., & Chiou, S. H. (2024, March 1). Cutting-Edge Therapies for Lung Cancer. [Internet]. *Cells*, *13*(5), 436. Retrieved November 2, 2025, from https://www.mdpi.com/2073-4409/13/5/436. DOI: 10.3390/cells13050436 PMID: 38474400

Li, C., Lei, S., Ding, L., Xu, Y., Wu, X., Wang, H., Zhang, Z., Gao, T., Zhang, Y., & Li, L. (2023, July 5). Global burden and trends of lung cancer incidence and mortality. *Chinese Medical Journal*, *136*(13), 1583–1590. DOI: 10.1097/CM9.0000000000002529 PMID: 37027426

Li, X., Krivtsov, V., Pan, C., Nassehi, A., Gao, R. X., & Ivanov, D. (2025, February). End-to-end supply chain resilience management using deep learning, survival analysis, and explainable artificial intelligence. [Internet]. *International Journal of Production Research*, *63*(3), 1174–1202. Retrieved November 1, 2025, from https://www.tandfonline.com/doi/full/10.1080/00207543.2024.2367685. DOI: 10.1080/00207543.2024.2367685

Liu, D., & Imai, N. (2025, September 24). Decision-Making Biomarkers Guiding Therapeutic Strategies in Hepatocellular Carcinoma: From Prediction to Personalized Care. [Internet]. *Cancers (Basel)*, *17*(19), 3105. Retrieved November 2, 2025, from https://www.mdpi.com/2072-6694/17/19/3105. DOI: 10.3390/cancers17193105 PMID: 41097634

Logullo, P., MacCarthy, A., Dhiman, P., Kirtley, S., Ma, J., Bullock, G., & Collins, G. S. (2023). Artificial intelligence in lung cancer diagnostic imaging: A review of the reporting and conduct of research published 2018-2019. *BJR Open*, *5*(1), 20220033. DOI: 10.1259/bjro.20220033 PMID: 37389003

Lu, C., Shiradkar, R., & Liu, Z. (2021, October 31). Integrating pathomics with radiomics and genomics for cancer prognosis: A brief review. *Chinese Journal of Cancer Research = Chung-Kuo Yen Cheng Yen Chiu*, *33*(5), 563–573. DOI: 10.21147/j.issn.1000-9604.2021.05.03 PMID: 34815630

Mikhael, P. G., Wohlwend, J., Yala, A., Karstens, L., Xiang, J., Takigami, A. K., Bourgouin, P. P., Chan, P. Y., Mrah, S., Amayri, W., Juan, Y.-H., Yang, C.-T., Wan, Y.-L., Lin, G., Sequist, L. V., Fintelmann, F. J., & Barzilay, R. (2023, April 20). Sybil: A Validated Deep Learning Model to Predict Future Lung Cancer Risk From a Single Low-Dose Chest Computed Tomography. *Journal of Clinical Oncology : Official Journal of the American Society of Clinical Oncology*, *41*(12), 2191–2200. DOI: 10.1200/JCO.22.01345 PMID: 36634294

Nguyen, H., Do, C., Tran, B., & Trinh, H. (2025, June 18). Rare sequential EGFR and ALK mutations in non-small cell lung cancer: A case report and literature review. *Oncology Letters*, *30*(2), 1–6. DOI: 10.3892/ol.2025.15144 PMID: 40606307

Nielsen, R. L., Monfeuga, T., Kitchen, R. R., Egerod, L., Leal, L. G., Schreyer, A. T. H., Gade, F. S., Sun, C., Helenius, M., Simonsen, L., Willert, M., Tahrani, A. A., McVey, Z., & Gupta, R. (2024, April 1). Data-driven identification of predictive risk biomarkers for subgroups of osteoarthritis using interpretable machine learning. [Internet]. *Nature Communications*, *15*(1), 2817. Retrieved November 1, 2025, from https://www.nature.com/articles/s41467-024-46663-4. DOI: 10.1038/s41467-024-46663-4 PMID: 38561399

Ogut, E. (2025, September 16). Artificial Intelligence in Clinical Medicine: Challenges Across Diagnostic Imaging, Clinical Decision Support, Surgery, Pathology, and Drug Discovery. [Internet]. *Clinics and Practice*, *15*(9), 169. Retrieved November 2, 2025, from https://www.mdpi.com/2039-7283/15/9/169. DOI: 10.3390/clinpract15090169 PMID: 41002784

Okenwa, C. D., Adeyemi, A. F., Orelaja, A., & Nasimbwa, R. (2024). Predictive Analytics In Financial Regulation: Advancing Compliance Models For Crime Prevention. *Iosr Journal Of Economics And Finance.*, *15*(4), 1–7.

Orstad, S., Fløtten, Ø., Madebo, T., Gulbrandsen, P., Strand, R., Lindemark, F., Fluge, S., Tilseth, R. H., & Schaufel, M. A. (2023, September). "The challenge is the complexity" – A qualitative study about decision-making in advanced lung cancer treatment. [Internet]. *Lung Cancer : Journal of the International Association for the Study of Lung Cancer*, *183*, 107312. Retrieved November 2, 2025, from https://linkinghub.elsevier.com/retrieve/pii/S0169500223008504. DOI: 10.1016/j.lungcan.2023.107312 PMID: 37481888

Pan, Y., Sheng, W., Shi, L., Jing, D., Jiang, W., Chen, J. C., Wang, H., & Qiu, J. (2023, June 1). Whole slide imaging-based deep learning to predict the treatment response of patients with non-small cell lung cancer. *Quantitative Imaging in Medicine and Surgery*, *13*(6), 3547–3555. DOI: 10.21037/qims-22-1098 PMID: 37284119

Parhi, S., Afrose, N., Rajendran, K., & Narayanasamy, D. (2025, March 4). Central composite design (CCD) based formulation, optimization, *in-vitro* and *ex-vivo* characterization of 5-fluorouracil-loaded emulgel for enhanced dermal penetration and psoriasis management. [Internet]. *Drug Development and Industrial Pharmacy*, *51*(3), 244–261. Retrieved November 2, 2025, from https://www.tandfonline.com/doi/full/10.1080/03639045.2025.2464782. DOI: 10.1080/03639045.2025.2464782 PMID: 39932381

Park, J., Brattoli, B., Shi, J., Senior, M., Qaiser, T., Rawson, J., Hwang, W., Shin, S., Lee, T., Ock, C.-Y., Pereira, S., Bannister, H., Riboni-Verri, E., Ali, S., & Moore, L. (2025, April 21). Abstract 2463: AI-based EGFR-mutation prediction from haematoxylin and eosin (H&E) images in non-small cell lung cancer (NSCLC): A global multi-cohort validation study. *Cancer Research*, *85*(8, Supplement_1), 2463–2463. DOI: 10.1158/1538-7445.AM2025-2463

Parra-Medina, R., Guerron-Gomez, G., Mendivelso-González, D., Gil-Gómez, J. H., Alzate, J. P., Gomez-Suarez, M., Polo, J. F., Sprockel, J. J., & Mosquera-Zamudio, A. (2025, May 30). Deep learning in histopathology images for prediction of oncogenic driver molecular alterations in lung cancer: A systematic review and meta-analysis. *Translational Lung Cancer Research*, *14*(5), 1756–1769. DOI: 10.21037/tlcr-2024-1196 PMID: 40535093

Penault-Llorca, F., & Socinski, M. A. (2025, March 10). Emerging molecular testing paradigms in non-small cell lung cancer management—Current perspectives and recommendations. *The Oncologist, 30*(3), oyae357. DOI: 10.1093/oncolo/oyae357 PMID: 40126879

Peters, A. A., Decasper, A., Munz, J., Klaus, J., Loebelenz, L. I., Hoffner, M. K. M., Hourscht, C., Heverhagen, J. T., Christe, A., & Ebner, L. (2021, May). Performance of an AI based CAD system in solid lung nodule detection on chest phantom radiographs compared to radiology residents and fellow radiologists. *Journal of Thoracic Disease, 13*(5), 2728–2737. DOI: 10.21037/jtd-20-3522 PMID: 34164165

Rankin, A. W., Duncan, B. B., Allen, C., Silbert, S. K., & Shah, N. N. (2025, March). Evolving strategies for addressing CAR T-cell toxicities. [Internet]. *Cancer Metastasis Reviews, 44*(1), 17. Retrieved November 2, 2025, from https://link.springer.com/10.1007/s10555-024-10227-1. DOI: 10.1007/s10555-024-10227-1 PMID: 39674824

Rodriguez, N. G., Mercado, N. C., Panjiyar, K., & Kunwor, R. (2025, October 8). Artificial intelligence versus radiologists in predicting lung cancer treatment response: A systematic review and meta-analysis. *Frontiers in Oncology, 15*, 1634694. DOI: 10.3389/fonc.2025.1634694 PMID: 41132734

Sartori, F., Codicè, F., Caranzano, I., Rollo, C., Birolo, G., Fariselli, P., & Pancotti, C. (2025, May 28). A Comprehensive Review of Deep Learning Applications with Multi-Omics Data in Cancer Research. [Internet]. *Genes, 16*(6), 648. Retrieved November 1, 2025, from https://www.mdpi.com/2073-4425/16/6/648. DOI: 10.3390/genes16060648 PMID: 40565540

Schreuder, A., Scholten, E. T., Van Ginneken, B., & Jacobs, C. (2021, May). Artificial intelligence for detection and characterization of pulmonary nodules in lung cancer CT screening: Ready for practice? *Translational Lung Cancer Research, 10*(5), 2378–2388. DOI: 10.21037/tlcr-2020-lcs-06 PMID: 34164285

Shahid, M. S., & Imran, A. (2025). Breast cancer detection using deep learning techniques: Challenges and future directions. *Multimedia Tools and Applications, 84*(6), 3257–3304.

Sharma, A., Lövgren, S. K., Eriksson, K. L., Wang, Y., Robertson, S., Hartman, J., & Rantalainen, M. (2024, August 14). Validation of an AI-based solution for breast cancer risk stratification using routine digital histopathology images. [Internet]. *Breast Cancer Research : BCR, 26*(1), 123. Retrieved November 1, 2025, from https://breast-cancer-research.biomedcentral.com/articles/10.1186/s13058-024-01879-6. DOI: 10.1186/s13058-024-01879-6 PMID: 39143539

Sharma, A., Verma, A., Malviya, R., & Yadav, S. (2022). Applications of Deep Learning in Radiation Therapy. In R. Malviya, G. Ghinea, R. K. Dhanaraj, B. Balusamy, & S. Sundram (Eds.), Deep Learning for Targeted Treatments [Internet]. 1st ed., pp. 247–288). Wiley., Available from https://onlinelibrary.wiley.com/doi/10.1002/9781119857983.ch9, Retrieved November 2, 2025, from. DOI: 10.1002/9781119857983.ch9

Sim, J. A., Huang, X., Horan, M. R., Stewart, C. M., Robison, L. L., Hudson, M. M., & Huang, I. C. (2023). Natural language processing with machine learning methods to analyze unstructured patient-reported outcomes derived from electronic health records: A systematic review. *Artificial Intelligence in Medicine, 146*, 102701.

Smolarz, B., Łukasiewicz, H., Samulak, D., Piekarska, E., Kołaciński, R., & Romanowicz, H. (2025, February 26). Lung Cancer—Epidemiology, Pathogenesis, Treatment and Molecular Aspect (Review of Literature). *International Journal of Molecular Sciences*, *26*(5), 2049. DOI: 10.3390/ijms26052049 PMID: 40076671

Song, A. H., Jaume, G., Williamson, D. F. K., Lu, M. Y., Vaidya, A., Miller, T. R., & Mahmood, F. (2023, October 2). Artificial intelligence for digital and computational pathology. [Internet]. *Nature Reviews Bioengineering*, *1*(12), 930–949. Retrieved November 2, 2025, from https://www.nature.com/articles/s44222-023-00096-8. DOI: 10.1038/s44222-023-00096-8

Sousa, R., Abelha, V., Peixoto, H., & Machado, J. (2024, December 17). Unlocking Healthcare Data Potential: A Comprehensive Integration Approach with GraphQL, openEHR, Redis, and Pervasive Business Intelligence. [Internet]. *Technologies*, *12*(12), 265. Retrieved November 2, 2025, from https://www.mdpi.com/2227-7080/12/12/265. DOI: 10.3390/technologies12120265

Tang, F. H., Fong, Y. W., Yung, S. H., Wong, C. K., Tu, C. L., Chan, M. T., & Radiomics-Clinical, A. I. (2023, July 25). Model with Probability Weighted Strategy for Prognosis Prediction in Non-Small Cell Lung Cancer. *Biomedicines*, *11*(8), 2093. DOI: 10.3390/biomedicines11082093 PMID: 37626590

Tayefi, M., Ngo, P., Chomutare, T., Dalianis, H., Salvi, E., Budrionis, A., ... Challenges and opportunities beyond structured data in analysis of electronic health records. WIREs Computational Stats [Internet]. 2021 Nov [cited 2025 Nov 2];13(6):e1549. Available from: https://wires.onlinelibrary.wiley.com/doi/10.1002/wics.1549

Terranova, N., & Venkatakrishnan, K. (2024). Machine learning in modeling disease trajectory and treatment outcomes: An emerging enabler for model-informed precision medicine. *Clinical Pharmacology and Therapeutics*, *115*(4), 720–726.

Tiwari, A., Mishra, S., & Kuo, T. R. (2025, June 2). Current AI technologies in cancer diagnostics and treatment. [Internet]. *Molecular Cancer*, *24*(1), 159. Retrieved November 2, 2025, from https://molecular-cancer.biomedcentral.com/articles/10.1186/s12943-025-02369-9. DOI: 10.1186/s12943-025-02369-9 PMID: 40457408

Tsimberidou, A. M., Fountzilas, E., Nikanjam, M., & Kurzrock, R. (2020, June). Review of precision cancer medicine: Evolution of the treatment paradigm. *Cancer Treatment Reviews*, *86*, 102019. DOI: 10.1016/j.ctrv.2020.102019 PMID: 32251926

Wang, H., Huang, J., Fang, X., Liu, M., Fan, X., & Li, Y. (2025, April 3). Advances in next-generation sequencing (NGS) applications in drug discovery and development. [Internet]. *Expert Opinion on Drug Discovery*, *20*(4), 537–550. Retrieved November 2, 2025, from https://www.tandfonline.com/doi/full/10.1080/17460441.2025.2481262. DOI: 10.1080/17460441.2025.2481262 PMID: 40099494

Wang, T. W., Wang, C. K., Hong, J. S., Chao, H. S., Chen, Y. M., & Wu, Y. T. (2025, February 12). Deep Learning in Thoracic Oncology: Meta-Analytical Insights into Lung Nodule Early-Detection Technologies. *Cancers (Basel)*, *17*(4), 621. DOI: 10.3390/cancers17040621 PMID: 40002216

Wang, Y., Zhang, W., Liu, X., Tian, L., Li, W., He, P., Huang, S., He, F., & Pan, X. (2025, January). Artificial intelligence in precision medicine for lung cancer: A bibliometric analysis. [Internet]. *Digital Health, 11*, 20552076241300229. Retrieved November 2, 2025, from https://journals.sagepub.com/doi/10.1177/20552076241300229. DOI: 10.1177/20552076241300229 PMID: 39758259

Wiegrebe, S., Kopper, P., Sonabend, R., Bischl, B., & Bender, A. (2024, February 19). Deep learning for survival analysis: A review. [Internet]. *Artificial Intelligence Review, 57*(3), 65. Retrieved November 1, 2025, from https://link.springer.com/10.1007/s10462-023-10681-3. DOI: 10.1007/s10462-023-10681-3

Wu, Y., & Xie, L. (2025). AI-driven multi-omics integration for multi-scale predictive modeling of genotype-environment-phenotype relationships. [Internet]. *Computational and Structural Biotechnology Journal, 27*, 265–277. Retrieved November 1, 2025, from https://linkinghub.elsevier.com/retrieve/pii/S2001037024004513. DOI: 10.1016/j.csbj.2024.12.030 PMID: 39886532

Xu, X., Li, J., Zhu, Z., Zhao, L., Wang, H., Song, C., Chen, Y., Zhao, Q., Yang, J., & Pei, Y. (2024, February 25). A Comprehensive Review on Synergy of Multi-Modal Data and AI Technologies in Medical Diagnosis. [Internet]. *Bioengineering (Basel, Switzerland), 11*(3), 219. Retrieved November 2, 2025, from https://www.mdpi.com/2306-5354/11/3/219. DOI: 10.3390/bioengineering11030219 PMID: 38534493

Xuyan, H., Meng, S., Chengxing, S., Haoxuan, L., & Jianlin, Z. (2025, October). Visual-language reasoning large language models for primary care: Advancing clinical decision support through multimodal AI. [Internet]. *The Visual Computer, 41*(13), 11327–11348. Retrieved November 2, 2025, from https://link.springer.com/10.1007/s00371-025-04109-y. DOI: 10.1007/s00371-025-04109-y

Ying, J., Yang, L., Yin, J. C., Xia, G., Xing, M., Chen, X., Pang, J., Wu, Y., Bao, H., Wu, X., Shao, Y., Zhu, L., & Cheng, X. (2021, September). Additive effects of variants of unknown significance in replication repair-associated DNA polymerase genes on mutational burden and prognosis across diverse cancers. *Journal for Immunotherapy of Cancer, 9*(9), e002336. DOI: 10.1136/jitc-2021-002336 PMID: 34479923

Young, A. L., Oxtoby, N. P., Garbarino, S., Fox, N. C., Barkhof, F., Schott, J. M., & Alexander, D. C. (2024, February). Data-driven modelling of neurodegenerative disease progression: Thinking outside the black box. [Internet]. *Nature Reviews. Neuroscience, 25*(2), 111–130. Retrieved November 1, 2025, from https://www.nature.com/articles/s41583-023-00779-6. DOI: 10.1038/s41583-023-00779-6 PMID: 38191721

Yu, Y., Cai, G., Lin, R., Wang, Z., Chen, Y., Tan, Y., He, Z., Sun, Z., Ouyang, W., Yao, H., & Zhang, K. (2024, December). Multimodal data fusion AI model uncovers tumor microenvironment immunotyping heterogeneity and enhanced risk stratification of breast cancer. [Internet]. *MedComm, 5*(12), e70023. Retrieved November 1, 2025, from https://onlinelibrary.wiley.com/doi/10.1002/mco2.70023. DOI: 10.1002/mco2.70023 PMID: 39669975

Zhang, N., Yang, X., Piao, M., Xun, Z., Wang, Y., Ning, C., Zhang, X., Zhang, L., Wang, Y., Wang, S., Chao, J., Lu, Z., Yang, X., Wang, H., & Zhao, H. (2024, February 14). Biomarkers and prognostic factors of PD-1/PD-L1 inhibitor-based therapy in patients with advanced hepatocellular carcinoma. [Internet]. *Biomarker Research, 12*(1), 26. Retrieved November 1, 2025, from https://biomarkerres.biomedcentral.com/articles/10.1186/s40364-023-00535-z. DOI: 10.1186/s40364-023-00535-z PMID: 38355603

Zhang, W., Guo, Y., & Jin, Q. (2023, September 27). Radiomics and Its Feature Selection: A Review. *Symmetry*, *15*(10), 1834. DOI: 10.3390/sym15101834

Zhong, Y., Yu, L., Zhao, J., Fang, Y., Yang, Y., Wu, Z., Wang, J., & Hu, W. (2021, June 24). Clinical Implementation of Automated Treatment Planning for Rectum Intensity-Modulated Radiotherapy Using Voxel-Based Dose Prediction and Post-Optimization Strategies. [Internet]. *Frontiers in Oncology*, *18(1)*, 697995. Retrieved November 2, 2025, from https://www.frontiersin.org/articles/10.3389/fonc.2021.697995/full. DOI: 10.3389/fonc.2021.697995 PMID: 34249757

Chapter 6

Omics-Nano-CRISPR Strategies for Combating Breast Cancer in the AI Age

Ashok Kumar Sah
http://orcid.org/0000-0002-7762-4351
A' Sharqiyah University, Oman

Sakshi Patel
http://orcid.org/0009-0000-2041-9325
Vidya University, India

Rahul Kumar
http://orcid.org/0009-0001-0237-3201
Vidya University, India

Prem Shankar Mishra
http://orcid.org/0000-0003-3797-1575
Vidya University, India

Shriram Sisodiya
Vidya University, India

Aryan Kumar
Vidya University, India

Rabab H. Elshaikh
http://orcid.org/0000-0003-0973-8661
A' Sharqiyah University, Oman

ABSTRACT

Breast cancer is a major global health challenge due to its molecular complexity and diverse clinical outcomes. This chapter explores the integration of omics technologies, nanotechnology, CRISPR-based genome editing, and artificial intelligence (AI) in breast cancer research and management. Omics approaches enable molecular profiling, biomarker discovery, and tumor classification, while nanotechnology improves targeted drug delivery and diagnostic precision. CRISPR technologies support functional genomics and therapeutic target identification. AI enhances analysis of complex biological and clinical data, improving diagnosis, treatment planning, and personalized care. The convergence of these technologies offers promising opportunities to advance precision oncology and transform breast cancer diagnosis, therapy, and prognosis.

DOI: 10.4018/979-8-3373-9715-3.ch006

1. INTRODUCTION

Breast cancer is the most frequently diagnosed cancer in women worldwide and is a significant cause of cancer morbidity and mortality. Although there has been impressive progress in screening, diagnosis and treatment, the disease still remains a very complex and challenging clinical problem, because it is molecularly complex, metastatic, resistant to therapy and prone to recurrence (Smolarz et al 2022). Although treatment with conventional methods (surgery, chemotherapy, radiotherapy, hormonal therapy or targeted therapies) has improved patient survival, treatment results may differ from person to person due to the complex nature of breast tumors. Hence, precision oncology strategies that could deliver more personalised and effective therapeutic interventions are increasingly needed (Kim et al, 2025).

Omics technologies have revolutionised breast cancer research, providing a global picture of the molecular mechanisms involved in breast tumour development and progression. The generation of genomic, transcriptomic, proteomic, metabolomic, and epigenomic high throughput data sets provide insight into the genetic and molecular changes that underlie disease initiation, prognosis, and response to therapy. The use of multi-omics data has also contributed to the understanding of breast cancer heterogeneity, and the discovery of new targets for drug development (Orsini et al, 2023, Zhang et al 2024).

Concurrently, nanotechnology has also found application as a powerful tool to enhance the diagnosis and treatment of cancer. The specificity of using nanoparticles allows for targeted drug delivery, controlled drug release, improved bioavailability, and decreased systemic toxicity. Furthermore, there are numerous applications of nanomaterials in cancer imaging, biosensing, and theragnostic uses, which offer potential for greater accuracy in disease monitoring and intervention (Mosleh-Shirazi et al 2022, Shizad et al 2025).

The other important advance in modern biomedical science is the use of genome editing technologies based on CRISPR. CRISPR-Cas systems have proven to be powerful tools for genomic sequence modification, and have been used to study the functions of genes, uncover vulnerabilities in cancer cells, and develop therapeutics targeting cancer genes. CRISPR technologies are used in the study of breast cancer for correcting disease-associated mutations, silencing oncogenic pathways, and overcoming therapeutic resistance (Palaz et al 2021, Iyer et al 2025).

Artificial Intelligence (AI) technology has also gained momentum, advancing the use of these technologies in oncology. AI and machine learning algorithms are able to process large and complex biological and clinical data with efficiency and precision and can be used for the discovery of biomarkers, patient stratification, predicting treatment response and for use in decision making (Vobugari et al 2022, Nafissi et al 2024). AI is ushering in a new age of precision medicine through the power of omics data, nanotechnology-based platforms, and CRISPR-based interventions.

Even though domains like multi-omics technologies, nanotechnology, artificial intelligence, and CRISPR-based gene editing have made tremendous strides, the present literature often discusses these developments separately. Comprehensive studies that examine these complimentary technologies in a cohesive precision oncology strategy for breast cancer are very rare. Furthermore, there aren't many research that critically assess the clinical validation, ethical issues, regulatory concerns, and translational impediments that influence their successful transfer to clinical practice. By providing a thorough and balanced view of the newest technologies, as well as their clinical potential and present limitations, this chapter seeks to close these knowledge gaps.

In light of the current advancements in multi-omics technologies, nanotechnology, CRISPR-based gene editing, and artificial intelligence (AI), the chapter provides an overview of the latest developments in precision oncology for breast cancer. The following keywords were used to search the major scientific databases (PubMed, Scopus, Web of Science) for relevant literature: breast cancer, multi-omics, genomics, proteomics, nanotechnology, CRISPR, artificial intelligence, and precision oncology. To guarantee a comprehensive overview of current developments, translational issues, and future research directions, a balance of recent peer-reviewed papers, clinical investigations, systematic reviews, and landmark pieces was sought. This chapter aims to critically analyze the many contributions made to precision oncology of breast cancer by multi-omics technologies, nanotechnology, genome editing based on CRISPR (clustered regularly interspaced short palindromic repeats), and artificial intelligence.

2. OMICS TECHNOLOGIES IN BREAST CANCER RESEARCH

The development of omics technologies has transformed the study of breast cancer by providing the ability to analyze all molecular changes that contribute to the initiation, progression, metastasis and drug responsiveness of the disease. Omics platforms are generating large-scale data sets, opposite to traditional methods that focus on a distinct set of biomarkers, such as genes, transcripts, proteins, metabolites and epigenetic modifications. These technologies have greatly contributed to the knowledge of breast cancer heterogeneity and the creation of precision medicine approaches (Neagu et al 2023, Ortiz and Andrechek 2023).

Omic technologies enable the identification of molecular signatures linked to breast cancer subtypes, enabling early diagnosis, prognosis, identification of new therapeutic targets and selection of cancer treatment. In addition, the introduction of next-generation sequencing, mass spectrometry and computational biology has quickened the coming together of several omics datasets, creating a systems-level perspective of tumour biology (Rossi et al 2022, Ran et al 2025). The major disciplines in the field of omics and their contributions to breast cancer research and clinical management are discussed in the following sections.

2.1 Genomics and Transcriptomics

Genomics and transcriptomics are the basis of molecular oncology and have significantly contributed to the understanding of the biology of breast cancer. The aim of genomics is to identify changes in the DNA level, such as single nucleotide variants, insertions, deletions, copy number variations and chromosomal rearrangements. Genes that are commonly mutated in hereditary breast cancer include BRCA1, BRCA2, TP53, and PIK3CA.BRCA1, BRCA2, TP53, and PIK3CA are genes that are strongly linked to breast cancer susceptibility, progression, and therapy (Li et al 2025). Modern sequencing technologies have allowed the sequencing and characterization of genome landscapes in various breast cancer subtypes (Yin et al, 2024, Panday et al 2021)

The study of expression patterns of genes via analysis of mRNA profiles is called transcriptomics, which is done through technologies like RNA sequencing and microarrays. A number of molecular subtypes of breast cancer have been identified using gene expression signatures: Luminal A, Luminal B, HER2-positive, Triple-Negative Breast Cancer (TNBC). These molecular classifications are very useful to know about the prognosis, treatment response and disease progression (Zhong et al 2024). Genomic and transcriptomic analyses provide key insights into the genetic and regulatory processes involved in the pathogenesis of breast cancer and can help guide the development of personalized therapeutic approaches.

2.2 Proteomics and Metabolomics

The translatability of data from genomics and transcriptomics to cellular function and the physiology of tumors, is being complemented by data from proteomics and metabolomics. Proteomics is the study of the expressions, structures, interactions and post-translational modification of proteins, which are the functional molecules in cells. Proteomic studies in breast cancer have revealed abnormal signaling pathways like PI3K/AKT/mTOR, MAPK and estrogen receptor signaling which play a role in the growth, metastasis and resistance to therapy of breast cancer (Panda et al 2025, Miziak et al 2023). Protein biomarkers are also critical in diagnosis and monitoring of diseases.

Metabolomics is a discipline that aims to systematically characterize all small-molecule metabolites that are generated from metabolic processes in cells. Cancer cells change their metabolic pathways to satisfy their new and greater needs for energy and biosynthetic pathways, including increased glucose uptake, changes in amino acid metabolism, and dysregulated lipid biosynthesis (Zaripov et al 2025). Such metabolic changes might be useful as biomarkers for early diagnosis and disease monitoring. Proteomics and metabolomics research offer much more than just genomic changes and play a crucial role in comprehending the behavior of tumors and in discovering new therapeutic targets in the management of breast cancer.

2.3 Multi-Omics Integration

Breast cancer cannot be studied by a single omics approach because it is too complex and heterogeneous. Multi-omics integration is the integration of data from genomics, transcriptomics, proteomics, metabolomics, and other platforms to provide a comprehensive view of tumor biology. This integrative approach allows for linking of alterations in the genome, changes in the pattern of gene expression, protein activity, and metabolic alterations in cancerous cells (Wang et al 2023).

Highly comprehensive computational methods and artificial intelligence algorithms support the analysis of large multi-dimensional datasets and the extraction of strong biomarkers and predictive signatures. Multi-omics investigations have identified new molecular mechanisms involved in therapeutic resistance, tumor progression and metastatic dissemination (Huang et al 2019). In addition, the integrated analyses enable patient stratification, target identification and precision treatment selection based on patient-specific molecular profiles.

2.3.1 Clinical Translation and Biomarker Validation

The molecular characterization of breast cancer has significantly progressed in recent years using multi-omics technologies, which have led to the discovery of a large number of potential diagnostic, prognostic and predictive breast cancer biomarkers. But there is a need to differentiate between those used in routine clinical practice and those that are still investigational. Genomic biomarkers that have been established, such as HER2 (ERBB2) amplification, estrogen receptor (ER) and progesterone receptor (PR) testing, mutations in the genes encoding BRCA1 and BRCA2, and multigene expression assays like Oncotype DX® and MammaPrint® have all shown clinical value and have been widely used for patient stratification and targeting therapy.

As compared to this, many transcriptomic, proteomic, metabolomic and integrated multi-omics biomarkers are still in the discovery or validation process and have not yet been widely used in the clinics.

The translation of their translation is hampered by a variety of factors such as cohort variability, non-standardized analytical methodology, insufficiently reproducible across independent studies, regulatory requirements, high implementation costs, and inadequate prospective clinical validation. In conclusion, while multi-omics will continue to provide increasing insight into the biology of breast cancer, additional large-scale multicenter studies are needed to support the analytical validity, clinical value and cost-effectiveness of introducing these technologies into routine precision oncology care.

Figure 1. Illustration of Omics Technologies in Breast Cancer Research.

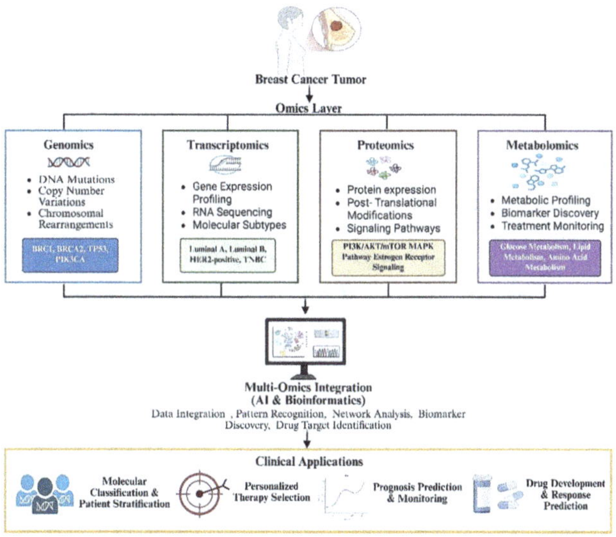

Multi-omics technologies will likely become more vital in the future to enable the development of personalized interventions and the improvement of breast cancer clinical outcomes, as the technologies evolve, shown in **table 1** (Liu et al 2024, Wu et al 222, Dai and Shen 2022)

Table 1. Major Omics Technologies and their Applications in Breast Cancer

Omics Technology	Biological Target	Common Analytical Methods	Key Applications in Breast Cancer
Genomics	DNA sequence alterations	Whole-genome sequencing (WGS), Whole-exome sequencing (WES), NGS	Detection of BRCA1/2, TP53, and PIK3CA mutations; risk assessment and target identification
Transcriptomics	RNA expression profiles	RNA-Seq, Microarray analysis	Molecular subtype classification, biomarker discovery, prognosis prediction
Proteomics	Protein abundance and modifications	Mass spectrometry, Protein arrays	Identification of dysregulated signaling pathways and therapeutic targets
Metabolomics	Cellular metabolites	LC-MS, GC-MS, NMR spectroscopy	Detection of metabolic biomarkers and monitoring treatment response
Multi-omics	Integrated molecular datasets	Bioinformatics, AI/ML-based integration	Precision oncology, biomarker discovery, patient stratification, and personalized treatment planning

3. NANOTECHNOLOGY- BASED APPROACHES FOR BREAST CANCER MANAGEMENT

The unique capability of manipulating materials at nanoscale (1–100 nm) has made nanotechnology a transformative tool in breast cancer diagnosis and treatment. The traditional strategies used to combat cancer present several challenges such as lack of solubility in the drug, lack of specificity for the tumor, rapid degradation of the drug, systemic toxicity, etc. Nanotechnology solves these problems because it allows a controlled release of a drug, better pharmacokinetics, more accumulation within a tumour, and less side effects (Tran et al 2022).

The physics and chemistry of nanomaterials can be designed to enable delivery of therapeutic agents to cancer cells, and little or no delivery to healthy tissues. Besides drug delivery, nanotechnology has made tremendous progress in cancer imaging, biosensing and real-time disease monitoring (Gupta et al, 2021). Recent developments have also resulted in the development of theragnostic platforms combining diagnostic and therapeutic functionality in the framework of a single nano system. Such innovations are helping to create more accurate, targeted and successful approaches to the treatment of breast cancer.

Figure 2. Nanotechnology-Based Approaches for Breast Cancer Diagnosis and Treatment.

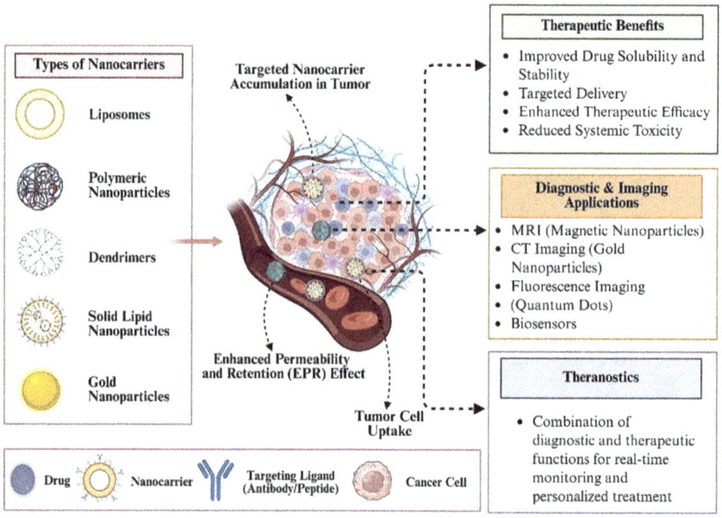

3.1 Nanocarriers for Drug Delivery

Breast cancer therapeutic nanotechnology systems include nanocarriers. These nanoscale delivery vehicles can enhance the solubility, stability, bioavailability and therapeutic efficacy of anticancer drugs and decrease the systemic toxicity. Nanocarriers are small and can be modified on their surface to accumulate preferentially in tumour tissue, which can lead to an increased concentration at the disease site, due to the enhanced permeability and retention (EPR) effect (Fatima et al 2023).

There are several nanocarrier platforms which have shown great promise for the treatment of breast cancer. Liposomes are vesicles composed of lipids and able to contain both hydrophilic and hydrophobic drugs. Polymeric nanoparticles can be engineered for controlled and sustained drug release; dendrimers can be engineered to be highly branched and can be engineered for multifunctional drug loading. The SLN's are both biocompatible and enhance the stability of the drug; the gold nanoparticles have unique optical and surface properties that can be exploited for therapeutic and diagnostic purposes (Lombardo et al, 2029). Together, these systems have the potential to boost treatment efficacy by increasing the efficiency of drug delivery, and reduce the damage to normal tissues, making them ideal candidates for next-generation breast cancer therapeutics.

3.2 Targeted Nanomedicine

The precision or targeted nanomedicine is a significant new development in precision oncology that allows targeted delivery of therapeutic agents to cancer cells. Unlike traditional chemotherapy, which also targets normal cells, targeted nanoparticles are designed to detect a certain marker that is expressed on breast cancer cells. This targeting is more selective, meaning that it is more effective at the site of the disease and less toxic and side effect producing in other parts of the body (Gote et al, 2021,Yang et al 2023).

Antibodies, peptides, aptamers or small molecule ligands which bind to tumor-associated receptors can be attached to the nanoparticles. The most frequently targeted biomarkers in breast cancer are the HER2 receptors, estrogen receptors (ER), and folate receptors. This could be, for instance, the case for the delivery of chemotherapeutic agents to HER2-overexpressing tumors by means of nanoparticles targeted to HER2, which would improve treatment efficacy, while simultaneously decreasing exposure of the body to the same agents (Dudhwala et al 2026). Likewise, the receptor specific nanocarriers enhance the uptake of drug within cells and therapeutic response. Targeting ligands can also be attached to nanocarriers to deliver other substances, such as nucleic acids, proteins, and gene editing tools (Sun et al 2022). Thus, personalized and highly efficient breast cancer treatment is achieved via the use of targeted nanomedicine, which is the new strategy

3.3 Nano- Diagnostics and Theragnostic

In addition to drug delivery, nanotechnology has been greatly utilized in the diagnosis of cancer with the development of highly sensitive imaging and biosensing platforms. Nanomaterials have special optical, magnetic and electronic properties which improve the early detection of tumors (Bhushan et al 2021, Dessale et al, 2022). The applications of the quantum dots, magnetic nanoparticles, and gold nanostructures in the field of breast cancer imaging and molecular diagnostics are broadly studied, as described in **table 2** (Torres Quintas et al 2024, Ashrafizadeh et al 2023, Kumar et al 2026).

They enhance the sensitivity and resolution of imaging modalities like magnetic resonance imaging (MRI), computed tomography (CT), fluorescence imaging, and biosensor-based detection systems. Early and accurate detection is crucial to enhancing the prognosis of the patient and treatment outcomes. In addition, the idea of theragnostic has become a great innovation in nanomedicine (Andoh et al 2024). Theragnostic nanoparticles combine diagnostic and therapeutic functions into a single entity, enabling the detection of a disease, targeted drug delivery and monitoring of treatment (Raheja et al 2026). These systems can deliver real-time data on the response to therapy and tumor progression, which enable personalized therapy decisions. In light of this, nano-diagnostics and theragnostic are urged to be of great importance to the future of precision oncology.

Table 2. Nanotechnology Platforms and their Applications in Breast Cancer Management

Nanotechnology Platform	Key Characteristics	Major Applications in Breast Cancer	Representative Examples
Liposomes	Biocompatible lipid vesicles capable of encapsulating drugs	Targeted chemotherapy and controlled drug release	Liposomal doxorubicin
Polymeric Nanoparticles	Biodegradable polymers with sustained release properties	Drug delivery and gene delivery	PLGA-based nanoparticles
Dendrimers	Highly branched nanoscale structures with multiple functional groups	Targeted drug delivery and imaging	PAMAM dendrimers
Solid Lipid Nanoparticles	Lipid-based carriers with high stability and low toxicity	Improved bioavailability of anticancer agents	Paclitaxel-loaded SLNs)
Gold Nanoparticles	Unique optical and photothermal properties	Imaging, photothermal therapy, and drug delivery	AuNP-mediated cancer therapy
Quantum Dots	Semiconductor nanocrystals with strong fluorescence	Tumor imaging and biomarker detection	Fluorescence-guided diagnostics
Magnetic Nanoparticles	Magnetic responsiveness and imaging capability	MRI contrast enhancement and targeted therapy	Iron oxide nanoparticles
Theragnostic Nanoparticles	Combined diagnostic and therapeutic functions	Real-time monitoring and personalized treatment	Multifunctional nano-theragnostic systems

3.4 Translational Challenges and Clinical Limitations of Nanotechnology

Nanotechnology has brought breath of new technologies in breast cancer diagnosis and therapy, in targeted delivery of drugs, as well as in improved cancer imaging and theragnostic applications. However, it is not yet widely adopted in clinical practice due to a number of issues. Due to the tumor vascularization, stromal composition, and heterogeneity of different patients, the enhanced permeability and retention (EPR) effect is variable, which leads to different amounts of nanoparticles accumulating in tumors and different levels of therapeutic efficacy. Also, the penetration in heterogenous tumours, manufacturing reproducibility, batch-to-batch variability, scalability of production, high costs, and high regulatory requirements, are obstacles to broad scale clinical use. Long term toxicity, biodistribution, and safety are also issues that must be investigated. While some nano-enabled platforms are approved for regulatory, a number of them still are in the preclinical or early clinical trial phase. Furthermore, few have been clearly superior in large prospective clinical trials compared with existing breast cancer treatments. Thus, ongoing clinical validation, manufacturing optimisation, and regulatory assessment are crucial for the successful incorporation of nanotechnology in the management of precision breast cancer.

4. CRISPR TECHNOLOGY FOR BREAST CANCER THERAPY

CRISPR (Clustered Regularly Interspaced Short Palindromic Repeats) technology is a groundbreaking advancement in genetic engineering and cancer research that allows for precise and efficient genome editing. CRISPR initially recognized as a bacterial adaptive immune system, has become a powerful molecular tool for the study of gene function, for the identification of therapeutic targets, and for the development of novel treatment strategies (Pont et al 2023, Banda et al 2024).

This CRISPR technology presents unparalleled opportunities for the study of the genes that regulate tumor initiation, progression, metastasis and drug resistance in research into breast cancer. In comparison with the traditional gene-editing methods, the CRISPR systems are simpler, less costly, and very versatile. They have helped to hasten the development of precision medicine strategies, which enable them to selectively modify disease-associated genes (Siva et al 2021, Ranga 2026). Moreover, the developments of the CRISPR based screening, gene regulation and cellular engineering are providing new opportunities for personalized breast cancer therapies. With ongoing research to tackle safety and delivery issues, CRISPR technology will likely be an increasingly key part of next-generation cancer treatment strategies.

4.1 Principles of CRISPR-Cas Systems

The CRISPR-Cas systems are genome editing tools that can create targeted changes in DNA sequences and can be programmed. The most popular system is CRISPR-Cas9, which comprises two elements: a guide RNA (gRNA) and the Cas9 nuclease. Cas9 recognizes a specific target sequence by binding to the guide RNA and then nuclease activity creates a double-stranded DNA break within the target sequence. The DNA repair mechanisms of cells then repair the break, producing disruption, correction, or alteration of the gene (Brezgin et al 2019, van Beljiouw et al, 2023).

Due to the versatility of CRISPR, there are multiple applications in cancer research. These include gene knockout for functional studies, gene correction of disease-causing mutations, activation of beneficial gene expression and repression of oncogenic pathways. More recently developed CRISPR variants, such as CRISPR interference (CRISPRi), CRISPR activation (CRISPRa), base editing, and prime editing, further enhance precision of editing while minimizing unintended changes to the genome (Wang et al 2022). These advances have made CRISPR a fundamental tool to gain a better understanding of the biology of breast cancer and create novel therapeutic strategies.

4.2 Identification of Therapeutic Targets

A significant utilization of the CRISPR technology in oncology is the screening of a large number of genes for novel therapeutic targets. Genes required for the survival and growth of cancer cells may be found using genome-wide CRISPR screens, which can simultaneously knock down or modify thousands of genes (Jiang et al 2020). The research provides valuable insights into the molecular mechanisms behind tumor growth and treatment resistance.

In breast cancer, CRISPR screening has led to the finding of genes that are connected to tumor development, metastatic spread, resistance to treatments and immune evasion. When treating aggressive diseases without targeted treatment, such as Triple-Negative Breast Cancer (TNBC), these evaluations may be very useful (Medina et al 2020, Maqbool et al 2023). By integrating the methods of functional genomics with CRISPR, significant pathways and genetic vulnerabilities have been discovered as prospective targets for therapeutic development. Additionally, the identification of clinically relevant biomarkers and treatment options is further enhanced by the integration of CRISPR screening data with omics data. As a result, the use of CRISPR technology is quickly emerging as a crucial instrument in the advancement of precision oncology research and target finding.

4.3 CRISPR-Based Therapeutic Strategies

Although the CRISPR system is able to modify genetic information to an unparalleled degree of accuracy, there is a lot of interest in employing it for therapeutic reasons in breast cancer. Repairing hereditary abnormalities in genes like BRCA1 and BRCA2 that are known to be vulnerable to breast and other cancers is a potential approach. Additionally, CRISPR may be used to activate tumor suppressor genes that have been deactivated in cancer or to stop the activation of oncogenes that promote tumor development (Ponti et al 2023, Yamamoto and Hirasawa 2022).

CRISPR is being used for cancer immunotherapy in addition to genome editing. T cells and natural killer (NK) cells are examples of immune cells that may be genetically altered to more effectively identify and eradicate tumor cells (Binnie et al 2021, Chehelgerdi et al 2024). Additionally, as **table 3** shows, efforts are being undertaken to create CRISPR-based techniques to achieve drug resistance and improve responses to therapeutic medications (Eskandari et al, 2024, Di Carlo and Sorrentino 2024, Alinejad et al 2023, 2023, Hou et al 2022). However, a number of obstacles must be addressed, including long-term safety issues, immunological response, off-target editing, and delivery efficiency. The standardization of CRISPR-based treatments for breast cancer will depend on future technology developments and clinical trials.

Table 3. CRISPR-Based Applications in Breast Cancer Research and Therapy

CRISPR Application	Mechanism	Breast Cancer Relevance	Potential Clinical Impact
Gene Knockout	Disruption of target gene function through DNA repair-induced mutations	Identification of oncogenes and essential survival genes	Discovery of novel drug targets
Gene Correction	Precise repair of disease-associated mutations	Correction of BRCA1/BRCA2 mutations	Prevention and treatment of hereditary breast cancer
Gene Activation (CRISPRa)	Upregulation of endogenous gene expression	Restoration of protective cellular pathways	Enhanced tumor suppression
Gene Repression (CRISPRi)	Silencing of target gene transcription	Inhibition of oncogenic signaling pathways	Reduction of tumor growth and progression
Genome-Wide CRISPR Screening	Systematic functional analysis of thousands of genes	Identification of genes involved in metastasis and drug resistance	Accelerated therapeutic target discovery
Immune Cell Engineering	Genetic modification of T cells or NK cells	Enhanced recognition and killing of tumor cells	Improved cancer immunotherapy
Base and Prime Editing	Precise nucleotide modification without double-strand breaks	Correction of pathogenic mutations with reduced off-target effects	Safer next-generation gene therapies

4.4 Challenges and Clinical Translation of CRISPR Technology

The technology of CRISPR-Cas has transformed the field of functional genomics and holds promise for precision oncology by enabling precise genome editing and the identification of therapeutic targets. In the field of breast cancer, however, therapeutic use of CRISPR is still very much in the preclinical stages, such as in vitro experiments and animal models. There are a number of biological and technical barriers to clinical translation. Targeting and delivery of CRISPR components to heterogeneous solid tumors is still challenging because of intratumoral heterogeneity and the complex tumor microenvironment.

Other issues to consider are off-target changes, editing efficiency that may differ from cell to cell, immune reactions, and the lack of long-term safety information for permanent genome editing. Extensive safety, efficacy, manufacturing quality and ethical compliance requirements are also required for regulatory approval. While CRISPR has tremendous potential to help progress breast cancer research, there is a significant translational hurdle. Before CRISPR-based therapies can be routinely used in precision breast cancer care, the delivery system would need to be further optimized, the specificity of the edited cells would need to be improved, long-term safety assessments would need to be conducted and clinical studies would need to be more thorough.

5. ARTIFICIAL INTELLIGENCE IN BREAST CANCER PRECISION MEDICINE

AI is playing a pivotal role in modern healthcare by making the analysis of big and complex data possible that is beyond the computational power of the traditional methods used. AI is used in breast cancer research and clinical care, including machine learning (ML), deep learning (DL), natural language processing, and predictive analytics, all of which enable data-driven decision-making. As more medical imaging, omics, health and clinical data becomes available, AI-driven techniques have the potential to enhance disease diagnosis, prognosis, and drug selection (Garrone and La Porta 2024, Sarhangi et al 2022)

AI algorithms have the potential to uncover patterns that go unnoticed, create predictive models, and assist physicians in reaching more informed and tailored choices. Perhaps most significantly, AI is also a bridge that merges molecular, clinical and imaging data, which is driving the advancement of precision oncology (Carini and Seyhan 2024, Sohrabei et al 2024). In the era of data-driven breast cancer management, AI is poised to become a key player in improving patient outcomes, minimizing diagnostic missteps, and driving faster advancements in treatment.

5.1 AI-Assisted Diagnosis

Early and accurate diagnosis is crucial to better breast cancer outcomes and AI has proved to be a helpful aid in improving diagnosis accuracy. AI and machine learning algorithms can process extensive amounts of medical imagery and recognize subtle abnormalities that can be hard for human eyes to spot reliably (Tiwari et al 2025). Their systems are now being used for mammography, ultrasound and magnetic resonance imaging (MRI), and histopathology image analysis.

Deep convolutional neural networks (CNNs), especially, have proven to be highly effective in breast lesion detection, breast tumor classification, and benign versus malignant classification. Using AI diagnostics systems can increase sensitivity and specificity and decrease false-positive and false-negative reading rates (Zhu et al 2023, Chougrad et al 2018). What's more, AI can assist radiologists and pathologists in automating image interpretation and flagging potentially suspicious samples for additional analysis. AI isn't a substitute for a clinician – it's a tool that supports them.AI is not meant to replace clinicians – instead, it should complement their work by increasing efficiency and consistency in diagnosis. With the continuous growth in imaging data, AI-powered diagnostics will play a crucial role in advancing earlier detection and better patient outcomes.

5.2 AI for Omics Data Analysis

With the development of genomics, transcriptomics, proteomics and metabolomics, biological data has been increasing at a rapid pace and needs advanced analytical methods. AI offers robust computational capabilities to glean valuable insights from these multidimensional datasets. Machine learning algorithms can detect highly complex molecular patterns, correlations and predictive signatures that may not be discernible by traditional statistical approaches (Ali and Mohammed 2024).

For breast cancer research, AI has continued to play an important role in the discovery of biomarkers, classification of molecular subtypes, identification of drug targets, and patient stratification. Supervised and unsupervised learning methods allow dealing with heterogeneous sets of data and assist in discovering networks of molecules associated with the disease (Batool et al 2024, Cheng et al 2026). Moreover, AI-driven integration of multi-omics data gives a global picture of the tumor biology and therapeutic response. These methods are useful in identifying novel diagnostic markers and personalized treatment strategies. AI has the potential to catalyze the adoption of precision medicine in breast cancer care by rapidly converting large-scale omics data into clinically relevant information.

5.3 Predictive Modeling and Personalized Treatment

The creation of predictive models to guide personalized oncology treatment is one of the most potential applications of AI. Not only can AI algorithms harness a variety of data sources like clinical records, imaging results, genomic information, treatment histories, and patient outcomes, but it can combine them all to produce customized predictions. These models help clinicians make the best use of therapies with specific patient populations (Jiang et al 2023).

AI-powered predictive systems are applied in breast cancer to help identify which patients will respond to treatments, determine the risk of recurrence, predict survival, and determine the likelihood of treatment response with targeted therapies. As more patient information is fed into the deep learning models, they have the ability to continually improve their accuracy, which will allow for adaptive and data-driven decision making in the clinical setting. AI plays a role in drug discovery and optimization, where it helps predict the therapeutic effects and side effects of drugs. These capabilities enable transitioning from the generic treatment strategy of current medicine to precision oncology, in which medicines are designed specifically for each patient based on their individual molecular and clinical profile, as presented in **table 4** (Modi et al 2020, Ming et al 2019, Doudican et al 2015, Dhillon et al 2023).

Table 4. Applications of Artificial Intelligence in Breast Cancer Precision Medicine

AI Application Area	Data Source	AI Techniques Used	Clinical Significance
Imaging-Based Diagnosis	Mammography, MRI, Ultrasound, Histopathology	Deep Learning, Convolutional Neural Networks (CNNs)	Improved detection of tumors and reduction of diagnostic errors
Biomarker Discovery	Genomics, Transcriptomics, Proteomics, Metabolomics	Machine Learning, Feature Selection Algorithms	Identification of novel diagnostic and prognostic biomarkers
Molecular Subtype Classification	Multi-omics datasets	Supervised and Unsupervised Learning	Accurate classification of breast cancer subtypes for treatment planning
Drug Target Prediction	Genomic and pathway datasets	Network Analysis, Deep Learning	Discovery of potential therapeutic targets
Patient Stratification	Clinical and molecular data	Clustering Algorithms, Predictive Modeling	Selection of patient-specific therapeutic strategies
Treatment Response Prediction	Clinical, imaging, and genomic data	Machine Learning and Deep Neural Networks	Prediction of therapy effectiveness and resistance
Survival and Recurrence Analysis	Longitudinal patient datasets	Predictive Analytics and Risk Modeling	Improved prognosis estimation and follow-up planning
AI-Driven Precision Oncology	Integrated multi-modal datasets	Multi-omics AI Platforms	Personalized treatment recommendations and clinical decision support

5.4 Challenges and Clinical Implementation of Artificial Intelligence

AI has proven to have great promise for improving breast cancer diagnosis, molecular profiling, treatment planning, and clinical decision support by analyzing imaging, pathological, and multi-omics data. It has, however, not yet been widely adopted in clinical practice due to a number of difficulties. The generalizability and training-data bias are inherent shortcomings of many AI models that are trained on retrospective data from high-income countries and specialized centers. Another challenge is the lack of explainability of deep learning models commonly known as the "black-box" problem which makes it more difficult for clinicians to have confidence and adopt deep learning models.

Moreover, there is a need for standardizing the data formats and interoperability when incorporating AI into hospital information systems and clinical processes. Agencies like FDA/EMA require strong safety, efficacy and reproducibility evidence for approval. Despite the reported success of retrospective studies, there is still limited clinical validation in prospective multicenter trials, warranting further studies to establish the benefits of this approach on patient outcomes, the development of transparent algorithms, as well as regulatory scrutiny.

6. CONVERGENCE OF OMICS, NANOTECHNOLOGY, CRISPR, AND AI

Breast cancer treatment of the future will involve the use of several emerging technologies, not just one. Omics technologies enable a comprehensive understanding of tumor biology; nanotechnology helps to deliver therapeutics to the target; CRISPR helps to perform precise genome editing; and artificial intelligence helps to integrate data to make clinical decisions (Mahato 2025, Sharma and Shing 2026). These technologies, taken together, form a synergic context for the field of precision oncology, which allows a more in-depth understanding of disease mechanisms and more personalized therapeutic approaches. These disciplines are combined to break from the one size fits all paradigm of interventions to the adaptive and patient-specific.

In addition, progress in the fields of computational biology and nanomedicine is helping to speed up the process of taking laboratory findings and applying them in the clinical world. The integration of Omics–Nano–CRISPR–AI technologies will undoubtedly transform the landscape of breast cancer diagnostics, prognosis, and therapy, ultimately enhancing the outcomes of breast cancer treatment and the quality of life for patients (Qureshi et al 2024, Sonalisha et al 2026).

Figure 3. Convergence of Omics Technologies, Nanotechnology, CRISPR Genome Editing, and Artificial Intelligence in Breast Cancer Precision Oncology.

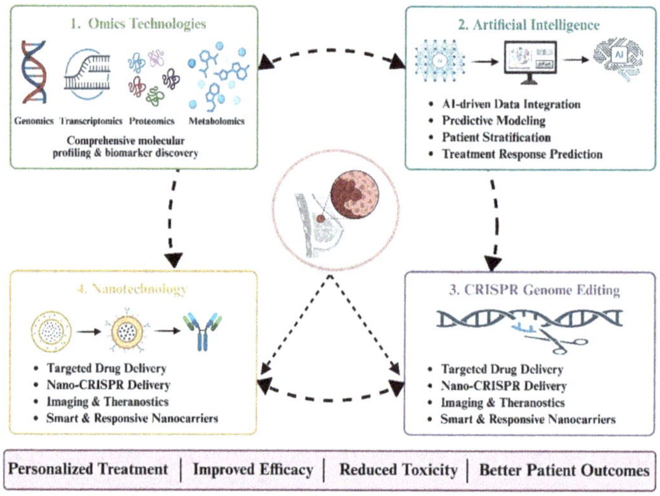

6.1 AI-Guided Multi-Omics Analysis

The explosion of genomics, transcriptomics, proteomics and metabolomics data has opened up unparalleled avenues to the understanding of breast cancer biology. But these datasets are complex and enormous and need sophisticated analytic capabilities that can uncover biological information. AI has proven to be a game-changer for integrating and interpreting multi-omics data (Kaur et al 2024).

Combination of machine learning and deep learning algorithms can detect hidden patterns, molecular signatures and interactions at different biological layers. These strategies enable identification of disease progression, treatment response and patient prognosis related biomarkers. AI-driven multi-omics analysis further enhances the understanding of tumor heterogeneity, allowing scientists to identify molecular subpopulations of breast cancer patients (Pant et al 2026, Kaur et al 2021). These insights are used for patient stratification and designing tailored therapeutic approaches. AI's integration of computational intelligence with expanding molecular data allows for a more precise identification of targets and aids in the development of more effective precision medicine strategies for breast cancer treatment.

6.2 Nano-Enabled CRISPR Delivery

While genome editing using CRISPR has great potential, delivery of CRISPR components to the desired cell is still a significant hurdle. One of the strategies that are proving to be a viable solution to these drawbacks is nanotechnology, which offers biocompatible and targeted delivery systems (Dubey et al 2022). The CRISPR-associated proteins and the guide RNAs or messenger RNAs can be wrapped in nanoparticles which can be delivered directly to tumor cells (Fang et al 2022).

There are several benefits for using nano-enabled delivery systems such as: protection of the CRISPR components from enzymatic degradation, increased cellular uptake, reduced immunogenicity, and efficient tumor-specific targeting. Currently, different nanocarriers including lipid nanoparticles, polymeric nanoparticles, and gold nanoparticles are explored for delivering CRISPR into cancer therapy (Hii and Wu 2024). Such systems can enhance the editing efficiency and reduce off-target effects and systemic toxicity. For breast cancer, the potential applications of nano-CRISPR platforms are highly promising in terms of disease-associated mutations correction, oncogene silencing and therapeutic precision (Garcia-Fernandez et al 2022, Hermain et al 2026). The potential for future gene-based cancer therapies becomes more than ever nano-enabled CRISPR systems.

6.3 Intelligent Precision Oncology Platforms

The incorporation of omics technologies, nanomedicine, CRISPR genome editing and artificial intelligence (AI) has allowed for the emergence of the concept of intelligent precision oncology platforms. Combining all of these technologies has led to the concept of intelligent precision oncology platforms. Their goal is to integrate molecular profiling, computational prediction, targeted delivery and gene editing tools into one platform to facilitate personalized cancer therapy (Gharti et al 2026, Bakhshi et al 2026).

In this kind of system, the patient individual molecular information generated by omics technologies can be analysed by AI algorithms to discover therapeutic targets and predicting treatment response. These delivery platforms are then based on nanotechnology and allow for the exact delivery of drugs or parts of the CRISPR system to tumour tissues, as indicated in **table 5** (Sabit et al 2025, Chen et al 2025, Srivastav et al 2025). Long-term follow-up with imaging, biosensors and computer modelling further aids in adaptive therapeutic decision-making. This holistic approach can maximize effectiveness of treatment and reduce toxicity and resistance. While there are still a number of technical and regulatory hurdles to overcome, intelligent precision oncology platforms hold great promise as a future paradigm of breast cancer management, providing highly personalized and data-driven therapeutic interventions (Khan et al 2025).

Table 5. Convergence of Omics, Nanotechnology, CRISPR, and AI in Breast Cancer Precision Oncology

Integrated Technology Component	Primary Function	Contribution to Breast Cancer Management	Future Potential
Multi-Omics Technologies	Molecular profiling of tumors	Identification of biomarkers, molecular subtypes, and therapeutic targets	Comprehensive patient-specific disease characterization
Artificial Intelligence	Data integration and predictive analytics	Biomarker discovery, patient stratification, treatment prediction	Real-time precision medicine and clinical decision support
Nanotechnology	Targeted delivery and diagnostics	Enhanced drug delivery, imaging, and reduced systemic toxicity	Smart and responsive nanomedicine platforms
CRISPR Genome Editing	Precise genetic modification	Gene correction, oncogene silencing, and functional genomics studies	Personalized gene-based therapies
AI-Guided Multi-Omics Platforms	Integration of heterogeneous biological data	Improved understanding of tumor heterogeneity and disease progression	Digital twin models for individualized treatment planning
Nano-CRISPR Systems	Delivery of genome-editing components	Enhanced editing efficiency and tumor-specific targeting	Clinical translation of precision genome editing (Lino et al., 2018)
Intelligent Precision Oncology Platforms	Unified therapeutic ecosystem	Combination of diagnostics, prediction, and therapy	Fully integrated adaptive cancer treatment systems (Ching et al., 2018)

6.4 Technological Promise Versus Clinical Reality

The combination of multi-omics technologies, nanotechnology, CRISPR-based genome editing, and artificial intelligence holds great promise for precision breast cancer research, as it contributes to improved molecular understanding, biomarker identification, therapeutic target selection, and clinical decision-making. Despite major technological developments, however, they are not in widespread use in day-to-day clinical practice, with the vast majority of the integrated platforms available only in pre-clinical studies, retrospective studies or in early phase clinical trials. The lack of standardization of multi-omics workflows, clinical validation and cost-effectiveness are some of the translational barriers that hinder broader use.

Training-data bias and lack of explainability are challenges for AI models, and manufacturing, targeted delivery, long-term safety, and regulatory approval are hurdles for nano-enabled therapies and CRISPR-based interventions. As a result, there are limited prospective multicenter studies to show significant improvements in patient outcomes. Hence, currently integrated Omics–Nano–CRISPR–AI platforms should be considered as evolutionary, not revolutionary technologies until they are further clinically validated and regulated.

7. CHALLENGES AND FUTURE PERSPECTIVES

Combination of omics technologies, nanotechnology, CRISPR genome editing and AI has provided unprecedented opportunities in breast cancer precision medicine. Although there have been great scientific advances, there are still a number of technical, clinical, ethical and regulatory hurdles preventing their more widespread use. It is crucial that these limitations are overcome to bring laboratory discoveries to safe, effective and clinically viable solutions (Alamri et al 2025, Ali 2023).

Moreover, future innovations will prioritize advancing data integration, boosting the precision of genome editing, designing smarter nanotherapeutics, and using AI to make predictive and personalized oncology. The barriers that presently stand in the way of the Omics–Nano–CRISPR–AI convergence are discussed, and current research directions are suggested that may aid in the future of breast cancer management (Vatta 2025, Gupta et al 2026).

All of these technologies have shown great promise, but there are several barriers to their clinical translation. The analyses that are based on "omics" technologies can yield vast amounts of data, which demands specialized expertise and expensive instrumentation and computational resources. Although the fusion of genomics, transcriptomics, proteomics and metabolomics data is still a challenge, the data formats, quality, and analysis methods do differ.

There are concerns about off-target genome editing effects, long term effects, and unintended genetic modification effects that are associated with CRISPR based therapies. Another challenge is efficient and tumor targeted delivery of the CRISPR components. Likewise, problems with the large-scale manufacture, reproducibility, pharmacokinetics, and regulatory approval of nanomedicine are also problematic. However, constructing dependable predictive models for AI demands large-scale, diverse, and high-quality datasets, which can be hard to access because of privacy issues and data-sharing constraints.

Ethical issues also complicate the use of gene editing and AI for health care, as illustrated in **table 6** (Parvi et al 2025, Alshorman et al 2026, Dhara et al2026) Therefore, addressing these scientific, technological and regulatory hurdles will be essential to realizing the true promise of precision oncology.

Future studies will likely aim to construct more comprehensive, precise, and clinically relevant precision oncology platforms. A potentially fruitful approach is the use of single-cell and spatial multi-omics technologies to gain insights into cellular heterogeneity and interactions with the tumor microenvironment that are otherwise impossible to obtain. All these methods have the potential to greatly facilitate biomarker discovery and therapeutic target identification.

AI will become increasingly relevant to the fields of drug discovery, prediction of drug treatment and clinical decision support. Real-time personalised treatment recommendations can be made possible by advanced machine learning models that are able to combine multi-modal data sets. At the same time, new generation of CRISPR technologies such as base editing and prime editing, plus epigenome editing technologies, will be expected to enhance the precision of the editing with reduced off-target effects (Villiger et al 2024, Saber sichani et al 2023).

Research on nanotechnology is progressing towards creating smart and stimuli-responsive nanocarriers that selectively release the therapeutics in response to signals from tumors. A close collaboration between molecular biologists, oncologists, bioengineers, data scientists and regulatory authorities will be needed for successful clinical translation of integrated Omics–Nano–CRISPR platforms (Suryawanshi et al 2025, Jayachandran 2026). These multi-disciplinary approaches will be critical to realizing breakthrough technologies into effective and accessible breast cancer therapies.

Table 6. Challenges and Future Research Directions for Omics- Nano- CRISPR-AI- Based Breast Cancer Precision Medicine

Technology Area	Current Challenges	Future Research Directions	Expected Impact
Omics Technologies	High costs, data complexity, limited standardization	Single-cell and spatial multi-omics integration	Improved understanding of tumor heterogeneity
Artificial Intelligence	Limited training datasets, algorithm bias, data privacy concerns	Explainable AI and multi-modal predictive models	Enhanced clinical decision-making and personalized treatment
CRISPR Genome Editing	Off-target effects, delivery challenges, safety concerns	Base editing, prime editing, and epigenome editing	Safer and more precise gene therapies
Nanotechnology	Regulatory hurdles, scalability, reproducibility issues	Smart and stimuli-responsive nanocarriers	Improved targeted delivery and reduced toxicity
Multi-Omics Integration	Computational complexity and heterogeneous datasets	AI-driven data fusion platforms	Better biomarker discovery and patient stratification
Clinical Translation	Limited clinical validation and regulatory uncertainty	Multi-center trials and standardized protocols	Faster adoption of precision oncology technologies
Interdisciplinary Collaboration	Fragmented expertise across disciplines	Integrated research and regulatory frameworks	Accelerated innovation and implementation in healthcare

7.1 Clinical Implementation Challenges

While there has been significant progress in integrating multi-omics, nanotechnology, CRISPR and artificial intelligence in the clinical management of breast cancer, several challenges remain to ensure such technologies are routinely used in the clinic. There is still a long way to go with the standardization of laboratory procedures, data analyses, and data interpretation, especially when dealing with multi-omics analyses that are produced on different platforms and in different laboratories.

In addition, molecular profiling is expensive, requires specific facilities, and requires multidisciplinary skills, which all these factors limit the widespread use. Moreover, many new technologies have been investigated in small cross-sectional or pilot studies, and the vast majority of prospective multicenter trials have failed to consistently show improvements in patient survival, quality of life or long-term clinical outcomes. Solving these translation hurdles will be a matter of standardizing methods, strong clinical validation, cost-effectiveness analyses and a close partnership between industry partners, regulatory agencies, and clinicians and researchers.

7.2 Ethical, Regulatory, and Data Governance Considerations

There are several ethical, legal, and regulatory considerations to address and discuss regarding the use of artificial intelligence, multi-omics technologies, and genome editing in precision oncology. The amount of genomic and clinical information that can be obtained from multi-omics analyses is vast, necessitating the need for strong consent procedures, privacy safeguards, secure data storage and responsible data sharing. Algorithmic bias can also occur due to the use of datasets that do not adequately reflect the diversity of patients which could lead to inequitable clinical decision-making with artificial intelligence algorithms. Transparency and explainability of algorithms are consequently critical for establishing trust in them, both with clinicians and patients.

Furthermore, AI systems for clinical decision-making, nanomedicine platforms and CRISPR therapeutics need to be thoroughly assessed by regulatory authorities before their broad clinical adoption. There are further ethical issues related to off-target effects, long-term safety and clinical regulation of genome-editing technologies. To provide a safe, ethical, and equitable use of emerging precision oncology technologies, continued development of internationally harmonized regulatory frameworks and responsible governance will be vital.

7.3 Health Equity and Global Accessibility

Precision oncology holds tremendous potential to revolutionize the treatment of breast cancer, but access to these cutting-edge technologies is limited and unevenly distributed in health care systems. The advanced nanotechnology-based therapies, comprehensive genomic sequencing, multi-omics profiling and advanced technical capacity are expensive and inaccessible in LMICs and resource-limited settings. If precision oncology technologies are only available in speciality academic centres, these differences could exacerbate the already significant inequalities in access to healthcare.

Future implementation strategies should focus on the development of low cost diagnostic platforms, standardized clinical protocols, international cooperation, staff training, and resource distribution to make sure advances in precision medicine are available to a diverse patient population, irrespective of their geographic or socioeconomic status

7.4 Future Research Priorities

Aiming to facilitate the widespread and sustainable use of precision oncology technologies in breast cancer care, future research should aim to produce high quality prospective clinical evidence. Large multicenter clinical trials, standardizing multi-omics workflows; creating explainable and generalizable artificial intelligence algorithms, optimizing the design and manufacturing of nanocarriers; and enhancing the delivery systems for CRISPR-based therapeutics are among the priorities. Successful clinical translation will also require both long term safety monitoring and regulatory harmonization and cost effectiveness evaluations. Moreover, the ability of the science and technology fields to work together – oncologists, molecular biologists, bioinformaticians, engineers, regulatory authorities, and policymakers – will be essential to making the promise of promising technological advances into safe and effective, equitable patient-centered care.

8. CONCLUSION

Breast cancer is still one of the most complicated and heterogeneous cancers and requires precision oncology interventions that can target its unique molecular features. The development of multi-omics technologies, nanotechnology, CRISPR-based genome editing, and artificial intelligence techniques, have led to a better understanding of breast cancer biology and generated novel opportunities to identify biomarkers, classify molecularly, deliver targeted drugs, engineer the genome, and make decisions using data in the clinic. All these technologies would offer a potential platform for more individualised diagnostic and therapeutic approaches.

Despite the progress made, however, it is essential to acknowledge that numerous precision oncology technologies are still in the research, development and clinical proof-of-concept phases. Although some multi-omics signatures, AI-assisted clinical tools, nano-enabled therapeutic platforms and CRISPR-based interventions are still in their investigational or early clinical testing stages, there are several that have been successfully integrated into clinical breast cancer care. Issues of analytical standardization, prospective clinical validation, manufacturing scalability, long-term safety, regulatory approval, data governance, algorithmic bias and equitable access still present a barrier to the widespread use.

The synergy of omics technologies, nanotechnology, CRISPR and artificial intelligence are significant strides in the direction of more precise and individualized breast cancer treatment. The present contribution of these integrated technologies, however, should be considered evolutionary and not revolutionary for their long-term clinical significance to depend on scientific rigor in their validation and demonstration of meaningful improvements in patient outcomes. Multidisciplinary collaboration, standardized methods, large prospective clinical studies, and responsible regulatory oversight are all necessary to ensure that progress toward safe, effective and equitable implementation. These new technologies will need to be further developed, and supported by strong clinical data, to bring these innovations into mainstream precision oncology care and enhance patient outcomes in breast cancer.

REFERENCES

Alamri, A. M., Assiri, A. A., Khan, B., & Khan, N. U. (2025). Next-generation oncology: Integrative therapeutic frontiers at the crossroads of precision genomics, immuno-engineering, and tumor micro-environment modulation. *Medical Oncology*, *42*(11), 482. DOI: 10.1007/s12032-025-03042-3 PMID: 40974488

Ali, A. M., & Mohammed, M. A. (2024). A comprehensive review of artificial intelligence approaches in omics data processing: Evaluating progress and challenges. *Int. J. Math. Stat. Comput. Sci.*, *2*, 114–167.

Ali, H. (2023). Artificial intelligence in multi-omics data integration: Advancing precision medicine, biomarker discovery and genomic-driven disease interventions. *Int J Sci Res Arch*, *8*(1), 1012–1030. DOI: 10.30574/ijsra.2023.8.1.0189

Alinejad, T., Modarressi, S., Sadri, Z., Hao, Z., & Chen, C. S. (2023). Diagnostic applications and therapeutic option of Cascade CRISPR/Cas in the modulation of miRNA in diverse cancers: Promises and obstacles. *Journal of Cancer Research and Clinical Oncology*, *149*(12), 9557–9575. DOI: 10.1007/s00432-023-04747-6 PMID: 37222810

Alshorman, J., Mehran, M. J., Miyanda Tembo, K., Bolideei, M., & Wang, Y. (2026). Applications of genome editing technologies in the treatment of human diseases. *Gene Therapy*, ●●●, 1–29. DOI: 10.1038/s41434-026-00627-z PMID: 42298089

Andoh, V., Ocansey, D. K. W., Naveed, H., Wang, N., Chen, L., Chen, K., & Mao, F. (2024). The advancing role of nanocomposites in cancer diagnosis and treatment. *International Journal of Nanomedicine*, *19*, 6099–6126. DOI: 10.2147/IJN.S471360 PMID: 38911500

Ashrafizadeh, M., Zarrabi, A., Bigham, A., Taheriazam, A., Saghari, Y., Mirzaei, S., Hashemi, M., Hushmandi, K., Karimi-Maleh, H., Nazarzadeh Zare, E., Sharifi, E., Ertas, Y. N., Rabiee, N., Sethi, G., & Shen, M. (2023, November). (Nano) platforms in breast cancer therapy: Drug/gene delivery, advanced nanocarriers and immunotherapy. *Medicinal Research Reviews*, *43*(6), 2115–2176. DOI: 10.1002/med.21971 PMID: 37165896

Bakhshi, A., Bakhshi, M., Hosseine, M., Hasannezhad, H., Rahdar, A., Fosso-Kankeu, E., & Pandey, S. (2026). Integrating Artificial Intelligence and Quantum Computing with Omics Data for Cancer Therapy and Diagnosis. *Intelligent Nanocarriers: AI-Based Tools in Cancer Diagnosis and Therapy*, 413-524.

Banda, A., Impomeni, O., Singh, A., Baloch, A. R., Hu, W., & Jaijyan, D. K. (2024). Precision in Action: The Role of Clustered Regularly Interspaced Short Palindromic Repeats/Cas in Gene Therapies. *Vaccines*, *12*(6), 636. Advance online publication. DOI: 10.3390/vaccines12060636 PMID: 38932365

Batool, Z., Kamal, M. A., & Shen, B. (2024). Advancements in triple-negative breast cancer sub-typing, diagnosis and treatment with assistance of artificial intelligence : A focused review. *Journal of Cancer Research and Clinical Oncology*, *150*(8), 383. DOI: 10.1007/s00432-024-05903-2 PMID: 39103624

Bhushan, A., Gonsalves, A., & Menon, J. U. (2021). Current State of Breast Cancer Diagnosis, Treatment, and Theranostics. *Pharmaceutics*, *13*(5), 723. Advance online publication. DOI: 10.3390/pharmaceutics13050723 PMID: 34069059

Binnie, A., Fernandes, E., Almeida-Lousada, H., de Mello, R. A., & Castelo-Branco, P. (2021). CRISPR-based strategies in infectious disease diagnosis and therapy. *Infection, 49*(3), 377–385. DOI: 10.1007/s15010-020-01554-w PMID: 33393066

Brezgin, S., Kostyusheva, A., Kostyushev, D., & Chulanov, V. (2019). Dead Cas Systems: Types, Principles, and Applications. *International Journal of Molecular Sciences, 20*(23), 6041. Advance online publication. DOI: 10.3390/ijms20236041 PMID: 31801211

Carini, C., & Seyhan, A. A. (2024). Tribulations and future opportunities for artificial intelligence in precision medicine. *Journal of Translational Medicine, 22*(1), 411. DOI: 10.1186/s12967-024-05067-0 PMID: 38702711

Chehelgerdi, M., Chehelgerdi, M., Khorramian-Ghahfarokhi, M., Shafieizadeh, M., Mahmoudi, E., Eskandari, F., Rashidi, M., Arshi, A., & Mokhtari-Farsani, A. (2024). Comprehensive review of CRISPR-based gene editing: Mechanisms, challenges, and applications in cancer therapy. *Molecular Cancer, 23*(1), 9. DOI: 10.1186/s12943-023-01925-5 PMID: 38195537

Chen, Y.-M., Hsiao, T.-H., Lin, C.-H., & Fann, Y. C. (2025). Unlocking precision medicine: Clinical applications of integrating health records, genetics, and immunology through artificial intelligence. *Journal of Biomedical Science, 32*(1), 16. DOI: 10.1186/s12929-024-01110-w PMID: 39915780

Y. Cheng, J. Kong, X. Liu, and S. Li, "Recent Advances and Emerging Directions in Machine Learning-Based Breast Cancer Drug Discovery: A Comprehensive Review," *Breast Cancer Targets Ther.*, vol. 18, no. null, p. 586786, Dec. 2026, .DOI: 10.2147/BCTT.S586786

Chougrad, H., Zouaki, H., & Alheyane, O. (2018). Deep Convolutional Neural Networks for breast cancer screening. *Computer Methods and Programs in Biomedicine, 157*, 19–30. DOI: 10.1016/j.cmpb.2018.01.011 PMID: 29477427

Dai, X., & Shen, L. (2022). Advances and trends in omics technology development. *Frontiers in Medicine, 9*, 911861. DOI: 10.3389/fmed.2022.911861 PMID: 35860739

Dessale, M., Mengistu, G., & Mengist, H. M. (2022). Nanotechnology: A promising approach for cancer diagnosis, therapeutics and theragnosis. *International Journal of Nanomedicine, 17*, 3735–3749. DOI: 10.2147/IJN.S378074 PMID: 36051353

Dhara, C., Sah, H., Gantayat, S., Rajput, M., & Mishra, S. (2026). CRISPR/Cas9 in perspective: Evaluating efficacy, delivery methods, and ethical challenges in genome editing. *Molecular Biology Reports, 53*(1), 719. DOI: 10.1007/s11033-026-11815-w PMID: 42090053

Dhillon, A., Singh, A., & Bhalla, V. K. (2023). A Systematic Review on Biomarker Identification for Cancer Diagnosis and Prognosis in Multi-omics: From Computational Needs to Machine Learning and Deep Learning. *Archives of Computational Methods in Engineering, 30*(2), 917–949. DOI: 10.1007/s11831-022-09821-9

Di Carlo, E., & Sorrentino, C. (2024). State of the art CRISPR-based strategies for cancer diagnostics and treatment. *Biomarker Research, 12*(1), 156. DOI: 10.1186/s40364-024-00701-x PMID: 39696697

Doudican, N. A., Kumar, A., Singh, N. K., Nair, P. R., Lala, D. A., Basu, K., Talawdekar, A. A., Sultana, Z., Tiwari, K., Tyagi, A., Abbasi, T., Vali, S., Vij, R., Fiala, M., King, J., Perle, M. A., & Mazumder, A. (2015). Personalization of cancer treatment using predictive simulation. *Journal of Translational Medicine*, *13*(1), 43. DOI: 10.1186/s12967-015-0399-y PMID: 25638213

Dubey, A. K., Gupta, V. K., Kujawska, M., Orive, G., Kim, N. Y., Li, C. Z., & Kaushik, A. (2022). Exploring nano-enabled CRISPR-Cas-powered strategies for efficient diagnostics and treatment of infectious diseases. *Journal of Nanostructure in Chemistry*, *12*(5), 833–864. DOI: 10.1007/s40097-022-00472-7 PMID: 35194511

Dudhwala, Y. D., Mehta, R. K., Kapoor, D. U., Shah, D. P., Shah, K. V., & Saini, V. (2026). Ligand-functionalized nanocarriers for targeted breast cancer therapy: Advances in receptor-mediated drug delivery. *Heal. Nanotechnol.*, *2*(1), 7. DOI: 10.1186/s44301-026-00029-5

Eskandari, F., Aali, M., Hadisadegh, S. N., & Azadeh, M. (2024). Advances in breast cancer research using CRISPR/Cas9 system. *Nano Select*, *5*(10), 2400015. DOI: 10.1002/nano.202400015

Fang, T., Cao, X., Ibnat, M., & Chen, G. (2022). Stimuli-responsive nanoformulations for CRISPR-Cas9 genome editing. *Journal of Nanobiotechnology*, *20*(1), 354. DOI: 10.1186/s12951-022-01570-y PMID: 35918694

Fatima Qizilbash, F., Sartaj, A., Qamar, Z., Kumar, S., Imran, M., Mohammed, Y., Ali, J., Baboota, S., & Ali, A. (2023, September 14). Nanotechnology revolutionises breast cancer treatment: Harnessing lipid-based nanocarriers to combat cancer cells. *Journal of Drug Targeting*, *31*(8), 794–816. DOI: 10.1080/1061186X.2023.2243403 PMID: 37525966

Garcia-Fernandez, A., Vivo-Llorca, G., Sancho, M., García-Jareño, A. B., Ramirez-Jimenez, L., Barber-Cano, E., Murguía, J., Orzáez, M., Sancenón, F., & Martinez-Manez, R. (2022). Nanodevices for the efficient codelivery of CRISPR-Cas9 editing machinery and an entrapped cargo: A proposal for dual anti-inflammatory therapy. *Pharmaceutics*, *14*(7), 1495. DOI: 10.3390/pharmaceutics14071495 PMID: 35890389

Garrone, O., & La Porta, C. A. M. (2024). Artificial Intelligence for Precision Oncology of Triple-Negative Breast Cancer: Learning from Melanoma. *Cancers (Basel)*, *16*(4), 692. Advance online publication. DOI: 10.3390/cancers16040692 PMID: 38398083

Gharti, L., Kaur, S. D., Vihal, S., Ramakrishna, K., Thakur, D., Kapoor, D., Sharma, L., & Kaurav, H. (2026). Recent advances in analytical techniques for cancer diagnostics and therapeutics: Combining state-of-the-art technologies for precision oncology. *Critical Reviews in Analytical Chemistry*, •••, 1–26. DOI: 10.1080/10408347.2026.2656439 PMID: 41983744

Gote, V., Nookala, A. R., Bolla, P. K., & Pal, D. (2021). Drug Resistance in Metastatic Breast Cancer: Tumor Targeted Nanomedicine to the Rescue. *International Journal of Molecular Sciences*, *22*(9), 4673. Advance online publication. DOI: 10.3390/ijms22094673 PMID: 33925129

Gupta, P., Neupane, Y. R., Parvez, S., & Kohli, K. (2021, December). Recent Advances in Targeted Nanotherapeutic Approaches for Breast Cancer Management. *Nanomedicine (London)*, *16*(29), 2605–2631. DOI: 10.2217/nnm-2021-0281 PMID: 34854336

Gupta, V., Kumar, D., Gupta, S., Tanwar, R., Jaiswal, N. K., Islam, M. M., Singh, S., Choudhary, N., Gowri, S., Webster, T., & Faiyazuddin, M. (2026). Nanotechnology-driven cancer therapies for precision oncology: Advances and clinical outlook. *International Journal of Nanomedicine*, *21*, 568254. DOI: 10.2147/IJN.S568254 PMID: 41869396

Hermain, S., Harshitha, K. S., & Naveen, N. R. (2026). Beyond chemotherapy: The rise of nucleic acid nanoformulations in personalized lung cancer therapy. *Cancer Pathogenesis and Therapy*, *4*, E16–E62. DOI: 10.1016/j.cpt.2026.04.002 PMID: 42597591

Hii, A. R. K., Qi, X., & Wu, Z. (2024, February). Advanced strategies for CRISPR/Cas9 delivery and applications in gene editing, therapy, and cancer detection using nanoparticles and nanocarriers. *Journal of Materials Chemistry. B*, *12*(6), 1467–1489. DOI: 10.1039/D3TB01850D PMID: 38288550

Hou, J., He, Z., Liu, T., Chen, D., Wang, B., Wen, Q., & Zheng, X. (2022). Evolution of molecular targeted cancer therapy: Mechanisms of drug resistance and novel opportunities identified by CRISPR-Cas9 screening. *Frontiers in Oncology*, *12*, 755053. DOI: 10.3389/fonc.2022.755053 PMID: 35372044

Huang, Z., Zhan, X., Xiang, S., Johnson, T. S., Helm, B., Yu, C. Y., Zhang, J., Salama, P., Rizkalla, M., Han, Z., & Huang, K. (2019, March 8). SALMON: Survival analysis learning with multi-omics neural networks on breast cancer. *Frontiers in Genetics*, *10*, 166. DOI: 10.3389/fgene.2019.00166 PMID: 30906311

Iyer, K. A., Tenchov, R., Lotti Diaz, L. M., Jain, P., Thite, T., Deng, Y., & Zhou, Q. A. (2025, November 24). CRISPR technology: Transforming the future of medicine and diagnostics. *Biochemistry*, *64*(24), 4628–4660. DOI: 10.1021/acs.biochem.5c00480 PMID: 41284414

Jayachandran, A., Ramakrishna, S., Shukla, R., Prithviraj, P., Kannourakis, G., & Poddar, A. (2026). *Synthetic Biology in Precision Medicine and Nanotechnology*. Handb. Synth. Biol. DOI: 10.1007/978-981-95-3957-4_37

Jiang, C., Meng, L., Yang, B., & Luo, X. (2020). Application of CRISPR/Cas9 gene editing technique in the study of cancer treatment. *Clinical Genetics*, *97*(1), 73–88. DOI: 10.1111/cge.13589 PMID: 31231788

Jiang, X., Hu, Z., Wang, S., & Zhang, Y. (2023). Deep Learning for Medical Image-Based Cancer Diagnosis. *Cancers (Basel)*, *15*(14), 3608. Advance online publication. DOI: 10.3390/cancers15143608 PMID: 37509272

Kaur, P., Singh, A., & Chana, I. (2021). Computational Techniques and Tools for Omics Data Analysis: State-of-the-Art, Challenges, and Future Directions. *Archives of Computational Methods in Engineering*, *28*(7), 4595–4631. DOI: 10.1007/s11831-021-09547-0 PMID: 33519179

Kaur, R., Gupta, S., Kulshrestha, S., Khandelwal, V., Pandey, S., Kumar, A., Sharma, G., Kumar, U., Parashar, D., & Das, K. (2024). Metabolomics-driven biomarker discovery for breast cancer prognosis and diagnosis. *Cells*, *14*(1), 5. DOI: 10.3390/cells14010005 PMID: 39791706

Khan, S. N., Danishuddin, M. W. A., Khan, M. W. A., Guarnera, L., & Akhtar, S. M. F. (2025). Khan, L. Guarnera, and S. M. F. Akhtar, "Multi-modal AI in precision medicine: Integrating genomics, imaging, and EHR data for clinical insights,". *Frontiers in Artificial Intelligence*, *8*, 1743921. DOI: 10.3389/frai.2025.1743921 PMID: 41573270

Kim, J., Harper, A., McCormack, V., Sung, H., Houssami, N., Morgan, E., Mutebi, M., Garvey, G., Soerjomataram, I., & Fidler-Benaoudia, M. M. (2025, April). Global patterns and trends in breast cancer incidence and mortality across 185 countries. *Nature Medicine*, *31*(4), 1154–1162. DOI: 10.1038/s41591-025-03502-3 PMID: 39994475

Kumar, S., Rathaur, S., Maity, D., Varshney, P., & Gayen, J. R. (2026). Nanotechnology based targeted drug delivery systems for breast cancer: Challenges, recent advancement and future perspectives. *Nanomedicine (London)*, *21*(7), 1027–1048. DOI: 10.1080/17435889.2026.2637591 PMID: 41773625

Li, X., Yang, M., Yang, L., Dang, X., Li, X., & Li, G. (2025). Sequencing of high-frequency mutated genes in breast cancer (BRCA) and associated-functions analysis. *International Journal of Clinical and Experimental Pathology*, *18*(2), 46–62. DOI: 10.62347/YODE5431 PMID: 40083350

Liu, X., Peng, T., Xu, M., Lin, S., Hu, B., Chu, T., Liu, B., Xu, Y., Ding, W., Li, L., Cao, C., & Wu, P. (2024, August 24). Spatial multi-omics: Deciphering technological landscape of integration of multi-omics and its applications. *Journal of Hematology & Oncology*, *17*(1), 72. DOI: 10.1186/s13045-024-01596-9 PMID: 39182134

Lombardo, D., Kiselev, M. A., & Caccamo, M. T. (2019). Smart Nanoparticles for Drug Delivery Application: Development of Versatile Nanocarrier Platforms in Biotechnology and Nanomedicine. *Journal of Nanomaterials*, *2019*(1), 3702518. DOI: 10.1155/2019/3702518

Mahato, T. K. (2025). Pharmaceutical Innovation and Integrative Research: Bridging Drug Discovery, Biotechnology, and Life Sciences for Next-Generation Therapies. *Biopress J. Comput. Life Sci.*, *1*(09), 14–17.

M. Maqbool, F. Bekele, and G. Fekadu, "Treatment Strategies Against Triple-Negative Breast Cancer: An Updated Review," *Breast Cancer Targets Ther.*, vol. 14, no. null, pp. 15–24, Dec. 2023, .DOI: 10.2147/BCTT.S348060

Medina, M. A., Oza, G., Sharma, A., Arriaga, L. G., Hernandez Hernandez, J. M., Rotello, V. M., & Ramirez, J. T. (2020). Triple-negative breast cancer: A review of conventional and advanced therapeutic strategies. *International Journal of Environmental Research and Public Health*, *17*(6), 2078. DOI: 10.3390/ijerph17062078 PMID: 32245065

Ming, C., Viassolo, V., Probst-Hensch, N., Chappuis, P. O., Dinov, I. D., & Katapodi, M. C. (2019). Machine learning techniques for personalized breast cancer risk prediction: Comparison with the BCRAT and BOADICEA models. *Breast Cancer Research : BCR*, *21*(1), 75. DOI: 10.1186/s13058-019-1158-4 PMID: 31221197

Miziak, P., Baran, M., Błaszczak, E., Przybyszewska-Podstawka, A., Kałafut, J., Smok-Kalwat, J., Dmoszyńska-Graniczka, M., Kiełbus, M., & Stepulak, A. (2023, September 23). Estrogen receptor signaling in breast cancer. *Cancers (Basel)*, *15*(19), 4689. DOI: 10.3390/cancers15194689 PMID: 37835383

Modi, N. D., Sorich, M. J., Rowland, A., Logan, J. M., McKinnon, R. A., Kichenadasse, G., Wiese, M. D., & Hopkins, A. M. (2020). A literature review of treatment-specific clinical prediction models in patients with breast cancer. *Critical Reviews in Oncology/Hematology*, *148*, 102908. DOI: 10.1016/j.critrevonc.2020.102908 PMID: 32109714

Mosleh-Shirazi, S., Abbasi, M., Moaddeli, M., Vaez, A., Shafiee, M., Kasaee, S. R., Amani, A. M., & Hatam, S. (2022, August 21). reza Moaddeli M, Vaez A, Shafiee M, Kasaee SR, Amani AM, Hatam S. Nanotechnology advances in the detection and treatment of cancer: An overview. *Nanotheranostics*, *6*(4), 400–423. DOI: 10.7150/ntno.74613 PMID: 36051855

Nafissi, N., Heiranizadeh, N., Shirinzadeh-Dastgiri, A., Vakili-Ojarood, M., Naseri, A., Danaei, M., Saberi, A., Aghasipour, M., Shiri, A., Yeganegi, M., & Rahmani, A. (2024). The application of artificial intelligence in breast cancer. *Eurasian Journal of Medicine and Oncology*, *8*(3), 235. DOI: 10.14744/ejmo.2024.45903

Neagu, A.-N., Whitham, D., Bruno, P., Morrissiey, H., Darie, C. A., & Darie, C. C. (2023). Omics-Based Investigations of Breast Cancer. *Molecules (Basel, Switzerland)*, *28*(12), 4768. Advance online publication. DOI: 10.3390/molecules28124768 PMID: 37375323

Orsini, A., Diquigiovanni, C., & Bonora, E. (2023). Omics Technologies Improving Breast Cancer Research and Diagnostics. *International Journal of Molecular Sciences*, *24*(16), 12690. Advance online publication. DOI: 10.3390/ijms241612690 PMID: 37628869

Ortiz, M. M. O., & Andrechek, E. R. (2023). Molecular Characterization and Landscape of Breast cancer Models from a multi-omics Perspective. *Journal of Mammary Gland Biology and Neoplasia*, *28*(1), 12. DOI: 10.1007/s10911-023-09540-2 PMID: 37269418

Palaz, F., Kalkan, A. K., Can, O., Demir, A. N., Tozluyurt, A., Ozcan, A., & Ozsoz, M. (2021, May 26). CRISPR-Cas13 system as a promising and versatile tool for cancer diagnosis, therapy, and research. *ACS Synthetic Biology*, *10*(6), 1245–1267. DOI: 10.1021/acssynbio.1c00107 PMID: 34037380

Panda, V. K., Mishra, B., Mahapatra, S., Swain, B., Malhotra, D., Saha, S., Khanra, S., Mishra, P., Majhi, S., Kumari, K., Nath, A. N., Saha, S., Jena, S., & Kundu, G. C. (2025, January 13). Molecular insights on signaling cascades in breast cancer: A comprehensive review. *Cancers (Basel)*, *17*(2), 234. DOI: 10.3390/cancers17020234 PMID: 39858015

Pandey, K., Lee, E., Park, N., Hur, J., Cho, Y. B., Katuwal, N. B., Kim, S. K., Lee, S. A., Kim, I., An, H. J., Hwang, S., & Moon, Y. W. (2021, January 25). Deregulated immune pathway associated with palbociclib resistance in preclinical breast cancer models: Integrative genomics and transcriptomics. *Genes*, *12*(2), 159. DOI: 10.3390/genes12020159 PMID: 33504001

Pant, S., Choudhari, J. K., Pathak, R. K., Yadav, A. K., & Singh, D. B. (2026). Biological big data analysis in the post-genomic era: Applications and limitations. *Network Modeling and Analysis in Health Informatics and Bioinformatics*, *15*(1), 62. DOI: 10.1007/s13721-026-00749-9

Parvin, N., Joo, S. W., Jung, J. H., & Mandal, T. K. (2025). Multimodal AI in Biomedicine: Pioneering the Future of Biomaterials, Diagnostics, and Personalized Healthcare. *Nanomaterials (Basel, Switzerland)*, *15*(12), 895. Advance online publication. DOI: 10.3390/nano15120895 PMID: 40559258

Pont, M., Marqués, M., Sorolla, M. A., Parisi, E., Urdanibia, I., Morales, S., Salud, A., & Sorolla, A. (2023). Applications of CRISPR technology to breast cancer and triple negative breast cancer research. *Cancers (Basel)*, *15*(17), 4364. DOI: 10.3390/cancers15174364 PMID: 37686639

Ponti, G., De Angelis, C., Ponti, R., Pongetti, L., Losi, L., Sticchi, A., Tomasi, A., & Ozben, T. (2023). Hereditary breast and ovarian cancer: From genes to molecular targeted therapies. *Critical Reviews in Clinical Laboratory Sciences*, *60*(8), 640–650. DOI: 10.1080/10408363.2023.2234488 PMID: 37455374

Qureshi, M. A., Khan, M. Y., Imran, A., Maqsood, Q., Hussain, N., & Ali, S. W. (2024). Revolutionizing Breast Cancer Care: Cutting-Edge Breakthroughs and Future Frontiers in Precision Medicine. In Rezaei, N. (Ed.), *Breast Cancer Treatment: An Interdisciplinary Approach* (pp. 115–141). Springer Nature Switzerland., DOI: 10.1007/16833_2024_254

Raheja, I. K., Kumar, P., Rajashekarappa, K. K., & Mahadevan, G. D. (2026). Nanobiosensors for Early Detection of Cancer: A Recent Update. *Biomedical Materials & Devices*, *4*(1), 78–96. DOI: 10.1007/s44174-024-00263-4

Ran, D., Li, J., Zhao, M., Du, L., Zhang, Y., & Zhu, J. "Artificial intelligence integrates multi-omics data for precision stratification and drug resistance prediction in breast cancer," *Front. Oncol.*, vol. Volume 15-2025, 2025, DOI: 10.3389/fonc.2025.1612474

Ranga, P. (2026). CRISPR and genome editing: precision tools for personalized therapies. In Ahmad, S. R. (Ed.), *The Gut Microbiome in Precision Medicine* (pp. 863–873). Academic Press., DOI: 10.1016/B978-0-443-40483-2.00047-9

Rossi, C., Cicalini, I., Cufaro, M. C., Consalvo, A., Upadhyaya, P., Sala, G., Antonucci, I., Del Boccio, P., Stuppia, L., & De Laurenzi, V. (2022, April 14). Breast cancer in the era of integrating "Omics" approaches. *Oncogenesis*, *11*(1), 17. DOI: 10.1038/s41389-022-00393-8 PMID: 35422484

Saber Sichani, A., Ranjbar, M., Baneshi, M., Torabi Zadeh, F., & Fallahi, J. (2023). A Review on Advanced CRISPR-Based Genome-Editing Tools: Base Editing and Prime Editing. *Molecular Biotechnology*, *65*(6), 849–860. DOI: 10.1007/s12033-022-00639-1 PMID: 36547823

Sabit, H., Arneth, B., Pawlik, T. M., Abdel-Ghany, S., Ghazy, A., Abdelazeem, R. M., Alqosaibi, A., Al-Dhuayan, I. S., Almulhim, J., Alrabiah, N. A., & Hashash, A. (2025). Leveraging single-cell multi-omics to decode tumor microenvironment diversity and therapeutic resistance. *Pharmaceuticals (Basel, Switzerland)*, *18*(1), 75. DOI: 10.3390/ph18010075 PMID: 39861138

Sarhangi, N., Hajjari, S., Heydari, S. F., Ganjizadeh, M., Rouhollah, F., & Hasanzad, M. (2022). Breast cancer in the era of precision medicine. *Molecular Biology Reports*, *49*(10), 10023–10037. DOI: 10.1007/s11033-022-07571-2 PMID: 35733061

Sharma, P., & Singh, S. (2026). AI-nanotech synergies: Recent advances in sustainable bio-manufacturing. *World Journal of Microbiology & Biotechnology*, *42*(7), 332. DOI: 10.1007/s11274-026-05067-w PMID: 42250160

Shirzad, M., Shaban, M., Mohammadzadeh, V., Rahdar, A., Fathi-karkan, S., Hoseini, Z. S., Najafi, M., Kharaba, Z., & Aboudzadeh, M. A. (2025, September). Artificial intelligence-assisted design of nanomedicines for breast cancer diagnosis and therapy: Advances, challenges, and future directions. *BioNanoScience*, *15*(3), 354. DOI: 10.1007/s12668-025-01980-w

N. Siva, S. Gupta, A. Gupta, J. N. Shukla, B. Malik, and N. Shukla, "Genome-editing approaches and applications: a brief review on CRISPR technology and its role in cancer," *3 Biotech*, vol. 11, no. 3, p. 146, 2021, .DOI: 10.1007/s13205-021-02680-4

Smolarz, B., Nowak, A. Z., & Romanowicz, H. (2022). Breast Cancer—Epidemiology, Classification, Pathogenesis and Treatment (Review of Literature). *Cancers (Basel)*, *14*(10), 2569. Advance online publication. DOI: 10.3390/cancers14102569 PMID: 35626173

Sohrabei, S., Moghaddasi, H., Hosseini, A., & Ehsanzadeh, S. J. (2024). Investigating the effects of artificial intelligence on the personalization of breast cancer management: A systematic study. *BMC Cancer*, *24*(1), 852. DOI: 10.1186/s12885-024-12575-1 PMID: 39026174

Sonalisha, S., Kirti, A., Asima, S. P., Parashar, R., Pal, S., Barik, D., Naser, S. S., Souto, E. B., & Verma, S. K. (2026). Translational paradigm of Advanced Nanoscale Strategies for Triple Negative Breast Cancer (TNBC): Mechanistic Insights, Metastatic Pathways, and emerging theragnosis. *Materials Advances*, *7*(6), 3072–3100. DOI: 10.1039/D5MA00866B

Srivastav, A. K., Mishra, M. K., Lillard, J. W.Jr, & Singh, R. (2025). Transforming Pharmacogenomics and CRISPR Gene Editing with the Power of Artificial Intelligence for Precision Medicine. *Pharmaceutics*, *17*(5), 555. Advance online publication. DOI: 10.3390/pharmaceutics17050555 PMID: 40430848

Sun, X., Liu, K., Lu, S., He, W., & Du, Z. (2022). Targeted Therapy and Immunotherapy for Heterogeneous Breast Cancer. *Cancers (Basel)*, *14*(21), 5456. Advance online publication. DOI: 10.3390/cancers14215456 PMID: 36358874

Suryawanshi, M. V., Bagban, I., & Patne, A. Y. (2025). Integrating Nanotechnology and Artificial Intelligence for Early Detection and Prognostication of Glioblastoma: A Translational Perspective. *Targets (Basel)*, *3*(4), 31. Advance online publication. DOI: 10.3390/targets3040031

Tiwari, A., Mishra, S., & Kuo, T.-R. (2025). Current AI technologies in cancer diagnostics and treatment. *Molecular Cancer*, *24*(1), 159. DOI: 10.1186/s12943-025-02369-9 PMID: 40457408

Torres Quintas, S., Canha-Borges, A., Oliveira, M. J., Sarmento, B., & Castro, F. (2024). Special Issue: Nanotherapeutics in Women's Health Emerging Nanotechnologies for Triple-Negative Breast Cancer Treatment. *Small*, *20*(41), 2300666. DOI: 10.1002/smll.202300666 PMID: 36978237

Tran, P., Lee, S.-E., Kim, D.-H., Pyo, Y.-C., & Park, J.-S. (2020). Recent advances of nanotechnology for the delivery of anticancer drugs for breast cancer treatment. *Journal of Pharmaceutical Investigation*, *50*(3), 261–270. DOI: 10.1007/s40005-019-00459-7

van Beljouw, S. P. B., Sanders, J., Rodríguez-Molina, A., & Brouns, S. J. J. (2023). RNA-targeting CRISPR–Cas systems. *Nature Reviews. Microbiology*, *21*(1), 21–34. DOI: 10.1038/s41579-022-00793-y PMID: 36171275

Vatta, S. (2025). *Role of Artificial Intelligence and Machine Learning in Advancing Nanomedicine for Breast Cancer*. Next-Gen Nanomedicine Breast Cancer From Bench to Bedside Beyond. DOI: 10.70593/978-93-7185-537-2_11

Villiger, L., Joung, J., Koblan, L., Weissman, J., Abudayyeh, O. O., & Gootenberg, J. S. (2024). CRISPR technologies for genome, epigenome and transcriptome editing. *Nature Reviews. Molecular Cell Biology*, *25*(6), 464–487. DOI: 10.1038/s41580-023-00697-6 PMID: 38308006

Vobugari, N., Raja, V., Sethi, U., Gandhi, K., Raja, K., & Surani, S. R. (2022). Advancements in Oncology with Artificial Intelligence—A Review Article. *Cancers (Basel)*, *14*(5), 1349. Advance online publication. DOI: 10.3390/cancers14051349 PMID: 35267657

Wang, S. W., Gao, C., Zheng, Y. M., Yi, L., Lu, J. C., Huang, X. Y., Cai, J.-B., Zhang, P.-F., Cui, Y.-H., & Ke, A. W. (2022). Current applications and future perspective of CRISPR/Cas9 gene editing in cancer. *Molecular Cancer*, *21*(1), 57. DOI: 10.1186/s12943-022-01518-8 PMID: 35189910

Wang, Z. Z., Li, X. H., Wen, X. L., Wang, N., Guo, Y., Zhu, X., Fu, S. H., Xiong, F. F., Bai, J., Gao, X. L., & Wang, H. J. (2023, March 21). Integration of multi-omics data reveals a novel hybrid breast cancer subtype and its biomarkers. *Frontiers in Oncology*, *13*, 1130092. DOI: 10.3389/fonc.2023.1130092 PMID: 37064087

Wu, X., Yang, X., Dai, Y., Zhao, Z., Zhu, J., Guo, H., & Yang, R. (2024). Single-cell sequencing to multi-omics: Technologies and applications. *Biomarker Research*, *12*(1), 110. DOI: 10.1186/s40364-024-00643-4 PMID: 39334490

Yamamoto, H., & Hirasawa, A. (2022). Homologous Recombination Deficiencies and Hereditary Tumors. *International Journal of Molecular Sciences*, *23*(1), 348. Advance online publication. DOI: 10.3390/ijms23010348 PMID: 35008774

Yang, F., He, Q., Dai, X., Zhang, X., & Song, D. (2023). The potential role of nanomedicine in the treatment of breast cancer to overcome the obstacles of current therapies. *Frontiers in Pharmacology*, *14*, 1143102. DOI: 10.3389/fphar.2023.1143102 PMID: 36909177

Yin, G., Liu, L., Yu, T., Yu, L., Feng, M., Zhou, C., Wang, X., Teng, G., Ma, Z., Zhou, W., Ye, C., Zhang, J., Ji, C., Zhao, L., Zhou, P., Guo, Y., Meng, X., Fu, Q., Zhang, Q., & Yu, Z. (2024). Genomic and transcriptomic analysis of breast cancer identifies novel signatures associated with response to neoadjuvant chemotherapy. *Genome Medicine*, *16*(1), 11. DOI: 10.1186/s13073-024-01286-8 PMID: 38217005

Zaripov, E. A., Khraibah, A., Kasyanchyk, P., Radchanka, A., Hüttmann, N., & Berezovski, M. V. (2025, May). CE–MS Metabolomic and LC–MS Proteomic Analyses of Breast Cancer Exosomes Reveal Alterations in Purine and Carnitine Metabolism. *Journal of Proteome Research*, *24*(5), 2505–2516. DOI: 10.1021/acs.jproteome.4c00795 PMID: 40036676

Zhang, C., Li, N., Zhang, P., Jiang, Z., Cheng, Y., Li, H., & Pang, Z. (2024). Advancing precision and personalized breast cancer treatment through multi-omics technologies. *American Journal of Cancer Research*, *14*(12), 5614–5627. DOI: 10.62347/MWNZ5609 PMID: 39803662

Zhong, H., Zhou, Z., Wang, H., Wang, R., Shen, K., Huang, R., & Wang, Z. (2024). The biological roles and clinical applications of the PI3K/AKT pathway in targeted therapy resistance in HER2-positive breast cancer: A comprehensive review. *International Journal of Molecular Sciences*, *25*(24), 13376. DOI: 10.3390/ijms252413376 PMID: 39769140

Zhu, Z., Wang, S.-H., & Zhang, Y.-D. (2023, March). A Survey of Convolutional Neural Network in Breast Cancer. *Computer Modeling in Engineering & Sciences*, *136*(3), 2127–2172. DOI: 10.32604/cmes.2023.025484 PMID: 37152661

Chapter 7
Omics-Nano-CRISPR Strategies for Combating Colorectal Cancer in the AI Age

B. Lokeshwari
http://orcid.org/0009-0005-6612-5134
Department of Microbiology, Sacred Heart College, Tirupattur, India

T. Elavarasi
http://orcid.org/0009-0003-0792-4484
Department of Microbiology, Sacred Heart College, Tirupattur, India

S. Indhu
http://orcid.org/0009-0005-9848-764X
Department of Microbiology, Sacred Heart College, Tirupattur, India

M. Manigandan
http://orcid.org/0000-0002-8115-2713
Department of Microbiology, Sacred Heart College, Tirupattur, India

ABSTRACT

Colorectal cancer is a molecularly heterogeneous disease driven by complex genetic, epigenetic, metabolic, inflammatory, microbial, and microenvironmental alterations that influence diagnosis, progression, therapeutic response, and resistance. The recent developments in omics techniques such as genomics, transcriptomics, proteomics, metabolomics, epigenomics, and microbiomics give a complete overview of tumor biology and help in biomarker discovery, molecular classifications, patient stratification, and precision medicine. This chapter explores the integrated application of omics, nanotechnology, CRISPR, and artificial intelligence in advancing colorectal cancer diagnosis, targeted treatment, personalized medicine, and future precision oncology.

DOI: 10.4018/979-8-3373-9715-3.ch007

1. INTRODUCTION

Colorectal Cancer (CRC) is one of the most molecularly heterogeneous cancers, as its initiation, progression, metastatic capabilities, and response to treatment are regulated by extensive molecular heterogeneity. While traditional pathological staging and histological classification are key to clinical decision making, genotypic, immunologic, metabolic, and therapeutic characteristics of tumors with similar anatomical patterns can often be quite distinct. CRC development typically is preceded by a multistep progression from normal epithelium through adenoma and ultimately to invasive carcinoma, with disruption of key pathways, including WNT–β-catenin, RAS–RAF–MEK, PI3K, transforming growth factor-β, DNA mismatch repair and p53-mediated cellular stress responses (Vasaikar et al., 2019). Moreover, epigenetic changes, chronic inflammation, gut microbial dysbiosis, changes in cellular metabolism and interactions between the tumor and the stromal compartment add to the intra- and intertumoral variability and disease evolution (Repici et al., 2020). While individual reviews on omics, nanotechnology, CRISPR, or artificial intelligence in colorectal cancer exist, this chapter provides a novel, integrated framework that systematically combines these technologies into a cohesive, stepwise precision oncology pipeline. We critically evaluate the translational readiness of each technology, moving from molecular discovery to clinical implementation, and propose a structured workflow for their integration. This comprehensive review aims to: (i) synthesize the current state of omics, nanotechnology, CRISPR, and AI in CRC research; (ii) critically assess the evidence base for each technology; and (iii) propose a framework for their clinical integration. This chapter addresses this gap by proposing a structured precision oncology pipeline that transitions from molecular discovery to clinical implementation, while critically evaluating the translational readiness and evidence quality of each technology.

Nevertheless, key to this shift in the omics technologies that provide opportunities for large-scale, high-throughput, and high-resolution analysis of the molecular makeup of colorectal cancers. The genomics technology identifies mutations and deletions, structural rearrangements, genomic instability, and genetic factors of susceptibility to cancer. Transcriptomics reveals gene expression profiles, signatures of immune infiltration, molecular subtypes, and states involved in the process of metastasis and treatment failure. Proteomics includes not only an abundance of proteins but also their functioning, localization in cells, signal transduction, and modifications. Metabolomics includes the study of the biochemical adaptation associated with rapid cell proliferation, oxidation-reduction potential, adaptation to nutrient availability, and immune suppression (Schlicker et al., 2012). Epigenomics involves the analysis of DNA methylation, histone modifications, chromatin accessibility, and non-coding RNAs regulation (Tie et al., 2022).

Artificial intelligence (AI), machine learning, deep learning, and multimodal computational systems provide powerful approaches for analysing the multidimensional data generated in colorectal cancer (CRC) research and clinical care. By integrating genomic, transcriptomic, digital pathology, radiological, electronic health record, microbiome, and treatment data, AI models can identify relationships among heterogeneous datasets and support diagnosis, prognosis, biomarker discovery, therapeutic decision-making, and long-term disease monitoring. Computer-assisted systems can facilitate real-time detection and characterization of colonic polyps during colonoscopy, thereby reducing the likelihood that clinically significant lesions will be overlooked (Peer et al., 2007). Deep learning techniques can also be employed for tissue classification, tumor segmentation, quantification of immune and stromal cells, and prediction of molecular properties such as microsatellite instability from whole slide histopathology images. Supervised learning can differentiate between disease states based on multi-omics data while unsupervised, semi-supervised, and self-supervised learning can identify novel molecular subtypes and their complex interactions (Grady & Carethers, 2008).

CRISPR technologies allow determining which genes are necessary for the tumor cell survival, proliferation, invasion, metastasis formation, immune escape and therapy resistance. Consequently, pooled CRISPR screening in standard cell lines, organoids, and tumors in vivo can uncover oncogene addictions, synthetic lethal interactions, vulnerabilities in signal transduction pathways, and drug resistance mechanisms. CRISPR-based genetic modification of T-cells and natural killer cells is studied as an approach to increase immune destruction of CRC cells; however, clinical implementation is currently hindered by low tumor delivery efficiency, off-target effects, genomic instability, immune response triggering, intratumor heterogeneity, and toxicity to normal tissue due to antigen expression (Le et al., 2017). Application of CRISPR along with omics, AI, and nanotechnology might contribute to prioritization of biomarkers, target validation, prediction of treatment, and targeted delivery of drugs, nucleic acids, contrast agents, immunomodulatory drugs, and genome editing components (Diaz et al., 2012; Hou et al., 2021).

This chapter gives an integrated applications of omics, nanotechnology, CRISPR, and artificial intelligence approaches in colorectal cancer. In order to be rigorous, we conducted an extensive literature search using the PubMed, Scopus, and Web of Science databases, where keywords such as 'colorectal cancer,' 'genomics,' 'transcriptomics,' 'proteomics,' 'metabolomics,' 'CRISPR-Cas9,' 'nanoparticles,' 'nanocarriers,' 'artificial intelligence,' 'machine learning,' 'precision oncology,' 'circulating tumor DNA,' and 'immunotherapy' were used. Peer-reviewed original research articles, systematic reviews, meta-analysis, clinical guidelines, and phase I-III clinical trials were selected and included from January 2015 to April 2026 to ensure translational significance. This chapter gives an extensive review of the integrated applications of omics, nanotechnology, CRISPR, and artificial intelligence approaches in colorectal cancer. The omics technologies (genomics, transcriptomics, proteomics, metabolomics, epigenomics, microbiomics) give the molecular level characterization whereas artificial intelligence helps in integrating the multidimensional data. This integrated precision oncology pipeline transitions from molecular discovery to clinical implementation, critically evaluating translational readiness and evidence quality for each technology,

2. OVERVIEW OF COLORECTAL CANCER

2.1 Types of Colorectal Cancer

Colorectal cancer comprises several different subtypes, with the classical adenocarcinoma constituting most of the diagnoses. This subtype is developed through the adenoma-carcinoma pathway, where there is progressive architectural distortion, cellular atypia, and stromal invasion. Grading is done according to the presence or absence of glands: well-differentiated cancers possess glands while poorly differentiated ones do not have glands, hence behaving more aggressively. Other types include mucinous adenocarcinoma, signet-ring cell carcinoma, medullary carcinoma, neuroendocrine tumors, gastrointestinal stromal tumors, and primary colorectal lymphomas, which are each diagnosed and treated differently. There have been molecular classifications of colorectal cancers in recent times, including the Consensus Molecular Subtypes (CMS). CMS is categorized into four types: immune-activated CMS1, canonical CMS2, metabolic CMS3, and mesenchymal/stromal CMS4 (Guinney et al., 2015).

2.2 Stages of Colorectal Cancer

The staging of colorectal cancers is based on the TNM system: Stage 0 (carcinoma in situ) is confined to the mucosa; Stage I involves submucosa or muscularis propria with no lymph node metastasis; Stage II penetrates the wall of the intestine without lymph node metastasis (IA, IB, IC depending on invasion); Stage III suggests regional lymph node metastasis (III, IIIB, IIIC depending on lymph node metastasis and T stage); Stage IV represents distant metastasis (IV, IVB, IVC depending on metastasis). Although the anatomical stage is still the main factor for prognosis, tumor biology, immune cell infiltration, and molecular profile are important factors which have an impact on the patient's survival. Therefore, modern staging includes information on the anatomical, histological, and molecular levels for better treatment planning (Kather et al., 2019).

3. MOLECULAR PATHOGENESIS OF COLORECTAL CANCER

3.1 Genetic Mutations in Colorectal Cancer

The genetic abnormalities initiating colorectal tumorigenesis generally involve disrupted epithelial regulation and commonly follow the adenoma–carcinoma sequence, beginning with *APC* inactivation. The loss of APC leads to the accumulation of β-catenin, contributing to an increase in cellular stemness and proliferation. Mutations in KRAS lead to the activation of signalling pathways that help to increase cell survival and therapy resistance. Chromosomal instability, especially with regard to 18q changes and the inactivation of TP53, which negatively impacts cellular stress response, often develops during tumour progression, allowing the survival of genetically unstable cells. While such a sequence offers a common pattern of tumour development, tumour evolution itself is affected not only by selection but also by genomic instability. Chromosomal instability and microsatellite instability are two main types of genomic instability in colorectal cancer (Dekker et al., 2019). This type of instability increases mutational load, hence affecting the immune recognition of the disease and its response to therapy. Another common pathway of colorectal carcinogenesis is serrated pathway, which features BRAF mutations, hypermethylation of the promoters, tumour aggression, and immune evasion strategies. In addition, interactions of tumours with their microenvironment and differences between primary and metastatic tumours need to be understood to create personalized therapy and make correct prognoses. The clinical application of these genetic insights requires a clear distinction between established biomarkers, emerging biomarkers, and experimental findings.

3.2 Epigenetic Changes and Tumor Progression

Epigenetic changes play a crucial role in colorectal carcinogenesis without changing the DNA sequence. Some of the mechanisms include methylation of DNA which can either activate or silence genes that play a role in tumor suppression, repair, and cell cycle regulation. CpG-island methylator phenotype is associated with certain clinical and genetic features. Other modifications of the histones and noncoding RNAs control other cancer pathways. Epigenomic change can also be caused by environmental influences. The phenotype adaptation caused by the plasticity of the malignant cell affects the tumor's progression.

Epigenetic alterations hold promise as targets for therapy and can serve as a biomarker for identification and surveillance but need to be carefully validated and standardized for clinical use (Lee et al., 2020).

3.3 Role of Inflammation and Gut Microbiota

The ability of chronic inflammation to increase the risk of colorectal tumorigenesis is mainly due to the constant interaction of the inflamed epithelial cells with multiple harmful factors including Reactive Oxygen Species and inflammatory cytokines, especially in conditions like Inflammatory Bowel Disease. This leads to the development of dysplasia and/or carcinoma, through the release of mediators from activated immune and stromal cells that affect key pathways involved in tumor survival and expansion. Further, chronic inflammation causes abnormal accumulation of cell clones and DNA damage and also weaken epithelial barriers, thereby allowing the penetration of microbes, making the tumor microenvironment even more complex (Hanahan, 2022). The gut microbiota also contributes to colorectal carcinogenesis through multiple mechanisms, including genotoxicity, chronic inflammation, and modulation of host immune responses. In particular, specific bacterial strains have been implicated in colorectal cancer pathogenesis. *Fusobacterium nucleatum* promotes tumor progression through FadA adhesin binding to E-cadherin, activating β-catenin signaling and recruiting myeloid-derived suppressor cells that suppress antitumor immunity (Kostic et al., 2012). Certain strains of *Escherichia coli* harboring the polyketide synthase (pks) genotoxic island produce colibactin, which induces DNA double-strand breaks and chromosomal instability in intestinal epithelial cells (Dejea et al., 2018). *Bacteroides fragilis* toxin (BFT) disrupts epithelial barrier function and activates oncogenic signaling pathways. Moreover, microbial metabolites, including secondary bile acids and short-chain fatty acids, can impact epithelial health, immune function, and therapeutic responses in cancer.

4. OMICS APPROACHES IN COLORECTAL CANCER

Omics approaches have revolutionized for molecular analysis of changes in the genome, transcriptomes, epigenome, and microbiome. Genomic profiling can detect recurrent mutations, copy-number changes and structural variants linked to the initiation, progression, hereditary predisposition and treatment response of tumors (Argelaguet et al., 2018). By the end of this chapter: (1) describe the key molecular pathways driving colorectal cancer pathogenesis; (2) explain the roles of genomics, transcriptomics, proteomics, metabolomics, epigenomics, and microbiomics in biomarker discovery and therapeutic targeting; (3) evaluate the clinical evidence supporting CRISPR-based gene editing and nanocarrier-mediated delivery in colorectal cancer; (4) critically assess AI applications in diagnosis, prognosis, and treatment selection; and (5) identify the regulatory, ethical, and translational barriers to implementing these technologies in clinical practices. Finally, the goal of colorectal cancer research using omics methods is to translate complex molecular data into reproducible tools to enhance screening, prognosis, treatment selection and longitudinal monitoring (Arthur et al., 2012). Table 1 summarizes the major omics technologies and their applications in colorectal cancer, highlighting their roles in biomarker discovery, molecular characterization, disease diagnosis, prognosis, therapeutic target identification, and precision medicine.

Table 1. Omics Technologies and Their Applications in Colorectal Cancer

Omics Technology	Major Biomarkers/Targets	Sample Type	Clinical Applications	Key References
Genomics	*APC, KRAS, NRAS, BRAF, TP53, PIK3CA, SMAD4*	Tissue, Blood (ctDNA)	Mutation detection, hereditary risk assessment, targeted therapy selection, prognosis	Cancer Genome Atlas Network, 2012; Dekker et al., 2019
Transcriptomics	Gene expression signatures (CMS subtypes), immune signatures, fusion transcripts	Tissue (Fresh/ Frozen, FFPE)	Molecular subtyping (CMS1-4), treatment response prediction, metastasis risk	Guinney et al., 2015; Dienstmann et al., 2017
Proteomics	Carcinoembryonic Antigen (CEA), CA19-9, signaling proteins, post-translational modifications	Tissue, Blood, Stool	Biomarker discovery, pathway analysis, monitoring, therapeutic target identification	Vasaikar et al., 2019; Zhang et al., 2014
Metabolomics	Lactate, amino acids (glutamine), lipids, nucleotides, TCA cycle intermediates	Tissue, Blood, Urine, Stool	Metabolic profiling, early detection, monitoring treatment response, understanding metabolic reprogramming	Hasin et al., 2017
Epigenomics	DNA methylation (CpG islands, CIMP), histone modifications, chromatin accessibility, noncoding RNAs	Tissue, Blood (ctDNA methylation)	Early diagnosis, molecular classification, monitoring (ctDNA methylation), biomarker identification	Lee et al., 2020; Pan et al., 2025
Microbiomics	*Fusobacterium nucleatum*, *Escherichia coli* (pks+), *Bacteroides fragilis*, microbial metabolites	Stool, Tissue	Cancer risk assessment, prognosis, immunotherapy response prediction, therapeutic modulation	Kostic et al., 2012; Wirbel et al., 2019; Sepich-Poore et al., 2021
Multi-Omics Integration	Combined molecular signatures (genomic, transcriptomic, proteomic, metabolomic)	Tissue, Blood, Multi-modal	Precision medicine, comprehensive patient stratification, personalized treatment, dynamic monitoring	Argelaguet et al., 2018; Vasaikar et al., 2019

4.1 Omics for Biomarker Discovery

Comprehensive measurement of biological molecules across different levels is essential in colorectal cancer analysis and can be realised by omics technologies. Genomic profiling is used to detect important mutations and biomarkers associated with the nature of the tumour and its response to treatment. Transcriptomics further contributes to the understanding by examining active gene expression, which is useful to identify different cancer sub-types. Early detection biomarkers derived from epigenomic knowledge of DNA methylation and chromatin states are vital (Barenholz, 2012). But the ability to differentiate between relevant biomarkers and age-related/benign changes is critical. Proteomics and metabolomics also reveal functional aspects of cancer, including changes in tumor behaviour, as determined by changes in protein activity and metabolic pathways. Microbiome profiling brings in an additional level and provides a novel pathway for noninvasive biomarker discovery. Multi-omics approaches involve analyzing various types of omics data together, and machine learning algorithms can be used to detect relevant patterns in the data associated with a disease. Circulating markers have potential as disease monitoring tools, but need to be well validated for clinical relevance and reliability. To be successful, multi-omics strategies need to be standardized and need to be able to show a benefit to patient care (Blanco et al., 2015).

4.2 Omics for Therapeutic Target Identification

The identification of oncogenic drivers that are required to maintain the survival of tumour cells in colorectal cancer can be achieved through genomic and transcriptomic profiling to help identify potential therapeutic targets. It's important to note that activation of mutations in genes such as RAS–RAF–MEK and ERBB2 are essential but not enough to consider them therapeutic targets, they must be functionally significant and have a sufficient percentage of cells present. Targeted inhibition could be affected by coexisting mutations. CRISPR and RNA interference are used in functional genomics to determine the key genes and processes underlying drug resistance. Single-cell methods offer an understanding of tumor heterogeneity and vulnerabilities. The analysis of omics is essential to identify targets associated with acquired drug resistance, including secondary mutations, signaling reactivation, etc (Brennan & Garrett, 2016). But the targets identified by the computational approach must be verified by different experimental techniques and normal tissue toxicity tests. Omics-based targets should ideally be supported by a variety of data types and be associated with patient identification biomarkers.

4.3 Omics-Based Precision Medicine

Precision Medicine for colorectal cancer can use the omics approach to provide a tailored treatment based on the molecular features of each patient, using biomarkers for diagnosis and therapy. Important mutations, such as RAS and BRAF, affect treatment effectiveness, and some biomarkers predict response to some treatments, such as immune checkpoint inhibition. Precision is improved with advanced techniques such as transcriptomic analyses, and patient-derived organoids, which assess tumor heterogeneity and treatment responsiveness. But, challenges in implementation include analytical quality, tumour diversity, and access that is equitable. Future precision medicine is likely to be multi-modal in nature and combine multiple biological and clinical data to achieve optimal results (Cancer Genome Atlas Network, 2012). Figure 1 presents a unified precision oncology workflow for colorectal cancer. The pipeline begins with multi-omics profiling (genomics, transcriptomics, proteomics, metabolomics, epigenomics, microbiomics) from tumor tissue or liquid biopsies. These data are then analyzed using AI/ML algorithms to identify molecular subtypes, prioritize driver alterations, and predict therapeutic vulnerabilities. Candidate targets identified by AI prioritization undergo functional validation using CRISPR-Cas9 screening in patient-derived organoids or in vivo models. Validated targets are then translated into therapeutic strategies using nanocarrier-mediated delivery of CRISPR components, chemotherapeutics, or immunomodulators. This integrated approach enables iterative refinement through longitudinal monitoring using ctDNA and imaging, creating a feedback loop for adaptive treatment.

Figure 1. Omics-Based Precision Medicine Workflow for Colorectal Cancer

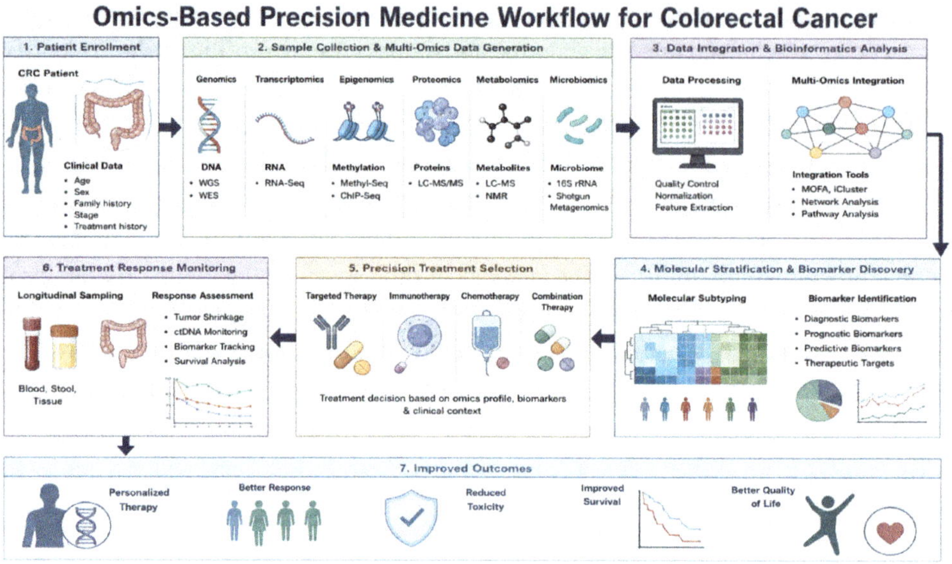

5. GENOMICS AND TRANSCRIPTOMICS IN COLORECTAL CANCER

5.1 Genomic Alterations and Driver Genes

Genomic profiling of colorectal tumors now routinely employs high-throughput platforms, including targeted gene panels, whole-exome sequencing (WES), and whole-genome sequencing (WGS). In targeted panel sequencing, (such as KRAS exons 2-4, NRAS, BRAF V600E and mismatch repair genes), making it economically efficient for diagnostic purposes. On the other hand, whole exome and whole genome sequencing will give you a larger landscape of rare variants, complex structural alterations and TMB, which have become more important in terms of choosing an immunotherapy. One of the most crucial analytical issues in cancer genomics concerns distinguishing driver mutations, which give tumor cells a selective advantage, from "passenger mutations". Another technical issue related to cancer sequencing is the problem of intratumoral heterogeneity, because conventional bulk sequencing is not able to identify subclonal mutations that represent only 10-20% of all tumor cells. The long-read sequencing technology (PacBio, Oxford Nanopore etc.) gives better resolution for complex structural alterations and repetitive genomic regions.

5.2 Next-Generation Sequencing in Colorectal Cancer

Next Generation Sequencing (NGS) has transformed colorectal cancer research and clinical testing using vast-scale genomic analyses (Dejea et al., 2018). These include targeted sequencing panels, whole-exome sequencing, or whole genome sequencing for more in-depth genomic information. RNA sequencing assesses gene expression and mutations; methylation sequencing and chromatin-accessibility assays yield epigenomic information. Preanalytical factors are important and particularly the integrity and contamination of the specimen. In the clinical setting, NGS can be used to help with the hereditary evaluation, treatment planning and tracking resistance. Liquid-biopsy sequencing provides information from the floating tumour DNA, helping when tissue samples are not available. However, challenges in bioinformatics and data interpretation of sequence data make a strong analytical pipeline essential to identify actual mutations from artifacts. Applications of AI for better variant prioritization and long-read sequencing are among the potential advancements that could be part of the future of NGS. Finally, the key is the correct identification and interpretation of molecular characteristics for improved patient care (Dienstmann et al., 2017).

5.3 Genomic and Transcriptomic Biomarkers

Genomic biomarkers are an important part of the diagnosis, prognosis and surveillance of colorectal cancer. Several markers are relevant to Lynch syndrome, prognosis and immunotherapy: mismatch-repair deficiency and microsatellite instability (MSI). Molecular classification is also aided by other markers like BRAF mutations and methylation patterns. New prognostic markers such as circulating tumor DNA have helped to identify molecular evidence for the persistence of disease after surgery and can also potentially predict recurrence earlier than imaging. Stool-based DNA has potential for a noninvasive detection method. Treatment efficacy is guided by predictive biomarkers such as KRAS and BRAF mutations. Single-cell RNA sequencing improvements in transcriptomic biomarkers will potentially improve personalised treatment strategies. To apply in the clinic, biomarker signatures should be reproducible, interpretable, and validated to ensure better patient outcomes (Drost et al., 2015).

6. PROTEOMICS, METABOLOMICS, AND EPIGENOMICS IN COLORECTAL CANCER

6.1 Proteomic Changes in Colorectal Tumors

Genomic mutations, transcriptional regulation and protein modification are all affecting proteomic changes in CRC, highlighting the functional impact on CRC biology. Proteomic profiling directly measures protein abundance and interactions that are directly associated with cell functions, whereas genomic analysis measures potential effects of mutations. Various proteins involved in cellular processes have been found to be different between malignant tissue and the normal tissue adjacent to the tumor by mass spectrometry. Key findings are that expression of proteins involved in proliferation is increased, while expression of proteins associated with intestinal differentiation is decreased (Fearon & Vogelstein, 1990). Furthermore, the amount of RNA does not necessarily correspond to the amount of protein, because of the efficiency of translation and modifications. Combined proteogenomic analysis provides a more comprehensive picture of the behaviour of the tumour than genomic profiling. Signaling pathways that are associated with cancer progression and resistance such as Wnt/β-catenin and RAS/RAF pathways are identified by proteomic analysis. The tumor microenvironment also contributes to proteomic changes, with various cell types impacting signaling and tumor dynamics. Proteomic remodeling helps in the metastatic process by altering patterns in protein expression, and proteins associated with this process could be biomarkers and therapeutic targets. But there are still issues with the translation of these findings into clinical applications, and the need for standardization in order to increase the reproducibility and specificity of the diagnostics (Guinney et al., 2015).

6.2 Protein Biomarkers and Signaling Pathways

Protein biomarkers for colorectal cancer can be found in a range of bio-samples and have the potential to help in diagnosis, prognosis, response to treatment and monitoring. Carcinoembryonic Antigen (CEA) is the most prevalent circulating marker, but it has limited sensitivity and is found in non-cancerous diseases that limit its ability to be used for screening purposes. Further markers and tissue analysis (including mismatch repair protein testing) helps to elucidate tumour biology. There are several signaling pathways that have effects on cancer progression, and the Wnt/β-catenin pathway plays a central role in tumorigenesis. Emerging technologies in multiplex analysis and proteomic measurements are instrumental in the discovery of useful markers and insight into therapeutic resistance. Future developments include integration of tissue and plasma proteomics for more specific therapeutic strategies that will address the complexities of colorectal cancer (Hart et al., 2015).

6.3 Metabolic Reprogramming in Colorectal Cancer

Metabolic reprogramming is an important aspect of colorectal cancer that allows the cancer cells to grow quickly and stay alive when they are stressed. Increased glucose uptake and lactate production are typical characteristics of colorectal tumors and are associated with the expression of a host of metabolic enzymes and transporters, which give rise to an aerobic glycolytic phenotype. There can also be metabolic heterogeneity within a tumour cell, with dynamic shifts between glycolysis and mitochondrial respiration depending on the environment. Tumour redox balance and growth is related to key metabolic pathways such as pentose-phosphate pathway and modified tricarboxylic-acid-cycle activity. Further, there is a significant remodeling of glutamine and lipid metabolism which supports nucleotide synthesis and energy production during stress conditions (Hasin et al., 2017). Metabolic cooperation is further enhanced by microbial metabolites and fibroblasts in the tumor microenvironment. Tumor cells also induce metabolism changes that impact antitumor immunity, through increased nutrient uptake in tumor cells that can impact T-cell function. Metabolomic analyses can provide biomarkers useful for early detection and monitoring of colorectal cancer, but metabolomic differences between populations make it difficult to use these data in a general way. The metabolic plasticity poses challenges for investigative strategies aimed at specific metabolic pathways, and combined strategies might be needed for effective treatment without compromising normal tissue function. Combination of metabolomics with other omics could help characterize different CRC groups, thereby increasing the potential for personalized treatments (Isella et al., 2015).

7. KEY BIOMARKERS AND THERAPEUTIC TARGETS

7.1 Diagnostic Biomarkers

Diagnostic biomarkers of colorectal neoplasia are markers that reveal the presence of disease before symptoms occur. Presently, screening options include feces immunochemical tests, fecal DNA tests, colonoscopy, and more. Testing stool is preferred because it is non-invasive. Fecal immunochemical testing useful but may fail to detect non-bleeding lesions. Multitarget stool DNA assays are more sensitive than hemoglobin testing alone because it uses genetic testing (Kather et al., 2020). New methods include blood tests for circulating tumor DNA (ctDNA) and methylation analysis, though blood tests are not currently as effective at detecting precancerous lesions. Biomarkers such as microbial signatures and microRNAs are being explored, but differences in the variation of microbial composition make their use difficult. Multimodal models and AI integration may enhance diagnostic accuracy, yet thorough validation and implementation considerations are crucial to ensure effectiveness and accessibility in colorectal cancer screening.

7.2 Prognostic Biomarkers

Prognostic biomarkers in colorectal cancer are valuable to obtain information on clinical outcome other than the established stage and grade. Improved evaluations using molecular biomarkers, which can detect differences between tumors at the same stage. For example, the presence of Mismatch-Repair (MMR) deficiency is a good prognosis, while the presence of BRAF V600E mutations is associated with highly aggressive disease. Circulating tumor DNA (ctDNA) is becoming an important prognostic marker and the presence of it after surgery indicates that there is still disease left behind and increased chances of recurrence. The DYNAMIC study indicates that ctDNA-driven approaches can help maximize the use of chemotherapy while maintaining the same survival benefits. Immune scoring, transcriptomic signatures, and metabolomic studies are other strategies to be mentioned, each pointing to a different tumor characteristic. This approach may be extended with the use of different data sources in future models, which would lead to further prospective trials to test the feasibility and accuracy of these methods for enhanced patient management (Kopetz et al., 2019).

7.3 Predictive Biomarkers

Predictive biomarkers play a key role in determining the response to therapy in metastatic colorectal cancer. It is important to perform extended RAS testing before starting treatment with an EGFR-targeted drug because patients with a KRAS or NRAS mutation are more likely to be dependent on MAPK signaling pathways than on EGFR signaling pathways. Various biological factors contribute to the better response of tumours on the left side compared to the right side (Kostic et al., 2012). BRAF V600E is a poor prognostic factor and combination therapy targeting both BRAF and EGFR is required to improve the effects of these inhibitors. In advanced cases, pembrolizumab outperforms chemotherapy, suggesting it has a good chance of working in patients who may have mismatch-repair deficiency and/or high microsatellite instability. Other markers such as HER2 amplification and NTRK gene fusion direct personalized treatments. The introduction of KRAS G12C inhibitors has altered therapeutic concepts and made it possible to combine it with EGFR blockade. It is critical to be able to monitor tumour heterogeneity, as resistance can occur during treatment. Future predictive models should include a wider variety of factors allowing for more accurate treatment decisions and reduce overtreatment (Le et al., 2015).

7.4 Biomarkers for Immunotherapy Response

Immunotherapy has transformed the treatment of a subset of colorectal cancers characterized by deficient DNA mismatch repair (dMMR) and/or high microsatellite instability (MSI-H). Mismatch repair is a critical DNA maintenance mechanism; when deficient (dMMR), DNA replication errors accumulate during cell division, generating a hypermutated phenotype with thousands of mutations, including microsatellite sequences. MSI-H status is closely associated with dMMR and is typically detected using immunohistochemistry (IHC) for MMR proteins (MLH1, MSH2, MSH6, PMS2) or polymerase chain reaction (PCR) for microsatellite markers (BAT25, BAT26, MONO27, NR21, NR24) (Marisa et al., 2013). The landmark KEYNOTE-177 trial demonstrated that pembrolizumab significantly improved median progression-free survival (16.5 months vs. 8.2 months; HR 0.60, 95% CI 0.45-0.80; p=0.0002) compared to chemotherapy in patients with MSI-H metastatic colorectal cancer, leading to FDA approval as first-line therapy (André et al., 2020). Objective response rates with pembrolizumab reached 43.8% compared to 33.1% with chemotherapy, with durable responses in a subset of patients. The CheckMate 142 trial further confirmed efficacy of nivolumab ± ipilimumab, with objective response rates of 31–55% in MSI-H metastatic CRC (Overman et al., 2017).

Notably, tumor mutational burden (TMB) is not a direct surrogate for MMR status, as some microsatellite-stable (MSS) tumors may have high TMB due to other mechanisms (e.g., *POLE/POLD1* exonuclease domain mutations) and may also respond to immunotherapy, though these represent rare cases. Emerging biomarkers include immune cell spatial organization (immunoscore), T-cell receptor clonality, tumor microenvironment features (e.g., CD8+ T-cell infiltration, PD-L1 expression), and circulating tumor DNA dynamics, though these require further validation in prospective trials. The majority of colorectal cancers (approximately 85%) are microsatellite-stable (MSS) and exhibit primary resistance to PD-1/PD-L1 inhibitors due to low immunogenicity, lack of neoantigen expression, immunosuppressive tumor microenvironment, and exclusion of CD8+ T cells from tumor beds. Ongoing research focuses on combination strategies to overcome this resistance, including dual checkpoint blockade (anti-PD-1 + anti-CTLA-4), targeted therapies combined with checkpoint inhibitors (e.g., BRAF/EGFR inhibition, KRAS G12C inhibitors, VEGF inhibition), oncolytic viruses, and epigenetic modulation. Early-phase trials show promising signals, but definitive Phase III data are awaited (Marisa et al., 2013).

7.5 Molecular Targets for Personalized Therapy

Molecular alterations, such as in the EGFR–RAS–RAF–MEK–ERK pathway, are personalised therapies in colorectal cancer. Mutations are a sign of resistance, a RAS wild-type tumor is a good candidate for EGFR-targeted therapy. Combined inhibition is necessary for BRAF V600E tumors because of feedback activation, as in the case of encorafenib and cetuximab. Another dual-targeting approach is available for KRAS G12C. Other targets of interest include HER2, NTRK fusions and angiogenic pathways. The latter have an important role in tumor vascularization, and do not have reliable predictive biomarkers. There are difficulties in targeting pathways such as Wnt/β-catenin because of the toxicity that this kind of approach would have on normal tissues. Future personalized therapies require full molecular profile and consider tumor heterogeneity. Adapting treatments on the basis of their responses and resistance, through techniques such as patient-derived organoids and liquid biopsies, are crucial. The overall strategy should be flexible and able to incorporate molecular and clinical data for better patient care (Matano et al., 2015).

8. CRISPR-BASED STRATEGIES IN COLORECTAL CANCER

8.1 CRISPR for Functional Gene Screening

CRISPR-Cas9 pooled loss-of-function screens have transformed functional genomics by enabling genome-wide identification of essential genes in cancer cells. In colorectal cancer, such screens have identified genes required for survival, proliferation, metastasis, and drug resistance (Hart et al., 2015; Shalem et al., 2014). For example, genome-scale CRISPR screens have identified novel dependencies such as *WRN* in microsatellite-unstable tumors, *SMARCA4* in specific CRC subtypes, and genes involved in resistance to chemotherapy and targeted agents (Hart et al., 2015). While these studies are robust in vitro, clinical translation requires independent validation in patient-derived models, as in vitro findings often fail to replicate in the complex tumor microenvironment. Importantly, candidate targets identified through CRISPR screens must demonstrate: (1) dependency in multiple CRC models (at least three independently derived models), (2) selective toxicity to tumor cells versus normal tissues (demonstrated in normal colonic epithelial cells), (3) evidence of target expression and functional significance in human tumors (using immunohistochemistry or transcriptomics), (4) resistance to the development of acquired resistance mechanisms (tested in long-term selection experiments), and (5) validation in patient-derived organoids or xenografts to confirm relevance *In vivo* (Overman et al., 2017). Three-dimensional organoids generated from patient tumors retain key histological and molecular features, preserving patient-specific genomic alterations, heterogeneity, and treatment response patterns. CRISPR screening in organoids has identified patient-specific dependencies that would be missed in standard cell lines (Matano et al., 2015). *In vivo* screening in tumor microenvironments further identifies genes involved in metastasis, immune evasion, and therapy resistance that are not captured in simplified in vitro systems. However, results from both platforms must be validated with care to confirm therapeutic relevance and to rule out context-specific and off-target effects. Future advances involve multiplexed CRISPR screens, single-cell resolution analyses, and integration with multi-omics data to identify individualized treatment plans based on comprehensive biological information (Rohart et al., 2017).

8.2 CRISPR in Immune Cell Engineering

CRISPR technology is helping to make immune cells, such as T lymphocytes and natural killer cells, more efficient at recognising and killing colorectal cancer cells. One of the main aims is to engineer Chimeric Antigen Receptor (CAR) T cells to Target Tumor-Associated Antigens (TAAs) such as carcinoembryonic antigen or HER2, which do not rely on major Histocompatibility Complex (MHC) presentation. One of the possibilities is to disrupt the genes of T cells that decrease their alloreactivity and to edit genes, e.g., PD-1, to increase the activation efficiency of T cells. The targeted nature of CRISPR's precision has enabled therapeutic insertion of receptors at predetermined sites in the genome, such as by tailoring T cells to target specific mutations within a tumor. But challenges such as the antigen heterogeneity of solid tumors and the potential for on-target, off-tumor toxicity make treatment difficult. Novel engineering strategies, such as logic-gated receptors and dual recognition systems, are being explored to enhance targeting specificity. Although the therapies could be successful, they are still in very early stages of research and require extensive testing for safety and production quality (Reinert et al., 2019).

9. NANOTECHNOLOGY IN COLORECTAL CANCER DIAGNOSIS AND THERAPY

9.1 Role of Nanotechnology in Cancer Medicine

Nanotechnology is based on the design of materials at the nanometer size scale, where properties are often significantly different from larger materials. For cancer therapy, especially colorectal cancer, nanoparticles can be used to deliver therapeutic agents and improve their therapeutic efficacy through a number of mechanisms, such as increasing their solubility, preventing their degradation, and controlling drug release from the nanoparticles. A wide range of administration options are discussed, such as oral, as well as systems that can withstand digestive issues. In addition to treatment, nanotechnology has also been used to facilitate cancer diagnosis, allowing particles to be used to concentrate imaging agents at the tumour site, making lesions easier to see and make better use of biomarkers to detect the presence of cancer. Several factors affect the effectiveness of nanoparticles such as biological interactions, and the complexities of the tumor environment. Issues include delivery problems and overcoming biological barriers, and improved technologies to be used for enhanced penetration and targeting. There is a need to solve these challenges and biological solutions to make nanotechnology a promising tool in oncology (Shalem et al., 2014).

9.2 Nanoparticles for Targeted Drug Delivery

There are three types of targeted drug delivery via nanoparticles: passive, active, and stimulus responsive. Passive targeting involves the use of pharmacokinetics of carriers and tumor vasculature defects to let long-circulating nanoparticles enter the tumor through permeable vessels. Active targeting involves attaching ligands to receptors on cancer cells or vasculature where ligands such as antibodies and peptides induce endocytosis. Optimization of the properties of the nanoparticle, in addition to ligand selection, is crucial for an effective targeting. The stimulus-responsive nanoparticles release the payload in response to the environmental cues, such as acidity and enzyme activity under the colorectal tumor microenvironment. Light or ultrasound can also be used to improve control over release, but this is difficult to achieve for deep tumors. Nanoparticles are engineered to target delivery of drugs, improve toxicity profile and allow co-delivery, but have limitations in terms of uptake by the liver, and immune reactions. It is imperative to carefully evaluate the delivery, intratumoral distribution and the pharmacodynamics of the tumor-cell delivered nanomedicine, and patient-specific imaging may allow for the individual optimization of nanomedicine formulations for tumor characteristics (Shi et al., 2017). Despite these advantages, nanoparticle delivery faces fundamental limitations: (1) the liver and reticuloendothelial system sequester the majority of systemically administered nanoparticles, limiting tumor accumulation to less than 5% of injected dose; (2) the enhanced permeability and retention (EPR) effect is heterogeneous in human tumors and may not reliably predict clinical efficacy; (3) active targeting ligands can be blocked by tumor heterogeneity or antigen loss; (4) stimulus-responsive systems may have insufficient tumor-specific triggers; and (5) manufacturing reproducibility and quality control remain challenging for clinical-scale production (Wilhelm et al., 2016; Shi et al., 2017). These barriers must be addressed through careful preclinical optimization and validated in large animal models before clinical translation.

9.3 Liposomes, Polymeric Nanoparticles, Dendrimers and Gold Nanoparticles

Liposomes are vesicles whose walls consist of phospholipid bilayers that are able to contain hydrophilic drugs in the aqueous compartment, and the hydrophobic drugs in the lipid membranes. They increase the solubility of drugs, stabilize the drug, and can be modified to have targeted drug release (ligands attached). Drug leakage and poor stability are challenges. Polymeric nanoparticles are made from synthetic or natural polymers that can be used to achieve controlled drug delivery, especially for small molecules and proteins, and have a flexible degradation rate. Branched macromolecules known as dendrimers have the property of having a very controlled drug attachment; however, they are toxic. The tunability of the properties and applications of gold nanoparticles is their hallmark; however, the lack of biodegradability and tissue retention is a concern. There is no single best drug delivery platform, it will depend on the drug, delivery system and release profile being sought. Hybrid systems are becoming a reality that will take advantage of these carriers' strengths (Stadtmauer et al., 2020).

9.4 Nanocarriers for Chemotherapy and Gene Therapy

For colorectal cancer, the addition of nanocarriers to the drugs in chemotherapy, such as 5-fluorouracil and oxaliplatin, increases the solubility, stability and targeted delivery of those drugs, which are known to have a high toxicity towards normal tissues. They can provide lower levels of circulating drug and localize the delivery to the vicinity of a tumour. Other innovations like sustained-release formulations and colon-targeted carriers help reduce early release of the drug, and co-encapsulation allows for co-release of two or more agents. But this needs to be studies under well designed conditions for the safety and efficacy to be established. The delivery of nucleic acids such as siRNA and CRISPR components with gene-therapy nanocarriers surmounts the difficulties of stability and cellular uptake. These carriers may have the potential to act as an inhibitor of oncogenes and to make tumours more sensitive to chemotherapy, however there is a concern that the carriers could have off-target effects and that the cellular editing may not be complete. If future developments allow for diagnostic and therapeutic capabilities to be combined in a single platform, specific to certain tumors, material safety and delivery efficiency will need to be seriously considered (Thomas et al., 2019).

10. NANO-CRISPR DELIVERY SYSTEMS

10.1 Lipid Nanoparticle-Based CRISPR Delivery

The lipid nanoparticles are the next-generation nonviral delivery system for CRISPR, which can be used to transiently edit the genome by encapsulating nucleic acids. They are made up of different lipids, which are used to promote cell uptake and reduce cell toxicity. In cancer models, lipid nanoparticles are efficient for delivering components such as Cas9 mRNA and guide RNAs, but the efficiency of delivering these to the tumor is problematic, with accumulation in the liver. Engineering modifications of the lipid structure and the use of surface ligands can improve targeting to colorectal tumors. The internalization is enhanced by systemic delivery methods, which, however, feature issues of cytosolic delivery and optimal RNA ratios. Possible applications are breaking open the mutated oncogenes, as well as combining other therapies such as immune stimulation or chemotherapy. While current evidence confirms that lipid nanoparticle-mediated CRISPR therapy is a possibility, challenges in targeting, immune response, and editing efficiency and safety in various tumor settings will need to be addressed for successful clinical application (Urban et al., 2018).

10.2 Polymeric Nanoparticle-Based CRISPR Delivery

Polymeric nanoparticles are multipurpose delivery vehicles for the components of the CRISPR system to be used in the treatment of colorectal tumors. Polyplexes or micelles can be formed by various materials such as polyethyleneimine and poly (lactic-co-glycolic acid) which are used for nucleic-acid and protein delivery using nanoscale delivery systems. These carriers offer the flexibility of structure in controlling the loading, stability and release of the cargo, especially biodegradable polymers, which reduce cell damage. What it takes to treat colorectal cancer in the future is found in novel technologies such as supramolecular systems that target mutant KRAS and gene editing nanoparticles taken by mouth. However, there are still some problems to be addressed, such as the potential negative impact of cationic polymers and the requirement to carefully formulate the product in order to preserve the integrity of the nucleic acids. Further research on machine learning could be used to develop and improve properties of the polymers and to design effective delivery systems, but therapeutic benefits would be confirmed with a lot of testing before clinical use (Vlachogiannis et al., 2018). The clinical application of nano-CRISPR systems requires rigorous safety evaluation: (1) off-target editing must be characterized using whole-genome sequencing; (2) immunogenicity of Cas9 protein and vector components requires assessment; (3) genotoxicity from DNA double-strand breaks must be quantified; (4) biodistribution and clearance must be determined in clinically relevant models; and (5) long-term effects of genomic modification must be monitored in longitudinal studies. Current evidence for nano-CRISPR in colorectal cancer is limited to preclinical models, and significant barriers—including inefficient tumor delivery, editing efficiency below therapeutic thresholds, and potential for genomic instability—must be overcome before clinical trials can be justified. Figure 2 illustrates the nano-CRISPR-based targeted drug delivery strategy for colorectal cancer, highlighting nanoparticle-mediated delivery of CRISPR/Cas components to tumor cells for precise gene editing and enhanced therapeutic efficacy.

Figure 2. Nano-CRISPR-Based Targeted Drug Delivery in Colorectal Cancer

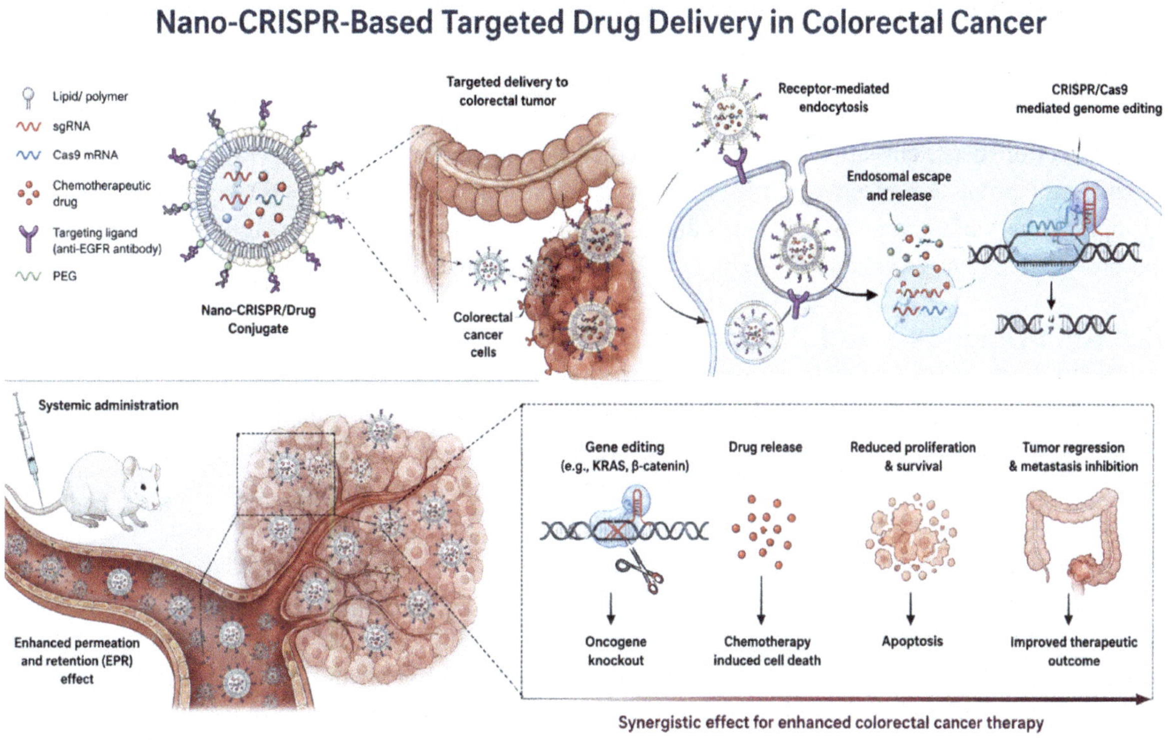

11. NANOTECHNOLOGY-BASED IMAGING AND THERANOSTICS

Nanotechnology based imaging involves using nanotechnology to augment conventional imaging techniques and thereby improve detection and monitoring of colorectal tumors. These nanoparticles can be loaded with imaging agents and are functionalized to target receptors that are present on tumors, thus enhancing the specificity and ability to detect small or heterogeneous lesions. This could be used in iron oxide nanoparticles in magnetic resonance imaging and in the generation of fluorescence signals with quantum dots. Further, theranostic systems integrate imaging and therapy, allowing the visualization of the delivery of the drug and the evaluation of the treatment response during its lifetime. However, there are obstacles to clinical translation: difficulties in tumour accumulation, toxicity and standardization of imaging protocols. Future developments should emphasize biodegradable materials, large-scale synthesis and the incorporation of artificial intelligence for better results in nanotheranostics to benefit patients (Vogelstein et al., 1988).

12. ARTIFICIAL INTELLIGENCE IN COLORECTAL CANCER RESEARCH

Artificial Intelligence (AI) has emerged as a promising computational tool for analyzing complex, high-dimensional data generated in colorectal cancer research and clinical care. However, the clinical impact of AI remains limited, with few FDA-cleared or CE-marked AI tools currently available for routine CRC management. Machine learning and deep learning algorithms have demonstrated potential in polyp detection, tissue classification, survival prediction, and biomarker discovery, yet most applications remain at the proof-of-concept or retrospective validation stage. This evaluates the current evidence for AI applications while acknowledging substantial limitations in data bias, generalizability, interpretability, and prospective clinical validation (Repici et al., 2020; Soleymanjahi et al., 2024).

12.1 AI in Diagnosis and Prognosis

The field of colorectal cancer diagnosis and prognosis is now of great value to Artificial Intelligence (AI). It helps in screening, confirmatory diagnosis and various stages. AI identifies polyps as they occur during procedures such as colonoscopy and classifies lesions. Deep-learning algorithms are used in histopathology to distinguish malignant and benign tissues, to segment tumors and classify tissue types. The use of comprehensive data sets has boosted the precision of AI applications with advanced systems. AI can also extract molecular characteristics from a histopathology image, aiding earlier detection of diseases like microsatellite instability. Prognostically, AI combines various patient and tumor data to deliver estimates of recurrence risk and survival. Successfully implementing these predictive models requires understanding how the models will affect clinical decisions, however, and effective risk stratification needs to consider false positive and false negative outcomes. Continuous evaluations are also important for improving AI tools and ensuring they are compatible with multi-disciplinary healthcare approaches (Wilhelm et al., 2016).

12.2 AI-Assisted Histopathology and Medical Imaging

AI-assisted histopathology uses a digitized colorectal tissue slide to generate quantitative maps of tumor morphology and microenvironments. WSI are huge in size and it creates computation challenges that require WSI to be split into smaller patches for analysis. Whereas, fully supervised systems demand a large amount of pathologist annotations, weakly supervised systems use a few annotations. Semi-supervised and self-supervised learning enhance efficiency by learning from unlabeled slides. These approaches are used to predict tissue type, to identify malignancy, and to provide an insight into molecular changes. While clinical AI must deliver interpretable results with which pathologists could confirm predictions, tissue artifacts and other obstacles could cause inaccuracies. AI analyzes information from different imaging modalities to forecast disease features in medical imaging. The combination of imaging methods and strong validation procedures is crucial to ensure consistency and accuracy. Histopathology provides both macro and micro views, when coupled with medical imaging provides a comprehensive picture of colorectal cancer. But problems such as missing data, spatial mapping issues must be solved. In practice, multimodal AI has been applied in clinical settings for tumor diagnosis and planning, but due to its complexity, the effectiveness of the systems needs to be carefully evaluated against simpler models to ensure reproducibility (Wirbel et al., 2019).

12.3 AI-Based Biomarker Discovery

AI biomarker discovery based on computational learning, instead of working with a small number of potential biomarkers, is able to identify features associated with colorectal cancer. Machine learning has the ability to analyze massive amounts of data, and supervised machine learning algorithms can classify the cancer state in different ways while unsupervised machine learning can reveal molecular subgroups. Various interactions and prognostic classifications have been studied using multi-omics approaches, and imaging data has been combined with genomic data to improve the prediction of biomarkers. Important advances include the ability to anticipate the presence of established molecular markers from common tissue images and discovering new markers using spatial analysis. Robust testing of AI-generated biomarkers across a variety of populations and methods is essential to validate them. Future developments might be based on the integration of multiple data sources and on the clinical utility, consistently aimed at enhancing patient outcomes (Yaeger et al., 2023).

13. CHALLENGES AND LIMITATIONS

Inter-individual and intra-tumoral biological and molecular heterogeneity is one of the main challenges in colorectal cancer research. The molecular pathways involved in the development of CRC can be heterogeneous, with different genomic alterations and sensitivity to treatment. The driver mutations (APC, KRAS, TP53) come in various combinations and may have different clinical relevance, depending on the tumour site and/or treatment history. It is also difficult to discover biomarkers in intratumoral heterogeneity, meaning that a single biopsy may not reflect all malignant subclones. Also, treatment could lead to the emergence of resistant sub-populations, affecting disease progression. While multi-omics approaches can improve molecular classification, there are challenges with integration because of the different features of the datasets. Numerous biomarkers found in small studies don't replicate across the population. Thus, the main issue is differentiating actionable molecular alterations from context-dependent ones. To address these, sampling along the whole longitudinal axis of the sample, multiregional tumor analysis, standardization of the biospecimen collection, and validation across various populations is necessary, as are other factors that could affect molecular patterns and outcomes, such as age and environmental factors (Yu et al., 2017).

Regulatory, ethical, economic and health-system issues are barriers to clinical use of new therapies. There is a need to establish the analytical comparability, clinical validity, utility, safety, cost-effectiveness and workflow integration of biomarker panels, gene editing products, and AI tools. Therapies that contain multiple components present regulatory challenges, and hence product characterization is essential. For permanent genome editing, long-term monitoring is essential because of the potential for adverse effects that may not occur until later in life. Issues related to ethics involve informed consent, incidental findings, ownership of biospecimens, and equitable access to treatments. Economic barriers include but are not limited to access to advanced diagnostics and the different reimbursement policies. There is also the lack of the skilled professionals, which can postpone progress. Effectively bringing innovations to diverse populations and helping to close health disparities requires collaboration on standards, data systems, and workforce development, as well as community engagement (André et al., 2020).

14. FUTURE PERSPECTIVES

Selective delivery systems and combination treatments involving resistant subclones, malignant cells and tumor microenvironment will be the targets for future therapeutic development. Other engineered carriers, such as next-generation lipid nanoparticles, are expected to be able to release drugs depending on certain conditions, allowing them to accumulate in the tumour and avoid toxicity. In addition, these nanocarriers can be used to co-deliver several therapeutic agents to break resistance pathways. Better gene-editing technology will increase precision and minimize DNA breaks, which will allow for the use of colorectal cancer applications to validate targets and re-establish tumor-suppressor action. Future immunotherapy strategies will involve a range of approaches to make non-responsive tumours responsive with the use of personalised treatments. Research needs to focus on integrated platforms to coordinate the delivery, editing, immune activation, and monitoring to tailor therapeutic strategies, with the possibility that these platforms can be theranostic, allowing for real-time visualization of treatment effectiveness (Peer et al., 2007).

AI and digital medicine will improve the care of colorectal cancer in multiple phases from screening to diagnosis, treatment and monitoring. By fusing clinical data, deep-learning systems could be used to enhance the detection of polyps, histopathological quantification, segmentation of metastases and prediction of prognosis. Future models are likely to integrate information from a variety of sources such as the imaging and genomics data into an integrated decision support system. Language models will help with clinical tasks but need to be managed. Patient privacy will be preserved in collaborative model training with federated learning, and digital twins might model disease progression to optimize treatment planning. Wearable tech will help to detect treatment side effects at an early stage. To achieve successful implementation, there is a need for clinical validation, smooth integration into healthcare systems, continued clinical trials, and flexible regulatory measures, all of which require collaboration between multiple disciplines. AI is not about replacing clinical reality; it's about supporting it and making decisions based on informed and evidence-based practices, alongside strong performance and safety monitoring (Tie et al., 2022). Practical implementation of precision oncology requires addressing several health-system challenges: (1) reimbursement models must align with the cost of comprehensive molecular profiling and targeted therapies; (2) healthcare infrastructure must support data integration across electronic health records, genomic platforms, and imaging systems; (3) workforce training in molecular oncology and bioinformatics is essential; (4) patient engagement and shared decision-making must be prioritized, particularly for experimental therapies with uncertain outcomes; (5) equitable access across socioeconomic, geographic, and demographic groups must be ensured to avoid exacerbating health disparities.

15. CONCLUSION

The convergence of omics technologies, nanotechnology, CRISPR genome editing, and artificial intelligence is transforming colorectal cancer research and clinical care. Multi-omics approaches now enable comprehensive molecular characterization, while CRISPR provides functional validation of therapeutic targets and nanocarriers offer improved drug and gene-editing delivery. AI enhances these advances by integrating complex biological and clinical datasets for biomarker discovery and clinical decision support. However, the evidence base for these technologies varies dramatically: immune checkpoint inhibitors and anti-EGFR therapies are well-established clinical standards supported by

Phase III evidence; ctDNA-guided therapy and AI-assisted endoscopy are emerging with prospective clinical validation; while CRISPR-engineered cell therapies, nano-CRISPR delivery, and multi-omics AI models remain experimental, supported primarily by preclinical or retrospective data. Significant challenges—tumor heterogeneity, data integration barriers, biosafety and off-target effects, manufacturing reproducibility, regulatory complexity, cost, and equitable access—must be systematically addressed through collaborative, multi-stakeholder initiatives. The translation of these promising technologies into routine clinical practice will require rigorous validation, standardized protocols, prospective clinical trials, health-economic evaluation, and thoughtful implementation strategies to ensure patients in diverse populations benefit from precision oncology.

REFERENCES

André, T., Shiu, K.-K., Kim, T. W., Jensen, B. V., Jensen, L. H., Punt, C., Smith, D., Garcia-Carbonero, R., Benavides, M., Gibbs, P., de la Fouchardiere, C., Rivera, F., Elez, E., Bendell, J., Le, D. T., Yoshino, T., Van Cutsem, E., Yang, P., Farooqui, M. Z. H., & Diaz, L. A.Jr. (2020). Pembrolizumab in microsatellite-instability-high advanced colorectal cancer. *The New England Journal of Medicine*, *383*(23), 2207–2218. DOI: 10.1056/NEJMoa2017699 PMID: 33264544

Argelaguet, R., Velten, B., Arnol, D., Dietrich, S., Zenz, T., Marioni, J. C., Buettner, F., Huber, W., & Stegle, O. (2018). Multi-Omics Factor Analysis—A framework for unsupervised integration of multi-omics data sets. *Molecular Systems Biology*, *14*(6), e8124. DOI: 10.15252/msb.20178124 PMID: 29925568

Arthur, J. C., Perez-Chanona, E., Mühlbauer, M., Tomkovich, S., Uronis, J. M., Fan, T.-J., Campbell, B. J., Abujamel, T., Dogan, B., Rogers, A. B., Rhodes, J. M., Stintzi, A., Simpson, K. W., Hansen, J. J., Keku, T. O., Fodor, A. A., & Jobin, C. (2012). Intestinal inflammation targets cancer-inducing activity of the microbiota. *Science*, *338*(6103), 120–123. DOI: 10.1126/science.1224820 PMID: 22903521

Barenholz, Y. (2012). Doxil®—The first FDA-approved nano-drug: Lessons learned. *Journal of Controlled Release : Official Journal of the Controlled Release Society*, *160*(2), 117–134. DOI: 10.1016/j.jconrel.2012.03.020 PMID: 22484195

Blanco, E., Shen, H., & Ferrari, M. (2015). Principles of nanoparticle design for overcoming biological barriers to drug delivery. *Nature Biotechnology*, *33*(9), 941–951. DOI: 10.1038/nbt.3330 PMID: 26348965

Brennan, C. A., & Garrett, W. S. (2016). Gut microbiota, inflammation, and colorectal cancer. *Annual Review of Microbiology*, *70*(1), 395–411. DOI: 10.1146/annurev-micro-102215-095513 PMID: 27607555

Cancer Genome Atlas Network. (2012). Comprehensive molecular characterization of human colon and rectal cancer. *Nature*, *487*(7407), 330–337. DOI: 10.1038/nature11252 PMID: 22810696

Dejea, C. M., Fathi, P., Craig, J. M., Boleij, A., Taddese, R., Geis, A. L., Wu, X., DeStefano Shields, C. E., Hechenbleikner, E. M., Huso, D. L., Anders, R. A., Giardiello, F. M., Wick, E. C., Wang, H., Wu, S., Pardoll, D. M., Housseau, F., & Sears, C. L. (2018). Patients with familial adenomatous polyposis harbor colonic biofilms containing tumorigenic bacteria. *Science*, *359*(6375), 592–597. DOI: 10.1126/science.aah3648 PMID: 29420293

Dekker, E., Tanis, P. J., Vleugels, J. L. A., Kasi, P. M., & Wallace, M. B. (2019). Colorectal cancer. *Lancet*, *394*(10207), 1467–1480. DOI: 10.1016/S0140-6736(19)32319-0 PMID: 31631858

Diaz, L. A.Jr, Williams, R. T., Wu, J., Kinde, I., Hecht, J. R., Berlin, J., Allen, B., Bozic, I., Reiter, J. G., Nowak, M. A., Kinzler, K. W., Oliner, K. S., & Vogelstein, B. (2012). The molecular evolution of acquired resistance to targeted EGFR blockade in colorectal cancers. *Nature*, *486*(7404), 537–540. DOI: 10.1038/nature11219 PMID: 22722843

Dienstmann, R., Vermeulen, L., Guinney, J., Kopetz, S., Tejpar, S., & Tabernero, J. (2017). Consensus molecular subtypes and the evolution of precision medicine in colorectal cancer. *Nature Reviews. Cancer*, *17*(2), 79–92. DOI: 10.1038/nrc.2016.126 PMID: 28050011

Drost, J., van Jaarsveld, R. H., Ponsioen, B., Zimberlin, C., van Boxtel, R., Buijs, A., Sachs, N., Overmeer, R. M., Offerhaus, G. J., Begthel, H., Korving, J., van de Wetering, M., Schwank, G., Logtenberg, M., Cuppen, E., Snippert, H. J. G., Medema, J. P., Kops, G. J. P. L., & Clevers, H. (2015). Sequential cancer mutations in cultured human intestinal stem cells. *Nature*, *521*(7550), 43–47. DOI: 10.1038/nature14415 PMID: 25924068

Fearon, E. R., & Vogelstein, B. (1990). A genetic model for colorectal tumorigenesis. *Cell*, *61*(5), 759–767. DOI: 10.1016/0092-8674(90)90186-I PMID: 2188735

Grady, W. M., & Carethers, J. M. (2008). Genomic and epigenetic instability in colorectal cancer pathogenesis. *Gastroenterology*, *135*(4), 1079–1099. DOI: 10.1053/j.gastro.2008.07.076 PMID: 18773902

Guinney, J., Dienstmann, R., Wang, X., de Reyniès, A., Schlicker, A., Soneson, C., Marisa, L., Roepman, P., Nyamundanda, G., Angelino, P., Bot, B. M., Morris, J. S., Simon, I. M., Gerster, S., Fessler, E., De Sousa, E., Melo, F., Missiaglia, E., Ramay, H., & Tejpar, S. (2015). The consensus molecular subtypes of colorectal cancer. *Nature Medicine*, *21*(11), 1350–1356. DOI: 10.1038/nm.3967 PMID: 26457759

Hanahan, D. (2022). Hallmarks of cancer: New dimensions. *Cancer Discovery*, *12*(1), 31–46. DOI: 10.1158/2159-8290.CD-21-1059 PMID: 35022204

Hart, T., Chandrashekhar, M., Aregger, M., Steinhart, Z., Brown, K. R., MacLeod, G., Mis, M., Zimmermann, M., Fradet-Turcotte, A., Sun, S., Mero, P., Dirks, P., Sidhu, S., Roth, F. P., Rissland, O. S., Durocher, D., Angers, S., & Moffat, J. (2015). High-resolution CRISPR screens reveal fitness genes and genotype-specific cancer liabilities. *Cell*, *163*(6), 1515–1526. DOI: 10.1016/j.cell.2015.11.015 PMID: 26627737

Hasin, Y., Seldin, M., & Lusis, A. (2017). Multi-omics approaches to disease. *Genome Biology*, *18*(1), 83. DOI: 10.1186/s13059-017-1215-1 PMID: 28476144

Hou, X., Zaks, T., Langer, R., & Dong, Y. (2021). Lipid nanoparticles for mRNA delivery. *Nature Reviews. Materials*, *6*(12), 1078–1094. DOI: 10.1038/s41578-021-00358-0 PMID: 34394960

Isella, C., Terrasi, A., Bellomo, S. E., Petti, C., Galatola, G., Muratore, A., Mellano, A., Senetta, R., Cassenti, A., Sonetto, C., Inghirami, G., Trusolino, L., Fekete, Z., De Ridder, M., Cassoni, P., Storme, G., Bertotti, A., & Medico, E. (2015). Stromal contribution to the colorectal cancer transcriptome. *Nature Genetics*, *47*(4), 312–319. DOI: 10.1038/ng.3224 PMID: 25706627

Kather, J. N., Heij, L. R., Grabsch, H. I., Loeffler, C., Echle, A., Muti, H. S., Krause, J., Niehues, J. M., Sommer, K. A. J., Bankhead, P., Kooreman, L. F. S., Schulte, J. J., Cipriani, N. A., Buelow, R. D., Boor, P., Ortiz-Brüchle, N., Hanby, A. M., Speirs, V., Kochanny, S., & Pearson, A. T. (2020). Pan-cancer image-based detection of clinically actionable genetic alterations. *Nature Cancer*, *1*(8), 800–810. DOI: 10.1038/s43018-020-0087-6 PMID: 33763651

Kather, J. N., Pearson, A. T., Halama, N., Jäger, D., Krause, J., Loosen, S. H., Marx, A., Boor, P., Tacke, F., Neumann, U. P., Grabsch, H. I., Yoshikawa, T., Brenner, H., Chang-Claude, J., Hoffmeister, M., Trautwein, C., & Luedde, T. (2019). Deep learning can predict microsatellite instability directly from histology in gastrointestinal cancer. *Nature Medicine*, *25*(7), 1054–1056. DOI: 10.1038/s41591-019-0462-y PMID: 31160815

Kopetz, S., Grothey, A., Yaeger, R., Van Cutsem, E., Desai, J., Yoshino, T., Wasan, H., Ciardiello, F., Loupakis, F., Hong, Y. S., Steeghs, N., Guren, T. K., Arkenau, H.-T., Garcia-Alfonso, P., Pfeiffer, P., Orlov, S., Lonardi, S., Elez, E., Kim, T.-W., & Tabernero, J. (2019). Encorafenib, binimetinib, and cetuximab in BRAF V600E-mutated colorectal cancer. *The New England Journal of Medicine*, *381*(17), 1632–1643. DOI: 10.1056/NEJMoa1908075 PMID: 31566309

Kostic, A. D., Gevers, D., Pedamallu, C. S., Michaud, M., Duke, F., Earl, A. M., Ojesina, A. I., Jung, J., Bass, A. J., Tabernero, J., Baselga, J., Liu, C., Shivdasani, R. A., Ogino, S., Birren, B. W., Huttenhower, C., Garrett, W. S., & Meyerson, M. (2012). Genomic analysis identifies association of *Fusobacterium* with colorectal carcinoma. *Genome Research*, *22*(2), 292–298. DOI: 10.1101/gr.126573.111 PMID: 22009990

Le, D. T., Durham, J. N., Smith, K. N., Wang, H., Bartlett, B. R., Aulakh, L. K., Lu, S., Kemberling, H., Wilt, C., Luber, B. S., Wong, F., Azad, N. S., Rucki, A. A., Laheru, D., Donehower, R., Zaheer, A., Fisher, G. A., Crocenzi, T. S., Lee, J. J., & Diaz, L. A.Jr. (2017). Mismatch repair deficiency predicts response of solid tumors to PD-1 blockade. *Science*, *357*(6349), 409–413. DOI: 10.1126/science.aan6733 PMID: 28596308

Le, D. T., Uram, J. N., Wang, H., Bartlett, B. R., Kemberling, H., Eyring, A. D., Skora, A. D., Luber, B. S., Azad, N. S., Laheru, D., Biedrzycki, B., Donehower, R. C., Zaheer, A., Fisher, G. A., Crocenzi, T. S., Lee, J. J., Duffy, S. M., Goldberg, R. M., de la Chapelle, A., & Diaz, L. A.Jr. (2015). PD-1 blockade in tumors with mismatch-repair deficiency. *The New England Journal of Medicine*, *372*(26), 2509–2520. DOI: 10.1056/NEJMoa1500596 PMID: 26028255

Lee, H.-O., Hong, Y., Etlioglu, H. E., Cho, Y. B., Pomella, V., Van den Bosch, B., Vanhecke, J., Verbandt, S., Hong, H., Min, J.-W., Kim, N., Eum, H. H., Qian, J., Boeckx, B., Lambrechts, D., Tsantoulis, P., De Hertogh, G., Chung, W., Lee, T., & Park, W.-Y. (2020). Lineage-dependent gene expression programs influence the immune landscape of colorectal cancer. *Nature Genetics*, *52*(6), 594–603. DOI: 10.1038/s41588-020-0636-z PMID: 32451460

Marisa, L., de Reyniès, A., Duval, A., Selves, J., Gaub, M. P., Vescovo, L., Etienne-Grimaldi, M.-C., Schiappa, R., Guenot, D., Ayadi, M., Kirzin, S., Chazal, M., Fléjou, J.-F., Benchimol, D., Berger, A., Lagarde, A., Pencreach, E., Piard, F., Elias, D., & Boige, V. (2013). Gene expression classification of colon cancer into molecular subtypes: Characterization, validation, and prognostic value. *PLoS Medicine*, *10*(5), e1001453. DOI: 10.1371/journal.pmed.1001453 PMID: 23700391

Matano, M., Date, S., Shimokawa, M., Takano, A., Fujii, M., Ohta, Y., Watanabe, T., Kanai, T., & Sato, T. (2015). Modeling colorectal cancer using CRISPR-Cas9-mediated engineering of human intestinal organoids. *Nature Medicine*, *21*(3), 256–262. DOI: 10.1038/nm.3802 PMID: 25706875

Overman, M. J., McDermott, R., Leach, J. L., Lonardi, S., Lenz, H.-J., Morse, M. A., Desai, J., Hill, A., Axelson, M., Moss, R. A., Goldberg, M. V., Cao, Z. A., Ledeine, J.-M., Maglinte, G. A., Kopetz, S., & André, T. (2017). Nivolumab in patients with metastatic DNA mismatch repair-deficient or microsatellite instability-high colorectal cancer (CheckMate 142): An open-label, multicentre, phase 2 study. *The Lancet. Oncology*, *18*(9), 1182–1191. DOI: 10.1016/S1470-2045(17)30422-9 PMID: 28734759

Peer, D., Karp, J. M., Hong, S., Farokhzad, O. C., Margalit, R., & Langer, R. (2007). Nanocarriers as an emerging platform for cancer therapy. *Nature Nanotechnology*, *2*(12), 751–760. DOI: 10.1038/nnano.2007.387 PMID: 18654426

Reinert, T., Henriksen, T. V., Christensen, E., Sharma, S., Salari, R., Sethi, H., Knudsen, M., Nordentoft, I., Wu, H.-T., Tin, A. S., Heilskov Rasmussen, M., Vang, S., Shchegrova, S., Frydendahl Boll Johansen, A., Srinivasan, R., Assaf, Z., Balcioglu, M., Olson, A., Dashner, S., & Lindbjerg Andersen, C. (2019). Analysis of plasma cell-free DNA by ultradeep sequencing in patients with stages I to III colorectal cancer. *JAMA Oncology*, *5*(8), 1124–1131. DOI: 10.1001/jamaoncol.2019.0528 PMID: 31070691

Repici, A., Badalamenti, M., Maselli, R., Correale, L., Radaelli, F., Rondonotti, E., Ferrara, E., Spadaccini, M., Alkandari, A., Fugazza, A., Anderloni, A., Galtieri, P. A., Pellegatta, G., Carrara, S., Di Leo, M., Craviotto, V., Lamonaca, L., Lorenzetti, R., Andrealli, A., & Wallace, M. B. (2020). Efficacy of real-time computer-aided detection of colorectal neoplasia in a randomized trial. *Gastroenterology*, *159*(2), 512–520.e7. DOI: 10.1053/j.gastro.2020.04.062 PMID: 32371116

Rohart, F., Gautier, B., Singh, A., & Lê Cao, K.-A. (2017). mixOmics: An R package for 'omics feature selection and multiple data integration. *PLoS Computational Biology*, *13*(11), e1005752. DOI: 10.1371/journal.pcbi.1005752 PMID: 29099853

Schlicker, A., Beran, G., Chresta, C. M., McWalter, G., Pritchard, A., Weston, S., Runswick, S., Davenport, S., Heathcote, K., Alferez Castro, D., Orphanides, G., French, T., & Wessels, L. F. A. (2012). Subtypes of primary colorectal tumors correlate with response to targeted treatment in colorectal cell lines. *BMC Medical Genomics*, *5*(1), 66. DOI: 10.1186/1755-8794-5-66 PMID: 23272949

Sepich-Poore, G. D., Zitvogel, L., Straussman, R., Hasty, J., Wargo, J. A., & Knight, R. (2021). The microbiome and human cancer. *Science*, *371*(6536), eabc4552. Advance online publication. DOI: 10.1126/science.abc4552 PMID: 33766858

Shalem, O., Sanjana, N. E., Hartenian, E., Shi, X., Scott, D. A., Mikkelson, T., Heckl, D., Ebert, B. L., Root, D. E., Doench, J. G., & Zhang, F. (2014). Genome-scale CRISPR-Cas9 knockout screening in human cells. *Science*, *343*(6166), 84–87. DOI: 10.1126/science.1247005 PMID: 24336571

Shi, J., Kantoff, P. W., Wooster, R., & Farokhzad, O. C. (2017). Cancer nanomedicine: Progress, challenges and opportunities. *Nature Reviews. Cancer*, *17*(1), 20–37. DOI: 10.1038/nrc.2016.108 PMID: 27834398

Soleymanjahi, S., Huebner, J., Elmansy, L., Rajashekar, N., Lüdtke, N., Paracha, R., Thompson, R., Grimshaw, A. A., Foroutan, F., Sultan, S., & Shung, D. L. (2024). Artificial intelligence-assisted colonoscopy for polyp detection: A systematic review and meta-analysis. *Annals of Internal Medicine*, *177*(12), 1652–1663. DOI: 10.7326/ANNALS-24-00981 PMID: 39531400

Stadtmauer, E. A., Fraietta, J. A., Davis, M. M., Cohen, A. D., Weber, K. L., Lancaster, E., Mangan, P. A., Kulikovskaya, I., Gupta, M., Chen, F., Tian, L., Gonzalez, V. E., Xu, J., Jung, I.-Y., Melenhorst, J. J., Plesa, G., Shea, J., Matlawski, T., Cervini, A., & June, C. H. (2020). CRISPR-engineered T cells in patients with refractory cancer. *Science*, *367*(6481), eaba7365. Advance online publication. DOI: 10.1126/science.aba7365 PMID: 32029687

Thomas, A. M., Manghi, P., Asnicar, F., Pasolli, E., Armanini, F., Zolfo, M., Beghini, F., Manara, S., Karcher, N., Pozzi, C., Gandini, S., Serrano, D., Tarallo, S., Francavilla, A., Gallo, G., Trompetto, M., Ferrero, G., Mizutani, S., Shiroma, H., & Segata, N. (2019). Metagenomic analysis of colorectal cancer datasets identifies cross-cohort microbial diagnostic signatures and a link with choline degradation. *Nature Medicine*, *25*(4), 667–678. DOI: 10.1038/s41591-019-0405-7 PMID: 30936548

Tie, J., Cohen, J. D., Lahouel, K., Lo, S. N., Wang, Y., Kosmider, S., Wong, R., Shapiro, J., Lee, M., Harris, S., Khattak, A., Burge, M., Harris, M., Lynam, J., Nott, L., Day, F., Hayes, T., McLachlan, S.-A., Lee, B., & Gibbs, P. (2022). Circulating tumor DNA analysis guiding adjuvant therapy in stage II colon cancer. *The New England Journal of Medicine, 386*(24), 2261–2272. DOI: 10.1056/NEJMoa2200075 PMID: 35657320

Urban, G., Tripathi, P., Alkayali, T., Mittal, M., Jalali, F., Karnes, W., & Baldi, P. (2018). Deep learning localizes and identifies polyps in real time with 96% accuracy in screening colonoscopy. *Gastroenterology, 155*(4), 1069–1078.e8. DOI: 10.1053/j.gastro.2018.06.037 PMID: 29928897

Vasaikar, S., Huang, C., Wang, X., Petyuk, V. A., Savage, S. R., Wen, B., Dou, Y., Zhang, Y., Shi, Z., Arshad, O. A., Gritsenko, M. A., Zimmerman, L. J., McDermott, J. E., Clauss, T. R., Moore, R. J., Zhao, R., Monroe, M. E., Wang, Y.-T., Chambers, M. C., & Zhang, B. (2019). Proteogenomic analysis of human colon cancer reveals new therapeutic opportunities. *Cell, 177*(4), 1035–1049.e19. DOI: 10.1016/j.cell.2019.03.030 PMID: 31031003

Vlachogiannis, G., Hedayat, S., Vatsiou, A., Jamin, Y., Fernández-Mateos, J., Khan, K., Lampis, A., Eason, K., Huntingford, I., Burke, R., Rata, M., Koh, D.-M., Tunariu, N., Collins, D., Hulkki-Wilson, S., Ragulan, C., Spiteri, I., Moorcraft, S. Y., Chau, I., & Valeri, N. (2018). Patient-derived organoids model treatment response of metastatic gastrointestinal cancers. *Science, 359*(6378), 920–926. DOI: 10.1126/science.aao2774 PMID: 29472484

Vogelstein, B., Fearon, E. R., Hamilton, S. R., Kern, S. E., Preisinger, A. C., Leppert, M., Nakamura, Y., White, R., Smits, A. M., & Bos, J. L. (1988). Genetic alterations during colorectal-tumor development. *The New England Journal of Medicine, 319*(9), 525–532. DOI: 10.1056/NEJM198809013190901 PMID: 2841597

Wilhelm, S., Tavares, A. J., Dai, Q., Ohta, S., Audet, J., Dvorak, H. F., & Chan, W. C. W. (2016). Analysis of nanoparticle delivery to tumours. *Nature Reviews. Materials, 1*(5), 16014. DOI: 10.1038/natrevmats.2016.14

Wirbel, J., Pyl, P. T., Kartal, E., Zych, K., Kashani, A., Milanese, A., Fleck, J. S., Voigt, A. Y., Palleja, A., Ponnudurai, R., Sunagawa, S., Coelho, L. P., Schrotz-King, P., Vogtmann, E., Habermann, N., Niméus, E., Thomas, A. M., Manghi, P., Gandini, S., & Zeller, G. (2019). Meta-analysis of fecal metagenomes reveals global microbial signatures that are specific for colorectal cancer. *Nature Medicine, 25*(4), 679–689. DOI: 10.1038/s41591-019-0406-6 PMID: 30936547

Yaeger, R., Weiss, J., Pelster, M. S., Spira, A. I., Barve, M., Ou, S.-H. I., Leal, T. A., Bekaii-Saab, T. S., Paweletz, C. P., Heavey, G. A., Christensen, J. G., Velastegui, K., Kheoh, T., Der-Torossian, H., Klempner, S. J., & Hong, D. S. (2023). Adagrasib with or without cetuximab in colorectal cancer. *The New England Journal of Medicine, 388*(1), 44–54. DOI: 10.1056/NEJMoa2212419 PMID: 36546659

Yu, J., Feng, Q., Wong, S. H., Zhang, D., Liang, Q. Y., Qin, Y., Tang, L., Zhao, H., Stenvang, J., Li, Y., Wang, X., Xu, X., Chen, N., Wu, W. K. K., Al-Aama, J., Nielsen, H. J., Kiilerich, P., Jensen, B. A. H., Yau, T. O., & Wang, J. (2017). Metagenomic analysis of faecal microbiome as a tool towards targeted non-invasive biomarkers for colorectal cancer. *Gut, 66*(1), 70–78. DOI: 10.1136/gutjnl-2015-309800 PMID: 26408641

Zhang, B., Wang, J., Wang, X., Zhu, J., Liu, Q., Shi, Z., Chambers, M. C., Zimmerman, L. J., Shaddox, K. F., Kim, S., Davies, S. R., Wang, S., Wang, P., Kinsinger, C. R., Rivers, R. C., Rodriguez, H., Townsend, R. R., Ellis, M. J. C., Carr, S. A., & Liebler, D. C. (2014). Proteogenomic characterization of human colon and rectal cancer. *Nature*, *513*(7518), 382–387. DOI: 10.1038/nature13438 PMID: 25043054

Chapter 8

Omics-Nano-CRISPR Strategies for Combating Prostate Cancer in the AI Age

Ashok Kumar Sah
http://orcid.org/0000-0002-7762-4351
A' Sharqiyah University, Oman

Rahul Kumar
http://orcid.org/0009-0001-0237-3201
Vidya University, India

Sakshi Patel
http://orcid.org/0009-0000-2041-9325
Vidya University, India

Prem Shankar Mishra
http://orcid.org/0000-0003-3797-1575
Vidya University, India

Aryan Kumar
Vidya University, India

Shriram Sisodiya
Vidya University, India

Rabab H. Elshaikh
http://orcid.org/0000-0003-0973-8661
A' Sharqiyah University, Oman

ABSTRACT

Prostate cancer remains a major global health challenge due to its molecular and phenotypic heterogeneity, disease progression, and therapeutic resistance. Advances in precision oncology integrating multi-omics technologies, artificial intelligence (AI), nanotechnology, and CRISPR-based genome editing are transforming prostate cancer diagnosis and treatment. Multi-omics approaches enable comprehensive molecular profiling for biomarker discovery, patient stratification, and therapeutic target identification, while AI integrates complex datasets to improve diagnosis, prognosis, and personalized treatment strategies. Nanotechnology enhances molecular imaging and targeted drug delivery, whereas CRISPR supports precise genome editing for functional validation and therapeutic development. This chapter examines the synergistic integration of these technologies and their potential to advance precision medicine and personalized prostate cancer management.

DOI: 10.4018/979-8-3373-9715-3.ch008

1. INTRODUCTION

Prostate cancer is the most common cancer in men globally, and is a leading cause of cancer-related morbidity and mortality. Although there has been major progress in screening, diagnosis and treatment, the disease is a significant clinical challenge because of its biological diversity, the different disease courses and the development of resistance to treatment. Localized prostate cancer can be treated effectively using surgery, radiotherapy or active surveillance, but if it does become advanced and metastatic, it is often not possible to treat and cure it using conventional therapies. These challenges underscore the need for more specific and tailored disease management strategies (Rawla 2019, Sekhoacha et al, 2022).

Precision medicine has revolutionized the understanding and management of prostate cancer by moving away from a one-size-fits-all approach towards each patient's therapy. Precision oncology is based on comprehensive molecular characterization of tumors to characterize the molecular changes at the genetic, epigenetic, transcriptomic, proteomic and metabolic level that underlie the initiation and progression of the disease (Pecoraro et al, 2026). High throughput omics technologies have advanced significantly since then and have allowed for global analysis of prostate cancer at various biological levels, which has resulted in the identification of biomarkers for early detection, prognosis, selection of therapy, and monitoring of disease. These insights are driving new ways to stratify patients and aid the development of targeted therapeutic interventions (Dimitrov and Popov 2026).

The use of nanotechnology has also proved to be a powerful tool for improving the diagnosis and treatment of prostate cancer, in addition to the use of the omics technologies, shown in **figure 1.** The sensitivity of diagnostic platforms can be enhanced, targeted drug delivery can be realized, and off-target toxicity minimized by controlled release of therapeutic agents using nanomaterials. These benefits are especially significant in prostate cancer where selective targeting of tumor cells is still an important goal (Jiang et al, 2024). At the same time, the arrival of genome editing using CRISPR has opened up new opportunities for studying genome function and precision therapeutics. CRISPR technologies provide novel opportunities to understand tumor biology and to devise next-generation therapies, as they can be used to modify genes that are associated with disease (Tabibian et al, 2024, Hashemi et al, 2026).

At the same time, artificial intelligence (AI) has also been transforming biomedical research with its sophisticated computational tools and approaches that tackle complex biomedical data (Naskar et al, 2025, Athanasopoulou et al, 2022). Simultaneously, advanced fields of artificial intelligence (AI) have been reshaping biomedical research by delivering powerful computational frameworks and methods to address intricate biomedical data (Silva 2024). AI-powered algorithms can combine various types of omics data, discover clinically-actionable molecular patterns, forecast drug response and aid in decision making (Tataru er al, 2021, Rabaan et al, 2022). Precision oncology is becoming a highly interconnected field due to the convergence of omics sciences, nanotechnology, genome engineering with CRISPR and AI (Jena et al, 2025). As these technologies continue to advance, they hold the promise of enhancing prostate cancer diagnosis, treatment, and prognosis, thereby contributing to the future of personalized medicine and improved healthcare outcomes in the AI era.

Figure 1. Illustrating the Convergence of Omics, Nanotechnology, CRISPR, and AI in Precision Prostate Cancer Management

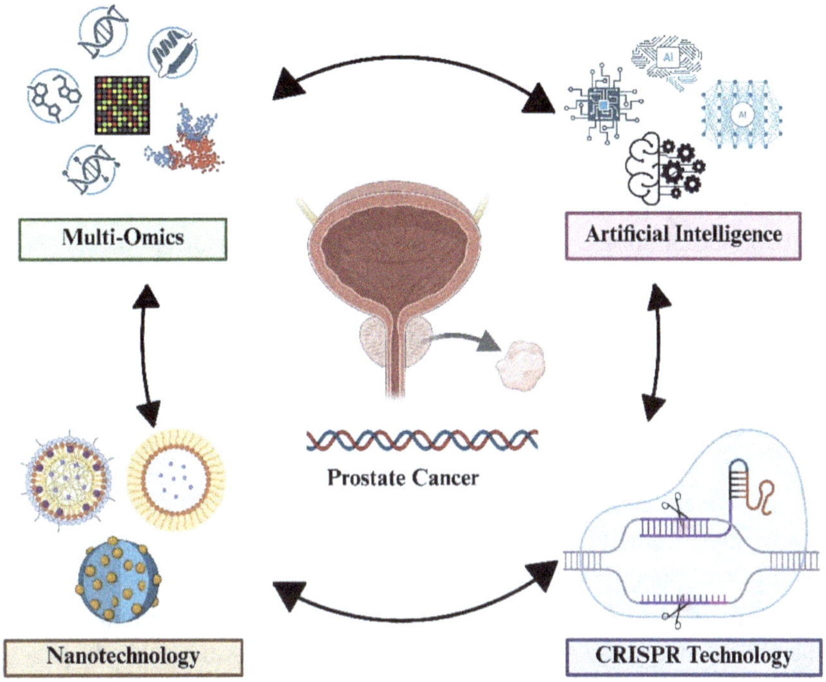

2. MULTI-OMICS LANDSCAPE OF PROSTATE CANCER

2.1 Genomics and Mutational Profiling

Genomic studies have made great advances to understanding prostate cancer, including the identification of genomic changes linked to the initiation of prostate cancer, its progression, and response to therapy. Recurrent mutations and structural variations of genes like TP53, PTEN, BRCA1, BRCA2, and AR (androgen receptor) have been identified using high-throughput sequencing technologies (Kwon and Joung 2025). These changes affect the regulation of important cellular processes such as DNA repair, cell cycle regulation and androgen signaling. Mutational profiling can be used to identify molecular subtypes of tumour cells, which can be used to assess risk level and choose appropriate treatment (Hachem et al, 2024, Ren et al 2023). In addition, genomic data can be used to identify patients who are likely to respond to targeted therapies, such as PARP inhibitors and other precision-based treatments, which can aid in personalized treatment approaches (Nevedomskaya and Haendler 2022).

2.2 Transcriptomics and Epigenomics

Transcriptomics looks at patterns of gene expression, which give insights to the functional state of the cancer cells, and is more than just static genomic alterations. Transcriptomic analyses in prostate cancer have identified patterns of expression that are dysregulated in pathways related to proliferation, metastasis, immune evasion and treatment resistance. Epigenomics looks at heritable changes in gene activity without changes in the DNA sequence, such as DNA methylation, histone modifications and chromatin remodeling, as well as other epigenetic changes (Li et al, 2025). Prostate cancer progression is marked by disruption of epigenetic regulation and this may affect the aggressiveness of the tumor (Pomerantz et al, 2025). Combining transcriptomic and epigenomic information can enhance our understanding of disease mechanisms and help identify new drug targets and clinically relevant biomarkers.

2.3 Proteomics and Metabolomics

Genomic and transcriptomic approaches allow to identify what kind of activities cells are able to perform, whereas proteomics and metabolomics help to assess what the real activity of tumours is. Proteomics involves the discovery and quantification of proteins, such as post-translation modifications that control signaling pathways and cell function. Metabolomics is the study of small molecule metabolites which are reflective of the changes in cellular metabolism involved in the growth and progression of cancer (Al-Daffaie et al, 2024, Salciccia et al, 2021). Prostate cancer has unique metabolic characteristics, such as changes in energy use and lipid metabolism. Such Omics technologies offer invaluable insight into the biology of tumors, their response to treatment, and progression of disease. They connect the dots between molecular changes and clinical manifestations and help to develop a more complete picture of how cancer unfolds.

2.4 Biomarker Discovery and Disease Stratification

Prostate cancer biomarkers for early detection, prognosis, therapy monitoring, and therapeutic decision making have been found more quickly because to the merging of many omics datasets. Combining biomarkers based on proteomic, metabolomic, transcriptomic, and genomic studies may improve illness detection accuracy and help predict clinical outcomes more precisely than traditional approaches. Multi-omics-based patient stratification may aid in defining individuals according to molecular characteristics, disease aggressiveness, and likely treatment outcomes. By enabling tailored treatment interventions and reducing unnecessary interventions, this approach contributes to the development of precision medicine (Marzec et al 2021, Kumar et al, 2022).

Additionally, artificial intelligence and computational techniques continue to advance, making it possible to find biomarkers that are more precise and clinically significant. Numerous omics methods provide complementary insights into the biology of prostate cancer, ranging from genetic alterations to metabolic reprogramming. Table 1 summarizes their several important uses in precision oncology (Testa et al, 2019, Sumiyoshi et al, 2021).

Table 1. Major Omics Technologies and their Applications in Prostate Cancer.

Omics Technology	Primary Target	Key Information Obtained	Applications in Prostate Cancer
Genomics	DNA sequence	Mutations, copy number variations, structural alterations	Identification of driver mutations and therapeutic targets
Transcriptomics	RNA transcripts	Gene expression profiles	Tumor classification and pathway analysis
Epigenomics	Epigenetic modifications	DNA methylation, histone modifications	Understanding gene regulation and disease progression
Proteomics	Proteins	Protein abundance and signaling networks	Biomarker discovery and therapeutic monitoring
Metabolomics	Metabolites	Cellular metabolic changes	Detection of metabolic signatures and treatment response

3. ARTIFICIAL INTELLIGENCE FOR OMICS-DRIVEN PROSTATE CANCER MANAGEMENT

3.1 AI and Machine Learning in Multi-Omics Integration

Genomic, transcriptomic, proteomic and metabolomic data are becoming more and more available, creating unprecedented opportunities for understanding prostate cancer biology. But these data sets are frequently so large or complex that traditional analysis is not able to handle them. The multi-omics data can be integrated and interpreted with powerful tools from artificial intelligence (AI) and machine learning (ML) algorithms (Yetgin 2025, Agarwal et al, 2024), shown in **figure 2.** These computational methods can uncover patterns, interactions and signatures of diseases at various biological levels. The ability of AI to integrate and analyze heterogeneous datasets in a single framework pays off in providing a systems-level understanding of prostate cancer behavior, which can help researchers identify novel therapeutic targets and enhance the precision of prostate cancer diagnosis and treatment (Cai et al, 2022).

Figure 2. AI-Assisted Integration of Multi-Omics and Clinical Data for Precision Prostate Cancer Management.

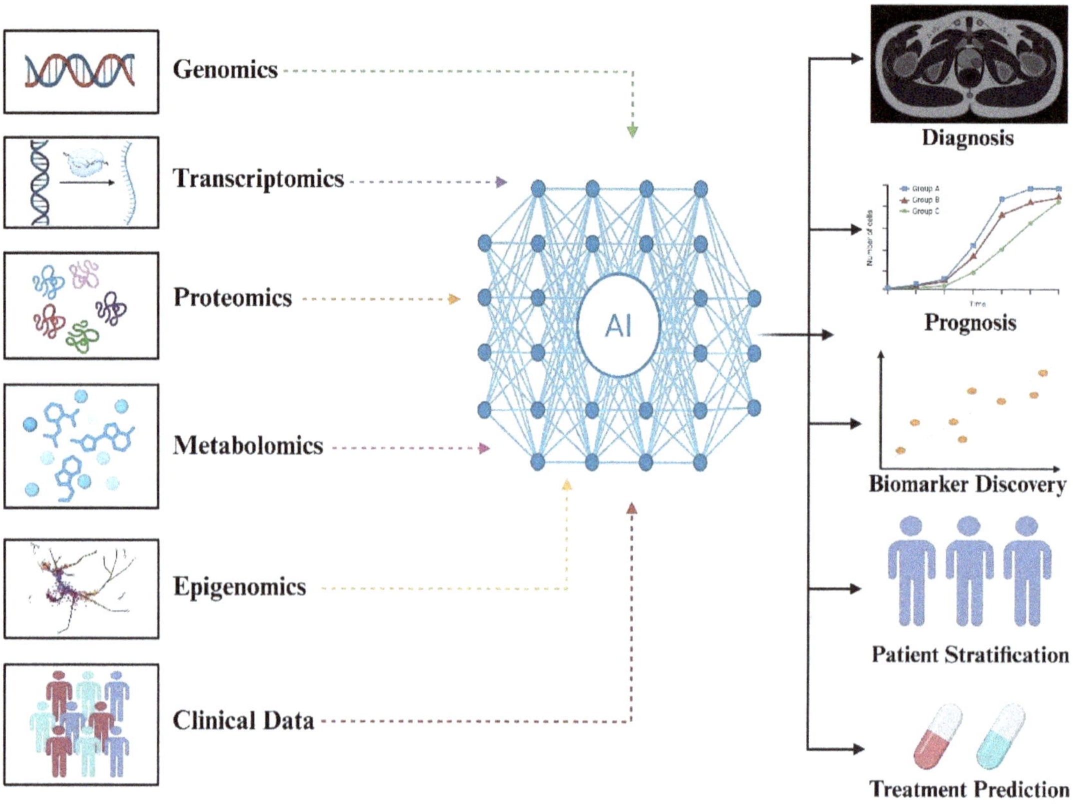

3.2 Predictive Modeling for Diagnosis and Prognosis

The use of AI algorithms to create predictive models is growing in prostate cancer, with the goal of improving diagnosis and prognosis (Baydoun et al 2024, Feretzakis et al, 2024). Machine learning algorithms are able to identify patterns in complex clinical, imaging, and omics data that are associated with disease but would be difficult to identify using a traditional statistical method. These models can provide early diagnosis of cancer, estimation of the aggressiveness of a tumour, prediction of the risk of recurrence, and estimation of the outcomes of treatment (Cai et al, 2022, Battineni and Sagaro 2020). AI systems can continually learn from vast amounts of data, enhancing prediction accuracy over time. These features can help healthcare providers better manage patient care, plan treatment more effectively, and determine if a patient needs more aggressive monitoring or treatment.

3.3 AI-Assisted Biomarker Identification and Patient Stratification

Developing effective biomarkers is crucial for precision oncology, and AI has proven to be a useful tool in the effort. High-dimensional omics data can be analyzed using advanced machine learning methods to find molecular features that are correlated with clinical outcome, therapeutic response, and disease progression (Dorado et al, 2025). These algorithms can use the biological relevance and predictive power of candidate biomarkers for prioritization (Jiang et al, 2026, Ali and Mohammed 2024). The AI also aids to classify patients into molecular subgroups based on risk types and treatment sensitivities. The stratification provides an extra level of personalized treatment, enabling physicians to choose the most effective therapy and reduce unnecessary treatment and enhance patient results.

3.4 Digital Pathology and Clinical Decision Support

Digital pathology involves producing high-resolution images of tissue and using AI-based analytical tools to enhance the analysis of prostate cancer specimens. Deep learning algorithms can also automatically analyze histopathology characteristics, identify tumor regions, estimate Gleason grades, and identify subtle morphological patterns related to disease progression. These technologies increase the consistency of diagnosis and decrease the observer variability (Rabilloud et al, 2023, Van et al, 2029). Outside of the realm of pathology, AI clinical decision support systems incorporate imaging, lab results, multi-omics data and patient records to help clinicians optimize treatment plans and assess risk. Digital pathology and decision-support systems help to improve clinical efficiency, achieve personalised care, and make accurate diagnosis. **Table 2** summarizes the various stages of prostate cancer management where AI can play a role (Zhu et al, 2024, Chan and Ding 2026, Eminaga et al, 2024).

AI has proven to be a powerful tool for prostate cancer diagnosis, prognosis, and clinical decision-making, but there are still some challenges that need to be addressed before it can be widely used in clinical practice. A large number of AI models are developed and tested on the basis of retrospective data collected mostly from Western populations and highly specialized academic institutions. Consequently, their efficacy may be limited to populations that are not underrepresented, in diverse healthcare settings, and in resource limited environments. Another drawback is that many of the machine learning and deep learning algorithms are "black box" models, meaning that they can make a prediction but can't tell you why.

This unexplainability might diminish clinician trust and regulatory acceptance. Moreover, the seamless integration of AI tools into existing hospital information systems, the pathologizing workflows, and electronic health records is technically complex and expensive. Regulatory processes for AI-driven medical innovations are also undergoing transformation, with a need for rigorous testing of safety, reliability, and clinical effectiveness before their widespread adoption. Most notably, although there are some promising studies that have examined the impact of AI-assisted methods in hindsight, there are relatively few prospective clinical trials that have consistently shown improved patient outcomes, survival, or quality of life. Therefore, AI is essentially a tool to support clinical practice and should be thought of as complementary to clinicians' expertise, not a replacement for it.

Table 2. Applications of Artificial Intelligence in Omics-Driven Prostate Cancer Management.

AI Application	Data Sources	Clinical/Reasearch Utility
Biomarker Discovery	Multi-omics datasets	Identification of diagnostic and prognostic biomarkers
Disease Diagnosis	Imaging and pathology data	Early and accurate cancer detection
Prognostic Prediction	Clinical and molecular data	Risk assessment and survival prediction

AI Application	Data Sources	Clinical/Reasearch Utility
Patient Stratification	Multi-omics profiles	Personalized treatment selection
Digital Pathology	Histopathological images	Automated grading and tumor characterization
Drug Discovery	Omics and pharmacological datasets	Identification of novel therapeutic targets

4. NANOTECHNOLOGY- BASED APPROACHES FOR PROSTATE CANCER DIAGNOSIS AND THERAPY

4.1 Nanomaterials and Nano Systems in Oncology

The materials with unique physicochemical properties generated by nanotechnology have opened up new possibilities in cancer research. Compared with traditional therapeutic platforms, nanomaterials like liposomes, polymeric nanoparticles, dendrimers, metallic nanoparticles, and lipid-based nanocarriers can more effectively interact with biological systems (Freitas et al, 2021). For prostate cancer, these nano systems can be engineered for greater stability of drugs, increased bioavailability, and selective targeting to tumor tissue (Kumar et al, 2024, Saadh et al, 2025). It has a large surface area and easy modifiability of its structures, which enable the attachment of targeting ligands, imaging agents, and therapeutic molecules. Therefore, nanotechnology offers several potential platforms to enhance the diagnosis and treatment of prostate cancer.

4.2 Nano-Enabled Imaging and Biosensing

The advancement of imaging agents and biosensors through nanotechnology has given tremendous impetus to the detection of cancer. The increasing signal intensity and contrast of imaging modalities including magnetic resonance imaging (MRI), computed tomography (CT), ultrasound, and fluorescence imaging, can be achieved with the use of nanoparticles (Han and Xu 2029, Li et al 2020). In prostate cancer, nano-enabled biosensors can detect disease associated biomarkers at very low concentration, thus allowing earlier and precise diagnosis (Patra et al, 2026, Chugh et al 2024).

Functionalized nanoparticles can also selectively target molecular targets found only in prostate cells with enhanced diagnostic accuracy (Lai et al 2024). These technologies enable ongoing surveillance of disease activity and treatment effectiveness, which facilitate better treatment management and timely treatment interventions.

4.3 Targeted Drug Delivery and Theragnostic

Targeted drug delivery is one of the exciting uses of nanotechnology for prostate cancer. Therapeutic agents can be packaged in nanocarriers and directed to the tumor cells, while sparing normal tissues from exposure, which helps to enhance therapeutic efficacy and decrease systemic toxicity (Deshmukh et al 2024). Prostate specific membrane antigen is a potential target or other markers that are specific to the tumor can be used to improve the selectivity of the target. Nanotechnology has also led to the creation of theragnostic platforms which incorporate diagnostic and therapeutic capabilities in a single system, besides its application in drug delivery (Fang et al, 2020, Barani et al 2020). The multifunctional nano systems could provide simultaneous disease detection, treatment monitoring, and controlled therapy, which could lead to a more personalized and efficient management of prostate cancer.

4.4 Challenges in Clinical Translation

Although their results in the lab have been promising, there are a number of challenges that remain to get nanotechnology-based treatments into general use clinically. Biological issues include the instability of nanoparticles, recognition by the immune system, variable penetration of tumors, and potential long-term toxicity. Manufacturing and regulatory issues also contribute to the complexity of translation, given that it is not easy to produce in large numbers and quality control standards are not standardized (Bolocan et al, 2025, Koo et al 2019).

Furthermore, the therapeutic performance in therapy is often different in experimental models and in the clinic because of differences between the models and the human tumors (Guo et al 2024). Before going to the clinic, full evaluation in terms of safety, efficacy, pharmacokinetics and biodistribution is crucial (Santis et al, 2020, Zaccariotto et al 2025). However, overcoming these challenges by developing new designs of nanomaterials, harmonizing regulations, and inter-disciplinary collaboration will be essential to fully harness the potential of nanotechnology in prostate cancer treatment.

Nanotechnology platforms have demonstrated great potential in the study of prostate cancer, but there are a number of challenges that must be overcome for successful clinical translation. Therapeutic efficacy may be limited by interindividual variation in the enhanced permeability and retention (EPR) effect, by problem of penetration in heterogeneous and multifocal tumors and by the concern of long-term nanotoxicity. Besides, manufacturing reproducibility, batch-to-batch consistency, scalability and high development cost impede large-scale implementation. Although promising preclinical and early clinical results have been achieved, few nanoformulations have been shown to be superior to existing drug treatments, and prospective randomized evidence that supports a survival benefit or a clinical benefit in terms of quality of life is still limited, thus warranting further clinical validation.

5. CRISPR Technologies for Precision Editing and Functional Genomics

5.1 CRISPR-Cas Systems and Mechanisms

CRISPR (Clustered Regularly Interspaced Short Palindromic Repeats) technology has revolutionized genome engineering by enabling precise and efficient modification of DNA sequences. The system is composed mainly of a Cas nuclease like Cas9 and a guide RNA (gRNA) that guides the Cas nuclease to a specific genomic target (Zhang et al 2021, Shademan et al, 2022). When the target sequence is detected, the Cas protein creates a break within the DNA sequence, which can be repaired by cellular repair processes, causing disruption or correcting or inserting the gene (Yang and Zhang 2023, Allemailem et al 2022). Recently, new CRISPR variants have been developed that achieve even higher levels of precision with minimal off-target modifications, such as base editing and prime editing systems, shown in **figure 3.** These properties render the CRISPR tool very useful to study and treat prostate cancer.

Figure 3. Mechanism and Applications of CRISPR-Cas Systems in Prostate Cancer Research.

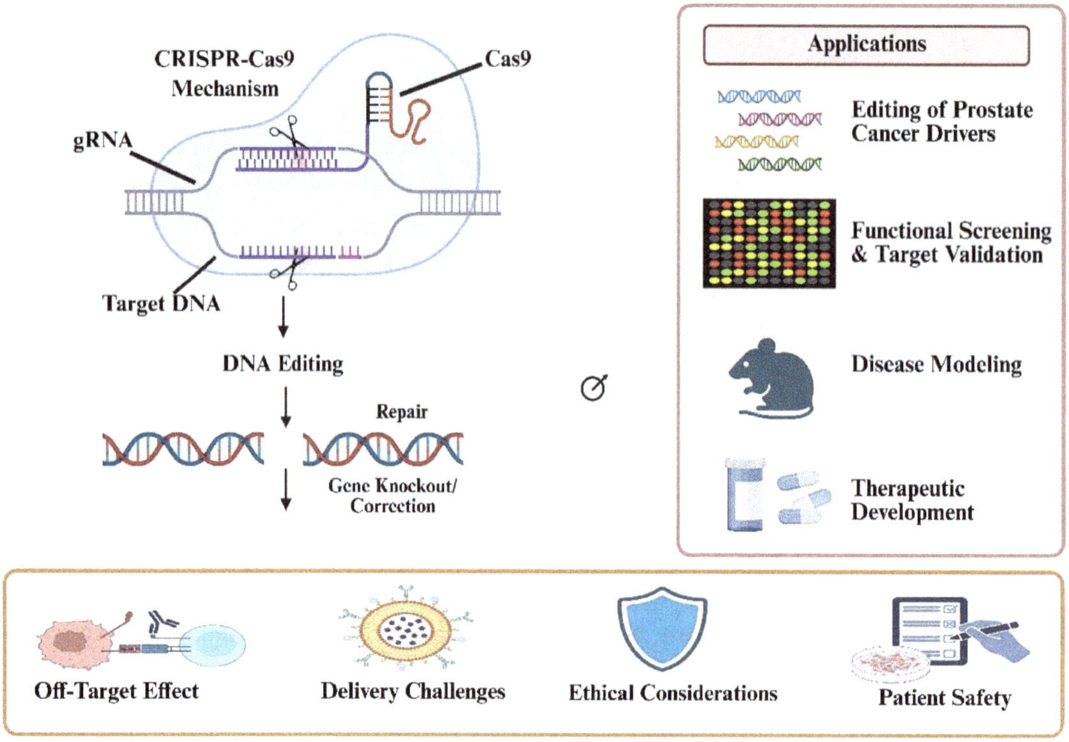

5.2 Gene Editing of Prostate Cancer Drivers

The discovery of important genetic triggers has led to opportunities for therapeutic interventions using the CRISPR technique in prostate cancer. Gene-editing techniques may be used to target oncogenes, tumor suppressor genes, and signaling pathways that play a role in the initiation, progression, and resistance to therapy of tumors. The androgen receptor (AR) signaling axis and genes involved in DNA repair (such as BRCA1 and 2) and the gene PTEN are important targets (Jillson et al 2021, Militaru et al 2023). Using CRISPR technology, these genes can be selectively altered, a process that can clarify the way they are used in living organisms and may help to restore the normal regulation of cells. These approaches can help to enhance the effectiveness of treatment, counteract drug resistance, and facilitate the creation of personalized treatment options based on the individual's molecular profile (Zhang 2021).

5.3 Functional Screening and Target Validations

The power of CRISPR is that it has become an indispensable tool for functional genomics, and is allowing researchers to conduct large-scale genetic screening. Genome-wide CRISPR screens enable researchers to systematically discover genes that participate in prostate cancer growth, metastasis, immune evasion and therapeutic resistance (Bock et al, 2022, Shi et al, 2015). By strategically modifying key genes by knocking them out, activating them, or repressing them, researchers can uncover the roles of individual molecular pathways and test the efficacy of drug targets. These screening techniques provide important information on the biology of the tumour and can highlight drug resistance that can be targeted for therapy (Zhou et al 2024). Combined with multi-omics data, functional studies using CRISPR technology expedite the identification of new biomarkers and therapeutic opportunities for precision oncology.

5.4 Safety, Specificity, and Ethical Considerations

Although the power of CRISPR is tremendous, there are critical safety and ethical concerns to be addressed before the technology can be used in the clinic. Off-target editing is a major issue since unintended changes to the genome could impact cell function and may cause unwanted effects (Emrani et al 2020). To further improve specificity and minimize off-target effects, researchers are working on better guide RNA design, more precise Cas variants, and better editing systems. Ethical issues range from access to gene editing treatments being balanced with equity to the responsible use of genetic information and monitoring of treated individuals over time. To implement effective and ethically acceptable prostate cancer management using the CRISPR-based approach, robust regulatory frameworks and comprehensive safety evaluations will be crucial (Piergentili et al, 2021, Panday et al 2026).

Despite the transformative potential of CRISPR-based genome editing, most prostate cancer applications remain confined to preclinical studies. Prospective clinical evidence demonstrating improvements in survival, disease control, or quality-of-life outcomes is currently lacking, and substantial work remains before CRISPR-based therapies can be routinely integrated into clinical practice.

6. INTEGRATED OMICS-NANO-CRISPR PLATFORMS IN THE AI AGE

6.1 AI-Guided Target Discovery and Therapeutic Design

The use of AI in multi-omics has revolutionized the discovery of therapeutic targets in prostate cancer. While traditional analyses of genomic, transcriptomic, proteomic, and metabolomic data are valuable, AI algorithms can process huge amounts of this data to identify genes, signaling pathways, and molecular interactions that drive disease processes – this process is not possible through traditional analysis (Verma and Kumar 2025, Bae et al 2025). These computational methods allow targets for action and the design of personalized treatment strategies to be prioritized. AI can also aid in the development of new drugs by predicting interactions between drugs and their targets, optimizing drug compounds for improving efficacy, and simulating drug response. AI can be a pivotal element in precision oncology 4.0, enabling quicker data analysis and decision-making (Sebastian and Peter 2022).

6.2 Nano-Carrier-Mediated CRISPR Delivery

One of the biggest hurdles for CRISPR therapeutic delivery is its efficiency and safety. Nanocarrier systems are a promising approach because they can shield the CRISPR components from degradation and deliver them to the target tumor site. Cas proteins, guide RNAs or gene-editing constructs can be encapsulated into lipid nanoparticles, polymeric nanoparticles and other engineered nanocarriers, which can deliver the constructs directly to prostate cancer cells (Vikal et al 2025, Wand et al 2025). Specific delivery results in increased efficiency of editing and decreases off-target effects and systemic toxicity. In addition, tumor-specific ligands can be attached to the nanocarriers to maximize cellular uptake and tissue selectivity. These breakthroughs are crucial to bringing CRISPR technologies into clinical use. As illustrated in table 3, the combination of nanotechnology and CRISPR has led to the development of new treatment approaches for prostate cancer (Seijas et al 2025, Xia et al 2021, Zhuang et al, 2025, Li et al 2020, Lee et al 2021).

Table 3. Nanotechnology and CRISPR-Based Strategies for Prostate Cancer Therapy

Strategy	Mechanism	Advantages	Potential Applications
Lipid Nanoparticles	Encapsulation of CRISPR components	High biocompatibility and delivery efficiency	Gene editing and targeted therapy
Polymeric Nanocarriers	Controlled release systems	Enhanced stability and prolonged circulation	Precision drug and gene delivery
Gold Nanoparticles	Targeted delivery and imaging	Multifunctional therapeutic platform	Theragnostic and tumor targeting
CRISPR-Cas9 Editing	Targeted DNA modification	Precise genome engineering	Functional genomics and therapeutic intervention
CRISPR Base Editing	Single nucleotide modification	Reduced DNA double-strand breaks	Correction of pathogenic mutations

6.3 Multi-Omics Informed Personalized Interventions

Multi-omics analysis generates a global picture of the molecular features of each tumor, which will allow the development of very personalized therapeutic strategies. Combining genomics, transcriptomics, proteomics and metabolomics enables more precise identification of disease drivers unique to individual patients, and a better understanding of their likely response to treatment. These molecular signatures can be used to inform targeted therapies, CRISPR-based strategies, and treatment platforms with nanotechnology (Chen et al, 2021, Johnston et al, 2025). AI also plays a key role in this process by analyzing complex data and providing clinical insights that can guide care. Multi-omics data and cutting-edge computational technologies enable precision medicine strategies that can deliver the maximum benefit from drugs and other interventions and reduce over-treatment and side effects (Ahmed 2020).

6.4 Emerging Synergistic and Translational Strategies

The intersection of omics technologies, nanotechnology, CRISPR genome editing, and artificial intelligence is generating extremely promising synergistic approaches to prostate cancer treatment (Farghama et al 2026, Kwon et al, 2025). Multi-omics platforms provide detailed molecular data, AI systems can analyze such data to identify clinically relevant patterns, genetic manipulation using CRISPR can be used to make precise changes, and targeted delivery and monitoring using nanotechnology (Qi et al 2024). All these technologies create integrated therapeutic ecosystems that are able to tackle the complexity and heterogeneity of prostate cancer (Vakili et al 2025). The current research on translation emphasizes creating and developing a platform of intelligent, patient-specific diagnostic, target identification, therapeutic intervention, and response monitoring platforms. While some obstacles exist, the integrated approaches have great promise for progressing precision oncology and for better long-term clinical outcomes.

6.5 Translational Reality Versus Technological Promise

Despite significant advancements, multi-omics technologies, artificial intelligence (AI), nanotechnology, and CRISPR-based gene editing have not yet revolutionized the daily treatment of prostate cancer. Despite the fact that these technologies have enhanced our understanding of disease biology and opened up new avenues for diagnosis and treatment, a number of translational obstacles still stand in the way of their widespread clinical use. Numerous prospective biomarkers that were found using multi-omics techniques have now been clinically confirmed and authorized for regular usage.

Similar to this, there are potential uses of AI in pathology, imaging, and predictive modeling; however, issues with data bias, interpretability, generalizability, and integration into clinical workflow have not yet been resolved. Scalability, manufacturing repeatability, tumor penetration, and long-term safety are only a few of the difficulties that still exist in the clinical translation of nanotechnology-based systems, despite their encouraging preclinical outcomes in the delivery of targeted medicines and theragnostics. Similar to how CRISPR has transformed functional genomics, delivery problems, a lack of long-term safety evidence, off-target effects, and ethical considerations remain obstacles to the practical use of this technique in prostate cancer. The evidence base is still biased toward retrospective studies, exploratory research, and preclinical studies, all of which are useful but not exactly the same thing, despite advancements in multi-omics, AI, nanotechnology, and CRISPR. It is comparatively uncommon to find well-designed prospective randomized studies with clinically meaningful evidence of patient benefits.

7. FUTURE PERSPECTIVE AND CLINICAL TRANSLATIONAL

7.1 Towards Precision and Personalized Prostate Cancer Care

The future of prostate cancer treatment is moving from population-based treatment models to precision and personalized treatment (Menon et al 2025, Rituraj et al 2025). The growing number of multi-omics technologies coupled with the development of artificial intelligence, nanomedicine and CRISPR-based genome editing are helping to gain a deeper understanding of tumour heterogeneity and individual disease characteristics (Rahman et al 2026, Gómez-Luque et al, 2026). These innovations enable the design of personalized diagnosis, prognosis and therapy as well as to increase the efficacy of therapy with fewer side effects. Combining molecular profiling with real-time clinical information is anticipated to help guide dynamic treatment planning and disease monitoring (Molla and Bitew 2025). With the advancement of these technologies, personalized medicine can have a profound impact on patient outcomes and quality of life.

7.2 Regulatory, Ethical, and Data-Governance Considerations

To realize the clinical application of new precision oncology technologies, a strong regulatory framework and ethical supervision are necessary. Multi-omic studies and AI systems generate vast amounts of patient information, which must be sensitive and subject to data privacy, security, informed consent, and data ownership concerns (Dupras and Bunnik 2021, Sabha 2025). Likewise, the ethical concerns of genetic manipulation, long-term safety, and equitable access to new treatments arise from the use of genome editing with CRISPR (Gupta et al, 2025, Schleidgen et al 2020). The regulatory bodies need to define clear guidelines to recognize, approve and monitor emerging technologies to ensure patient safety and clinical reliability. Clear governance processes and multi-disciplinary working will be key to building public trust and responsible innovation in prostate cancer care (Nahata et al 2026).

7.3 Current Limitations and Research Opportunities

While significant advances have been made in recent years, there are still several challenges that hinder the broad clinical use of the integrated omics–nano–CRISPR technologies. These include the high costs of implementation, data quality variability, lack of standardisation and difficulty in translating results from the laboratory to routine clinical practice, biological complexity (Kang 2026). There are also the issues of interpretability of AI models, dataset bias, and testing across various patient populations. Better integration of data, increased precision in gene-editing, safer nanocarriers, and large-scale clinical studies should be explored in future studies. Solving these challenges will speed the delivery of innovative technologies to the field of prostate cancer care where they can become effective, accessible, and evidence-based solutions. Although there have been significant advances in technology, there are still some issues to be addressed in the scientific and translational aspects. **Table 4** identifies and outlines key limitations and opportunities for the future (Liu et al 2026, Zack et al 2025, Tafirenyika 2023, Han et al 2020, Sharma et al 2021).

7.3.1 Tumor Heterogeneity and Overdiagnosis Challenges

Prostate cancer exhibits substantial biological heterogeneity, including intratumoral variability, multifocal disease, and dynamic molecular evolution during treatment, which complicate biomarker discovery, validation, and reproducibility. Molecular signatures identified in one tumor region or patient population may not consistently predict outcomes across diverse clinical settings. Additionally, highly sensitive multi-omics, molecular diagnostic, and imaging technologies may increase the detection of clinically indolent lesions, raising concerns regarding overdiagnosis and overtreatment. Future precision oncology strategies must address both temporal and spatial tumor heterogeneity while balancing early detection with avoidance of unnecessary interventions to improve biomarker reliability, patient stratification, and clinical translation

Table 4. Challenges and Future Directions of Omics-Nano-CRISPR approaches in the AI Age.

Challenge	Impact	Future Direction
Multi-omics Data Complexity	Difficult data integration and interpretation	Advanced AI-driven analytical frameworks
AI Model Bias and Interpretability	Reduced clinical reliability	Development of explainable AI systems
CRISPR Off-Target Effects	Safety concerns	High-fidelity editing systems
Nanoparticle Toxicity	Potential adverse biological effects	Biocompatible and biodegradable nanomaterials
Regulatory Uncertainty	Delayed clinical translation	Harmonized regulatory guidelines
High Treatment Costs	Limited accessibility	Cost-effective precision medicine platforms

7.4 Equity and Global Accessibility Challenges

Precision oncology technologies have great potential to enhance prostate cancer treatment, but these benefits are not necessarily uniformly shared across populations in the world. The technology required to implement comprehensive genomic sequencing, advanced molecular profiling, AI-powered analytical platforms and nanotechnology-based therapeutics can be expensive and demand specialized infrastructure. These requirements could be challenging, especially in low- and middle-income countries and in resource-limited health care systems. Moreover, the widespread machine learning models tend to be trained on data from high-income countries, which might limit their applicability and performance in lower-resourced areas. If not targeted to make it more affordable, build infrastructure to help, and diversify the population of biomedical researchers, precision oncology could inadvertently exacerbate current disparities in health care. Access to emerging technologies should, therefore, continue to be a critical issue in future research, policy production and implementation strategies.

8. CONCLUSION

The molecular heterogeneity, complexity of disease progression and frequent development of therapeutic resistance make prostate cancer one of the biggest health challenges worldwide. In recent years, precision medicine has revolutionized the knowledge and management of this disease, moving beyond a blanket approach to more personalized treatment. The advent of multi-omics technologies, artificial intelligence (AI), nanotechnology and CRISPR-based genome editing has been a major catalyst for the transition. Together, these technologies have contributed to a better understanding of the genetic, transcriptomic, epigenetic, proteomic, and metabolic pathways involved in prostate cancer, facilitating the discovery of new biomarkers, molecular classification of prostate cancers, patient stratification, and the identification of potential therapeutic targets. The integration and interpretation of complex biological, clinical and imaging data have been further enhanced with AI, enabling precision oncology.

Likewise, the development of nanotechnology has enabled breakthroughs in disease detection, targeted delivery of drugs and theragnostic applications, and CRISPR technologies have led to significant progress in functional genomics research and targeted genome engineering. These strategies can combine to create a more powerful system for developing individualized diagnostic and therapeutic strategies for prostate cancer. These encouraging advances, however, should be viewed with proper circumspection in the light of their clinical results. A number of the biomarkers, decision support systems based on artificial intelligence, nano-mediated therapeutics, and CRISPR-based interventions derived from the omics are still in early developmental and/or translation phases. Data standardization, reproducibility, interpretability of models, scalability of manufacturing, long-term safety, regulatory clearance, cost-effectiveness and availability of access remain challenges to widespread clinical implementation.

Moreover, problems with tumour heterogeneity, overtreatment of non-aggressive disease and the lack of prospective randomized data showing survival and quality of life benefits of routine clinical use of these technologies further underscore the disconnect between technology promise and practice. Finally, the combination of omics sciences, AI, nanotechnology, and CRISPR is a promising and developing paradigm for precision prostate cancer treatment. At the moment its effect is more evolutionary than revolutionary, however. Continued technological innovation and rigorous clinical validation, standardisation, responsible regulation and equitable access to ensure meaningful and sustainable benefits to patients globally will be required for future progress.

REFERENCES

Agarwal, N., & Nupur, P. K. (2024). Paul, and S. K. Mishra, "Artificial Intelligence and Machine Learning for Analysis of Multi-omics. In Mani, I., & Singh, V. (Eds.), *Multi-Omics Analysis of the Human Microbiome: From Technology to Clinical Applications* (pp. 339–354). Springer Nature Singapore., DOI: 10.1007/978-981-97-1844-3_16

Ahmed, Z. (2020). Practicing precision medicine with intelligently integrative clinical and multi-omics data analysis. *Human Genomics, 14*(1), 35. DOI: 10.1186/s40246-020-00287-z PMID: 33008459

Al-Daffaie, F. M., Al-Mudhafar, S. F., Alhomsi, A., Tarazi, H., Almehdi, A. M., El-Huneidi, W., Abu-Gharbieh, E., Bustanji, Y., Alqudah, M. A., Abuhelwa, A. Y., Guella, A., Alzoubi, K. H., & Semreen, M. H. (2024, May 7). Metabolomics and proteomics in prostate cancer research: Overview, analytical techniques, data analysis, and recent clinical applications. *International Journal of Molecular Sciences, 25*(10), 5071. DOI: 10.3390/ijms25105071 PMID: 38791108

Ali, A. M., & Mohammed, M. A. (2024). A comprehensive review of artificial intelligence approaches in omics data processing: Evaluating progress and challenges. *Int. J. Math. Stat. Comput. Sci., 2*, 114–167.

Allemailem, K. S., Alsahli, M. A., Almatroudi, A., Alrumaihi, F., Alkhaleefah, F. K., Rahmani, A. H., & Khan, A. A. (2022, December). Current updates of CRISPR/Cas9-mediated genome editing and targeting within tumor cells: An innovative strategy of cancer management. *Cancer Communications, 42*(12), 1257–1287. DOI: 10.1002/cac2.12366 PMID: 36209487

Athanasopoulou, K., Daneva, G. N., Adamopoulos, P. G., & Scorilas, A. (2022). Artificial intelligence: The milestone in modern biomedical research. *BioMedInformatics, 2*(4), 727–744. DOI: 10.3390/biomedinformatics2040049

Bae, H., Ji, H., Konstantinov, K., Sluyter, R., Ariga, K., Kim, Y. H., & Kim, J. H. (2025, October). Artificial intelligence-driven nanoarchitectonics for smart targeted drug delivery. *Advanced Materials, 37*(42), e10239. DOI: 10.1002/adma.202510239 PMID: 40772340

Barani, M., Sabir, F., Rahdar, A., Arshad, R., & Kyzas, G. Z. (2020). Nanotreatment and nanodiagnosis of prostate cancer: Recent updates. *Nanomaterials (Basel, Switzerland), 10*(9), 1696. DOI: 10.3390/nano10091696 PMID: 32872181

Battineni, G., Sagaro, G. G., Chinatalapudi, N., & Amenta, F. (2020, March 31). Applications of Machine Learning Predictive Models in the Chronic Disease Diagnosis. *Journal of Personalized Medicine, 10*(2), 21. DOI: 10.3390/jpm10020021 PMID: 32244292

Baydoun, A., Jia, A. Y., Zaorsky, N. G., Kashani, R., Rao, S., Shoag, J. E., Vince, R. A.Jr, Bittencourt, L. K., Zuhour, R., Price, A. T., Arsenault, T. H., & Spratt, D. E. (2024, March). Artificial intelligence applications in prostate cancer. *Prostate Cancer and Prostatic Diseases, 27*(1), 37–45. DOI: 10.1038/s41391-023-00684-0 PMID: 37296271

Bock, C., Datlinger, P., Chardon, F., Coelho, M. A., Dong, M. B., Lawson, K. A., Lu, T., Maroc, L., Norman, T. M., Song, B., Stanley, G., Chen, S., Garnett, M., Li, W., Moffat, J., Qi, L. S., Shapiro, R. S., Shendure, J., Weissman, J. S., & Zhuang, X. (2022, February 10). High-content CRISPR screening. *Nature Reviews. Methods Primers*, *2*(1), 8. DOI: 10.1038/s43586-021-00093-4 PMID: 37214176

Bolocan, V.-O., Mitoi, A., Nicu-Canareica, O., Băean, M.-L., Medar, C., & Popa, G.-A. (2025). Artificial Intelligence in Prostate MRI: Current Evidence and Clinical Translation Challenges—A Narrative Review. *Journal of Imaging*, *11*(10), 335. Advance online publication. DOI: 10.3390/jimaging11100335 PMID: 41150011

Cai, Z., Poulos, R. C., Liu, J., & Zhong, Q. (2022, January 22). Machine learning for multi-omics data integration in cancer. *iScience*, *25*(2), 103798. DOI: 10.1016/j.isci.2022.103798 PMID: 35169688

Chan, G. J., & Ding, C. C. (2026, March). Advances in artificial intelligence in prostate cancer pathology. *Seminars in Diagnostic Pathology*, *43*(2), 150995. DOI: 10.1016/j.semdp.2026.150995 PMID: 41807227

Chen, F., Alphonse, M., & Liu, Q. (2020). Strategies for nonviral nanoparticle-based delivery of CRISPR/Cas9 therapeutics. *Wiley Interdisciplinary Reviews. Nanomedicine and Nanobiotechnology*, *12*(3), e1609. DOI: 10.1002/wnan.1609 PMID: 31797562

Chugh, V., Basu, A., Kaushik, A., Bhansali, S., & Basu, A. K. (2024). Employing nano-enabled artificial intelligence (AI)-based smart technologies for prediction, screening, and detection of cancer. *Nanoscale*, *16*(11), 5458–5486. DOI: 10.1039/D3NR05648A PMID: 38391246

da Silva, R. G. L. (2024, May 21). The advancement of artificial intelligence in biomedical research and health innovation: Challenges and opportunities in emerging economies. *Globalization and Health*, *20*(1), 44. DOI: 10.1186/s12992-024-01049-5 PMID: 38773458

De Santis, M., Mottet, N., Cornford, P., & Gillessen, S. (2020). Precision oncology for metastatic prostate cancer: Translation into practice. *European Urology*, *78*(6), 771–774. DOI: 10.1016/j.eururo.2020.08.021 PMID: 32863055

Deshmukh, R., Singh, V., Harwansh, R. K., Agrawal, R., Garg, A., Singh, S., Elossaily, G. M., Ansari, M. N., Ali, N., & Prajapati, B. G. (2024, February 20). Emerging trends of nanomedicines in the management of prostate cancer: Perspectives and potential applications. *Pharmaceutics*, *16*(3), 297. DOI: 10.3390/pharmaceutics16030297 PMID: 38543191

Dimitrov, G., & Popov, E. (2026). In Vaschetto, L. M., & Srivastava, S. (Eds.), *Precision Medicine in Prostate Cancer BT - Molecular Precision Medicine in Cancer* (pp. 105–135). Springer Nature Switzerland., DOI: 10.1007/978-3-032-24071-2_4

Dorado, C. Q., Ramírez, A. S., Pérez, M. P., Centeno, M. S., Mici, L. P., Güemez, C. M., Albers Acosta, E., Luis, G. C., Costal, M., Diez, P. T., Laorden, N. R., Díaz, R. J., Velasco Balanza, C., & Manso, L. S. J. (2025, November 24). Redefining Prostate Cancer Precision: Radiogenomics, Theragnostics, and AI-Driven Biomarkers. *Cancers (Basel)*, *17*(23), 3747. DOI: 10.3390/cancers17233747 PMID: 41374949

Dupras, C., & Bunnik, E. M. (2021). Toward a framework for assessing privacy risks in multi-omic research and databases. *The American Journal of Bioethics : AJOB*, *21*(12), 46–64. DOI: 10.1080/15265161.2020.1863516 PMID: 33433298

Eminaga, O., Abbas, M., Kunder, C., Tolkach, Y., Han, R., Brooks, J. D., Nolley, R., Semjonow, A., Boegemann, M., West, R., Long, J., Fan, R. E., & Bettendorf, O. (2024, March 4). Critical evaluation of artificial intelligence as a digital twin of pathologists for prostate cancer pathology. *Scientific Reports*, *14*(1), 5284. DOI: 10.1038/s41598-024-55228-w PMID: 38438436

Emrani, J., Ahmed, M., Newman, R. H., Thomas, M. D., Mowa, C. N., & Teleha, J. (2020, July). Introducing Chemistry Students to Emerging Technologies in Gene Editing, Their Applications, and Ethical Considerations. *Journal of Chemical Education*, *97*(7), 1931–1943. DOI: 10.1021/acs.jchemed.9b01154

Fang, Y., Lin, S., Yang, F., Situ, J., Lin, S., & Luo, Y. (2020). Aptamer-Conjugated Multifunctional Polymeric Nanoparticles as Cancer-Targeted, MRI-Ultrasensitive Drug Delivery Systems for Treatment of Castration-Resistant Prostate Cancer. *BioMed Research International*, *2020*(1), 9186583. DOI: 10.1155/2020/9186583 PMID: 32420382

Farghama, K., Khan, M. S., Wajid, M., & Shaffique, S. (2026). AI Driven Computational Nanomedicine for Personalized Cancer Therapy. *BioNanoScience*, *16*(7), 447. DOI: 10.1007/s12668-026-02681-8

Feretzakis, G., Juliebø-Jones, P., Tsaturyan, A., Sener, T. E., Verykios, V. S., Karapiperis, D., Bellos, T., Katsimperis, S., Angelopoulos, P., Varkarakis, I., Skolarikos, A., Somani, B., & Tzelves, L. (2024, February 16). Emerging trends in AI and radiomics for bladder, kidney, and prostate cancer: A critical review. *Cancers (Basel)*, *16*(4), 810. DOI: 10.3390/cancers16040810 PMID: 38398201

Freitas, L. F., Ferreira, A. H., Thipe, V. C., Varca, G. H., Lima, C. S., Batista, J. G., Riello, F. N., Nogueira, K., Cruz, C. P., Mendes, G. O., Rodrigues, A. S., Sousa, T. S., Alves, V. M., & Lugão, A. B. (2021, September 30). The state of the art of theranostic nanomaterials for lung, breast, and prostate cancers. *Nanomaterials (Basel, Switzerland)*, *11*(10), 2579. DOI: 10.3390/nano11102579 PMID: 34685018

Gómez-Luque, M. Á., Rodríguez-Marcos, P., Campanario-Pérez, R., Medina-González, A., Cárdenas-Fernández, A., Orlandi-Oliveira, W., & Ruibal-Moldes, M. (2026, February). Towards precision oncology: A new predictive machine learning model for early progression to castration resistant prostate cancer. *The Prostate*, *86*(3), 320–328. DOI: 10.1002/pros.70088 PMID: 41186168

Guo, H., Xu, X., Zhang, J., Du, Y., Yang, X., He, Z., Zhao, L., Liang, T., & Guo, L. (2024, August 9). The pivotal role of preclinical animal models in anti-cancer drug discovery and personalized cancer therapy strategies. *Pharmaceuticals (Basel, Switzerland)*, *17*(8), 1048. DOI: 10.3390/ph17081048 PMID: 39204153

Gupta, U., Pranav, A., Modi, R. K., & Dubey, A. (2025). Ethical implications of AI in CRISPR: responsible genome editing. In *Bioinformatics and Beyond* (pp. 256–275). CRC Press. DOI: 10.1201/9781003508403-13

Hachem, S., Yehya, A., El Masri, J., Mavingire, N., Johnson, J. R., Dwead, A. M., Kattour, N., Bouchi, Y., Kobeissy, F., Rais-Bahrami, S., Mechref, Y., Abou-Kheir, W., & Woods-Burnham, L. (2024, September 25). Contemporary update on clinical and experimental prostate cancer biomarkers: A multi-omics-focused approach to detection and risk stratification. *Biology (Basel)*, *13*(10), 762. DOI: 10.3390/biology13100762 PMID: 39452071

Han, H. A., Pang, J. K. S., & Soh, B.-S. (2020). Mitigating off-target effects in CRISPR/Cas9-mediated in vivo gene editing. *Journal of Molecular Medicine : Official Organ of the "Gesellschaft Deutscher Naturforscher und Ärzte"*, *98*(5), 615–632. DOI: 10.1007/s00109-020-01893-z PMID: 32198625

Han, X., Xu, K., Taratula, O., & Farsad, K. (2019, January 17). Applications of nanoparticles in biomedical imaging. *Nanoscale*, *11*(3), 799–819. DOI: 10.1039/C8NR07769J PMID: 30603750

Hashemi, M., Khosravinejad, A., Alimohammadi, M., Jamshidian, F., Shirazi, R. M., Ghannad Zadeh, R., Jamali, B., Esfahani, K. S., Shirazi, A., Zali, N., & Farahani, N. 2026. CRISPR/Cas9 in the Treatment of Prostate Cancer. In *INNOVATIVE THERAPEUTIC APPROACHES BY CRISPR/CAS9 IN CANCER THERAPY* (pp. 226-253).

Jena, R., Samal, H. B., Sharma, J., Suresh, P., Mishra, A. P., & Nigam, M. (2025). Biotechnology in drug discovery and development for cancer. In *Biotechnology and cancer therapeutics* (pp. 447–478). Springer. DOI: 10.1007/978-981-96-4959-4_14

Jiang, S., Yang, C., Huang, Z., Guo, Z., Lu, F., Nian, X., Chen, Z., Luo, P., Jiang, J., Gao, X., Li, M., & Liu, F. (2026, April). Multicentre Evaluation of an AI-Assisted Urine Test for Clinically Significant Prostate Cancer in Men Undergoing Initial Biopsy. *Journal of Extracellular Vesicles*, *15*(4), e70233. DOI: 10.1002/jev2.70233 PMID: 41877634

Jiang, Y., Wang, C., Zu, C., Rong, X., Yu, Q., & Jiang, J. (2024, October 2). Synergistic Potential of Nanomedicine in Prostate Cancer Immunotherapy: Breakthroughs and Prospects. *International Journal of Nanomedicine*, *19*, 9459–9486. DOI: 10.2147/IJN.S466396 PMID: 39371481

Jillson, L. K., Yette, G. A., Laajala, T. D., Tilley, W. D., Costello, J. C., & Cramer, S. D. (2021). Androgen Receptor Signaling in Prostate Cancer Genomic Subtypes. *Cancers (Basel)*, *13*(13), 3272. Advance online publication. DOI: 10.3390/cancers13133272 PMID: 34208794

Johnston, M., Dissanayake-Perera, S., Collins, J. J., Stevens, M. M., & Dincer, C. (2025). Convergence of nanotechnology and CRISPR-based diagnostics. *Nature Nanotechnology*, *20*(10), 1365–1373. DOI: 10.1038/s41565-025-02018-8 PMID: 41039096

J. Kang, X. Ke, Y. Zhao, and Y. Zhou, "Nanotechnology-Assisted Molecular Profiling: Emerging Advances in Circulating Tumor DNA Detection," *Int. J. Nanomedicine*, vol. 21, no. null, p. 585118, Dec. 2026, .DOI: 10.2147/IJN.S585118

Koo, K. M., Mainwaring, P. N., Tomlins, S. A., & Trau, M. (2019). Merging new-age biomarkers and nanodiagnostics for precision prostate cancer management. *Nature Reviews. Urology*, *16*(5), 302–317. DOI: 10.1038/s41585-019-0178-2 PMID: 30962568

Kumar, A., Lunawat, A. K., Kumar, A., Sharma, T., Islam, M. M., Kahlon, M. S., Mukherjee, D., Narang, R. K., & Raikwar, S. (2024, March 6). Recent trends in nanocarrier-based drug delivery system for prostate cancer. *AAPS PharmSciTech*, *25*(3), 55. DOI: 10.1208/s12249-024-02765-2 PMID: 38448649

Kumar, V., Randhawa, P., Bilodeau, R., Mercola, D., McClelland, M., Agrawal, A., Nguyen, J., Castro, P., Ittmann, M. M., & Rahmatpanah, F. (2022, October 8). Spatial profiling of the prostate cancer tumor microenvironment reveals multiple differences in gene expression and correlation with recurrence risk. *Cancers (Basel)*, *14*(19), 4923. DOI: 10.3390/cancers14194923 PMID: 36230846

Kwon, W. A., & Joung, J. Y. (2025, April 27). Precision Targeting in Metastatic Prostate Cancer: Molecular Insights to Therapeutic Frontiers. *Biomolecules*, *15*(5), 625. DOI: 10.3390/biom15050625 PMID: 40427518

W.-A. Kwon, H. Kim, and Y. S. Song, "Prostate Cancer at the Microbial Crossroads: Illuminating a New Frontier in Precision Medicine," *World J. Mens. Health*, vol. 43, 2025.

Lai, J., Luo, Z., Chen, L., & Wu, Z. (2024). Advances in nanotechnology-based targeted-contrast agents for computed tomography and magnetic resonance. *Science Progress*, *107*(1), 00368504241228076. DOI: 10.1177/00368504241228076 PMID: 38332327

Lee, H. K., Oh, Y., Hong, J., Lee, S. H., & Hur, J. K. (2021, February). Development of CRISPR technology for precise single-base genome editing: A brief review. *BMB Reports*, *54*(2), 98–105. DOI: 10.5483/BMBRep.2021.54.2.217 PMID: 33298245

Li, H., Yang, Y., Hong, W., Huang, M., Wu, M., & Zhao, X. (2020). Applications of genome editing technology in the targeted therapy of human diseases: Mechanisms, advances and prospects. *Signal Transduction and Targeted Therapy*, *5*(1), 1. DOI: 10.1038/s41392-019-0089-y PMID: 32296011

Li, L., Cho, K. H., Yu, X., & Cheng, S. (2025, May). Systematic multi-omics investigation of androgen receptor driven gene expression and epigenetics changes in prostate cancer. *Computers in Biology and Medicine*, *189*, 110000. DOI: 10.1016/j.compbiomed.2025.110000 PMID: 40056843

Li, L., Guan, Y., Xiong, H., Deng, T., Ji, Q., Xu, Z., Kang, Y., & Pang, J. (2020, September). Fundamentals and applications of nanoparticles for ultrasound-based imaging and therapy. *Nano Select*, *1*(3), 263–284. DOI: 10.1002/nano.202000035

Liu, F., Beck, S., Yang, L., Luo, H., & Zhang, K. (2026). Advancing AI for multi-omics and clinical data integration in basic and translational cancer research. *Nature Reviews. Cancer*, *26*(7), 1–16. DOI: 10.1038/s41568-026-00922-2 PMID: 42014628

Marzec, J., Ross-Adams, H., Pirrò, S., Wang, J., Zhu, Y., Mao, X., Gadaleta, E., Ahmad, A. S., North, B. V., Kammerer-Jacquet, S. F., Stankiewicz, E., Kudahetti, S. C., Beltran, L., Ren, G., Berney, D. M., Lu, Y.-J., & Chelala, C. (2021, January 19). The transcriptomic landscape of prostate cancer development and progression: An integrative analysis. *Cancers (Basel)*, *13*(2), 345. DOI: 10.3390/cancers13020345 PMID: 33477882

Menon, A. V., Song, B., Chao, L., Sriram, D., Chansky, P., Bakshi, I., Ulianova, J., & Li, W. (2025, April 11). Unraveling the future of genomics: CRISPR, single-cell omics, and the applications in cancer and immunology. *Frontiers in Genome Editing*, *7*, 1565387. DOI: 10.3389/fgeed.2025.1565387 PMID: 40292231

Militaru, F. C., Militaru, V., Crisan, N., Bocsan, I. C., Udrea, A. A., Catana, A., Kutasi, E., & Militaru, M. S. (2023, October 1). Molecular basis and therapeutic targets in prostate cancer: A comprehensive review. *Biomolecules and Biomedicine*, *23*(5), 760–771. DOI: 10.17305/bb.2023.8782 PMID: 37021836

Molla, G., & Bitew, M. (2025). The Future of Cancer Diagnosis and Treatment: Unlocking the Power of Biomarkers and Personalized Molecular-Targeted Therapies. *Journal of Molecular Pathology*, *6*(3), 20. Advance online publication. DOI: 10.3390/jmp6030020

Nahata, L., Khare, V., Madnoorkar, N., Khare, S., & Jain, R. (2026). Multidisciplinary Cooperation: Combining Viewpoints to Provide the Best Possible Patient Care. In *Individualized Treatment Strategies in the Radiotherapy Era* (pp. 136–150). CRC Press.

Naskar, S., Sharma, S., Kuotsu, K., Halder, S., Pal, G., Saha, S., Mondal, S., Biswas, U. K., Jana, M., & Bhattacharjee, S. (2025, May 28). The biomedical applications of artificial intelligence: An overview of decades of research. *Journal of Drug Targeting*, *33*(5), 717–748. DOI: 10.1080/1061186X.2024.2448711 PMID: 39744873

Nevedomskaya, E., & Haendler, B. (2022, June 3). From Omics to Multi-Omics Approaches for In-Depth Analysis of the Molecular Mechanisms of Prostate Cancer. *International Journal of Molecular Sciences*, *23*(11), 6281. DOI: 10.3390/ijms23116281 PMID: 35682963

Pandey, A., Dey, S., & Gayen, D. (2026). Advances in CRISPR-Cas technology, mechanisms, applications, and ethical challenges. *Discover Plants*, *3*(1), 243. DOI: 10.1007/s44372-026-00705-x

Patra, S., Shand, H., & Ghorai, S. (2026). Nanobiosensor-Based Strategies for Early Cancer Detection and Therapeutic Monitoring. *Biomedical Materials & Devices*, •••. Advance online publication. DOI: 10.1007/s44174-026-00688-z

Pecoraro, M., Messina, E., Novelli, S., Laschena, L., Blasilli, G., Tronci, E., & Panebianco, V. (2026, June 19). Digital twin technologies in prostate cancer as a frontier for precision medicine. *European Radiology Experimental*, *10*(1), 93. DOI: 10.1186/s41747-026-00753-8 PMID: 42319650

Piergentili, R., Del Rio, A., Signore, F., Umani Ronchi, F., Marinelli, E., & Zaami, S. (2021). CRISPR-Cas and Its Wide-Ranging Applications: From Human Genome Editing to Environmental Implications, Technical Limitations, Hazards and Bioethical Issues. *Cells*, *10*(5), 969. Advance online publication. DOI: 10.3390/cells10050969 PMID: 33919194

Pomerantz, M. M., Qiu, X., Zhu, Y., Takeda, D. Y., Pan, W., Baca, S. C., Gusev, A., Korthauer, K. D., Severson, T. M., Ha, G., Viswanathan, S. R., Seo, J.-H., Nguyen, H. M., Zhang, B., Pasaniuc, B., Giambartolomei, C., Alaiwi, S. A., Bell, C. A., O'Connor, E. P., & Freedman, M. L. (2020, August). Prostate cancer reactivates developmental epigenomic programs during metastatic progression. *Nature Genetics*, *52*(8), 790–799. DOI: 10.1038/s41588-020-0664-8 PMID: 32690948

Qi, L., Li, Z., Liu, J., & Chen, X. (2024). Omics-enhanced nanomedicine for cancer therapy. *Advanced Materials*, *36*(50), 2409102. DOI: 10.1002/adma.202409102 PMID: 39473316

Rabaan, A. A., Bakhrebah, M. A., AlSaihati, H., Alhumaid, S., Alsubki, R. A., Turkistani, S. A., Al-Abdulhadi, S., Aldawood, Y., Alsaleh, A. A., Alhashem, Y. N., Almatouq, J. A., Alqatari, A. A., Alahmed, H. E., Sharbini, D. A., Alahmadi, A. F., Alsalman, F., Alsayyah, A., & Mutair, A. A. (2022, November 14). Artificial intelligence for clinical diagnosis and treatment of prostate cancer. *Cancers (Basel)*, *14*(22), 5595. DOI: 10.3390/cancers14225595 PMID: 36428686

Rabilloud, N., Allaume, P., Acosta, O., De Crevoisier, R., Bourgade, R., Loussouarn, D., Rioux-Leclercq, N., Khene, Z. E., Mathieu, R., Bensalah, K., Pecot, T., & Kammerer-Jacquet, S.-F. (2023, August 14). Deep learning methodologies applied to digital pathology in prostate cancer: A systematic review. *Diagnostics (Basel)*, *13*(16), 2676. DOI: 10.3390/diagnostics13162676 PMID: 37627935

Rahman, S., Arun, A. S., Kim, I. Y., Oh, W. K., Kim, J. W., & Kim, W. J. (2026). Navigating the winding road toward precision prostate cancer care. *Gene, 984*, 149966. DOI: 10.1016/j.gene.2025.149966 PMID: 41419150

Rawla, P. (2019, April). Epidemiology of Prostate Cancer. *World Journal of Oncology, 10*(2), 63–89. DOI: 10.14740/wjon1191 PMID: 31068988

Ren, S., Li, J., Dorado, J., Sierra, A., Gonzalez-Diaz, H., Duardo, A., & Shen, B. (2023, December 18). From molecular mechanisms of prostate cancer to translational applications: Based on multi-omics fusion analysis and intelligent medicine. *Health Information Science and Systems, 12*(1), 6. DOI: 10.1007/s13755-023-00264-5 PMID: 38125666

Rituraj, R. S., Pal, R. S., Wahlang, J., Pal, Y., Chaitanya, M. V. N. L., & Saxena, S. (2025). Pal, J. Wahlang, Y. Pal, M. Chaitanya, and S. Saxena, "Precision oncology: Transforming cancer care through personalized medicine,". *Medical Oncology, 42*(7), 246. DOI: 10.1007/s12032-025-02817-y PMID: 40488843

Saadh, M. J., Khidr, W. A., Alfarttoosi, K. H., Bishoyi, A. K., Ganesan, S., Shankhyan, A., Gayathri, S., Rizaev, J., Taher, W. M., Alwan, M., Jawad, M. J., & Al-Nuaimi, A. M. A. (2025, February 23). Metal nanoparticles as a promising therapeutic approach for prostate cancer diagnosis and therapy: A comprehensive review. *Medical Oncology, 42*(4), 83. DOI: 10.1007/s12032-025-02633-4 PMID: 39987535

Sabha, S. U. (2025). Artificial Intelligence and Omics in Health and Diseases. In Macha, M. A., Assad, A., & Bhat, M. R. (Eds.), *Artificial Intelligence in Human Health and Diseases* (pp. 75–92). Springer Nature Singapore., DOI: 10.1007/978-981-96-8176-1_5

Salciccia, S., Capriotti, A. L., Lagana, A., Fais, S., Logozzi, M., De Berardinis, E., Busetto, G. M., Di Pierro, G. B., Ricciuti, G. P., Del Giudice, F., Sciarra, A., Carroll, P. R., Cooperberg, M. R., Sciarra, B., & Maggi, M. (2021, April 22). Biomarkers in prostate cancer diagnosis: From current knowledge to the role of metabolomics and exosomes. *International Journal of Molecular Sciences, 22*(9), 4367. DOI: 10.3390/ijms22094367 PMID: 33922033

Schleidgen, S., Dederer, H. G., Sgodda, S., Cravcisin, S., Lüneburg, L., Cantz, T., & Heinemann, T. (2020, September 11). Human germline editing in the era of CRISPR-Cas: Risk and uncertainty, intergenerational responsibility, therapeutic legitimacy. *BMC Medical Ethics, 21*(1), 87. DOI: 10.1186/s12910-020-00487-1 PMID: 32912206

Sebastian, A. M., & Peter, D. (2022). Artificial intelligence in cancer research: Trends, challenges and future directions. *Life (Basel, Switzerland), 12*(12), 1991. DOI: 10.3390/life12121991 PMID: 36556356

Seijas, A., Cora, D., Novo, M., Al-Soufi, W., Sánchez, L., & Arana, Á. J. (2025). CRISPR/Cas9 Delivery Systems to Enhance Gene Editing Efficiency. *International Journal of Molecular Sciences, 26*(9), 4420. Advance online publication. DOI: 10.3390/ijms26094420 PMID: 40362657

Sekhoacha, M., Riet, K., Motloung, P., Gumenku, L., Adegoke, A., & Mashele, S. (2022, September 5). Prostate Cancer Review: Genetics, Diagnosis, Treatment Options, and Alternative Approaches. *Molecules (Basel, Switzerland), 27*(17), 5730. DOI: 10.3390/molecules27175730 PMID: 36080493

Shademan, B., Masjedi, S., Karamad, V., Isazadeh, A., Sogutlu, F., Rad, M. H., & Nourazarian, A. (2022, October). CRISPR technology in cancer diagnosis and treatment: Opportunities and challenges. *Biochemical Genetics*, *60*(5), 1446–1470. DOI: 10.1007/s10528-022-10193-9 PMID: 35092559

Sharma, S., Parveen, R., & Chatterji, B. P. (2021). Toxicology of Nanoparticles in Drug Delivery. *Current Pathobiology Reports*, *9*(4), 133–144. DOI: 10.1007/s40139-021-00227-z PMID: 34840918

Shi, J., Wang, E., Milazzo, J. P., Wang, Z., Kinney, J. B., & Vakoc, C. R. (2015). Discovery of cancer drug targets by CRISPR-Cas9 screening of protein domains. *Nature Biotechnology*, *33*(6), 661–667. DOI: 10.1038/nbt.3235 PMID: 25961408

Sumiyoshi, T., Chi, K. N., & Wyatt, A. W. (2021, June). Clinical implications of genomic alterations in metastatic prostate cancer. *Prostate Cancer and Prostatic Diseases*, *24*(2), 310–322. DOI: 10.1038/s41391-020-00308-x PMID: 33452452

Tabibian, M., Moghaddam, F. S., Motevaseli, E., & Ghafouri-Fard, S. (2024, October 17). Targeting mRNA-coding genes in prostate cancer using CRISPR/Cas9 technology with a special focus on androgen receptor signaling. *Cell Communication and Signaling : CCS*, *22*(1), 504. DOI: 10.1186/s12964-024-01833-1 PMID: 39420406

Tafirenyika, S. (2023). *AI in healthcare: Predictive modeling, explainability, and clinical impact.* World J Adv Res Rev.

Tătaru, O. S., Vartolomei, M. D., Rassweiler, J. J., Virgil, O., Lucarelli, G., Porpiglia, F., Amparore, D., Manfredi, M., Carrieri, G., Falagario, U., Terracciano, D., de Cobelli, O., Busetto, G. M., Giudice, F. D., & Ferro, M. (2021, February 20). Artificial intelligence and machine learning in prostate cancer patient management—Current trends and future perspectives. *Diagnostics (Basel)*, *11*(2), 354. DOI: 10.3390/diagnostics11020354 PMID: 33672608

Testa, U., Castelli, G., & Pelosi, E. (2019, July 30). Cellular and Molecular Mechanisms Underlying Prostate Cancer Development: Therapeutic Implications. *Medicines (Basel, Switzerland)*, *6*(3), 82. DOI: 10.3390/medicines6030082 PMID: 31366128

Vakili, S., Beheshti, I., Barzegar Behrooz, A., Łos, M. J., Vitorino, R., & Ghavami, S. (2025). Transforming Prostate Cancer Care: Innovations in Diagnosis, Treatment, and Future Directions. *International Journal of Molecular Sciences*, *26*(11), 5386. Advance online publication. DOI: 10.3390/ijms26115386 PMID: 40508192

van Wijk, Y., Halilaj, I., van Limbergen, E., Walsh, S., Lutgens, L., Lambin, P., & Vanneste, B. G. (2019). Decision support systems in prostate cancer treatment: An overview. *BioMed Research International*, *2019*(1), 4961768. DOI: 10.1155/2019/4961768 PMID: 31281840

Verma, V., & Kumar, D. (2025). Artificial intelligence and machine learning in drug discovery: From lead discovery to clinical validation (2020–2025). *Letters in Drug Design & Discovery*, *22*(12), 100341. DOI: 10.1016/j.lddd.2026.100341

Vikal, A., Maurya, R., Bhowmik, S., Patel, P., Singh, R., Das Gupta, G., & Das Kurmi, B. (2025, April). Precision Genome Editing: The Synergy of CRISPR and Nanotechnology in Cancer Treatment. *Current Cancer Therapy Reviews*, *21*(3), 265–277. DOI: 10.2174/0115733947290474240315062214

Wang, M., Liu, H., Huang, J., Cai, T., Xu, Z. P., & Zhang, L. (2025). Advancing cancer gene therapy: The emerging role of nanoparticle delivery systems. *Journal of Nanobiotechnology*, *23*(1), 362. DOI: 10.1186/s12951-025-03433-8 PMID: 40394591

Xia, W., Tao, Z., Zhu, B., Zhang, W., Liu, C., Chen, S., & Song, M. (2021, August 24). Targeted delivery of drugs and genes using polymer nanocarriers for cancer therapy. *International Journal of Molecular Sciences*, *22*(17), 9118. DOI: 10.3390/ijms22179118 PMID: 34502028

Yang, X., & Zhang, B. (2023). A review on CRISPR/Cas: A versatile tool for cancer screening, diagnosis, and clinic treatment. *Functional & Integrative Genomics*, *23*(2), 182. DOI: 10.1007/s10142-023-01117-w PMID: 37231285

Yetgin, A. (2025, April 7). Revolutionizing multi-omics analysis with artificial intelligence and data processing. *Quantitative Biology*, *13*(3), e70002. DOI: 10.1002/qub2.70002 PMID: 41675959

Zaccariotto, G. D., Bistaffa, M. J., Zapata, A. M., Rodero, C., Coelho, F., Quitiba, J. V., Lima, L., Sterman, R., Cardoso, V. M., & Zucolotto, V. (2025, April 9). Cancer nanovaccines: Mechanisms, design principles, and clinical translation. *ACS Nano*, *19*(17), 16204–16223. DOI: 10.1021/acsnano.4c15765 PMID: 40202241

Zack, M., Stupichev, D. N., Moore, A. J., Slobodchikov, I. D., Sokolov, D. G., Trifonov, I. F., & Gobbs, A. (2025, September 1). Artificial intelligence and multi-omics in pharmacogenomics: A new era of precision medicine. *Mayo Clinic Proceedings. Digital Health*, *3*(3), 100246. DOI: 10.1016/j.mcpdig.2025.100246 PMID: 40881104

Zhang, B. (2021). CRISPR/Cas gene therapy. *Journal of Cellular Physiology*, *236*(4), 2459–2481. DOI: 10.1002/jcp.30064 PMID: 32959897

Zhang, H., Qin, C., An, C., Zheng, X., Wen, S., Chen, W., Liu, X., Lv, Z., Yang, P., Xu, W., Gao, W., & Wu, Y. (2021, October 1). Application of the CRISPR/Cas9-based gene editing technique in basic research, diagnosis, and therapy of cancer. *Molecular Cancer*, *20*(1), 126. DOI: 10.1186/s12943-021-01431-6 PMID: 34598686

Zhou, H., Ye, P., Xiong, W., Duan, X., Jing, S., He, Y., Zeng, Z., Wei, Y., & Ye, Q. (2024). Genome-scale CRISPR-Cas9 screening in stem cells: Theories, applications and challenges. *Stem Cell Research & Therapy*, *15*(1), 218. DOI: 10.1186/s13287-024-03831-z PMID: 39026343

Zhu, M., Sali, R., Baba, F., Khasawneh, H., Ryndin, M., Leveillee, R. J., Hurwitz, M. D., Lui, K., Dixon, C., & Zhang, D. Y. (2024, August 25). Artificial intelligence in pathologic diagnosis, prognosis and prediction of prostate cancer. *American Journal of Clinical and Experimental Urology*, *12*(4), 200–215. DOI: 10.62347/JSAE9732 PMID: 39308594

Zhuang, L., Lian, Y., & Zhu, T. (2025). Multifunctional gold nanoparticles: Bridging detection, diagnosis, and targeted therapy in cancer. *Molecular Cancer*, *24*(1), 228. DOI: 10.1186/s12943-025-02431-6 PMID: 40993729

Chapter 9
Omics–Nano–CRISPR Strategies for Combating Skin Cancer in the AI Age

Ashish Singh Chauhan
http://orcid.org/0009-0004-7154-2584
Uttaranchal Institute of Pharmaceutical Sciences, Uttaranchal University, India

Sumit Durgapal
Uttaranchal Institute of Pharmaceutical Sciences, Uttaranchal University, India

Jyoti Pandit
School of Pharmaceutical Sciences, Jigaysa University, India

Arvind Raghav
Shri Ram Murti Smarak College of Engineering and Technology, India

ABSTRACT

Skin cancer, including melanoma and non-melanoma skin cancer, remains a major global health challenge due to molecular heterogeneity, treatment resistance, and variable therapeutic responses. This chapter explores the integration of multi-omics technologies, nanotechnology, CRISPR-based genome editing, and artificial intelligence to advance precision dermato-oncology. Multi-omics approaches reveal driver mutations, dysregulated signalling pathways, and tumour immune microenvironment dynamics, while lipid- and polymer-based nanocarriers enable targeted topical, transdermal, and intratumoural drug delivery. CRISPR-Cas9 expands therapeutic opportunities through precise editing of oncogenic drivers and resistance pathways, although efficient delivery remains a key limitation. AI serves as the unifying computational framework by integrating omics data, optimizing nanocarrier design, prioritizing therapeutic targets, and supporting image-based diagnosis. Together, these technologies provide a promising framework for personalized, adaptive, and minimally invasive skin cancer management.

DOI: 10.4018/979-8-3373-9715-3.ch009

1. INTRODUCTION

Skin cancer occupies a paradoxical place in global oncology: it is simultaneously one of the most common and one of the most survivable malignancies when caught early, yet it accounts for a disproportionate share of cancer-related disfigurement, healthcare expenditure, and, for melanoma, mortality (Bray et al., 2024; Díaz de la Guardia-Bolívar et al., 2022; Sung et al., 2021; Whiteman et al., 2016). Melanoma ranks as the seventeenth most common cancer worldwide, with an estimated 331,722 new cases and roughly 58,667 deaths in the most recent GLOBOCAN cycle, while NMSC ranks fifth most common, with over 1.2 million new diagnoses annually. Incidence is markedly skewed toward fair-skinned, heavily sun-exposed populations such as Australia, New Zealand, North America, and parts of Europe, but rising incidence in traditionally lower-burden regions signals changing exposures that will keep skin cancer a growing public-health priority for decades to come.

Chronic and intermittent UV exposure, fair skin phenotype, immunosuppression, and inherited susceptibility loci together define the principal risk architecture of skin cancer. UV radiation is a direct mutagen, generating cyclobutane pyrimidine dimers and C-to-T transition "UV signatures" detectable in the genomes of both melanoma and NMSC that frequently strike the tumour-suppressor and oncogenic hotspots driving malignant transformation. This mechanistic directness is part of why skin cancer is such a productive model system for precision oncology: the causal chain from exposure to specific, targetable mutations (Brash, 2015; *Cadet J, Douki T. Formation of UV-Induced DNA Damage. Photochem Photobiol Sci. 2018;17:1816–1841*, n.d.; Pfeifer et al., 2005) is unusually well characterised compared with many internal malignancies.

Skin malignancies are broadly divided into melanoma, arising from melanocytes, and non-melanoma skin cancer, principally basal cell carcinoma (BCC) and cutaneous squamous cell carcinoma (cSCC) (Akbani et al., 2015a; Göppner & Leverkus, 2011; Pellegrini et al., 2017), both arising from keratinocytes. BCC, the most common human cancer overall, only rarely metastasises, yet its slow, relentless local invasion can produce severe tissue destruction if treatment is delayed; cSCC carries somewhat higher metastatic potential, with chronic UV exposure and immunosuppression as major risk amplifiers; and melanoma, though far less common, causes the substantial majority of skin-cancer deaths because of its early tendency toward vertical growth-phase transition and lymphatic/haematogenous spread.

Current treatment paradigms, including surgical excision, radiotherapy, chemotherapy, targeted inhibitors, and immune checkpoint blockade, have measurably improved outcomes over the past fifteen years (Luke et al., 2017; Ribas & Wolchok, 2018; Stratigos et al., 2020) Surgical excision with clear margins remains curative for most localised BCC and cSCC, but cannot address disseminated disease, and repeated excisions in patients with extensive field-cancerised skin impose a cumulative burden of scarring and morbidity. In advanced melanoma, BRAF/MEK-targeted inhibitor combinations and checkpoint inhibitors against CTLA-4 and PD-1 have transformed survival expectations, yet many patients fail to respond (primary resistance) or relapse after an initial response (acquired resistance), and combination checkpoint blockade, while more effective, carries a substantially higher burden of adverse events.

Several structural limitations recur across these modalities. Intratumoral heterogeneity allows resistant subclones, present even before treatment begins, to expand under therapeutic pressure, and no single biomarker reliably separates responders from non-responders; PD-L1 expression, mutational burden, and interferon-gamma signatures each carry predictive value, but none is sufficient alone. Systemic toxicity restricts many regimens, and cutaneous delivery is complicated by the stratum corneum's barrier function. It is this gap, between molecular understanding of why therapies fail and the practical tools to act on it, that motivates the integrated omics, nanotechnology, CRISPR, and AI approach discussed throughout this chapter: omics identifies what to target and in which patient; nanotechnology and CRISPR provide the means to intervene precisely and locally; and AI supplies the computational scaffolding to integrate heterogeneous data, prioritise candidates, and adapt therapy as a tumour evolves (*Lim SY, Rizos H. Single-Cell Sequencing Reveals Melanoma Cell-State Plasticity and Therapeutic Resistance. Nat Rev Clin Oncol. 2024*, n.d.; Manguso et al., 2017a; Zuo et al., 2021a).

Skin cancer is an unusually favourable proving ground for this approach: the organ is directly accessible for imaging, biopsy, and topical or intratumoral delivery without systemic exposure; its mutational landscape is unusually well characterised, with UV-signature mutagenesis providing a well-validated causal narrative linking exposure to druggable lesions; and it already possesses a substantial nanomedicine and immuno-oncology literature (Borgheti-Cardoso et al., 2020; Diaz et al., 2023; Gupta et al., 2022; Song et al., 2021) The chapter proceeds accordingly: Section 2 establishes the molecular landscape any precision strategy must target; Section 3 details the omics technologies used to characterise it; Sections 4 and 5 describe the nanotechnology and CRISPR tools available to act on it; Section 6 describes the AI methods that make integration tractable; Section 7 proposes an integrated framework; and Section 8 assesses the barriers that still separate this framework from routine practice.

2. MOLECULAR LANDSCAPE OF SKIN CANCER

2.1 Melanoma

Cutaneous melanoma is driven by a relatively small number of recurrently mutated oncogenes acting through the MAPK and PI3K-AKT pathways. BRAF is the single most frequently altered gene, with activating mutations, overwhelmingly V600E and, less commonly, V600K, reported in approximately 40-50% of cutaneous melanomas. NRAS mutations, concentrated at codon Q61, represent the second most frequent driver class, while KIT mutations, though less frequent overall, become disproportionately important in acral and mucosal subtypes where BRAF mutations are rare. BRAF-V600 status is now a mandatory molecular test determining eligibility for BRAF/MEK inhibitor combination therapy, illustrating how directly melanoma's molecular landscape has already been translated into clinical decision-making. (Akbani et al., 2015a; Luke et al., 2017; Zuo et al., 2021a).

UV-induced genomic instability is the dominant mutational process shaping the melanoma genome. Cutaneous melanomas carry among the highest tumour mutational burdens of any solid cancer, dominated by C-to-T transitions at dipyrimidine sites, the classic signature of UV-induced cyclobutane pyrimidine dimer formation and imperfect excision repair. This high mutational load is double-edged: it increases the likelihood of generating targetable driver mutations and immunogenic neoantigens, partly explaining melanoma's relative responsiveness to checkpoint blockade, but it also raises the probability that a resistant subclone already exists before treatment begins (Brash, 2015; *Lim SY, Rizos H. Single-Cell Sequencing Reveals Melanoma Cell-State Plasticity and Therapeutic Resistance. Nat Rev Clin Oncol. 2024*, n.d.; Pfeifer et al., 2005)).

The immune microenvironment is now understood as a co-determinant of melanoma behaviour rather than a passive bystander. A "T-cell-inflamed" microenvironment, characterised by dense CD8+ T-cell infiltration, an interferon-gamma-responsive signature, and PD-L1 expression, correlates with superior response to anti-PD-1 therapy, whereas an immunologically "cold" tumour predicts primary resistance.

Because no single biomarker captures this complexity, biomarker research increasingly favours multi-parameter panels combining mutational burden, immune signatures, and even gut-microbiome composition (Manguso et al., 2017a; Ribas & Wolchok, 2018).

2.2 Non-Melanoma Skin Cancer

Non-melanoma skin cancer follows a different, though equally well-characterised, molecular logic. Basal cell carcinoma is overwhelmingly driven by aberrant Hedgehog-pathway activation, most commonly through inactivating mutations in the tumour-suppressor gene PTCH1 (roughly 70-90% of BCCs) or, less often, activating mutations in the downstream effector SMO; both converge on constitutive, ligand-independent GLI-mediated transcription. TP53 mutations are the second most common alteration in BCC (Göppner & Leverkus, 2011; Pellegrini et al., 2017), found in roughly 44-65% of tumours, and frequently co-occur with rather than substitute for Hedgehog pathway mutations, indicating cooperating, largely independent hits. This mechanistic understanding directly enabled smoothened (SMO) inhibitors such as vismodegib and sonidegib for locally advanced and metastatic BCC, a clear precedent for how molecular characterisation translates into targeted therapy.

Cutaneous squamous cell carcinoma shares the UV-signature mutational backdrop of melanoma and BCC but carries its own recurrent driver set, most consistently TP53, NOTCH1, and NOTCH2, alongside FAT1 alterations (Pellegrini et al., 2017; Stratigos et al., 2020) several of these same mutations are also detectable in histologically normal, chronically sun-exposed skin, implying a substantial "field" of pre-malignant clonal mutation typically precedes overt tumour formation. This has direct clinical relevance for recurrence patterns after excision and for field-directed topical therapies rather than single-lesion treatment alone.

2.3 Subtype Diversity and Clinical Staging Relevance

Beyond the canonical BRAF-NRAS-KIT triad, recurrent non-coding TERT promoter mutations are now recognised as among the most frequent alterations in cutaneous melanoma, elevating telomerase expression and conferring replicative immortality; these mutations arise early and often co-occur with BRAF or NRAS mutations. Acral and mucosal melanomas, arising on non-sun-exposed palmar, plantar, subungual, and mucosal surfaces, present a genetically distinct picture dominated by KIT mutations and copy-number alterations rather than UV-signature point mutations, a distinction with direct treatment implications since KIT-mutant melanomas may respond to KIT-targeted tyrosine kinase inhibitors that have no role in BRAF- or NRAS-driven disease. This subtype-specific stratification (summarised in Figure 1) illustrates why a single "melanoma" treatment algorithm is increasingly inadequate.(Luke et al., 2017; Zuo et al., 2021).

This molecular landscape's clinical staging relevance extends beyond initial treatment selection. Breslow thickness, ulceration status, and mitotic rate remain the histopathological cornerstones of melanoma staging, but molecular features increasingly refine prognosis: BRAF-mutant tumours have historically shown differential responses to first-line therapy, and co-occurring mutations (NF1 loss, CDKN2A inactivation) modify both prognosis and durability of response. For NMSC, high-risk features including perineural invasion, TP53 mutation burden, and NOTCH pathway alterations are increasingly used alongside staging to identify patients who need more aggressive management (Diaz et al., 2023;

Lim SY, Rizos H. Single-Cell Sequencing Reveals Melanoma Cell-State Plasticity and Therapeutic Resistance. Nat Rev Clin Oncol. 2024, n.d.; Ribas & Wolchok, 2018).

Figure 1. Principal molecular drivers and downstream signalling in melanoma versus non-melanoma skin cancer (basal cell carcinoma and squamous cell carcinoma), converging on a shared tumour immune microenvironment that shapes treatment response.

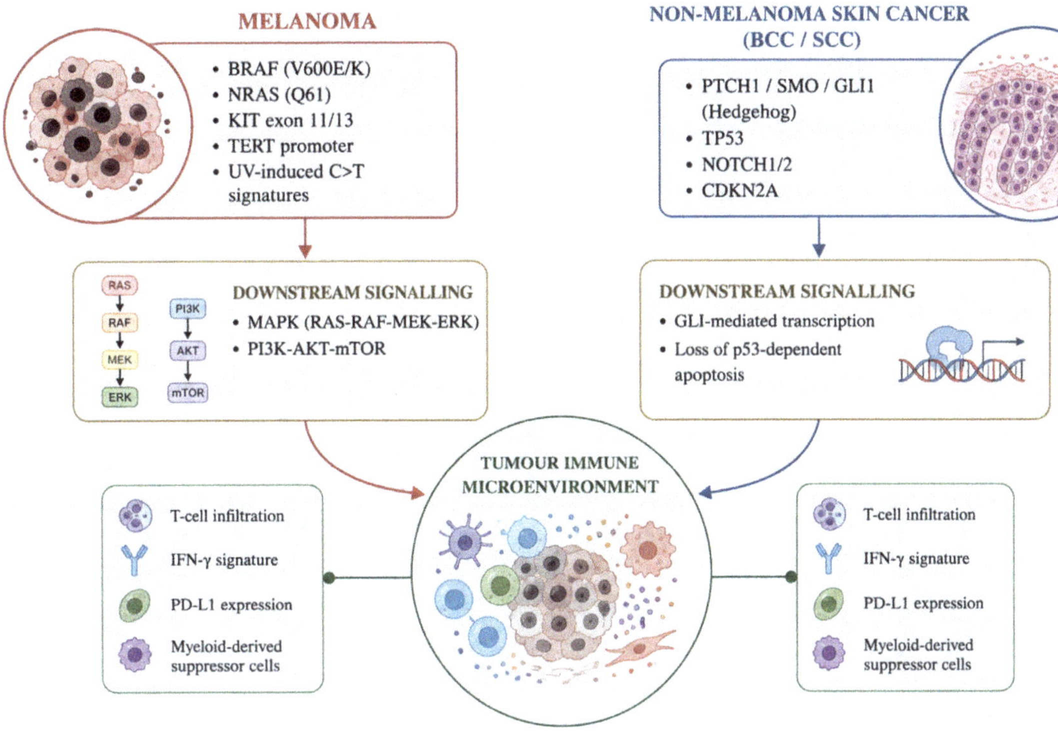

Table 1. Comparative Overview of Melanoma, BCC, and cSCC

Feature	Melanoma	Basal Cell Carcinoma (BCC)	Cutaneous Squamous Cell Carcinoma (cSCC)
Cell of origin	Melanocytes	Basal keratinocytes	Suprabasal/differentiated keratinocytes
Approx. share of skin cancer deaths	Majority of deaths despite lower incidence	Rarely fatal	Low, but higher than BCC
Principal driver genes	BRAF (V600E/K), NRAS (Q61), KIT (acral/mucosal)	PTCH1, SMO, GLI1 (Hedgehog); TP53	TP53, NOTCH1/2, FAT1

continued on following page

Table 1. Continued

Feature	Melanoma	Basal Cell Carcinoma (BCC)	Cutaneous Squamous Cell Carcinoma (cSCC)
Metastatic potential	High (lymphatic & haematogenous)	Very low	Moderate, higher with immunosuppression
Key targeted therapy precedent	BRAF/MEK inhibitors; anti-PD-1/CTLA-4	Smoothened (SMO) inhibitors (vismodegib, sonidegib)	Immune checkpoint inhibitors (advanced disease)

Table 1 makes explicit why a single "skin cancer" treatment algorithm is insufficient: the three principal subtypes differ not only in cell of origin but in the specific molecular lesions that drive them, in their metastatic behaviour, and in which targeted therapy, if any, has established clinical precedent. This divergence is precisely what an omics-informed, patient-specific characterisation strategy is designed to resolve at the individual-patient level rather than the population level.

3. OMICS IN SKIN CANCER

3.1 Genomics and Mutation Profiling

Next-generation sequencing has become the workhorse technology for cataloguing the somatic mutations, copy-number alterations, and structural variants that distinguish skin cancer subtypes. Panel-based and whole-exome sequencing now routinely inform BRAF-inhibitor eligibility in melanoma while uncovering co-occurring resistance-associated alterations, such as NF1 loss. Genome-scale sequencing has also revealed that BCC pathogenesis extends beyond the canonical Hedgehog-TP53 axis to include Hippo-YAP effectors such as PTPN14, LATS1, and MYCN, together with recurrent non-coding promoter mutations indicating a more complex genetic network than previously assumed (Akbani et al., 2015b; Hayward et al., 2017; Hodis et al., 2012; Shain & Bastian, 2016).

3.2 Transcriptomics and Immune Signatures

RNA sequencing adds a dynamic, functional layer on top of static mutational data, capturing pathway activation states and immune-infiltration signatures. Transcriptomic immune signatures, particularly interferon-gamma response modules, are informative correlates of checkpoint-inhibitor responsiveness, helping stratify patients likely to benefit from single-agent PD-1 blockade from those who may need combination immunotherapy. Because transcriptional states are plastic under therapeutic pressure, monitoring during treatment, not merely at diagnosis, is increasingly recognised as necessary to track emerging resistance (Hugo et al., 2016; Jerby-Arnon et al., 2018; Tirosh et al., 2016).

3.3 Proteomics and Signalling Cascades

Mass-spectrometry-based and antibody-array proteomic platforms provide a functional readout that complements genomic and transcriptomic data by characterising signalling-network activation at the level of protein phosphorylation. Because oncogenic signalling is ultimately executed through protein modification rather than gene expression alone, proteomic profiling can reveal mechanisms of adaptive resistance, such as MAPK reactivation through alternative RAF isoforms, that are not reliably predictable from DNA sequencing alone, particularly in BRAF-inhibitor-resistant melanoma (Avanzato et al., 2020; Mertins et al., 2016; Zhang et al., 2014).

3.4 Metabolomics and Oxidative Stress

Skin cancer cells frequently undergo metabolic reprogramming toward aerobic glycolysis and altered lipid metabolism to sustain rapid proliferation, while UV-generated reactive oxygen species contribute both to mutagenesis and to selective pressure on antioxidant defence pathways. Metabolomic profiling of tumour tissue and biofluids can reveal these shifts, offering candidate disease-activity biomarkers, and identifying metabolic vulnerabilities, such as dependence on specific glycolytic or lipogenic intermediates, that may be therapeutically exploitable, particularly alongside nanocarrier systems that co-deliver metabolic inhibitors with conventional cytotoxic or targeted agents (Faubert et al., 2020; Pavlova & Thompson, 2016).

3.5 Single-Cell Sequencing

Bulk sequencing methods average signals across a heterogeneous mixture of malignant, immune, and stromal cells, obscuring rare subpopulations disproportionately responsible for treatment failure. Single-cell RNA sequencing resolves this by profiling individual cells, revealing that melanoma cells exist along a spectrum of transcriptional states, commonly described as melanocytic, transitory, neural-crest-like, and undifferentiated, that differ markedly in sensitivity to targeted therapy and immunotherapy Neural-crest-like and undifferentiated states are consistently enriched in immunotherapy-resistant tumours, and studies of matched pre- and post-treatment biopsies show resistance frequently emerges through selection of pre-existing minority subclones rather than acquisition of new mutations under treatment. Comparable approaches applied to NMSC are beginning to reveal analogous heterogeneity, reinforcing that intratumoural heterogeneity, not a single driver mutation, ultimately determines durable treatment response (Rambow et al., 2018).

3.6 Liquid Biopsy and Circulating Tumour DNA

Liquid biopsy approaches, particularly circulating tumour DNA (ctDNA) analysis, are increasingly incorporated as a complementary, minimally invasive omics layer. Because melanoma and, to a lesser extent, advanced NMSC shed detectable tumour-derived DNA into peripheral blood, serial ctDNA sampling offers a practical route to longitudinal monitoring of tumour burden and emerging resistance mutations without repeated tissue biopsy. Integrating ctDNA data with tissue-based omics is itself an AI-dependent task, since distinguishing genuine low-frequency tumour-derived variants from background noise and clonal haematopoiesis requires sophisticated machine-learning filtering rather than simple variant-calling thresholds.

Considered together, these five omics layers (Figure 2), genomic, transcriptomic, proteomic, metabolomic, and single-cell/liquid-biopsy, are complementary rather than redundant. A tumour's DNA sequence establishes what is structurally possible; its transcriptome and proteome reveal what is being executed at a given moment; its metabolome reflects downstream physiological consequences; and single-cell/liquid- biopsy approaches reveal how these states are distributed across a heterogeneous population and change over time. The central challenge, addressed in Section 6, is that no human analyst can integrate five such high-dimensional layers within a clinically actionable timeframe without substantial computational assistance (Lee et al., 2017; Wan et al., 2017).

Figure 2. Multi-omics data flow in skin cancer research, from genomic, transcriptomic, proteomic, metabolomic, and single-cell/spatial layers into an AI-mediated integration step supporting biomarker discovery and patient stratification.

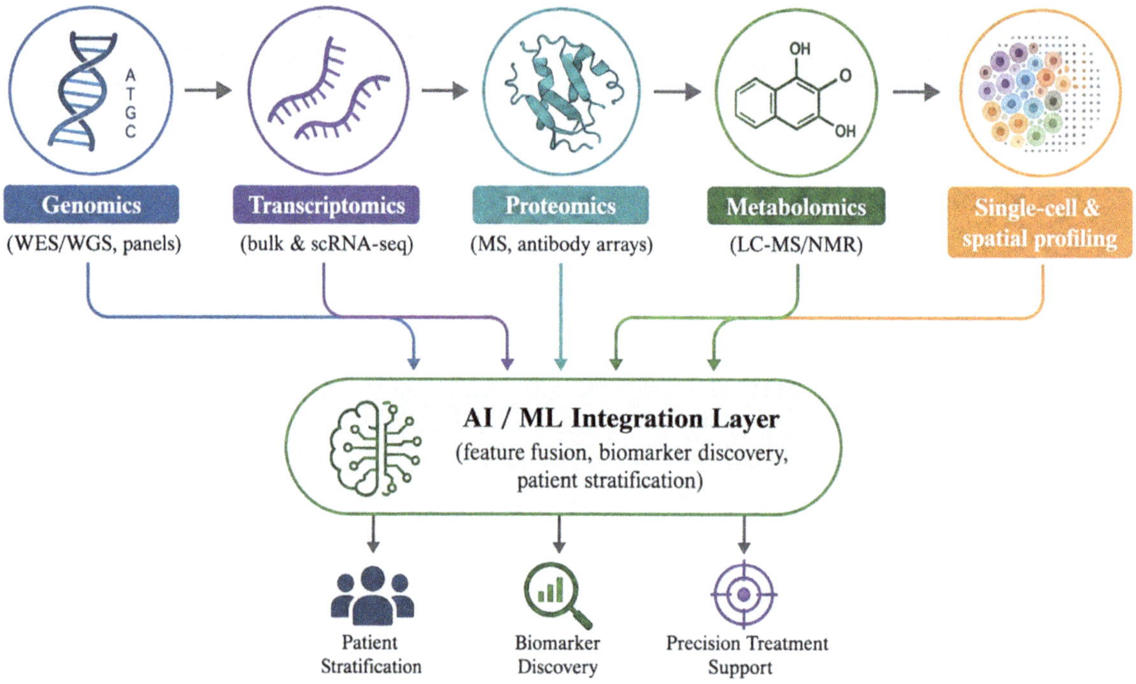

Table 2. Omics Platforms in Skin Cancer Research

Omics layer	Core technology	Principal application in skin cancer
Genomics	Whole-exome/whole-genome sequencing, targeted panels	Driver-mutation identification; BRAF-inhibitor eligibility; UV-signature analysis
Transcriptomics	Bulk RNA-seq, immune gene panels	Pathway activation state; interferon-gamma/checkpoint-response signatures
Proteomics	Mass spectrometry, antibody arrays, phosphoproteomics	Signalling network activation; adaptive resistance mechanism discovery

continued on following page

Table 2. Continued

Omics layer	Core technology	Principal application in skin cancer
Metabolomics	LC-MS, NMR spectroscopy	Metabolic reprogramming and oxidative-stress biomarker discovery
Single-cell/spatial	scRNA-seq, spatial transcriptomics	Intratumoural heterogeneity; cell-state plasticity; resistant subclone tracking

Table 2 summarises how each omics layer contributes a distinct, non-substitutable perspective on tumour biology; the practical challenge, taken up again in Section 6, is combining outputs from all five layers into a single, clinically interpretable patient profile within a timeframe compatible with treatment decision-making.

4. NANOTECHNOLOGY IN SKIN CANCER THERAPY

Nanotechnology addresses two problems that conventional formulations handle poorly in skin cancer: penetration through the stratum corneum for topical therapy, and selective accumulation at tumour sites for both topical and systemic therapy. A 2023 systematic review of controlled trials concluded that nanoscale drug-delivery systems demonstrate significant antineoplastic and anti-metastatic properties in skin cancer models, including increased bioavailability, reduced systemic toxicity, and enhanced permeability-and-retention-driven tumour accumulation, while noting that nanoformulations for skin cancer have generally lagged behind those for other malignancies. Closing that translational gap is a central motivation for the strategies discussed in this section.

4.1 Topical Nanocarriers

Topical nanocarriers exploit the fact that the stratum corneum, while a formidable barrier, is partially compromised by tumour-associated inflammation, ulceration, or hyperproliferation, giving appropriately engineered nanoparticles preferential access to diseased over healthy tissue. A comprehensive review of localised topical delivery systems for skin cancer catalogued liposomes, niosomes, solid lipid nanoparticles, nanostructured lipid carriers, polymeric nanoparticles, dendrimers, and nanoemulsions as the principal platforms under investigation, each offering a distinct balance of penetration enhancement, drug loading, and stability. Phospholipid-based liposomes containing cholesterol, ceramides, and cholesterol sulphate, for example, show enhanced accumulation in the epidermis and dermis, while flexible, deformable vesicle modifications (incorporating agents such as oleic acid) further increase skin deposition and drug diffusion (Torchilin, 2014) .

4.2 Lipid Nanoparticles

Lipid nanoparticles, including solid lipid nanoparticles (SLNs), nanostructured lipid carriers (NLCs), and ionizable-lipid nanoparticle (LNP) systems, offer enhanced drug solubility and bioavailability while reducing off-target toxicity. Reviews of LNP applications in melanoma report their use for delivering temozolomide and other cytotoxic and repurposed agents, with murine studies showing extended survival relative to free-drug controls, and silica-based nanoparticles loaded with agents such as verteporfin, cisplatin, or resveratrol have separately been shown to reduce lymphangiogenesis, systemic toxicity, and tumour growth in preclinical melanoma models. A representative example is a dacarbazine-loaded nanoparticle and nanocream topical delivery system developed specifically to overcome the poor aqueous solubility, short circulating half-life, and systemic toxicity that limit conventional dacarbazine therapy in melanoma (Bobo et al., 2016), demonstrating the principle that nanocarrier reformulation can rehabilitate even long-established but pharmacokinetically limited chemotherapeutic agents for topical use.

4.3 Polymeric Nanoparticles

Polymeric nanoparticles constructed from biodegradable, biocompatible materials such as poly(lactic-co-glycolic acid) (PLGA), chitosan, and polylactic acid (PLA) offer versatile, tunable platforms for both hydrophilic and hydrophobic anticancer agents. Beyond passive encapsulation, polymeric systems are increasingly engineered with tumour-targeting ligands and stimulus-responsive (pH- or enzyme-sensitive) release mechanisms that concentrate drug release at the acidic, protease-rich tumour microenvironment while sparing surrounding healthy tissue, an approach with particular relevance for combination delivery of chemotherapeutic and immunomodulatory agents to melanoma and NMSC lesions.

4.4 Photothermal and Photodynamic Nanosystems

Photothermal and photodynamic nanosystems add an activatable, spatially controllable dimension to nanocarrier-based therapy. Gold, carbon-based, and other plasmonic or photosensitiser-loaded nanoparticles can be engineered to accumulate preferentially at the tumour site and then be activated by an external light source, generating localised heat (photothermal therapy) or reactive oxygen species (photodynamic therapy) that destroy malignant cells while sparing adjacent normal skin, a strategy with obvious appeal for cutaneous tumours that are directly accessible to external light delivery, unlike most internal malignancies.

4.5 Transdermal and Localised Delivery Advantages

Collectively, these nanocarrier platforms confer several advantages for cutaneous oncology relative to conventional formulations: enhanced penetration across the stratum corneum barrier, prolonged and controlled drug release that reduces dosing frequency, preferential tumour accumulation through both passive (enhanced permeability and retention) and active (ligand-mediated) targeting mechanisms, and materially reduced systemic toxicity when a topical or localised route is feasible. These properties are precisely what is required to make CRISPR genome-editing components, discussed in the next section, deliverable to cutaneous tumours in a form that is both effective and clinically tolerable.

4.6 Combination Nanoparticle-Immunotherapy Co-Delivery

Combination co-delivery strategies, in which a single nanocarrier encapsulates both a cytotoxic or targeted agent and an immunomodulatory payload, represent a further active area of development directly relevant to the immune-microenvironment considerations discussed in Section 2. Because a substantial fraction of melanoma treatment failure is attributable to an immunologically “cold” tumour microenvironment rather than to inadequate cytotoxic killing alone, nanocarriers designed to simultaneously deliver a targeted or chemotherapeutic agent alongside an immune-adjuvant, checkpoint-modulating small molecule, or even a CRISPR construct directed at an immune-evasion gene, offer a rational route to converting a cold tumour microenvironment into one more (Danaei et al., 2018) receptive to systemic checkpoint blockade, without requiring two entirely separate treatment courses. Co-delivery formulations of this kind remain earlier in development than single-agent nanocarrier systems, but they follow directly from the mechanistic logic established by the CRISPR checkpoint-screening literature discussed in Section 5.

4.7 Clinical Translation Status

Clinical translation status (Figure 3) varies considerably by carrier type. Liposomal and, to a lesser degree, polymeric nanoparticle formulations of established cytotoxic agents already have regulatory approval in other tumour types, a precedent that skin-cancer-specific formulations can draw upon. A completed phase 1/2 trial of a topical nanoparticle paclitaxel ointment in cutaneous breast-cancer metastases demonstrated acceptable safety, illustrating that topical nanocarrier delivery is already feasible in humans, even though dedicated large-scale trials in primary melanoma and NMSC remain few. Photothermal and photodynamic nanosystems occupy an earlier stage of development for skin cancer specifically, though non-nanoparticle photodynamic therapy is already approved for superficial BCC and actinic keratosis in several jurisdictions.

Figure 3. Principal nanocarrier platforms investigated for cutaneous drug delivery in skin cancer, and the therapeutic advantages they confer relative to conventional topical or systemic formulations.

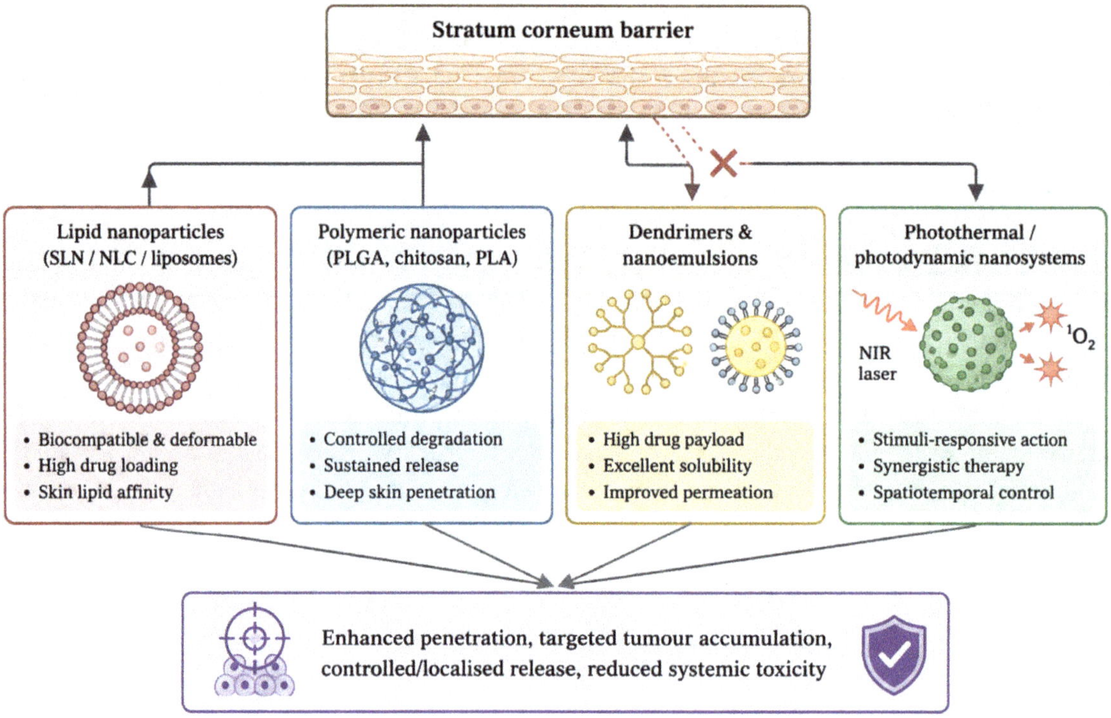

Table 3. Nanocarrier Systems for Cutaneous Delivery

Nanocarrier type	Representative composition	Key advantage for cutaneous delivery
Liposomes	Phospholipids, cholesterol, ceramides	High epidermal/dermal accumulation; flexible vesicle penetration
Solid lipid NP / NLC	Solid and liquid lipid matrix	Improved drug solubility and stability; controlled release
Polymeric NP	PLGA, chitosan, PLA	Tunable, stimulus-responsive release; ligand-mediated targeting
Dendrimers / nanoemulsions	Branched polymers; oil-in-water systems	High drug loading; enhanced skin permeation
Photothermal/photodynamic NP	Gold, carbon-based, or photosensitiser-loaded particles	Externally activatable, spatially confined tumour ablation

Table 3 highlights that no single nanocarrier chemistry dominates across all clinical needs: lipid-based systems currently offer the most established manufacturing and regulatory precedent, polymeric systems offer the greatest flexibility for stimulus-responsive and ligand-targeted designs, and photothermal/photodynamic systems uniquely exploit the external accessibility of cutaneous tumours for spatially confined, light-activated therapy.

5. CRISPR-BASED STRATEGIES

CRISPR-Cas9 genome editing has moved from a laboratory discovery tool to a platform with direct translational relevance in skin cancer research, serving three broadly distinct functions: correcting or disrupting oncogenic drivers, modulating immune-checkpoint biology, and systematically screening for the genes that confer resistance to existing therapies.

5.1 Targeting BRAF-Mutant Melanoma

Targeting BRAF-mutant melanoma with CRISPR has been used both to validate resistance mechanisms and, prospectively, to engineer isogenic cell models that precisely recreate clinically observed resistance mutations. CRISPR/Cas9-mediated introduction of specific point mutations into RAS and MEK1 genes has been used to generate isogenic BRAF- and MEK-inhibitor-resistant (Chiappetta et al., 2021; Duttke et al., 2015; Zuo et al., 2021b) melanoma cell lines that differ from their parental line by only the engineered mutation, allowing resistance phenotypes to be attributed unambiguously to a single genetic change rather than to the confounding effects of prolonged drug selection, and providing a controlled platform for testing candidate combination therapies against defined resistance mechanisms.

5.2 Immune Checkpoint Gene Modulation

Epigenetic reprogramming is an increasingly important CRISPR application, using catalytically dead Cas9 (dCas9) (Gilbert et al., 2013; Konermann et al., 2015) fused to transcriptional activators or repressors to modulate gene expression without permanent double-strand breaks. This is particularly attractive for reactivating silenced tumour-suppressor genes; genome-scale CRISPR-Cas9 transcriptional activation screening has, for example, identified long non-coding RNAs associated with BRAF-inhibitor (dabrafenib) resistance in melanoma, expanding the therapeutic target space beyond protein-coding genes (Manguso et al., 2017b; Patel et al., 2017).

5.3 Epigenetic Reprogramming

Epigenetic reprogramming represents an increasingly important CRISPR application, using catalytically dead Cas9 (dCas9) (Gilbert et al., 2013; Konermann et al., 2015) fused to transcriptional activators or repressors to modulate gene expression without introducing permanent double-strand breaks. This approach is particularly attractive for reactivating silenced tumour-suppressor genes or transiently modulating resistance-associated long non-coding RNAs; genome-scale CRISPR-Cas9 transcriptional activation screening has, for example, been used to identify long non-coding RNAs associated with BRAF-inhibitor (dabrafenib) resistance in melanoma, expanding the therapeutic target space beyond conventional protein-coding genes.

5.4 CRISPR Screening for Resistance Pathways

CRISPR screening for resistance pathways more broadly has become a standard unbiased discovery method across skin cancer research. A genome-wide CRISPR-Cas9 knockout screen using the high-efficiency Brunello sgRNA library (Doench, 2018; Sanson et al., 2018) in BRAF-V600E-mutant A375 melanoma cells identified 33 genes regulating vemurafenib resistance, 14 of which had not previously been reported, with gene-ontology enrichment analysis implicating histone modification, transcriptional regulation, and cell-cycle control as recurrent resistance mechanisms alongside the expected MAPK-pathway reactivation genes. Findings of this kind illustrate both the discovery power of genome-scale functional screening and the practical challenge it creates: each newly identified resistance gene is a candidate therapeutic target, but converting that target into a deliverable, tumour-restricted therapy still requires a delivery vehicle capable of reaching cutaneous tumour cells with acceptable safety, precisely the role that nanocarrier-based CRISPR delivery is beginning to fill.

5.5 Delivery Vector Considerations

The choice of delivery vector fundamentally shapes what is achievable with CRISPR in skin cancer. Viral vectors, particularly AAV and lentivirus, offer high transduction efficiency but carry meaningful drawbacks: limited cargo capacity for the relatively large Cas9 protein alongside guide RNA, pre-existing immunity that can neutralise the vector, and a small but non-zero risk of insertional mutagenesis. Non-viral nanocarrier-based delivery, particularly ionizable lipid nanoparticles, has consequently become the platform of greatest interest for in vivo CRISPR delivery, since it avoids pre-existing viral immunity, can be manufactured at scale, and can package CRISPR components as plasmid DNA, mRNA encoding Cas9 alongside guide RNA, or pre-assembled ribonucleoprotein complexes, the last offering transient Cas9 exposure that narrows the window for off-target editing. Selective tissue-targeting lipid nanoparticle designs represent a further refinement relevant to restricting CRISPR activity to cutaneous tumour tissue rather than systemic distribution (Finn et al., 2018; Gillmore et al., 2021; Lino et al., 2018).

5.6 Base and Prime Editing as Safer Alternatives

Base and prime editors offer a further refinement over conventional CRISPR-Cas9 nuclease editing that is particularly relevant to the safety concerns raised throughout this section (illustrated in the overall CRISPR delivery workflow in Figure 4). Because conventional Cas9 editing introduces a double-strand break that is then repaired, often imprecisely, it carries an inherent risk of unintended insertions, deletions, or larger rearrangements at both on-target and off-target sites. Base editors, which directly convert one DNA base to another without cutting both strands, and prime editors, which use a reverse-transcriptase fusion protein to write new sequence without a double-strand break, substantially reduce this risk and are increasingly favoured where the desired edit is a precise, single-nucleotide change (such as correcting a specific BRAF or TP53 mutation) rather than a knockout. Their larger size and more complex machinery, however, place additional demands on the nanocarrier packaging systems discussed in Section 4.6, illustrating how tightly the CRISPR and nanotechnology pillars are mechanically, not merely conceptually, linked (Anzalone et al., 2019; Komor et al., 2016).

Figure 4. Nanoparticle-enabled CRISPR-Cas9 workflow for skin cancer, from target selection and guide-RNA design through nanocarrier packaging, cutaneous or intratumoural delivery, and functional validation of edited cells.

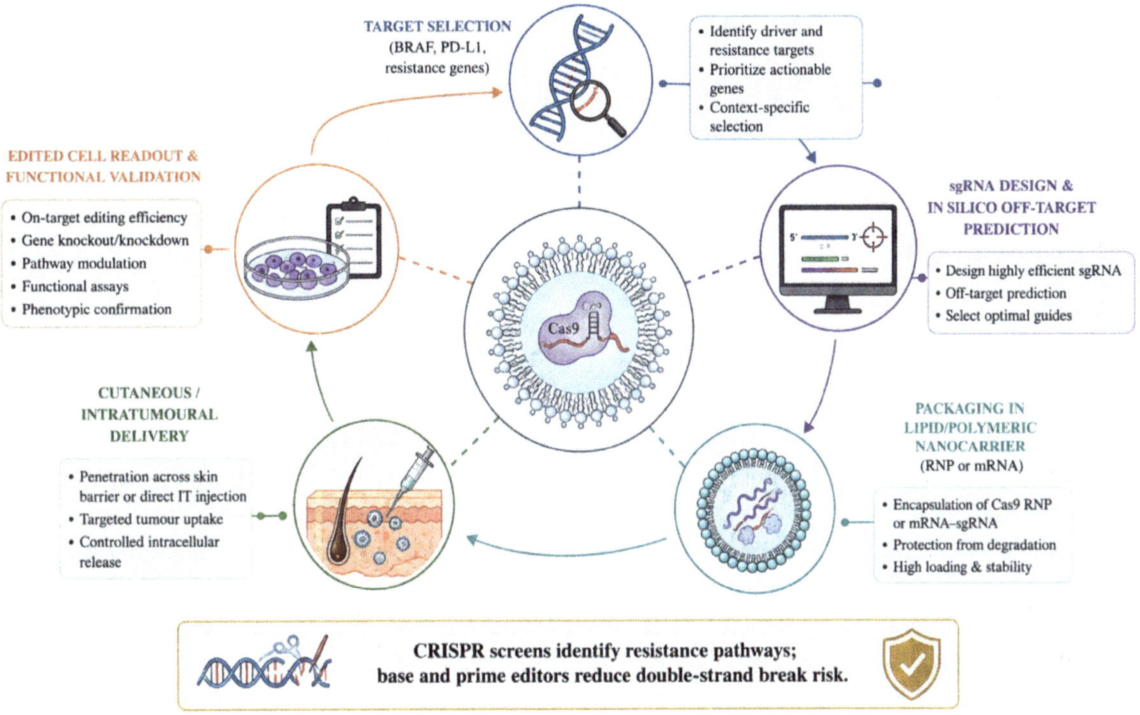

Table 4. CRISPR-Based Strategies in Skin Cancer

CRISPR application	Representative target/approach	Reported or intended outcome
Isogenic resistance modelling	CRISPR-engineered RAS/MEK1 point mutations	Defined, drug-pressure-independent models of BRAF/ MEK-inhibitor resistance
Immune-checkpoint sensitisation	Pooled in vivo screen (PD-L1, CD47, PTPN2)	Identification of PTPN2 loss as a checkpoint-therapy sensitiser
Epigenetic reprogramming	dCas9-based transcriptional activation of lncRNAs	Identification of lncRNAs driving BRAF-inhibitor resistance
Genome-wide resistance screening	Brunello sgRNA library knockout screen (A375 cells)	33 vemurafenib-resistance genes identified, 14 novel

Table 4 illustrates that CRISPR's contribution to skin cancer research to date has been overwhelmingly at the discovery and mechanistic-validation stage, using genome-scale screening and isogenic modelling to identify resistance and immune-sensitising genes, rather than at the stage of direct therapeutic gene editing delivered to patients, a distinction that matters for calibrating expectations about how soon CRISPR-based treatment, as opposed to CRISPR-based target discovery, is likely to reach the clinic for this disease group.

6. AI IN SKIN CANCER RESEARCH AND THERAPY

Artificial intelligence functions in this chapter's framework not as a single application but as the computational layer that makes the other three domains, omics, nanotechnology, and CRISPR, practically integrable at the scale and speed that clinical translation requires.

6.1 AI-Based Dermoscopy Analysis

AI-based dermoscopy analysis is the most clinically mature AI application in skin cancer. Convolutional neural networks trained on large dermoscopic image datasets, most notably HAM10000, have repeatedly matched or exceeded the diagnostic accuracy of practising dermatologists in distinguishing melanoma from benign lesions. Reported performance commonly falls in the 85-98% accuracy range across benchmarks such as ISIC and PH2, and head-to-head comparisons have found deep-learning classifiers achieving accuracies (86-87%) that exceed board-certified dermatologists (~80%) and non-expert clinicians (~63%). Because standard dermoscopic examination alone can yield melanoma-detection sensitivity as low as 40% in complex cases, and the number of biopsies needed to find one true melanoma can approach 30 to 1, AI-assisted triage carries substantial potential to reduce unnecessary biopsies while improving early detection (Esteva et al., 2017; Tschandl et al., 2018).

6.2 Predictive Mutation Modelling

Predictive mutation modelling extends AI beyond image classification into genomic and clinical data, using models trained on sequencing and outcome data to predict BRAF, NRAS, or KIT mutation status, anticipate resistance-mutation emergence, or estimate immunotherapy response probability directly from combined imaging, histopathology, and molecular features (Haenssle et al., 2018)), reducing dependence on sequential, single-modality testing (Topol, 2019) .

6.3 AI-Guided Nanoformulation Optimisation

AI-guided nanoformulation optimisation is a rapidly maturing application with direct relevance to the nanocarrier platforms discussed earlier. Machine-learning approaches are now applied across essentially every stage of nanoparticle drug-delivery development, from predicting optimal synthesis parameters and quality attributes such as particle size, zeta potential, and encapsulation efficiency, to modelling nanoparticle behaviour in biological fluids, including protein corona formation and tumour extravasation. Kernel-based and other regression models represent the markedly non-linear relationship between nanoparticle physicochemical properties and tumour delivery efficiency, offering a route to rational, computationally guided formulation design that reduces the empirical trial-and-error burden repeatedly cited as a key reason skin cancer nanoformulations have lagged behind those for other tumour types

(*Wang Y, et al. Artificial Intelligence for Nanomedicine: Machine Learning in Nanoparticle Design and Optimization. Adv Drug Deliv Rev. 2025;214:115133*, n.d.) .

6.4 Integration of Omics Datasets

Integration of omics datasets is, in many respects, the application area where AI is least optional and most essential, because the high dimensionality, heterogeneity, and sparsity of combined genomic, transcriptomic, proteomic, metabolomic, and single-cell datasets exceed what conventional statistical methods can reliably handle. Deep-learning approaches, including graph neural networks, transformer-based architectures, and generative models such as variational autoencoders, are increasingly used to perform intelligent imputation of missing data, non-linear batch correction across datasets generated on different platforms, and multi-omics fusion for tasks such as molecular subtyping, survival stratification, and immunotherapy-response prediction. Explainable AI methods are simultaneously being developed to address the clinical-trust barrier that opaque "black box" models otherwise create, since a prediction that cannot be interpreted or audited is difficult to act on responsibly in a treatment-decision context, particularly where the underlying data are subject to real limitations of dimensionality, missingness, and cross-platform heterogeneity that no algorithm can fully eliminate.

6.5 Explainability and Regulatory Trust

Explainability methods are an increasingly necessary companion to raw diagnostic accuracy for AI-based dermoscopy tools intended for real clinical use. Techniques such as saliency mapping and Shapley-value-based feature attribution let a clinician see which regions of a dermoscopic image, or which input features in a multi-omics prediction, most strongly influenced a given AI output, allowing outputs inconsistent with recognised diagnostic criteria such as the ABCD rule to be flagged for closer review rather than accepted uncritically. This is not merely a matter of clinician comfort: regulatory frameworks for AI-based diagnostic software increasingly require interpretability evidence as part of approval and post-market surveillance, making explainability a practical prerequisite for regulatory clearance rather than an optional enhancement.

6.6 Federated Learning and Data Equity

Data privacy and cross-institutional collaboration present a further constraint on AI development, since the large, diverse datasets required to train robust models typically exceed what any single institution can assemble alone, particularly for rarer subtypes such as acral or mucosal melanoma. Federated learning, in which model training occurs locally at each institution and only aggregated parameters, not patient-level data, are shared centrally, is increasingly explored as a means of pooling statistical power without centralising sensitive data, with particular relevance for building dermoscopic classifiers that perform reliably across skin tones, since models trained predominantly on lighter-skinned populations have repeatedly performed less reliably on darker skin tones, a well-documented equity concern that any responsible deployment of AI-based dermoscopy must address.

Table 5. AI Applications Across the Integrated Framework

AI application domain	Representative method	Role in the integrated framework
Dermoscopic image analysis	Convolutional neural networks (e.g., DenseNet, HRNet-based CAD)	Automated melanoma/BCC/cSCC classification; biopsy triage
Mutation/response prediction	Supervised ML on genomic + clinical features	Predicting BRAF/NRAS/KIT status and immunotherapy response probability
Nanoformulation optimisation	Kernel regression, generative and graph-based ML	Predicting particle size, zeta potential, encapsulation efficiency, biodistribution
Multi-omics integration	Graph neural networks, transformers, variational autoencoders	Data fusion, imputation, batch correction, and biomarker discovery

Table 5 underscores that AI's contribution spans the full pipeline from diagnosis to formulation to data integration, rather than being confined to the image-classification role most commonly associated with "AI in dermatology," and it is this breadth of application that justifies treating AI as the coordinating layer of the integrated framework proposed in Section 7 rather than as merely one tool among several.

7. INTEGRATED OMICS–NANO–CRISPR–AI FRAMEWORK

The preceding sections have deliberately treated omics, nanotechnology, CRISPR, and AI as distinct domains, each with its own literature, methods, and translational maturity. The central proposition of this chapter is that their combination, rather than any single domain in isolation, is what closes the gap between mechanistic understanding and deliverable therapy identified in the introduction.

7.1 Precision Targeting

Precision targeting in this integrated model begins with a patient-specific molecular profile, genomic, transcriptomic, and, where feasible, single-cell data, that AI models process to identify the dominant driver mutations, resistance-associated alterations, and immune microenvironment phenotype present in an individual patient's tumour. This profile then determines which nanocarrier-CRISPR construct is most appropriate: a BRAF-V600E-driven, immunologically "cold" melanoma might be prioritised for a nanoparticle-delivered CRISPR construct targeting a resistance-conferring gene identified through genome-wide screening, administered alongside checkpoint blockade, whereas a Hedgehog-pathway-driven BCC with an accessible cutaneous location might instead be prioritised for topical nanocarrier-delivered Hedgehog-pathway modulation without requiring genome editing at all (Hasin et al., 2017) .

7.2 Personalised Topical Genome Editing

Personalised topical genome editing extends this logic specifically to cutaneous tumours' unique accessibility advantage: unlike most internal malignancies, skin lesions can, in principle, be treated with topically or intratumourally delivered nanocarrier-CRISPR constructs that never require systemic circulation, substantially reducing off-target tissue exposure and the associated safety concerns that dominate discussion of in vivo CRISPR delivery for other cancer types. This accessibility is precisely the advantage that nanocarrier-mediated CRISPR delivery to cutaneous tumours is positioned to exploit, though, as the next section discusses, exploiting it safely still requires solving nontrivial delivery and off-target risk problems (Karczewski & Snyder, 2018) .

7.3 Adaptive Therapy Models

Adaptive therapy models complete the framework by treating skin cancer management as a continuous, monitored process rather than a single treatment decision. AI-integrated omics and imaging monitoring during treatment can, in principle, detect the earliest transcriptional or immune-microenvironment signatures of emerging resistance before it becomes clinically apparent, allowing the nanocarrier-CRISPR-immunotherapy combination to be re-optimised in response, an approach conceptually analogous to adaptive antibiotic stewardship but applied to a tumour's continuously evolving clonal architecture. This adaptive vision remains substantially aspirational relative to the individual component technologies already in clinical or near-clinical use, and its realisation depends directly on resolving the delivery, safety, data-integration, and regulatory challenges discussed in the final section (Hood & Friend, 2011) .

7.4 Technology Readiness Across the Four Pillars

It is worth being explicit about the differing maturity of the four pillars this framework combines, since conflating them risks overstating how close the fully integrated model is to clinical reality. Omics profiling and AI-based dermoscopy classification are the most mature, with genomic testing for BRAF status already standard of care and AI-assisted image triage entering early clinical deployment. Nanocarrier-based topical and photodynamic delivery occupies an intermediate position, with individual formulations in trials or, for non-nanoparticle photodynamic therapy, already approved, but comparatively few nanocarrier-specific trials in skin cancer. CRISPR-based intervention, particularly in vivo, remains the least mature pillar; genome-scale screening has already delivered clinically relevant target discoveries, but no nanocarrier-CRISPR construct has yet reached clinical trial for melanoma or NMSC, and the field is still resolving delivery and off-target safety questions in more tractable organ systems, such as the liver, before extending to solid cutaneous tumours. Framing the model honestly, as one clinically established pillar, one clinically emerging pillar, and one largely preclinical pillar, coordinated by an increasingly capable but not yet fully validated computational layer, is essential for setting realistic expectations (Ashley, 2016) .

7.5 An Illustrative Composite Case Scenario

To make the integrated framework more concrete, consider a hypothetical composite patient: a 58-year-old presents with a 2.1 mm-thickness cutaneous melanoma on the lower leg. Sequencing confirms a BRAF-V600E mutation with a co-occurring TERT promoter mutation; transcriptomic profiling shows a "cold" immune signature; and single-cell sequencing of the pre-treatment biopsy reveals a small resistant subclone already present. An AI model integrating these layers flags a moderate probability of early resistance to single-agent BRAF/MEK inhibition and identifies the cold microenvironment as a likely driver of suboptimal checkpoint response. The team elects combination BRAF/MEK inhibition with checkpoint blockade, considers a nanocarrier-delivered topical adjunct targeting the resistant subclone's dependency pathway (identified by analogy to the CRISPR screening data reviewed in Section 5), and uses serial ctDNA monitoring to detect relapse earlier than imaging alone would allow. This scenario is deliberately aspirational, since nanocarrier-delivered adjuvant therapy guided by CRISPR screening remains preclinical (Section 7.4), but each step draws on methods already validated in the literature reviewed here, illustrating that the gap to this integrated vision is one of coordination and clinical validation rather than missing science.

8. CHALLENGES AND FUTURE DIRECTIONS

8.1 Delivery Barriers

Delivery barriers remain the single most consistently cited obstacle across the nanotechnology and CRISPR literature reviewed in this chapter. The stratum corneum, while partially compromised at tumour sites, still limits the depth and consistency of topical nanocarrier penetration, and systemically administered nanoparticles continue to accumulate disproportionately in the liver and spleen rather than at cutaneous or subcutaneous tumour sites, raising both efficacy and off-target safety concerns. For CRISPR components specifically, rapid enzymatic degradation, inefficient targeting of the intended tissue, and limited packaging capacity of both viral and non-viral vectors continue to constrain in vivo delivery, even as ionizable lipid nanoparticles have emerged as the most clinically advanced non-viral platform, exemplified by the first successful human in vivo CRISPR therapy (an LNP-delivered treatment for hereditary transthyretin amyloidosis), which targeted the comparatively accessible liver rather than a solid tumour (Mitchell et al., 2021) .

8.2 Off-Target Gene Editing

Off-target gene editing is both a technical and an ethical concern that scales with the ambition of in vivo CRISPR applications. Because no single computational or experimental method reliably predicts every bona fide off-target editing site, current best practice for genomic safety assessment relies on a multilayered strategy combining high-sensitivity in silico guide-RNA design tools, unbiased in vitro cleavage mapping, whole-genome sequencing of edited cells, and targeted deep amplicon sequencing at candidate off-target loci. This concern is amplified for in vivo, as opposed to ex vivo, editing: an ex vivo-edited cell population (as used, for example, in hereditary blood-disorder therapies) can be fully characterised before re-infusion, whereas an in vivo-edited tumour cannot be biopsied and screened with the same completeness before the edit takes effect, meaning that nanocarrier-CRISPR systems for skin cancer must be designed with off-target risk mitigation, high-fidelity Cas variants, precisely tuned guide RNAs, and base or prime editors that avoid double-strand breaks where feasible, built in from the outset rather than addressed retrospectively (Tsai & Joung, 2016) .

8.3 Data Integration Complexity

Data integration complexity, already introduced in the AI section, deserves explicit treatment as a challenge rather than only as an opportunity. Multi-omics datasets relevant to skin cancer are generated on heterogeneous platforms, at different institutions, with inconsistent metadata standards and substantial missingness, particularly for rarer data types such as single-cell and spatial transcriptomics. Even sophisticated deep-learning integration architectures cannot fully compensate for genuinely absent data or for systematic batch effects that are indistinguishable, without careful experimental design, from true biological signal. Explainability remains a further practical barrier: a treatment-relevant AI prediction generated by an opaque model is difficult for a treating clinician to act on with confidence, and difficult for a regulator to evaluate, regardless of the model's measured accuracy on retrospective data (Huang et al., 2017) .

8.4 Regulatory and Ethical Considerations

Regulatory and ethical considerations for combined nanocarrier-CRISPR-AI approaches in skin cancer sit at the intersection of several previously separate regulatory frameworks: gene-therapy regulation (with its established requirements for genomic safety characterisation), nanomedicine-specific regulatory guidance (which remains less mature and less harmonised across jurisdictions than small-molecule drug regulation), and, increasingly, AI-specific regulatory expectations around model validation, transparency, and post-market monitoring. In vivo genome editing raises particular ethical scrutiny regarding the possibility, however remote for a topically or intratumourally restricted cutaneous application, of unintended germline transmission or off-tissue effects, and clinical translation pathways for combined nanoparticle-CRISPR therapeutics remain considerably less well-trodden than those for either nanomedicines or gene therapies individually. For lower- and middle-income countries, where the burden of NMSC is substantial even if lower than in fair-skinned, high-UV-exposure populations, the cost and manufacturing complexity of nanocarrier-CRISPR (*Human Genome Editing*, 2017) systems raise additional equity considerations that will need to be addressed if these technologies are to benefit patients beyond a small number of high-resource health systems.

8.5 Standardisation and Reproducibility

Standardisation and reproducibility across the nanoparticle characterisation and CRISPR off-target assessment methods discussed in this chapter constitute a further, more technical but no less consequential challenge. Nanoparticle size, zeta potential, polydispersity, and encapsulation efficiency are measured using a range of instruments and protocols that are not always directly comparable across laboratories, complicating efforts to build the large, consistent training datasets that the AI-guided formulation-optimisation methods described in Section 6.3 depend upon. Similarly, off-target CRISPR editing assessment currently relies on a multilayered combination of in silico prediction, in vitro cleavage mapping, and sequencing-based validation for which no single, universally adopted protocol yet exists, making cross-study comparison of reported off-target rates for different guide RNAs or delivery systems difficult. Coordinated efforts toward standardised characterisation and reporting protocols, of the kind that have gradually matured in adjacent fields such as nanomedicine regulatory science and clinical genomics, would materially accelerate progress toward the integrated framework proposed in this chapter by making results from different research groups genuinely comparable rather than merely individually informative.

8.6 Cost, Scalability, and Equity

Cost, manufacturing scalability, and workforce readiness constitute a final, often underappreciated set of challenges that will determine whether an integrated Omics-Nano-CRISPR-AI approach reaches broad clinical practice or remains confined to a small number of specialist centres. Multi-omics profiling, nanocarrier manufacturing under good manufacturing practice conditions, and CRISPR reagent production each carry substantial per-patient cost relative to conventional excision or standard systemic therapy, and integrating them into a single coordinated treatment pathway requires clinical and laboratory infrastructure, including sequencing capacity, controlled nanoparticle manufacturing, and computational expertise for AI model deployment and interpretation, that is unevenly distributed even within high-income health systems and substantially scarcer in the low- and middle-income countries that bear a meaningful share of the global NMSC burden. Realising the equity potential of this framework, rather than inadvertently widening existing disparities in cancer-care access, will require deliberate attention to cost reduction, simplified manufacturing platforms suitable for decentralised production, and international collaborative infrastructure for data sharing and model training, considerations that are as integral to the translational roadmap for this technology as the underlying biological and engineering science itself (Collins & Varmus, 2015).

9. IMPLICATIONS

9.1 Implications for Research

For researchers, the clearest near-term priority is generating skin-cancer-specific clinical evidence for nanocarrier-delivered therapeutics and, eventually, nanocarrier-delivered CRISPR constructs, since the great majority of supporting evidence reviewed in this chapter for these two pillars currently comes from preclinical models or from clinical experience in other tumour types rather than from dedicated melanoma or NMSC trials. Prospective studies that pair nanocarrier or CRISPR interventions with the multi-omics stratification strategies described in Section 3, rather than treating molecular profiling and delivery innovation as separate research programmes, would directly test the integrated framework's central premise.

9.2 Implications for Clinical Practice

For clinicians and health systems, the more immediate implication is that AI-assisted dermoscopy and genomic biomarker testing are already sufficiently mature to inform practice today, provided that deployment includes explicit validation across the full range of skin tones and clinical settings in which a tool will be used, and provided that any AI diagnostic support tool is treated, and regulated, as a decision-support aid that augments rather than replaces dermatologist judgement. Regulatory bodies in several jurisdictions now formally classify such tools as software-as-a-medical-device, requiring evidence of analytical and clinical validity, algorithmic transparency, and post-market performance monitoring before clinical deployment, a framework that is considerably more mature for image-based diagnostic AI than for the nanocarrier-CRISPR combination products discussed elsewhere in this chapter, and clinicians adopting these tools should expect to engage with that regulatory evidence base directly rather than treating AI-based triage as a purely technical, pre-validated black box.

10. CONCLUSION

The convergence of multi-omics profiling, nanotechnology-enabled delivery, CRISPR genome editing, and artificial intelligence represents a coherent, evidence-supported direction for the future of dermato-oncology rather than a speculative aggregation of unrelated technologies. Each component already carries meaningful independent clinical or preclinical traction: BRAF-mutation testing and Hedgehog-pathway-targeted therapy are established standards of care; nanocarrier-based topical and photodynamic therapies are progressing through preclinical and early clinical development; genome-scale CRISPR screening has already identified clinically relevant resistance and immunotherapy-sensitising genes; and AI-based dermoscopy classifiers are approaching or exceeding specialist-level diagnostic accuracy. What remains genuinely unresolved, and what this chapter has tried to represent honestly rather than optimistically, is the integration step: safely and efficiently delivering CRISPR payloads to cutaneous tumours in vivo, building multi-omics AI pipelines robust enough for clinical rather than purely research use, and establishing regulatory pathways equal to the hybrid nature of nanoparticle-gene-editing-AI combination products. Skin cancer, by virtue of its direct accessibility, its unusually well-characterised UV-driven mutational landscape, and its already-substantial nanomedicine and immunotherapy literature, is arguably better positioned than most solid tumours to serve as the proving ground where this integrated Omics-Nano-CRISPR-AI paradigm is first realised in routine clinical practice.

REFERENCES

Akbani, R., Akdemir, K. C., Aksoy, B. A., Albert, M., Ally, A., Amin, S. B., Arachchi, H., Arora, A., Auman, J. T., Ayala, B., Baboud, J., Balasundaram, M., Balu, S., Barnabas, N., Bartlett, J., Bartlett, P., Bastian, B. C., Baylin, S. B., Behera, M., & Zou, L. (2015, June). Genomic Classification of Cutaneous Melanoma. *Cell, 161*(7), 1681–1696. DOI: 10.1016/j.cell.2015.05.044 PMID: 26091043

Akbani, R., Akdemir, K. C., Aksoy, B. A., Albert, M., Ally, A., Amin, S. B., Arachchi, H., Arora, A., Auman, J. T., Ayala, B., Baboud, J., Balasundaram, M., Balu, S., Barnabas, N., Bartlett, J., Bartlett, P., Bastian, B. C., Baylin, S. B., Behera, M., & Zou, L. (2015, June). Genomic Classification of Cutaneous Melanoma. *Cell, 161*(7), 1681–1696. DOI: 10.1016/j.cell.2015.05.044 PMID: 26091043

Anzalone, A. V., Randolph, P. B., Davis, J. R., Sousa, A. A., Koblan, L. W., Levy, J. M., Chen, P. J., Wilson, C., Newby, G. A., Raguram, A., & Liu, D. R. (2019, December). Search-and-replace genome editing without double-strand breaks or donor DNA. *Nature, 576*(7785), 149–157. DOI: 10.1038/s41586-019-1711-4 PMID: 31634902

Ashley, E. A. (2016, September). Towards precision medicine. *Nature Reviews. Genetics, 17*(9), 507–522. DOI: 10.1038/nrg.2016.86 PMID: 27528417

Avanzato, V. A., Matson, M. J., Seifert, S. N., Pryce, R., Williamson, B. N., Anzick, S. L., Barbian, K., Judson, S. D., Fischer, E. R., Martens, C., Bowden, T. A., de Wit, E., Riedo, F. X., & Munster, V. J. (2020, December). Case Study: Prolonged Infectious SARS-CoV-2 Shedding from an Asymptomatic Immunocompromised Individual with Cancer. *Cell, 183*(7), 1901–1912.e9. DOI: 10.1016/j.cell.2020.10.049 PMID: 33248470

Bobo, D., Robinson, K. J., Islam, J., Thurecht, K. J., & Corrie, S. R. (2016, October). Nanoparticle-Based Medicines: A Review of FDA-Approved Materials and Clinical Trials to Date. *Pharmaceutical Research, 33*(10), 2373–2387. DOI: 10.1007/s11095-016-1958-5 PMID: 27299311

Borgheti-Cardoso, L. N., Viegas, J. S. R., Silvestrini, A. V. P., Caron, A. L., Praça, F. G., Kravicz, M., & Bentley, M. V. L. B. (2020, January). Nanotechnology approaches in the current therapy of skin cancer. *Advanced Drug Delivery Reviews, 153*, 109–136. DOI: 10.1016/j.addr.2020.02.005 PMID: 32113956

Brash, D. E. (2015, January). '<scp>UV</scp> Signature Mutations'. *Photochemistry and Photobiology, 91*(1), 15–26. DOI: 10.1111/php.12377 PMID: 25354245

Bray, F., Laversanne, M., Sung, H., Ferlay, J., Siegel, R. L., Soerjomataram, I., & Jemal, A. (2024, May). Global cancer statistics 2022: GLOBOCAN estimates of incidence and mortality worldwide for 36 cancers in 185 countries. *CA: a Cancer Journal for Clinicians, 74*(3), 229–263. DOI: 10.3322/caac.21834 PMID: 38572751

Cadet, J., & Douki, T. (2018). Formation of UV-induced DNA damage contributing to skin cancer development. *Photochemical & Photobiological Sciences, 17*(12), 1816–1841.

Chiappetta, M., Lococo, F., Leuzzi, G., Sperduti, I., Bria, E., Petracca Ciavarella, L., Mucilli, F., Filosso, P. L., Ratto, G., Spaggiari, L., Facciolo, F., & Margaritora, S. (2021, March). Survival Analysis in Single N2 Station Lung Adenocarcinoma: The Prognostic Role of Involved Lymph Nodes and Adjuvant Therapy. *Cancers (Basel)*, *13*(6), 1326. DOI: 10.3390/cancers13061326 PMID: 33809513

Collins, F. S., & Varmus, H. (2015, February). A New Initiative on Precision Medicine. *The New England Journal of Medicine*, *372*(9), 793–795. DOI: 10.1056/NEJMp1500523 PMID: 25635347

Danaei, M., Dehghankhold, M., Ataei, S., Hasanzadeh Davarani, F., Javanmard, R., Dokhani, A., Khorasani, S., & Mozafari, M. R. (2018, May). Impact of Particle Size and Polydispersity Index on the Clinical Applications of Lipidic Nanocarrier Systems. *Pharmaceutics*, *10*(2), 57. DOI: 10.3390/pharmaceutics10020057 PMID: 29783687

Diaz, M. J., Natarelli, N., Aflatooni, S., Aleman, S. J., Neelam, S., Tran, J. T., Taneja, K., Lucke-Wold, B., & Forouzandeh, M. (2023, July). Nanoparticle-Based Treatment Approaches for Skin Cancer: A Systematic Review. *Current Oncology (Toronto, Ont.)*, *30*(8), 7112–7131. DOI: 10.3390/curroncol30080516 PMID: 37622997

Díaz de la Guardia-Bolívar, E., Barrios-Rodríguez, R., Zwir, I., Jiménez-Moleón, J. J., & del Val, C. (2022, July). Identification of novel prostate cancer genes in patients stratified by Gleason classification: Role of antitumoral genes. *International Journal of Cancer*, *151*(2), 255–264. DOI: 10.1002/ijc.33988 PMID: 35234293

Doench, J. G. (2018, February). Am I ready for CRISPR? A user's guide to genetic screens. *Nature Reviews. Genetics*, *19*(2), 67–80. DOI: 10.1038/nrg.2017.97 PMID: 29199283

Duttke, S. H. C., Lacadie, S. A., Ibrahim, M. M., Glass, C. K., Corcoran, D. L., Benner, C., Heinz, S., Kadonaga, J. T., & Ohler, U. (2015, November). Perspectives on Unidirectional versus Divergent Transcription. *Molecular Cell*, *60*(3), 348–349. DOI: 10.1016/j.molcel.2015.10.014 PMID: 26545075

Esteva, A., Kuprel, B., Novoa, R. A., Ko, J., Swetter, S. M., Blau, H. M., & Thrun, S. (2017, February). Dermatologist-level classification of skin cancer with deep neural networks. *Nature*, *542*(7639), 115–118. DOI: 10.1038/nature21056 PMID: 28117445

[] B. Faubert, A. Solmonson, and R. J. DeBerardinis, 'Metabolic reprogramming and cancer progression', *Science (1979).*, vol. 368, no. 6487, Apr. 2020, .DOI: 10.1126/science.aaw5473

Finn, J. D., Smith, A. R., Patel, M. C., Shaw, L., Youniss, M. R., van Heteren, J., Dirstine, T., Ciullo, C., Lescarbeau, R., Seitzer, J., Shah, R. R., Shah, A., Ling, D., Growe, J., Pink, M., Rohde, E., Wood, K. M., Salomon, W. E., Harrington, W. F., & Morrissey, D. V. (2018, February). A Single Administration of CRISPR/Cas9 Lipid Nanoparticles Achieves Robust and Persistent In Vivo Genome Editing. *Cell Reports*, *22*(9), 2227–2235. DOI: 10.1016/j.celrep.2018.02.014 PMID: 29490262

Gilbert, L. A., Larson, M. H., Morsut, L., Liu, Z., Brar, G. A., Torres, S. E., Stern-Ginossar, N., Brandman, O., Whitehead, E. H., Doudna, J. A., Lim, W. A., Weissman, J. S., & Qi, L. S. (2013, July). CRISPR-Mediated Modular RNA-Guided Regulation of Transcription in Eukaryotes. *Cell*, *154*(2), 442–451. DOI: 10.1016/j.cell.2013.06.044 PMID: 23849981

Gillmore, J. D., Gane, E., Taubel, J., Kao, J., Fontana, M., Maitland, M. L., Seitzer, J., O'Connell, D., Walsh, K. R., Wood, K., Phillips, J., Xu, Y., Amaral, A., Boyd, A. P., Cehelsky, J. E., McKee, M. D., Schiermeier, A., Harari, O., Murphy, A., & Lebwohl, D. (2021, August). CRISPR-Cas9 In Vivo Gene Editing for Transthyretin Amyloidosis. *The New England Journal of Medicine*, *385*(6), 493–502. DOI: 10.1056/NEJMoa2107454 PMID: 34215024

Göppner, D., & Leverkus, M. (2011). Basal Cell Carcinoma: From the Molecular Understanding of the Pathogenesis to Targeted Therapy of Progressive Disease. *Journal of Skin Cancer*, *2011*, 1–8. DOI: 10.1155/2011/650258 PMID: 21253551

Gupta, N., Gupta, G. D., & Singh, D. (2022, October). Localized topical drug delivery systems for skin cancer: Current approaches and future prospects. *Frontiers in Nanotechnology*, *4*, 1006628. Advance online publication. DOI: 10.3389/fnano.2022.1006628

Haenssle, H. A., Fink, C., Schneiderbauer, R., Toberer, F., Buhl, T., Blum, A., Kalloo, A., Hassen, A. B. H., Thomas, L., Enk, A., Uhlmann, L., Alt, C., Arenbergerova, M., Bakos, R., Baltzer, A., Bertlich, I., Blum, A., Bokor-Billmann, T., Bowling, J., & Zalaudek, I. (2018, August). Man against machine: Diagnostic performance of a deep learning convolutional neural network for dermoscopic melanoma recognition in comparison to 58 dermatologists. *Annals of Oncology : Official Journal of the European Society for Medical Oncology*, *29*(8), 1836–1842. DOI: 10.1093/annonc/mdy166 PMID: 29846502

Hasin, Y., Seldin, M., & Lusis, A. (2017, December). Multi-omics approaches to disease. *Genome Biology*, *18*(1), 83. DOI: 10.1186/s13059-017-1215-1 PMID: 28476144

Hayward, N. K., Wilmott, J. S., Waddell, N., Johansson, P. A., Field, M. A., Nones, K., Patch, A.-M., Kakavand, H., Alexandrov, L. B., Burke, H., Jakrot, V., Kazakoff, S., Holmes, O., Leonard, C., Sabarinathan, R., Mularoni, L., Wood, S., Xu, Q., Waddell, N., & Mann, G. J. (2017, May). Whole-genome landscapes of major melanoma subtypes. *Nature*, *545*(7653), 175–180. DOI: 10.1038/nature22071 PMID: 28467829

Hodis, E., Watson, I. R., Kryukov, G. V., Arold, S. T., Imielinski, M., Theurillat, J.-P., Nickerson, E., Auclair, D., Li, L., Place, C., DiCara, D., Ramos, A. H., Lawrence, M. S., Cibulskis, K., Sivachenko, A., Voet, D., Saksena, G., Stransky, N., Onofrio, R. C., & Chin, L. (2012, July). A Landscape of Driver Mutations in Melanoma. *Cell*, *150*(2), 251–263. DOI: 10.1016/j.cell.2012.06.024 PMID: 22817889

Hood, L., & Friend, S. H. (2011, March). Predictive, personalized, preventive, participatory (P4) cancer medicine. *Nature Reviews. Clinical Oncology*, *8*(3), 184–187. DOI: 10.1038/nrclinonc.2010.227 PMID: 21364692

Huang, S., Chaudhary, K., & Garmire, L. X. (2017, June). More Is Better: Recent Progress in Multi-Omics Data Integration Methods. *Frontiers in Genetics*, *8*, 84. Advance online publication. DOI: 10.3389/fgene.2017.00084 PMID: 28670325

Hugo, W., Zaretsky, J. M., Sun, L., Song, C., Moreno, B. H., Hu-Lieskovan, S., Berent-Maoz, B., Pang, J., Chmielowski, B., Cherry, G., Seja, E., Lomeli, S., Kong, X., Kelley, M. C., Sosman, J. A., Johnson, D. B., Ribas, A., & Lo, R. S. (2016, March). Genomic and Transcriptomic Features of Response to Anti-PD-1 Therapy in Metastatic Melanoma. *Cell*, *165*(1), 35–44. DOI: 10.1016/j.cell.2016.02.065 PMID: 26997480

Human Genome Editing. (2017). National Academies Press., DOI: 10.17226/24623

Jerby-Arnon, L., Shah, P., Cuoco, M. S., Rodman, C., Su, M.-J., Melms, J. C., Leeson, R., Kanodia, A., Mei, S., Lin, J.-R., Wang, S., Rabasha, B., Liu, D., Zhang, G., Margolais, C., Ashenberg, O., Ott, P. A., Buchbinder, E. I., Haq, R., & Regev, A. (2018, November). A Cancer Cell Program Promotes T Cell Exclusion and Resistance to Checkpoint Blockade. *Cell*, *175*(4), 984–997.e24. DOI: 10.1016/j.cell.2018.09.006 PMID: 30388455

Karczewski, K. J., & Snyder, M. P. (2018, May). Integrative omics for health and disease. *Nature Reviews. Genetics*, *19*(5), 299–310. DOI: 10.1038/nrg.2018.4 PMID: 29479082

Komor, A. C., Kim, Y. B., Packer, M. S., Zuris, J. A., & Liu, D. R. (2016, May). Programmable editing of a target base in genomic DNA without double-stranded DNA cleavage. *Nature*, *533*(7603), 420–424. DOI: 10.1038/nature17946 PMID: 27096365

Konermann, S., Brigham, M. D., Trevino, A. E., Joung, J., Abudayyeh, O. O., Barcena, C., Hsu, P. D., Habib, N., Gootenberg, J. S., Nishimasu, H., Nureki, O., & Zhang, F. (2015, January). Genome-scale transcriptional activation by an engineered CRISPR-Cas9 complex. *Nature*, *517*(7536), 583–588. DOI: 10.1038/nature14136 PMID: 25494202

Lee, J. H., Long, G. V., Boyd, S., Lo, S., Menzies, A. M., Tembe, V., Guminski, A., Jakrot, V., Scolyer, R. A., Mann, G. J., Kefford, R. F., Carlino, M. S., & Rizos, H. (2017, May). Circulating tumour DNA predicts response to anti-PD1 antibodies in metastatic melanoma. *Annals of Oncology : Official Journal of the European Society for Medical Oncology*, *28*(5), 1130–1136. DOI: 10.1093/annonc/mdx026 PMID: 28327969

Lim, S. Y., Lin, Y., Lee, J. H., Pedersen, B., Stewart, A., Scolyer, R. A., & Rizos, H. (2024). Single-cell RNA sequencing reveals melanoma cell state-dependent heterogeneity of response to MAPK inhibitors. *EBioMedicine*, •••, 107.

Lino, C. A., Harper, J. C., Carney, J. P., & Timlin, J. A. (2018, January). Delivering CRISPR: A review of the challenges and approaches. *Drug Delivery*, *25*(1), 1234–1257. DOI: 10.1080/10717544.2018.1474964 PMID: 29801422

Luke, J. J., Flaherty, K. T., Ribas, A., & Long, G. V. (2017, August). Targeted agents and immunotherapies: Optimizing outcomes in melanoma. *Nature Reviews. Clinical Oncology*, *14*(8), 463–482. DOI: 10.1038/nrclinonc.2017.43 PMID: 28374786

Manguso, R. T., Pope, H. W., Zimmer, M. D., Brown, F. D., Yates, K. B., Miller, B. C., Collins, N. B., Bi, K., LaFleur, M. W., Juneja, V. R., Weiss, S. A., Lo, J., Fisher, D. E., Miao, D., Van Allen, E., Root, D. E., Sharpe, A. H., Doench, J. G., & Haining, W. N. (2017, July). In vivo CRISPR screening identifies Ptpn2 as a cancer immunotherapy target. *Nature*, *547*(7664), 413–418. DOI: 10.1038/nature23270 PMID: 28723893

Manguso, R. T., Pope, H. W., Zimmer, M. D., Brown, F. D., Yates, K. B., Miller, B. C., Collins, N. B., Bi, K., LaFleur, M. W., Juneja, V. R., Weiss, S. A., Lo, J., Fisher, D. E., Miao, D., Van Allen, E., Root, D. E., Sharpe, A. H., Doench, J. G., & Haining, W. N. (2017, July). In vivo CRISPR screening identifies Ptpn2 as a cancer immunotherapy target. *Nature*, *547*(7664), 413–418. DOI: 10.1038/nature23270 PMID: 28723893

Mertins, P., Mani, D. R., Ruggles, K. V., Gillette, M. A., Clauser, K. R., Wang, P., Wang, X., Qiao, J. W., Cao, S., Petralia, F., Kawaler, E., Mundt, F., Krug, K., Tu, Z., Lei, J. T., Gatza, M. L., Wilkerson, M., Perou, C. M., Yellapantula, V., & Carr, S. A. (2016, June). Proteogenomics connects somatic mutations to signalling in breast cancer. *Nature*, *534*(7605), 55–62. DOI: 10.1038/nature18003 PMID: 27251275

Mitchell, M. J., Billingsley, M. M., Haley, R. M., Wechsler, M. E., Peppas, N. A., & Langer, R. (2021, February). Engineering precision nanoparticles for drug delivery. *Nature Reviews. Drug Discovery*, *20*(2), 101–124. DOI: 10.1038/s41573-020-0090-8 PMID: 33277608

Patel, S. J., Sanjana, N. E., Kishton, R. J., Eidizadeh, A., Vodnala, S. K., Cam, M., Gartner, J. J., Jia, L., Steinberg, S. M., Yamamoto, T. N., Merchant, A. S., Mehta, G. U., Chichura, A., Shalem, O., Tran, E., Eil, R., Sukumar, M., Guijarro, E. P., Day, C.-P., & Restifo, N. P. (2017, August). Identification of essential genes for cancer immunotherapy. *Nature*, *548*(7669), 537–542. DOI: 10.1038/nature23477 PMID: 28783722

Pavlova, N. N., & Thompson, C. B. (2016, January). The Emerging Hallmarks of Cancer Metabolism. *Cell Metabolism*, *23*(1), 27–47. DOI: 10.1016/j.cmet.2015.12.006 PMID: 26771115

Pellegrini, C., Maturo, M., Di Nardo, L., Ciciarelli, V., Gutiérrez García-Rodrigo, C., & Fargnoli, M. (2017, November). Understanding the Molecular Genetics of Basal Cell Carcinoma. *International Journal of Molecular Sciences*, *18*(11), 2485. DOI: 10.3390/ijms18112485 PMID: 29165358

Pfeifer, G. P., You, Y.-H., & Besaratinia, A. (2005, April). Mutations induced by ultraviolet light. *Mutation Research. Fundamental and Molecular Mechanisms of Mutagenesis*, *571*(1–2), 19–31. DOI: 10.1016/j.mrfmmm.2004.06.057 PMID: 15748635

Rambow, F., Rogiers, A., Marin-Bejar, O., Aibar, S., Femel, J., Dewaele, M., Karras, P., Brown, D., Chang, Y. H., Debiec-Rychter, M., Adriaens, C., Radaelli, E., Wolter, P., Bechter, O., Dummer, R., Levesque, M., Piris, A., Frederick, D. T., Boland, G., & Marine, J.-C. (2018, August). Toward Minimal Residual Disease-Directed Therapy in Melanoma. *Cell*, *174*(4), 843–855.e19. DOI: 10.1016/j.cell.2018.06.025 PMID: 30017245

[] A. Ribas and J. D. Wolchok, 'Cancer immunotherapy using checkpoint blockade', *Science (1979).*, vol. 359, no. 6382, pp. 1350–1355, Mar. 2018, .DOI: 10.1126/science.aar4060

Sanson, K. R., Hanna, R. E., Hegde, M., Donovan, K. F., Strand, C., Sullender, M. E., Vaimberg, E. W., Goodale, A., Root, D. E., Piccioni, F., & Doench, J. G. (2018, December). Optimized libraries for CRISPR-Cas9 genetic screens with multiple modalities. *Nature Communications*, *9*(1), 5416. DOI: 10.1038/s41467-018-07901-8 PMID: 30575746

Shain, A. H., & Bastian, B. C. (2016, June). From melanocytes to melanomas. *Nature Reviews. Cancer*, *16*(6), 345–358. DOI: 10.1038/nrc.2016.37 PMID: 27125352

Song, M., Liu, C., Chen, S., & Zhang, W. (2021, February). Nanocarrier-Based Drug Delivery for Melanoma Therapeutics. *International Journal of Molecular Sciences*, *22*(4), 1873. DOI: 10.3390/ijms22041873 PMID: 33668591

Stratigos, A. J., Garbe, C., Dessinioti, C., Lebbe, C., Bataille, V., Bastholt, L., Dreno, B., Fargnoli, M. C., Forsea, A. M., Frenard, C., Harwood, C. A., Hauschild, A., Hoeller, C., Kandolf-Sekulovic, L., Kaufmann, R., Kelleners-Smeets, N. W. J., Malvehy, J., del Marmol, V., Middleton, M. R., & Grob, J.-J. (2020, March). European interdisciplinary guideline on invasive squamous cell carcinoma of the skin: Part 1. epidemiology, diagnostics and prevention. *European Journal of Cancer : Official Journal for European Organization for Research and Treatment of Cancer (EORTC) [and] European Association for Cancer Research (EACR)*, *128*, 60–82. DOI: 10.1016/j.ejca.2020.01.007 PMID: 32113941

Sung, H., Ferlay, J., Siegel, R. L., Laversanne, M., Soerjomataram, I., Jemal, A., & Bray, F. (2021, May). Global Cancer Statistics 2020: GLOBOCAN Estimates of Incidence and Mortality Worldwide for 36 Cancers in 185 Countries. *CA: a Cancer Journal for Clinicians*, *71*(3), 209–249. DOI: 10.3322/caac.21660 PMID: 33538338

Tirosh, I., Izar, B., Prakadan, S. M., Wadsworth, M. H., Treacy, D., Trombetta, J. J., & Garraway, L. A. (2016). Dissecting the multicellular ecosystem of metastatic melanoma by single-cell RNA-seq. *Science*, *352*(6282), 189–196.

Topol, E. J. (2019, January). High-performance medicine: The convergence of human and artificial intelligence. *Nature Medicine*, *25*(1), 44–56. DOI: 10.1038/s41591-018-0300-7 PMID: 30617339

Torchilin, V. P. (2014, November). Multifunctional, stimuli-sensitive nanoparticulate systems for drug delivery. *Nature Reviews. Drug Discovery*, *13*(11), 813–827. DOI: 10.1038/nrd4333 PMID: 25287120

Tsai, S. Q., & Joung, J. K. (2016, May). Defining and improving the genome-wide specificities of CRISPR–Cas9 nucleases. *Nature Reviews. Genetics*, *17*(5), 300–312. DOI: 10.1038/nrg.2016.28 PMID: 27087594

Tschandl, P., Rosendahl, C., & Kittler, H. (2018, August). The HAM10000 dataset, a large collection of multi-source dermatoscopic images of common pigmented skin lesions. *Scientific Data*, *5*(1), 180161. DOI: 10.1038/sdata.2018.161 PMID: 30106392

Wan, J. C. M., Massie, C., Garcia-Corbacho, J., Mouliere, F., Brenton, J. D., Caldas, C., Pacey, S., Baird, R., & Rosenfeld, N. (2017, April). Liquid biopsies come of age: Towards implementation of circulating tumour DNA. *Nature Reviews. Cancer*, *17*(4), 223–238. DOI: 10.1038/nrc.2017.7 PMID: 28233803

Wang, Q., Liu, Y., Li, C., Xu, B., Xu, S., & Liu, B. (2025). Machine learning-enhanced nanoparticle design for precision cancer drug delivery. *Advancement of Science*, *12*(30), e03138.

Whiteman, D. C., Green, A. C., & Olsen, C. M. (2016, June). The Growing Burden of Invasive Melanoma: Projections of Incidence Rates and Numbers of New Cases in Six Susceptible Populations through 2031. *The Journal of Investigative Dermatology*, *136*(6), 1161–1171. DOI: 10.1016/j.jid.2016.01.035 PMID: 26902923

Zhang, B., Wang, J., Wang, X., Zhu, J., Liu, Q., Shi, Z., Chambers, M. C., Zimmerman, L. J., Shaddox, K. F., Kim, S., Davies, S. R., Wang, S., Wang, P., Kinsinger, C. R., Rivers, R. C., Rodriguez, H., Townsend, R. R., Ellis, M. J. C., Carr, S. A., & Liebler, D. C. (2014, September). Proteogenomic characterization of human colon and rectal cancer. *Nature*, *513*(7518), 382–387. DOI: 10.1038/nature13438 PMID: 25043054

Zuo, S. Q., Li, C., Liu, Y. L., Tan, Y. H., Wan, X., Xu, T., Li, Q., Wang, L., Wu, Y. L., Deng, F. M., & Tang, B. (2021, October). Cordycepin inhibits cell senescence by ameliorating lysosomal dysfunction and inducing autophagy through the AMPK and mTOR–p70S6K pathway. *FEBS Open Bio*, *11*(10), 2705–2714. DOI: 10.1002/2211-5463.13263 PMID: 34448542

Zuo, S. Q., Li, C., Liu, Y. L., Tan, Y. H., Wan, X., Xu, T., Li, Q., Wang, L., Wu, Y. L., Deng, F. M., & Tang, B. (2021, October). Cordycepin inhibits cell senescence by ameliorating lysosomal dysfunction and inducing autophagy through the AMPK and mTOR–p70S6K pathway. *FEBS Open Bio*, *11*(10), 2705–2714. DOI: 10.1002/2211-5463.13263 PMID: 34448542

Chapter 10
AI-Enhanced Precision Medicine for Leukemia

Manoj Mani
http://orcid.org/0009-0008-6601-0339
Bharathiar University, India

Praveenkumar Ravi
http://orcid.org/0009-0004-7527-4097
Bharathiar University, India

Thangapandi Pandiyan
http://orcid.org/0009-0007-1879-5170
Bharathiar University, India

Preethi Amardeep Ray
Bharathiar University, India

Antony Prabhu Jeyabal Philomenathan
http://orcid.org/0009-0001-6500-9538
Bharathiar University, India

Nalini Devi Dharumaiyan
Bharathidasan University, India

Velayuthaprabhu Shanmugam
Bharathiar University, India

Vijaya Anand Arumugam
http://orcid.org/0000-0001-7485-1586
Bharathiar University, India

ABSTRACT

Artificial intelligence (AI) is transforming management of leukemia through integration of computation with precision medicine. The integration of machine learning, deep learning technologies has gained momentum in diagnostic, prognostic, and therapeutic applications. AI-enabled platforms boost the accuracy of subtyping, track minimal residual disease, and revise treatment through predictive modeling. Additional technologies like natural language processing, liquid biopsies, and federated learning boost the value of patient data while enhancing privacy and equity. In addition to diagnostic applications, AI accelerates drug development pipelines, informs adaptive clinical trials, and reinforces immunotherapy and gene editing. Prospective technologies such as quantum computing and synthetic biology emphasizes AI's transformational potential in the treatment of leukemia. This transdisciplinary integration reveals AI-enabled precision oncology, patient-specific plans to counter disease heterogeneity, therapeutic resistance, and socioeconomic disparities in management of leukemia.

DOI: 10.4018/979-8-3373-9715-3.ch010

1. INTRODUCTION

For the last two decades, the worldwide incidence and mortality rates for leukemia have exhibited an overall decrease, though the absolute number of cases remains increasing, mostly due to population growth. Specifically, the age-standardized incidence rate (ASIR) for leukemia declined from 6.89 per 100,000 in 1990 to 5.63 in 2021, and the age-standardized mortality rate (ASMR) decreased from 5.56 to 3.89 per 100,000 during the same period (Yun *et al.,* 2024). It was equaled in the United States by negative trends such that ASIR decreased by 27.33% and ASMR reduced by 25.82% between 1990 and 2021. Despite such encouraging global trends, acute myeloid leukemia (AML) is a deviation in rising incidence and mortality rates, thereby pointing towards persistent problems in its management (Saadeh *et al.,* 2024). Further, there are some regional differences; while high-income countries have generally experienced declining leukemia burden, others among the lower- and middle-income countries have experienced opposite trends (Huang & Zhang, 2025). In young people, acute lymphoid leukemia (ALL) continues to be an ominous threat, as one of the leading causes of cancer death for people under the age of 20 years, despite therapeutic progress (Durga *et al.,* 2024). Socioeconomic determinants, including Human Development Index and gross domestic product per capita, significantly affect leukemia burden, with higher development in regions. Together, these trends highlight the twofold difficulty of narrowing worldwide healthcare disparities and enhancing treatment for subtypes such as AML and ALL (Huang *et al.,* 2022).

Evolution of cancer precision medicine, particularly in leukemia, has transformed strategies of treatment by enabling biomarker-driven and enhanced stratification of patients. Molecular diagnostics and targeted therapies have been pivotal, and there is no example better than the discovery of the Breakpoint Cluster Region-Abelson (BCR-ABL) fusion gene in chronic Myeloid leukemia (CML), which directly led to the creation of tyrosine kinase inhibitors and revolutionized disease management (Berger & Van Allen, 2016). In AML, advances in disease classification and risk stratification incorporate novel molecular markers, refining diagnostic and predictive accuracy of outcome. Furthermore, the application of measurable residual disease (MRD) monitoring is now a sensitive measure of response to therapy, permitting dynamic risk evaluation and early therapeutic intervention (Hernández-Sánchez & Bullinger, 2023). Biomarkers, therefore, remain at the forefront of precision oncology as pivotal prognosticators for therapy selection based on individualized molecular and genetic profiles, providing more efficiency with fewer side effects (Mahmood *et al.,* 2024).

At the same time, innovative technologies such as liquid biopsies have also gained popularity as non-invasive alternatives to traditional tissue sampling, enabling real-time monitoring of disease progression and adaptive treatment planning (Hays, 2021). Moreover, the integration of multi-omics platforms-e.g., genomics, proteomics, and transcriptomics-has unveiled a comprehensive perspective on leukemia pathogenesis, its new drug targets, as well as furthered the precision of clinical decision-making. These developments come at a cost, however, with inherent challenges including continued high implementation costs, demand for sophisticated bioinformatics infrastructure, and ethical barriers of unequal access, for which continued research and investment are required (Asrina, 2024).

In this evolving paradigm, artificial intelligence (AI) has come to symbolize a groundbreaking revolution in managing leukemia, complementing precision medicine by improving diagnosis, prognosis, individualized treatment, and monitoring of the patient. Machine learning (ML) algorithms and image analysis have significantly improved the detection and classification of leukemia, increasing speed and accuracy compared to traditional methods. Convolutional neural networks (CNNs), for example, have proved to be extremely effective in hematological imaging and clinical data analysis for predicting CML and AML prognosis (Achir *et al.,* 2025). In CML, AI has made it possible to design predictive models for early diagnosis as well as optimized therapy inputs, resulting in improved patient outcomes (Ram *et al.,* 2024). Although still in nascent stages, AI-based tools for CML management are becoming increasingly capable of converging biochemical and biomolecular data to refine personalized treatment plans (Bernardi *et al.,* 2024). Similarly, AI and computational pathology have enhanced AML prognostic prediction and therapeutic stratification, accelerating the imperative to modify treatment to individual patient profiles (Venunadh & Joy, 2024). Moreover, AI platforms that integrate molecular data sets into practice have continued to fuel the move from "one size fits all" to more tailored treatment protocols and improved outcomes (Kubasch *et al.,* 2022). Despite such advances, the integration of AI into leukemia treatment is also faced with issues like heterogeneity of data quality, scalability of models across populations of differing diversity, as well as ensuring equitable global access (Achir *et al.,* 2025).

Though some achievements have already been reached in the use of artificial intelligence (AI) in oncology, AI in leukemia has not been systematically implemented because of the heterogeneity of the disease, lack of multi-center validation, and lack of combination of genomic, clinical, and imaging datasets. In most studies published till now, the focus has been on algorithm development without sufficient consideration of model explainability, clinical translation, regulatory approval, and cost-effectiveness. Moreover, there is not enough data to prove the necessity of the use of AI-assisted decision-making systems in the treatment of leukemia in the clinical setting, especially in under-resourced healthcare systems. This chapter aims to fill the gap by combining the most recent advancements in AI-assisted diagnostic, predictive, precision, immunotherapy, drug discovery, and clinical translation of leukemia.

2. TYPES OF LEUKEMIA AND THEIR CLINICAL PROFILES

ALL is a blood cancer involving the uncontrollable growth of immature lymphoid progenitors, with a strong predilection in children, having peak prevalence from 2 to 6 years of age (Pillai & Carroll, 2022). Clinical symptoms of marrow failure, including anemia, frequent infections, and bleeding manifestations, are accompanied by extramedullary disease, including lymphadenopathy and splenomegaly due to leukemic infiltrates (Malard & Mohty, 2020). At the molecular level, ALL involves heterogeneous genetic abnormalities from simple chromosomal translocations and fusion oncogenes to non-coding mutations and point mutations, which have been cataloged in sequence through high-throughput sequencing (Kimura & Mullighan, 2024). One high-risk variant, Philadelphia chromosome-like (Ph-like) ALL, involves mutations in cytokine receptor and tyrosine kinase-related pathways, highlighting the molecular variability of the disease (Yokota & Kanakura, 2016). Of further interest are *IKZF1* mutations, which compromise the function of the transcription factor *IKAROS*, powerfully linking with treatment refractoriness and poor prognosis, especially in B-cell ALL (Mullighan, 2012). Although ALL covers about 80% of pediatric leukemias, adult patients still have poor cure rates, reflecting biological and therapeutic differences with age (Saghir, 2023). Targeted therapy with tyrosine kinase inhibitors in *BCR::ABL1*-positive ALL has enhanced survival; however, the complexity of the disease persists and requires integrated diagnostic approaches, including next-gen sequencing and optical genome mapping, to further narrow risk classification and therapy (El Mahdaoui *et al.,* 2024). Challenges, however, persist in clinical translation of genomic profiling, highlighting future research needs in the area of epigenomic, proteomic, and microenvironmental causes of disease (Kimura & Mullighan, 2024; Saghir, 2023).

AML, on the other hand, is the most frequent type of acute leukemia in the adult population and results from clonal proliferation of myeloid progenitors with dysfunctional differentiation, with subsequent marrow failure and peripheral blood buildup of immature blasts (Becker *et al.*, 2024). While therapeutic developments have been substantial, targeted therapy resistance is one of the greatest clinical challenges. Resistance to *FLT3* inhibitors is commonly due to on-target mutations or dysregulated activation of downstream pathways, while *IDH* inhibitors like ivosidenib and enasidenib encounter secondary mutations or metabolic changes, including *IDH1/IDH2* isotype switching (Mecklenbrauck & Heuser, 2023; Zhuang *et al.*, 2022). While targeted therapies of *FLT3*, *IDH1/2*, and *BCL2* have evolved with novel therapeutic options, they all too often achieve short-term responses with evidence of residual leukemic stem cells (LSCs) and protective crosstalk with the marrow niche (Jones & Smith, 2024). Emerging modalities such as immunotherapeutic modalities and nanotechnology-based drug delivery continue to be explored to increase bioavailability, bypass resistance, and engage leukemic-specific antigens (Uddin *et al.*, 2021). Present clinical trials more commonly incorporate combinatorial regimens including cytotoxic, targeted, and cellular therapeutic modalities with an extension of survival results in high-risk AML patients in mind (Lachowiez *et al.*, 2023).

Chronic lymphocytic leukemia (CLL) is the most prevalent adult leukemia within Western populations and is the result of clonal proliferation of CD5+ B cells (Zhang & Kipps, 2014). Disease course is markedly influenced by genetic mutations and leukemic microenvironmental influences, respectively enhancing CLL cell proliferation and survival. Central to disease biology are interactions involving B cell receptor (BCR)-mediated and chemokine receptor-facilitated homing to lymphoid organs and immune dysregulation (Zhang & Kipps, 2014). Prognostic classification in CLL is significantly dependent on molecular biomarkers such as *TP53* disruption, *NOTCH1* mutations, mutated *IGHV* status, and *CD49d* expression, all individually predicting disease course and therapeutic responsiveness (Cohen *et al.*, 2020). Other markers, such as *ZAP-70* expression and *IgVH* gene mutational status, also possess independent prognosis (Stilgenbauer, 2007). Refinements in understanding these molecular pathways resulted in novel targeted agents such as inhibitors of *BCR* signaling and *BCL2*, which have improved clinical outcomes through effective alternatives to traditional chemoimmunotherapy (Crombie & Brown, 2019). However, CLL heterogeneity continues to bedevil single-pronged approaches to treatment, forcing adaptive, biomarker-guided approaches.

CML is a myeloproliferative neoplasm fueled by the seminal *BCR-ABL* fusion gene, formed from the Philadelphia chromosome translocation involving chromosomes 9 and 22. This creates constitutive tyrosine kinase activity, which accelerates cellular growth while inhibiting apoptosis in hematopoietic stem cells. Tyrosine kinase inhibitors (TKIs), such as imatinib, dasatinib, nilotinib, bosutinib, and ponatinib, have changed CML to a mostly manageable disease, with many patients attaining near-normal life expectancy (Wu *et al.*, 2024; El-Tanani *et al.*, 2024; Sholikah, 2017; Khalid & Riasat, 2023; Sorel *et al.*, 2017). Nevertheless, drug resistance through kinase domain mutations and residual presence of quiescent leukemic stem cells are major clinical challenges. Therefore, second- and third-generation TKIs were generated to bypass resistance, in addition to continuous attempts to eliminate LSCs through innovative therapeutic options. Standard molecular monitoring of *BCR-ABL* transcript levels continues to play a pivotal role in determining therapy direction and treatment benchmarks, but an absolute cure is still lacking.

Mixed-phenotype acute leukemia (MPAL) is a rare and diagnostically controversial disease, responsible for 2–4% of acute leukemias. MPAL is characterized by co-expression of definitional markers of ALL and AML blasts and requires strong immunophenotypic and genomic analysis. Lineage attribution according to WHO guidelines is determined with markers like myeloperoxidase in the case of myeloid blasts and *CD19* in the case of B lymphoid blasts. The genomic composition of MPAL is heterogeneous, with frequent mutations in *TP53*, *RUNX1*, and *FLT3*, and cytogenetic abnormalities like *BCR-ABL1* Philadelphia chromosome translocation (Awasthi *et al.*, 2019; Aggarwal & Weinberg, 2021; Hennawi *et al.*, 2022; Alexander & Orgel, 2021). Therapeutic management is controversial, but retrospective data attest that ALL-like induction therapy with subsequent allogeneic stem cell transplantation in first complete remission can achieve superior results, but poor prognosis persists with a median survival of around 19.5 months. Biological heterogeneity and the rare incidence of MPAL require more prospective research to improve classification and define therapeutic targets more accurately.

In parallel with biological and clinical developments, AI has been an effective adjuvant for subtyping leukemias through integration of clinical, genomic, and imaging data sets. CNNs and transfer learning-based frameworks like deep learning (DL) networks have shown unprecedented accuracies in the classification of subtypes, with Inception-ResNet achieving 95.52% classification of five subtypes of leukemias based on images (Hasan *et al.*, 2024). Advanced networks like Long Short-Term Memory (LSTM) architecture-based frameworks have shown over 98% accuracy in subtyping leukemias through gene expression profiles (Fatima *et al.*, 2024). Explainable AI (XAI) frameworks like Single Cell Based Explainable Multiple Instance Learning Algorithm (SCEMILA) have also improved interpretability through classification of subtypes of AML directly from smears of peripheral blood with high concordance with expert grading and possible recognition of subtype-specific features without annotation of single-cell data (Hehr *et al.*, 2023). Classical ML software like logistic regression and multilayer perceptrons have also proved robust through the classification of AML from ALL through transcriptomic data, with accuracies reaching 91% (Raut *et al.*, 2024). Even with current challenges about data heterogeneity, generalizability of AI-based frameworks, and equitable accessibility, AI-based technologies have immense potential through early detection augmentation, accurate subtyping classification, and individualized therapeutic approaches across leukemic subtypes (Achir *et al.*, 2025).

3. MOLECULAR, GENOMIC, AND MULTI-OMICS LANDSCAPE OF LEUKEMIA

Multi-omics approaches combining genomics, transcriptomics, proteomics, and metabolomics have greatly facilitated the classification and therapeutic targeting of leukemia, especially AML. Conventional genomic profiling has largely been centered on the detection of mutations and cytogenetic abnormalities; these changes, however, do not necessarily reflect gene and protein expression levels, which more directly relate to phenotypic disease manifestations (Zhang *et al.*, 2024). In order to address these limitations, multi-omics platforms have evolved as an integrated framework to probe the molecular complexity of leukemia (**Figure 1**). Proteogenomic research, like research from the Clinical Proteomic Tumor Analysis Consortium, has demonstrated proteogenomic clusters in AML with unique proteogenomic profiles, revealing molecular drivers and phenotypic heterogeneities undetectable through genomic data alone. These clusters not only define AML hierarchies of differentiation but also define mechanistic correlations to cancer-specific metabolic reprogramming and therapeutic vulnerabilities, guiding precision oncology strategies (Tyner *et al.*, 2018). Complementary to this, proteomics and phosphoproteomics have shed light on complex networks of signaling, revealing novel targets like mitochondrial dependencies invisible through DNA or RNA sequencing alone (Joshi *et al.,* 2022). Concurrent with this, metabolomics has provided insightful information on the biochemical reprogramming of leukemia cells through the capture of dynamic changes in metabolites during disease progression, providing functional insight into cellular metabolic dependencies.

Figure 1. Molecular Profiling of Leukemia Enhanced by AI

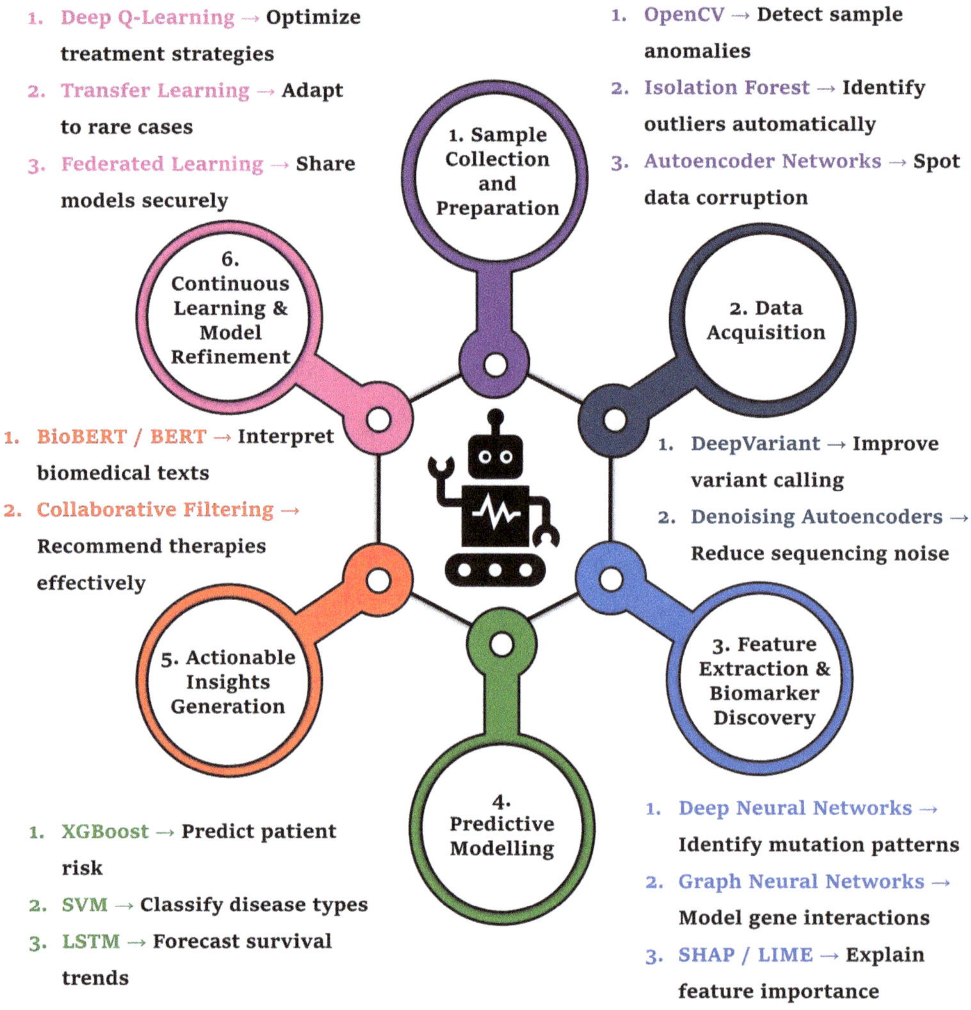

Inclusion of functional profiling, such as gene expression tests and ex vivo drug sensitivity screening, provides additional complementarity to multi-omics information. These approaches reinforce understanding of leukemogenesis, clonal heterogeneity, and drug resistance mechanisms and consequently aid in developing patient-specific therapeutic regimes (Iyer *et al.,* 2024). These methodologies, together, refine molecular classification frameworks and generate predictive biomarkers and therapeutic targets, solidifying the transition to precision medicine in leukemia (Turk *et al.,* 2024). Recent advances with single-cell RNA sequencing (scRNA-seq) and spatial transcriptomics have added an extra dimension of resolution, enabling high-dimensional mapping of cellular heterogeneity and the leukemic microenvironment. These technologies have already revolutionized understanding of neurological and oncological diseases, including Alzheimer's disease, gastric, prostate, and pancreatic cancers, through the expression of cell-state diversity, tumour–immune interactions, and spatially resolved signalling programmes (He *et al.,* 2024; Liu *et al.,* 2024b; Zhu & Qin, 2023; Li *et al.,* 2024b; Zhu *et al.,* 2024). In leukemia, scRNA-seq and spatially resolved omics should dissect out microenvironmental signals controlling leukemic stem cell survival, immune evasiveness, and therapeutic insensitivity and consequently refine stratification of prognosis and optimisation of therapy.

4. AI ARCHITECTURES AND METHODS IN LEUKEMIA CARE

The use of ML, DL, transformer-based networks, and graph neural networks (GNNs) is acquiring more prominence in leukemia research, diagnostics, and optimization of therapeutic plans, with marked improvement over traditional methods being achieved. DL, especially with the aid of CNNs, has shown considerable promise in increasing diagnostic precision and operational efficiency and hence overcoming the fundamental limitations of routine flow cytometry and microscopy (Rodriguez *et al.,* 2024). In ALL, traditional ML algorithms such as support vector machines and random forests, and state-of-the-art DL models such as AlexNet and ResNet50, have been applied with success to achieve earlier detection, which is essential to achieving desirable clinical outcomes (Faiz *et al.,* 2024).

In line with these developments, GNNs are turning out to be revolutionary tools in the evaluation of structured biomedical data in oncology. Their strength lies in integrating multimodal data such as molecular structures, gene expression networks, and protein–protein interactions, hence making possible predictive modeling and treatment stratification methods (Gogoshin & Rodin, 2023). This potential integrates well with the rising emphasis on precision oncology since GNNs present an architecture to handle biological complexity better compared to conventional ones. Individually, ML, DL, and GNN-based methods indicate the translatability of AI in terms of enhancing screening, diagnosis, and treatment planning. More work in predictive modeling with patient-specific features such as symptoms and hematologic profiles is, however, necessary to customize treatment paths (El Alaoui *et al.,* 2022).

One of the main enablers of these AI-based methodologies is federated learning (FL), which efficiently resolves problems of privacy surrounding sensitive clinical data. FL architectures allow distributed model training across diverse healthcare establishments without the need to store data centrally (Muthalakshmi *et al.,* 2024). In this setup, local models are trained locally, and only the model's parameters are summed in a master location, so the original data's confidentiality is maintained (Raza, 2023). For increased security, FL incorporates data protection methods like differential privacy, which ensures individual data points cannot be inferred from the model's output, and homomorphic encryption, which allows computations over encrypted data without decryption. The addition of distributed database structures further fortifies FL as it corrects data heterogeneity and provides data synchronization across establishments in real-time (Bandla, 2024). Performance metrics attest that FL attains levels of accuracy and strength comparable to central methodologies, while simultaneously maintaining strong protection of privacy (Kumari *et al.,* 2023). For these reasons, this framework is a critical methodological shift in AI's clinical application in treating leukemia.

Even with the application of AI-based privacy protection enabling safe uses, the value remains in clinical applications. Techniques such as Local Interpretable Model-agnostic Explanations (LIME), Shapley Additive Explanations (SHAP), and counterfactual reasoning from XAI enable interpretability and transparency of applied models to data sets from leukemia. SHAP, which uses cooperative game theory, provides Shapley values to individual features and hence estimates their contributions to predictions as numerical values, while LIME provides locally interpretable approximations to explain selected cases' behavior during modeling (Swathi & Challa, 2023; Rao *et al.,* 2022). Aside from these approaches, counterfactual reasoning defines hypothetical alternative scenarios unveiling underlying causal relationships and highlighting selected features' necessity or sufficiency in determining predictions (Presciuttini *et al.,* 2024; Kommiya Mothilal *et al.,* 2021). Through the provision of transparent and interpretable results to clinicians, XAI promotes AI-based recommendation trust and ease of clinical integration. These frameworks of interpretability collectively ensure the models achieve high performance while also being clinically dependable (Presciuttini *et al.,* 2024).

In addition to its diagnostic potential, reinforcement learning (RL) has also been an efficient approach to optimize sequential decision-making in treatment and therapeutic alterations in the management of leukemia. Through iterative incorporation of learning from patient responses and adaptations in disease evolution, RL allows adaptive treatment schedules individualized according to unique patient trajectories (Saranya & Sivaraman, 2024). Earlier applications in oncology have already established the potential of RL to optimize chemotherapy policies, with improved survival rates and decreases in side effects through data-based dynamic changes in treatment (Dixit, 2025). Incorporation with pharmacokinetic/pharmacodynamic (PK/PD) modeling best enhances dosing accuracy, which is particularly critical because of the narrow therapeutic margins typical of leukemia therapy (Tosca *et al.,* 2024). Regardless of their great potential, limitations like limited clinical data, high computational intensity, and the need for transparency in decision-making pose major challenges (Jayaraman *et al.,* 2024). Overcoming the challenges through multi- and transdisciplinary research and intensive clinical verification will ease the realization of the potential of RL to realize precision and patient-specific care in patients with leukemia (Ekundayo, 2024).

5. AI IN DIAGNOSIS AND EARLY DETECTION

DL architectures have shown great potential in allowing the automation of the diagnosis of leukemias across modalities such as blood smears, bone marrow biopsies, and radiologic images. CNNs in particular stand out with high diagnostic accuracies in ALLs and other subtype leukemias, with some selected models attaining accuracies of 100% in diagnosing leukemia from images of the bone marrow (Rodriguez *et al.,* 2024). These frameworks typically utilize advanced methodologies such as Cat-Boosting, XG-Boosting, and transfer learning, which increase performance considerably so as to outclass traditional diagnostic processes (Elsayed *et al.,* 2023). Nevertheless, challenges abound with respect to dataset variability, computational expense, as well as generalizability of clinical applicability (Faiz *et al.,* 2024). In order to allay the "black-box" aspect inherent in these frameworks, XAI methodologies are increasingly being incorporated so as to introduce greater transparency and clinical trust, thereby allowing the results to reliably be interpreted from the healthcare professional's perspective. Though still regarded as the diagnostic gold standard, DL frameworks utilized on peripheral blood smears have also been able to exhibit significant diagnostic potential, with the sub-discipline having attracted relatively humble interest in research-based publications (Achir *et al.,* 2025).

Along with image-based diagnostics, Natural Language Processing (NLP) has emerged as a critical AI approach to information organization from unstructured clinical information sources, including electronic health records (EHRs), pathology reports, and physician notes. Although rich in patient-specific information, these records are often sparsely utilized in a meaningful way because of their unstructured nature. NLP corrects this limitation through the extraction, categorization, and formatting of clinically relevant features, enhancing patient care, research endeavors, and administrative processes. Basic processes like tokenization, named entity recognition (NER), and relation extraction allow the systematic recognition of diagnostic predictors, treatment procedures, and prognostic factors (Gautam, 2024). Complex NLP frameworks like Bidirectional Encoder Representations from Transformers (BERT) and spaCy have been applied to study both structured and unstructured clinical data sets and have yielded patterns directly benefiting clinical decision-making (Hossain *et al.,* 2024). Challenges like the paucity of marked data sets, problems with integration with existing healthcare infrastructure, and maintenance of interpretability continue to prevail (Kondra *et al.,* 2024). Addressing these challenges requires domain-specific approaches like transfer learning and active learning, tuned according to the unique properties of medical text (Khosravi *et al.,* 2022). Thus, NLP has an integral role to play in supplying a level of diagnostic intelligence, consequently supporting the comprehensive clinical value of textual records.

In parallel, liquid biopsy technologies, including circulating tumor DNA (ctDNA), circulating tumor cells (CTCs), and exosomes, are gaining recognition as non-invasive approaches to early detection of leukemia. ctDNA is a particularly mature biomarker with the capability to distinguish rare genetic mutations critical in the recognition of early-stage disease and monitoring of disease development (Jiang, 2024). Techniques such as digital droplet PCR (ddPCR) and next-generation sequencing (NGS) have enhanced the sensitivity and specificity of ctDNA analysis to support early detection and patient-specific classification of treatment (Prabhakar *et al.,* 2024). CTCs provide intact tumor cells with morphologic and molecular description, while capture efficiency is a significant barrier; exosomes reveal abundant material but have analytical complexity due to heterogeneity and nonspecific markers (Yan *et al.,* 2020). Importantly, liquid biopsies circumvent challenges of invasive sampling of tissue, capturing tumor heterogeneity, and enabling longitudinal monitoring. Nevertheless, clinical application requires standardization of methods and validation to achieve specificity and sensitivity with potential application in routine use (Prabhakar *et al.,* 2024; Yan *et al.,* 2020). As refinement continues, liquid biopsy-based testing has the potential to transform early detection and precision medicine in leukemia (Anitha *et al.,* 2024).

The integration of novel molecular approaches has been further augmented by AI-based methods developed to increase sensitivity on ddPCR and NGS platforms, leading to previously unparalleled detection capability for MRD monitoring. ddPCR has been validated in the recognition of clinically relevant mutations, such as *FLT3-TKD* in AML, exhibiting as low as 0.01% variant allelic fraction of detection (Li *et al.,* 2024a). Similarly, NGS approaches, particularly those involving error-corrected sequencing (ECS), allow recognition of point mutations with 0.001% sensitivity and accommodate thorough gene fusions and structural variant examination across a variety of disease settings (Crowgey *et al.,* 2020). ddPCR assays centered on oncogenic *KMT2A* fusions in therapy-related AML also highlight this potential through achieving recognition thresholds with as much precision as 1 in 1,000,000 cells (Young *et al.,* 2023). Furthermore, the integration of DL and ensemble learning within these platforms improves diagnostic accuracy with classification performances of 100% precision and an F1 score of 0.997, outperforming conventional approaches (Syed *et al.,* 2024).

Several clinically tested AI-based platforms illustrate the real-world applications of AI-based leukemia diagnosis. Deep learning algorithms based on peripheral blood smears have proven to achieve accuracies greater than 95% for diagnosing acute leukemia under conditions of rigorous validation studies. Likewise, AI-based MRD detection using flow cytometry and next-generation sequencing has outperformed their respective control groups in terms of sensitivity in predicting relapse. However, these positive findings are based largely on retrospective studies, whereas prospective multicenter clinical trials examining patient outcomes are relatively sparse.

6. AI IN RISK STRATIFICATION AND PROGNOSIS

Integrating multi-omics data with clinical features has been an essential method of refining leukemic risk prediction and patient stratification. For AML, the model of Universal Analysis of Methylation Omics for Complex Systems (UAMOCS) is an illustration of integrating DNA methylation, genomics, transcriptomics, and ex vivo drug sensitivity data to differentiate patients in three subtypes with variable survival destinies and drug response profiles (Song *et al.,* 2023). Beyond projecting traditional cytogenetic risk groups, UAMOCS predicts actionable targets, such as receptor tyrosine kinase inhibitors in the UAMOCS1 subtype, with translation potential. Addressing the challenge of partial omics data, the IntegrAO framework employs GNNs to reliably differentiate AML patients with partial profiles, unveiling underlying biological heterogeneities and refined clinical stratification. Complementary to these frameworks, the Canonical Correlation Analysis with Recursive Feature Elimination Selector (CCARS) algorithm integrates canonical correlation analysis and recursive feature elimination, enhancing feature selection and classification accuracy considerably for early leukemic detection, superior to classical methods such as PCA and Lasso Regression (Nikshya *et al.,* 2024). Extending to B-cell ALL, the COMBAT system demonstrates proof of concept of integrated models through combining transcriptomic, epigenetic, and genomic features to differentiate subtypes with variable prognostic and therapeutic potential, with significant emergence of subgroups benefiting maximally from allogeneic hematopoietic stem cell transplantation (Song *et al.,* 2025).

With the incorporation of multi-omics data, AI-based prognostic models have been developed to predict treatment response, relapse incidence, and survival rates in diverse subtypes of leukemia. In the case of CML, predictive models integrated with morphometric diagnostic variables have increased the accuracy of predicting complete cytogenetic response (CCgR) with imatinib treatment beyond conventional scoring methods like EUTOS (Ćojbašić *et al.,* 2018). Similarly, in AML, decision forest-based models have determined relapse-related genes and pathways, like *MCL1* and *BAD.p112*, and have encouraged individualized therapeutic plans. For CLL, XAI methods have increased the IPI's prognostic potential through the elicitation of clinically meaningful immune cell populations, like CD4+ T cells, correlated with clinical outcomes (Hoffmann *et al.,* 2023). Pediatric ALL has also benefited from AI-augmented prognostic tools, wherein Random Forest computational methods have been highly accurate in relapse prediction using a modest set of clinical predictors, thus supporting stratified therapeutic plans through their attempts to elicit therapeutic efficacy without undue toxicity (Pan *et al.,* 2017). These developments reflect the capability of AI to not just refine hazard appraisal but also generate clinically meaningful prognostic data.

In this context, the recognition and tracking of MRD are influential predictors of prognosis in leukemia and cognate malignancies. The intersection of AI with computational approaches, coupled with sequencing technologies, is transforming the evaluation of MRD. In ALL, MRD monitoring is pivotal in therapeutic guidance, employing flow cytometry, PCR, and NGS with varying sensitivities, necessitating laboratory standardization to facilitate valid interpretation (Kruse *et al.,* 2020; Correia *et al.,* 2021). In myeloid malignancies, liquid biopsies (LBs) are undergoing evaluation for MRD detection, with recommended AI integration intended to support the interrogation of intricate molecular signatures, although clinical application awaits further standardization and substantiation (Allam *et al.,* 2023). In multiple myeloma, MRD assessment with NGS and flow cytometry has long exceeded traditional complete response evaluation, solidifying its position as a surrogate endpoint for survival outcomes (Mina *et al.,* 2020).

Beyond laboratory-based testing, wearable-based patient monitoring systems are revolutionizing prognostic evaluation through continuous monitoring of disease course and therapeutic response. Physiological signals like ECG, temperature, pulse, and SpO_2 are captured through wearable devices and IoT-enabled sensors and yield continuous cardiovascular and systemic health evaluation (Short *et al.,* 2018). The movement towards patient-centered monitoring is also bolstered through mobile communication technologies, enabling data transmission from a distance, minimizing human error, clinical workload, and healthcare expenses while optimizing early detection and proactive management (Naddeo *et al.,* 2017). Two-way communication between patients and physicians is also enabled through these systems, a particularly worthwhile capability in resource-poor regions of healthcare (Khanzode *et al.,* 2024). Integration of ML algorithms with wearable devices like smartwatches and pedometers has also furthered uses for continuous disease monitoring, like early recognition and intensity estimation of COVID-19 (Mary & Raj, 2024). Likewise, IoT-based health monitoring systems aided through cloud analytics allow clinicians to derive actionable intelligence that facilitates timely intervention and individualized advice to ultimately yield better patient outcomes (Prakash & Harish, 2024). Through integration of their real-time patient monitoring with AI-based predictive modeling, these monitoring systems also have potential for dynamic and adaptive risk stratification during care for leukemia.

Various AI models have specific strengths in predicting leukemia prognosis. Standard machine learning models such as random forest and XGboost are highly interpretable and can work very well with clinical data sets. Deep learning models excel in high-dimensional data sets such as genomics and imaging data, although they tend to demand much more extensive training cohorts and computational power than the aforementioned machine learning models. Graph neural networks show great potential in incorporating interaction network information but are mostly used in research settings.

7. FUNCTIONAL PRECISION MEDICINE AND PERSONALIZED THERAPEUTICS

In silico approaches to drug repurposing and patient-specific target discovery in leukemia treatment have drawn significant interest, presenting novel routes to overcoming drug resistance and enhancing clinical efficacy. Drug repurposing offers a time- and cost-saving alternative to established drug development pipelines, an essential consideration in leukemia, wherein therapeutic resistance so commonly imperils long-term remission. Through systematic screening of approved drugs for new oncological indications, the approach facilitates the extension of therapeutic options, especially when monotherapies are inadequate for refractory patients (Kenmogne *et al.,* 2022). In silico investigations targeting molecular drivers of resistance have registered significant success, exemplified through the discovery of inhibitors of the BCR kinase domain in CML and disabling *Grb-2* protein interactions to restore tyrosine kinase inhibitor sensitivity (Natarajan *et al.,* 2019). Through the integration of genomic profiling and ML, drug repurposing methodologies have matured from basic ligand screening to encompass structural biology, mutational signatures, and network-based models, consequently optimizing candidate prioritization and translation potential (Cousins *et al.,* 2024). Sophisticated pipelines like DrugReSC, through single-cell RNA sequencing to attack disease-critical subpopulations of cells, reflect an increasing focus on precision methodologies enabling context-specific therapy identification (Liu *et al.,* 2024a). Likewise, the DCPT platform, through drug-target interaction networks and polypharmacology-based modeling, reflects how patient-specific predictions can inform synergistic drug combinations, exemplified in T-cell prolymphocytic leukemia (T-PLL) (He *et al.,* 2018).

In an attempt to expand computational predictions, preclinical functional models such as organoids, patient-derived xenografts (PDX), and microfluidic simulation devices (organs-on-chips) are finding increasing application to justify therapeutic hypotheses and tailor individualized treatment strategies. Organoids generated from patient tumor tissue offer three-dimensional in vitro platforms with retention of the molecular and cellular complexities of the host tumor. This retention of complexity enables application in high-throughput screening of therapeutic agents, identification of the mechanisms of resistance, and identification of biomarkers (Kim *et al.,* 2022; Swain & Maddi, 2024). Concomitantly, PDX models with retention of the genetic and phenotypic integrity of the donor tumors in vivo permit longitudinal assessment of therapeutic responses and metastatic behaviors and consequently act as efficient platforms to formulate patient-specific targeted therapies (Gao *et al.,* 2023). Organs-on-chips introduce an additional dimension through a mimicry of the microenvironmental conditions, such as stromal interactions, vascular flow, and modulations of the immune system necessary to appropriately envision systemic treatment efficacy (Zhu *et al.,* 2023). Although each has distinct strengths, the selection has to match clinical aims and the unique diseased context, with an appreciation of the importance of integrative approaches to best capture translational outcomes (He *et al.,* 2024).

Extending these experimental platforms, AI-enabled platforms are transforming therapeutic individualization through integration with multi-omics data and clinical variables. ML- and DL-based algorithms are able to interrogate rich, heterogeneous data sets, including genomic, transcriptomic, and proteomic data, with the ultimate aim of identifying predictive biomarkers, delineating patient subgroups, and making individualized therapeutic regimen recommendations (Heuser *et al.,* 2020; Vegesna, 2024). These predictors enable precision chemotherapy, targeted therapy, and immunotherapy through the prediction of patient-specific molecular vulnerabilities, consequently reducing toxicity while attaining maximum therapeutic efficacy. For immuno-oncology, AI hastens the identification of novel therapeutic targets, such as neoantigens, amenable to the development of personalized vaccines or adaptive T-cell therapies (Olawade *et al.,* 2025). Further integration of AI recommendations with EHRs and contemporaneous clinical outcomes provides adaptive feedback loops with consequent improvement in the vision of learning healthcare systems (Terzi, 2024).

Aside from conventional static therapeutic guidelines, adaptive therapeutic strategies via deep reinforcement learning (DRL) also enable ongoing personalization of leukemia therapy. Through successive integration of longitudinal clinical information, DRL algorithms adaptively optimize drug schedules and dosages in real-time while preventing drug resistance from breaching conventional treatment protocols. Preclinical experiments with metastatic tumors have established DRL's capability to optimize progression-free survival via individualized treatment schedules adapted to tumor behavior, including variable monitoring conditions (Gallagher *et al.,* 2024). More generally, reinforcement learning strategies in oncology also enable cumulative outcome optimization across several modalities like chemotherapy, radiation, and immunotherapy, an expression of their adaptability in clinical decision-making (Ekundayo, 2024).

Furthermore, adjunct AI-augmented decision support systems complement these adaptive treatments via explanatory discovery of mechanisms of treatment response and recommendation of synergistic combinations best adapted to the individualized needs of the individual patients (Molefi *et al.,* 2023).

8. AI IN IMMUNOTHERAPY AND NEXT-GENERATION TREATMENTS

Next-generation immunotherapies have revolutionized the practice of oncology, with novel therapeutic options for patients who in the past had few alternatives. A vital role of the immune system is the elimination of malignant cells via ongoing immune surveillance, primarily employing cytotoxic T lymphocytes (CTLs) and natural killer (NK) cells. A sophisticated approach in this context involves the utilization of bispecific antibodies to eradicate cancer cells (**Figure 2**) (Garg *et al.,* 2024). Recent developments in immunotherapy have utilized this native surveillance potential, moving forward from early experimental vaccines and cytokine-based approaches to innovative modalities like immune checkpoint inhibitors (ICIs) and engineered cell therapies. Therapeutic agents targeting inhibitory receptors, namely *PD-1* and *CTLA-4*, have shown persistent clinical responses in malignancies like melanoma and lung cancer, hence securing ICIs as standard components of current cancer therapy (Naimi *et al.,* 2022). The pivotal finding by Allison and colleagues that blocking CTLA-4 might unleash antitumor immunity led to a shift in therapeutic approaches, culminating in the FDA licensure of ipilimumab in 2011 and marking the first checkpoint inhibitor to achieve such an honor (Goswami *et al.,* 2024).

Figure 2. AI-Driven Workflow for Bispecific Antibody Development

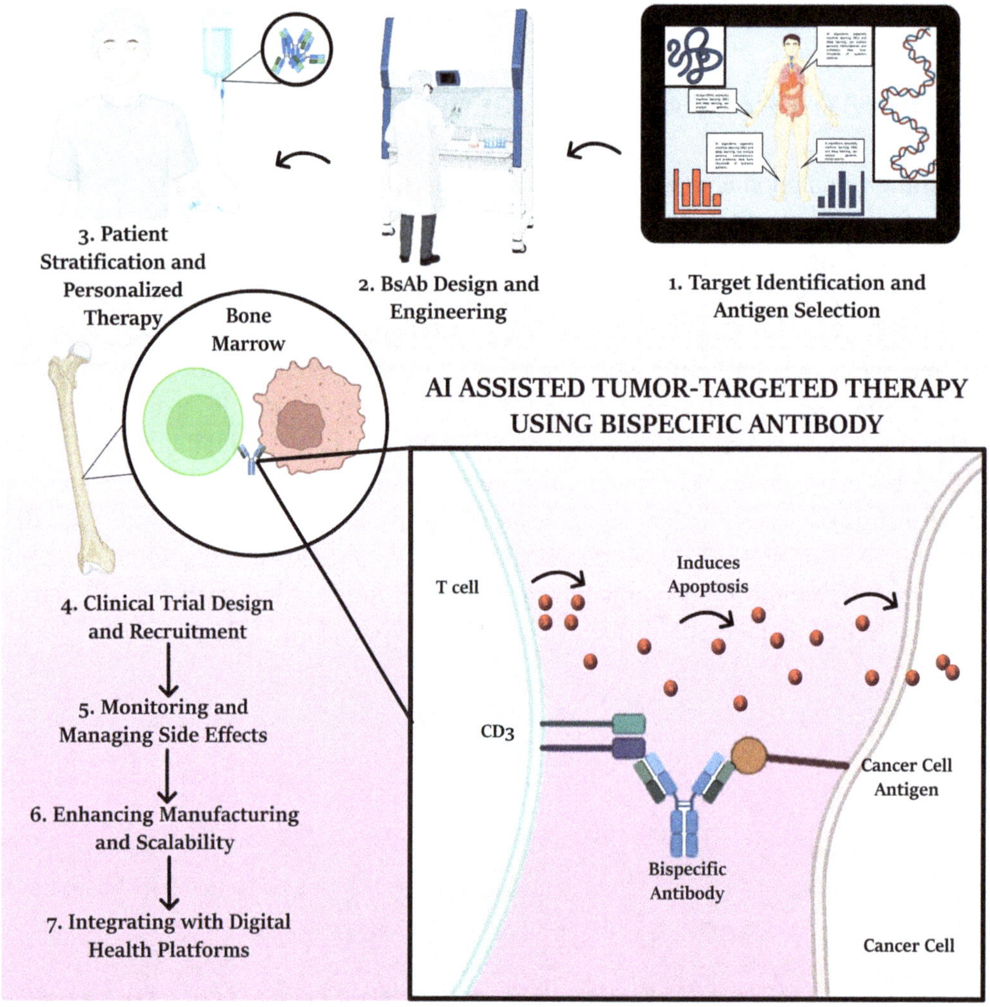

Chimeric antigen receptor (CAR)-T cell therapy has been confirmed as an efficient development in the management of hematological cancers, especially relapsed or refractory ALL and large B-cell lymphoma (Naimi *et al.,* 2022). Despite these achievements, it is ineffective against solid tumors because of weakened cytotoxicity, inefficient tumor infiltration, and the presence of highly suppressive microenvironments (Ai *et al.,* 2024; Zeng & Zhang, 2022). AI-driven tumor-specific antigen discovery and knowledge of the various stages of tumor development, like proliferation, invasion, angiogenesis, and metastasis, are essential in increasing the potential of CAR-T cells to target tumors. Proteins like LunX, which execute their functions in stimulating tumor growth and metastasis through ERK and JNK pathways, are promising targets in increasing T cell cytotoxicity (**Figure 3**) (Zheng *et al.,* 2015). Overcoming these biological and technical challenges remains essential in increasing the clinical applicability of CAR-T therapy for hematologic malignancies.

Figure 3. AI-Driven Optimization of CAR-T Cell Therapies in Leukemia Care

In parallel, individualized cancer vaccines have gained traction through the use of patient-specific neoantigens. Nevertheless, limitations in identifying neoantigens, designing vaccines, and clinical confirmation continue to impede progress. AI has been an enabling tool to break through these hurdles by combining ML and DL across several phases of vaccine development. AI algorithms enhance neo-antigen discovery through proper peptide–MHC binding prediction, antigen processing modeling, and modeling of T cell receptor interaction (Kong, 2025). Beyond the realm of epitope selection, AI-based codon optimization and untranslated region (UTR) engineering boost mRNA stability and translational efficiency, while DL-based structural modeling facilitates accelerated nanoparticle formation and delivery platforms. These computational technologies collectively enrich the design of individualized vaccines, making them promising next generations of immunotherapies. AI also has a critical role in the rationale and optimization of the design of cell-based therapies. Engineered cells are naturally multidimensional platforms requiring the integration of heterogeneous molecular, genomic, and environmental data. ML strategies, such as already established successful platforms in protein structure prediction and enzyme engineering, continue to expand their utility to forecast cell behavior, optimize receptor design, and predict patient-specific therapeutic outcomes (Capponi & Daniels, 2023). Such prediction platforms, coupled with ensuing advanced experimental platforms, promise to systematically optimize the safety, efficacy, and scalability of engineered cell therapies.

9. AI IN DRUG DISCOVERY AND CLINICAL TRIALS

AI is a groundbreaking driver of change in cancer drug development and discovery, holding the promise to speed up processes that otherwise are labor-intensive and expensive. AI approaches-such as ML and DL-allow extensive integration of intricate data sets, hence making it easier to discover new drug targets, lead compound optimization, and therapeutic outcome prediction. In lead optimization and de novo drug design, generative models driven by AI can generate structurally diverse molecular scaffolds and fine-tune the physicochemical properties to optimize efficacy and safety. Preclinical development is enhanced by AI-driven in silico simulations, especially in the prediction of toxicity and modeling of pharmacokinetics, allowing for more effective prioritization of drug candidates. Additionally, AI expedites biomarker discovery through the recognition of molecular signatures that facilitate patient stratification as well as precision therapy development (Sarvepalli & Vadarevu, 2025).

The clinical application of AI is most prominent in hematological neoplasms like leukemia. Leukemia is distinguished by the uncontrolled growth of aberrant hematopoietic cells, resulting in systemic spread through the circulatory and lymphatic systems. Subtypes like acute monocytic leukemia (AMoL) remain formidable clinical challenges, with poor prognosis, especially in the elderly (Shehadi *et al.,* 2020). In CML, the *BCR-ABL* oncogene induces leukemogenesis by virtue of constitutive tyrosine kinase activity. First-generation TKIs like imatinib, which inhibit ATP binding to the BCR-ABL fusion protein, changed therapy with a big step; nonetheless, resistance following kinase domain mutations made it important to develop second-generation inhibitors like bosutinib, nilotinib, dasatinib, ponatinib, and bafetinib (Kumar *et al.,* 2016). AI-based computational models are now being used to predict mutation-driven mechanisms of resistance, refine inhibitor design, and suggest synergistic therapeutic approaches based on patient-specific molecular profiles.

Whereas most solid tumors do not offer constant access to peripheral blood and bone marrow samples, leukemia can easily facilitate longitudinal AI analysis and therapy optimization. Recently, there have been several efforts to use AI to predict resistance to FLT3 inhibitors, BCL-2 inhibitors, and tyrosine kinase inhibitors, and at the same time identify patients who will be good candidates for combined therapy. These are just a few examples of the potential of AI in leukemia that demonstrate its application in therapy beyond drug development.

In the clinical trial design space, AI has emerged as a potent driver of improved efficiency, accuracy, and responsiveness. AI-powered platforms accelerate patient recruitment by quickly aligning participants with eligibility criteria, thereby avoiding enrollment delays and enhancing the representativeness of populations. Real-time patient data are tracked through predictive analytics based on ML to identify safety signals and adverse events, permitting real-time course corrections and enhancing trial safety profiles (Gupta *et al.,* 2025). Integration of real-world evidence from EHRs also strengthens trial design, providing information regarding patient histories and aiding in the selection of potential therapeutic candidates. However, challenges like ethical issues related to maintaining patient privacy, non-uniform regulatory alignment on the validation of AI, and algorithmic biases with non-diverse training datasets are still significant considerations. These are addressed through standardized data formats, strict governance, and representativeness in the dataset (Sedano *et al.,* 2025). Adaptive trial designs are another area where AI integration is highly valuable. By using cumulative data, adaptive designs can adjust important features like dosing, regimen, and patient stratification dynamically. Bayesian modeling and longitudinal biomarker incorporation allow for more accurate predictions of outcomes and optimization of trials. A notable example is the I-SPY2 platform, which employs an adaptive phase II design to evaluate multiple investigational agents concurrently while seamlessly transitioning into confirmatory stages (Berry, 2012). Such approaches exemplify how AI-powered adaptive methodologies accelerate therapeutic development while maximizing trial informativeness.

AI's impact also extends to diagnostic technologies underpinning trial enrollment and monitoring. Leukemia subtype identification automated systems through peripheral blood smear analysis have been highly accurate in classifying white blood cells, aiding in early diagnosis and personalized therapeutic intervention (Haque *et al.,* 2024). The exponential rise of molecular technologies such as next-generation sequencing (NGS) and mass spectrometry has further enriched the data environment that can be used by AI for insights. NGS allows for the full profiling of structural variants, translocations, copy number variations (CNVs), single-nucleotide variants, DNA methylation, and transcriptome signatures, while mass spectrometry adds proteomic characterization of tumor tissues (Benard *et al.,* 2019). Combining these multi-dimensional data with AI analytics not only speeds up drug discovery pipelines but also increases the accuracy and responsiveness of clinical trials.

10. AI FOR MONITORING, WEARABLES, AND REMOTE CARE

Biomarkers-biological markers that are measurable signatures-are at the heart of health status assessment and medical decision-making. The U.S. Food and Drug Administration (FDA) BEST document defines a biomarker as "a characteristic that is measured as an indicator of normal biological processes, pathogenic processes, or biological responses to an exposure or intervention, including therapeutic interventions." Digital biomarkers, on the other hand, are less well-defined. Instead of being a direct statistical measure, they are composites of streams of physiological, behavioral, and environmental data captured using sensors and wearable technologies. .

The relevance of early diagnosis is evidenced by the elevated mortality of multifactorial conditions such as cancer, cardiovascular disease, mental illness, and infectious diseases, whereby delayed diagnosis results in diminishing treatment outcomes and increasing healthcare expenditure. Since Clark introduced an electrochemical glucose electrode in the 1950s, biosensor technologies have seen immense progress, resulting in noninvasive glucometers that enable continuous monitoring through smartphones (Wasilewski *et al.,* 2024). These advances illustrate the deployment of biosensors within digital health systems, setting the stage for AI-based monitoring platforms. AI has catalyzed the shift toward automated, scalable disease detection from traditional, manual screening techniques (e.g., Pap smear, HPV testing, breast exam) towards early detection of disease. Utilizing multimodal datasets, AI-based systems allow for real-time monitoring, predictive modeling, and tailored intervention protocols. In oncology, business platforms like CureMetrix's cmAssist, Qlarity Imaging's QuantX, and BUCAD utilize AI algorithms in all mammography, MRI, and ultrasound to enhance the sensitivity and specificity of cancer detection (Goel *et al.,* 2025).

In addition to image-based detection, AI is being more heavily embedded in multimodal monitoring platforms that integrate liquid biopsies, wearable sensors, and digital pathology. NGS strategies applied to circulating tumor DNA and exosomes, with the support of ML and DL algorithms, reach AUCs >0.90 across cancer types. CancerSEEK illustrates the convergence of multi-omic characteristics for pan-cancer detection, while AI-powered CNNs like Optellum's LCP-CNN and U-Net-based MRI models surpass traditional risk scores in pulmonary and neurological cancer evaluation. These systems not only enable early diagnosis but also allow for recurrence prediction and longitudinal monitoring, in which radiomics-based and "Deep Profiler" strategies guide personalized treatment dosing (Hunter *et al.,* 2022).

The function of imaging tools in AI-supported remote monitoring is especially pivotal in resource-poor settings, where cost-efficient, transportable technology fills the gaps in diagnosis. High-resolution microendoscopy delivers real-time cellular imaging to identify precancerous and cancerous lesions, and optical coherence tomography provides high-resolution 3D visualization ideal for point-of-care oncology settings. In the same manner, low-cost ultrasound equipment, in conjunction with AI-powered computer-aided detection and diagnosis systems, improves diagnostic accuracy while diminishing reliance on highly specialized expertise (Haney *et al.,* 2017; Huang *et al.,* 2004). Complementary imaging, like optical imaging using non-ionizing visible, ultraviolet, or infrared light, extends even safer and more repeatable monitoring performance in oncology and provides the ability for exact non-invasive real-time monitoring of tumor development and treatment outcome (Taruttis & Ntziachristos, 2012).

11. ETHICAL, LEGAL, AND SOCIAL CONSIDERATIONS

Algorithmic equity in leukemia treatment, as a component of general healthcare AI implementations, is an area of research of high significance due to the potential to perpetuate existing health inequality (Bhatia *et al.,* 2024). Biases can arise due to imbalanced datasets, poor feature representation, or contextual narrowness, causing overrepresentation among vulnerable groups (Bhatia *et al.,* 2024). Decentralized learning, disentanglement, and explainability of AI models are necessary to achieve fair healthcare provision. Individual fairness-treating comparable persons similarly-has gained prominence, but implementation in healthcare is in an embryonic stage (Anderson & Visweswaran, 2025). Embedding fairness measures and bias avoidance methodologies in AI workflows is crucial to achieving inclusivity and equity reduction (Xu *et al.,* 2022). Correspondingly, participatory design and stakeholder input are becoming increasingly advised to reflect the sociotechnical processes of healthcare provision, developing AI solutions attuned to diverse patient needs (Griffin *et al.,* 2024).

The implementation of AI-based healthcare systems also leads to urgent data protection, patient permission, and cybersecurity issues. The "Integrated Security and Ethics Model" of Gawankar et al. stresses integrating ethical principles with security processes through effective governance and risk management approaches (Gawankar *et al.,* 2024). Gupta stresses that intelligent healthcare technologies require strong security to safeguard sensitive data while meeting cost-effectiveness and ethical obligations (Gupta, 2025). Google's Project Nightingale case study highlights AI-based healthcare data-governance complexities with a need for openness, public deliberation, and adaptive regulation to meet shifting privacy needs (Arigbabu *et al.,* 2024). As a parallel, Jeelani et al. suggest contemporary algorithmic strategies and technologies to maintain data confidentiality, stressing innovative methodologies in the area (Jeelani *et al.,* 2024).

Developments of AI-based diagnostic and therapeutic regulatory paths pose challenges of validation, approval, and monitoring. Lopez-Perez et al. put forward an external validation framework consisting of four phases to maintain reliability and efficacy of AI models in varying healthcare settings (Lopez-Perez *et al.,* 2024). In the United States, the FDA has also started codification of guidelines on AI-enabled in silico trials and healthcare devices in an attempt to maintain the safety and efficacy of predictive algorithms (Opderbeck, 2019; Schulz *et al.,* 2019). Of interest, FDA clearances of cardiology devices with AI-enabled capability reflect regulatory sternness in obtaining assurance and permitting clinical application of AI/ML technologies (Maher & Chang, 2024).

Trust among clinicians, patients, and AI systems is founded on explainability, transparency, and ethical governance. XAI techniques are necessary to help clinicians interpret results from algorithms and integrate them in evidence-based decision-making (Sîrbu & Mercioni, 2024). Training of clinicians also has an equally pivotal role, allowing healthcare professionals to possess an understanding of AI's capabilities and limitations so as to better realize patient outcomes (Asan *et al.,* 2020). Patient education also fortifies trust through explaining data use concerning personal data, resolving privacy challenges, and acquiring informed consent (Albalawi *et al.,* 2024). Governance frameworks-supplemented with effective data stewardship, regulatory scrutiny, and multidisciplinary working-provide structural protection in aid of responsible AI implementation. Furthermore, actions such as diversified clinical recruitment trials and augmented privacy-preserving methods are also necessary in realizing equitable and trustworthy AI-based health care (Sîrbu & Mercioni, 2024).

12. INTEGRATION WITH CLINICAL WORKFLOWS

Incorporation of AI in Clinical Decision Support Systems (CDSS) in the management of leukemia requires strategic planning, enhancing diagnostic precision, personalization of treatments, and improved patient outcomes. AI systems are capable of processing high-volume clinical data in real time, producing individualized insights to maximize diagnostic pathways and therapeutic advice. This aspect is most significant in leukemia, in which early recognition and accurate treatment stratification are critical. For instance, in AML, AI-powered CDSS can advise on personalization of therapy protocols through prognostic modeling, providing clinicians with decision support in the intermediate-risk groups, usually tricky to evaluate accurately (Tupsakhare, 2023; Castro *et al.,* 2024). Through the incorporation of multimodal data comprising genomic, laboratory, and imaging data, AI-based CDSS minimize diagnostic confusion, reduce mistakes, and enhance patient management (Mitra & Rehman, 2025; Atul, 2025).

Inclusion of AI in clinical workflows extends beyond diagnostic support to workflow automation, with AI-based recommendations and alerts streamlining routine processes. By combining ML algorithms with EHRs, AI-powered CDSS draw on rich collections of similar patient cases to support clinicians with evidence-informed decision-making (Pindi, 2019). Automation of laboratories with AI cements healthcare delivery through reduction of human error, better diagnostic accuracy, and increased test reporting velocity. Enhancements have an immediate positive impact on efficiency and clinical outcomes. Moreover, application of Robotic Process Automation (RPA) with AI in administrative spheres-scheduling, billing, and resource allocation-reduces operational overhead and limits error propagation, freeing more time for patient-oriented care (Nimkar *et al.,* 2024). AI-powered workflow decision automation with heterogeneous patient data sources generates priority indicators and clinical advice, navigating physicians through pivotal decision-making points (Chen *et al.,* 2021).

Clinical application of AI technology calls for unimpeded integration into hospital information systems (HIS), laboratory information management systems (LIMS), and electronic health records (EHRs). Instead of substituting hematologists, the use of AI should be a decision support system helping in diagnostics, risk prediction, treatment choice, and patient monitoring. The establishment of multidisciplinary cooperation among doctors, bioinformaticians, laboratory scientists, and regulatory experts will be crucial for the appropriate implementation of AI technology in leukemia treatment.

But unresolved challenges-office data protection, bias in algorithms, explainability, and astronomical implementation costs-continue to impede widespread implementation. Focusing on rising above them with standard frameworks, collaborative proof-of-concept studies, and networking with stakeholders will facilitate the realization of the full potential of AI-based CDSS and their integration with clinical workflows (Sopruchi & Rashid, 2024).

13. CHALLENGES AND LIMITATIONS

In AI-powered leukemia research, long-standing issues of data standardization, interoperability, and reproducibility persist in restricting translational impact. Data standardization is the heart of having comparable documentation and comparative capability across healthcare infrastructures. As an exemplar, the development of common data elements in myeloid leukemia led to an 86% consensus from professionals, evidencing the potential of aligned data frameworks (Varghese *et al.,* 2016). Parallel issues abound, with ONCO-FAIR projects integrating the HL7 FHIR standard to facilitate data exchange in oncology chemotherapy treatments as an exemplar of the benefit of upstream interoperability across clinical care and research infrastructures (Guilbert *et al.,* 2024). Similarly, the EuCanImage consortium advocates crisp clinical data standards permitting interoperability in precision oncology but also notes the methodological and ethico-legal issues inhibiting implementation across practical settings. As an aid to enhancing reproducibility, the National Cancer Institute Imaging Data Commons has tried to standardize large, heterogeneous data collections according to industry guidelines, permitting reproducible AI processing pipelines across institutional sites (Fedorov *et al.,* 2023). Complementary projects such as ReproSchema offer schema-centric frameworks standardizing data extracted from surveys to boost the quality of reproduction through enforced compliant behavior across varying research environments (Chen *et al.,* 2025).

Apart from data infrastructure, technical challenges like overfitting, poor validation, and lack of clinical trust hamper the implementation of AI-based diagnostic and prognostic instruments in leukemia. Overfitting is a widespread challenge, especially in research with small sample sizes and high-dimensional data, where models tend to memorize noise instead of biologically relevant patterns (Aliferis & Simon, 2024). This challenge is compounded by suboptimal selection of models and error estimation methods. To avoid these pitfalls, sound validation protocols are essential.

Systematic reviews, however, reflect that still a large percentage of ML prediction models retain a high risk of bias, oftentimes attributed to poor management of overfitting and deficient validation methodologies (Navarro *et al.,* 2021). Trust in AI technologies adds a further challenge, due to their transparent "black box" characteristics, which may erode clinician trust and hinder implementation in high-stakes decision-making processes. XAI methods-comprising SHAP and LIME-are gaining increasing utility in increasing interpretability, so gaining insight into contributions of individual features to model predictions (Tuan, 2024). These methods do more than pave the way to clinical acceptability but also comply with ethical and regulatory demands for transparency in clinical AI. Closing the gap between the computational complexity of algorithms and clinical interpretability is an essential facet of translating computational developments into trustworthy instruments in the diagnosis and planning of the treatment of leukemia.

14. FUTURE DIRECTIONS AND EMERGING TRENDS

Quantum computing is rising as a transformative technology in speeding up intricate genomic analysis and AI-based model-building for leukemia. Through its ability to cut processing time and computational resource needs exceedingly, it holds transformative potential for DNA analysis, paving the way for high-resolution genomic research essential for precision medicine and AI-facilitated drug discovery (Döhner *et al.,* 2022). In research on leukemia, quantum computing improves AI research methods through more effective feature selection via Quantum-Inspired Optimization Algorithms (QIOA), enhancing the predictive power of ML-based detection of leukemia. This blending is enabling the detection of genetic mutations and biomarkers while also expediting drug discovery and the development of patient-oriented therapeutic plans. Notably, the exponential scalability capability of quantum computing over traditional architecture facilitates genomic analysis breakthroughs and predictive modeling essential in personalizing therapy plans and optimizing clinical trials. In AML, quantum-based modeling of molecular markers uses qubits to define genetic frequencies and pathways, achieving more accurate diagnostic results (Tiwari, 2023; Khoury *et al.,* 2022).

In tandem with the development of quantum computing, AI-powered advancements in synthetic biology and gene editing technologies are revolutionizing the landscape of next-gen therapeutic modalities in leukemias. Platforms such as BE-DICT and DeepCRISPR assist in the prediction and prevention of off-target effects to increase the precision and efficacy of CRISPR-based therapies (Lin *et al.,* 2024). In leukemic therapy, CRISPR science makes possible targeted knockout or insertion of genes intended to check aberrant cell growth with promising potential for personalized therapeutic modalities. The integration of AI with genome editing best optimizes guide RNA designs and solidifies the precision of next-gen editing platforms such as base and prime editing, which make DNA changes to the very base without involving standard repair mechanisms. This integration accelerates the discovery of novel therapeutic targets and the creation of individualized medicines founded on genetic blueprints (Dixit *et al.,* 2024). Though all such developments have been accomplished, convergence involving AI, synthetic biology, and genome editing also presents considerable challenges, such as ethical controversies, increased regulation, and the need to devise effective governing structures securing efficient and equitable clinical translation. Even so, multidisciplinary collaboration involving these convergences has transformative potential and provides avenues to very individualized and effective therapeutic modalities in leukemias (Liv, 2024).

The areas for future research include:

1. Creation of extensive, multicenter leukemia data sets, encompassing ethnically diverse populations.
2. Clinical trials to determine if AI-enhanced decision support can increase patient survival rate and enhance quality of care.
3. Increased usage of explainable AI approaches to enhance clinicians' confidence and facilitate regulation.
4. Standardization of multi-omics data integration and validation processes.

5. Assessment of cost-effectiveness and implementation feasibility, especially in resource-limited healthcare systems, where the application of AI can significantly increase the availability of diagnosis.

15. CONCLUSION

The incorporation of AI in leukemia research and clinical care represents a transformative move in precision oncology. Through the integration of genomic, proteomic, and clinical data, AI-based models improved risk stratification, enabled earlier detection, and guided adaptive treatment planning. These computational advancements are supplemented with liquid biopsies, XAI, and federated learning, enhancing diagnostic precision, interpretability, and safe data sharing across networks. At the therapeutic level, AI hastens drug repurposing, guides individualized chemotherapy and immunotherapy, and empowers adaptive clinical trials dynamically responding to patient outcomes. Reinforcement learning, with the potential of simultaneous tracking with wearables and biosensors in real time, further exemplifies AI's potential in the provision of individualized, patient-centric care. Though artificial intelligence has shown immense potential in the diagnosis, prognosis, treatment, and decision-making of leukemia, its clinical application is currently limited by deficiencies in data quality, external validation, transparency of algorithms, and acceptance from the regulatory authorities. Further developments in the field will be facilitated through collaborations from different disciplines, standardization of validation protocols, and clinical trials focusing on outcome measures in patients instead of just the algorithm's efficiency. With time, AI is expected to develop from a supplementary analysis tool into an essential part of precision leukemia care.

DATA AVAILABILITY

The authors confirm that the data supporting the findings of this study are available within the book/ chapter.

ACKNOWLEDGEMENT

All the authors are thankful to their respective Universities and Institutions for their valuable supports.

REFERENCES

Achir, A., Debbarh, I., Zoubir, N., Battas, I., Medromi, H., & Moutaouakkil, F. (2025). Advances in leukemia detection and classification: A systematic review of AI and image processing techniques. *F1000 Research, 13*, 1536. DOI: 10.12688/f1000research.159318.2 PMID: 41200664

Aggarwal, N., & Weinberg, O. K. (2021). Update on acute leukemias of ambiguous lineage. *Clinics in Laboratory Medicine, 41*(3), 453–466. DOI: 10.1016/j.cll.2021.03.016 PMID: 34304775

Ai, K., Liu, B., Chen, X., Huang, C., Yang, L., Zhang, W., & Lai, P. (2024). Optimizing CAR-T cell therapy for solid tumors: Current challenges and potential strategies. *Journal of Hematology & Oncology, 17*(1), 105. DOI: 10.1186/s13045-024-01625-7 PMID: 39501358

Albalawi, A. F., Yassen, M. H., Almuraydhi, K. M., Althobaiti, A. D., Alzahrani, H. H., & Alqahtani, K. M. (2024). Ethical obligations and patient consent in the integration of artificial intelligence in clinical decision-making. *JOHS, 04*(12), 957–963. DOI: 10.52533/JOHS.2024.41241

Alexander, T. B., & Orgel, E. (2021). Mixed phenotype acute leukemia: Current approaches to diagnosis and treatment. *Current Oncology Reports, 23*(2), 22. DOI: 10.1007/s11912-020-01010-w PMID: 33544265

Aliferis, C., & Simon, G. (2024). Overfitting, underfitting and general model overconfidence and under-performance pitfalls and best practices in machine learning and AI. *Artificial intelligence and machine learning in health care and medical sciences: Best practices and pitfalls*, 477-524.

Allam, S., Nasr, K., Khalid, F., Shah, Z., Suheb, M. Z. K., Mulla, S., & Chaulagain, C. P. (2023). Liquid biopsies and minimal residual disease in myeloid malignancies. *Frontiers in Oncology, 13*, 1164017. DOI: 10.3389/fonc.2023.1164017 PMID: 37213280

Anderson, J. W., & Visweswaran, S. (2025). Algorithmic individual fairness and healthcare: A scoping review. *JAMIA Open, 8*(1), ooae149. DOI: 10.1093/jamiaopen/ooae149 PMID: 39737346

Anitha, K., Posinasetty, B., Kumari, K. N., Chenchula, S., Padmavathi, R., Prakash, S., & Radhika, C. (2024). Liquid biopsy for precision diagnostics and therapeutics. *Clinica Chimica Acta; International Journal of Clinical Chemistry, 554*, 117746. DOI: 10.1016/j.cca.2023.117746 PMID: 38151071

Arigbabu, A. T., Olaniyi, O. O., Adigwe, C. S., Adebiyi, O. O., & Ajayi, S. A. (2024). Data governance in AI-enabled healthcare systems: A case of the project nightingale. *Asian Journal of Research in Computer Science, 17*(5), 85–107. DOI: 10.9734/ajrcos/2024/v17i5441

Asan, O., Bayrak, A. E., & Choudhury, A. (2020). Artificial intelligence and human trust in healthcare: Focus on clinicians. *Journal of Medical Internet Research, 22*(6), e15154. DOI: 10.2196/15154 PMID: 32558657

Asrina, A. (2024). Precision medicine approaches in oncology: Current trends and future directions. *Advances in Healthcare Research, 2*(1), 36–48. DOI: 10.60079/ahr.v2i1.370

Atul, R. B. (2025). Transforming healthcare delivery: AI-powered clinical decision support systems. *INTERNATIONAL JOURNAL, 11*(1), 339–347.

Awasthi, N. P., Singh, P., & Agrawal, N. (2019). Mixed phenotype acute leukemia, B/myeloid with t (9; 22): Diagnostic and therapeutic challenges. *Indian Journal of Hematology & Blood Transfusion : An Official Journal of Indian Society of Hematology and Blood Transfusion, 35*(1), 179–181. DOI: 10.1007/s12288-018-1020-3 PMID: 30828170

Bandla, C. (2024). Distributed Database Architectures for Federated Medical Training. *International Journal of Advanced Research in Science* [IJARSCT]. *Tongxin Jishu, 4*(2), 633–640.

Becker, M., Farina, K. A., & Mascarenhas, J. (2024). Acute myeloid leukemia: Current understanding and management. *JAAPA : Official Journal of the American Academy of Physician Assistants, 37*(1), 34–39. DOI: 10.1097/01.JAA.0000995680.52352.b5 PMID: 38128137

Benard, B., Gentles, A. J., Köhnke, T., Majeti, R., & Thomas, D. (2019). Data mining for mutation-specific targets in acute myeloid leukemia. *Leukemia, 33*(4), 826–843. DOI: 10.1038/s41375-019-0387-y PMID: 30728456

Berger, M. F., & Van Allen, E. M. (2016). Delivering on the promise of precision cancer medicine. *Genome Medicine, 8*(1), 110. DOI: 10.1186/s13073-016-0373-1 PMID: 27782852

Bernardi, S., Vallati, M., & Gatta, R. (2024). Artificial Intelligence-based management of adult chronic myeloid leukemia: Where are we and where are we going? *Cancers (Basel), 16*(5), 848. DOI: 10.3390/cancers16050848 PMID: 38473210

Berry, D. A. (2012). Adaptive clinical trials in oncology. *Nature Reviews. Clinical Oncology, 9*(4), 199–207. DOI: 10.1038/nrclinonc.2011.165 PMID: 22064459

Bhatia, S., Kumar, A., & Tandon, S. (2024). Uncovering the challenges from algorithmic bias affecting the marginalized patient groups in healthcare. *Anuj and Tandon, Stuti, Uncovering the Challenges From Algorithmic Bias Affecting the Marginalized Patient Groups in Healthcare (May 30, 2024).*

Capponi, S., & Daniels, K. G. (2023). Harnessing the power of artificial intelligence to advance cell therapy. *Immunological Reviews, 320*(1), 147–165. DOI: 10.1111/imr.13236 PMID: 37415280

Castro, G. A., Almeida, J. M., Machado-Neto, J. A., & Almeida, T. A. (2024). A decision support system to recommend appropriate therapy protocol for AML patients. *Frontiers in Artificial Intelligence, 7*, 1343447. DOI: 10.3389/frai.2024.1343447 PMID: 38510471

Chen, B., Jing, Z., Kshirsagar, A., Gkanatsios, N., & Chui, H. (2021). *U.S. Patent Application No. 16/944,995.*

Chen, Y., Jarecka, D., Abraham, S. A., Gau, R., Ng, E., Low, D. M., & Ghosh, S. S. (2025). Standardizing survey data collection to enhance reproducibility: Development and comparative evaluation of the ReproSchema ecosystem. *Journal of Medical Internet Research, 27*, e63343. DOI: 10.2196/63343 PMID: 40644691

Cohen, J. A., Bomben, R., Pozzo, F., Tissino, E., Härzschel, A., Hartmann, T. N., & Gattei, V. (2020). An updated perspective on current prognostic and predictive biomarkers in chronic lymphocytic leukemia in the context of chemoimmunotherapy and novel targeted therapy. *Cancers (Basel), 12*(4), 894. DOI: 10.3390/cancers12040894 PMID: 32272636

Ćojbašić, Ž., Ćojbašić, I., Marković, N., & Trajanović, M. (2018). Machine Learning Leukemia Clinical Outcome Prediction. *pathways, 1*, 2.

Correia, R. P., Bento, L. C., de Sousa, F. A., Barroso, R. D. S., Campregher, P. V., & Bacal, N. S. (2021). How I investigate minimal residual disease in acute lymphoblastic leukemia. *International Journal of Laboratory Hematology, 43*(3), 354–363. DOI: 10.1111/ijlh.13463 PMID: 33423385

Cousins, H. C., Nayar, G., & Altman, R. B. (2024). Computational approaches to drug repurposing: Methods, challenges, and opportunities. *Annual Review of Biomedical Data Science, 7*(1), 7. DOI: 10.1146/annurev-biodatasci-110123-025333 PMID: 38598857

Crombie, J. L., & Brown, J. R. (2019). Advances in drug-based therapies in chronic lymphocytic leukemia and future prospects. *Advances in Cell and Gene Therapy, 2*(3), e51. DOI: 10.1002/acg2.51

Crowgey, E. L., Mahajan, N., Wong, W. H., Gopalakrishnapillai, A., Barwe, S. P., Kolb, E. A., & Druley, T. E. (2020). Error-corrected sequencing strategies enable comprehensive detection of leukemic mutations relevant for diagnosis and minimal residual disease monitoring. *BMC Medical Genomics, 13*(1), 32. DOI: 10.1186/s12920-020-0671-8 PMID: 32131829

Dixit, R. R. (2025). Applying reinforcement learning to optimize cancer treatment protocols in machine learning frameworks. In *Artificial Intelligence Revolutionizing Cancer Care* (pp. 20–35). CRC Press.

Dixit, S., Kumar, A., Srinivasan, K., Vincent, P. D. R., & Ramu Krishnan, N. (2024). Advancing genome editing with artificial intelligence: Opportunities, challenges, and future directions. *Frontiers in Bioengineering and Biotechnology, 11*, 1335901. DOI: 10.3389/fbioe.2023.1335901 PMID: 38260726

Döhner, H., Wei, A. H., Appelbaum, F. R., Craddock, C., DiNardo, C. D., Dombret, H., & Löwenberg, B. (2022). Diagnosis and management of AML in adults: 2022 recommendations from an international expert panel on behalf of the ELN. *Blood, 140*(12), 1345–1377. DOI: 10.1182/blood.2022016867 PMID: 35797463

Durga, D., Gadhiya, D., Suhani, F., Singh, N. P., Lakkimsetti, M., Syed, S., & Desai, H. D. (2024). Epidemiological Trends of Acute Lymphoid Leukemia in Individuals Younger Than 20 Years across 204 Countries and Territories from 1990-2021: A Benchmarking Global Analysis. *Blood, 144*(Supplement 1), 1428. DOI: 10.1182/blood-2024-212326

Ekundayo, F. (2024). Reinforcement learning in treatment pathway optimization: A case study in oncology. *International Journal of Science and Research Archive, 13*(02), 2187–2205. DOI: 10.30574/ijsra.2024.13.2.2450

El Alaoui, Y., Elomri, A., Qaraqe, M., Padmanabhan, R., Yasin Taha, R., El Omri, H., & Aboumarzouk, O. (2022). A review of artificial intelligence applications in hematology management: Current practices and future prospects. *Journal of Medical Internet Research, 24*(7), e36490. DOI: 10.2196/36490 PMID: 35819826

El Mahdaoui, C., Dehbi, H., & Cherkaoui, S. (2024). A Review of Emerging Cytogenetic and Molecular Approaches in Identifying Abnormalities in Acute Lymphoblastic Leukemia.

El-Tanani, M., Nsairat, H., Matalka, I. I., Lee, Y. F., Rizzo, M., Aljabali, A. A., & Tambuwala, M. M. (2024). The impact of the BCR-ABL oncogene in the pathology and treatment of chronic myeloid leukemia. *Pathology, Research and Practice, 254*, 155161. DOI: 10.1016/j.prp.2024.155161 PMID: 38280275

Elsayed, B., Elhadary, M., Elshoeibi, R. M., Elshoeibi, A. M., Badr, A., Metwally, O., & Yassin, M. (2023). Deep learning enhances acute lymphoblastic leukemia diagnosis and classification using bone marrow images. *Frontiers in Oncology, 13*, 1330977. DOI: 10.3389/fonc.2023.1330977 PMID: 38125946

Faiz, M., Mounika, B. G., Akbar, M., & Srivastava, S. (2024). Deep and machine learning for acute lymphoblastic leukemia diagnosis: A comprehensive review. *ADCAIJ: Advances in Distributed Computing and Artificial Intelligence Journal, 13*, e31420–e31420. DOI: 10.14201/adcaij.31420

Fatima, S. S., Sultana, S., Perwaiz, N., Lashari, T. A., & Jabeen, F. (2024). Experimental Analysis of Deep Learning Techniques for Leukemia Cancer Classification using Micro Gene Array Dataset. In *2024 International Conference on Frontiers of Information Technology (FIT)* (pp. 1-6). IEEE. DOI: 10.1109/FIT63703.2024.10838426

Fedorov, A., Longabaugh, W. J., Pot, D., Clunie, D. A., Pieper, S. D., Gibbs, D. L., & Kikinis, R. (2023). National cancer institute imaging data commons: Toward transparency, reproducibility, and scalability in imaging artificial intelligence. *Radiographics : A Review Publication of the Radiological Society of North America, Inc., 43*(12), e230180. DOI: 10.1148/rg.230180 PMID: 37999984

Gallagher, K., Strobl, M. A., Park, D. S., Spoendlin, F. C., Gatenby, R. A., Maini, P. K., & Anderson, A. R. (2024). Mathematical model-driven deep learning enables personalized adaptive therapy. *Cancer Research, 84*(11), 1929–1941. DOI: 10.1158/0008-5472.CAN-23-2040 PMID: 38569183

Gao, J., Lan, J., Liao, H., Yang, F., Qiu, P., Jin, F., & Zhu, Y. (2023). Promising preclinical patient-derived organoid (PDO) and xenograft (PDX) models in upper gastrointestinal cancers: Progress and challenges. *BMC Cancer, 23*(1), 1205. DOI: 10.1186/s12885-023-11434-9 PMID: 38062430

Garg, P., Pareek, S., Kulkarni, P., Horne, D., Salgia, R., & Singhal, S. S. (2024). Next-Generation Immunotherapy: Advancing Clinical Applications in Cancer Treatment. *Journal of Clinical Medicine, 13*(21), 6537. DOI: 10.3390/jcm13216537 PMID: 39518676

Gautam, L. (2024). Natural language processing-based structured data extraction from unstructured clinical notes. *Notes (Philadelphia, Pa.), 6*, 9.

Gawankar, S., Nair, S., Pawar, V., Vhatkar, A., & Chavan, P. (2024). Patient privacy and data security in the era of AI-driven healthcare. In *2024 8th International Conference on Computing, Communication, Control and Automation (ICCUBEA)* (pp. 1-6). IEEE. DOI: 10.1109/ICCUBEA61740.2024.10775004

Goel, I., Bhaskar, Y., Kumar, N., Singh, S., Amanullah, M., Dhar, R., & Karmakar, S. (2025). Role of AI in empowering and redefining the oncology care landscape: Perspective from a developing nation. *Frontiers in Digital Health, 7*, 1550407. DOI: 10.3389/fdgth.2025.1550407 PMID: 40103737

Gogoshin, G., & Rodin, A. S. (2023). Graph neural networks in cancer and oncology research: Emerging and future trends. *Cancers (Basel), 15*(24), 5858. DOI: 10.3390/cancers15245858 PMID: 38136405

Goswami, S., Pauken, K. E., Wang, L., & Sharma, P. (2024). Next-generation combination approaches for immune checkpoint therapy. *Nature Immunology, 25*(12), 2186–2199. DOI: 10.1038/s41590-024-02015-4 PMID: 39587347

Griffin, A. C., Wang, K. H., Leung, T. I., & Facelli, J. C. (2024). Fairness and Inclusion in Biomedical Artificial Intelligence Research and Clinical Use: Technical and Social Perspectives. *Journal of Biomedical Informatics, 157*, 104693. DOI: 10.1016/j.jbi.2024.104693 PMID: 39019301

Guilbert, B., Rieu, T., Delamarre, D., Laurent, F., Serre, F., Leroy, S., & Cuggia, M. (2024). ONCO-FAIR project: Improving data interoperability in oncology chemotherapy treatments for data reuse. *Studies in Health Technology and Informatics, 316*, 1373–1377. DOI: 10.3233/SHTI240667 PMID: 39176636

Gupta, S. (2025). Ethical, privacy, and security issues in smart healthcare. In *Driving Global Health and Sustainable Development Goals With Smart Technology* (pp. 83–98). IGI Global Scientific Publishing.

Gupta, Y., Srivastava, V., & Singh, R. K. (2025). AI-enhanced patient-centric clinical trial design. In. AIP Conference Proceedings: Vol. 3254. *No. 1* (p. 020020). AIP Publishing LLC.

Haney, K., Tandon, P., Divi, R., Ossandon, M. R., Baker, H., & Pearlman, P. C. (2017). The role of affordable, point-of-care technologies for cancer care in low-and middle-income countries: A review and commentary. *IEEE Journal of Translational Engineering in Health and Medicine, 5*, 1–14. DOI: 10.1109/JTEHM.2017.2761764 PMID: 29204328

Haque, R., Al Sakib, A., Hossain, M. F., Islam, F., Ibne Aziz, F., Ahmed, M. R., & Hasan, M. J. (2024). Advancing early leukemia diagnostics: A comprehensive study incorporating image processing and transfer learning. *BioMedInformatics, 4*(2), 966–991. DOI: 10.3390/biomedinformatics4020054

Hasan, J., Hasan, K., Al Noman, A., Hasan, S., Sultana, S., Arafat, M. A., & Ahmed, M. R. (2024). Transforming leukemia classification: A comprehensive study on deep learning models for enhanced diagnostic accuracy. In *2024 IEEE International Conference on Power, Electrical, Electronics and Industrial Applications (PEEIACON)* (pp. 1-6). IEEE. DOI: 10.1109/PEEIACON63629.2024.10800693

Hays, P. (2021). Great strides in precision medicine: personalized oncology, immunotherapies, and molecular diagnostics. In *Advancing Healthcare Through Personalized Medicine* (pp. 141–417). Springer International Publishing. DOI: 10.1007/978-3-030-80100-7_5

He, L., Tang, J., Andersson, E. I., Timonen, S., Koschmieder, S., Wennerberg, K., & Aittokallio, T. (2018). Patient-customized drug combination prediction and testing for T-cell prolymphocytic leukemia patients. *Cancer Research, 78*(9), 2407–2418. DOI: 10.1158/0008-5472.CAN-17-3644 PMID: 29483097

He, Y., Lu, W., Zhou, X., Mu, J., & Shen, W. (2024). Unraveling Alzheimer's disease: Insights from single-cell sequencing and spatial transcriptomic. *Frontiers in Neurology, 15*, 1515981. DOI: 10.3389/fneur.2024.1515981 PMID: 39741706

Hehr, M., Sadafi, A., Matek, C., Lienemann, P., Pohlkamp, C., Haferlach, T., & Marr, C. (2023). Explainable AI identifies diagnostic cells of genetic AML subtypes. *PLOS Digital Health*, *2*(3), e0000187. DOI: 10.1371/journal.pdig.0000187 PMID: 36921004

Hennawi, M., Pakasticali, N., Tashkandi, H., & Hussaini, M. (2022). Genomic landscape of mixed-phenotype acute leukemia. *International Journal of Molecular Sciences*, *23*(19), 11259. DOI: 10.3390/ijms231911259 PMID: 36232559

Hernández-Sánchez, A., & Bullinger, L. (2023). Recent advances in precision medicine for acute myeloid leukemia. *Current Opinion in Oncology*, *35*(6), 581–588. DOI: 10.1097/CCO.0000000000000965 PMID: 37621173

Heuser, M., Ofran, Y., Boissel, N., Mauri, S. B., Craddock, C., & Janssen, J.ESMO Guidelines Committee. (2020). Acute myeloid leukaemia in adult patients: ESMO Clinical Practice Guidelines for diagnosis, treatment and follow-up. *Annals of Oncology : Official Journal of the European Society for Medical Oncology*, *31*(6), 697–712. DOI: 10.1016/j.annonc.2020.02.018 PMID: 32171751

Hoffmann, J., Eminovic, S., Wilhelm, C., Krause, S. W., Neubauer, A., Thrun, M. C., & Brendel, C. (2023). Prediction of clinical outcomes with explainable artificial intelligence in patients with chronic lymphocytic leukemia. *Current Oncology (Toronto, Ont.)*, *30*(2), 1903–1915. DOI: 10.3390/curroncol30020148 PMID: 36826109

Hossain, M. R., Mahabub, S., Al Masum, A., & Jahan, I. (2024). Natural language processing (nlp) in analyzing electronic health records for better decision making. *Journal of Computer Science and Technology Studies*, *6*(5), 216–228. DOI: 10.32996/jcsts.2024.6.5.18

Huang, J., Chan, S. C., Ngai, C. H., Lok, V., Zhang, L., Lucero-Prisno, D. E.III, & Wong, M. C. (2022). Disease burden, risk factors, and trends of leukaemia: A global analysis. *Frontiers in Oncology*, *12*, 904292. DOI: 10.3389/fonc.2022.904292 PMID: 35936709

Huang, P., & Zhang, J. (2025). Global leukemia burden and trends: A comprehensive analysis of temporal and spatial variations from 1990-2021 using GBD (Global Burden of Disease) data. *BMC Public Health*, *25*(1), 262. DOI: 10.1186/s12889-025-21428-w PMID: 39838416

Huang, Y. L., Chen, D. R., & Liu, Y. K. (2004). Breast cancer diagnosis using image retrieval for different ultrasonic systems. In *2004 International Conference on Image Processing, 2004. ICIP'04.* (Vol. 5, pp. 2957-2960). IEEE.

Hunter, B., Hindocha, S., & Lee, R. W. (2022). The role of artificial intelligence in early cancer diagnosis. *Cancers (Basel)*, *14*(6), 1524. DOI: 10.3390/cancers14061524 PMID: 35326674

Iyer, P., Jasdanwala, S. S., Wang, Y., Bhatia, K., & Bhatt, S. (2024). Decoding acute myeloid leukemia: A clinician's guide to functional profiling. *Diagnostics (Basel)*, *14*(22), 2560. DOI: 10.3390/diagnostics14222560 PMID: 39594226

Jayaraman, P., Desman, J., Sabounchi, M., Nadkarni, G. N., & Sakhuja, A. (2024). A primer on reinforcement learning in medicine for clinicians. *NPJ Digital Medicine*, *7*(1), 337. DOI: 10.1038/s41746-024-01316-0 PMID: 39592855

Jeelani, O. F., Njie, M., & Korzhuk, V. (2024). Methods and algorithms of ensuring data privacy in ai-based healthcare systems and technologies. *Digital Health and Informatics Innovations for Sustainable Health Care Systems*, 1260-1261.

Jiang, H. (2024). Latest research progress of liquid biopsy in tumor-a narrative review. *Cancer Management and Research, 16*, 1031–1042. DOI: 10.2147/CMAR.S479338 PMID: 39165347

Jones, M. F., & Smith, C. C. (2024). Mechanisms of resistance to targeted therapies in AML. *Annual Review of Cancer Biology*, *8*(1), 81–96. DOI: 10.1146/annurev-cancerbio-062822-025055

Joshi, S. K., Tognon, C. E., Druker, B. J., & Rodland, K. D. (2022). Oncoproteomic profiling of AML: Moving beyond genomics. *Expert Review of Proteomics*, *19*(7-12), 283–287. DOI: 10.1080/14789450.2023.2176757 PMID: 36734985

Kenmogne, V. L., Nweke, E. E., Takundwa, M. M., Fru, P. N., & Thimiri Govinda Raj, D. B. (2022). *Application of drug repurposing-based precision medicine platform for leukaemia patient treatment. Cell Biology and Translational Medicine* (Vol. 19). Perspectives in Diverse Human Diseases and Their Therapeutic Options.

Khalid, R., & Riasat, S. (2023). Molecular pathogenesis and treatment strategies of Chronic Myeloid Leukemia (CML). *Sudan Journal of Medical Sciences*, *18*(4), 525–538. DOI: 10.18502/sjms.v18i4.14741

Khanzode, P. C., Mohod, A. P., & Raut, P. P. (2024). A real-time patient health monitoring system. In *Predictive Data Modelling for Biomedical Data and Imaging* (pp. 189–204). River Publishers. DOI: 10.1201/9781003516859-9

Khosravi, B., Rouzrokh, P., & Erickson, B. J. (2022). Getting more out of large databases and EHRs with natural language processing and artificial intelligence: The future is here. *The Journal of Bone and Joint Surgery. American Volume*, *104*(Suppl 3), 51–55. DOI: 10.2106/JBJS.22.00567 PMID: 36260045

Khoury, J. D., Solary, E., Abla, O., Akkari, Y., Alaggio, R., Apperley, J. F., & Hochhaus, A. (2022). The 5th edition of the World Health Organization classification of haematolymphoid tumours: myeloid and histiocytic/dendritic neoplasms. *leukemia, 36*(7), 1703-1719.

Kim, Y. S., Hsieh, A. C., & Lam, H. M. (2022). Bladder cancer patient-derived organoids and avatars for personalized cancer discovery. *European Urology Focus*, *8*(3), 657–659. DOI: 10.1016/j.euf.2022.07.006 PMID: 35915037

Kimura, S., & Mullighan, C. G. (2024). Molecular basis of acute lymphoblastic leukemia. *Molecular Hematology*, 67-81.

Kommiya Mothilal, R., Mahajan, D., Tan, C., & Sharma, A. (2021). Towards unifying feature attribution and counterfactual explanations: Different means to the same end. In *Proceedings of the 2021 AAAI/ACM Conference on AI, Ethics, and Society* (pp. 652-663). DOI: 10.1145/3461702.3462597

Kondra, S., Xu, W., & Raghavan, V. V. (2024). Unlocking the power of clinical notes: natural language processing in healthcare. *Acta Scientific MEDICAL SCIENCES (ISSN: 2582-0931), 8*(6).

Kong, H. (2025). Advances in personalized cancer vaccine development: AI applications from neoantigen discovery to mRNA formulation. *Biochem*, *5*(2), 5. DOI: 10.3390/biochem5020005

Kruse, A., Abdel-Azim, N., Kim, H. N., Ruan, Y., Phan, V., Ogana, H., & Kim, Y. M. (2020). Minimal residual disease detection in acute lymphoblastic leukemia. *International Journal of Molecular Sciences, 21*(3), 1054. DOI: 10.3390/ijms21031054 PMID: 32033444

Kubasch, A. S., Grieb, N., Oeser, A., Haferlach, C., & Platzbecker, U. (2022). RETRACTED ARTICLE: Use of artificial intelligence in the management of acute leukemia. *Die Onkologie, 28*(8), 731–732. DOI: 10.1007/s00761-022-01153-4

Kumar, H., Raj, U., Gupta, S., & Varadwaj, P. K. (2016). In-silico identification of inhibitors against mutated BCR-ABL protein of chronic myeloid leukemia: A virtual screening and molecular dynamics simulation study. *Journal of Biomolecular Structure & Dynamics, 34*(10), 2171–2183. DOI: 10.1080/07391102.2015.1110046 PMID: 26479578

Kumari, R., Sah, D. K., Gupta, S., Cengiz, K., & Ivković, N. (2023). Advancing medical recommendations with federated learning on decentralized data: A roadmap for implementation. *IEEE Transactions on Consumer Electronics, 70*(1), 2666–2674. DOI: 10.1109/TCE.2023.3334159

Lachowiez, C. A., DiNardo, C. D., & Loghavi, S. (2023). Molecularly targeted therapy in acute myeloid leukemia: Current treatment landscape and mechanisms of response and resistance. *Cancers (Basel), 15*(5), 1617. DOI: 10.3390/cancers15051617 PMID: 36900407

Li, E. W., Tran, N. Y. K., McCulloch, D., Krigstein, M., Catalano, A., Othman, J., & Iland, H. (2024a). FLT3-TKD measurable residual disease detection using droplet digital PCR and clinical applications in acute myeloid leukemia. *International Journal of Molecular Sciences, 25*(11), 5771. DOI: 10.3390/ijms25115771 PMID: 38891959

Li, Z., Li, Z., Luo, Y., Chen, W., Fang, Y., Xiong, Y., & Zhu, J. (2024b). Application and new findings of scRNA-seq and ST-seq in prostate cancer. *Cell Regeneration (London, England), 13*(1), 23. DOI: 10.1186/s13619-024-00206-w PMID: 39470950

Lin, L., Cui, Z., Wang, S., Chen, Y., & Zong, Y. (2024). AI-driven protein engineering for DNA sequence modification. *Journal of Theory and Practice of Engineering Science, 4*(03), 183–190. DOI: 10.53469/jtpes.2024.04(03).17

Liu, C., Zhang, Y., Liang, Y., Zhang, T., & Wang, G. (2024a). DrugReSC: Targeting disease-critical cell subpopulations with single-cell transcriptomic data for drug repurposing in cancer. *Briefings in Bioinformatics, 25*(6), bbae490. DOI: 10.1093/bib/bbae490 PMID: 39350337

Liu, W., Puri, A., Fu, D., Chen, L., Wang, C., Kellis, M., & Yang, J. (2024b). Dissecting the tumor microenvironment in response to immune checkpoint inhibitors via single-cell and spatial transcriptomics. *Clinical & Experimental Metastasis, 41*(4), 313–332. DOI: 10.1007/s10585-023-10246-2 PMID: 38064127

Liv, B. (2024). *Unlocking the future of biotechnology: The transformative potential of CRISPR and emerging gene-editing technologies*. Next Generation Journal for The Young Researchers.

Lopez-Perez, L., Merino, B., Rujas, M., Maccaro, A., Guillén, S., Pecchia, L., & Fico, G. (2024). Regulatory frameworks and validation strategies for advancing artificial intelligence in healthcare. In *European Medical and Biological Engineering Conference* (pp. 260-265). Cham: Springer Nature Switzerland. DOI: 10.1007/978-3-031-61628-0_28

Maher, K., & Chang, A. C. (2024). Regulatory frameworks for artificial intelligence in cardiovascular medicine and surgery. In *Intelligence-Based Cardiology and Cardiac Surgery* (pp. 421–426). Academic Press. DOI: 10.1016/B978-0-323-90534-3.00026-3

Mahmood, A. A. R., Jha, A. M., & Manivannan, K. (2024). Precision medicine: personalizing the fight against cancer. *International Journal of Trends in OncoScience*, 10-18.

Malard, F., & Mohty, M. (2020). Acute lymphoblastic leukaemia. *Lancet, 395*(10230), 1146–1162. DOI: 10.1016/S0140-6736(19)33018-1 PMID: 32247396

Mary, L. W., & Raj, S. A. A. (2024). Smart wearable devices for prediction using machine learning techniques: depth exploration. In *2024 7th International Conference on Circuit Power and Computing Technologies (ICCPCT)* (Vol. 1, pp. 1522-1526). IEEE. DOI: 10.1109/ICCPCT61902.2024.10672922

Mecklenbrauck, R., & Heuser, M. (2023). Resistance to targeted therapies in acute myeloid leukemia. *Clinical & Experimental Metastasis, 40*(1), 33–44. DOI: 10.1007/s10585-022-10189-0 PMID: 36318439

Mina, R., Oliva, S., & Boccadoro, M. (2020). Minimal residual disease in multiple myeloma: State of the art and future perspectives. *Journal of Clinical Medicine*, *9*(7), 2142. DOI: 10.3390/jcm9072142 PMID: 32645952

Mitra, U., & Rehman, S. U. (2025). Leveraging AI and machine learning for next-generation clinical decision support systems (CDSS). In *AI-Driven Innovation in healthcare data analytics* (pp. 83-112). IGI Global Scientific Publishing.

Molefi, T., Marima, R., Demetriou, D., Basera, A., & Dlamini, Z. (2023). Employing AI-Powered decision support systems in recommending the most effective therapeutic approaches for Individual Cancer patients: Maximising Therapeutic Efficacy. In *Artificial Intelligence and Precision Oncology: bridging Cancer Research and clinical decision support* (pp. 259–275). Springer Nature Switzerland. DOI: 10.1007/978-3-031-21506-3_13

Mullighan, C. G. (2012). The molecular genetic makeup of acute lymphoblastic leukemia. *Hematology 2010, the American Society of Hematology Education Program Book, 2012*(1), 389-396.

Muthalakshmi, M., Jeyapal, K., & Vinoth, M. PS, D., Murugan, N. S., & Sheela, K. S. (2024). Federated learning for secure and privacy-preserving medical image analysis in decentralized healthcare systems. In *2024 5th international conference on electronics and sustainable communication systems (ICESC)* (pp. 1442-1447). IEEE.

Naddeo, S., Verde, L., Forastiere, M., De Pietro, G., & Sannino, G. (2017). A real-time m-health monitoring system: An integrated solution combining the use of several wearable sensors and mobile devices. In *Special Session on Smart Medical Devices-From Lab to Clinical Practice* (Vol. 6, pp. 545-552). SciTePress.

Naimi, A., Mohammed, R. N., Raji, A., Chupradit, S., Yumashev, A. V., Suksatan, W., & Razeghian, E. (2022). Tumor immunotherapies by immune checkpoint inhibitors (ICIs); the pros and cons. *Cell Communication and Signaling : CCS, 20*(1), 44. DOI: 10.1186/s12964-022-00854-y PMID: 35392976

Natarajan, A., Thangarajan, R., & Kesavan, S. (2019). Repurposing drugs by in silico methods to target BCR kinase domain in chronic myeloid leukemia. *Asian Pacific Journal of Cancer Prevention : APJCP, 20*(11), 3399–3406. DOI: 10.31557/APJCP.2019.20.11.3399 PMID: 31759365

Navarro, C. L. A., Damen, J. A., Takada, T., Nijman, S. W., Dhiman, P., Ma, J., & Hooft, L. (2021). Risk of bias in studies on prediction models developed using supervised machine learning techniques: systematic review. *bmj, 375*.

Nikshya, J. E., Karthikeyan, M. S., Prasad, S., Krishnan, R. S., Balamurugan, S., & Raj, J. R. F. (2024). A Machine Learning Framework for Integrating Multi-Omics Data for Early Leukemia Detection. In *2024 8th International Conference on I-SMAC (IoT in Social, Mobile, Analytics and Cloud)(I-SMAC)* (pp. 1348-1356). IEEE. DOI: 10.1109/I-SMAC61858.2024.10714596

Olawade, D. B., Clement David-Olawade, A., Adereni, T., Egbon, E., Teke, J., & Boussios, S. (2025). Integrating AI into cancer immunotherapy-A narrative review of current applications and future directions. *Diseases (Basel, Switzerland), 13*(1), 24. DOI: 10.3390/diseases13010024 PMID: 39851488

Opderbeck, D. W. (2019). Artificial intelligence in pharmaceuticals, biologics, and medical devices: Present and future regulatory models. *Fordham Law Review*, *88*, 553.

Pan, L., Liu, G., Lin, F., Zhong, S., Xia, H., Sun, X., & Liang, H. (2017). Machine learning applications for prediction of relapse in childhood acute lymphoblastic leukemia. *Scientific Reports*, *7*(1), 7402. DOI: 10.1038/s41598-017-07408-0 PMID: 28784991

Pillai, P. M., & Carroll, W. L. (2022). Acute lymphoblastic leukemia. In *Lanzkowsky's Manual of Pediatric Hematology and Oncology* (pp. 413–438). Academic Press. DOI: 10.1016/B978-0-12-821671-2.00004-0

Pindi, V. (2019). A Ai-assisted clinical decision support systems: Enhancing diagnostic accuracy and treatment recommendations. *International Journal of Innovations in Engineering Research and Technology*, *6*(10), 1–10. DOI: 10.26662/ijiert.v6i10.pp1-10

Prabhakar, P. K., Paliya, N., Malik, V., & Sharma, K. (2024). Liquid biopsies in oncology: the future of non-invasive cancer detection. *International Research Journal of Medicine and Surgery, 1*(3),15-19.

Prakash, G., & Harish, S. (2024). Health monitoring system by using IoT. In *2024 International Conference on Power, Energy, Control and Transmission Systems (ICPECTS)* (pp. 1-9). IEEE.

Presciuttini, A., Cantini, A., & Portioli-Staudacher, A. (2024). From explanations to actions: leveraging shap, lime, and counterfactual analysis for operational excellence in maintenance decisions. In *2024 4th International Conference on Electrical, Computer, Communications and Mechatronics Engineering (ICECCME)* (pp. 1-6). IEEE. DOI: 10.1109/ICECCME62383.2024.10796106

Ram, M., Afrash, M. R., Moulaei, K., Parvin, M., Esmaeeli, E., Karbasi, Z., & Sabahi, A. (2024). Application of artificial intelligence in chronic myeloid leukemia (CML) disease prediction and management: A scoping review. *BMC Cancer*, *24*(1), 1026. DOI: 10.1186/s12885-024-12764-y PMID: 39164653

Rao, S., Mehta, S., Kulkarni, S., Dalvi, H., Katre, N., & Narvekar, M. (2022). *A study of LIME and SHAP model explainers for autonomous disease predictions. In 2022 ieee bombay section signature conference (ibssc)*. IEEE.

Raut, R., Lokare, S., Gavkare, S., & Narute, M. (2024). Leukemia subtype classification via gene expression using machine learning algorithms. In *2024 IEEE International Conference on Information Technology, Electronics and Intelligent Communication Systems (ICITEICS)* (pp. 1-7). IEEE. DOI: 10.1109/ICITEICS61368.2024.10625370

Raza, A. (2023). *Secure and privacy-preserving federated learning with explainable artificial intelligence for smart healthcare system*. University of Kent.

Rodriguez, J. R., Fernandez, S., Swartz, N., Alonge, A., Bhullar, F., Betros, T., & Jacobs, R. J. (2024). A chronological overview of using deep learning for leukemia detection: A scoping review. *Cureus, 16*(5). Advance online publication. DOI: 10.7759/cureus.61379 PMID: 38846204

Saadeh, A., Al-Bitar, F., Al Ta'ani, O., Abdulelah, A. A., Usmani, A. A., Jibrin, D., & Scott, K. (2024). The burden of leukemia in the united states from 1990 to 2021. *Blood, 144*(Supplement 1), 2421. DOI: 10.1182/blood-2024-208967

Saghir, S. A. M. (2023). A new insight updates in diagnosis and management of acute lymphoblastic leukemia, cytogenetics, immunophenotyping, and proteomic profile. *Electronic Journal of General Medicine, 20*(5), em519. DOI: 10.29333/ejgm/13386

Saranya, G., & Sivaraman, K. (2024). Applying reinforcement learning to optimize treatment plans for chronic disease management. In *2024 International Conference on Advances in Computing, Communication and Applied Informatics (ACCAI)* (pp. 1-6). IEEE. DOI: 10.1109/ACCAI61061.2024.10601732

Sarvepalli, S., & Vadarevu, S. (2025). Role of artificial intelligence in cancer drug discovery and development. *Cancer Letters, 627*, 217821. DOI: 10.1016/j.canlet.2025.217821 PMID: 40414522

Schulz, W. L., Durant, T. J., & Krumholz, H. M. (2019). Validation and regulation of clinical artificial intelligence.

Sedano, R., Solitano, V., Vuyyuru, S. K., Yuan, Y., Hanžel, J., Ma, C., & Jairath, V. (2025). Artificial intelligence to revolutionize IBD clinical trials: A comprehensive review. *Therapeutic Advances in Gastroenterology, 18*, 17562848251321915. DOI: 10.1177/17562848251321915 PMID: 39996136

Shehadi, I. A., Rashdan, H. R., & Abdelmonsef, A. H. (2020). Homology modeling and virtual screening studies of antigen MLAA-42 protein: Identification of novel drug candidates against leukemia-An in silico approach. *Computational and Mathematical Methods in Medicine, 2020*(1), 8196147. DOI: 10.1155/2020/8196147 PMID: 32256683

Sholikah, T. A. (2017). Fusion gene bcr-abl: from etiopathogenesis to the management of Chronic Myeloid Leukemia. *JKKI: Jurnal Kedokteran dan Kesehatan Indonesia*, 29-37.

Short, N. J., Rytting, M. E., & Cortes, J. E. (2018). Acute myeloid leukaemia. *Lancet, 392*(10147), 593–606. DOI: 10.1016/S0140-6736(18)31041-9 PMID: 30078459

Sîrbu, C. L., & Mercioni, M. A. (2024). Fostering trust in AI-driven healthcare: a brief review of ethical and practical considerations. In *2024 International Symposium on Electronics and Telecommunications (ISETC)* (pp. 1-4). IEEE. DOI: 10.1109/ISETC63109.2024.10797264

Song, Y., Liu, T., Hao, Q., Fang, Q., Gong, X., Li, Y., & Mi, Y. (2025). Comprehensive omics-based classification system in adult patients with B-cell acute lymphoblastic leukemia. *Molecular Oncology*, *19*(12), 3578–3595. DOI: 10.1002/1878-0261.70053 PMID: 40388565

Song, Y., Wang, Z., Zhang, G., Hou, J., Liu, K., Wei, S., & Mi, Y. (2023). Integrative multi-omic analysis for prognosis stratification in acute myeloid leukemia. *Blood*, *142*(Supplement 1), 5984. DOI: 10.1182/blood-2023-173211

Sopruchi, A. D., & Rashid, A. (2024). The Integration of AI-Driven Decision Support Systems in Healthcare: Enhancements, Challenges, and Future Directions. *IDOSR Journal of Computer and Applied Sciences*, *9*(2), 17–25. DOI: 10.59298/JCAS/2024/92.1725

Sorel, N., Cayssials, É., Brizard, F., & Chomel, J. C. (2017). Treatment and molecular monitoring update in chronic myeloid leukemia management. *Annales de Biologie Clinique*, *75*(2), 129–145. DOI: 10.1684/abc.2017.1233 PMID: 28377326

Stilgenbauer, S. (2007). Progress and prognosis in chronic lymphocytic leukaemia. In *Hematology Meeting Reports,1*(5), 3-11.

Swain, B., & Maddi, S. (2024). Patient-derived Tumor Organoids: Generation and Applications in Disease Modeling and Personalized Therapy. *Nature Cell and Science*, *2*(3), 185–197. DOI: 10.61474/ncs.2024.00008

Swathi, Y., & Challa, M. (2023). A comparative analysis of explainable AI techniques for enhanced model interpretability. In *2023 3rd international conference on pervasive computing and social networking (ICPCSN)* (pp. 229-234). IEEE.

Syed, S. A., Nisar, H., Jaffari, R., Hum, Y. C., Jen, L. Y., & Qaisar, M. (2024). Enhancing leukemia detection: an automated approach using deep learning and ensemble techniques. http://hdl.handle.net/20.500.14131/1460

Taruttis, A., & Ntziachristos, V. (2012). Translational optical imaging. *AJR. American Journal of Roentgenology*, *199*(2), 263–271. DOI: 10.2214/AJR.11.8431 PMID: 22826386

Terzi, T. A. (2024). Using artificial intelligence for personalized cancer treatment. *Next Front. Life Sci. AI*, *8*(1), 133–136. DOI: 10.62802/mpfsy492

Tiwari, A. (2023). Simulating molecular markers in acute myeloid leukemia using quantum computing. In *2023 IEEE MIT Undergraduate Research Technology Conference (URTC)* (pp. 1-4). IEEE. DOI: 10.1109/URTC60662.2023.10534988

Tosca, E. M., De Carlo, A., Ronchi, D., & Magni, P. (2024). Model-informed reinforcement learning for enabling precision dosing via adaptive dosing. *Clinical Pharmacology and Therapeutics*, *116*(3), 619–636. DOI: 10.1002/cpt.3356 PMID: 38989560

Tuan, D. A. (2024). Bridging the gap between black box AI and clinical practice: Advancing explainable AI for trust, ethics, and personalized healthcare diagnostics. DOI: 10.20944/preprints202409.1974.v2

Tupsakhare, P. (2023). Improving clinical decision support in health care through ais. progress in medical sciences. PMS-E118. *Prog Med Sci*, *7*(3), 1–4.

Turk, A., Čeh, E., Calin, G. A., & Kunej, T. (2024). Multiple omics levels of chronic lymphocytic leukemia. *Cell Death Discovery*, *10*(1), 293. DOI: 10.1038/s41420-024-02068-2 PMID: 38906881

Tyner, J. W., Tognon, C. E., Bottomly, D., Wilmot, B., Kurtz, S. E., Savage, S. L., & Druker, B. J. (2018). Functional genomic landscape of acute myeloid leukaemia. *Nature*, *562*(7728), 526–531. DOI: 10.1038/s41586-018-0623-z PMID: 30333627

Uddin, R., Darwish, N. H., & Mousa, S. A. (2021). Acute myeloid leukemia mutations and future mechanistic target to overcome resistance. *Current Treatment Options in Oncology*, *22*(9), 76. DOI: 10.1007/s11864-021-00880-x PMID: 34213682

Varghese, J., Holz, C., Neuhaus, P., Bernardi, M., Boehm, A., Ganser, A., & Dugas, M. (2016). Key data elements in myeloid leukemia. In *Exploring Complexity in Health: An Interdisciplinary Systems Approach* (pp. 282–286). IOS Press.

Vegesna, V. V. (2024). Medicine: a frame work for tailored cancer treatment. *International journal of innovative research in advanced engineering, 6*(11), 747-752.

Venunadh, A., & Joy, S. P. (2024). A comprehensive study on technological advancements in treatment and prognosis of acute myeloid leukemia. In *2024 International Conference on Intelligent and Innovative Technologies in Computing, Electrical and Electronics (IITCEE)* (pp. 1-7). IEEE. DOI: 10.1109/IITCEE59897.2024.10467781

Wasilewski, T., Kamysz, W., & Gębicki, J. (2024). AI-assisted detection of biomarkers by sensors and biosensors for early diagnosis and monitoring. *Biosensors (Basel)*, *14*(7), 356. DOI: 10.3390/bios14070356 PMID: 39056632

Wu, A., Liu, X., Fruhstorfer, C., & Jiang, X. (2024). Clinical insights into structure, regulation, and targeting of ABL kinases in human leukemia. *International Journal of Molecular Sciences*, *25*(6), 3307. DOI: 10.3390/ijms25063307 PMID: 38542279

Xu, J., Xiao, Y., Wang, W. H., Ning, Y., Shenkman, E. A., Bian, J., & Wang, F. (2022). Algorithmic fairness in computational medicine. *EBioMedicine*, *84*, 104250. DOI: 10.1016/j.ebiom.2022.104250 PMID: 36084616

Yan, X., Yeh, C., & Zou, L. (2020). Clinical Applications of Circulating Tumor DNA, Circulating Tumor Cells, and Exosomes as Liquid Biopsy-Based Tumor Biomarkers. *Journal of Applied Bioanalysis*, *6*(3), 107–130. DOI: 10.17145/jab.20.013

Yokota, T., & Kanakura, Y. (2016). Genetic abnormalities associated with acute lymphoblastic leukemia. *Cancer Science*, *107*(6), 721–725. DOI: 10.1111/cas.12927 PMID: 26991355

Young, A. L., Davis, H. C., & Challen, G. A. (2023). Droplet digital PCR for oncogenic KMT2A fusion detection. *The Journal of Molecular Diagnostics : JMD*, *25*(12), 898–906. DOI: 10.1016/j.jmoldx.2023.09.006 PMID: 37813299

Yun, Z., Zhang, Y., Wang, J., Xue, C., Wu, J., Han, X., & Hou, L. (2024). Global, regional and national burden of leukemia, 1990-2021: Analysis of data from the global burden of disease study 2021. *Blood*, *144*(Supplement 1), 7907. DOI: 10.1182/blood-2024-205478

Zeng, W., & Zhang, P. (2022). Resistance and recurrence of malignancies after CAR-T cell therapy. *Experimental Cell Research, 410*(2), 112971. DOI: 10.1016/j.yexcr.2021.112971 PMID: 34906583

Zhang, S., & Kipps, T. J. (2014). The pathogenesis of chronic lymphocytic leukemia. *Annual Review of Pathology, 9*(1), 103–118. DOI: 10.1146/annurev-pathol-020712-163955 PMID: 23987584

Zhang, Z., Huang, J., Zhang, Z., Shen, H., Tang, X., Wu, D., & Chen, S. (2024). Application of omics in the diagnosis, prognosis, and treatment of acute myeloid leukemia. *Biomarker Research, 12*(1), 60. DOI: 10.1186/s40364-024-00600-1 PMID: 38858750

Zheng, X., Cheng, M., Fu, B., Fan, X., Wang, Q., Yu, X., & Wei, H. (2015). Targeting LUNX inhibits non–small cell lung cancer growth and metastasis. *Cancer Research, 75*(6), 1080–1090. DOI: 10.1158/0008-5472.CAN-14-1831 PMID: 25600649

Zhu, C. X., & Qin, J. J. (2023). Single-cell RNA sequencing and spatial transcriptomic technologies and applications in exploring gastric cancer: A review. *Oncology Advances, 1*(1), 6–16. DOI: 10.14218/OnA.2023.00039

Zhu, J., Ji, L., Chen, Y., Li, H., Huang, M., Dai, Z., & Chu, X. (2023). Organoids and organs-on-chips: Insights into predicting the efficacy of systemic treatment in colorectal cancer. *Cell Death Discovery, 9*(1), 72. DOI: 10.1038/s41420-023-01354-9 PMID: 36813783

Zhu, J., Zhang, K., Chen, Y., Ge, X., Wu, J., Xu, P., & Yao, J. (2024). Progress of single-cell RNA sequencing combined with spatial transcriptomics in tumour microenvironment and treatment of pancreatic cancer. *Journal of Translational Medicine, 22*(1), 563. DOI: 10.1186/s12967-024-05307-3 PMID: 38867230

Zhuang, X., Pei, H. Z., Li, T., Huang, J., Guo, Y., Zhao, Y., & Chen, Y. (2022). The molecular mechanisms of resistance to IDH inhibitors in acute myeloid leukemia. *Frontiers in Oncology, 12*, 931462. DOI: 10.3389/fonc.2022.931462 PMID: 35814406

APPENDIX

Abbreviations

AI - Artificial intelligence
ASIR - Age-Standardized Incidence Rate
ASMR - Age-Standardized Mortality Rate
AML - Acute Myeloid Leukemia
ALL - Acute Lymphoid Leukemia
CML - Chronic Myeloid Leukemia
BCR-ABL - Breakpoint Cluster Region-Abelson
MRD - Measurable Residual Disease
CNNs - Convolutional neural networks
CLL - Chronic Lymphocytic Leukemia
LSCs - Leukemic Stem Cells
BCR - B Cell Receptor
TKIs - Tyrosine Kinase Inhibitors
MPAL - Mixed-Phenotype Acute Leukemia
LSTM - Long Short-Term Memory
SCEMILA - Single Cell-Based Explainable Multiple Instance Learning Algorithm
XAI - Explainable AI
scRNA-seq - Single-Cell RNA Sequencing
ML - Machine learning
DL - Deep Learning
GNNs - Graph Neural Networks
FL - Federated Learning
LIME - Local Interpretable Model-agnostic Explanations
SHAP - Shapley Additive Explanations
RL - Reinforcement Learning
PK/PD - Pharmacokinetic/Pharmacodynamic
NLP - Natural Language Processing
EHRs - Electronic Health Records
NER - Named Entity Recognition
BERT - Bidirectional Encoder Representations from Transformers
ctDNA - Circulating Tumor DNA
CTCs - Circulating Tumor Cells
ddPCR - Digital Droplet PCR
NGS - Next-Generation Sequencing
ECS - Error-Corrected Sequencing
UAMOCS - Universal Analysis of Methylation Omics for Complex Systems
CCARS - Canonical Correlation Analysis with Recursive Feature Elimination Selector
CCgR - Complete Cytogenetic Response
NSCLC - Non-Small Cell Lung Cancer

WGS - Whole-Genome Sequencing
LBs - Liquid Biopsies
T-PLL- T-cell prolymphocytic leukemia
PDX - Patient-Derived Xenografts
DRL - Deep Reinforcement Learning
CTLs - Cytotoxic T Lymphocytes
NK - Natural Killer
ICIs - Immune Checkpoint Inhibitors
CAR - Chimeric Antigen Receptor
UTR - Untranslated Region
CNVs - Copy Number Variations
FDA - Food and Drug Administration
TGF-β - Transforming Growth Factor-β
CDSS - Clinical Decision Support Systems
QIOA - Quantum-Inspired Optimization Algorithms

Chapter 11
Nanofibers in Oncology Theranostics:
Omics-Nano-CRISPR

Vidyashree Kuballi
Sri Sathya Sai University for Human Excellence, India

Veda Paragonnavar
Sri Sathya Sai University for Human Excellence, India

Nagashri Nanjundeshwara
Sri Sathya Sai University for Human Excellence, India

Ramya Raghavan
http://orcid.org/0000-0002-9953-543X
Sri Sathya Sai University for Human Excellence, India

ABSTRACT

The convergence of omics technologies, nanofiber-based nanotechnology, and CRISPR-Cas systems is enabling precision diagnostics, targeted therapy, and real-time monitoring. These integrated platforms are advancing the management of benign tumors, tumor grading, metastasis, imaging, and theranostics, with demonstrated efficacy in both preclinical and clinical settings. Current review literature shows that electrospun nanofibers already support local chemotherapy, combinatorial therapy, cancer detection, cell capture, 3D tumor modeling, metastasis studies, and theranostics. At the same time, newer work extends this platform toward smarter, biomaterial-fabricated systems and CRISPR-enabled delivery concepts. Key challenges remain in optimizing delivery efficiency, minimizing off-target effects, and standardizing clinical translation. Future research will focus on expanding multiplexing capabilities, refining single-cell and spatial omics integration, and developing robust, patient-specific theranostic platforms to further advance the promise of precision oncology.

DOI: 10.4018/979-8-3373-9715-3.ch011

INTRODUCTION

Modern oncology increasingly interprets cancer not as a static disease entity but as a dynamic ecological process involving developmental programming, metabolic plasticity, immune remodeling, epigenetic drift, and age-associated genomic instability. Electrospun biomimetic nanofibers have emerged as critical interfaces linking omics-informed diagnostics with CRISPR-enabled therapeutics and theranostic imaging systems. The convergence of multi-omics, nanofiber engineering, and CRISPR-based genome editing has turned translational oncology into a more molecularly directed discipline, where diagnosis, modeling, delivery, and therapy can be linked within one workflow. Nanofibers functionalized with CRISPR-Cas systems enable spatially targeted editing of oncogenic loci, delivery of siRNA/miRNA cargos, localized immunomodulation, and real-time imaging through fluorescent or magnetic nanoparticles. Nanofibers can simultaneously support sustained release, biomarker sensing, and cell-interactive function, which is essential for localized therapy with lower toxicity. Nanofiber drug delivery supports localized, sustained release as a central advantage over conventional systemic administration. Omics can identify subtype-specific targets, resistance pathways, and metastatic signatures. (Kanaujia et al., 2026) Electrospun nanofibers recapitulate extracellular matrix topology and tumor biomechanical architecture, enabling superior cancer modeling compared with conventional two-dimensional culture systems. Nanofiber matrices regulate epithelial-mesenchymal transition, cancer stemness, immune infiltration, and angiogenesis through nanoscale mechanical signaling. Omics analyses embedded into systems permit longitudinal characterization of metastatic evolution, clonal selection, and therapeutic resistance. At the same time, CRISPR can validate or silence those targets, and nanofibers can deliver tools locally or present them in a biomimetic 3D format that better reflects real tumor biology. Multifunctional fibers embedded with gold nanoparticles, quantum dots, iron oxide nanocrystals, or near-infrared fluorophores permit multimodal MRI, PET, fluorescence, and photoacoustic imaging. Chemotherapeutics, immunomodulators, or CRISPR constructs respond to pH, hypoxia, redox stress, or ultrasound activation. The translation of CRISPR-based oncology interventions has been hampered by three persistent delivery challenges: (i) efficient intracellular delivery of large Cas9 ribonucleoprotein (RNP) or mRNA/sgRNA cargoes, (ii) targeting of the specific intratumoral or tissue-niche cellular subpopulation (LGR5+ CSCs at the crypt base), and (iii) timing of release to the circadian window of maximum CRISPR activity and minimum off-target toxicity. (Feng et al., 2026) Nanofiber platforms address all three challenges through their inherent capacity for sustained, tunable release, surface functionalization with targeting ligands, and integration of clock-responsive polymer chemistries. Omics-guided nanofiber systems now allow molecular stratification of benign lesions such as adenomas, fibroadenomas, meningiomas, papillomas, and dysplastic nevi. Electrospun scaffolds loaded with biosensors detect mutation signatures including KRAS, APC, BRAF, PTEN, and TP53 alterations before overt malignancy. The key design principle unifying all nanofiber platforms is that the release trigger should be aligned with a circadian physiological or molecular signal predictable by wearable biosensors or organoid characterization. This synergistic combination enables targeted, multimodal support to advance tumor grading, metastasis modeling, imaging, and theranostics. (Fareed et al., 2026)

Multi-omics improves tumor stratification by genomic, transcriptomic, proteomic, and metabolomic signals into a more complete disease map. In nanofiber-based oncology, ligands, drugs, CRISPR guides, and imaging agents can target tumor phenotypes, rather than relying on one-size-fits-all treatment. This makes the combined platforms relevant for benign-to-malignant progression, grading, metastasis monitoring, and imaging-guided therapy. (Nie et al., 2026) Spatial biology is emerging as a central component of oncology theranostics. Spatial transcriptomics, spatial proteomics, multiplex immunofluorescence, imaging mass cytometry, and digital spatial profiling are highly multiplexed. Immunohistochemistry enables simultaneous visualization of tumor architecture, immune cell localization, stromal remodeling, vascular organization, and therapeutic target expression within intact tissue. Artificial intelligence reconstructs tumor ecosystems by quantifying immune cell interactions, tertiary lymphoid structures, cancer-associated fibroblasts, hypoxic regions, angiogenic niches, and intratumoral heterogeneity. (M et al., 2026) Spatial biomarkers are under investigation for predicting responses to immune checkpoint inhibitors, antibody-drug conjugates, antiangiogenic therapies, and adoptive cell therapies. Electrospun nanofibers have already been used to engineer pre-metastatic niches and bone microenvironments for metastasis studies, which makes them relevant not only for therapy but also for mechanism discovery. (Yavuz et al., 2026) Tumor grading can benefit from nanofiber-enabled imaging and capture platforms that enrich circulating tumor cells or detect biomarkers associated with aggressiveness. When coupled with omics, grading becomes more than histology, it becomes a multi-layered functional readout that can include molecular heterogeneity, metastatic potential, and therapy resistance signatures (Sánchez et al., 2026). Three-dimensional computational pathology is advancing through reconstruction of serial histological sections into volumetric tumor models. Integration with optical tissue clearing, light-sheet microscopy, micro-computed tomography, and spatial molecular imaging permits quantitative characterization of invasive fronts, vascular architecture, lymphatic dissemination, immune infiltration, and metastatic niches. Volumetric biomarkers are being investigated for surgical margin assessment, tumor evolution, and treatment response prediction.

Label-free optical imaging technologies including stimulated Raman scattering microscopy, coherent anti-Stokes Raman scattering microscopy, multiphoton microscopy, quantitative phase imaging, second harmonic generation microscopy, fluorescence lifetime imaging, and hyperspectral imaging generate biochemical fingerprints without exogenous labels. (Lakshmaiah & Raghavan, 2026) Artificial intelligence extracts metabolic, lipidomic, collagen, and mitochondrial signatures associated with malignant transformation, treatment response, and therapy-induced remodeling. Technologies are under evaluation for rapid intraoperative diagnosis and real-time surgical guidance. Nanofiber systems for cancer detection, sensing, cancer cell imaging, and capture of circulating tumor cells toward smart theranostic designs. Most nanofiber oncology systems remain preclinical, and scale-up manufacturing, reproducibility, burst release, immune compatibility, CRISPR cargo stability, and imaging or sensing signals remain clinically interpretable. Another issue is that sophisticated multimodal systems can become too complex for regulatory approval when combining editing, delivery, imaging, and responsive devices. (Wadan et al., 2026) For 50 years the central unsolved problem of early cancer diagnosis has been that in stage 1 disease, there may be only 1-10 mutant DNA fragments, exosomes or circulating tumor cells in an entire milliliters of blood. (Table 1) This chapter examines the mechanistic basis of omics technologies applicable to tumor characterization across grades, CRISPR platforms employed for intervention, and nanofiber scaffold systems as delivery and theranostic vehicles.

Table 1. Integration of Omics-Nano-CRISPR in Oncology Theranostics

Multiomics	Nanomedicine	CRISPR Bioengineering	Theranostic Objective
Multiomic biomarker discovery	Imaging nanoparticles	Functional target validation	Patient stratification Precision Companion Diagnostics
Spatial multiomics	Targeted nanodelivery	Local genome editing	Tissue-specific therapy Spatial Precision Oncology
Longitudinal omics monitoring	Stimuli-responsive drug release	Dynamic gene editing	Real-time treatment optimization Adaptive Cancer Therapy
Neoantigen identification	mRNA/lipid nanoparticle delivery	Immune cell engineering	Precision immunotherapy Personalized Cancer Vaccines
Single-cell immune profiling	Nanoparticle immune modulation	CAR-T/CAR-NK engineering	Durable antitumor immunity Next-Generation Cell Therapy
Spatial immune profiling	Stromal-targeted nanoparticles	Immune checkpoint editing	Overcoming therapeutic resistance Tumor Microenvironment Reprogramming
ctDNA, cfRNA, exosomal biomarkers	Image-guided nanocarriers	Mutation-specific editing	Early intervention Liquid Biopsy-Guided Therapy
Organoid multiomics	Drug screening nanoparticles	Genome-engineered organoids	Individualized drug selection Functional Precision Oncology
Integrated longitudinal multiomics	Pharmacokinetic nanomodeling	Virtual genome intervention	Predictive treatment simulation Digital Twin Oncology
Continuous multimodal biomarker integration	Intelligent nanorobotic delivery	Programmable genome engineering	Closed-loop precision oncology Autonomous Theranostic Ecosystems

METHODOLOGY

A comprehensive literature search was performed from January 2021 to May 2026 across five major databases: PubMed/MEDLINE, Scopus, Web of Science Core Collection, Embase, and IEEE Xplore for AI/biosensor integration. This review was conducted following the scale for narrative reviews to ensure comprehensive coverage, transparency, and reproducibility of the integrated omics-nano-CRISPR-nanofiber literature. (Figure 1)

Figure 1. Methodology framework allowed for cross-comparison of cancer-specific barriers and shared nano-CRISPR solutions.

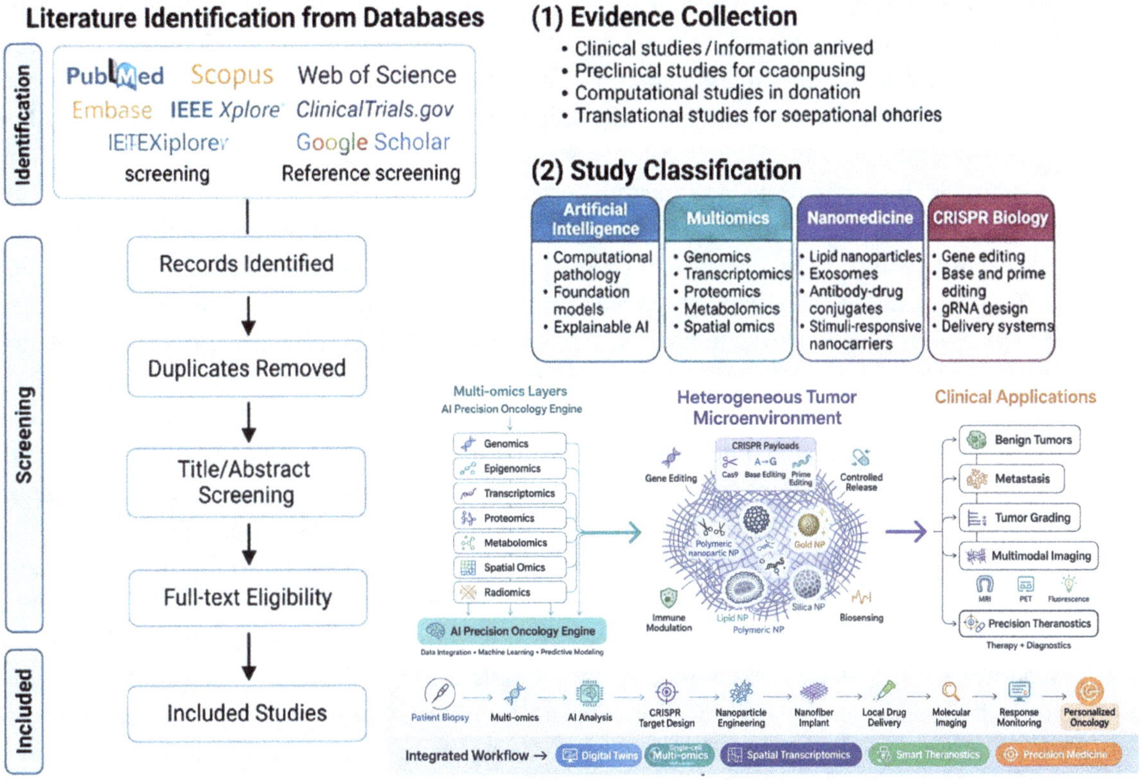

Inclusion crieria: (i) Original research, systematic reviews, and high-impact reviews focused on integration of ≥2 omics layers with nano-delivery or CRISPR editing in cancer, (ii) Studies reporting electrospun nanofiber applications for CRISPR, drug delivery, biosensing, or molecular imaging (MRI, PET, fluorescence), (iii) Cancer-specific studies detailing molecular targets (HER2/BRCA1/2 in breast, EGFR/KRAS in lung, BBB penetration in glioblastoma, dense stroma/YAP in pancreatic), (iv) English language. Exclusion criteria: (i) Non-cancer CRISPR applications without translational relevance, (ii) Nanofiber studies without omics or gene-editing integration, (iii) Editorials, conference abstracts without full data, (iv) Duplicate records.

Data were extracted using a standardized form capturing: (1) Omics type and platform, (2) Nanofiber composition (PCL, PLGA, chitosan, gelatin, composite) and fabrication method, (3) Nanoparticle type (AuNP, Fe3O4, quantum dots, LNP), (4) CRISPR payload (Cas9, dCas9, Cas12/Cas13, siRNA, mRNA, miRNA) and editing strategy, (5) Cancer type and hallmark targeted, (6) AI/biosensor/imaging modality, (7) Therapeutic outcome and theranostic potential.

CONCEPTUAL FRAMEWORK

Traditional oncology often treats tumors by tissue of origin and stage. Integrated precision oncology now views cancer as a dynamic system. Genomic lesions, epigenetic states, transcriptional programs, proteomic networks define tumors. Metabolic rewiring and epigenetic configurations arising during developmental windows cause heritable alterations. Mutations and gene expression changes promote tumorigenesis, progressing from benign growths to high-grade metastatic carcinomas. Omics profiling has greatly improved the distinction between benign and malignant tumors. In uterine tumors, comprehensive genomic, transcriptomic, and methylation analyses differentiate benign leiomyomas from malignant leiomyosarcomas, identifying cell cycle gene overexpression and chromosomal abnormalities as hallmarks of malignancy. (Zhu et al., 2026) Liquid biopsy integration represents a rapidly expanding area of theranostic pathology. Artificial intelligence combines histopathological features with circulating tumor DNA (ctDNA), circulating tumor cells, extracellular vesicles, circulating RNA, methylation signatures, tumor-educated platelets, and plasma proteomics to develop composite biomarkers for early diagnosis, minimal residual disease detection, recurrence surveillance, and longitudinal therapeutic monitoring. Multimodal integration improves sensitivity compared with individual biomarker platforms.

Single-cell and spatial multiomics are redefining tumor characterization. Artificial intelligence integrates single-cell RNA sequencing, single-cell ATAC sequencing, spatial transcriptomics, spatial metabolomics, and digital pathology to reconstruct clonal evolution, lineage plasticity, immune escape, therapeutic resistance, and tumor ecosystem dynamics. approaches facilitate identification of rare cellular populations responsible for relapse and metastatic progression. Neonatal retinoblastoma driven by biallelic RB1 loss, adolescent EWSR1-FLI1 Ewing sarcoma, middle-age KRAS-mutant colorectal carcinoma, and geriatric AML arising on clonal hematopoiesis of indeterminate potential share uncontrolled cellular proliferation but differ profoundly in their epigenetic landscape, developmental context, and immune microenvironment. (Iyer et al., 2026) Similar approaches in thyroid tumors using nanoparticle-assisted proteomics have revealed protein signatures that distinguish benign adenomas from carcinomas and correlate with tumor grade and metastatic potential. (Fiorentino et al., 2026) Proteogenomics and spatial omics enhance tumor grading by molecular alterations. Omics data combined with nanotechnology allow biomarker detection and molecular grading through nanoparticle-assisted proteomics and spatial omics platforms.

Multi-omics platforms (genomics, transcriptomics, proteomics, metabolomics, and epitranscriptomics) with next-generation CRISPR-based epigenome editors have many applications. None of these technologies work particularly well alone. The revolution is in how they are now being combined into fully integrated stacked hybrid biosensors:

First, magnetic nanoparticles isolate and concentrate the target biomarker from 1ml of whole blood. Next, CRISPR scans the concentrated sample with single nucleotide precision, and provides 10,000x biological signal amplification. Finally, plasmonic, quantum dot or graphene nanomaterials convert that biological signal into an extremely bright, measurable optical or electrical output. This stacked amplification results in systems that now routinely achieve attomolar detection limits in laboratory testing.

Neonatal and early-pediatric cancer genomes have fewer somatic point mutations than adult malignancies but a higher burden of large-scale structural variants, gene fusions, and oncogenic amplifications. The ETV6-RUNX1 fusion in childhood ALL is detectable as a preleukemic clone in cord blood in 1% of newborns, showing the somatic hit occurs prenatally. MYCN amplification in neuroblastoma, H3K27M oncohistone mutations in diffuse intrinsic pontine glioma (DIPG), and SMARCB1 biallelic loss in AT/RT are examples of pediatric cancer driven by epigenetic rather than mutational mechanisms for CRISPR strategy selection.(Pan et al., 2026) Knowledge graph-based oncology is emerging as a systems biology framework for precision therapeutics. Artificial intelligence constructs interconnected networks linking genes, signaling pathways, protein interactions, immune checkpoints, histological phenotypes, imaging biomarkers, drug targets, adverse events, and clinical outcomes. Graph neural networks prioritize therapeutic combinations, identify mechanisms of resistance, and generate mechanistic hypotheses for biomarker-guided interventions. approaches are increasingly incorporated into precision oncology decision-support platforms. Digital pathology is increasingly integrated with patient-derived organoids, tumor-on-chip systems, ex vivo tissue cultures, and functional precision medicine platforms. Histological and molecular data guide establishment of individualized experimental tumor models that are subsequently exposed to targeted therapies, immunotherapies, antibody-drug conjugates, radiopharmaceuticals, and combination regimens. Artificial intelligence correlates ex vivo drug responses with tissue morphology and molecular signatures to identify patient-specific therapeutic strategies.

Nanofiber delivery architectures present an unprecedented opportunity to dissect, monitor, and therapeutically reverse developmental epigenetic lesions. Adult cancers accumulating over decades of somatic mutagenesis display the highest mutational burdens. KRAS G12D/V/C mutations (CRC, PDAC, NSCLC), EGFR exon 19 deletions and L858R substitutions (NSCLC), TP53 hotspot gain-of-function mutations (breast, ovarian, CRC), and HER2 amplifications (breast, gastric) each represent high-priority CRISPR targets that base and prime editing can now address with sub-5% off-target frequency.(Xiu et al., 2026) Multi-omics gives a layered view of tumor heterogeneity, therapy resistance, immune escape, and metastatic potential, which is essential because tumors evolve under treatment pressure and frequently diverge from the primary lesion to metastatic sites.

Epigenomic landscape analysis of adult malignancies consistently identifies hypermethylation of PER1, PER2, and BMAL1 promoters as a recurrent feature of aggressive, MDR-enriched tumor subtypes across CRC, lung, breast, and hepatocellular carcinoma. (Han et al., 2025) As described, circadian clock silencing drives constitutive MEX3A expression, ABCB1 upregulation, and LGR5+ CSC expansion the molecular constellation that Omics-Nano-CRISPR strategies must target.(Q. Zhao et al., 2026)

Multi-omics technologies have enabled the identification of metastatic biomarkers and signaling pathways, EMT markers, MMP3, WNT11, SAA1, and LRRN4CL, across cancer types. (Zang et al., 2026) Biomarkers inform the design of CRISPR-based gene editing interventions delivered via nanofiber platforms. For instance, a Nano-CRISPR scaffold targeting haem oxygenase-1 (HO-1) has demonstrated effective suppression of metastasis and durable immune memory in melanoma models when combined with immune checkpoint blockade.(Zhu et al., 2026) Dendrimer lipid nanoparticles co-delivering siRNA against FAK and CRISPR/Cas9 targeting PD-L1 have shown enhanced tumor penetration and significant inhibition of tumor growth and metastasis in multiple animal models. (Majeed et al., 2026) Nanofiber-based systems deliver CRISPR tools to suppress metastasis-driving genes and modulate the tumor microenvironment. (Table 2) CRISPR then becomes the causal instrument that can test whether a candidate driver is truly actionable. At the same time, nanofiber systems provide localized, structured delivery or *in vitro* tumor-like niches for these experiments.

Table 2. Multiomics Innovations in Human Oncology Theranostics

Multiomics Technology	Biological Component	AI Integration	Theranostic Application
Whole-Genome Sequencing (WGS)	Genomic variants	Variant prioritization, predictive modeling	Precision targeted therapy selection
Whole-Exome Sequencing (WES)	Coding mutations	Driver mutation identification	Companion diagnostics
Bulk RNA Sequencing	Transcriptome	Molecular subtype classification	Treatment stratification
Single-Cell RNA Sequencing (scRNA-seq)	Cellular heterogeneity	Cell-state mapping	Resistance mechanism discovery
Spatial Transcriptomics	Tissue-specific gene expression	Spatial AI mapping	Tumor microenvironment characterization
Single-Cell Multiomics	DNA, RNA, chromatin	Cell lineage reconstruction	Clonal evolution analysis
Proteomics	Protein expression	Biomarker discovery	Therapeutic target identification
Phosphoproteomics	Signaling pathways	Kinase network modeling	Drug response prediction
Metabolomics	Cancer metabolism	Metabolic pathway analysis	Precision metabolic therapy
Lipidomics	Lipid metabolism	Lipid biomarker profiling	Immunometabolism-guided therapy
Glycoproteomics	Protein glycosylation	Glycan pattern recognition	Immunotherapy optimization
Epigenomics	DNA methylation, histone modification	Epigenetic signature prediction	Epigenetic therapeutics
Chromatin Accessibility (scATAC-seq)	Regulatory chromatin	Gene regulation inference	Resistance prediction
Microbiomics	Tumor-associated microbiota	Host-microbiome interaction modeling	Immunotherapy response prediction
Spatial Multiomics	Morphology + genomics + proteomics	Multimodal foundation models	Precision tumor ecosystem mapping

NANOFIBER ENGINEERING ROLE

Nanofiber-based platforms are electrospun polymeric scaffolds, self-assembling peptide nanofibers, DNA origami structures, carbon nanotubes, and lipid-polymer hybrids. The programmable delivery of molecular cargoes enables sustained-release kinetics for diagnostic imaging. Nanofibers in the extracellular matrix have high surface area for biomolecule loading, with spatial control over release kinetics and cell interactions. Review literature shows that electrospun nanofibers have been used for 3D cancer models, diagnostics, cell capture, localized therapy, and cancer imaging, making them one of the most versatile scaffold classes in oncology. (Yavuz et al., 2026) Self-assembling peptide nanofibers (SAPNs) functionalized with bone-marrow homing sequences DWIVA fibronectin-derived sequence recognized by VLA-4 on HSPCs achieve selective accumulation in marrow sinusoids after intravenous administration. The SAPN scaffold undergoes β-sheet assembly at physiological pH and disassembles under the mildly acidic TME (pH 6.5–6.8) of malignant marrow, releasing the CRISPR payload preferentially in the leukemic niche. (Shokoya et al., 2026) Tumor TME is strongly influenced by stiffness, porosity, adhesiveness, and geometry. In the oncological context, electrospun PLGA (poly-lactic-co-glycolic acid)

Electrospun nanofibers have been used for 3D cancer models, diagnostics, cell capture, localized therapy, and cancer imaging. The Gliadel wafer, an FDA-approved carmustine-loaded polymer implant for glioblastoma, established for locoregional nanofiber delivery. Next-generation electrospun constructs co-deliver CRISPR ribonucleoprotein (RNP) complexes alongside chemotherapy or gene editing and cytotoxicity. Chitosan and hyaluronic acid nanofibers, graphene quantum dots exploit CD44 receptor overexpression on cancer stem cells for active targeting. (Vashani et al., 2026) At the same time, coaxial electrospinning enables core-shell architectures that provide spatially and temporally controlled sequential release of distinct therapeutic agents. Histological biomarkers are being investigated to select patients suitable for nanoparticle-mediated chemotherapy, photothermal therapy, photodynamic therapy, magnetic hyperthermia, ultrasound-responsive nanocarriers, and image-guided targeted drug delivery. Artificial intelligence predicts nanoparticle penetration, vascular permeability, stromal barriers, and intratumoral distribution based on tissue architecture and spatial biomarkers.

Digital twins represent one of the most ambitious translational innovations. Individualized computational cancer models integrate longitudinal histopathology, radiology, genomics, circulating biomarkers, treatment history, pharmacokinetics, pharmacodynamics, immune monitoring, and clinical outcomes into continuously updated simulations. Digital twins are being explored for virtual therapeutic testing, optimization of combination therapies, adaptive dosing strategies, prediction of resistance evolution, and personalized surveillance schedules.

Carbon Nanotube Fibers and Photothermal Theranostics

Carbon nanotubes (CNTs) are cylindrical graphene allotropes with extraordinary optical, electrical, and mechanical properties. CNTs occupy a unique position in oncological nanofiber photothermal transduction, enabling light-triggered ablation of cancer cells. CNTs are irradiated in the near-infrared window (700–1000 nm). Functionalized single-walled CNTs (SWCNTs) conjugated to polyethylene glycol, target ligands folic acid, transferrin, anti-EGFR antibodies. CNTs act as therapeutic payloads for doxorubicin, siRNA, and CRISPR components to achieve antitumor activity in melanoma, lung carcinoma, and pancreatic cancer models. The photothermal ablation induces hyperthermia that transiently increases tumor vascular permeability, creating a thermal EPR enhancement into poorly vascularized tumor cores.(Sandoval et al., 2026) Furthermore, CNT Raman spectroscopy enables label-free, multiplexed molecular imaging of tumor biomarkers, transforming CNT nanofibers into dual-purpose photothermal-theranostic platforms.

Nanofibers serve as *in vitro* tumor analogs for studying invasion, metastasis, and drug response. Second, they function as local depots for chemotherapeutics, genes, or biologicals. Nanofibers support theranostic sensing, imaging, and therapy. (Shokoya et al., 2026) Nano-in-nano approaches use nanoparticles within nanofibers to overcome tumor microenvironment barriers and strengthen therapeutic payload delivery. (Table 3)

Silk Fibroin and ECM-Mimetic Nanofibers for Geriatric TME

The geriatric TME is distinguished by elevated MMP-9 secretion from SASP-activated senescent stroma, fibrotic ECM remodeling, and reduced vascular permeability compared to the more proliferative adult TME. Silk fibroin nanofibers, crosslinked with MMP-9 cleavable GPLGIAGQ peptide linkers, are ideally matched to this environment. Silk fibroin nanofibers remain intact in normal, low-MMP tissue but dissolve in the MMP-9-rich geriatric TME, releasing CRISPR payloads precisely at the sites of senescent stromal activation.(Tomljenovic-Hanic & Khalid, 2026) SASP-triggered release mechanism for aged, senescence-rich microenvironments. It also provides a passive targeting strategy in active clock-phase timing PNIPAM or REV-ERB-gated systems.(Abo Qoura et al., 2026)

Table 3. Nanomedicine Platforms in Oncology Theranostics

Nanotechnology Platform	Therapeutic Cargo	Imaging Modality	Theranostic Function
Lipid Nanoparticles	siRNA, mRNA	MRI, fluorescence	Gene delivery
Polymeric Nanoparticles	Chemotherapeutics	MRI, PET	Controlled drug delivery
Exosome Nanocarriers	RNA, proteins	Optical imaging	Precision drug delivery
Biomimetic Nanoparticles	Cell membrane-coated drugs	Multimodal imaging	Immune evasion
Gold Nanoparticles	Photothermal agents	CT, photoacoustic	Photothermal therapy
Magnetic Iron Oxide Nanoparticles	Chemotherapy	MRI	Hyperthermia + imaging
Mesoporous Silica Nanoparticles	Multi-drug payloads	Fluorescence	Stimuli-responsive delivery
Quantum Dots	Imaging probes	Fluorescence	Molecular imaging
Upconversion Nanoparticles	Photosensitizers	NIR imaging	Deep tissue photodynamic therapy
Metal-Organic Framework Nanoparticles	Multiple therapeutics	MRI, CT	Controlled multifunctional delivery
Nanogels	Biologics	Fluorescence	Controlled release
DNA Origami Nanostructures	CRISPR, siRNA	Optical imaging	Programmable nanomedicine
Stimuli-Responsive Nanocarriers	Chemotherapeutics	Multimodal imaging	pH, ROS, enzyme-responsive release
Theranostic Nanoplatforms	Drug + imaging probe	PET/MRI/CT	Simultaneous diagnosis and therapy
Immune Nanoparticles	Immune checkpoint inhibitors	Molecular imaging	Cancer immunotherapy

CRISPR AS A FUNCTIONAL TOOL

In parallel, CRISPR-Cas9 genome-editing tools for epigenome editing target genomic loci for methylation, demethylation, histone remodeling, or transcriptional activation and repression. The therapeutic potential is critically dependent on tissue-targeted delivery systems. (Pan et al., 2026) CRISPR/Cas systems have shifted cancer research from correlation to causation. They allow investigators to disrupt oncogenes, validate candidate resistance pathways, model disease, and reprogram immune cells. At the same time, newer work also explores more selective strategies such as editing amplified oncogenes to induce tumor cell death. A 2026 study reported that CRISPR editing of amplified oncogenes reduced tumor growth and triggered immunogenic cell death in preclinical models. (Zang et al., 2026)

CRISPR components can be delivered as plasmid DNA, mRNA, or ribonucleoprotein complexes. Challenges related to stability, endosomal escape, nuclear access, off-target activity, and immunogenicity remain.(Pallavi et al., 2026) Nanotechnology-based delivery systems and CRISPR delivery with nanocarriers can be paired with a delivery platform that is localized, tunable, and potentially image-guided.

Major CRISPR Diagnostic Platforms

SHERLOCK (Specific High-Sensitivity Enzymatic Reporter UnLOCKing)

- Utilizes Cas13a protein that exhibits collateral RNA cleavage activity upon target RNA recognition
- Combined with recombinase polymerase amplification (RPA) for target preamplification
- Capable of detecting cancer-associated mutations, viral oncogenes, and fusion transcripts at attomolar concentrations
- Has been successfully applied for detecting KRAS mutations, BCR-ABL fusion transcripts, and HPV-related cancer biomarkers

DETECTR (DNA Endonuclease-Targeted CRISPR Trans Reporter)

- Employs Cas12a protein that exhibits collateral single-stranded DNA cleavage upon target dsDNA recognition
- Combined with loop-mediated isothermal amplification (LAMP) for signal enhancement
- Enables rapid detection of cancer-related DNA mutations, HPV subtypes, and ctDNA alterations
- Provides results within 30–60 minutes without sophisticated instrumentation

CasRx and Other Novel CRISPR Effectors

- Next-generation CRISPR effectors with enhanced collateral cleavage activities
- Enable detection of cancer-associated RNA biomarkers with improved sensitivity and reduced background signals
- Facilitate multiplexed detection of multiple cancer targets simultaneously

Self-Assembling Peptide Nanofibers: Biological Mimicry for Tumor Penetration

Self-assembling peptide nanofibers (SAPNs) are formed through the hierarchical organization of short peptide sequences bearing alternating hydrophilic and hydrophobic residues into beta-sheet-rich fibrous networks. Canonical SAPN sequences RADA16-I form hydrogel scaffolds at physiological pH with mechanical properties matching soft tumor tissue, enabling minimally invasive injection and in situ gelling of therapeutic payload to the tumor mass.(Nazar Saeed et al., 2026) For CRISPR delivery, SAPNs have been functionalized with nuclear localization sequences, cell-penetrating peptides, and endosome-disrupting lipid moieties to achieve efficient cytosolic and nuclear delivery of Cas9 RNP complexes without viral vectors. In glioblastoma models, RADA16-functionalized nanofibers loaded with Cas9-sgRNA complexes targeting EGFR penetrated blood-brain barrier surrogates and achieved significant tumor volume reduction in orthotopic murine experiments.(Neves et al., 2026) The epigenomic profiling confirms on-target H3K27 demethylation consistent with CRISPR-mediated TSG reactivation.

Table 4. Nanofiber Delivery Platforms for CRISPR Oncology Payloads

Nanofiber Fabrication	Responsive CRISPR Payload	Application
PCL/PLA electrospun scaffold Coaxial electrospinning	LCST-thermoresponsive shell (PNIPAM) LNP-encapsulated Cas9 RNP + sgRNA (KRAS G12D)	ZT16–20 timed release matched to ABCB1 trough
Self-assembling peptide nanofiber (SAPN) pH/redox-triggered β-sheet assembly	REV-ERBα-degradable crosslinker dissolves during active phase Base editor (ABE8e) targeting ETV6-RUNX1 junction	Pediatric ALL intra-marrow sustained delivery
Collagen-mimetic nanofiber hydrogel Electrospinning + crosslinking	Melatonin-receptor-decorated surface, CRISPRa-BMAL1 dCas9-VPR activation construct	Geriatric clock amplitude restoration in CHIP/AML stem cells
PLGA-chitosan core–shell fiber Emulsion electrospinning	pH-responsive chitosan shell (TME acidosis trigger) Prime editor (PE3b) targeting TP53 R175H	Adult solid tumors — depot implant, clock-timed burst via PLGA crystallinity
Silk fibroin nanofiber mat Wet-electrospinning + methanol annealing	Enzymatic degradation by MMP-9 (SASP-upregulated in aged TME) CRISPR-dCas9-DNMT3A (epigenome editor) — KRAS enhancer methylation	Geriatric PDAC, SASP-activated, clock-neutral slow release
Lipid-coated carbon nanofiber CVD + lipid wrapping	Optogenetic surface receptor (Cry2-LOV2), NIR-activated release at clinician-set ZT MEX3A siRNA + sgRNA (LGR5 promoter CRISPRi)	Adult/geriatric CRC CSC ablation, wearable-biosensor-triggered NIR pulse
Decellularized ECM nanofiber Tissue decellularization + electrospinning	Native circadian ECM remodeling cues (fibronectin rhythms) Lentiviral PER2 re-expression cassette	Pediatric/adult solid tumors, PER2 silencing reversal, MDR resensitization.

DNA Origami Nanofibers: Programmable Molecular Architectures

DNA origami technology involves the folding of long single-stranded DNA scaffolds using complementary 'staple' oligonucleotides into precise three-dimensional nanostructures. Nanofiber morphologies include tubes, rods, barrels, and ribbons, each with sub-nanometer-scale programmable surface chemistry. (Freedman et al., 2026) DNA origami nanofibers are uniquely suited to CRISPR delivery because the same scaffold chemistry enables co-assembly of sgRNA sequences, Cas9 protein attachment domains, targeting aptamers, and imaging probes into a single integrated nanoplatform with spatially defined cargo organization. Their resistance to nuclease degradation, achieved through phosphorothioate backbones or PEG coating, allows systemic delivery. In hematological malignancies, DNA origami tubes bearing anti-MUC1 aptamers and loaded with Cas9-DNMT3A epigenomic editors demonstrated leukemia cell specificity and targeted methylation installation at oncogenic promoters with minimal off-target activity. (Zhang et al., 2026)

Lipid-Polymer Hybrid Nanofibers and Systemic CRISPR Delivery

The systemic delivery of CRISPR Cas9 mRNA, sgRNA, or RNP complexes to solid tumors and metastatic lesions requires nanocarriers that combine the endosomal escape efficiency of cationic lipids with the serum stability and sustained-release characteristics of polymer cores. (Korrapati et al., 2026)

Lipid-polymer hybrid nanoparticle (LPHN) platforms. The PLGA or polycaprolactone (PCL) polymer core is surrounded by lipid shells of DOTAP, DOPE, and PEG-lipid, forming fibrous morphologies via coaxial electrospinning or microfluidic fiber generation. (X. Zhao et al., 2026) Hybrid nanofibers achieve lipid-mediated membrane fusion-dependent endosomal escape and the polymer core-mediated protection of nucleic acid cargo from enzymatic degradation. In breast cancer and colorectal cancer models, LPHN nanofibers delivering mRNA and BRCA1-targeting sgRNAs demonstrated site-specific CpG demethylation at the BRCA1 promoter, restored BRCA1 protein expression, and sensitized cancer cells to PARP inhibition. (He et al., 2025)

MULTI-OMICS IN CANCER

Multi-omics has changed how cancers are classified and treated by linking molecular profiles to prognosis and response. (Table 4) In glioblastoma, for example, genomic and epigenomic biomarkers such as IDH mutation status and MGMT promoter methylation have already altered the taxonomy of disease and the downstream treatment plan. (Giercuszkiewicz-Haśnik et al., 2026) A major strength of multi-omics is that it can reveal tumor states that histology alone cannot resolve. Plasmonic nanoparticles exhibit highly sensitive shifts in resonance wavelength in response to changes in the local refractive index upon target biomarker binding, enabling label-free and real-time cancer biomarker detection. Single-cell and spatial approaches capture intratumoral heterogeneity. Metastatic niches often acquire distinct molecular programs compared with the primary tumor. Proteomic and metabolomic layers expose functional phenotypes that may not be obvious from DNA sequencing alone. (Kaźmierczak-Siedlecka et al., 2026) Beyond CNS tumors, integrated omics can identify actionable vulnerabilities in solid and hematologic malignancies, oncogene amplification, immune evasion signatures, and signaling pathways associated with resistance. The future trajectory of oncology theranostics is converging toward multimodal precision oncology ecosystems, where digital pathology, spatial biology, liquid biopsy, functional tumor models, molecular imaging, artificial intelligence, and computational systems biology operate as integrated clinical decision-support platforms. (Korrapati et al., 2026) Rather than functioning solely as diagnostic tools, technologies are being developed as predictive, prognostic, and therapeutic companion systems capable of identifying actionable biomarkers, guiding targeted interventions, monitoring treatment response, and enabling adaptive precision oncology. (Table 5) Many of these approaches are already in early clinical implementation or prospective validation studies, while others remain in translational research with substantial potential for future clinical integration.

Table 5. CRISPR and Genome Bioengineering Innovations in Oncology Theranostics

CRISPR Technology	Engineering Strategy	Oncology Application	Theranostic Potential
CRISPR-Cas9	Gene knockout	Oncogene disruption	Precision genome editing
CRISPR Base Editing	Single nucleotide correction	Driver mutation correction	Personalized therapeutics
Prime Editing	Precise DNA editing	Mutation repair	Precision correction
CRISPRa	Gene activation	Tumor suppressor restoration	Functional reprogramming
CRISPRi	Gene repression	Oncogene silencing	Target validation

continued on following page

Table 5. Continued

CRISPR Technology	Engineering Strategy	Oncology Application	Theranostic Potential
Epigenome Editing	DNA methylation modification	Gene regulation	Epigenetic therapy
RNA Editing (Cas13)	RNA modification	Transcript correction	Transient therapeutic editing
Multiplex CRISPR	Multi-gene engineering	Combination cancer therapy	Polygenic targeting
CRISPR Functional Genomics	Genome-wide screening	Drug target discovery	Biomarker identification
CRISPR Lineage Tracing	Cellular evolution mapping	Tumor evolution	Resistance monitoring
CRISPR Synthetic Lethality	Dual pathway targeting	Precision oncology	Combination therapy
CRISPR Immune Engineering	CAR-T/CAR-NK modification	Adoptive immunotherapy	Personalized immunotherapy
In vivo CRISPR Delivery	Nanoparticle-mediated editing	Solid tumors	Targeted genome therapy
Organoid Genome Engineering	Patient-derived organoids	Functional precision medicine	Drug response prediction
CRISPR Molecular Diagnostics	CRISPR-based biosensors	Liquid biopsy	Early cancer detection

Benign Tumors: Epigenomic Prevention and Early Intervention

Precision oncology integrates molecular biomarkers with imaging for early disease detection. Uterine leiomyomas, thyroid follicular adenomas, colorectal adenomas, neurofibromas, and hepatic hemangiomas collectively affect hundreds of millions of patients worldwide and consume substantial healthcare resources through surveillance, intervention, and management of mass-effect symptoms. The epigenomic benign entities carry traces of developmental programming. Focal CpG island methylation alters histone modification patterns at developmental enhancers and miRNA expression, creating a permissive chromatin environment for clonal expansion without malignant transformation.(Iyer et al., 2026)

Magnetoresistance-Based Biosensors: Changes in magnetic resistance signals upon biomarker binding to MNP surfaces enable ultrasensitive and label-free detection of cancer-associated analytes.

Nuclear Magnetic Resonance (NMR) Biosensing: NMR biosensors can detect cancer biomarkers in unprocessed biological samples with high sensitivity and specificity.

Magnetic Relaxation Switching mechanism enables colorimetric-like detection of cancer biomarkers through measurable changes in T2 relaxation times upon MNP aggregation or dispersion. Multimodal Imaging and Sensing with other nanomaterials create multifunctional theranostic platforms capable of simultaneous cancer detection and treatment. Multimodal biomarker fusion by combining CTG-derived physiological biomarkers with serum biomarkers, circulating extracellular vesicles, cell-free fetal DNA, placental transcriptomics, metabolomics, lipidomics, proteomics, epigenomics, microbiome profiling, and exposome measurements. (Byalahunasishetter et al., 2026) AI enables hierarchical integration of heterogeneous datasets through multimodal transformers, Bayesian graphical models, probabilistic causal inference, graph representation learning, and knowledge graphs capable of identifying developmental signatures.

The application of omics technologies to benign tumors has revealed molecular subtypes with distinct developmental epigenomic signatures that correlate with growth rate, symptom burden, and malignant potential. In uterine leiomyomas, for example, MED12 mutation-positive tumors exhibit a distinct methylome from MED12 wild-type lesions, with the latter showing epigenomic features reminiscent of smooth muscle developmental progenitors. (T A et al., 2026) CRISPR-nanofibers used in benign tumors are for epigenomic normalization. It helps suppress aberrant growth-promoting chromatin states more than for cytotoxic tumor elimination. Silk fibroin nanofiber implants loaded with CRISPRi constructs targeting MED12-driven enhancers, delivered via minimally invasive hysteroscopy or laparoscopy, are being evaluated in preclinical leiomyoma models as alternatives to medical or surgical management. (Pandey et al., 2026) Magnetic nanoparticles (MNPs), particularly iron oxide nanoparticles (Fe_3O_4 and γ-Fe_2O_3), possess unique superparamagnetic properties that make them highly valuable in cancer biosensing, separation, and targeted drug delivery applications. Similarly, in colorectal adenomas identified at colonoscopy, endoscopically delivered PLGA nanofibers carrying CRISPRa constructs targeting APC promoter demethylation represent a conceptual framework for endoscopic epigenomic intervention to prevent adenoma-to-carcinoma progression. (Shi et al., 2026)

Figure 2. Cancer-Type-Specific Applications Linking Molecular Targets To Nanofiber-CRISPR Solutions.

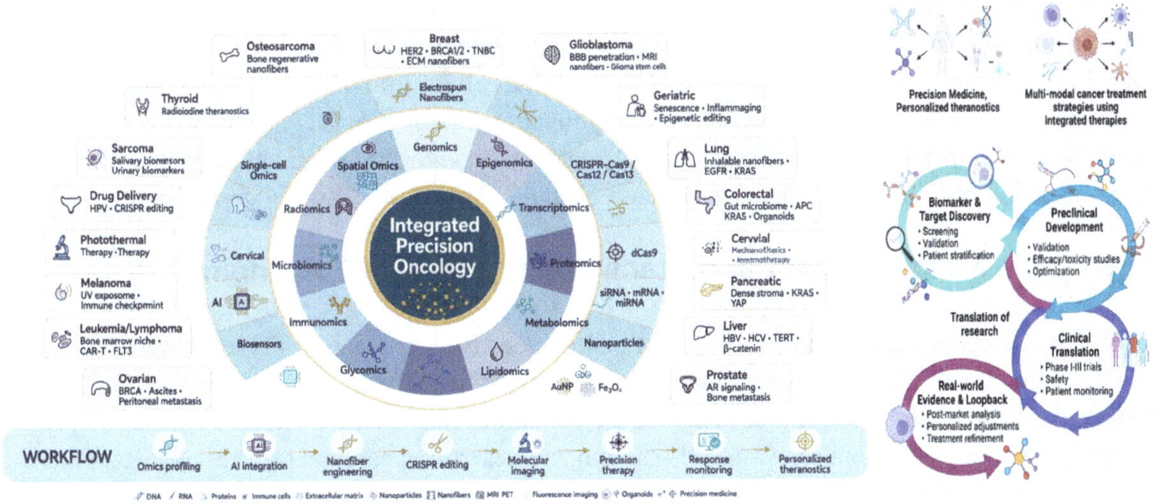

Low-Grade Tumors: Precision Epigenomic Profiling

Low-grade (grade I and II) gliomas, well-differentiated thyroid carcinoma, low-grade serous ovarian carcinoma, and ductal carcinoma in situ of the breast are characterized by relatively indolent biology. The extended clinical timelines and sensitivity carry the perpetual risk of dedifferentiation to high-grade malignancy.(Figure 2) Omics profiling of low-grade tumors has delineated key developmental epigenomic features that distinguish entities at high risk of malignant transformation from IDH-mutant lower-grade gliomas. For example, the G-CIMP phenotype, characterized by genome-wide CpG island hypermethylation downstream of the oncometabolite 2-HG, is associated with improved prognosis compared with IDH wild-type tumors. (Bai et al., 2026) The mechanical flexibility of graphene makes it an ideal material for developing wearable biosensing devices capable of continuous real-time monitoring of cancer biomarkers in sweat, saliva, or interstitial fluids. Graphene-based nano-biosensors have demonstrated exceptional sensitivity in detecting miRNA panels, cancer-derived extracellular vesicles, and circulating tumor cells (CTCs), offering promising tools for liquid biopsy-based cancer diagnostics.

Theranostic oncology combines diagnosis and treatment within integrated clinical workflows. Prenatal theranostics support antioxidant modulation, inflammatory pathway regulation. PLGA electrospun nanofiber wafers, analogous to the Gliadel wafer but incorporating CRISPR epigenomic payloads alongside temozolomide, represent the most clinically proximate nanofiber intervention strategy for low-grade glioma. Post-resection implantation of constructs at the tumor cavity margin enables sustained local delivery of dCas9-TET1 constructs targeting CDKN2A promoter demethylation, restoring p16 expression in residual tumor cells and suppressing recurrence in syngeneic murine data. Liquid biopsy of circulating tumor DNA (ctDNA) methylation from plasma is a minimally invasive method for monitoring epigenomic editing efficiency. (Ventelou et al., 2026) Early detection of tumor recurrence creates a closed-loop management of low-grade tumor surveillance. Digital pathology transformed oncology through whole-slide imaging, computational histopathology, and foundation models trained on large image repositories. Equivalent innovation is emerging through digital placental pathology integrated with prenatal imaging and CTG analysis. Quantification of histological architecture, vascular remodeling, trophoblast invasion, spiral artery transformation, immune cell infiltration, extracellular matrix remodeling, and placental angiogenesis through computational pathology provides mechanistic explanations for physiological abnormalities detected by fetal heart rate monitoring. Foundation AI models trained across placental histology, fetal imaging, physiological monitoring, and molecular datasets.

High-Grade Malignancies: Reprogramming Aggressive Epigenomes

High-grade glioblastoma multiforme (GBM), triple-negative breast cancer (TNBC), pancreatic ductal adenocarcinoma (PDAC), hepatocellular carcinoma (HCC), and small-cell lung cancer (SCLC) are therapeutically recalcitrant. High-grade oncological conditions show genetic heterogeneity and hyperdynamic epigenomes, with immune evasion. These epigenomes exhibit cardinal features of developmental reprogramming, including H3K27me3-mediated TSG silencing and super-enhancer-driven oncogene amplification. Lineage-inappropriate chromatin accessibility and extensive ncRNA dysregulation sustain stem-like transcriptional programs and phenotypic plasticity in response to therapeutic pressure.(Ke & Cardoso, 2026) Functionalized graphene platforms combined with aptamers enable highly selective and sensitive detection of cancer cells, exosomes, and protein biomarkers.

Multi-omics characterization of high-grade tumors using integrated scRNA-Seq, ATAC-Seq, and Hi-C has revealed epigenomic heterogeneity. A small population of epigenetically plastic cancer stem cells (CSCs) can repopulate the tumor following therapy. Self-assembling peptide nanofibers (SAPNs) carrying Cas9 RNP complexes and equipped with CSC-targeting ligands, anti-CD44, anti-CD133, and anti-EpCAM antibodies. (Zhou et al., 2026) SAPNs are being developed as CSC-specific epigenomic intervention platforms for high-grade malignancies. In GBM models, intratumoral injection of RADA16 SAPNs delivered multiplexed CRISPR targeting of EGFR amplification and the H3K27M mutation via prime editing. EZH2 CRISPRi demonstrated synergistic antitumor activity, extending median survival by 47% in orthotopic GBM models with sustained epigenomic reprogramming, as confirmed by longitudinal methylome profiling via ctDNA.(Z. Li et al., 2025)

Metastasis: Targeting the Epigenomic Metastatic Cascade

Metastasis involves epithelial-to-mesenchymal transition or intravasation, survival in circulation as circulating tumor cells (CTCs), extravasation, and colonization of the metastatic niche. Each metastasis stage is governed by dynamic epigenomic reprogramming of aspects of developmental programs. EMT in the epigenomic state of mesenchymal progenitors during embryogenesis gains H3K27ac at the SNAI1, TWIST1, and ZEB1 enhancers and loses EpCAM-associated epithelial chromatin signatures. (Table 6) Graphene-based electrodes exhibit exceptional electron transfer capabilities, significantly improving the sensitivity and detection limits of electrochemical cancer biosensors.

Lipid-polymer hybrid nanofiber platforms enable anti-metastatic CRISPR epigenomics via intravenous systemic delivery.(Vu et al., 2026) Nanofibers functionalized with anti-EpCAM aptamers and loaded with Cas13d target HOTAIR. lncRNAs recruit PRC2 to silence epithelial gene loci during EMT. In colorectal cancer models, RNA sequencing of circulating tumor cell counts and hepatic metastatic colonization confirms re-expression of EMT-suppressive epithelial genes. Integration of cfDNA methylation sequencing as a liquid biopsy readout with nanofiber-delivered epigenomic editing creates a pharmacodynamic monitoring framework. (G. Li et al., 2026) Graphene oxide effectively quenches fluorescence signals, which forms the basis of highly sensitive FRET-based biosensors for detecting cancer-related nucleic acids and proteins. The anti-metastatic epigenomic real-time assessment of EMT reversal links to metastatic burden reduction. Additionally, CRISPR-mediated editing of the metastatic epigenomic programs in hepatic stellate cells, pulmonary fibroblasts, and bone marrow stromal cells creates pre-metastatic niches. Graphene-based FET devices can detect cancer biomarkers at extremely low concentrations by measuring changes in electrical conductance upon biomarker binding.

Table 6. Examples of Cancer Therapies Show Omics-Nano-CRISPR Applications

Life Stage	Cancer Molecular Targets	Intervention
Neonatal (0–1 yr)	Retinoblastoma, hepatoblastoma, sacrococcygeal teratoma, congenital leukemia Germline RB1/TP53, high fetal proteome persistence, immature DNA repair	CRISPR base editing of RB1 germline, ECM nanofiber intraocular depot, multi-omics neonatal atlas
Pediatric (1–12 yr)	ALL, AML, medulloblastoma, nephroblastoma, neuroblastoma, DIPG Oncohistone mutations (H3K27M), ETV6-RUNX1 fusion, MYCN amplification, ALK/ NTRK fusions	Oncohistone prime editing (DIPG), SAPN bone-marrow nanofiber for ALL, scMulti-omics atlas
Adolescent Young Adult (13–25 yr)	Ewing sarcoma, osteosarcoma, germ cell tumors, Hodgkin lymphoma, CML EWSR1-FLI1 fusion, BRD4/MYC axis, HR pathway competency high, immunogenic TME	CRISPRi EWSR1-FLI1 enhancer, checkpoint inhibitor chronotherapy at ZT14–18, nanofiber scaffold
Adult (26–55 yr)	CRC, lung, breast, pancreatic, gastric, ovarian cancers, GBM KRAS/TP53/EGFR/BRCA mutations, MDR via ABCB1/ABCG2, clock-gated TME	Base editing KRAS G12D, ZT-timed nanofiber oxaliplatin, MEX3A CRISPRi at ZT8, AI chronotherapy
Midlife (56–65 yr)	CRC, NSCLC, HCC, RCC, prostate, bladder, myeloma Epigenetic drift, CDKN2A/PER2 methylation, MDR enrichment, clock amplitude declining	dCas9-TET1 PER2/BMAL1 demethylation nanofiber, ZT-resolved proteomics, CRY stabilizer delivery

continued on following page

Table 6. Continued

Life Stage	Cancer Molecular Targets	Intervention
Early Geriatric (66–75 yr)	CRC, NSCLC, prostate, AML/MDS, myeloma, bladder, melanoma CHIP, SASP, senescent stroma, MDR, clock disruption, inflammaging	CRISPRa BMAL1 clock restoration, silk fibroin MMP-9-triggered nanofiber, CHIP editing in HSPCs
Advanced Geriatric (76+ yr)	AML/MDS, CLL, prostate, NSCLC, CRC, thyroid (indolent) Proteostasis collapse, severe clock arrhythmicity, immunosenescence, polypharmacy interactions	Synthetic clock reactivation (CRISPRa BMAL1+RORα+CRY2 multiplex), wearable phase induction, minimal-toxicity chrono-nanofiber regimens

TRANSLATIONAL INTEGRATION

The translational value of combining three domains is that each one solves a limitation of the others. Multi-omics identifies the right target and the right patient, CRISPR provides the intervention, and nanofibers provide a controllable environment for delivery or testing. For example, omics can identify a pathway associated with metastasis or immune evasion, CRISPR can validate that target in tumor cells, and a nanofiber scaffold can release the editor locally or model the tumor microenvironment under realistic conditions. Quantum dots (QDs) are semiconductor nanocrystals typically ranging from 2 to 10 nanometers in size that exhibit exceptional and tunable optical and fluorescence properties. Their unique size-dependent photoluminescence, broad absorption spectra, narrow emission bandwidths, and remarkable photostability make them highly advantageous over conventional fluorescent dyes in biosensing applications.

Nanofiber-Enhanced MRI and Nuclear Imaging

Molecular imaging of tumors using nanofiber-based contrast or biomarker agents provides spatial resolution and molecular specificity that conventional contrast agents cannot achieve. (N. Li et al., 2026) Superparamagnetic iron oxide nanoparticle (SPION)-loaded PLGA nanofibers provide T2/T2 contrast enhancement on MRI while simultaneously delivering therapeutic payloads, enabling co-registration of tumor imaging data with therapeutic drug distribution.(Caf et al., 2026) In brain tumors, the blood-brain barrier limits conventional contrast agent penetration, and tumor margins are difficult to delineate. In positron emission tomography (PET), 64Cu- cerium, or 18F-labeled peptide nanofibers bearing tumor-targeting ligands enable whole-body metabolic and molecular imaging of metastatic lesions with pico-molar sensitivity. (Hamato et al., 2026) Recent innovations have demonstrated QD-based nano-biosensors capable of detecting circulating tumor DNA (ctDNA) and exosomal microRNAs in blood samples with extraordinary sensitivity, offering a non-invasive liquid biopsy approach for early cancer detection.

Photoacoustic and Fluorescence Imaging for Intraoperative Guidance

Photoacoustic imaging (PAI), thermoelastic expansion of light-absorbing nanoagents to generate ultrasonic contrast. It offers high spatial resolution in deep tissue without the depth limitations of optical fluorescence imaging. Carbon nanotube nanofibers and gold nanorod-polymer composites are among the most efficient photoacoustic contrast agents, providing 5–10 mm depth penetration with sub-100 μm spatial resolution. (Duarah et al., 2026) CNTs enable visualization of tumor microarchitecture, invasion fronts, and lymphatic channels. Near-infrared-II (NIR-II) fluorescence imaging using quantum dot-silk nanofiber composites helps in surgical oncology. CNTs achieve clear resection margins in breast, head and neck, and colorectal cancers. (Vashani et al., 2026) CRISPRa-activated fluorescent reporter systems are embedded within nanofibers. TSG re-expression drives GFP or near-infrared fluorescent protein (iRFP) expression with in situ confirmation of epigenomic editing success at the cellular level during intraoperative procedures. Tunable Fluorescence by simply adjusting the size of QDs, emission wavelength can be precisely controlled across the visible and near-infrared spectrum, enabling multiplexed detection of multiple cancer biomarkers simultaneously. QDs exhibit significantly higher quantum yields compared to organic fluorophores, resulting in enhanced signal sensitivity. Bioconjugation in QDs can be readily functionalized with antibodies, aptamers, peptides, or nucleic acids to specifically target cancer-associated biomarkers such as prostate-specific antigen (PSA), carcinoembryonic antigen (CEA), and HER2. Fluorescence resonance energy transfer (FRET) mechanisms using QDs enable highly sensitive detection of cancer-related nucleic acids, proteins, and miRNAs at femtomolar concentrations.

Closed-Loop Theranostic Platforms: Toward Autonomous Tumor Management

The convergence of omics diagnostics, CRISPR epigenome editing, nanofiber delivery, and molecular imaging within a single integrated theranostic platform represents the frontier of precision oncology. (Table 7) Gold nanoparticles (AuNPs) are among the most extensively studied nanomaterials in biosensing due to their remarkable optical, electrical, and chemical properties. Their localized surface plasmon resonance (LSPR), ease of surface functionalization, biocompatibility, and excellent electrochemical activity make them ideal candidates for cancer biomarker detection. Such platforms are conceptually designed to execute a closed-loop therapeutic cycle. Molecular imaging of tumor epigenomic state using biomarker-targeted nanofibers provides diagnostic information on chromatin accessibility and gene expression. (Iyer et al., 2026) AI-assisted multi-omics CRISPR intervention targets epigenomic editing strategies. The nanofiber platform delivers the therapeutic CRISPR payload to the tumor with spatiotemporal real-time imaging of therapeutic reporter gene activation, metabolic normalization, or structural tumor regression. (Banerjee et al., 2026) The pharmacodynamic nanobiosensor refines the therapeutic strategy in the microbubble-nanofiber hybrid constructs. (Lee et al., 2026) Ultrasound-triggered microbubble cavitation enables sonoporation-enhanced drug delivery and real-time contrast-enhanced ultrasound (CEUS) imaging of tumor vasculature, providing a delivery mechanism and imaging nanobiosensor within a single acoustic procedure.(Rajput et al., 2026) Gold nanoparticles serve as powerful SERS substrates that amplify Raman signals by several orders of magnitude, enabling ultrasensitive detection of cancer biomarkers at single-molecule levels.

AuNP-modified electrodes significantly enhance electron transfer kinetics and electrochemical signal amplification for detecting cancer-related proteins, nucleic acids, and metabolites. Lateral Flow Assays using Gold nanoparticle-based lateral flow strips provide rapid, cost-effective, and point-of-care cancer biomarker detection without the need for sophisticated instrumentation.

Table 7. CRISPR Editing Modalities Mapped to Cancer Therapies

CRISPR	Molecular Mechanism	Pediatric Cancer	Adult Cancer
Cas9 nuclease (DSB)	Blunt DSB → NHEJ/HDR	MYCN amplicon deletion, ALK exon 23 frameshift	KRAS G12D allele-specific KO, EGFR exon 19 del
Base editing (ABE/CBE)	Deaminase-mediated A>G or C>T transition, no DSB	ETV6-RUNX1 splice donor disruption, NRAS Q61K reversion	BRAF V600E C>T correction, PIK3CA H1047R inactivation
Prime editing (PE3b)	Pegylated reverse transcriptase, precise substitution/insertion/deletion	H3K27M (K27M→K27) oncohistone correction in DIPG organoids	BRCA2 c.5946del frameshift repair, MLH1 promoter demethylation
CRISPRa (dCas9-VPR/SAM)	Transcriptional activation via VP64-p65-Rta or SAM complex	RASSF1A tumor suppressor activation, miR-34a restoration	PER1/PER2 re-expression (clock restoration + MDR reversal)
CRISPRi (dCas9-KRAB)	Transcriptional silencing via KRAB-mediated H3K9me3 deposition	MYCN/MYC enhancer silencing, LIN28B repression	ABCB1 ZT-peak suppression, MEX3A overexpression silencing
Epigenome editing (dCas9-TET1/DNMT3A)	Targeted CpG methylation/demethylation, no sequence change	H3K27M-driven hypermethylation reversal at TSS clusters in DIPG	PER2/CDKN2A promoter demethylation, enhancer reactivation
CRISPR base-excision (nCas9-APOBEC)	Cytidine deaminase, C>U>T, nicks opposite strand	Oncofusion splice site disruption (SS18-SSX)	CDKN2A promoter hypermethylation reversion (C-to-T at key CpGs)

TRANSLATIONAL CHALLENGES AND THE PATH TO CLINICAL IMPLEMENTATION

The clinical translation of CRISPR-nanofiber epigenomic platforms confronts a set of interconnected safety challenges that must be systematically addressed through preclinical and early-phase clinical investigation. CRISPR off-targets unintended sequence cleavage or epigenomic modification at genomic loci with partial sgRNA complementarity. Epigenomic editors that do not generate DSBs mean off-target epigenomic changes may persist. High-fidelity Cas9 variants (eSpCas9-1.1, HiFi Cas9, HypaCas9) have off-target cleavage rates. (Zang et al., 2026) Regulatory agencies require GUIDE-Seq and CIRCLE-Seq whole-genome unbiased off-target profiling. The immunogenicity of Cas9 protein from bacterial pathogens, against adaptive immune responses in clinical serological studies and nanofiber demands immunological characterization. (Pan et al., 2026) Complement activation assays, cytokine release assays, and *in vivo* inflammatory profiling of PEGylation and zwitterionic surface coatings for immunological tolerance are under active investigation. AuNP-based nano-biosensors have demonstrated remarkable performance in detecting tumor-derived exosomes, ctDNA mutations, and protein biomarkers associated with lung, breast, colorectal, and prostate cancers.

Delivery Barriers and Systemic Pharmacokinetics

The tumor microenvironment presents barriers to nanofiber penetration via extracellular collagen, fibronectin, and hyaluronan. Interstitial fluid pressure (IFP) opposes convective transport, and leaky vasculature creates heterogeneous perfusion in the acidic, hypoxic TME milieu. Matrix metalloproteinase

(MMP)-responsive nanofibers exploit the tumor-specific proteolytic environment to enable in situ size reduction and penetration enhancement. (B. Zhao et al., 2026) Further ultrasound-mediated acoustic cavitation for mechanical ECM disruption, co-administration of hyaluronidase or losartan to deplete stromal barriers, and active targeting of nanofibers to tumor vasculature receptors (VEGFR2, αvβ3 integrin) for transcytosis-mediated tumor entry. Systemic pharmacokinetics of nanofibers govern size, surface charge, and protein corona formation, with optimal tumor accumulation achieved by nanoparticles in the 50–200 nm size range with near-neutral to slightly negative surface charge and PEG corona to minimize mononuclear phagocyte system (MPS) clearance. (Iminova et al., 2026)

Regulatory Pathways and Clinical Trial Design

The regulatory classification of CRISPR-nanofiber theranostic platforms is complex, as they simultaneously implicate gene therapy and drug-device nanomedicine. The FDA's Center for Biologics Evaluation and Research (CBER), Center for Devices and Radiological Health (CDRH), and Center for Drug Evaluation and Research (CDER) regulate the application for CRISPR-nanofiber therapeutics. Future research needs to minimize genotoxicity (per ICH S2 guidelines), carcinogenicity potential, reproductive toxicity, and the long-term biodistribution of nanofiber degradation products. Clinical trial omics profiling panels select patients most likely to benefit from epigenomic editing. PMA or 510(k) regulatory approval creates a logistically complex co-development pathway in early-phase clinical trial designs. Upcoming platforms will likely employ adaptive seamless Phase I/II designs with expansion cohorts stratified by omics-defined tumor epigenomic subtypes.

Table 8. Applications of Omics-Nano-CRISPR Pipeline: From Biopsy to Oncotherapy

Stage	Technology	AI Wearable Input	Output
Molecular Profiling	scRNA-seq + scATAC-seq + spatial omics (Xenium/MERFISH) + clock reporter biopsy organoid	Wearable DLMO estimation provides approximate phase prior for organoid entrainment timing	Personalized omics atlas + circadian phenotype (τ, phase, amplitude, BMAL1 methylation status)
Vulnerability Identification	Multi-omics integration (MOFA+, DIABLO), clock-gene co-expression network (WGCNA)	AI (transformer) maps wearable time-series to organoid ZT-phase alignment	Ranked list of CRISPR targets with ZT-priority windows, ABCB1 trough, MEX3A peak, BMAL1 nadir
CRISPR Design Validation	Off-target prediction (CRISPOR, CHOPCHOP), base editor efficiency modeling, chrono-organoid editing trial	Neural ODE model predicts editing efficiency vs. clock phase from organoid training data.	Lead CRISPR modality + sgRNA + optimal ZT delivery window per patient
Nanofiber Fabrication	Clock-responsive nanofiber selected per TME/ life-stage profile, payload loaded (RNP or mRNA + sgRNA)	RL agent confirms nanofiber release lag calibration with wearable phase estimate.	Patient-specific nanofiber depot with pre-programmed circadian release trigger
Therapeutic Delivery	Intratumoral/intravenous / implanted nanofiber depot, wearable-gated NIR trigger if optogenetic fiber	Wearable streams real-time phase, RL agent adjusts NIR pulse timing or IV infusion start	First CRISPR-nanofiber dose delivered at patient-specific optimal ZT with <30 min precision
Monitoring Adaptation	Liquid biopsy ctDNA (editing confirmation), SASP/CHIP panel, organoid re-profiling at cycle 3	LSTM model ingests updated wearable + ctDNA data, re-optimizes subsequent dosing ZT windows	Adaptive chronotherapy schedule updated every treatment cycle, therapy resistance detected early.

FUTURE DIRECTION

Spatial multi-omics data with three-dimensional nanofibers have been employed as implantable local drug delivery systems at tumor resection sites. (Table 8) The Gliadel wafer, an FDA-approved carmustine-loaded polymer implant for glioblastoma, or chitosan and hyaluronic acid nanofibers, exploit CD44 receptor overexpression on cancer stem cells for active targeting. At the same time, coaxial electrospinning enables core-shell architectures that provide spatially and temporally controlled sequential release of distinct therapeutic agents. (B. Zhao et al., 2026) The most advanced working prototypes today include:

- CRISPR-Graphene FET chips that provide a digital yes/no result for 16 common cancer mutations in 12 minutes, with no sample preparation
- Plasmonic-CRISPR tests that produce a visible red/blue color change detectable by a smartphone camera, suitable for point of care screening in any setting
- Quantum dot multiplexed assays that can simultaneously measure 20 different protein and nucleic acid biomarkers, and predict the primary origin of an unknown tumor

Figure 3. Multiomics, Nanomedicine, and CRISPR in Precision Oncology Applications

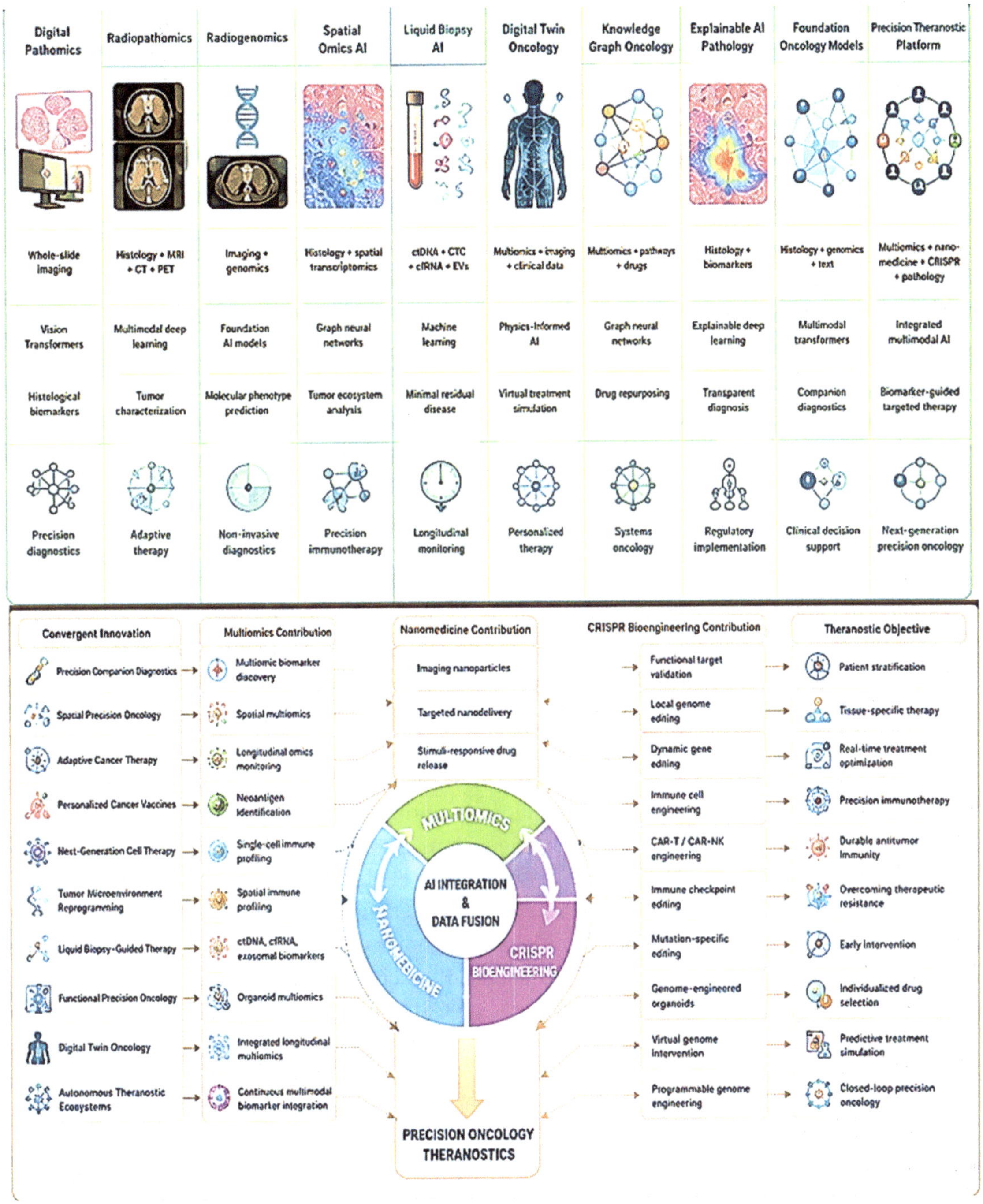

Artificial intelligence-supported histological and molecular data integration is evolving toward theranostic pathology. Nano-biosensors incorporating quantum dots, gold nanoparticles, graphene-related materials, magnetic nanoparticles, and plasmonic nanoparticles platforms for early cancer diagnosis.

allow tissue morphology, molecular alterations, spatial biology. (Figure 3) Therapeutic biomarkers are integrated into unified computational platforms for precision cancer management. Multi-omics technologies provide the analytical resolution required to characterize developmental epigenetic landscapes and identify clinically relevant environmental toxicities and parental factors. The evidence presented demonstrates that developmentally programmed epigenetic alterations constitute foundational oncogenic events that operate across the full spectrum of tumor biology, from pre-malignant benign lesions through low- and high-grade malignancies to metastatic dissemination. (Ramesh et al., 2026) The One Health functionalities aim to promote wellness within nanofiber-based theranostics and epigenomic disease for all, closing the loop between tumor characterization and therapeutic intervention. (J et al., 2026; Jangama et al., 2026) Translational challenges in safety, delivery efficiency, immunogenicity, and regulatory complexity are tractable. Ethical, legal, and informed preclinical evidence base depends on multidisciplinary AI, materials science, and synthetic biology. Similar to emerging human oncology, digital pathology is transitioning from static diagnosis to dynamic prediction of therapeutic response, resistance mechanisms, minimal residual disease, and longitudinal treatment monitoring. Whole-slide imaging (WSI) is becoming the digital backbone for integrating histopathology with genomics, transcriptomics, proteomics, metabolomics, radiomics, liquid biopsy, and clinical metadata through multimodal deep learning architectures.

Computational pathology is increasingly incorporating multimodal foundation models capable of learning generalized histomorphological representations across diverse cancer types. (Table 9) Models integrate histological images with genomic alterations, RNA expression profiles, immune signatures, and clinical outcomes to identify molecular subtypes directly from routine hematoxylin and eosin sections. Current studies have demonstrated prediction of microsatellite instability (MSI), tumor mutational burden (TMB), homologous recombination deficiency (HRD), EGFR mutations, KRAS mutations, BRAF alterations, TP53 status, IDH mutations, and HER2 expression from digital pathology, providing potential companion diagnostics without immediate molecular testing. It will enable the design of spatially precise nanofiber delivery strategies that differentially target distinct tumor compartments. Stem cell niches, invasion fronts, and immune-excluded zones require appropriate CRISPR epigenomic editors. Artificial intelligence, deep learning, and machine learning algorithms, as well as graph neural networks, are trained on multi-omics cancer datasets (TCGA, ICGC, GEO). (Lee et al., 2026)Transformer architectures will accelerate the identification of developmentally programmed epigenomic targets and the design of sgRNA sequences that maximize therapeutic specificity. The choice of CRISPR modality for the oncogenic lesion, the developmental stage of the target tissue, and the circadian phase-sensitivity of the editing machinery itself. (Ventelou et al., 2026) rDSB-inducing Cas9 nuclease editing carries genotoxic risk in rapidly proliferating neonatal and pediatric tissues with high replicative stress, base editing and prime editing, which operate without double-strand breaks, offer substantially improved safety profiles in these contexts. Geriatric cancers with global epigenetic dysregulation are targeted with dCas9-based epigenome editors, TET1 demethylases, and DNMT3A methyltransferases rather than sequence-altering nucleases.

In vivo base editing via lipid nanoparticle (LNP) systemic delivery: Ongoing clinical trials using BEAM-101 for sickle cell disease, VERVE-101 for PCSK9 editing are establishing the safety and efficacy of systemically delivered LNP-base editor systems. Incorporating clock-responsive polymer shells into LNPs. LNP hybrid nano-CRISPR will extend clock-gated oncology indications without requiring implanted depot fibers.(Vu et al., 2026)

Spatial CRISPR: Combining Visium/Xenium spatial transcriptomics with CRISPR functional screens in chrono-organoids will enable spatial-circadian dissection of which tumor subregions (core vs. invasive front vs. immune-excluded border) are clock-competent and therefore CRISPR-accessible at which ZT windows. The spatially targeted, temporally gated CRISPR delivery matches tumor heterogeneity at single-cell and single-voxel resolution.(Pan et al., 2026)

Table 9. Integrated AI-Driven Oncology Theranostic Platforms

Integrated Platform	Data Components	AI Methodology	Clinical Output
Digital Pathomics	Whole-slide imaging	Vision Transformers	Histological biomarkers
Radiopathomics	Histology + MRI + CT + PET	Multimodal deep learning	Tumor characterization
Radiogenomics	Imaging + genomics	Foundation AI models	Molecular phenotype prediction
Spatial Omics AI	Histology + spatial transcriptomics	Graph neural networks	Tumor ecosystem analysis
Liquid Biopsy AI	ctDNA + CTC + cfRNA + EVs	Machine learning	Minimal residual disease
Digital Twin Oncology	Multiomics + imaging + clinical data	Physics-informed AI	Virtual treatment simulation
Knowledge Graph Oncology	Multiomics + pathways + drugs	Graph neural networks	Drug repurposing
Explainable AI Pathology	Histology + biomarkers	Explainable deep learning	Transparent diagnosis
Foundation Oncology Models	Histology + genomics + text	Multimodal transformers	Companion diagnostics
Precision Theranostic Platform	Multiomics + nanomedicine + CRISPR + pathology	Integrated multimodal AI	Biomarker-guided targeted therapy

Synthetic biology clock circuits: CRISPR-mediated insertion of synthetic clock gene circuits. The oscillating gene networks generate user-programmable ~24 h oscillations independent of endogenous clock disruption. The tumor cells could provide autonomous intracellular ZT signals that gate CRISPR payload or sgRNA expression under a synthetic clock promoter, independent of the intact endogenous clock, in advanced geriatric or treatment-refractory cancers. (Iyer et al., 2026)

Organoid-on-chip chrono-CRISPR screening: Microfluidic chips combining entrained chrono-organoids with integrated electrochemical biosensors for ZT monitoring, coupled to automated nanofiber payload delivery. The programmable ZT windows will enable fully automated, high-throughput ZT-resolved CRISPR screens across patient-derived organoid biobanks. (Zhou et al., 2026)

Epigenetic clock reversal: BMAL1 CRISPRa + RORα CRISPRa multiplex clock restoration decelerates GrimAge/PhenoAge methylation clocks in clinical pilot studies. (Han et al., 2025) The framework becomes applicable to preventing age-associated oncogenesis in high-risk geriatric populations. (B. Zhao et al., 2026)

Liquid biopsy: Multi-omics ctDNA methylation profiling, circulating miRNA, extracellular vesicle proteomics, and cell-free RNA with nanofiber-delivered CRISPR therapy will create epigenomic editing pharmacodynamics and therapeutic response. Synthetic biology toggle switches that maintain durable silencing or activation states through self-reinforcing chromatin circuits. (Ventelou et al., 2026)

Nanofiber-based contrast agents and biosensors have been engineered for MRI, fluorescence, photoacoustic, and Raman imaging. The high surface area and tunable chemical sensitivity and specificity in CRISPR-based diagnostic platforms, such as SHERLOCK, SHERLOCKv2, DETECTR, and CRISPR/Cas13a-gold nanoarrays, enable nucleic acid methods. Rapid, programmable, and ultrasensitive detection of tumor biomarkers supports multiplexed detection of tumor-associated nucleic acids, proteins, and metabolites, enabling early cancer detection, grading, and treatment monitoring. (Sherlock et al., 2023) Organoid and patient-derived xenograft (PDX) models will use biobanked multi-omics data and a CRISPR-nanofiber preclinical epigenomic editing strategy. Selection of strategies and collaborative consortia, International Cancer Genome Consortium (ICGC) will be essential to generate the scale of multi-omics data required to train and validate AI models for epigenomic vulnerability prediction. (Zhou et al., 2026) Theranostic platforms at the omics-nano-CRISPR interface are designed to co-deliver imaging agents and therapeutic payloads. Electrospun and coaxial nanofibers, CRISPR-Cas9 RNPs, siRNAs, chemotherapeutics, and imaging agents enable controlled, stimuli-responsive release triggered by pH, redox, NIR, or ROS. (Hamato et al., 2026) Omics-guided personalization, as exemplified by the DSCN platform and MSC membrane-coated nanofibrils, allows for individualized therapy design based on patient-specific molecular profiles. (Rajput et al., 2026)

Finally, the emerging field of *in vivo* base edits, or base editor constructs, is delivered via nanofibers to correct founding oncogenic mutations in situ within primary tumors directly. The convergence of advanced nanomaterials, including quantum dots, gold nanoparticles, graphene-related materials, magnetic nanoparticles, and plasmonic nanostructures, with the revolutionary CRISPR-based molecular diagnostic technologies represents a paradigm shift in early cancer detection. These integrated platforms offer extraordinary sensitivity, specificity, speed, and versatility that far surpass conventional diagnostic approaches. By enabling the detection of cancer biomarkers at ultralow concentrations from minimally invasive liquid biopsies, these technologies hold immense promise for transforming cancer diagnostics toward earlier detection, improved patient stratification, real-time treatment monitoring, and ultimately, significantly improved cancer survival rates. The curative intent ranges from surgical resection to pharmacological epigenomic normalization and is likely stage- and indication-specific. The most promising studies will likely integrate spatial omics, responsive nanofiber scaffolds, and programmable genome editing into a single translational pipeline.(Shokoya et al., 2026) CRISPR still faces safety, off-target, and delivery hurdles, and nanofiber systems can be difficult to standardize for scale-up, reproducibility, and regulatory approval.(Yavuz et al., 2026) Multi-omics adds further complexity in data integration, interpretation, privacy, and health equity.

CONCLUSION

The convergence of multi-omics, nanofiber engineering, and CRISPR-based genome editing has transformed translational oncology from a histopathological discipline into a systems-level theranostics. Modern oncology increasingly interprets cancer as a dynamic ecological process involving developmental programming, metabolic plasticity, immune remodeling, epigenetic drift, and age-associated genomic instability. Electrospun biomimetic nanofibers serve as critical interfaces that link omics-informed diagnostics with CRISPR-enabled therapeutics and theranostic imaging systems.

REFERENCES

Abo Qoura, L., Churov, A. V., Maltseva, O. N., Eruslanova, K. A., & Zeng, G. (2026). Senescent cells in systemic aging: SASP heterogeneity, immune escape, and endocrine modulation. *Biogerontology, 27*(3), 103. DOI: 10.1007/s10522-026-10449-6 PMID: 42113304

Bai, S., Li, W., Yin, N., Cao, Y., Zhang, S., Wang, Z., Qu, D., Wang, B., Li, Y., Liu, X., Jiang, Y., & Wang, Y. (2026). Bi-Target Nanomodulator for Glioma-Associated Myeloid Cells Polarization by Inducing Pyroptosis and Regulating Tryptophan Metabolism to Reshape Immunosuppressive Microenvironment of IDH-Mutant Gliomas. *Advanced Materials, 23021*(22), e23021. Advance online publication. DOI: 10.1002/adma.202523021 PMID: 41863418

Banerjee, S., Mariya, A., Vangapally, S., & Paidimuddala, B. (2026). Nanobody-drug conjugates as versatile tools for improving therapeutic potential. *Biochemical Engineering Journal, 228*, 110072. DOI: 10.1016/j.bej.2026.110072

Byalahunasishetter, B. I., Chowdareddy, S., & Raghavan, R. (2026). Cancer Risk Prediction Based on Gut Microbiota Analysis. *Smart Microbiomes for Personalized Medicine*, 307–342. DOI: 10.4018/979-8-3373-8172-5.ch009

Caf, M., Esmaeilnejad-Ahranjani, P., Kolosnjaj-Tabi, J., Sabotič, J., Berlec, A., Zaveršek, N., Pajk, S., Zahirović, A., Golzio, M., Milosevic, I., & Kralj, S. (2026). Magnetic Field-Driven Strategies for Biofilm Disruption: From Iron Oxide Nanoparticles to Adaptive Swarms of Magnetic Microrobots. *ACS Nano, 20*(1), 34–58. DOI: 10.1021/acsnano.5c14390 PMID: 41477707

Duarah, R., Torné-Morató, H., Zhang, G., Amin, Y., Pabast, M., Sharma, N., He, S., Das, M. R., & Pompa, P. P. (2026). Next-Generation Wearable Optical Sensors for Personalized Health and Point-of-Care Diagnostics—A Systematic Review. *Advanced Healthcare Materials, 15*(10), e04419. DOI: 10.1002/adhm.202504419 PMID: 41486722

Fareed, M. A., Zhang, B., Memon, K., Hao, Y., Liu, Z., & Zhao, G. (2026). Recent advances in electrospun nanofiber-based strategies for diabetic wound healing application. *Materials Today. Chemistry, 51*, 103259. DOI: 10.1016/j.mtchem.2025.103259

Feng, L., Yu, P., He, N., Zhang, Q., Tang, Y., Geng, J., Lu, Q., Song, C., & Chen, F. (2026). Label-Free MicroRNA Diagnostics: From CRISPR Nucleases to Nanomaterial-Enhanced Transducers. *ACS Synthetic Biology, 15*(3), 915–954. DOI: 10.1021/acssynbio.5c00868 PMID: 41709870

Fiorentino, V., Giordano, W., Pizzimenti, C., Zuccalà, V., Ieni, A., Molinario, C., Cannavò, S., Campennì, A., Tralongo, P., Martini, M., Giuffrè, G., Larocca, L. M., Fadda, G., & Rossi, E. D. (2026). Molecular profiling of thyroid nodules on cytologic samples: Findings from an Italian multi-institutional cohort. *Cancer Cytopathology, 134*(1), e70065. DOI: 10.1002/cncy.70065 PMID: 41370117

Freedman, B. S., Bulte, J. W. M., Conklin, B. R., Judge, L. M., Dwinell, M. R., Geurts, A. M., Sitton, M. J., Mahajan, V., Kiani, S., Gersbach, C. A., Ebrahimkhani, M. R., Kelly, J. J., Ronald, J. A., Morizane, R., Gupta, N., Shakeri-Zadeh, A., Vo, N., Saha, K., Saxena, S., & Tsai, S. Q. (2026). Monitoring biological effects of somatic cell genome editing. *Nature Reviews. Genetics, 27*(4), 1–20. DOI: 10.1038/s41576-025-00916-0 PMID: 41530266

Giercuszkiewicz-Haśnik, K., Morak-Młodawska, B., & Jeleń, M. (2026). Molecularly Targeted Therapies in Oncology: Mechanisms, Resistance, and Combination Strategies. *Molecules (Basel, Switzerland)*, *31*(7), 1195. DOI: 10.3390/molecules31071195 PMID: 41976235

Hamato, A., Lee, D., Ota, R., Hemmati, H., Muller, M., Wang, Y., Pandian, L. S., Kim, S., Yamaya, T., Cherry, S. R., Glodo, J., & Kwon, S. I. (2026). Ultra-dense lutetium oxide ceramic scintillators for positron emission tomography. *Physics in Medicine and Biology*, *71*(4), 045014. DOI: 10.1088/1361-6560/ae456f PMID: 41678960

Han, J., Xiong, Z., Liu, Y., Li, J., Li, C., Zhang, W., Gao, J., Shuai, X., Wang, Z., Wang, G., Shu, X., Tao, K., & Cai, M. (2025). Circadian rhythm in gastrointestinal cancer: Clinical applications and future perspectives. *Annals of Medicine*, *57*(1), 2529571. DOI: 10.1080/07853890.2025.2529571 PMID: 40616824

He, Y., Tu, X., Xue, Y., Chen, Y., Ye, B., Li, X., Li, D., Zhong, Z., & Zhong, Q. (2025). CRISPR screening redefines therapeutic target identification and drug discovery with precision and scalability. *Journal of Pharmaceutical Analysis*, *16*(2), 101357. DOI: 10.1016/j.jpha.2025.101357 PMID: 41732321

Iminova, R., Berganayeva, G., Zhurtbayeva, A., Abdurakhmanova, L., Almabekova, A., Shepilov, D., Vassilina, G., Nurmahanova, A., Kairalapova, G., & Dyusebaeva, M. (2026). From PEGylation to Next-Generation Polymers: Overcoming Biological Barriers-A Review. *Molecules (Basel, Switzerland)*, *31*(4), 675–675. DOI: 10.3390/molecules31040675 PMID: 41752452

Iyer, K. A., Koynova-Tenchov, R., Sasso, J. M., Thite, T., Deng, Y., & Zhou, Q. A. (2026). The Future of Epigenetics: Emerging Technologies and Clinical Applications. *ACS Pharmacology & Translational Science*, *9*(3), 506–544. DOI: 10.1021/acsptsci.5c00729 PMID: 41852637

J, S. P., Moger, S. S., Byalahunasishetter, B. I., & Raghavan, R. (2026). The One Health Framework and Antimicrobial Resistance: Role of Herbal Medicine. *Herbal Medicine Approaches to Antimicrobial Resistance and One Health*, 1–36. DOI: 10.4018/979-8-3373-7810-7.ch001

Jangama, R., Suresh, C., Korrapati, C., & Raghavan, R. (2026). Hybrid Coating Systems: Bioactive Molecules and Nanoparticles for Cancer Therapy. *Biomedical Applications of Laser-Processed Coatings*, 99–134. DOI: 10.4018/979-8-3373-4104-0.ch003

Kanaujia, K. A., Yadav, V. K., Yadav, S. S., Talha, M., Saraf, S. A., & Kumar, S. (2026). 4D Printing in Healthcare: Innovations, Challenges, and Future Directions. *ACS Applied Bio Materials*, *9*(2), 493–528. DOI: 10.1021/acsabm.5c01165 PMID: 41474416

Kaźmierczak-Siedlecka, K., Kucharski, R., Stachowska, E., Pelikant-Małecka, I., Marano, L., Makarewicz, W., Kalinowska, M., Dambrauskas, Ž., & Kalinowski, L. (2026). Intra-tumoural microenvironment and bugs-based drug design: Foreseeable future in oncology and immuno-oncology. *Frontiers in Pharmacology*, *17*, 1732712. Advance online publication. DOI: 10.3389/fphar.2026.1732712 PMID: 41693784

Ke, X., & Cardoso, W. V. (2026). Transitional cell states at the crossroads of development, disease, and repair-regeneration. *Physiological Reviews*, *106*(3), 2059–2078. DOI: 10.1152/physrev.00041.2025 PMID: 42012134

Korrapati, C., Chowdareddy, S., Suresh, C., & Raghavan, R. (2026). Cancer Cell Membrane-Coated Nanoparticles: Therapeutic Advancements and Hybrid Systems. *Applications of Nanomaterials for Oncology*, 131–160. DOI: 10.4018/979-8-3373-3765-4.ch005

Lakshmaiah, A., & Raghavan, R. (2026). Deep Learning-Enhanced Hyperspectral Imaging: Applications in Oncology Cancer Diagnosis and Tumor Margin Identification. *Biomedical Applications in Deep Learning-Enhanced Hyperspectral Imaging*, 335–368. DOI: 10.4018/979-8-3373-7183-2.ch011

Lee, H., Menezes, V., Zeng, S., Kim, C., Baseman, C. M., Kim, J. H., Padmanabhan, S., Premdas, P., Djeddar, N., Bryksin, A., Pandey, N., Anastasiadis, P., Kim, A. J., MacDonald, T. J., Bettegowda, C., Woodworth, G. F., Herrmann, F. J., & Arvanitis, C. (2026). Data-Driven Feedback Identifies Focused Ultrasound Exposure Regimens for Improved Nanotheranostic Targeting of the Brain. *Advanced Science (Weinheim, Baden-Württemberg, Germany)*, *13*(13), e17834. DOI: 10.1002/advs.202517834 PMID: 41498583

Li, G., Xu, J., Tian, X., Xiao, J., Long, J., Chen, Y., Shen, W., & Zhao, S. (2026). Intrapatient variation in PD-L1 expression and tumor mutational burden and implications for outcomes to immune checkpoint inhibitor therapy in patients with non-small-cell lung cancer. *Journal for Immunotherapy of Cancer*, *14*(1), e013328. Advance online publication. DOI: 10.1136/jitc-2025-013328 PMID: 41571296

Li, N., Xiao, L., Li, S., Zhang, L., & Chen, S. (2026). β-Galactosidase-Activated Near-Infrared Fluorescence and 1H/19F MRI Trimodal Probe for Real-Time In Vivo Imaging of Ovarian Cancer. *Analytical Chemistry*, *98*(7), 5261–5270. DOI: 10.1021/acs.analchem.5c05864 PMID: 41672934

Li, Z., Du, L., Du, B., Ullah, Z., Zhang, Y., Tu, Y., Zhou, Y., & Guo, B. (2025). Inorganic and hybrid nanomaterials for NIR-II fluorescence imaging-guided therapy of Glioblastoma and perspectives. *Theranostics*, *15*(12), 5616–5665. DOI: 10.7150/thno.112204 PMID: 40365286

M, M. K., P, D. A., J, S. P., Moger, S. S., & Raghavan, R. (2026). Spatial Multi-Omics in Cancers: Integrating Molecular Landscapes for Precision Oncology. *Spatially Resolved Single-Cell and Multiple Omics in Human Diseases*, 217–248. DOI: 10.4018/979-8-3373-4849-0.ch007

Majeed, H., Irfan, M., Irfan, M., & Iftikhar, T. (2026). Intelligent multi omics and industry 6.0 and 7.0 enabled technologies for deciphering the tumor immune microenvironment and advancing cancer immunotherapy. *Toxicology Research*, *15*(1), tfag002. Advance online publication. DOI: 10.1093/toxres/tfag002 PMID: 41623584

Nazar Saeed, T., Al-Hussainy, A. F., Sanghvi, G., Ballal, S., Singh, A., Sudhakar, T., Mishra, S., Kubaev, A., Taher, S. G., Alwan, M., Jawad, M., & Mushtaq, H. (2026). Tunable silk fibroin platforms for controlled drug delivery: A biocompatible approach. *Journal of Drug Delivery Science and Technology*, *115*, 107649. DOI: 10.1016/j.jddst.2025.107649

Neves, A. R., Mineiro, R., Quintela, T., & Costa, D. (2026). The Potential of Chronotherapy and Nanotherapy-Based Strategies for Glioblastoma Treatment. *Pharmaceutics*, *18*(2), 235. DOI: 10.3390/pharmaceutics18020235 PMID: 41754976

Nie, C., Shi, Z., Song, Z., Wang, F., & Min, J. (2026). Artificial intelligence and multiomics for longevity across species. *Element*, *1*(1), n/a, n/a. DOI: 10.20517/element.2026.01

Pallavi, V. N., Chowdareddy, S., Byalahunasishetter, B. I., & Raghavan, R. (2026). CRISPR-Cas9 Technology for Precision Cancer Medicine: Target Discovery, Disease Modeling, and Therapeutics. *CRISPR-Cas9 Technology for Precision Medicine*, 35–72. DOI: 10.4018/979-8-3373-4857-5.ch002

Pan, J., Li, B., Wang, Y., Han, Y., Liu, G., & Sun, S.-K. (2026). Physical stimuli-responsive CRISPR-Cas9 systems for spatiotemporally precise control of genome engineering. *Theranostics, 16*(6), 2984–3010. DOI: 10.7150/thno.122238 PMID: 41510162

Pandey, G. A., Kashikar, P. M., Nathani, K., Mangrulkar, S., Sawarkar, S. P., & Omri, A. (2026). Biomimetic Scaffolds and Extracellular Matrix-Based Strategies for Myofiber Regeneration in Volumetric Muscle Loss. *Drug Design, Development and Therapy*, *20*(0), 544862. DOI: 10.2147/DDDT.S544862 PMID: 41924553

Rajput, S., Khandia, R., Gaur, T., Anant, A., Kamal, M. A., Gurjar, P., Al-Hussain, S. A., & Zaki, M. E. A. (2026). Advancements in nanobiosensors for early cancer detection, challenges, treatment, and future prospects: A comprehensive review. *Discover Nano*, *21*(1), 83. DOI: 10.1186/s11671-026-04478-w PMID: 41874885

Ramesh, R. N., Palled, S. B., Shivaraju, A., & Raghavan, R. (2026). Reproductive Toxicities and Transgenerational Cancer Risk From Climate-Related Occupational Hazards. *One Health Perspectives on Climate Change and Emerging Occupational Health Risks*, 191–230. DOI: 10.4018/979-8-3373-8470-2.ch007

Sandoval, S., Gonçalves, G., Barrio, J. P., Kharlamova, M. V., & Tobías-Rossell, G. (2026). A Comprehensive Review on Filled Carbon Nanotubes: Synthesis, Properties and Applications. *Chemical Reviews*, *126*(4), 2283–2390. DOI: 10.1021/acs.chemrev.5c00219 PMID: 41636552

Sherlock, L., Martin, B. R., Behsangar, S., & Mok, K. H. (2023). Application of novel AI-based algorithms to biobank data: Uncovering of new features and linear relationships. *Frontiers in Medicine*, *10*, 1162808. DOI: 10.3389/fmed.2023.1162808 PMID: 37521348

Shi, P., Ren, C., & Tong, H. (2026). Research progress of stem cells in the treatment of atherosclerosis. *Frontiers in Cell and Developmental Biology*, *13*, 1722416. DOI: 10.3389/fcell.2025.1722416 PMID: 41552013

Shokoya, M. M., Szabó, E., Nagy, Z. K., Sebe, I., & Zelkó, R. (2026). Therapeutic nanofibrous scaffolds in precision drug delivery: Formulation design, challenges and regulatory frameworks. *Expert Opinion on Drug Delivery*, *23*(5), 823–839. DOI: 10.1080/17425247.2026.2619643 PMID: 41549842

T A, A., Vallazhath, A., Maniyal, V., & Menon, D. (. (2026). Advances in therapeutic strategies for atherosclerosis: From pharmacologics to stents and stent coatings. *Journal of Materials Chemistry. B*, *14*(8), 2407–2437. DOI: 10.1039/D5TB02293B PMID: 41668608

Tomljenovic-Hanic, S., & Khalid, A. (2026). Silk Fibroin for Biomedical Applications with Emphasis on Bioimaging, Biosensing and Regenerative Systems: A Review. *Molecules (Basel, Switzerland)*, *31*(7), 1142. DOI: 10.3390/molecules31071142 PMID: 41976184

Vashani, D., Paul, H., Nguyen, S., Topkiran, U. C., Valimukhametova, A. R., Dorsky, A., Sottile, O., Durmaz, L., Gonzalez-Rodriguez, R., & Naumov, A. V. (2026). Precursor-dependent optical and structural properties of eleven NIR-emissive graphene quantum dots for bioimaging applications. *2D Materials, 13*(2), 025017. DOI: 10.1088/2053-1583/ae4e41

Ventelou, L., Ginzac, A., Ferreira, M.-C., Canetti, L., Philippe, S., Pinard, C., Durando, X., & Bernadach, M. (2026). Circulating tumour DNA in head and neck squamous-cell carcinomas: A literature review. *Cancer Treatment Reviews, 142*, 103070. Advance online publication. DOI: 10.1016/j.ctrv.2025.103070 PMID: 41435530

Vu, S., Ebah, K., Zaibi, F., Qaiss, A., Siaj, M., & Izquierdo, R. (2026). Wet-Spun Graphene-Enhanced PVDF Fibers for Flexible Nanocomposites. *Materials (Basel), 19*(7), 1376. DOI: 10.3390/ma19071376 PMID: 41976663

Wadan, A.-H. S., Ali, M. H., & Ellakwa, D. E.-S. (2026). Sustainable nanomaterials for precision dental medicine: Green synthesis, therapeutic applications, and future directions. *Journal of Nanobiotechnology, 24*(1), 190. Advance online publication. DOI: 10.1186/s12951-025-04008-3 PMID: 41652613

Xiu, X., Wu, Y., Li, J., Lin, D., Sun, X., Su, X., Weng, Z., Zuo, X., Yang, X., Fan, C., Wang, Y., Zhang, D. Y., & Song, P. (2026). Anchored random reverse primer sequencing for quantitative detection of novel gene fusions. *Nature Biomedical Engineering*, •••, 1–16. DOI: 10.1038/s41551-025-01564-9 PMID: 41482592

Yavuz, Y., Kartal, I., Cesur, S., Kanli, Z., Kaya, E., Tinaz, G., & Gunduz, O. (2026). Electrospun PVA/CS/HA/BA Nanofiber Scaffolds with Enhanced Mechanical Stability and Antifungal Activity for Bone Tissue Engineering. *Materials (Basel), 19*(2), 412. DOI: 10.3390/ma19020412 PMID: 41598123

Zang, Z., Chen, J., Dong, Y., Chen, L., Yang, M., Mu, M., Zhou, L., Zhang, W., Zou, G., & Liu, C. (2026). CRISPR-Cas12a biosensing technology advances and applications in precision diagnostics and cancer research. *Talanta, 301*, 129298. DOI: 10.1016/j.talanta.2025.129298 PMID: 41500123

Zhang, Z., Tian, W., Gu, J., Li, J., Liu, M., Soleymani, L., & Li, Y. (2026). A general strategy to enhance aptamer affinity by suppressing dissociation through symmetric assembly. *Nucleic Acids Research, 54*(6), gkag268. Advance online publication. DOI: 10.1093/nar/gkag268 PMID: 41873762

Zhao, B., Tian, Y., Yu, M., Han, J., Chen, Z., Zhang, J., Liu, J., Geng, L., & Lei, Y. (2026). The Protective Effect of N, Ce-doped Carbon Dots Against H2O2-Induced Oxidative Damage in MOVAS Cells. *Applied Biochemistry and Biotechnology, 198*(3), 1855–1868. DOI: 10.1007/s12010-025-05540-6 PMID: 41504846

Zhao, Q., Yu, Y., Pu, C., Zheng, S., Chen, L., Zeng, F., Liu, L., & Li, D. (2026). Research progress on the spatiotemporal dynamics of therapy-induced senescence in remodeling the tumor microenvironment. *Frontiers in Immunology, 17*, 1727142. DOI: 10.3389/fimmu.2026.1727142 PMID: 41694369

Zhao, X., Liu, D., Li, G., Xu, W., & Wu, G. (2026). Nanomedicines Reshape the Tumor Microenvironment: Multidimensional Strategies from Modulating "Barriers" to Metabolic Intervention. *International Journal of Nanomedicine, 21*(0), 1–24. DOI: 10.2147/IJN.S570411 PMID: 41858588

Zhou, J., Abasolo, S. A., Yang, Y., Guo, F., Xing, M., & Zhu, D. (2026). Biomimetic Woven Bone Organoids with Multifocal Centers for Retaining Stemness and Prolonging Osteogenic Activity. *Advanced Functional Materials*, *36*(36), e27855. Advance online publication. DOI: 10.1002/adfm.202527855

Zhu, G., He, J., Shi, N., Cai, Z., Zhang, J., Wang, S. C. Y., Li, J., Zhang, M., & Wang, Y. (2026). Prognostic, chemotherapy and immunotherapy roles of GPR37/GPR37L1 in pan-cancer. *Medicine*, *105*(10), e47813. DOI: 10.1097/MD.0000000000047813 PMID: 41790623

Chapter 12
Nanoparticle-Mediated Gene Editing and Immunotherapy in Liver and Prostate Cancers

S. Mayilazhagan
http://orcid.org/0009-0007-4750-6905
Department of Microbiology, Sacred Heart College, Tirupattur, India

S. Dayana
http://orcid.org/0009-0008-7135-5606
Department of Microbiology, Sacred Heart College, Tirupattur, India

T. Elavarasi
http://orcid.org/0009-0003-0792-4484
Department of Microbiology, Sacred Heart College, Tirupattur, India

M. Manigandan
http://orcid.org/0000-0002-8115-2713
Department of Microbiology, Sacred Heart College, Tirupattur, India

K. Gayathri
http://orcid.org/0009-0007-3786-2009
Department of Microbiology, Sacred Heart College, Tirupattur, India

ABSTRACT

Nanoparticle-mediated gene editing and immunotherapy have emerged as transformative approaches for the treatment of liver and prostate cancers, addressing limitations associated with conventional therapies such as chemotherapy, radiotherapy, and surgical interventions. The integration of nanotechnology with advanced gene-editing tools, particularly CRISPR-Cas systems, enables precise modification of cancer-associated genes while minimizing off-target effects. Nanoparticles serve as efficient delivery vehicles for gene-editing components, therapeutic nucleic acids, and immunomodulatory agents, enhancing cellular uptake, target specificity, and therapeutic efficacy. In liver cancer, nanoparticle-based platforms facilitate targeted disruption of oncogenic pathways, restoration of tumour suppressor genes,

DOI: 10.4018/979-8-3373-9715-3.ch012

and modulation of the tumour microenvironment to suppress tumour progression. Similarly, in prostate cancer, nanoparticle-assisted gene editing enables the regulation of androgen receptor signalling and other molecular drivers responsible for tumour growth and metastasis.

1. INTRODUCTION

Cancer remains a huge disease burden and cause of death globally; particularly liver and prostate cancers account for a significant burden of cancer. The majority of liver cancers are primary liver cancer (Hepatocellular Carcinoma (HCC)) often diagnosed at late stages and characterized by heterogeneous tumours, high recurrence and limited treatment options, which result in poor prognosis (Davodabadi et al., 2023). Likewise, prostate cancer is the second most prevalent cancer in the world in men and at its most advanced stages can transform into the less susceptible Castration-Resistant Prostate Cancer (CRPC) and have poor clinical outcomes. Surgery, chemotherapy, radiotherapy, targeted therapy and hormone therapy have enhanced patient survival, but therapeutic resistance, systemic toxicity, off-target and immune evasion still hinder long-term treatment efficacy (Emran et al., 2022). Precision oncology is a transformative approach in recent advances in molecular medicine, which has been established through gene editing and cancer immunotherapy. The CRISPR-Cas9 system is unique due to its simplicity, high editing efficiency, programmability and capacity to make highly precise genetic modifications, and has garnered considerable interest among genome-editing technologies (Boehnke et al., 2022). CRISPR-Cas9 offers the potential to disrupt oncogenes, restore tumour suppressor genes, correct disease-associated mutations and modulate immune-related pathways, which could open new approaches to liver and prostate cancer therapy. But delivering the CRISPR components into cells efficiently and safely continues to be a big hurdle, since naked nucleic acids are easily degraded by enzymes, have low cell-uptake and rapidly clear from circulation (Sun et al., 2023).

Nanotechnology has become a very useful platform to penetrate these delivery hurdles. Nanoparticles, specifically Lipid Nano Particles (LNPs), have been shown to possess outstanding delivery properties for messenger RNA (mRNA), Target tissues with small interfering RNA (siRNA), plasmid DNA or CRISPR-Cas9 components (Liu et al., 2023). They are highly biocompatible, biodegradable, have a high loading capacity and shield the nucleic acids from degradation, which will significantly enhance therapeutic efficiency and decrease the systemic toxicity. To increase the selective accumulation in tumours, enhance cellular uptake and reduce off-target effects, the surface is modified with targeting ligands, antibodies, peptides or polymers. Lipid nanoparticle-based nucleic acid delivery systems have become successful in the clinical setting, thus boosting the use of these systems in cancer gene therapy and precision medicine (Yoo et al., 2024). At the same time, cancer immunotherapy has transformed Immune checkpoint inhibitors, cancer vaccines, cytokine therapies and adoptive cell therapies are the tools being used to bring the body's immune system into the arena of oncology, where it can target and kill cancer cells. However, the immunosuppressive tumour microenvironment, lack of immune cell infiltration, suboptimal delivery of immunotherapeutic drugs to the tumour and immune escape mechanisms often hinder the therapeutic success, especially in solid tumours like liver and prostate cancer (Ding et al., 2024). The delivery systems based on nanoparticles offer some of the best solutions: by controlled delivery of immunomodulatory molecules, improving antigen presentation, promoting immune-cell activation, they also allow delivery of checkpoint inhibitors and cytokines, while keeping their side effects in the local area. In addition, co-delivery of gene-editing tools along with immunotherapeutic agents by nanoparticles provide a synergistic approach that can correct oncogenic alterations and simultaneously boost the immune response against the tumour (Mitsogiannis et al., 2022). The combination of nanotechnology, genome editing and immunotherapy is a great leap forward in precision cancer treatment. Future advancements in the engineering of nanoparticles, their targeted delivery, design of biomaterials and personalized medicine should further increase the therapeutic specificity, safety and clinical translation. Such integrated strategies are attractive candidates to overcome therapeutic resistance, inhibit tumour expansion, minimize toxicity and enhance survival in patients with liver and prostate cancers (Meng et al., 2023).

2. OVERVIEW OF LIPID NANOPARTICLES

2.1. Lipid Nanoparticles (LNPs)

Lipid Nanoparticles (LNPs) are one of the most promising and powerful non-viral delivery platforms for nucleic acids, such as messenger RNA (mRNA), small interfering RNA (siRNA), plasmid DNA and CRISPR/Cas9. They are biocompatible, biodegradable and highly encapsulating, which makes them promising carriers for gene editing and cancer immunotherapies (Buschmann et al., 2021). The basic components of a typical LNP formulation are the ionizable lipids, phospholipids, cholesterol, and Poly-Ethylene Glycol (PEG)-lipids. Ionizable lipids are electrically neutral in physiological conditions, and consequently not toxic throughout the body, but become positively charged in acidic endosomal compartments and enable endosomal escape and release of the therapeutic cargo. Phospholipids maintain membrane integrity, while cholesterol provides stability and PEG-lipids minimize protein adsorption and quick clearance by MPS (Wong et al., 2022). For liver and prostate cancers, LNPs can be used to deliver a wide range of gene editing systems and therapeutics directly to the tumor tissue with CRISPR/Cas9 technology. They are also very small in size, allowing them to efficiently enter cells via endocytosis, and targeted cell binding occurs via ligand, antibody, or peptide modifications (Mohamed Isa et al., 2021). The highly fenestrated liver vasculature readily traps LNPs, facilitating efficient delivery of LNPs to the liver in HCC. In the case of prostate cancer, the ligand-functionalized LNPs are more tumor-specific, and it has been shown that they have fewer off-target effects. LNPs offer higher loading capacity of genetic material, lower immunogenicity, and no risk of insertional mutagenesis, as opposed to viral vectors. In terms of the importance of these properties, LNPs are a valuable platform for precision genome editing and targeted immunotherapy in the treatment of cancer patients (Murugesan & Raman, 2022).

2.2. Characterization and Manufacturing

There are several different ways to prepare lipid nanoparticles, of which microfluidic mixing is the most common. This technique has become popular because it is so easy to scale up, repeat and control operational parameters (Murugesan & Raman, 2022). The microfluidic mixing is achieved by co-injecting the lipids in an ethanolic (organic) phase and the nucleic acids in an aqueous (buffered) phase. These phases come together in microchannels where chaotic mixing takes place, allowing self-assembly and complexation of components to form LNPs. The parameters of the mixing process, such as the pH of the buffer, the L/NA ratio, and the flow rate affect the properties of LNPs (Guimarães et al., 2023). Characterization is critical for the reproducibility of LNPs and to predict their performance. Some of the most important parameters are the particle size and the Polydispersity Index (PDI) which is usually determined by either Dynamic Light Scattering (DLS) or nanoparticle tracking analysis (NTA). Zeta potential is used to determine the colloidal stability of the formulations (Costa et al., 2024). The encapsulation efficiency and nucleic acid content are usually determined by measuring using fluorescence-based assays like Ribo Green (RNA) and Pico Green (DNA) assays. To analyses the morphology and internal structure of LNPs, the following techniques are used: Transmission electron microscopy (TEM) or cryo- TEM. One of the key parameters used to check quality control of LNP is stability under different storage conditions, which can affect the various characteristics of the LNP, such as particle size, structural integrity, nucleic acid retention and transfection activity. Sterility control and the number of endotoxins is also important to avoid unwanted immune reactions (He et al., 2023).

Meanwhile, Machine Learning (ML) is developing into a powerful tool for the design of LNP, allowing relationships to be extracted from large datasets that can be predictive (Tian et al., 2023). The development of mRNA vaccines has been predicted based on compositional data using ML methods, and it was found that key substructural components of ionizable lipids play key roles in immunogenicity. The potential of DOE and ML to be integrated together is especially interesting, as it involves the use of statistically sound experimentation and adaptive predictive modelling (Tenchov et al., 2023).

2.3. LNP Pharmacokinetics and Pharmacodynamics

After administration, lipid nanoparticles are absorbed, distributed, metabolized and eliminated based on their pharmacokinetic behavior. After intravenous injection, the LNPs move through the bloodstream as a protected form of nucleic acid delivery that protects against enzymatic degradation (Wang et al., 2023). Particles with different surface properties, PEG content, lipid composition and size vary in their circulation time. Generally, small and evenly distributed nanoparticles have long circulation times and are more penetrating to tissues. Liver is also a major deposition site as apolipoproteins in circulation bind to the surface of LNP, facilitating uptake through hepatocyte receptors. Surface modifications and targeting ligands can be used to further target nanoparticles to specific tumor tissues to increase delivery efficacy and reduce systemic exposure (Balasubramanian & Sundrud, 2023).

Pharmacodynamically, LNPs are taken up by endocytosis, then escape from the endosome with the help of ionizable lipids. After entering the cytoplasm, the mRNA is translated to generate therapeutic proteins, while the components of CRISPR/Cas9 are used to promote precise genome editing by targeting specific DNA sequences (Chen et al., 2024). Effective delivery into the cell, editing efficiency and long-term expression are all important to the biological response. In liver cancer, efficient uptake in liver leads to high therapeutic activity, while targeted LNP formulations enhance liver delivery to prostate tumors. To get the most out of the treatment, increase the therapeutic efficacy, decrease off-target editing, reduce toxicity, and maximize long-term treatment, optimizing the pharmacokinetic and pharmacodynamic properties is essential (AnishRuban et al., 2025).

2.4. mRNA vs DNA Delivery

The selection of genetic cargo that will be enclosed in LNPs needs to consider the nature and restrictions that come along with each of the molecules based on the desired use of the delivery system (Gambacorta et al., 2023). For the therapy to be successful, nucleic acids need to pass through various biological hurdles, both intracellular and extracellular. Nucleic acid, especially mRNA, must be shielded against the action of endonucleases when administered in vivo. Subsequently, the LNPs must reach the target tissue, enter the target cells and escape from the endosomes (Ichihara et al., 2023). Translation happens immediately upon delivery of mRNA into the cytoplasm; the expression peaks between 6 to 24 hours, but it is transient, declining gradually with time as mRNA gets degraded. This is good because it helps to regulate the effects of cell-based therapy like CAR-T and prevents systemic effects of cytokines (Dhatchinamoorthy et al., 2021).

In comparison, pDNA has slow expression kinetics, since it must first travel into the nucleus and undergo transcription to mRNA before translation occurs in the cytoplasm, taking 24 to 72 hours. Prolonged antigen expression occurs using pDNA, thus making it a better choice for vaccines, since prolonged exposure results in better and lasting immune response (El-Ashmawy et al., 2025). Whereas, concerning stability, pDNA is more stable chemically and structurally than mRNA. Consequently, it can be stored under less severe conditions or even at room temperature. On the contrary, mRNA is highly sensitive to hydrolysis, hence, its storage should be done in much more severe conditions, that is, between −20 and −80°C (Gibson & Deweese, 2022). These differences greatly affect the logistics, storage, and the cost of the nucleic acid therapy, hence, their availability on the global scale. The knowledge about the delivery of mRNA and pDNA helps in understanding the gene delivery in general and the strengths and weaknesses of LNPs in comparison with other delivery systems (O'Leary et al., 2024).

2.5. LNPs vs Other Gene Delivery Systems

Gene delivery can be achieved using viral vectors, electroporation, and LNPs. Viral vectors, including lentiviruses and adenoviruses, are popular gene delivery techniques since they have high transduction efficiency and capacity for sustained expression (Garcia et al., 2024). Yet, while viral vectors are highly efficient and can maintain sustained expression of genes permanent in case of lentiviruses because of genomic integration and sustained episomal expression in case of adenoviruses these techniques can pose problems of insertional mutagenesis, high immunogenicity, lack of repetitive administration possibilities, and expensive production costs. On the other hand, LNPs pose no risk of genomic integration, are less likely to provoke unwanted immune response, provide scalability and cheaper production, yet LNPs-based transgene expression is a transient process that reaches maximal protein production only 24 hours after administration and subsides to below therapeutic concentrations within seven days (Mu et al., 2021).

Electroporation is a common technique utilized for *ex vivo* cell manipulation including, for example, CAR-T cell generation. Though electroporation proves to be highly effective, it causes great cytotoxic effects and reduction of cell viability, which is an obstacle for its broad clinical use. In contrast to electroporation, LNPs have been shown to have extremely low toxicity and can be applied for both *ex vivo* and *vivo* applications (Dai et al., 2021).

Another class of non-viral vectors is cationic polymers, e.g. Polyethyleneimine (PEI), possessing features such as low manufacturing cost and flexibility of optimization. Yet, these vectors are toxic and have limited in *vivo* efficiency that makes LNPs a better choice. Although polymeric nanocarriers have some advantages in relation to formulation and modification, they are characterized by lower gene delivery efficiency and unpredictable *in vivo* behavior. Regarding the immune response, LNP-based vectors cause a powerful innate immune response, which can be useful for vaccination, whereas in case of polymeric vectors it will depend on their charge and composition (Xie et al., 2023).

To conclude, while evaluating various platforms for delivering genetic materials, LNP has become an extremely promising technology that combines safety, versatility, and efficiency (Runnels et al., 2023).

As a result, it represents one of the leading technologies that can substitute the role of viral vectors, competing even with such non-viral platforms as electroporation and polymer delivery systems. This delivery method has gained increasing attention due to its efficiency and safety. Also, the technology provides high versatility and can be used for both *in vivo* and *ex vivo* applications. Lipid nanoparticles provide an effective balance between delivery efficiency, safety, and scalability, making them one of the leading non-viral platforms for clinical gene-editing applications are shown in table 1 (Wang et al., 2023).

Table 1. Comparison of Major Gene Delivery Platforms for Cancer Therapy

Delivery Platform	Advantages	Limitations	Clinical Relevance
Lipid Nanoparticles (LNPs)	Biocompatible, biodegradable, repeat dosing, high nucleic acid loading	Transient gene expression	Approved for nucleic acid delivery and widely investigated in cancer therapy
Viral Vectors	High transfection efficiency, long-term expression	Immunogenicity, insertional mutagenesis, costly production	Gene therapy applications
Electroporation	Efficient ex vivo gene transfer	Cell toxicity, reduced viability	Mainly CAR-T cell preparation
Polymeric Nanoparticles	Flexible formulation, low production cost	Lower in vivo efficiency, higher toxicity	Experimental cancer delivery systems

2.6. Advantages of *in vivo* Delivery over *ex vivo* Manipulation

The *in vivo* delivery of the genes means that the CRISPR/Cas9 system contained in the lipid nanoparticles is directly injected into the patient, where the therapeutic nucleic acids can be delivered to the target tissues without the need for removing cells from the patients, expanding them *in vitro* and then reintroducing back to the patients (Liu et al., 2022). In this way, the *in vivo* delivery reduces the complexity of the process and makes it much safer since there is no danger of contaminating the therapeutic nucleic acids with pathogens when working with the cells. LNPs can protect the genetic content of nanoparticles in circulation, help in their cellular uptake *via* endocytosis, and ensure the release inside the cells using ionizable lipids (Knaggs et al., 2025). The main advantage of *in vivo* delivery compared to *ex vivo* manipulation is the ability of therapeutic nanoparticles to reach the tumor tissue directly from the systemic blood flow. Repeatable administration is possible without the need to perform repeated cell harvesting and transplantation since there is an extended period of action of the therapy (Jörgensen et al., 2023). The structure of the liver vessels makes it easier for LNP to accumulate in the tumor tissue, providing enhanced CRISPR/Cas9 delivery and efficiency of genome editing. For the prostate tumor, the modification of the LNPs surface provides improved selective targeting of tumor cells without affecting healthy tissues. Figure 1 illustrates lipid nanoparticles encapsulate Cas9 mRNA and guide RNA, protect them during circulation, promote cellular uptake through endocytosis, facilitate endosomal escape, and enable precise genome editing at the target site off-target effects (Ho et al., 2021).

Figure 1. Lipid Nanoparticle-Mediated Delivery of CRISPR-Cas9 for Targeted Cancer Therapy

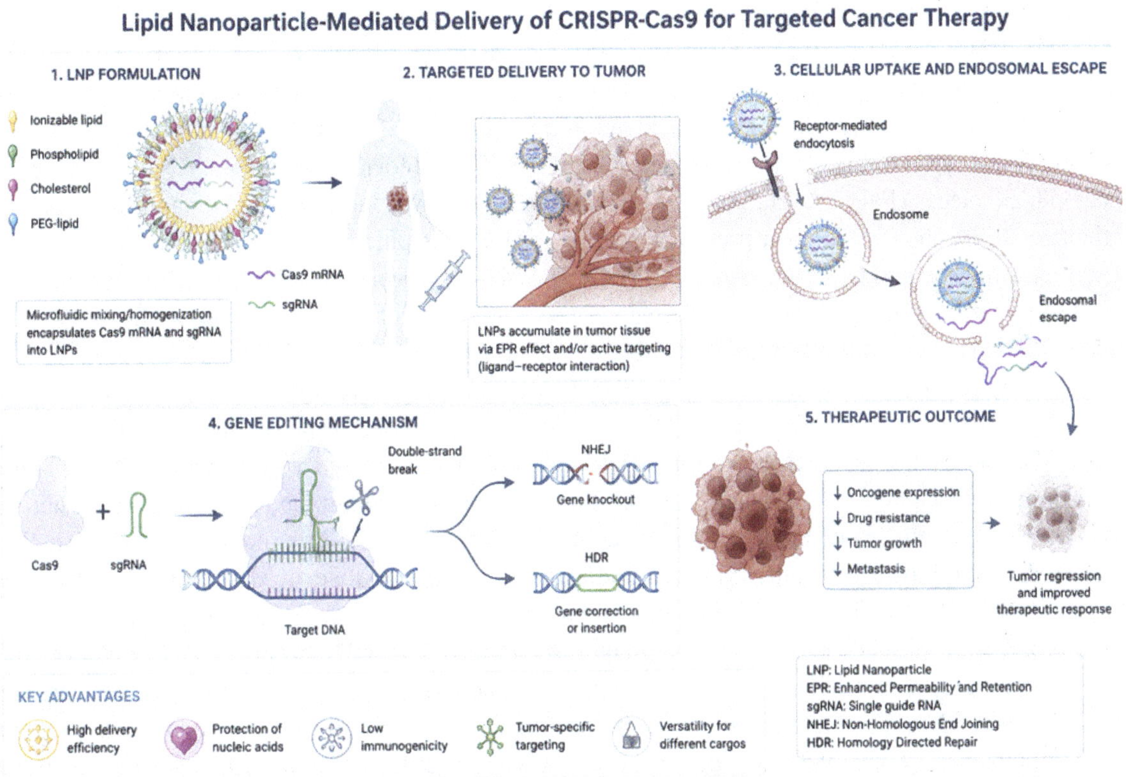

3. MECHANISMS OF CRISPR-CAS9 FOR GENOME EDITING

Genetic engineering in which the sequence of the DNA of living cells is inserted, deleted or altered is referred to as genome editing. CRISPR is an acronym for clustered regularly interspaced short palindromic repeat, which are short and partially repeating DNA sequences found in the genomes of prokaryotes. CRISPR and the related proteins act as the defense system of prokaryotes from viruses and bacteriophages (Sebastiani et al., 2021). CRISPR-Cas9 is made up of three components. A) Single guide RNA (sgRNA) crRNA array is a sequence of 20nt which will bind to the target DNA. B) The trans-activating crRNA (tracrRNA) is a short non-coding RNA that can base pair with the crRNA (gRNA) to form the functional guide RNA. PAM (Protospacer Adjacent Motif) is the Cas9 nuclease which cuts the DNA strand after three to four bases from the protospacer neighbouring motif. PAM is a short and conserved sequence motif (2-5bp) that is found in the invading DNA and is very important in selecting and cutting the target DNA. It is located next to the crRNA targeted sequence. There are two main domains in the Cas9 nuclease, namely, the RuvC-like domain and the HNH domain. The part of the tracrRNA in the guide is the handle for Cas9, and crRNA is used to guide the whole complex to the target infectious sequences. In total, there are six types of CRISPR-Cas system found in a wide range of bacteria and archaea, such as, class 1 (types I, III, and IV) and class 2 (types II, V, and VI) (Wu et al., 2020). While class 1 systems require more than one Cas proteins, class 2 systems utilize only a single Cas protein and thus are easy to use and more efficient. CRISPR-Cas9 Type II, among the Class 2 systems, is quite popular and is highly studied. As a result of the immune system's reactions to the foreign genetic material, foreign sections of the DNA become part of the host's genome during the process of immunisation. Thus, the host can defend against the same pathogen due to its genetic memory because of the previous infection. The CRISPR locus is then transcribed and processed into mature CRISPR RNAs (crRNAs). In the process of silencing, trac rRNA binds to the repeat regions of crRNA and forms unique dual RNA structure like jigsaw puzzle pieces (Fajrial et al., 2020). The two RNA guide will facilitate once bound to the crRNA by the trac rRNA, the Cas9 scissor protein will cleave any DNA containing a compatible 20 nucleotides (nt) target sequence and Cas9 accessory molecule (PAM) sequence. Cas9 cleaves the double-stranded DNA through the action of two nuclease domains; HNH and RuvC, ending up producing a Double-Strand Break (DSB). The nuclease HNH domain will cut the complementary strand of the DNA (target strand) to the RNA guide. The non-complementary DNA strand (non-target strand) will be cut by the nuclease RuvC. The Cas9 of S. pyogenes is a big DNA endonuclease with multiple domains and functions. A mutant of either of the two nuclease domains will cause the Cas9 to cleave one strand of DNA. This is because the HNH catalytic domain can cut the other strand of a single guide RNA (sgRNA), when the RuvC catalytic domain is mutated and inactivated. The name given to this mutation is Nickase Cas9 (nCas9). When two nCas9 proteins bind to two adjacent Protospacer Adjacent Motifs (PAM) at the locus of interest, Double-Strand Break (DSB) will happen with high specificity (Eftekhari et al., 2023). Non-Homologous End Joining (NHEJ) and Homology-Directed Repair (HDR) represent two distinct types of DNA repair mechanisms. The former results in small and random InDels at the site of cleavage, and this may cause DNA frame-shift mutations and/or creation of premature stop codons, finally leading to gene knockout. Alternatively, the latter provides a way to replace a mutated or defective sequence with the corresponding wild-type sequence, giving rise to targeted genome modifications. The use of donor DNA templates is required for introducing a target sequence into the genome through the HDR process. In general, while HDR represents a relatively dependable DNA repair pathway, this method is usually less efficient than the NHEJ one. Targeted genome editing, different from random mutagenesis approaches such as EMS or radiation, allows for precise modifications of genomes at certain genomic locations (Chen et al., 2020).

The CRISPR-Cas9 system outperforms other genome editing systems, such as ZFNs and TALENs, which use the nuclease domain of the FokI endonuclease for generating double strand DNA breaks (Mollanoori et al., 2021). For instance, ZFN is a series of Cys2-His2 ZF domains, which bind to a particular protospacer adjacent motif (PAM). This characteristic of it makes it hard to select a suitable target sequence as it will depending on the different PAMs. If two ZFNs are bound together to recognize a particular DNA sequence of 18-24 bp, it forms a dimer. However, the application of ZFN and TALEN systems is limited in terms of possible off-target activity (Aldosari et al., 2021).

4. APPLICATIONS OF CRISPR-CAS9

4.1. Role in Gene Therapy

Gene therapy has been developed as a new revolutionary treatment modality that is used for the correction, replacement, or silencing of the genes associated with disease development. Concerning liver and prostate cancers, gene therapy is aimed at manipulating genetic abnormalities causing the onset, progression, metastases, and drug resistance of tumors (Buschmann et al., 2021). The application of CRISPR-Cas9 technology for genome editing in gene therapy has considerably contributed to the progress in this field owing to high accuracy of targeting and modifying specific DNA segments. In contrast to traditional gene substitution techniques, CRISPR-Cas9 technique can be utilized for disrupting the activity of oncogenes, correcting disease-causing mutations, activating tumor suppressors, and regulating immune responses. Nonetheless, the success of CRISPR-based gene therapy strongly depends on the delivery systems that can be safely applied for the introduction of genome-editing components into the desired cells. One of the most promising approaches to delivering CRISPR components into the cells is represented by lipid nanoparticles (LNPs) as they allow for the efficient encapsulation and protection of Cas9 mRNA, gRNA, plasmid DNA, or RNP complexes (Scarpini et al., 2021). Following administration, the LNPs enhance cellular uptake by endocytosis and facilitate endosomal escape, thus enhancing intracellular delivery of genome editing elements. In case of hepatocellular carcinoma, accumulation of LNPs in the liver tissues is naturally increased due to the presence of fenestrations and lipid metabolism in liver vasculature, which leads to higher efficiency of gene delivery. In case of prostate cancer, LNPs modified with ligands specifically target tumor associated receptors including the prostate-specific membrane antigen (PSMA). It makes LNP-mediated CRISPR delivery an excellent tool for precise gene therapy in liver and prostate cancers (Knezevic et al., 2021).

Combination of gene editing based on nanoparticles with cancer immunotherapy has significantly expanded possibilities of gene therapy for liver and prostate cancers. CRISPR-Cas9 could be used for modification of genes regulating immune system, inhibition of immune checkpoint pathways, enhancement of antigen presentation, or increase of cytotoxicity of immune cells against tumor cells (Chiang et al., 2021). At the same time, LNPs can serve as an effective tool to co-deliver genome-editing tools along with therapeutic nucleic acids, mRNA encoding cytokines, or cancer vaccines antigens for creating a multimodal therapeutic approach where several strategies could be united into one delivery system. Such an approach not only enables cancer growth via genome editing but also promotes a creation of long-lasting antitumor immune response that decreases chances of cancer recurrence and spreading (Mitsogiannis et al., 2022). Unlike viral vectors, LNPs have low immunogenicity, no risks of insertional mutagenesis, can carry larger genetic cargo, and allow repeat dosing in cases of prolonged therapy. Continuous development of ionizable lipids chemistry, functionalization of nanoparticles' surface, and targeting properties of LNPs have made LNP-mediated gene delivery even more efficient and safer. Thus, nanoparticle-mediated CRISPR-Cas9 delivery can become a next generation therapeutic platform where genome editing and immunotherapy will be united, providing high potential for treating liver and prostate cancer patients (Buschmann et al., 2021).

4.2. Role in the Generation of Animal Models

The gene targeting technique using homologous recombination and embryonic stem cell has been a conventional way of genome modification of animals for a while now. This technique has been highly important in proving the correlation between genetic mutations and developmental phenotypes or diseases. However, in some species, the use of gene targeting is restricted owing to its complex technique and lack of embryonic stem cells (Murugesan & Raman, 2022). Recent research has shown that CRISPR-Cas9 system is able to produce targeted modifications in the germline of a variety of model species at a fast pace, hence contributing positively to functional genomics. Gene modification in one-cell stage zebrafish embryos can be done successfully by injecting Cas9 expressing mRNAs together with adaptable sgRNAs (Davodabadi et al., 2023). In addition, the co-injection of Cas9 mRNAs/sgRNAs with mutant oligos has helped to produce gene knock-in mice with precise point mutations in two genes. Others have reported the generation of reporters and conditional mutants in a single step by introducing into mouse zygotes a Cas9 mRNA, several different sgRNAs and multiple size DNA vectors. Moreover, deletion mice have been developed using two different target sequences in a gene by using sgRNAs (Emran et al., 2022).

5. LNP-MEDIATED GENE DELIVERY FOR THERAPEUTIC APPLICATIONS

5.1. LNP Delivery of CRISPR/Cas9 Genome Editing

Delivery of CRISPR/Cas9 genome editing components using Lipid Nanoparticles (LNPs) can be regarded as the most effective non-viral method of delivery of such components for use in treatment of cancer (Boehnke et al., 2022). All these genome editing components are enclosed within LNPs so that they can be protected against enzymatic degradation when traveling in the bloodstream. Ionizable lipids found in LNP formulation are electrically neutral under physiological pH levels, thus making them less toxic and increasing their circulation time, but get positively charged in acidic endosomal compartments. Thus, they can escape from the endosomes for efficient intracellular delivery of genome editing components (Sun et al., 2023). On entering the cell through endocytosis, Cas9 mRNA gets translated to Cas9 protein that binds to sgRNA and moves to the nucleus where it identifies and cleaves the target DNA sequence located near Protospacer Adjacent Motif (PAM). The double-strand DNA cut is then repaired through Non-Homologous End Joining (NHEJ) or Homology-Directed Repair (HDR). When compared with viral vectors, there are some key benefits of LNP delivery system such as reduced immunogenicity, absence of insertional mutagenesis, flexibility in choice of cargo, increased cargo loading capacity, and the ability for repeated application (Liu et al., 2023).

LNPs have been proven to be effective for delivery of CRISPR/Cas9 systems in cases of liver and prostate cancers. The liver is a perfect candidate for LNP-based delivery due to its natural targeting by nanoparticles through the fenestrated vessels and increased lipid metabolism. As a result, LNP delivery allows for high efficiency editing of genetic material of the hepatocellular carcinoma, involving oncogenes, tumor suppressors, and immune regulators (Yoo et al., 2024). In prostate cancer, surface-modified LNPs with the use of antibodies, peptides or receptor ligands allow for specific recognition of PSMA and other tumor-associated receptors. These multimodal delivery systems help to efficiently kill the tumor cells, activate the immune system, minimize the chances of therapeutic resistance, and prevent tumor relapse. The ongoing developments in ionizable lipid chemistry, surface modification of nanoparticles, and targeted delivery of LNPs towards tissues only add up to safety, precision, and applicability of LNP-based CRISPR/Cas9 gene editing technology in clinical settings (Ding et al., 2024).

5.2. Stimuli-Responsive Nanoparticles

Stimuli-sensitive NPs are synthesized in such a way that they deliver their therapeutic payload in response to certain stimuli in the TME or externally applied stimuli. Such stimuli may include changes in pH, redox potential, enzymatic reactions, hypoxia or application of various physical stimuli. Stimuli-sensitive NPs react to these stimuli provided by the tumour and enable targeted delivery of drugs (Mitsogiannis et al., 2022).

5.2.1. pH Sensitivity

The modification of the release kinetics of the NP cargo becomes crucial to overcome MDR and to provide efficient delivery of the drug within the TME, reducing potential toxicity of the drug and improving its efficiency (Meng et al., 2023). There are many different stimuli used in the development of actively targeted NPs; however, the change in pH value has gained significant attention due to the presence of multiple pH gradients in the human body. Specifically, the TME is usually characterized by lower (pH ~6.8) pH value than that in healthy tissues outside tumours (extracellular pH ~7.4) both in epithelial and non-epithelial tumours. This low pH is mainly due to the presence of the anaerobic glycolysis and oxidative phosphorylation in the hypoxic conditions leading to the formation of lactic acid. To use the pH gradient in tumour tissues, a pH-sensitive liposome nano liposomal formulation of doxorubicin was created for the glioma therapy. Further evidence of pH-responsiveness and stability was provided by incorporation of paclitaxel within the liposomal bilayer of PSL. In fact, at physiological pH 7.4 the liposomes were stable, whereas changes in supramolecular structure have been observed at lower pH values, indicating pH-responsive nature of these nanoparticles (Buschmann et al., 2021).

5.2.2. Redox Responsiveness

High concentrations of GSH have been observed in tumour cells, particularly in those that display resistance towards drugs. Disulfide bonds in redox-responsive nanoparticles become unstable in the reduced intracellular environment and undergo cleavage to enable the release of the drug cargo such as doxorubicin or siRNA directly into the cytoplasm. This method allows delivery of drugs without the involvement of efflux transporters (Murugesan & Raman, 2022).

5.2.3. Hypoxia Responsiveness

It is one of the features of solid tumours and contributes towards drug resistance of cancer by promoting angiogenesis through HIF-1α pathway, epithelial-mesenchymal transition, invasion, metastasis and stem-like properties as well as increasing glycolysis. Presence of tumour hypoxia in solid tumours and its contribution towards resistance makes it a great target for drug delivery systems (Sung et al., 2021). Hypoxia-responsive nanoparticles are engineered by using compounds including nitroimidazoles, nitro benzyl alcohols and azo compounds predominantly. They show significant changes in terms of hydrophobicity and surface charge of nanoparticles under hypoxic conditions and cause changes in behaviour of nanoparticles including structural destabilization. Numerous hypoxia-responsive nanoparticles have been prepared till date. Though no hypoxia-responsive nanoparticles have gained clinical approval so far, many of them have shown good prospects in preclinical studies. For instance, hypoxia-responsive polymer micelles loaded with dicoumarol (NAD(P)H: quinone oxidoreductase 1 inhibitor) and sorafenib (anti-cancer drug) inhibited the expression of HIF-1α and thus sensitized ovarian cancer cells to apoptosis caused by sorafenib. By providing a targeted release of drugs in resistant region of hypoxic tumours, the nanoparticulate system can provide a promising solution (Guimarães et al., 2023).

6. CANCER IMMUNOTHERAPIES FOR LIVER AND PROSTATE CANCERS

6.1. CAR-T Cell Therapy

Chimeric Antigen Receptor T-cell (CAR-T) Therapy is a highly evolved type of Adoptive Cell Immunotherapy involving genetic modification of T cells of a particular patient, enabling them to identify and kill cancerous cells harboring certain types of tumor-specific antigens (Costa et al., 2024). The CAR structure includes an antigen-binding region outside the cell, a transmembrane section, and signal domains of intracellular receptors that lead to activation of cytotoxicity of the T cells in response to the antigen interaction. CAR-T cell therapy has shown exceptional efficacy in treating hematological cancers and is being explored for application in solid tumors, including liver and prostate cancers. In liver cancer, GPC3 and Alpha-Fetoprotein (AFP) can be the target of the CAR-T cells, while in prostate cancer, one of the most widely studied targets is the Prostate-Specific Membrane Antigen (PSMA) (He et al., 2023).

Even though CAR-T therapy has a great capacity for treating cancer patients, some obstacles like poor tumor infiltration, immunosuppressive TME, antigen heterogeneity, and inadequate persistence of T cells after their transfer should be considered (Tian et al., 2023). Recently, Lipid Nanoparticles (LNPs) were proven as an efficient means of optimizing CAR-T therapy due to the delivery of mRNAs, gene-editing agents, or immunomodulatory factors into immune cells or tumors. Delivery through LNP provides transient CAR expression without permanent genome editing and allows for repeatable doses if necessary. In addition, delivery of CRISPR/Cas9 via LNPs helps to disrupt genes of inhibitory immune checkpoints or to increase CAR-T cell activity and improve anti-tumor effects in cases of liver and prostate cancers (Tenchov et al., 2023).

6.2. Cancer Vaccines

Another strategy involves presenting peptides that have been processed by APCs in order to stimulate antigen-specific cytotoxic T cells.10 Immunotherapy involving cancer vaccines represents another method and has found its application via cell-based delivery platforms like the one represented by Sipuleucel-T (dendritic cell vaccine), showing encouraging results against castration-resistant prostate cancer. Nevertheless, other cell-based delivery platforms have demonstrated low efficiency in clinical trials. Novel cancer vaccines as a vaccine platform could benefit immensely from new delivery approaches such as LNPs, which may become a replacement for existing delivery strategies without using cells or viruses. Via such platforms, nucleic acids will be delivered in a targeted manner to APCs and result in better stimulation of antigen-specific cytotoxic T cells (Wang et al., 2023).

Moreover, LNPs can be chemically modified to increase their uptake by APCs. Such a strategy has been examined by BioNTech in a phase I/II clinical trial where lipoplexes containing RNA of Claudin 6 (a tumor-associated antigen) are applied together with CAR-T cells targeting the same antigen (Balasubramanian & Sundrud, 2023). The application of LNPs in improving the efficiency of cancer therapies has already been broadly reported in the works on the use of the HPV models of cancer, including oropharyngeal cancer. Thus, the success of RNA and sRNA delivery resulted in the decrease of tumour volume. It was also possible to observe the tumour volume decrease in B16F10-OVA mouse models after vaccination with OVA either as mRNA or with the protein in the context of LNP. In general, vaccines can be viewed as rather versatile methods in the case of their efficient use for antigen presentation. Moreover, the delivery of nucleic acids (DNA and RNA) via LNPs is a well-proven strategy of fast vaccine development, as illustrated in the strategy applied in the COVID-19 vaccines. Hence, the LNP platform can become a practical way of developing personalized cancer vaccines based on the sequencing data of patients' biopsies (Chen et al., 2024).

6.3. Cytokine-Based Immunotherapy

Cytokine-based immunotherapy makes use of naturally occurring proteins that play an important role in immune regulation such as proliferation, activation, differentiation and destruction of cancerous cells via cytotoxicity. Cytokines like IL-2, IL-12, IFN-α, and IFN-γ have been used to treat liver and prostate cancers owing to the ability of these cytokines to stimulate T-cells and Natural Killer (NK) cells. Cytokines aid in destruction of malignant cells, anti-tumor angiogenesis and antigen presentation in the tumor environment (AnishRuban et al., 2025). Lipid nanoparticles are efficient delivery vehicles for cytokines in the form of mRNA or other therapeutic nucleic acids encoding these cytokines, resulting in cytokine production in tumors and immune cells. Furthermore, genome editing using CRISPR/Cas9 system can target either immune cells or cancer cells for enhancing their response to cytokines, blocking their immune-suppressive mechanisms, or regulating cytokine production. For example, in hepatocellular carcinoma and prostate cancers, the delivery of cytokine through LNP in conjunction with genome editing may have an additional advantage of boosting anti-tumoral immune response, increasing the infiltration of immune cells, and overcoming the immune resistance of tumors (Gambacorta et al., 2023).

6.4. Bispecific Antibodies and LNP Functionalization

BsAbs provide unprecedented advances in the field of therapeutic antibodies with unique abilities in disease targeting and therapy. BsAbs can be considered an exciting new form of immunotherapy drug with great possibilities to tackle various kinds of diseases like cancer and autoimmune conditions. Contrary to classical monoclonal antibodies, which bind only to one antigen, BsAbs are engineered to bind to two different antigens or epitopes. Dual binding capacity of BsAbs allows bridging cell-to-cell interactions (Ichihara et al., 2023). The main application of the significance of bispecific antibodies is in the treatment of cancer. BsAbs simultaneously interact with the tumor associated antigen and CD3 subunit on T cells. This binding results in recruitment of T cells to the tumor site, whereupon activation of T cells takes place followed by secretion of cytolytic substances and elimination of tumor cells (Dhatchinamoorthy et al., 2021).

Bispecific Antibodies can be functionalized by making modifications on them, through the application of covalent or non-covalent binding of particles or chemical substances to the surface of the antibody. The process of Functionalization helps in fine-tuning the biological activity and efficacy of BsAbs. Through advanced antibody engineering techniques, BsAbs can be developed for improved stability, low immunogenicity, and better pharmacokinetics (Gibson & Deweese, 2022). Another strategy where BsAbs may play a key role in increasing the efficiency of LNP delivery is the functionalization of nanoparticles. Figure 2 shows that lipid nanoparticles enhance the delivery of CAR-T modulators, cancer vaccines, cytokines, immune checkpoint inhibitors, and bispecific antibodies, resulting in improved immune activation, tumour targeting, and therapeutic efficacy (Garcia et al., 2024).

7. SYNERGISTIC APPROACH AND COMBINATION THERAPY

7.1 Combining Nanomedicine with Other Therapies

Reasons behind the integration of nanomedicine with other methods in the treatment of prostate cancer are associated with the multi-faceted and complex TME of the disease. The use of nanomedicine results in target-oriented delivery of drugs and higher bioavailability of medication, which enhances the mechanisms of other traditional treatments. As an illustration, a study conducted on polymer-based nanoparticles specifically developed for combination chemotherapy in tumour targeting has shown synergy in genotoxicity of the treatment in prostate cancer by delivering chemotherapeutics such as doxorubicin and 5-fluorouracil (Xie et al., 2023).

7.1.1 Chemotherapy

Traditional chemotherapy still plays an essential role in treating prostate cancer by utilizing toxic chemicals that kill cancer cells. Used in clinical treatment, traditional chemotherapeutic drugs have a low half-life in the plasma and fast distribution throughout healthy tissue, which causes many side effects. Moreover, due to a lack of drug retention and accumulation in the tumours, together with the Multi-Drug Resistance (MDR) caused by the chemotherapy, cancer cells become resistant to the similar chemical structures of the drugs, thus leading to treatment inefficiency. Consequently, other approaches should be found in order to achieve better results. Due to the progress in nanomedicine field, the use of multi-functional nanocarriers with radiotherapy became a promising strategy (Runnels et al., 2023). The ways to enhance the efficiency of chemotherapeutic drugs in nanomaterials involve targeted chemotherapy to the sub organelles, enhancement of chemotherapeutic drugs in tumours, overcoming drug resistance mechanisms and combination chemotherapy. The encapsulation of the chemotherapeutic drugs in nano-carriers provides protection from the efflux pumps. These nanocarriers surpass conventional chemotherapeutic agents by virtue of their action on tumour cells or blood vessels with less toxicity, overcoming multidrug resistance and increasing the efficacy of the drugs as well as lowering side effects. Materials like Borle toxins, doxorubicin, and actinomycin are used in the development of nanomedical agents.80 Furthermore, certain liposomal chemotherapeutic drugs consisting of paclitaxel and doxorubicin have been used clinically with few side effects (Wang et al., 2023).

7.1.2 Radiotherapy

Radiotherapy is one of the major treatment approaches to prostate cancer which could be employed either as monotherapy or combined therapy with surgical and chemotherapy. Even though it is quite effective, it has some adverse effects as it does not recognize the difference between healthy and disease tissue. Thus, there were numerous studies on defining the dose which would kill cancer cells but would not cause extra damages to healthy tissues (Liu et al., 2022). There were numerous studies on the advantages of using nanomedicine delivery for improving the efficiency of Chemoradiotherapy (CRT) and improving delivery for inducing direct DNA damages to minimize side effects of drugs and doses of radiation. Nanomaterials tagged with Imaging capabilities can be enhanced by using magnetic resonance imaging (MRI), positron emission tomography (PET) or computed tomography (CT) contrast agents. In addition, nanotechnology allows developing nanoparticles to decrease tumour hypoxia, regulate the immunosuppressive TME, and dramatically improve radiotherapy efficacy. Lipid nanoparticles improve the targeted delivery of immune-modulating agents and gene-editing components, enhancing therapeutic efficacy while reducing systemic toxicity and immune escape that were given in table 2 (Knaggs et al., 2025).

Table 2. Nanoparticle-Based Immunotherapeutic Approaches in Liver and Prostate Cancer

Immunotherapy Strategy	Nanoparticle Role	Therapeutic Benefit	Target Cancer
CAR-T Cell Therapy	Delivery of mRNA and CRISPR components	Improves T-cell activity and persistence	Liver, Prostate
Cancer Vaccines	Delivery of tumor antigens or mRNA	Enhances antigen presentation and immune memory	Liver, Prostate
Cytokine Therapy	Targeted cytokine gene delivery	Reduces systemic toxicity and increases immune activation	Liver, Prostate
Bispecific Antibodies	Surface functionalization of LNPs	Improves tumour-specific targeting and T-cell recruitment	Liver, Prostate
Combination Therapy	Co-delivery of CRISPR and immunomodulators	Synergistic tumour suppression and reduced resistance	Liver, Prostate

8. FUTURE PERSPECTIVES OF NANOPARTICLE-MEDIATED GENE EDITING AND CANCER IMMUNOTHERAPY

The utilization of nanoparticles in gene editing and cancer immunotherapy will bring about a revolution in the future of precision oncology due to the ability to employ highly targeted, personalized and minimally invasive approaches in the treatment of liver and prostate cancers (Jörgensen et al., 2023). Further progress in the field of nanotechnology will lead to the emergence of advanced multifunctional nanoparticles that can deliver at the same time CRISPR/Cas gene editing systems, therapeutic genes, immune regulators, and diagnostics tools into the tumors with very high selectivity. The emerging intelligent nanocarriers will have improved biocompatibility and prolonged circulation times along with increased tumor targeting by virtue of surface modification with ligands, antibodies, peptides, and aptamers recognizing cancer-specific biomarkers. It is worth noting that stimuli-responsive nanoparticles whose drug release depends on the local microenvironment conditions such as pH, enzyme level, hypoxia, oxidative stress, and exposure to external stimuli (light, ultrasound, magnetism) will help gene editing and activation of immune system (Ho et al., 2021).

However, the future of cancer treatment using nanoparticles will be based on a combination of genome editing and state-of-the-art immunotherapy to counteract the effects of tumor heterogeneity and immune resistance. Editing immune checkpoint molecules, oncogenes, tumor suppressor genes, and tumor microenvironment-related genes through CRISPR technology will greatly increase anti-tumor immunity while decreasing immune escape mechanisms that are usually seen in liver and prostate cancers (Sebastiani et al., 2021). In this case, nanoparticles will be used to deliver gene-editing products together with immune checkpoint inhibitors, cancer vaccines, cytokines, CAR-T cell modulators, and other immune-modulating nucleic acids on one therapy platform. This approach is projected to reprogram the tumor microenvironment and restore the balance by increasing T cell infiltration, activation of dendritic cells, suppression of regulatory T cells and myeloid-derived suppressor cells, and induction of immunological memory. The development of biomarker-based personalized cancer treatment will help select individual patients for treatment according to their genetic mutations and immune characteristics (Wu et al., 2020).

Novel technologies such as organ-on-a-chip, patient-derived organoids, digital twins, and imaging technologies will offer more precise preclinical models for evaluation of the therapy before clinical studies. With further development of collaboration between nanotechnologists, molecular biologists, immunologists, clinicians, and regulatory agencies, the nanoparticle-mediated gene editing and immunotherapy have all chances to become a basis of precision medicine of the future which will ensure safe and efficient treatment of liver and prostate cancer and improvement of patients' survival and quality of life (Chen et al., 2020).

9. CONCLUSION

Nanoparticle-mediated gene editing and immunotherapy represent a major advancement in the treatment of liver and prostate cancers by integrating the precision of genome engineering with the specificity of targeted drug delivery and immune modulation. Lipid nanoparticles have proven to be extremely effective non-viral delivery systems, and are now enabling the efficient delivery of Cas9 and guide RNAs in a safe manner, mRNA, siRNA, and immunotherapeutic agents while minimized systemic toxicity and off-target effects. Their favourable biocompatibility, biodegradability, and capacity for surface functionalization allow selective tumour targeting and enhanced intracellular delivery, thereby improving therapeutic efficacy. In parallel, advances in cancer immunotherapy, including CAR-T cell therapy, cancer vaccines, cytokine-based therapies, and bispecific antibodies, have demonstrated significant potential for overcoming immune evasion and promoting durable antitumor responses.

The convergence of these technologies offers a synergistic therapeutic strategy capable of simultaneously correcting oncogenic alterations, remodelling the tumour microenvironment, and enhancing immune-mediated tumour elimination. Furthermore, stimuli-responsive nanoparticles and combination therapies with chemotherapy and radiotherapy offer repeat chances to defeat drug resistance and enhance clinical results. Although these are positive advances, there are delivery efficiency, long term biosafety, manufacturing scalability challenges, regulatory approval, and clinical translation remain to be addressed. Continued progress in nanoparticle engineering, biomarker-guided precision medicine, artificial intelligence-assisted nanocarrier design, and patient-specific therapeutic strategies is expected to accelerate the transition of these technologies from experimental research to routine clinical practice. Overall, nanoparticle-mediated gene editing combined with the potential of immunotherapy to improve survival rates, decrease recurrence and create safer and more effective and individualized treatment is enormous, and enhance the quality of life of patients with liver and prostate cancers.

REFERENCES

Aldosari, B. N., Alfagih, I. M., & Almurshedi, A. S. (2021). Lipid nanoparticles as delivery systems for RNA-based vaccines. *Pharmaceutics, 13*(2), 206. DOI: 10.3390/pharmaceutics13020206 PMID: 33540942

AnishRuban. (1880). S., Raj, F. J., & Thangaraj, P. (2025). Phytochemical intervention in BCRP-driven cancer drug resistance: A comprehensive review. Biochimica et Biophysica Acta (BBA)-. *Revisiones en Cáncer*, (3), 189349.

Balasubramanian, A., & Sundrud, M. S. (2023). ATP-dependent transporters: Emerging players at the crossroads of immunity and metabolism. *Frontiers in Immunology, 14*, 1286696. DOI: 10.3389/fimmu.2023.1286696 PMID: 38022644

Boehnke, N., Straehla, J. P., Safford, H. C., Kocak, M., Rees, M. G., Ronan, M., Rosenberg, D., Adelmann, C. H., Chivukula, R. R., Nabar, N., Ebrahimkhani, M. R., & Hammond, P. T. (2022). Massively parallel pooled screening reveals genomic determinants of nanoparticle delivery. *Science, 377*(6604), eabm5551. Advance online publication. DOI: 10.1126/science.abm5551 PMID: 35862544

Buschmann, M. D., Carrasco, M. J., Alishetty, S., Paige, M., Alameh, M.-G., & Weissman, D. (2021). Nanomaterial Delivery Systems for mRNA Vaccines. *Vaccines, 9*(1), 65. DOI: 10.3390/vaccines9010065 PMID: 33478109

Buschmann, M. D., Carrasco, M. J., Alishetty, S., Paige, M., Alameh, M.-G., & Weissman, D. (2021). Nanomaterial delivery systems for mRNA vaccines. *Vaccines, 9*(1), 65. DOI: 10.3390/vaccines9010065 PMID: 33478109

Buschmann, M. D., Carrasco, M. J., Alishetty, S., Paige, M., Alameh, M.-G., & Weissman, D. (2021). Nanomaterial Delivery Systems for mRNA Vaccines. *Vaccines, 9*(1), 65. DOI: 10.3390/vaccines9010065 PMID: 33478109

Chen, F., Alphonse, M., & Liu, Q. (2020). Strategies for nonviral nanoparticle-based delivery of CRISPR/Cas9 therapeutics. *Wiley Interdisciplinary Reviews. Nanomedicine and Nanobiotechnology, 12*(3), e1609. DOI: 10.1002/wnan.1609 PMID: 31797562

Chen, X.-Y., Wu, Z.-X., Wang, J.-Q., Teng, Q.-X., Tang, H., Liu, Q., Chen, Z.-S., & Chen, W. (2024). Multidrug resistance transporters P-gp and BCRP limit the efficacy of ATR inhibitor ceralasertib in cancer cells. *Frontiers in Pharmacology, 15*, 1400699. DOI: 10.3389/fphar.2024.1400699 PMID: 38756373

Chiang, C. L.-L., Rovelli, R., Sarivalasis, A., & Kandalaft, L. E. (2021). Integrating cancer vaccines in the standard-of-care of ovarian cancer: Translating preclinical models to human. *Cancers (Basel), 13*(18), 4553. DOI: 10.3390/cancers13184553 PMID: 34572778

Costa, P. A. C., da Silva, W. N., Moura Prazeres, P. H. D., Lima, A. S., Oliveira, V. P., Silva, J. B., Santos, R. R., Birbrair, A., Azevedo, V., & Guimarães, P. P. G. (2024). siRNA lipid nanoparticles for CXCL12 silencing modulate brain immune response during Zika infection. *Biomedicine & Pharmacotherapy = Biomédecine & Pharmacothérapie, 170*, 115981. DOI: 10.1016/j.biopha.2023.115981 PMID: 38091634

Dai, L., Shen, G., Wang, Y., Yang, P., Wang, H., & Liu, Z. (2021). PSMA-targeted melanin-like nanoparticles as a multifunctional nanoplatform for prostate cancer theranostics. *Journal of Materials Chemistry. B*, *9*(4), 1151–1161. DOI: 10.1039/D0TB02576C PMID: 33434248

Davodabadi, F., Sajjadi, S. F., Sarhadi, M., Mirghasemi, S., Nadali Hezaveh, M., Khosravi, S., Kamali Andani, M., Cordani, M., Basiri, M., & Ghavami, S. (2023). Cancer chemotherapy resistance: Mechanisms and recent breakthrough in targeted drug delivery. *European Journal of Pharmacology*, *958*, 176013. DOI: 10.1016/j.ejphar.2023.176013 PMID: 37633322

Dhatchinamoorthy, K., Colbert, J. D., & Rock, K. L. (2021). Cancer Immune Evasion Through Loss of MHC Class I Antigen Presentation. *Frontiers in Immunology*, *12*, 636568. DOI: 10.3389/fimmu.2021.636568 PMID: 33767702

Ding, L., Agrawal, P., Singh, S. K., Chhonker, Y. S., Sun, J., & Murry, D. J. (2024). Polymer-Based Drug Delivery Systems for Cancer Therapeutics. *Polymers*, *16*(6), 843. DOI: 10.3390/polym16060843 PMID: 38543448

Eftekhari, A., Kryschi, C., Pamies, D., Gulec, S., Ahmadian, E., Janas, D., Davaran, S., & Khalilov, R. (2023). Natural and synthetic nanovectors for cancer therapy. *Nanotheranostics*, *7*(3), 236–257. DOI: 10.7150/ntno.77564 PMID: 37064613

El-Ashmawy, N. E., Al-Ashmawy, G. M., Hamada, O. B., & Khedr, N. F. (2025). The role of ABCG2 in health and disease: Linking cancer therapy resistance and other disorders. *Life Sciences*, *360*, 123245. DOI: 10.1016/j.lfs.2024.123245 PMID: 39561874

Emran, T. B., Shahriar, A., Mahmud, A. R., Rahman, T., Abir, M. H., Siddiquee, M. F.-R., Ahmed, H., Rahman, N., Nainu, F., Wahyudin, E., & Sarker, M. R. (2022). Multidrug Resistance in Cancer: Understanding Molecular Mechanisms, Immunoprevention and Therapeutic Approaches. *Frontiers in Oncology*, *12*, 891652. DOI: 10.3389/fonc.2022.891652 PMID: 35814435

Fajrial, A. K., He, Q. Q., Wirusanti, N. I., Slansky, J. E., & Ding, X. (2020). A review of emerging physical transfection methods for CRISPR/Cas9-mediated gene editing. *Theranostics*, *10*(12), 5532–5549. DOI: 10.7150/thno.43465 PMID: 32373229

Gambacorta, N., Gasperi, V., Guzzo, T., Di Leva, F. S., Ciriaco, F., Sánchez, C., Tullio, V., Rozzi, D., Marinelli, L., Topai, A., Nicolotti, O., & Maccarrone, M. (2023). Exploring the 1,3-benzoxazine chemotype for cannabinoid receptor 2 as a promising anti-cancer therapeutic. *European Journal of Medicinal Chemistry*, *259*, 115647. DOI: 10.1016/j.ejmech.2023.115647 PMID: 37478557

Garcia, G. G., Schmidt, C. J., & Hajal, C. (2024). The tumor microenvironment in therapy resistance. *Frontiers in Lab on a Chip Technologies*, *3*, 1420233. DOI: 10.3389/frlct.2024.1420233

Gibson, E. G., & Deweese, J. E. (2022). Structural and Biochemical Basis of Etoposide-Resistant Mutations in Topoisomerase IIα. *Symmetry*, *14*(7), 1309. DOI: 10.3390/sym14071309

Guimarães, P. P. G., Figueroa-Espada, C. G., Riley, R. S., Lammers, T., Dreaden, E. C., Sempkowski, M., Bhatia, S. N., Hammond, P. T., Langer, R., Jaklenec, A., & Blankschtein, D. (2023). In vivo bone marrow microenvironment siRNA delivery using lipid–polymer nanoparticles for multiple myeloma therapy. *Proceedings of the National Academy of Sciences of the United States of America*, *120*(25), e2215711120. DOI: 10.1073/pnas.2215711120 PMID: 37310997

He, M., Cao, Y., Chi, C., Yang, M., Zhang, Y., Chen, W., Zhang, H., Li, W., Wang, J., & Liu, Z. (2023). Unleashing novel horizons in advanced prostate cancer treatment: Investigating the potential of prostate specific membrane antigen-targeted nanomedicine-based combination therapy. *Frontiers in Immunology*, *14*, 1265751. DOI: 10.3389/fimmu.2023.1265751 PMID: 37795091

Ho, W., Gao, M., Li, F., Li, X., Zhang, X. Q., & Xu, X. (2021). Next-Generation Vaccines: nanoparticle-Mediated DNA and mRNA Delivery. *Advanced Healthcare Materials*, *10*(8), 2001812. DOI: 10.1002/adhm.202001812 PMID: 33458958

Ichihara, G., Katsumata, Y., Sugiura, Y., Matsuoka, Y., Maeda, R., Endo, J., Anzai, A., Shirakawa, K., Moriyama, H., Kitakata, H., & Fukuda, K. (2023). MRP1-Dependent Extracellular Release of Glutathione Induces Cardiomyocyte Ferroptosis After Ischemia-Reperfusion. *Circulation Research*, *133*(10), 861–876. DOI: 10.1161/CIRCRESAHA.123.323517 PMID: 37818671

Jörgensen, A. M., Wibel, R., & Bernkop-Schnürch, A. (2023). Biodegradable Cationic and Ionizable Cationic Lipids: A Roadmap for Safer Pharmaceutical Excipients. *Small*, *19*(17), 2206968. DOI: 10.1002/smll.202206968 PMID: 36610004

Knaggs, K. L. M., Sun, Y., Walz, B. A., Pang, J., & Khan, O. F. (2025). The role of excipients in lipid nanoparticle metabolism: Implications for enhanced therapeutic effect. *Therapeutic Delivery*, *16*(7), 687–700. DOI: 10.1080/20415990.2025.2506977 PMID: 40438965

Knezevic, I., Liu, M. A., Peden, K., Zhou, T., & Kang, H.-N. (2021). Development of mRNA vaccines: Scientific and regulatory issues. *Vaccines*, *9*(2), 81. DOI: 10.3390/vaccines9020081 PMID: 33498787

Liu, H., Zhu, X., Wei, Y., Song, C., & Wang, Y. (2023). Recent advances in targeted gene silencing and cancer therapy by nanoparticle-based delivery systems. *Biomedicine & Pharmacotherapy = Biomédecine & Pharmacothérapie*, *157*, 114065. DOI: 10.1016/j.biopha.2022.114065 PMID: 36481408

Liu, J. Q., Zhang, C., Zhang, X., Zhao, F., Li, J., Yang, Y., Chen, S., Wang, H., Zhang, Y., & Liu, W. (2022). Intratumoral delivery of IL-12 and IL-27 mRNA using lipid nanoparticles for cancer immunotherapy. *Journal of Controlled Release : Official Journal of the Controlled Release Society*, *345*, 306–313. DOI: 10.1016/j.jconrel.2022.03.021 PMID: 35301053

Meng, L., Teng, Z., Yang, S., Wang, X., Li, J., Liu, W., Zhang, Y., Wang, X., Chen, Q., & Zhang, Z. (2023). Biomimetic nanoparticles for DC vaccination: A versatile approach to boost cancer immunotherapy. *Nanoscale*, *15*(14), 6432–6455. DOI: 10.1039/D2NR07071E PMID: 36916703

Mitsogiannis, I., Tzelves, L., Dellis, A., Issa, H., Papatsoris, A., & Moussa, M. (2022). Prostate cancer immunotherapy. *Expert Opinion on Biological Therapy*, *22*(5), 577–590. DOI: 10.1080/14712598.2022.2027904 PMID: 35037527

Mitsogiannis, I., Tzelves, L., Dellis, A., Issa, H., Papatsoris, A., & Moussa, M. (2022). Prostate cancer immunotherapy. *Expert Opinion on Biological Therapy*, *22*(5), 577–590. DOI: 10.1080/14712598.2022.2027904 PMID: 35037527

Mohamed Isa, E. D., Ahmad, H., Abdul Rahman, M. B., & Gill, M. R. (2021). Progress in Mesoporous Silica Nanoparticles as Drug Delivery Agents for Cancer Treatment. *Pharmaceutics*, *13*(2), 152. DOI: 10.3390/pharmaceutics13020152 PMID: 33498885

Mollanoori, H., Rahmati, Y., Hassani, B., Mehr, M. H., & Teimourian, S. (2021). Promising therapeutic approaches using CRISPR/Cas9 genome editing technology in the treatment of Duchenne muscular dystrophy. *Genes & Diseases*, *8*(2), 146–156. DOI: 10.1016/j.gendis.2019.12.007 PMID: 33997161

Mu, X., Zhang, M., Wei, A., Yang, F., Yin, J., Wang, X., Li, Y., Liu, Y., Chen, H., & Zhang, H. (2021). Doxorubicin and PD-L1 siRNA co-delivery with stem cell membrane-coated polydopamine nanoparticles for the targeted chemoimmunotherapy of PCa bone metastases. *Nanoscale*, *13*(19), 8998–9008. DOI: 10.1039/D0NR08024A PMID: 33973580

Murugesan, R., & Raman, S. (2022). Recent Trends in Carbon Nanotubes Based Prostate Cancer Therapy: A Biomedical Hybrid for Diagnosis and Treatment. *Current Drug Delivery*, *19*(2), 229–237. DOI: 10.2174/18755704MTE08NDki3 PMID: 33655834

Murugesan, R., & Raman, S. (2022). Recent Trends in Carbon Nanotubes Based Prostate Cancer Therapy: A Biomedical Hybrid for Diagnosis and Treatment. *Current Drug Delivery*, *19*(2), 229–237. DOI: 10.2174/18755704MTE08NDki3 PMID: 33655834

O'Leary, B., Skinner, H., Schoenfeld, J. D., Licitra, L., Le Tourneau, C., Esdar, C., Schroeder, A., Salmio, S., & Psyrri, A. (2024). Evasion of apoptosis and treatment resistance in squamous cell carcinoma of the head and neck. *Cancer Treatment Reviews*, *129*, 102773. DOI: 10.1016/j.ctrv.2024.102773 PMID: 38878677

Runnels, J., Bloom, J. R., Hsieh, K., Chen, Y., Wang, Z., Li, X., Zhang, M., Liu, W., Yang, C., & Wang, X. (2023). Combining Radiotherapy and Immunotherapy in Head and Neck Cancer. *Biomedicines*, *11*(8), 2097. DOI: 10.3390/biomedicines11082097 PMID: 37626594

Scarpini, S., Morigi, F., Betti, L., Dondi, A., Biagi, C., & Lanari, M. (2021). Development of a vaccine against human cytomegalovirus: Advances, barriers, and implications for the clinical practice. *Vaccines*, *9*(6), 551. DOI: 10.3390/vaccines9060551 PMID: 34070277

Sebastiani, F., Yanez Arteta, M., Lerche, M., Porcar, L., Lang, C., Bragg, R., Elmore, C. S., Krishnamurthy, R., Cárdenas, M., Nylander, T., & Dabkowska, A. P. (2021). Apolipoprotein E Binding Drives Structural and Compositional Rearrangement of mRNA-Containing Lipid Nanoparticles. *ACS Nano*, *15*(4), 6709–6722. DOI: 10.1021/acsnano.0c10064 PMID: 33754708

Sun, X., Zhao, P., Lin, J., Chen, K., & Shen, J. (2023). Recent advances in access to overcome cancer drug resistance by nanocarrier drug delivery system. *Cancer Drug Resistance (Alhambra, Calif.)*, *6*(2), 390–415. DOI: 10.20517/cdr.2023.16 PMID: 37457134

Sung, H., Ferlay, J., Siegel, R. L., Laversanne, M., Soerjomataram, I., Jemal, A., & Bray, F. (2021). Global Cancer Statistics 2020: GLOBOCAN Estimates of Incidence and Mortality Worldwide for 36 Cancers in 185 Countries. *CA: a Cancer Journal for Clinicians*, *71*(3), 209–249. DOI: 10.3322/caac.21660 PMID: 33538338

Tenchov, R., Sasso, J. M., & Zhou, Q. A. (2023). PEGylated Lipid Nanoparticle Formulations: Immunological Safety and Efficiency Perspective. *Bioconjugate Chemistry*, *34*(6), 941–960. DOI: 10.1021/acs.bioconjchem.3c00174 PMID: 37162501

Tian, Y., Liu, Z., Wang, J., Huang, W., Li, S., Zhang, N., Wang, H., Xu, Y., Cao, Y., & Liu, Q. (2023). Nanomedicine for Combination Urologic Cancer Immunotherapy. *Pharmaceutics*, *15*(2), 546. DOI: 10.3390/pharmaceutics15020546 PMID: 36839868

Wang, Q., Atluri, K., Tiwari, A. K., & Babu, R. J. (2023). Exploring the Application of Micellar Drug Delivery Systems in Cancer Nanomedicine. *Pharmaceuticals (Basel, Switzerland)*, *16*(3), 433. DOI: 10.3390/ph16030433 PMID: 36986532

Wang, Z., Xing, H., Liu, A., Zhang, J., Chen, L., Zhu, W., Li, X., Wang, Y., Zhang, X., & Liu, Q. (2023). Multifunctional nano-system for multi-mode targeted imaging and enhanced photothermal therapy of metastatic prostate cancer. *Acta Biomaterialia*, *166*, 581–592. DOI: 10.1016/j.actbio.2023.05.014 PMID: 37172637

Wong, K. H., Yang, D., Chen, S., He, C., & Chen, M. (2022). Development of nanoscale drug delivery systems of dihydroartemisinin for cancer therapy: A review. *Asian Journal of Pharmaceutical Sciences*, *17*(4), 475–490. DOI: 10.1016/j.ajps.2022.04.005 PMID: 36105316

Wu, Z., Wang, Y., Zhang, Y., Chen, W., Wang, Y., & Ji, Q. (2020). Strategies for developing CRISPR-based gene editing methods in bacteria. *Small Methods*, *4*(2), 1900560. DOI: 10.1002/smtd.201900560

Xie, L., Li, J., Wang, L., & Dai, Y. (2023). Engineering metal-phenolic networks for enhancing cancer therapy by tumor microenvironment modulation. *Wiley Interdisciplinary Reviews. Nanomedicine and Nanobiotechnology*, *15*(3), e1864. DOI: 10.1002/wnan.1864 PMID: 36333962

Yoo, H., Kim, Y., Kim, J., Cho, H., & Kim, K. (2024). Overcoming Cancer Drug Resistance with Nanoparticle Strategies for Key Protein Inhibition. *Molecules (Basel, Switzerland)*, *29*(17), 3994. DOI: 10.3390/molecules29173994 PMID: 39274842

Chapter 13

Artificial Intelligence in Drug Manufacturing, Quality Control, and Post-Market Surveillance

Kazi Asraf Ali
http://orcid.org/0000-0002-1753-7504
Maulana Abul Kalam Azad University of Technology West Bengal, India

Sabyasachi Choudhuri
http://orcid.org/0000-0002-7736-8490
Maulana Abul Kalam Azad University of Technology West Bengal, India

Sachin Kuila
Maulana Abul Kalam Azad University of Technology West Bengal, India

Paramita Paul
University of North Bengal, India

Rituparna Das
Bengal School of Technology, India

Sanjit Kumar Roy
Maulana Abul Kalam Azad University of Technology West Bengal, India

ABSTRACT

Artificial Intelligence (AI) is now widely used not just in early drug discovery but also in drug manufacturing, quality control, and post market surveillance. This chapter looks at how AI is being used throughout the pharmaceutical product lifecycle, with a focus on new developments in manufacturing, better quality assurance, and real-time monitoring of drugs already on the market. The chapter starts by looking at how AI technologies like machine learning, deep learning, computer vision, and robotics are affecting pharmaceutical manufacturing. It covers topics such as predictive maintenance, real-time process monitoring, digital twins, and autonomous process control, which all help make manufacturing smarter as part of Industry 4.0. The chapter also covers the regulatory environment, concerns, and ethics of AI in the production of pharmaceuticals and their monitoring and provides a vision of how these technologies will ensure safety, effectiveness, and efficiency in the evolving AI-enabled pharmaceuticals sector.

DOI: 10.4018/979-8-3373-9715-3.ch013

1. INTRODUCTION

1.1 Overview of AI in the Pharmaceutical Lifecycle

Artificial intelligence is transforming the pharmaceutical life cycle by bringing greater efficiency, accuracy, and innovation to every stage, from early drug discovery to post-market surveillance (Oualikene-Gonin et al., 2024; Sampene & Nyirenda, 2024). In drug discovery, AI systems can rapidly analyze vast amounts of biological and chemical data to identify promising drug candidates, predict how molecules may behave, and optimize their structures for better performance (Gupta et al., 2021). During preclinical development, AI models simulate complex biological interactions, helping researchers reduce reliance on animal testing while accelerating the evaluation of drug safety and effectiveness (Gangwal & Lavecchia, 2025). In clinical trials, AI improves patient recruitment through predictive analytics, enables real-time monitoring of patient data, and supports earlier identification of potential adverse events, ultimately making trials more efficient and patient-centered (Chopra et al., 2023b; Harrer et al., 2019). Beyond development, AI-driven tools also assist regulatory processes by organizing and analyzing large datasets, ensuring accuracy, and supporting compliance with global regulatory standards. Together, these applications illustrate how AI is reshaping pharmaceutical development into a faster, smarter, and more responsive process. RiskWatch, DocShifter, Freyr SubmitPro, Veeva Vault, cortical.io, Litera Microsystems, and other AI-based tools already in use by regulatory professionals are examined, along with the advantages and challenges of incorporating these cutting-edge systems into regulatory procedures (Ajmal et al., 2025). In the manufacturing process, AI enhances production efficiency, quality control, and supply chain management by predicting demand fluctuations and optimizing inventory (Pasupuleti et al., 2024). Finally, in post-market surveillance, AI continuously monitors real-world data to detect adverse drug reactions and improve pharmacovigilance efforts (Nagar et al., 2025a) . Overall, AI is not only expediting the development of new treatments but also improving patient outcomes and making the pharmaceutical industry more agile and data-driven.

Key Digital Technologies Driving the Transformation

A range of digital technologies is driving the transformation of the pharmaceutical sector, each contributing to smarter, more connected, and more efficient operations (Sandner et al., 2020). The Internet of Things enables real-time monitoring of equipment and supports predictive maintenance, helping prevent unexpected failures and reduce downtime. Artificial intelligence and machine learning play a central role in analyzing complex datasets, generating predictive insights, and supporting activities such as drug discovery and process optimization. Blockchain technology adds a layer of security and transparency by enabling reliable tracking of pharmaceutical products across the supply chain, thereby reducing the risk of counterfeiting. Cloud computing provides scalable data storage and collaborative platforms that allow teams to access, share, and analyze information efficiently. Robotics and automation enhance manufacturing performance by increasing speed, consistency, and precision while minimizing human error. Big data analytics brings together information from clinical trials, market trends, and real-world evidence, turning large and complex datasets into actionable insights that support better decision-making across the pharmaceutical lifecycle.

1.2 Need for Digital Transformation in Pharmaceutical Manufacturing and Surveillance

Digital transformation has become increasingly important in pharmaceutical manufacturing and surveillance as the industry faces growing complexity, stricter regulatory expectations, and a constant demand for efficiency and transparency (Subha et al., 2024). Traditional manufacturing and monitoring systems often rely on manual processes, paper-based documentation, or isolated data silos, making it difficult to keep pace with global supply chains, personalized therapies, and rising quality standards (Hammad et al., 2026). By adopting digital technologies such as IoT sensors, AI-driven analytics, and automated production systems, manufacturers can monitor processes in real time, maintain better control over operations, and quickly detect factors that may compromise product quality (Sarkar & Paul, 2025). In the context of pharmaceutical surveillance, digital transformation enables seamless integration of diverse data sources, ranging from manufacturing records to real-world patient data. This connected approach supports more timely, accurate, and proactive pharmacovigilance, ultimately improving patient safety and regulatory compliance (Shamim et al., 2024). There are offensive reasons why digital transformation in pharmaceutical manufacturing and monitoring is a growing necessity. Both industries are confronting changing needs whose changes can be relieve through the incorporation of emerging digital technologies. A few of the critical requirements or benefits of digital transformation in these areas are as follows:

For the Improved Efficiency and Productivity

Digital transformation enables better process automation and smart manufacturing, significantly increasing efficiency and productivity. Pharmaceutical sector is accustomed to high precision and productivity. Production can be made leaner through digital transformation by automating menial tasks, lessening human errors and improving operational efficiency (Rajesh & Elumalai, 2025). Digital Equipment & Real-time machine/ Latest technology tools like IoT Sensors, Ai based analytics can monitor the equipment and process and smart manufacturing. This enables predictive maintenance, minimizing downtime, and optimizing production schedules, which is critical in meeting demand without compromising quality (Rakholia et al., 2025; Soori et al., 2023).

Regulatory Compliance and Traceability

The industry is regulated with stringent data validities accuracy, transparency and traceabilities. Digital technology can also provide guarantee for data integrity that is the data collected, kept and auditable from drug's beginning until its end (from development to distribution) using ALCOA+ Principles (Sembiring & Novagusda, 2024). Digital platforms are able to track the manufacturing conditions (temperature, humidity, etc.) in real time so that compliance with Good Manufacturing Practices (GMP) is maintained on-line and automatically documented for audits (Kodumuru et al., 2025a).

Enhanced Drug Safety and Surveillance

In the area of pharmacovigilance, digital solutions can be used to facilitate faster and more efficient collection and analysis of post-market safety information. AI models/ML can play a role in safety signal detection, sifting through pools of patient reports from clinical trials or medical databases to help identify potential problems faster and more efficiently than burgundy book team analyses (Nagar et al., 2025b). Ever-faster real world evidence Digital transformation really catalyses that process in accelerated and broader acquisition of real-time information from the field, which are often essential to oversee long-term safety and efficacy of pharmaceutical drugs after they go off the market (W. C. Lee et al., 2025).

Cost Reduction and Resource Optimization

Pharmaceuticals Reduce operational costs by automating periodic tasks, and collaborate over predictive maintenance or regulatory compliance management processes (Rajesh & Elumalai, 2025). Digital solutions can maximize human resources, production capacity and raw material utilization leading to the efficient use of resources for the complete manufacturing process (Jagtap et al., 2021).

Faster Time-to-Market

Next digitalization Digitization of the economy makes possible new more agile production processes that enables faster adjustment to both market demands and regulation changes (Gunasekaran & Yusuf, 2002). This holds even more relevance in situations like fast-track drug development and production, characteristic of health emergencies (like the COVID-19 pandemic). AI and data analytics are now being more widely deployed in drug discovery and development, which has the potential to expedite identification of candidate therapies for clinical trials, reducing the time needed to bring a new drug to market thus resulted in accelerated R&D (Jarallah et al., 2025).

Improved Patient-Centric Models

Personalized Medicines: Personalized drug treatment for patients based upon patient data, use of genomic information, and real-world evidence (Heesen et al., 2024). This can result in better patient outcomes and fewer wasteful treatments. Remote patient monitoring can also help patients adhere to their prescribed medications and provide continuous data to healthcare professionals on the effectiveness of treatment (Rakholia et al., 2025).

Increased Competitive Advantage: The Advantages are as Follows

Innovation and Market Leadership: Enterprises embracing digital transformation will innovate faster, more informed by market insight and customer feedback. This can be a decisive competitive advantage in an industry where innovation is what keeps you in the lead.

Digital Platforms for Customer Engagement: Digital tools can enable pharmaceutical companies to deepen their relationships with healthcare professionals and consumers by disseminating more relevant, timely information on products, clinical trials, and safety updates.

Enhanced Data Analytics and AI/ML Integration

The Use of AI, machine learning, and big data analytics in pharma manufacturing provides a better understanding of production processes and quality control, as well as market behavior. Predictive and prescriptive analytics, for example, can anticipate potential disruptions or forecast equipment breakdowns, enabling preventive measures to avoid unnecessary risks and maintain uninterrupted service.

Global Collaboration and Knowledge Sharing

Like manufacturers, regulators, suppliers, and healthcare providers working in the pharmaceutical value chain can facilitate real-time collaboration. This supports decision-making and enables the transmission of crucial intelligence. The use of digital platforms can foster partnerships among research institutions, start-ups, and major pharmaceutical companies to accelerate the discovery and development of drugs.

So, Digital transformation is no longer a nice-to-have for pharmaceutical manufacturers and regulatory agencies — it's the difference between being competitive, compliant, and responsive at speed in an increasingly dynamic global environment. By embracing digital technologies, pharmaceutical companies can further improve efficiency, enhance drug safety, ensure compliance, and better serve both healthcare professionals and patients.

1.3 Scope of AI Applications Beyond Discovery and Design

AI has already been involved as revolutionizing a wide variety of fields far beyond drug discovery and drug design. AI is no longer limited to a few specialized applications; its influence continues to grow across diverse sectors, often in ways that were once difficult to imagine. In healthcare, in particular, AI has emerged as a powerful tool with far-reaching impact. AI models can analyze medical images such as MRI scans and X-rays to support early disease detection and help predict potential health risks based on a patient's medical history (Ghaffar Nia et al., 2023). Beyond diagnosis, AI enables the development of personalized treatments by integrating genetic information with lifestyle and clinical data, allowing therapies to be tailored to individual patients (Ghanem et al., 2024). AI-powered chatbots and virtual assistants further enhance patient care by providing real-time health guidance, supporting the management of chronic conditions, and answering health-related questions around the clock (Kurniawan et al., 2024). In drug development, AI contributes to faster and more efficient approval processes by optimizing clinical trial design and closely monitoring patient responses (Chopra et al., 2023a). AI also forms the foundation of autonomous technologies, enabling robots, drones, and vehicles to operate independently. In manufacturing, warehouses, and even surgical settings, AI-driven robots perform complex tasks with increasing precision and reliability (Deeba et al., 2025).

Beyond healthcare, AI plays a significant role in finance and risk management. It supports real-time fraud detection by continuously monitoring transaction patterns and flagging suspicious activities (Upadhyaya, 2025). AI is also widely used in stock market prediction, automated trading, and portfolio optimization, helping investors make data-driven decisions (Sutiene et al., 2024). In customer service, AI-powered virtual assistants and chat systems enhance user experience through rapid responses and personalized interactions (Chaturvedi & Verma, 2023). Collectively, these examples highlight the broad and transformative impact of AI-driven technologies, including their growing influence across all stages of the pharmaceutical life cycle (Huanbutta et al., 2024a).

2. AI IN DRUG MANUFACTURING

2.1 Predictive Modeling and Process Optimization

Predictive modeling and process optimization work hand-in-hand to enhance decision-making and operational efficiency across industries. Predictive modeling uses statistical algorithms and machine learning techniques to analyze historical and real-time data, identifying patterns that help forecast future outcomes such as demand levels, equipment failures, or quality deviations (Strielkowski et al., 2023). These predictions feed directly into process optimization, where mathematical optimization, simulation or AI-based methods are used to find the best possible actions or configurations in order to meet certain objectives—be it minimising costs, waste, time spent or improving quality of output (Krzywanski et al., 2024). Together, they provide a proactive, data-driven strategy that allows businesses to better anticipate impact and allocate resources more efficiently while remaining agile in adapting operations for increased efficiency and resilience.

2.2 Process Analytical Technologies (PAT) Enhanced by AI

AI-driven PAT is revolutionizing drug production through continuous, intelligent monitoring and control of manufacturing processes. Models based on AI can make sense of data from sensors, spectroscopic tools and inline measurements in real time to detect subtle trends or deviations that may not be noticed by humans (Sharma et al., 2024). This can result in predictive corrections on production which make quality more consistent and prevent batch failures. Furthermore, machine learning can facilitate adaptive process control where parameters and process conditions are updated with varying inputs. Connecting PAT and AI, pharmaceutical manufacturers are provided a more powerful, efficient and robust production setup to enable continuous manufacturing, reduce costs and increase regulatory compliance through deeper process understanding.

2.3 Real-Time Process Control Using ML and Deep Learning

AI, ML, and deep learning (DL) are transforming **real-time process control in pharmaceutical drug manufacturing.** And AI is also changing how drugs are made, thanks you ML and ISIL real-time-control process (Kodumuru et al., 2025a). State of-the-art models monitor the time series of sensor like temperature, pressure, pH or spectral to identify PVT except for human perception error (H. Zhou et al., 2025). Before deviations occur, these systems can anticipate them and modify equipment parameters automatically or determine the best reaction conditions to achieve consistent product quality. DL solution can be effective to handle complex and high-dimensional data generated by PAT tools, accelerating decision making and minimizing batch failure. In this context, the AI powered real-time monitoring increases efficiency, compliance with regulations and speed of providing safe and high-quality pharmaceuticals.

2.4 Robotics and Automation in Formulation and Packaging

With the development of artificial intelligence, drug production is changing as robotics and automation are incorporated into formulation/formulation process and packaging. Using artificial intelligence (AI)-powered systems to analyze large data sets enables the calculation of optimal mixing times, ingredient proportions and conditions so that coherent, high-quality recipes can be consistently developed. In the production line, an intelligent robot based on machine-vision algorithms can achieve better precision and faster performance for dosing, tablet inspection and sterile filling compared to manual operation (Xia et al., 2021).AI automatically detects defects on packing lines, adjusts labeling design on the spot and traces products for compliance with all regulations. In combination these technologies increase efficiencies, reduce errors, improve safety and allow for more versatile data-driven pharmaceutical production (Huanbutta et al., 2024b).

2.5 Use of Digital Twins for Manufacturing Simulation

AI is transforming drug manufacturing by enabling smarter, more efficient, and highly controlled production processes. One of the most impactful applications is the use of **digital twins**, virtual replicas of physical manufacturing systems that allow companies to simulate, predict, and optimize operations in real time (Maharjan et al., 2025). When AI-based analytics are combined with these digital models, equipment performance can be predicted and process changes can be tested without affecting real production; potential quality issues may be identified early and product purity can stay more consistent.This combination of AI and digital twins supports faster scale-up, reduces waste, lowers costs, and strengthens compliance with regulatory standards, ultimately accelerating the delivery of safe and effective medicines(Chaparro-Cárdenas et al., 2025).

2.6 Case Studies: Pfizer, Novartis, and GSK Smart Factory Initiatives

Major pharmaceutical companies such as Pfizer, Novartis and GlaxoSmithKline (GSK) have deployed “smart-factory” or digital / Industry 4.0 manufacturing initiative by applying digital technologies to their operations (Phiri et al., 2025). “Smart manufacturing” or “Pharma 4.0” refers to using digital technologies such as IoT sensors, automation, robotics, data analytics, AI, ML, digital twins, real-time monitoring, continuous manufacturing, to modernize traditional batch-based and manual pharmaceutical manufacturing (Arden et al., 2021). GSK improved production efficiency and reduced downtime using IoT and Big Data to get consistent quality, while minimizing human error. Pfizer used AI-driven analytics ‘digital twin’ to optimize to optimize processes, quality and its supply chain as per the market demands. Novartis adopted AI for predictive maintenance, IoT for real-time monitoring, and PAT to control of manufacturing quality and ensuring consistency which helped lower production costs and increase efficiency (Phiri et al., 2025).

3. AI IN QUALITY CONTROL

3.1 AI Detecting Defects in Tablets, Capsules and Vials

Artificial intelligence is a breakthrough in pharmaceutical industry in quality analysis of the products. The images of tablets are identified and analysed by AI algorithms and computer system to identify possible flaws like chips, cracks, discolouration, changes in shape and size. AI algorithms are identified as high-level precision and recall which can simply identify faults by massive data set of interpreted photographs (Vora et al., 2023a). Traditional methods like X-ray computed tomography used for quality analysis of interior surface of tablets are time consuming and is a hindrance to demand in rapid production. Hence, deep learning is implemented along with X-ray tomography for physiochemical quality analysis of tablets. AI powered defect detections come with more accuracy as well as identification speed that made it took over the manual inspection thereby reduced the human errors and subjective judgement. AI systems' real-time monitoring capabilities guarantee the early identification of flaws, enabling rapid intervention and averting the distribution of defective tablets (Andrianandrianina Johanesa et al., 2024; Broday, 2022; Divyansh Jain, 2024).

Quality of pharmaceutical products directly impacts human health so this marks the importance of ensuing improved techniques for production of superior standard products and minimizing health concerns. Despite their relevance in healthcare, capsule-based formulations frequently have a number of flaws because of production technological restrictions (Bian et al., 2024; J. Zhou et al., 2020). The flaws include bubble entrapment, shrinkage, deformities leading to irregular dosing, reduced effectiveness which all together impacts the potency and shelf life of the drug. Surface morphology analysis of capsule formulation can be both external and internal and defects detection emphasize on image classification and computer-based vision analysis. Recently computer vision-based defect detection solutions are adopted mostly by the medical sectors to detection of defect in capsule formulations. To accomplish online capsule defect identification, a scanning-based area growth technique was used, which was based on conventional image segmentation methods (J. Zhou et al., 2020).

Making sterile injectables is a difficult and intricate procedure and by hand it has always been regarded as a risky endeavour. As per GMP (good manufacturing practice) regulations, proposed by the FDA regulatory to establish maximum manufacturing standards. The predictive analytics has been used in machine learning to process data for parenteral product sterility, which is the key attribute for product safety, using computer algorithms. Thus, the goal of pharmaceutical development is to create products with a predetermined level of quality through the use of a specified manufacturing procedure. Since the use of sterile injectable preparations has been steadily increasing in recent years, the pharmaceutical industry has decided to automate production. This can lower the possibility of human error by producing thorough and systematic high-dimensional data at a high input with computerized algorithmic language, which is necessary for future generations. Automation allows for increased safety, decreased exposure to dangerous substances, high reproducibility, and traceability of every stage of the manufacturing process. Since the manual procedure can be safely replaced by automation, parenteral production department of pharmaceutical industries are now using it to formally identify preventive measures and predict issues (Miązek & Świątowiec-Szczepańska, 2020; Mohammady et al., 2021).

3.2 Pattern Recognition in Process Deviation

A fundamental component of AI is pattern recognition, a methodological discovery for identification and analysis of patterns of data. Group of tasks which were once done by realms of human can now be done automatically by machine learning technologies making decision making much easier. Statistical techniques use ideas from statistical inference and probability theory to offer a probabilistic foundation for pattern recognition. The Bayes theorem is used by Bayesian classifiers, a well-known family of statistical models, to determine the likelihood that a data point belongs to a specific class based on likelihood functions and prior probabilities. K-means and hierarchical algorithms are part of clustering algorithms that helps in grouping similar data points and uncover underlying patterns and structures.

Prediction and connection between independent and dependent variables are done by regression analysis (C. Kalaiselvi & G. Arunasenbagam, 2017).

By highlighting the rules and connections between parts, syntactic approaches concentrate on the structural elements of patterns. Syntactic pattern modelling is made possible by formal grammars, a mathematical framework for characterizing languages. The computational model known as finite automata, which can identify regular languages, is employed in lexical analysis and pattern matching applications.

Pattern recognition has been transformed by neural networks, which are modelled after the interconnected network of neurons found in the human brain. Through several levels of abstraction, these models are able to learn hierarchical representations of data, identifying intricate patterns and characteristics that are challenging to identify with conventional techniques. By taking advantage of the spatial correlation between pixels, Convolutional Neural Networks (CNNs) are highly effective at image and video recognition tasks (A. Kumar, 2003; Suzuki, 2002).

Features Consisting of Pattern Recognition

Pattern recognition techniques comprehend several features which separates it from other AI and ML (machine learning) standards (A. Kumar, 2003).

Classification of data: The process of classifying a data sample involves placing it into one of several predetermined groups. Applications like picture recognition, spam filtering, and handwriting analysis rely on this prevalent type of pattern recognition (A. Kumar, 2025).

Clustering: Without the use of labelled data, related data points can be grouped by an unsupervised learning technique called clustering. It comes in very handy when working with big, unstructured datasets (A. Kumar, 2003).

Feature extraction: The process of turning unprocessed data into a more practicable collection of features that best captures the most important information is known as feature extraction. For pattern recognition to be effective, raw data frequently contains too much noise or irrelevant information. Feature extraction highlights the essential elements of the data that aid in finding patterns and helps to minimize dimensionality (A. Kumar, 2025).

Reduction of Dimensionality: Data may contain hundreds or thousands of features in many pattern recognition applications, which can cause calculations to be sluggish and ineffective. In order to minimize the number of features while maintaining the maximum amount of information, dimensionality reduction techniques like Principal Component Analysis (PCA) and t-SNE (t-distributed Stochastic Neighbor Embedding) are employed (Vora et al., 2023a).

Supervised and unsupervised data: There are two types of pattern recognition: supervised and unsupervised. The system is trained using labelled data in supervised learning, which means that every training example has a label or result attached to it. When the objective of a classification task is to predict a predetermined class label, this is typical.

Deep learning integration: Pattern recognition has been transformed in recent years by deep learning. In applications including speech recognition, picture identification, and natural language processing, deep neural networks (DNNs), in particular convolutional neural networks (CNNs) and recurrent neural networks (RNNs), have shown impressive performance (A. Kumar, 2003).

Utility of Pattern Recognition in Health Care

The pattern recognition is reforming the health service industry, when it comes to disease prediction, image classification and medical diagnosis applications. In radiology, for example, machines that recognize patterns can detect anomalies in pictures of MRIs or X-rays, helping identify diseases like cancer, fractures or heart disease. To uncover mutations or markers associated with particular diseases, genomics applies machine learning-based pattern recognition to huge data sets of genes. Compared to human radiologists, AI-driven models have shown greater accuracy in certain tasks, providing earlier diagnosis and more individualized treatment options (Singh et al., 2009; Suzuki, 2002).

3.3 Real-Time Release Testing Enabled by AI Models

To ensure and maintain quality standards of the production parameters in pharmaceutical industry PAT technique is adopted, a system that develops, evaluates, and controls pharmaceutical production processes using real-time measurements of critical quality attributes (CQAs). PAT is enhanced by AI and ML since they provide advanced data analysis tools. For instance, PAT regularly tracks CQAs using spectroscopy. Real-time spectrophotometric data can be analysed by machine learning algorithms to spot departures from the desired quality standards. Quick modifications to the manufacturing process based on this real-time feedback can guarantee consistent product quality (Singh et al., 2009).

3.4 Predictive Quality Modelling Using Historical Manufacturing Data

With applications ranging from yield optimization to predictive maintenance, predictive analytics has been extensively researched in broader manufacturing fields. Predictive models have been created in pharmaceutical quality assurance to anticipate batch variances, allowing for pre-emptive actions prior to nonconformities occurring. For example, time-series models have been utilized to identify microbial development in cleanroom settings, while regression-based models have been utilized to forecast dissolving results for solid dosage forms (Singh et al., 2009).

3.5 Machine Learning Algorithms for Root-Cause Analysis of Batch Failures

Root cause analysis in the context of manufacturing process is the method of identifying errors or quality deviations in manufactured products. Real root cause in unwanted quality deviation in manufacturing process can finally decrease the effectiveness of the product. One of the challenging factors in the traditional way of root cause analysis is identifying the real root cause of the process. This suggests look up with the expert knowledge to define the casual factors involved in manufacturing line. The drawbacks in this process includes – lack of clarification of the root cause source and solve among people, expert knowledge may not be stored, saved or transferred for future reference and lastly, no guarantee of accuracy in the data or expert knowledge (Khinvasara et al., 2024; Olatunde Omisola et al., 2024; Singh et al., 2009).

Automatic root cause analysis or failure diagnostics have been used previously within software and server diagnostics. Machine learning method is widely suitable because the data is of large size, digital and easily accessible. These applications being suitable for large sized manufacturing duties are extensively accessed by chemical processing units as well. A branch of artificial intelligence called "machine learning" is focused on researching and creating methods that allow computers to learn from data. Basically, the goal is to train a data-driven model that can be used to make predictions and judgments. There are strengths in other areas, particularly statistics and mathematical optimization. A probable graph that illustrates the conditional dependence structure between random variables is known as a probabilistic graphic model. Finding structured distributions and providing a concise description are made possible by a graph-based representation. In this context, the random variables are represented as nodes in a graph, and the edges show the direct, probabilistic relationships between them. There are two primary families of graphical representations of distributions: the more general framework of Markov random fields, which uses undirected graphs, and Bayesian networks, which employ directed, acyclic graphs. The joint probability distribution of fall variables is factorized in the model (Olatunde Omisola et al., 2024).

3.6 Integration of AI With Laboratory Information Management Systems

The laboratory uses laboratory information management systems (LIS), a new application technology that combines computer technology, automatic control technology, quality management science, and analysis technology, to accomplish automatic test result detection, quality control, release, and archiving. Better medical services for patients, online services for ward medical personnel, guaranteed test result correctness and dependability, and a reduction in needless human error are all possible with the launch and deployment of LIS systems. In Japan, Hitachi links over a dozen inspection instruments via a network and distributes specimens among them by combining the information management system, inspection tools, and auxiliary equipment into a sizable integrated system. Although LIS was developed relatively late in China, there aren't many differences between laboratory information management in China and other countries due to the country's rapid information technology development. This study topic creates and implements a collection of laboratory information systems (LISs) to achieve the automation, standardization, and informatization of clinical laboratory work. The current system network structure design scheme is learned via the study of the inspection department staff's needs in order to develop the function that this system requires (Q. Fu et al., 2023).

4. AI IN REGULATORY COMPLIANCE AND VALIDATION

4.1 Regulatory Considerations

The legal landscape surrounding AI and ML technologies in the pharmaceutical business is rapidly evolving, and health authorities are having a hard time adjusting existing rules to account for the unique characteristics and challenges that these technologies present. The raising acceptance of AI/ML technologies for pharmacovigilance, regulatory evaluation and drug development by health authorities is also motivating the establishment of protocols/norms for these entities. For instance, FDA has published a Digital Health Innovation Action Plan and a proposed framework for AI/ML-based medical devices to spur innovation, clarity from the regulatory standpoint and patient safety & effectiveness (E. Zhou et al., 2024).

4.2 GxP Compliance and AI Model Validation Strategies

The pharmaceutical production industry has been able to keep up with public health, thanks to the manufacture of safe and effective drug products. This industry operates under stringent rules the so-called Good Practice (GxP) guidelines to which quality requirements such as GMP for Good Manufacturing Practice, GLP for Good Laboratory Practice and GCP for Good Clinical Practice are assigned. These regulations aim to ensure patient safety and ensure regulatory compliance by ensuring that pharmaceutical products are produced and controlled in a consistent manner according to quality standards. But the added complexity of manufacturing processes, and rapid adoption of digital technologies, has raised awareness about security and integrity within the system. One key way to address these problems is by leveraging state-of-the-art technologies, including AI. The application of AI (Artificial intelligence, AI) seems to show promise and potential for enhancing the security of pharmaceutical manufacturing systems, with perhaps even greater benefits to be realized in machine learning and predictive analytics.

AI can identify patterns and anomalies in the sheer magnitude of data collected during manufacturing that may be evidence of potential security risks or compliance problems. Not only does it improve operational integrity, but this proactive approach ensures adherence to GxP standards that reduce the risk of regulatory violations and related penalties. In addition, AI is employed for reasons other than just the monitoring of regulatory compliance in pharmaceutical production.

AI presents the industry with a game changing opportunity due to its ability to enhance data quality and reliability as well as facilitate processes. The use of AI technologies does raise concerns, however, including data privacy and ethics as well as the need for personnel to interpret insights generated by AI (Sadiya Inamdar, 2024; E. Zhou et al., 2024).

4.3 AI Transparency and Explainability for Regulatory Approval

Artificial intelligence is rapidly reshaping industries worldwide by enabling greater efficiency, innovation, and data-driven decision-making. However, as AI becomes more deeply embedded in highly regulated sectors such as healthcare, banking, and law, concerns around accountability, transparency, and regulatory compliance have grown significantly. One of the key challenges is understanding how complex AI systems arrive at their decisions, especially when those decisions directly affect human lives. In this context, Explainable AI (XAI) has emerged as a critical solution. XAI focuses on making AI outputs clear, interpretable, and understandable to human stakeholders, including clinicians, regulators, and patients. As AI systems grow more sophisticated and autonomous, the ability to explain their decision-making processes is essential for building trust and ensuring responsible use. In healthcare, regulatory frameworks such as the Food and Drug Administration (FDA) guidelines and the Health Insurance Portability and Accountability Act (HIPAA) play a central role in governing AI adoption. HIPAA emphasizes the protection and transparency of patient data, requiring AI systems that handle sensitive health information to maintain strict data privacy and security standards. At the same time, the FDA oversees AI-enabled medical devices and mandates detailed documentation and clear explanations of how AI systems function and make decisions, ensuring that patient safety remains the top priority (Balasubramaniam et al., 2023).

4.4 Digital Documentation and Automated Audit Trails

A thorough comprehension and conceptualization of a study problem are made possible by qualitative research investigations. Nonetheless, there is ongoing discussion on the validity of qualitative research. Researchers that lean toward quantitative research paradigms frequently question its validity, criticizing it for lacking scientific rigor, representing a collection of subjective opinions and anecdotes, and being prone to researcher bias. Numerous data sources and a substantial amount of researcher-generated data, such as notes regarding the study's context, methodological choices, data analysis techniques, and the researcher's own self-awareness, are frequently used in qualitative research. In order to solve what refers to as the "intellectual puzzle" to comprehend and explain social phenomena by logical inference, the researcher actively immerses themselves in this body of information. Through iterative procedures of data collecting and analysis, the research design develops and evolves, necessitating frequent decisions from the researcher that may alter the study's direction (Houghton et al., 2013).

5. POST-MARKET SURVEILLANCE AND PHARMACOVIGILANCE

Post-market surveillance (PMS) and pharmacovigilance (PV) represent essential phases in the pharmaceutical product lifecycle, focused on continuous safety monitoring, risk detection, and quality assurance after drug approval. The integration of Artificial Intelligence (AI)—including machine learning (ML), deep learning (DL), and natural language processing (NLP)—has transformed these systems from passive reporting mechanisms into proactive, predictive, and intelligent safety networks (Barbieri et al., 2025).

AI-powered tools are increasingly transforming pharmacovigilance by automating many tasks that were traditionally time-consuming and manual. These technologies now support key activities such as the detection of adverse drug reactions, case processing, signal identification, and causality assessment. For example, systems like vigiMatch, developed by the Uppsala Monitoring Centre, help identify duplicate reports and detect safety signals more efficiently within large pharmacovigilance databases. Automated methods for causality assessment, including those based on Bayesian network models and decision-tree algorithms, further enhance the consistency and speed of safety evaluations. By reducing variability in assessments and shortening response times, these approaches support faster and more reliable decision-making. Importantly, the structured and interpretable outputs generated by such AI models make the results easier to understand and apply in both regulatory review and clinical practice, ultimately strengthening drug safety monitoring (Algarvio et al., 2025).

A holistic risk-based and human-centered approach to the application of AI in pharmacovigilance. This model focuses on human involvement, algorithm disclosure, privacy, fairness, and equity to facilitate the ethical and responsible use of AI. Such systems would require constant validation for reliability and trust or they may not be worthy of the label of AI- assisted safety systems (Pasas-Farmer & Jain, 2025).

AI-driven methods go beyond the traditional pharmacovigilance, covering only spontaneous reporting and embrace other real-world evidence (RWE) sources: electronic health records (EHRs), patient registries, claims databases, and social media data. Multi-modal approaches trained on such sources can be used as early warning systems to identify emerging safety signals and safety concerns earlier than traditional methodologies, enabling near real-time pharmacovigilance and public health interventions (Dimitsaki et al., 2024).

With the development of large language models (LLMs) and generative AI, PV has been further pushed by being able to automatically translate, create intelligent summarization and extract structured information from global safety reports. These solutions facilitate regulatory reporting and cross-border data sharing, thereby reinforcing global pharmacovigilance networks (Praveen et al., 2023).

5.1 NLP for Automated Adverse Event Detection from Unstructured Sources (FAERS, EHRs, and Social Media)

Natural Language Processing (NLP) has emerged as a transformative tool in pharmacovigilance (PV) for extracting and analyzing adverse drug events (ADEs) from unstructured data sources such as FDA's Adverse Event Reporting System (FAERS), Electronic Health Records (EHRs), and various social media platforms. These sources contain vast amounts of valuable but heterogeneous textual data—ranging from clinical narratives and case reports to patient-generated content—that traditional data mining methods cannot efficiently process(Golder et al., 2025).

NLP allows for automatic recognition and structuring of free text information into entities of interest (such as drugs, adverse events, time) and their relationships to support causality assessment and signal detection. In NLP is used for the deduplication of ICSRs and linking structured and unstructured data in FAERS, which results in an increase in accuracy when evaluating safety signal(Botsis & Kreimeyer, 2023). In EHRs, NLP models such as MADEx have shown good performance in extracting drug-event associations from clinical notes to support near real-time monitoring and signaling. In social media studies, patient-reported ADRs have been detected with the use of NLP and deep learning models (e.g., BiLSTM, BERT, RoBERTa); however, challenges including linguistic variability, data noise, and reliability persist(Desai et al., 2025).

The adoption of multi-modal AI frameworks that fuse information contained in NLP with other data modalities such as knowledge graphs and multimodal deep learning has even further increased the accuracy and detection timeliness across data sources. Nevertheless, complete clinical adoption is still hindered by the requirement of data standardization, validation and interpretability on different datasets(Nagar et al., 2025c).

5.2 AI Algorithms for Signal Detection and Causality Assessment in Pharmacovigilance

Artificial Intelligence (AI) has become a transformative force in signal detection and causality assessment within pharmacovigilance (PV)—the science of detecting, assessing, and preventing adverse drug reactions (ADRs) during post-market surveillance. By leveraging machine learning (ML) and natural language processing (NLP), AI enhances the ability to identify meaningful safety signals and establish causal relationships between drugs and adverse events across massive, complex datasets(L. Fu et al., 2025).

5.2.1 AI for Signal Detection

The incorporation of AI into social surveillance of signals can be a substantial improvement upon classical signal detection methods (e.g., PRR, ROR in frequentist disproportionality analyses, BCPNN in Bayesian methods). Research reviewed in Pharmaceutical Medicine (2025) outline the application of tree based ensembled methods such as Random Forests or GBM, and clustering techniques (k-means) to be more effective than traditional statistical approaches in finding safety signals. These models outperform in detecting emerging or rare ADRs as they capture multi-way signals in the large safety databases, including FAERS, VigiBase and EHR repositories. Additionally, deep learning and reinforcement learning techniques are being investigated to enhance early signal prediction and trend detection utilizing data from different sources (Warner et al., 2025).

AI-enhanced signal detection systems can also operate in real-time, integrating diverse data streams from structured reports and unstructured sources (e.g., social media, patient forums). NLP models, including transformer-based architectures (GPT, BERT, MedLEE), enable text mining of adverse event narratives, standardizing terminology (e.g., MedDRA, ATC) for more consistent and interpretable analysis (Fardale et al., 2024).

5.2.2 AI for Causality Assessment

Signal causality assessment—establishing whether an observed signal reflects a genuine causal relationship between a drug and an adverse event—is usually made using expert medical opinion. In news, recently published in CIOMS Working Group XIV (2025) draft, reports how the combination of decision tree and Bayesian Network models were being use for predictive causality analysis. One such model, the CASCADE model, was developed by training on >70K annotated safety cases from which it has reported a high sensitivity (0.90) and positive predictive value (0.78) as compared to physician judgment. Its interpretable nature enables causal reasoning based on, for example, temporal relationships, rechallenge outcomes and labeledness parameters—fulfilling the main regulatory requirement of transparency and explainability in AI-based PVA systems (Pande, 2018).

5.3 Risk Prediction Models Using Real-World Evidence (RWE)

Real-world data (RWD), including electronic health records, registries, insurance claims, and wearable devices, is a key source for real world evidence (RWE), the latter of which has become increasingly important in our efforts to develop risk prediction models pertaining to drug and medical device safety. These models integrate advanced analytics—particularly artificial intelligence (AI) and machine learning (ML)—to predict adverse outcomes, optimize regulatory decisions, and improve patient safety across the product lifecycle (Dang, 2023).

As Alice (2025) stated in The Role of Real-World Evidence in Regulatory Decision-Making for Medical Device Safety, the steady stream of data from connected health technologies enables real-time and forward-looking risk assessment. Using AI-based algorithms, it is possible to scan a wide spectrum of datasets to determine safety signals, predict uncommon adverse events and significant profiles in risk that can be modelled more accurately compared with traditional trial-gathered data. The integration of AI and big data analytics enables the early detection of potential hazards, while also accounting for complex factors such as patient comorbidities and environmental influences.

Complementing this, Ioakeim-Skoufa et al. (2025) note in Optimizing Real-World Evidence Studies for Regulatory Decision-Making and Pharmacovigilance that RWE-based risk models also perform best in heterogeneous populations (elderly, pediatric, multimorbid patients), where randomized controlled trials (RCTs) are often not feasible. Their LPRQ (Legal, Provenance, Relevance and Quality) framework helps to guarantee the integrity, interoperability and compliance of data sources used for predictive modelling.

These AI-enhanced RWE models support regulatory decision-making by:

- Enabling real-time signal detection and causal risk prediction through continuous learning algorithms.
- Integrating patient-reported outcomes and digital biomarkers for personalized risk estimation.
- Improving post-market surveillance through adaptive models that evolve as new data become available.
- Enhancing global harmonization, as regulatory bodies like the FDA, EMA, and Health Canada increasingly adopt frameworks that encourage the use of RWE in safety evaluations.

Together, these developments represent a paradigm shift from retrospective risk assessment to predictive, data-driven safety management, enabling regulators and manufacturers to anticipate and mitigate safety issues before they impact patients (Ioakeim-Skoufa et al., 2025).

5.4 Artificial Intelligence in Detecting Substandard or Counterfeit Medicines

Artificial Intelligence (AI) has become a transformative tool in combating the global challenge of substandard and counterfeit medicines, integrating advanced sensing, data analytics, and pattern recognition to ensure pharmaceutical quality and patient safety. Counterfeit and falsified medicines not only threaten public health but also undermine trust in healthcare systems, particularly in low- and middle-income countries where regulatory oversight and analytical capacity are often limited (Huanbutta et al., 2024c).

According to Kumar et al. (2025). Artificial Intelligence in Pharmacovigilance (Current Computer Science) AI-powered systems can provide real-time anomaly detection on drug formulations and packaging with the help of ML, DL and computer vision algorithms. These use visual, chemical and spectral measurements to separate the real thing from fakes. For example, CNN-based models can perceive minute packaging irregularities, label variances, and microstructure flaws that are often missed by human inspection (D. Kumar et al., 2025).

The CIOMS Working Group XIV (2025) report on AI in pharmacovigilance further considers the role of post-market quality assurance with AI, where AI technology for integrated supply chain surveillance, IoT-enabled sensors and predictive analytics may be used to track the integrity of drugs along their lifecycle. Using data from manufacturing, storage and retailer points, AI algorithms can highlight potential quality deviations or risks of falsification before those products make it to patients.

In addition, AI-assisted spectroscopic techniques (e.g., Raman, near-infra-red (NIR) and hyperspectral imaging), combined with ML classifiers including support vector machines (SVMs) and random forests (RFs), demonstrate high sensitivity in the examination of counterfeit or adulterated drugs. These AI-based analytical systems allow for portable, non-invasive and cost-effective quality control — which is crucially important if tests are to be carried out on site at borders, pharmacies, or resource-poor locations (S. Lee et al., 2025).

Collectively, these advances demonstrate how AI is enhancing pharmaceutical quality control and monitoring by combining chemical informatics, digital authentication and predictive intelligence. AI speeds up detection AND contributes to global health security by establishing adaptive data driven systems which will thwart the flow of counterfeit medicines in the supply chain (Vora et al., 2023b).

5.5 Mobile Apps and Wearable Devices for Patient-Reported Outcomes and Monitoring

Mobile health (mHealth) technologies—such as smartphone apps and wearables are changing the game for post-market surveillance and pharmacovigilance by allowing real-time, patient centered data collection. Such tools provide patients the means to report ADRs, monitor responses to treatments and disseminate health outcomes, all of which supply regulators and manufacturers with a continuous stream of real-world evidence (RWE) that complements traditional data sources (Dubale et al., 2024).

According to the BMC Medical Informatics and Decision Making (2022) study, the use of mobile applications, such as the WHO-endorsed Med Safety app, has dramatically increased ADR reporting to a higher quality and quantity level especially in LMICs. These applications cut reporting delays that are innate to paper-based techniques and standardize submissions in E2B-compliant formats, thereby allowing them to be interoperable with global PV databases such as VigiBase. More than 80% of assessed ADR apps showed increasing reporting activity post-implementation, highlighting their effectiveness in expanding drug safety monitoring (Fukushima et al., 2022).

Complementing these apps, wearable devices—as described in *The Role of Real-World Evidence in Regulatory Decision-Making for Medical Device Safety (2025)*—generate continuous physiological and behavioral data, allowing regulators to monitor patient-reported outcomes (PROs) such as treatment adherence, side effects, and quality-of-life measures in real time. This integration of wearable-derived and app-based data supports dynamic safety signal detection, enhances post-market responsiveness, and contributes to more personalized and proactive pharmacovigilance (Valla et al., 2023).

The International Journal of Trend in Scientific Research and Development (2023) also highlights that the combination of mobile apps, wearables, and AI based analytics makes it possible for automated pattern recognition and anomaly identification from the patient health data. This model is intended to enable the early detection of potential safety issues leading to a data-driven, collaborative surveillance network in which patients are both collector and creator of observations relating to their drugs safety.

5.6 Case Study: AI in COVID-19 Vaccine Safety Monitoring

The COVID-19 pandemic catalyzed an unprecedented integration of Artificial Intelligence (AI) into global pharmacovigilance systems, particularly for vaccine safety surveillance and adverse event monitoring. As detailed in the attached sources, AI-driven tools were leveraged to enhance real-time monitoring, causal assessment, and signal detection for vaccine-related adverse events following immunization (AEFIs).

Technology The AI-based monitoring systems employed machine learning (ML), deep learning (DL) and big data analytics for analysing huge sets of data from electronic health records/mobility/social media/mobile applications/and global reporting systems (eg, VigiBase, VAERS, EudraVigilance). Such models could help to quickly detect potential safety signals and quantify modelling relationship with demographic and clinical factors, thereby accelerating the response time for risk assessment with greater accuracy. For example, ML algorithms analyzed post-vaccination patterns to detect rare adverse events like myocarditis or thrombosis, significantly enhancing sensitivity compared to traditional passive reporting methods (Arora et al., 2021).

Concurrently, AI platforms like α-Satellite integrated mobile, demographic and epidemiological information to calculate risk for COVID-19 along with potential vaccination coverage so as to enable focus-based interventions and public health policy. Artificial intelligence also facilitated automatic causality assessment which combined Bayesian inference, WHO-UMC systems and real-world data to distinguish between coincidental and vaccine-related cases.

Furthermore, AI-enabled mobile apps and dashboards enhanced the capture of patient-reported outcomes and underreported AEFIs, particularly in low- and middle-income countries, addressing one of pharmacovigilance's most persistent challenges (Hodel et al., 2024).

6. INTEGRATION FRAMEWORKS AND INTEROPERABILITY

The potential of transforming the pharmaceutical production, quality control and post market surveillance by AI relies on a robust integration framework. These platforms will need to effectively connect the relatively siloed world of today with real time, continuous data inputs necessary for advanced analytics (balancing high throughput with intense regulatory requirements).

6.1. Data Interoperability Across Systems and Platforms

Pharmaceutical and healthcare environments are characterized by a complex, fragmented data landscape, where data originates from disparate sources and formats (Eichler et al., 2019). Achieving interoperability—the functional ability for these diverse systems to effectively communicate and exchange data—is a foundational requirement for efficient operations and coordinated public health responses. AI serves as a central technology for overcoming this challenge (Dhandayuthapani V., 2024).

AI algorithms significantly improve data harmonization and semantic alignment by automatically detecting patterns, aligning varied data formats, and standardizing semantics across systems (Dutta et al., 2021). Specifically, machine learning (ML) models classify, cluster, and normalize data schemas, transforming fragmented source data into consistent, structured formats while substantially reducing the necessity for manual preprocessing. Semantic reasoning tools contribute to consistent data interpretation by matching concepts across disparate technical vocabularies, such as identifying synonyms for complex medical concepts. Furthermore, Natural Language Processing (NLP) is critical for extracting actionable insights from the vast amounts of unstructured text data inherent in the industry, including clinical narratives, adverse event reports, and complex regulatory documentation (Derek & Collings, 2025).

A critical architectural solution for leveraging diverse datasets while mitigating risks associated with data confidentiality and intellectual property (IP) is the implementation of federated architectures (Farahani & Monsefi, 2023a). Federated learning (FL) allows multiple institutions to collaboratively train robust AI models without requiring the physical transfer or centralization of raw, sensitive data (Farahani & Monsefi, 2023a). Instead, only model updates are exchanged. By distributing the computational training effort, FL ensures data privacy and security, addressing regulatory concerns related to confidentiality agreements and trade secrets. This approach provides a scalable mechanism for leveraging the statistical power of large, diverse datasets while maintaining the physical and legal security of proprietary data (Table 1).

Table 1. Interoperability Techniques and Benefits

AI Technique	Primary Function	Application to Pharmaceutical Data	Source
Machine Learning (ML)	Data Harmonization & Schema Alignment	Transforms fragmented data into consistent, structured formats across varied databases, reducing manual preprocessing effort.	(Dutta et al., 2021)
Semantic Reasoning/Ontology Matching	Concept Harmonization	Ensures consistent interpretation of technical and medical concepts (e.g., identifying synonyms) across different systems.	(Dutta et al., 2021)
Federated Learning (FL)	Decentralized Collaboration	Enables collaborative training of models using broad datasets without compromising data confidentiality or IP security.	(Farahani & Monsefi, 2023b)
Natural Language Processing (NLP)	Unstructured Data Interpretation	Interprets clinical narratives, adverse event reports, and unstructured regulatory text.	(Derek & Collings, 2025)

6.2. Cloud-Based AI Ecosystems and Pharma-Tech Partnerships

The scalability and real time processing capability required for advanced AI are fundamentally dependent on supporting infrastructure, notably cloud computing, which enables the integration of AI with the Internet of Things (IoT) (Kodumuru et al., 2025b). This infrastructure is essential for handling the continuous, massive data streams generated by IoT sensors that monitor critical process parameters in real time.

The complexity of specialized AI applications often necessitates deep collaboration through pharma tech partnerships. Core technical competencies, such as Computer Vision (CV) for automating visual quality checks or NLP for accelerated detection of safety signals in pharmacovigilance datasets (Khemani et al., 2025), require expert development. Crucially, any AI service deployed in a Good Manufacturing Practice (GMP) environment must align its technical framework with GxP-critical processes, ensuring rigorous validation according to established standards such as GAMP.

6.3. Integrating AI With Enterprise Resource Planning (ERP) and Manufacturing Execution Systems (MES)

The strategic deployment of AI within pharmaceutical manufacturing relies on its seamless integration into the hierarchical structure of enterprise control systems, exemplified by the ISA-95 standard (McGinnis, 2019). This integration elevates operational platforms from mere static recorders to proactive, self-learning intelligence systems.

Traditional Enterprise Resource Planning (ERP) systems primarily functioned as repositories for centralized business data (Bahssas et al., 2015). With AI integration, ERPs evolve into dynamic platforms that optimize workflows, improve resource utilization, and increase business agility. Machine learning algorithms process vast amounts of structured and unstructured data to automate complex functions such as demand forecasting, inventory management, and stringent regulatory tracking. Similarly, the Manufacturing Execution System (MES), which operates at Level 3 of the ISA-95 standard, aligns operations with Pharma 4.0 initiatives by providing real time visibility into production data, tracking batch genealogy, and ensuring adherence to GMP (Yue et al., 2019). Integrating AI into the MES facilitates predictive quality control by analysing historical data to predict the likelihood of a batch meeting specification.

Furthermore, AI analyzes sensor data for predictive maintenance, anticipating equipment failures and significantly reducing production downtime.

The critical connection, standardized by ISA-95, links AI driven analytics (often housed at Level 4) with the tactical factory floor operations (Level 3 MES) (McGinnis, 2019). This linkage enables a continuous closed loop of innovation and compliance. Predictive demand forecasts generated by the ERP inform MES scheduling, while the MES's real time performance data is fed back into the ERP for continuous optimization of resource utilization. This continuous feedback loop allows pharmaceutical companies to adapt rapidly to market needs and process variations while maintaining strict regulatory oversight.

7. CHALLENGES AND LIMITATIONS

The Implementation of AI in GxP environments brings special challenges due to the data integrity, algorithm transparency, regulatory maturity and readiness of human workforce.

7.1. Data Quality, Availability, and Biases

The performance and reliability of an AI model are fundamentally determined by the quality and representativeness of its training data (Saini et al., 2025). When datasets are biased or non-representative, the resulting AI models may exhibit bias, potentially amplifying existing health inequalities (Nwebonyi & McKay, 2024). This can manifest in reduced diagnostic precision or unfair outcomes for specific racial or gender groups.

While bias pathways are frequently discussed in the context of upstream processes (research and development), the risk of bias can also emerge in downstream medicine production, particularly with complex targeted therapies. This risk is pronounced when utilizing sophisticated predictive tools, such as Digital Twins (DTs), for manufacturing processes like CAR T cell therapy (Aghamiri & Amin, 2025). If the underlying models are trained on synthetic datasets or cellular data derived from "healthy donors," they may not accurately represent the characteristics of the actual patient population receiving treatment. The failure to model the specific biology of the target population means that the manufacturing process parameters, adjusted automatically by the AI, will be optimized for the wrong biological profile. This results in products that are technically compliant but clinically suboptimal, directly compromising the efficacy of personalized medicine and generating disparities in treatment outcome (Nwebonyi & McKay, 2024). Therefore, data bias transitions from an ethical consideration to a critical quality attribute risk that impacts measurable product performance.

7.2. Lack of Standardization in AI Implementation

The advanced manufacturing of the pharmaceutical industry often integrates areas of technological novelty or application of existent methods in forms where any best practice is not defined (*Artificial Intelligence in Drug Manufacturing*, n.d.). It's this high tempo of innovation that stands a stark contrast to the regulated GxP-stable constraints. Although GAMP 5 remains the core framework for validating automated systems 2, the inherently dynamic and non-deterministic nature of modern AI systems requires evolving validation strategies that move beyond static, one-time qualification (Niazi, 2025).

7.3. Black-Box Algorithms and Regulatory Hesitance

A significant barrier to AI adoption is the complexity of non-deterministic AI/ML algorithms, which can create a "black-box" scenario (Niazi, 2025). It is challenging to provide a clear explanation of how specific changes in model inputs lead to particular model outputs, complicating traditional validation and audit processes (*Artificial Intelligence in Drug Manufacturing*, n.d.). Regulatory bodies, including the FDA, remain cautious due to the difficulties in assessing the safety and efficacy of such opaque systems.1

Explainable AI (XAI) is considered necessary to mitigate this transparency deficit. XAI requires models to explain in an understandable and accurate way, why they believe they should output one predicted value over another, capturing the decision process. 1 Drug applications need to establish criteria that substantiate the performance of the model and mechanisms for maintaining its explicability with respect to product quality over time. To proactively navigate this technological shift, the FDA's Center for Drug Evaluation and Research (CDER) established the Emerging Technology Program (ETP) in 2014 to collaborate with companies and inform the development of a necessary future regulatory framework (*Artificial Intelligence in Drug Manufacturing*, n.d.).

Acknowledging the continuous, adaptive nature of AI/ML Systems, the regulatory discourse has introduced a notion of progressive validation. This perspective is a revolutionary change in the GxP philosophy, moving from the validation of an individual instance of a system towards validating a managed process of change within it (Niazi, 2025). Progressive validation structures changes into categories and delineates the validation activities required to implement that change so AI models can learn and improve over time without losing regulatorily validated state.

7.4. Workforce Training and Change Management

The shift toward smart manufacturing necessitates comprehensive workforce transformation. Manufacturing environments are becoming increasingly complex, demanding that workers possess high levels of digital literacy to effectively solve problems and interact with AI driven systems (Pandit et al., 2024). This technological evolution has highlighted a critical global shortage of skilled, digitally literate personnel.

In short, closing the skills gap requires an organization to: Invest resources intelligently in upskilling/reskilling efforts Foster collaboration Human-Robotic workforces Overcome cultural issues (e.g., fear of job loss) Documented research has shown that powerful training programs in data science and specific AI tools, such as machine learning and computer vision, can significantly boost workforce productivity.

In addition, effective deployment will also demand robust org anizational change management — which is a requirement in any GxP environment for reviewing and managing anything that could impact the validated state of systems, processes, or equipment (Niazi, 2025). Systemic treatment for organizing, validating, and finalizing AI-introduced changes Digital transformation necessitates a systemic approach for the organization, validation and solidification of changes introduced by Artificial Intelligence. As a result, the procedural control (referred to as GxP change management in this paper) implementation should be integrated with the innovative validation model for managing evolving AI/ML models.This ensures that technical changes are reviewed by qualified representatives before implementation, maintaining both operational effectiveness and regulatory control (Niazi, 2025). **Table 2** demonstrates the challenges and mitigation strategies.

Table 2. Critical Challenges and Mitigation Strategies

Challenge Area	Specific Risk/ Limitation	Impact on Product Quality/Equity	Technical/Procedural Mitigation	Source
Data Integrity	Bias from unrepresentative datasets	Leads to unreliable predictions, reduced diagnostic precision, or amplification of health inequalities	Rigorous data governance; federated learning for diversity; bias detection algorithms	(Saini et al., 2025)
Advanced Manufacturing	Downstream bias in Digital Twins (targeted therapy)	Prediction models based on "healthy donor" data may be unrepresentative of patient population, compromising personalized treatment efficacy	Careful selection of training data; model calibration specific to patient cohorts	(Nwebonyi & McKay, 2024)
Regulatory Compliance	Black-Box Algorithms	Inability to explain model output prevents validation and auditability, leading to regulatory hesitancy	Mandatory use of Explainable AI (XAI) to provide evidence/ reasoning for decisions	(Saini et al., 2025)
Workforce/Operations	Digital Literacy Gap/ Job Fear	Critical shortage of staff capable of interacting with, validating, and maintaining AI/ML systems	Investment in upskilling/ reskilling programs; implementing structured change management (GxP compliant)	(Saini et al., 2025)

8. FUTURE PERSPECTIVES

The future of AI in pharmaceutical manufacturing centres around the autonomous, agile production system and it is here that the concept of The Digital Twin (DT) acts as the architectural catalyst.

Digital Twins Digital Twins are high-fidelity digital reproductions that emulate the states, properties and behaviour of their physical equivalents (Mihai et al., 2022). A mature DT is defined by three fundamental characteristics: a physical system, its digital twin and continuous bidirectional information flow between them. DTs synthesize advancements in Process Analytical Technology (PAT), complex process modelling, and advanced data integration to enable high-fidelity predictive simulation and optimization (Vora et al., 2023c) as shown in **Table 3**. For example, the AI algorithms functioning within a DT may exploit PAT data and critical material attributes (CMAs) such as particle size distribution to predict drug release and dissolution profiles with precision. Although the pharmaceutical industry has generally embraced QbD systems, the use of integrated DTs throughout biopharmaceutical production is still a nascent possibility as opposed to standard practice in the industry now.

The integration of AI, DTs, and advanced Manufacturing Execution Systems (MES) is enabling autonomous production planning and execution. To do that effectively MES architectures of the future will harness AI to quickly create an initial production plan and keep on looking for more efficient ones in real time (Duarte et al., 2025). This enables rapid replanning, and thus is equipped for the "on-the-fly" transition from an original plan to a newly optimized one—albeit while batch production may already be in progress.This dynamic responsiveness drastically increases manufacturing flexibility and resilience against unforeseen events, such as equipment malfunction or material shortages (Vyskočil et al., 2023).

The fusion of PAT/IoT (continuous monitoring), AI/ML (predictive quality control) and DTs (high-fidelity process modelling) provide technical reassurances that power Real Time Release. If a stand-alone MES is performing validated process changes driven from DT simulations, QA become seamlessly integrated into the production process (Vora et al., 2023c). This ability to prove quality on-the-fly and to manage process dynamism is the milestone that regulators need to meet in order for RTR will eventually become widespread.

Table 3. The Digital Twin Architecture

DT Component	Description	AI/PAT Integration Role	Strategic Outcome
Physical Component	Real-world manufacturing equipment and processes	Monitored by IoT sensors and PAT for real-time data streaming	Provides operational context and input data
Virtual Component (Model)	High-fidelity simulation of dynamics and behavior	AI algorithms perform process modelling and predictive quality control	Enables predictive analysis and optimization simulation
Information Exchange	Continuous bidirectional communication	Supports autonomous decision making and "on-the-fly replanning" by the MES	Facilitates agility and responsiveness to changes/faults

CONCLUSION

Artificial intelligence is transforming the pharmaceutical lifecycle by enabling smarter manufacturing, robust quality control, and proactive post-market surveillance. The integration of AI-driven analytics enhances process efficiency, ensures consistent product quality, and supports early detection of safety signals, ultimately contributing to safer, more effective, and patient-centric medicines while strengthening regulatory compliance and decision-making.

REFERENCES

Aghamiri, S. S., & Amin, R. (2025). The Potential Use of Digital Twin Technology for Advancing CAR-T Cell Therapy. *Current Issues in Molecular Biology*, *47*(5), 321. DOI: 10.3390/cimb47050321 PMID: 40699720

Ajmal, C. S., Yerram, S., Abishek, V., Nizam, V. P. M., Aglave, G., Patnam, J. D., Raghuvanshi, R. S., & Srivastava, S. (2025). Innovative Approaches in Regulatory Affairs: Leveraging Artificial Intelligence and Machine Learning for Efficient Compliance and Decision-Making. *The AAPS Journal*, *27*(1), 22. DOI: 10.1208/s12248-024-01006-5 PMID: 39776314

Algarvio, R. C., Conceição, J., Rodrigues, P. P., Ribeiro, I., & Ferreira-da-Silva, R. (2025). Artificial intelligence in pharmacovigilance: A narrative review and practical experience with an expert-defined Bayesian network tool. *International Journal of Clinical Pharmacy*, *47*(4), 932–944. DOI: 10.1007/s11096-025-01975-3 PMID: 40736603

Andrianandrianina Johanesa, T. V., Equeter, L., & Mahmoudi, S. A. (2024). Survey on AI Applications for Product Quality Control and Predictive Maintenance in Industry 4.0. *Electronics (Basel)*, *13*(5), 976. DOI: 10.3390/electronics13050976

Arden, N. S., Fisher, A. C., Tyner, K., Yu, L. X., Lee, S. L., & Kopcha, M. (2021). Industry 4.0 for pharmaceutical manufacturing: Preparing for the smart factories of the future. *International Journal of Pharmaceutics*, *602*, 120554. DOI: 10.1016/j.ijpharm.2021.120554 PMID: 33794326

Arora, G., Joshi, J., Mandal, R. S., Shrivastava, N., Virmani, R., & Sethi, T. (2021). Artificial Intelligence in Surveillance, Diagnosis, Drug Discovery and Vaccine Development against COVID-19. *Pathogens (Basel, Switzerland)*, *10*(8), 1048. DOI: 10.3390/pathogens10081048 PMID: 34451513

Artificial Intelligence in Drug Manufacturing. (n.d.). US FDA. Retrieved November 3, 2025, from https://www.fda.gov/media/165743/download

Bahssas, D. M., AlBar, A. M., & Hoque, M. (2015). Enterprise Resource Planning (ERP) Systems: Design, Trends and Deployment. *The International Technology Management Review*, *5*(2), 72. DOI: 10.2991/itmr.2015.5.2.2

Balasubramaniam, N., Kauppinen, M., Rannisto, A., Hiekkanen, K., & Kujala, S. (2023). Transparency and explainability of AI systems: From ethical guidelines to requirements. *Information and Software Technology*, *159*, 107197. DOI: 10.1016/j.infsof.2023.107197

Barbieri, M. A., Battini, V., Carnovale, C., Cocco, M., Papoutsi, D. G., Heckmann, N. S., Dong, G., Rossi, A., Peker, S., Van Manen, R. P., Thapar, S., & Sessa, M. (2025). Artificial intelligence in pharmacovigilance signal management: A review of tools, implementations, research, and regulatory landscape. *Expert Opinion on Drug Safety*, •••, 1–16. DOI: 10.1080/14740338.2025.2545926 PMID: 40778768

Bian, W., Huang, Y., Shan, B., & Chang, Y. (2024). Capsule Defect Detection and Classification Using Enhanced MobileNet Model. *American Journal of Biochemistry and Biotechnology*, *20*(4), 356–364. DOI: 10.3844/ajbbsp.2024.356.364

Botsis, T., & Kreimeyer, K. (2023). Improving drug safety with adverse event detection using natural language processing. *Expert Opinion on Drug Safety*, *22*(8), 659–668. DOI: 10.1080/14740338.2023.2228197 PMID: 37339273

Broday, E. E. (2022). The evolution of quality: From inspection to quality 4.0. *International Journal of Quality and Service Sciences*, *14*(3), 368–382. DOI: 10.1108/IJQSS-09-2021-0121

Chaparro-Cárdenas, S. L., Ramirez-Bautista, J.-A., Terven, J., Córdova-Esparza, D.-M., Romero-Gonzalez, J.-A., Ramírez-Pedraza, A., & Chavez-Urbiola, E. A. (2025). A Technological Review of Digital Twins and Artificial Intelligence for Personalized and Predictive Healthcare. *Health Care*, *13*(14), 1763. DOI: 10.3390/healthcare13141763 PMID: 40724790

Chaturvedi, R., & Verma, S. (2023). Opportunities and Challenges of AI-Driven Customer Service. In Sheth, J. N., Jain, V., Mogaji, E., & Ambika, A. (Eds.), *Artificial Intelligence in Customer Service* (pp. 33–71). Springer International Publishing., DOI: 10.1007/978-3-031-33898-4_3

Chopra, H., Annu, , Shin, D. K., Munjal, K., Priyanka, , Dhama, K., & Emran, T. B. (2023a). Revolutionizing clinical trials: The role of AI in accelerating medical breakthroughs. *International Journal of Surgery*, *109*(12), 4211–4220. DOI: 10.1097/JS9.0000000000000705 PMID: 38259001

Dang, A. (2023). Real-World Evidence: A Primer. *Pharmaceutical Medicine*, *37*(1), 25–36. DOI: 10.1007/s40290-022-00456-6 PMID: 36604368

Deeba, K., Chinnpa Prabu Shankar, K., Gnanavel, S., & Elsisi, M. (2025). Artificial Intelligence, Computer Vision, and Robotics for Industry 5.0. In Tyagi, A. K. (Ed.), *Next Generation Data Science and Blockchain Technology for Industry 5.0* (1st ed., pp. 295–324). Wiley., DOI: 10.1002/9781394245215.ch12

Derek, V., & Collings, P. (2025). *Natural Language Processing (NLP) in Healthcare AI: Enhancing Clinical Insight Extraction from Unstructured Patient Data. Preprints*. DOI: 10.22541/au.174845034.41890518/v1

Desai, K. S., Tiyyala, V. M., Tiyyala, P., Yeola, A., Gallegos-Rangel, A., Montiel-Torres, A., Allen, M. R., Dredze, M., Vandrey, R. G., Thrul, J., Leas, E. C., Hogarth, M., Smith, D. M., & Ayers, J. W. (2025). Waldo: Automated discovery of adverse events from unstructured self reports. *PLOS Digital Health*, *4*(9), e0001011. DOI: 10.1371/journal.pdig.0001011 PMID: 41026671

Dhandayuthapani, V. B. (2024). *Advancing Healthcare: AI Integration, Interoperability and Sustainability Challenges*. DOI: 10.25929/J3XXXY63

Dimitsaki, S., Natsiavas, P., & Jaulent, M.-C. (2024). Applying AI to Structured Real-World Data for Pharmacovigilance Purposes: Scoping Review. *Journal of Medical Internet Research*, *26*, e57824. DOI: 10.2196/57824 PMID: 39753222

Duarte, J. G., Duarte, M. G., Piedade, A. P., & Mascarenhas-Melo, F. (2025). Rethinking Pharmaceutical Industry with Quality by Design: Application in Research, Development, Manufacturing, and Quality Assurance. *The AAPS Journal*, *27*(4), 96. DOI: 10.1208/s12248-025-01079-w PMID: 40389714

Dubale, A. T., Tareke, A. A., Butta, F. W., Shibabaw, A. A., Eniyew, E. B., Ahmed, M. H., Kassie, S. Y., Demsash, A. W., Chereka, A. A., Dube, G. N., Walle, A. D., & Kitil, G. W. (2024). Healthcare professionals' willingness to utilize a mobile health application for adverse drug reaction reporting in a limited resource setting: An input for digital health, 2023. *European Journal of Obstetrics & Gynecology and Reproductive Biology: X, 23*, 100324. DOI: 10.1016/j.eurox.2024.100324 PMID: 39050924

Dutta, A., Deb, T., & Pathak, S. (2021). Automated Data Harmonization (ADH) using Artificial Intelligence (AI). *OPSEARCH, 58*(2), 257–275. DOI: 10.1007/s12597-020-00467-4

Eichler, H., Bloechl-Daum, B., Broich, K., Kyrle, P. A., Oderkirk, J., Rasi, G., Santos Ivo, R., Schuurman, A., Senderovitz, T., Slawomirski, L., Wenzl, M., & Paris, V. (2019). Data Rich, Information Poor: Can We Use Electronic Health Records to Create a Learning Healthcare System for Pharmaceuticals? *Clinical Pharmacology and Therapeutics, 105*(4), 912–922. DOI: 10.1002/cpt.1226 PMID: 30178490

Farahani, B., & Monsefi, A. K. (2023a). Smart and collaborative industrial IoT: A federated learning and data space approach. *Digital Communications and Networks, 9*(2), 436–447. DOI: 10.1016/j.dcan.2023.01.022

Farahani, B., & Monsefi, A. K. (2023b). Smart and collaborative industrial IoT: A federated learning and data space approach. *Digital Communications and Networks, 9*(2), 436–447. DOI: 10.1016/j.dcan.2023.01.022

Fardale, S., Goyal, A., Palav, M. R., Palav, S. M., Kalaiarasan, C., & Namdeo, A. K. (2024). Real-Time Market Analysis With AI-Enhanced Signal Processing. In Pandey, D., Muniandi, B., Pandey, B. K., & George, A. S. (Eds.), (pp. 45–60). Advances in Computational Intelligence and Robotics. IGI Global., DOI: 10.4018/979-8-3693-7367-5.ch004

Fu, L., Jia, G., Liu, Z., Pang, X., & Cui, Y. (2025). The applications and advances of artificial intelligence in drug regulation: A global perspective. *Acta Pharmaceutica Sinica. B, 15*(1), 1–14. DOI: 10.1016/j.apsb.2024.11.006 PMID: 40041903

Fu, Q., Lai, J., Zhong, T., & Ran, L. (2023). Design and Implementation of Clinical LIS360 Laboratory Management System Based on AI Technology. *International Journal of Computational Intelligence Systems, 16*(1), 33. DOI: 10.1007/s44196-023-00207-8

Fukushima, A., Iessa, N., Balakrishnan, M. R., & Pal, S. N. (2022). Smartphone-based mobile applications for adverse drug reactions reporting: Global status and country experience. *BMC Medical Informatics and Decision Making, 22*(1), 118. DOI: 10.1186/s12911-022-01832-7 PMID: 35501745

Gangwal, A., & Lavecchia, A. (2025). Artificial intelligence in preclinical research: Enhancing digital twins and organ-on-chip to reduce animal testing. *Drug Discovery Today, 30*(5), 104360. DOI: 10.1016/j.drudis.2025.104360 PMID: 40252989

Ghaffar Nia, N., Kaplanoglu, E., & Nasab, A. (2023). Evaluation of artificial intelligence techniques in disease diagnosis and prediction. *Discover Artificial Intelligence, 3*(1), 5. DOI: 10.1007/s44163-023-00049-5 PMID: 40478140

Ghanem, M., Ghaith, A. K., & Bydon, M. (2024). Artificial intelligence and personalized medicine: Transforming patient care. In *The New Era of Precision Medicine* (pp. 131–142). Elsevier., DOI: 10.1016/B978-0-443-13963-5.00012-1

Golder, S., Xu, D., O'Connor, K., Wang, Y., Batra, M., & Hernandez, G. G. (2025). Leveraging Natural Language Processing and Machine Learning Methods for Adverse Drug Event Detection in Electronic Health/Medical Records: A Scoping Review. *Drug Safety*, *48*(4), 321–337. DOI: 10.1007/s40264-024-01505-6 PMID: 39786481

Gunasekaran, A., & Yusuf, Y. Y. (2002). Agile manufacturing: A taxonomy of strategic and technological imperatives. *International Journal of Production Research*, *40*(6), 1357–1385. DOI: 10.1080/00207540110118370

Gupta, R., Srivastava, D., Sahu, M., Tiwari, S., Ambasta, R. K., & Kumar, P. (2021). Artificial intelligence to deep learning: Machine intelligence approach for drug discovery. *Molecular Diversity*, *25*(3), 1315–1360. DOI: 10.1007/s11030-021-10217-3 PMID: 33844136

Hammad, M., Panaousis, M., Ali, H., & Khan, W. A. (2026). Industry 4.0 in Pharmaceuticals: A New Era of Smart Manufacturing. In M. Hammad, M. Panaousis, H. Ali, & W. A. Khan, *Smart Manufacturing Blueprint: Navigating Industry 4.0 Across Diverse Sectors* (pp. 165–211). Springer Nature Switzerland. DOI: 10.1007/978-3-032-00214-3_6

Harrer, S., Shah, P., Antony, B., & Hu, J. (2019). Artificial Intelligence for Clinical Trial Design. *Trends in Pharmacological Sciences*, *40*(8), 577–591. DOI: 10.1016/j.tips.2019.05.005 PMID: 31326235

Heesen, P., Schelling, G., Birbaumer, M., Jäger, R., Bode, B., Studer, G., & Fuchs, B. (2024). Real-World-Time Data and RCT Synergy: Advancing Personalized Medicine and Sarcoma Care through Digital Innovation. *Cancers (Basel)*, *16*(14), 2516. DOI: 10.3390/cancers16142516 PMID: 39061155

Hodel, K. V. S., Fiuza, B. S. D., Conceição, R. S., Aleluia, A. C. M., Pitanga, T. N., Fonseca, L. M. D. S., Valente, C. O., Minafra-Rezende, C. S., & Machado, B. A. S. (2024). Pharmacovigilance in Vaccines: Importance, Main Aspects, Perspectives, and Challenges—A Narrative Review. *Pharmaceuticals (Basel, Switzerland)*, *17*(6), 807. DOI: 10.3390/ph17060807 PMID: 38931474

Houghton, C., Casey, D., Shaw, D., & Murphy, K. (2013). Rigour in qualitative case-study research. *Nurse Researcher*, *20*(4), 12–17. DOI: 10.7748/nr2013.03.20.4.12.e326 PMID: 23520707

Huanbutta, K., Burapapadh, K., Kraisit, P., Sriamornsak, P., Ganokratanaa, T., Suwanpitak, K., & Sangnim, T. (2024a). Artificial intelligence-driven pharmaceutical industry: A paradigm shift in drug discovery, formulation development, manufacturing, quality control, and post-market surveillance. *European Journal of Pharmaceutical Sciences : Official Journal of the European Federation for Pharmaceutical Sciences*, *203*, 106938. DOI: 10.1016/j.ejps.2024.106938 PMID: 39419129

Ioakeim-Skoufa, I., Atkins, K., & Hernández-Rodríguez, M. Á. (2025). Optimizing real-world evidence studies for regulatory decision-making and impact assessment in pharmacovigilance. *British Journal of Clinical Pharmacology*, *91*(4), 1092–1095. DOI: 10.1111/bcp.16393 PMID: 39821103

Jagtap, S., Garcia-Garcia, G., & Rahimifard, S. (2021). Optimisation of the resource efficiency of food manufacturing via the Internet of Things. *Computers in Industry, 127*, 103397. DOI: 10.1016/j.compind.2021.103397

Jain, D. (2024). Artificial Intelligence in Quality Control Systems: A Cross-Industry Analysis of Applications, Benefits, and Implementation Frameworks. *International Journal of Scientific Research in Computer Science. Engineering and Information Technology, 10*(6), 1321–1333. DOI: 10.32628/CSEIT241061162

Jarallah, S. J., Almughem, F. A., Alhumaid, N. K., Fayez, N. A., Alradwan, I., Alsulami, K. A., Tawfik, E. A., & Alshehri, A. A. (2025). Artificial intelligence revolution in drug discovery: A paradigm shift in pharmaceutical innovation. *International Journal of Pharmaceutics, 680*, 125789. DOI: 10.1016/j.ijpharm.2025.125789 PMID: 40451590

Kalaiselvi, C., & Arunasenbagam, G. (2017). An Overview of Statistical Pattern Recognition—A Review. *Journal of Environmental Nanotechnology, 6*(1), 75–78. DOI: 10.13074/jent.2017.03.171235

Khemani, D., Malave, D. S., Shinde, S., Shukla, M., Shikalgar, R., & Talwar, H. (2025). AI-driven pharmacovigilance: Enhancing adverse drug reaction detection with deep learning and NLP. *MethodsX, 15*, 103460. DOI: 10.1016/j.mex.2025.103460 PMID: 40678458

Khinvasara, T., Ness, S., & Shankar, A. (2024). Leveraging AI for Enhanced Quality Assurance in Medical Device Manufacturing. *Asian Journal of Research in Computer Science, 17*(6), 13–35. DOI: 10.9734/ajrcos/2024/v17i6454

Kodumuru, R., Sarkar, S., Parepally, V., & Chandarana, J. (2025a). Artificial Intelligence and Internet of Things Integration in Pharmaceutical Manufacturing: A Smart Synergy. *Pharmaceutics, 17*(3), 290. DOI: 10.3390/pharmaceutics17030290 PMID: 40142954

Krzywanski, J., Sosnowski, M., Grabowska, K., Zylka, A., Lasek, L., & Kijo-Kleczkowska, A. (2024). Advanced Computational Methods for Modeling, Prediction and Optimization—A Review. *Materials (Basel), 17*(14), 3521. DOI: 10.3390/ma17143521 PMID: 39063813

Kumar, A. (2003). Neural network based detection of local textile defects. *Pattern Recognition, 36*(7), 1645–1659. DOI: 10.1016/S0031-3203(03)00005-0

Kumar, A. (2025). *AI Driven Precision Oncology: Predictive Biomarker Discovery and Personalized Treatment Optimization Using Genomic Data.* DOI: 10.5281/ZENODO.15037946

Kumar, D., Kaur, A., Shruti, , & Kaur, D. (2025). Artificial Intelligence in Pharmacovigilance. *Current Computer Science, 04*, e29503779340108. DOI: 10.2174/0129503779340108241226103521

Kurniawan, M. H., Handiyani, H., Nuraini, T., Hariyati, R. T. S., & Sutrisno, S. (2024). A systematic review of artificial intelligence-powered (AI-powered) chatbot intervention for managing chronic illness. *Annals of Medicine, 56*(1), 2302980. DOI: 10.1080/07853890.2024.2302980 PMID: 38466897

Lee, S., Park, R., & Jung, H. S. (2025). AI-Enhanced Surface-Enhanced Raman Scattering for Accurate and Sensitive Biomedical Sensing. *Advanced Intelligent Discovery, 202500030.* Advance online publication. DOI: 10.1002/aidi.202500030

Lee, W. C., Blanchette, C., Pokras, S., Shaikh, J., & Miller, J. (2025). The evolution and future of integrated evidence planning. *Expert Review of Pharmacoeconomics & Outcomes Research, 25*(6), 855–862. DOI: 10.1080/14737167.2025.2497876 PMID: 40266581

Maharjan, R., Kim, N. A., Kim, K. H., & Jeong, S. H. (2025). Transformative roles of digital twins from drug discovery to continuous manufacturing: Pharmaceutical and biopharmaceutical perspectives. *International Journal of Pharmaceutics: X, 10*, 100409. DOI: 10.1016/j.ijpx.2025.100409 PMID: 41089758

McGinnis, L. F. (2019). *Formalizing ISA-95 level 3 control with smart manufacturing system models* (NIST GCR 19-022; p. NIST GCR 19-022). National Institute of Standards and Technology. DOI: 10.6028/NIST.GCR.19-022

Miązek, A., & Świątowiec-Szczepańska, J. (2020). The application of content analysis as a research method in management sciences. In Michałkiewicz, A., & Mierzejewska, W. (Eds.), *Contemporary organisation and management. Challenges and trends* (pp. 159–172). Wydawnictwo Uniwersytetu Łódzkiego., DOI: 10.18778/8220-333-2.10

Mohammady, M., Hadidi, M., Iman Ghetmiri, S., & Yousefi, G. (2021). Design of ultra-fine carvedilol nanococrystals: Development of a safe and stable injectable formulation. *European Journal of Pharmaceutics and Biopharmaceutics : Official Journal of Arbeitsgemeinschaft für Pharmazeutische Verfahrenstechnik e.V, 168*, 139–151. DOI: 10.1016/j.ejpb.2021.08.015 PMID: 34481906

Nagar, A., Gobburu, J., & Chakravarty, A. (2025a). Artificial intelligence in pharmacovigilance: Advancing drug safety monitoring and regulatory integration. *Therapeutic Advances in Drug Safety, 16*, 20420986251361435. DOI: 10.1177/20420986251361435 PMID: 40756610

Niazi, S. K. (2025). Regulatory Perspectives for AI/ML Implementation in Pharmaceutical GMP Environments. *Pharmaceuticals (Basel, Switzerland), 18*(6), 901. DOI: 10.3390/ph18060901 PMID: 40573297

Nwebonyi, N., & McKay, F. (2024). Exploring bias risks in artificial intelligence and targeted medicines manufacturing. *BMC Medical Ethics, 25*(1), 113. DOI: 10.1186/s12910-024-01112-1 PMID: 39415204

Olatunde Omisola, J., Oluwasegun Shiyanbola, J., & Omotunde Osho, G. (2024). A Predictive Quality Assurance Model Using Lean Six Sigma: Integrating FMEA, SPC, and Root Cause Analysis for Zero-Defect Production Systems. *International Journal of Advanced Multidisciplinary Research and Studies, 4*(6), 1481–1497. DOI: 10.62225/2583049X.2024.4.6.4051

Oualikene-Gonin, W., Jaulent, M.-C., Thierry, J.-P., Oliveira-Martins, S., Belgodère, L., Maison, P., & Ankri, J.The Scientific Advisory Board of ANSM. (2024). Artificial intelligence integration in the drug lifecycle and in regulatory science: Policy implications, challenges and opportunities. *Frontiers in Pharmacology, 15*, 1437167. DOI: 10.3389/fphar.2024.1437167 PMID: 39156111

Pande, S. (2018). Causality or Relatedness assessment in adverse drug reaction and its relevance in dermatology. *Indian Journal of Dermatology, 63*(1), 18. DOI: 10.4103/ijd.IJD_579_17 PMID: 29527021

Pandit, S., Sarkar, S. K., Barik, S., & Sahu, S. (2024). AI in Skill Development, Critical Thinking, Digital Literacy, and AI-Driven Job Preparation. In Mobo, F. D. (Ed.), (pp. 23–76). Advances in Educational Technologies and Instructional Design. IGI Global., DOI: 10.4018/979-8-3693-8191-5.ch002

Pasas-Farmer, S., & Jain, R. (2025). From discovery to delivery: Governance of AI in the pharmaceutical industry. *Green Analytical Chemistry*, *13*, 100268. DOI: 10.1016/j.greeac.2025.100268

Pasupuleti, V., Thuraka, B., Kodete, C. S., & Malisetty, S. (2024). Enhancing Supply Chain Agility and Sustainability through Machine Learning: Optimization Techniques for Logistics and Inventory Management. *Logistics (Basel)*, *8*(3), 73. DOI: 10.3390/logistics8030073

Phiri, V. J., Battas, I., Semmar, A., Medromi, H., & Moutaouakkil, F. (2025). Towards enterprise-wide pharma 4.0 adoption. *Scientific African*, *28*, e02771. DOI: 10.1016/j.sciaf.2025.e02771

Praveen, J., Kumar Cm, K., & Channappa, A. H. (2023). Transforming Pharmacovigilance Using Gen AI: Innovations in Aggregate Reporting, Signal Detection, and Safety Surveillance. *Journal of Multidisciplinary Research*, •••, 9–16. DOI: 10.37022/tjmdr.v3i3.484

Rajesh, M. V., & Elumalai, K. (2025). The transformative power of artificial intelligence in pharmaceutical manufacturing: Enhancing efficiency, product quality, and safety. *Journal of Holistic Integrative Pharmacy*, *6*(2), 125–135. DOI: 10.1016/j.jhip.2025.03.007

Rakholia, R., Suárez-Cetrulo, A. L., Singh, M., & Carbajo, R. S. (2025). Integrating AI and IoT for Predictive Maintenance in Industry 4.0 Manufacturing Environments: A Practical Approach. *Information (Basel)*, *16*(9), 737. DOI: 10.3390/info16090737

Sadiya Inamdar, T. K. (2024). *Harnessing AI And Machine Learning in Pharmaceutical Quality Assurance*. DOI: 10.5281/ZENODO.14186540

Saini, J. P. S., Thakur, A., & Yadav, D. (2025). AI-driven innovations in pharmaceuticals: Optimizing drug discovery and industry operations. *RSC Pharmaceutics*, *2*(3), 437–454. DOI: 10.1039/D4PM00323C

Sampene, A. K., & Nyirenda, F. (2024). Evaluating the effect of artificial intelligence on pharmaceutical product and drug discovery in China. *Future Journal of Pharmaceutical Sciences*, *10*(1), 58. DOI: 10.1186/s43094-024-00632-2

Sandner, P., Gross, J., & Richter, R. (2020). Convergence of Blockchain, IoT, and AI. *Frontiers in Blockchain*, *3*, 522600. DOI: 10.3389/fbloc.2020.522600

Sarkar, B., & Paul, R. K. (2025). AI-Driven Manufacturing Processes. In B. Sarkar & R. K. Paul, *AI for Advanced Manufacturing and Industrial Applications* (pp. 19–59). Springer Nature Switzerland. DOI: 10.1007/978-3-031-86091-1_2

Sembiring, M., & Novagusda, F. (2024). Enhancing Data Security Resilience in AI-Driven Digital Transformation: Exploring Industry Challenges and Solutions Through ALCOA+ Principles. *Acta Informatica Medica*, *32*(1), 65. DOI: 10.5455/aim.2024.32.65-70 PMID: 38585596

Shamim, M. A., Shamim, M. A., Arora, P., & Dwivedi, P. (2024). Artificial intelligence and big data for pharmacovigilance and patient safety. *Journal of Medicine, Surgery, and Public Health*, *3*, 100139. DOI: 10.1016/j.glmedi.2024.100139

Sharma, S., Sharma, K., & Grover, S. (2024). Real-Time Data Analysis with Smart Sensors. In Gulati, S. (Ed.), *Application of Artificial Intelligence in Wastewater Treatment* (pp. 127–153). Springer Nature Switzerland., DOI: 10.1007/978-3-031-69433-2_5

Singh, R., Johnsten, T., Raghavan, V., & Xie, Y. (2009). An efficient algorithm for discovering positive and negative patterns. *2009 IEEE International Conference on Granular Computing*, 507–512. DOI: 10.1109/GRC.2009.5255068

Soori, M., Arezoo, B., & Dastres, R. (2023). Internet of things for smart factories in industry 4.0, a review. *Internet of Things and Cyber-Physical Systems, 3*, 192–204. DOI: 10.1016/j.iotcps.2023.04.006

Strielkowski, W., Vlasov, A., Selivanov, K., Muraviev, K., & Shakhnov, V. (2023). Prospects and Challenges of the Machine Learning and Data-Driven Methods for the Predictive Analysis of Power Systems: A Review. *Energies, 16*(10), 4025. DOI: 10.3390/en16104025

Subha, S., Shanmugathai, M., Prasanth, A., Varagi, S. S., & Dhanashree, V. (2024). Digital Transformation in the Pharmaceutical and Biotech Industry: Challenges and Research Directions. In Dhanaraj, R. K., Balusamy, B., Samuel, P., Bashir, A. K., & Kadry, S. (Eds.), *Digital Twins in Industrial Production and Smart Manufacturing* (1st ed., pp. 297–324). Wiley., DOI: 10.1002/9781394195336.ch13

Sutiene, K., Schwendner, P., Sipos, C., Lorenzo, L., Mirchev, M., Lameski, P., Kabasinskas, A., Tidjani, C., Ozturkkal, B., & Cerneviciene, J. (2024). Enhancing portfolio management using artificial intelligence: Literature review. *Frontiers in Artificial Intelligence, 7*, 1371502. DOI: 10.3389/frai.2024.1371502 PMID: 38650961

Suzuki, E. (2002). UNDIRECTED DISCOVERY OF INTERESTING EXCEPTION RULES. *International Journal of Pattern Recognition and Artificial Intelligence, 16*(08), 1065–1086. DOI: 10.1142/S0218001402002155

Upadhyaya, P. (2025). The Role of AI in Identifying and Preventing Fraudulent Activities. In Gupta, S. K., Rosak-Szyrocka, J., Rena, R., Shieh, C.-S., & Erkol Bayram, G. (Eds.), *The Impact of Artificial Intelligence on Finance: Transforming Financial Technologies* (Vol. 53, pp. 367–381). Springer Nature Switzerland., DOI: 10.1007/978-3-031-92916-8_18

Valla, V., Tzelepi, K., Charitou, P., Lewis, A., Polatidis, B., Koukoura, A., Karapatsia, A., Antonopoulou, K., Prodromidou, K., Papadaki, E., & Vassiliadis, E. (2023). Use of Real-World Evidence for International Regulatory Decision Making in Medical Devices. *International Journal of Digital Health, 3*(1), 1. DOI: 10.29337/ijdh.50

Vora, L. K., Gholap, A. D., Jetha, K., Thakur, R. R. S., Solanki, H. K., & Chavda, V. P. (2023a). Artificial Intelligence in Pharmaceutical Technology and Drug Delivery Design. *Pharmaceutics, 15*(7), 1916. DOI: 10.3390/pharmaceutics15071916 PMID: 37514102

Vyskočil, J., Douda, P., Novák, P., & Wally, B. (2023). A Digital Twin-Based Distributed Manufacturing Execution System for Industry 4.0 with AI-Powered On-The-Fly Replanning Capabilities. *Sustainability (Basel), 15*(7), 6251. DOI: 10.3390/su15076251

Warner, J., Prada Jardim, A., & Albera, C. (2025). Artificial Intelligence: Applications in Pharmacovigilance Signal Management. *Pharmaceutical Medicine, 39*(3), 183–198. DOI: 10.1007/s40290-025-00561-2 PMID: 40257538

Xia, K., Saidy, C., Kirkpatrick, M., Anumbe, N., Sheth, A., & Harik, R. (2021). Towards Semantic Integration of Machine Vision Systems to Aid Manufacturing Event Understanding. *Sensors (Basel), 21*(13), 4276. DOI: 10.3390/s21134276 PMID: 34206522

Yue, L., Niu, P., & Wang, Y. (2019). Guidelines for defining user requirement specifications (URS) of manufacturing execution system (MES) based on ISA-95 standard. *Journal of Physics: Conference Series, 1168*, 032065. DOI: 10.1088/1742-6596/1168/3/032065

Zhou, E., Shen, Q., & Hou, Y. (2024). Integrating artificial intelligence into the modernization of traditional Chinese medicine industry: A review. *Frontiers in Pharmacology, 15*, 1181183. DOI: 10.3389/fphar.2024.1181183 PMID: 38464717

Zhou, H., Zhang, H., Zhang, R., Yuan, X., & Chang, H. (2025). AI-driven photonic noses: From conventional sensors to cloud-to-edge intelligent microsystems. *Microsystems & Nanoengineering, 11*(1), 209. DOI: 10.1038/s41378-025-01058-3 PMID: 41198617

Zhou, J., He, J., Li, G., & Liu, Y. (2020). Identifying Capsule Defect Based on an Improved Convolutional Neural Network. *Shock and Vibration, 2020*, 1–9. DOI: 10.1155/2020/8887723

Chapter 14

Biosafety, Bioethics, and Regulatory Challenges in Omics-Nano-CRISPR Oncology

S. Dayana
http://orcid.org/0009-0008-7135-5606
Department of Microbiology, Sacred Heart College, Tirupattur, India

K. Kesavardhini
http://orcid.org/0009-0005-8235-042X
Department of Microbiology, Sacred Heart College, Tirupattur, India

P. Ragavi
Department of Microbiology, Sacred Heart College, Tirupattur, India

B. Lokeshwari
http://orcid.org/0009-0005-6612-5134
Department of Microbiology, Sacred Heart College, Tirupattur, India

P. Saranraj
http://orcid.org/0000-0002-5052-6532
Department of Microbiology, Sacred Heart College, Tirupattur, India

ABSTRACT

Omics-Nano-CRISPR technologies are pioneering precision oncology by integrating multi-omics data, nanotechnology, and genome editing to improve cancer diagnosis and treatment. In addition to their therapeutic value, these technologies confront significant biosafety, bioethical and regulatory issues that must be addressed carefully, both scientifically and ethically within society. The future of Omics-Nano-CRISPR technologies in oncology is expected to be supported by ongoing advancements in explainable AI, enhanced bio-safety surveillance systems, and the development and harmonisation of international regulations and ethical governance structures. Finally, addressing bio-safety, bioethical and regulatory challenges will be crucial for the advancement of the safe, effective and socially responsible precision cancer medicine.

DOI: 10.4018/979-8-3373-9715-3.ch014

1. INTRODUCTION

Cancer remains one of the leading causes of death worldwide and a healthcare, research and policy challenge. Although disease diagnostics, treatments and precision medicine have made great strides, there is still a significant number of cancer patients with high rates of recurrence, drug resistance and treatment related toxicity (Arriaga-Canon et al., 2022). Omics technologies, nanotechnology and genome editing using CRISPR have transformed oncology by providing a detailed molecular characterization, targeted drug delivery, and precise genetic modifications. Tumours are studied using genomics, transcriptomics, proteomics, metabolomics and other omics platforms, that provide valuable insights into tumour biology, disease progression and patient-specific therapeutic responses. Combined with nanotechnology and the CRISPR-system, these methods have unlocked unprecedented possibilities for the development of personalized diagnostics and treatments of cancer. While all three technologies are advancing rapidly, and are being clinically applied, they also present important biosafety, bioethical and regulatory challenges that must be carefully addressed to assure safe and responsible innovation and patient protection (Wang et al., 2022).

Despite remarkable advances, significant barriers remain before integrated omics–nano–CRISPR platforms become routine in clinical oncology. Current limitations include incomplete understanding of long-term nanoparticle toxicity, unpredictable genome-editing outcomes, insufficient validation of AI algorithms across diverse populations, lack of standardized multi-omics integration pipelines, and limited regulatory harmonization. Future research should prioritize clinically validated AI-assisted decision systems, safer genome-editing platforms, biodegradable nanocarriers, and internationally accepted biosafety standards for precision oncology (Satam et al., 2023).

The genomics revolution in oncology generates a wealth of biological and genetic data, which can be leveraged for biomarker discovery, prognosis and therapy development (Casati, 2018). Combining multi-omics data with artificial intelligence and machine learning has greatly improved the understanding of the heterogeneity of cancer and the identification of new drug targets. Nanotechnology, on the other hand, is a promising technology for enhancing cancer diagnosis and treatment by creating nanoscale drug delivery devices, imaging agents, biosensors and theragnostic platforms (Sadek et al., 2023). Nanocarriers can specifically target the tumour tissues, increase drug bioavailability and decrease systemic toxicity, thus increasing the efficacy of the drugs. However, due to the growing utilization of nanoparticles in clinical applications, there are concerns about the potential for unintended biological interactions, environmental persistence, biodistribution and long-term toxicity. The ability to understand and manage these risks is key to the safe introduction of nanotechnology innovations into daily oncology care. Advances in genome editing have also led to the development of novel CRISPR-Cas-based technologies that allow for the precise modification of cancer-related genes, thereby revolutionizing cancer research and therapeutic development (Shamjetsabam et al., 2024).

CRISPR systems have a tremendous potential for gene editing, correcting genetic mutations, engineering of immune cells, identification of new drug targets and personalized cancer treatment (Kumar et al., 2013). The sensitivity and specificity of cancer detection technologies have also been enhanced with recent advances in CRISPR-diagnostics. The use of CRISPR-technology in clinical settings, however, poses a serious biosafety issue, such as off-target mutations, unwanted genomic changes, immune reactions, and long-term genetic effects. These issues are made even more complicated when incorporating platforms for delivery of CRISPR-systems using nanotechnology along with sensitive omics data. Therefore, strong risk assessment methodologies and comprehensive safety evaluations are necessary to limit the possible adverse effects and to ensure safe clinical use. Along with biosafety concerns, omics, nanotechnology, and CRISPR-based interventions present many bioethical questions. There are privacy and data ownership, informed consent, and equitable access to advanced healthcare technologies concerns associated with the collection, storage and analysis of such vast genomic and molecular data sets. Ethical issues also involve the possibility of genetic engineering misuse, of genome-editing being discriminatory, of genetic information being used for discrimination, and of personalised cancer treatment being unaffordable (Song et al., 2011).

Although the integration of omics technologies, nanotechnology, CRISPR genome editing, and artificial intelligence has transformed precision oncology, several scientific and translational challenges remain unresolved (Yadav et al., 2023). Long-term biosafety of engineered nanomaterials, unintended CRISPR off-target effects, algorithmic bias in AI-driven clinical decision-making, and limited interoperability of multi-omics datasets continue to restrict widespread clinical implementation. Future research should prioritize safer genome-editing strategies, standardized multi-omics workflows, explainable AI models, biodegradable nanocarriers, and harmonized international regulatory frameworks to facilitate responsible clinical translation (Babu et al., 2021).

2. MOLECULAR FOUNDATIONS OF OMICS-NANO-CRISPR SYSTEMS

2.1 Genomics, Proteomics, and Transcriptomics

The modern oncology tools of the omics, genomics, proteomics and transcriptomics are all necessary to give an overview of the mechanisms of cancer development and progression (Maqsood et al., 2023). Genomics involves the study of the entire genome of cancer cells, including mutations, abnormalities of chromosomes, copy number differences and predispositions of the genome that help initiate and develop tumors. New sequencing methods have allowed for the identification of cancer-associated genes in a few hours and the development of molecular markers to early detection of cancer and the tailoring of individual treatments. Transcriptomics not only offers the opportunity to gain insight into the expression patterns and the transcripts produced by cancer cells, but it can also help researchers to understand cell responses, signaling pathways and the progression of the disease itself (Sharma et al., 2013). These strategies could help to explain tumor heterogeneity and to identify target areas for precision medicine.Proteomics can also provide insights into the structure, function, interactions and expression of proteins in cancerous cells and tissues. Protein molecules, which are the molecules involved in cellular activities, are the molecules that reflect disease mechanisms, therapeutic response and resistance pathway in proteomics. A combined approach to genomics, transcriptomics and proteomics provides a multi-dimensional perspective of cancer biology, which helps identify novel cancer markers and develop personalized cancer therapies. These omics technologies in concert with state-of-the-art computation and artificial intelligence will be essential for precision cancer diagnostics and targeted therapeutics and for the data-driven oncology practice (Green and Sambrook, 2018).

2.2 CRISPR-Cas Systems and Gene Editing Mechanisms

CRISPR-Cas systems have transformed genome editing, in both molecular biology and cancer research, offering a precise and effective way of genome editing CRISPR-Cas is an adaptive immune system in bacteria and archaea, which is used to modify DNA sequences by incorporating guide RNA sequences into the CRISPR array that guide Cas nucleases to the DNA sequence to be modified (Sheeja et al., 2021). The most popular system is CRISPR-Cas9 which is a system that introduces site-specific double-stranded DNA breaks that can be repaired by cell mechanisms to produce gene disruption, correction or insertion. In oncology, CRISPR technologies are used to study genes in cancer, find therapeutic targets, model cancer progression and create novel treatment strategies that will fix the genetic defects causing cancer (Veryaskina et al., 2022).

These are some of the latest advances in CRISPR and the creation of next generation gene-editing technologies that further improve precision and minimize the off-target effects of gene editing, including base editing, prime editing and CRISPR interference systems (Yadav et al., 2023). These innovations allow the modification of specific nucleotides without causing a lot of DNA damages, thus enhancing the safety and therapeutic potential of these modifications. In cancer research, researchers have figured out how to use the CRISPR to make killer cells more effective, and to create treatments tailored to an individual's unique cancer. However, there are still numerous challenges to be addressed, including efficient delivery, genomic stability, ethical and bio-safety issues. Addressing these concerns will be essential for the successful clinical translation of oncology applications with CRISPR (Satam et al., 2023).

2.3 Smart Nanocarriers for Therapeutic Delivery

Smart nanocarriers are high-tech drug delivery systems which may improve the precision of cancer treatment.These nanoscale systems are designed to deliver therapeutic agents to tumor tissues and avoid damage to normal tissues (Capone et al., 2024). Different nanocarrier technologies have been developed, such as liposomes, polymeric nanoparticles, dendrimers, nanogels, lipid nanoparticles, and inorganic nanomaterials, which improve the solubility, stability, circulation time and bioavailability of the drug. It is possible to take advantage of these differences in targeted drug delivery by using smart nanocarriers that take advantage of tumor-specific characteristics, including enhanced permeability and retention effects, pH changes, enzyme activity, and the expression pattern of receptors. This highly specific therapy will not cause as much toxicity in the body as regular chemotherapy will, and will also be more effective in the treatment (Meaburn and Schulz, 2012).

The use of smart nanocarriers combined with the advanced molecular technologies has further broadened the range of their applications in precision oncology (Reardon et al., 2023). Multiple therapeutic agents such as chemotherapeutic drugs, nucleic acids, proteins, and CRISPR components can be co-delivered with nanocarriers to allow for combination therapies specifically for individual patients. There is added control over therapeutic delivery with the use of stimuli-responsive nanocarriers that release cargo upon exposure to specific environmental stimuli, including temperature, pH, light or magnetic field. Furthermore, theranostic systems based on nanocarriers can also be designed to be diagnostic and therapeutic, thus enabling real-time monitoring of the response to treatment. These developments could mean that smart nanocarriers are key ingredients in the development of safer and more effective cancer treatments (Thomas et al., 2019).

2.4 AI Integration in Precision Oncology

In precision oncology, Artificial Intelligence (AI) is becoming an essential tool for understanding and analyzing complex biological and clinical information (Lynch et al., 2024). Cancer research today generates huge amounts of data on genomes, transcriptomes, proteomes, medical imaging, medical records and clinical trials. These datasets are valuable for understanding the preferences and habits of consumers, and can be analyzed with AI technologies like machine learning, deep learning, natural language processing, and predictive analytics. AI can assist in cancer diagnosis, risk stratification, biomarker identification, treatment decision making, and prognosis assessment, by integrating multi-omics data and clinical parameters. The ability to do this allows tailored approaches to medicine where treatment is decided by the molecular uniqueness of the patient. The integration of AI with CRISPR technologies and nanomedicine is ushering in a new era of innovations in oncology (Steel et al., 2025).

The applications of genome editing can be made safer and more effective with the use of AI algorithms which can simulate the outcomes of genome editing, identify the best regions of the genome to target with CRISPR and evaluate possible off target effects (Ewim et al., 2023). Likewise, the design and optimization of smart nanocarriers can be facilitated by AI, and its behaviour, biodistribution and therapeutic effects can be predicted. AI-powered decision support systems can help healthcare providers in the medical field to analyse complex diagnostic data and formulate customized treatment strategies (Song et al., 2011). As the power of computing continues to grow, the future of AI in precision oncology looks bright for further improving the precision, efficiency, and accessibility of the treatment, which will result in better outcomes for the patient and a more effective management of cancer (Sharma et al., 2023).

3. BIOSAFETY CONCERNS IN CRISPR-BASED CANCER THERAPIES

3.1 Off-Target Gene Editing Effects

Off-target gene editing is one of the biggest biosafety problems of CRISPR cancer therapies. CRISPR-Cas systems are designed to recognize and modify a specific sequence of DNA, but sometimes they can bind to parts of DNA with similar sequences (Kuppan et al., 2024). Such accidental changes could involve insertions, deletions, substitutions or chromosomal rearrangements in non-target genes, which could interfere with their normal functions. For use in oncology, where accuracy is paramount, off-target activity may be detrimental to efficacy and pose unexpected risks. The off-target events frequency and severity depend on many factors including guide RNA design, complexity of target sequence, delivery method, and the cellular environment. Therefore, much optimization and validation is needed before interventions using CRISPR are safe to use in clinical practice. The newly developed high fidelity Cas enzymes, engineering strategies of guide RNA, and computational tools for predicting off-target activity have advanced genome editing technologies (Fatumo et al., 2020).

Whole genome sequencing, targeted deep sequencing and bioinformatics analyses are increasingly being used to identify and characterize unwanted genomic modifications (Sadek et al., 2023). The technologies allow for a full assessment of editing specificity and aid in the optimization of CRISPR systems for therapeutic application. Additionally, recent gene-editing techniques like base editing and prime editing have shown increased accuracy by minimizing the frequency of double-stranded DNA breaks and thus minimizing the potential for unintended alterations of the genome. For improving the safety profile of CRISPR-based cancer therapies, it is crucial to make continuous technological advancements (Olesen, 2018).

Rigorous preclinical testing, standardized safety assessment protocols and long-term monitoring of patients treated is required for the management of off-target effects (Satam et al., 2023). Prior to the introduction of genome editing therapies into the clinic, regulatory agencies highlight the need for extensively evaluating risks. Furthermore, ethical problems can occur, because unwanted effects of genetic modification can be harmful to a patient or have unwanted biological effects. AI and machine learning are also being used to predict and detect off-target events, helping to ensure the reliability and safety of therapeutic strategies. The continual evolution of CRISPR technologies demands reducing off-target effects of precision oncology interventions as a key priority (Viergever and Hendriks, 2016).

3.2 Genomic Instability and Mutagenesis

The primary concern with genome-editing technologies in oncology is the potential for genomic instability and mutagenesis (Ezeanolue et al., 2019). The CRISPR-induced modification of DNA usually results in the generation of double-stranded breaks in the DNA, which trigger innate cellular DNA repair processes including non-homologous end joining and homology-directed repair. These processes allow for the targeted alteration of cells' genetic material, but can also produce unwanted mutations, chromosomal translocation, big deletions and structural genomic rearrangements. These changes can affect crucial genes, trigger oncogenes or silence tumor suppressor genes, leading to elevated possibilities of malignant mutation or worsening. One of the challenges to the safe use of CRISPR techniques in cancer therapy is the potential for the induction of additional genomic changes (Ward et al., 2018).

Several factors such as editing efficiency, target site selection, activity of repair pathways and cellular context influence the extent of genomic instability (Low and Tagle, 2017). Genome editing frequently can have unintended genetic effects, especially on cancer cells, which may also have pre-existing genomic abnormalities and defective DNA repair mechanisms. To reduce mutagenesis, researchers are creating new strategies such as the use of high-precision editing systems, transient delivery and other editing methods that do not involve significant DNA damage. Advanced molecular characterization techniques enable detailed evaluation of the genomic integrity after editing procedures, to determine potential safety concerns prior to clinical implementation. The genomic instability is a serious issue that must be addressed through a complete biosafety assessment during the entire course of therapeutic development (Sakolish et al., 2018).

Genome-editing induced mutations need to be investigated in long-term studies to determine their persistence and biological effects (Song et al., 2011). The importance of regulation is growing exponentially and detailed genomic analysis and risk assessment are becoming more and more crucial to ensure patient safety. In addition, there are ethical issues with the potential for unwanted hereditary and somatic mutations and responsible research practice and good clinical management are important. New breakthroughs in precision genome editing, molecular diagnostics, and computational biology are likely to further diminish risks of genomic instability and to enable the creation of safer gene-based cancer therapeutics.

3.3 Immune Responses to CRISPR Components

Immunity to components of these CRISPR technologies is a significant hurdle to clinical translation for cancer therapeutics.Several CRISPR systems are based on Cas proteins from bacterial pathogens (e.g., Cas9, Cas12), which might be identified as foreign by the human immune system. Exposure to these proteins may lead to immune and inflammatory response, immune-mediated toxicity and less efficacy (Abinaya and Viswanathan, 2021). Treatment can be further complicated by pre-existing immunity due to previous exposure to bacteria with similar proteins. Such immune responses can negatively affect the editing efficiency, cause damage to normal tissues and hinder the durability of CRISPR therapies. Current research is focused on ways to minimize the immunogenicity and make the CRISPR systems more compatible with the human immune system. These strategies can involve engineering less immunogenic variants of Cas proteins, delivery of Cas proteins *via* a transient delivery system, and using a nanoparticle delivery system to shield CRISPR components from immune uptake. *Ex vivo* gene-editing techniques, which involve editing cells out of the body and then returning them to the patient, also minimize *in vivo* immune interactions (Alsulami, 2021).

In addition, sophisticated immunological screening techniques are being utilized to identify patients with a higher risk of immunological complications and tailor treatment regimen as required (Anwar et al., 2020). To make oncology interventions with CRISPR technology safe and effective, it is critical to understand and control immune responses. In order to properly evaluate the immune risk and more importantly to devise proper mitigation strategies, extensive testing in the preclinical and clinical settings is necessary. To gain approval of gene editing therapies, regulatory authorities request detailed immunogenicity testing. Furthermore, it is important to have long-term follow-up of immune response to identify late adverse events and ensure patient safety. As genome editing technologies keep developing, immunological barriers will be critical to clear the way for more widespread adoption in precision oncology (Ashique et al., 2023). Figure 1 illustrates the CRISPR-Cas9 delivery process, the activation of innate and adaptive immune responses, and the factors affecting the efficacy and safety of CRISPR-based cancer therapy, including immune recognition and anti-Cas9 immunity.

Figure 1. Immune Responses and Challenges in CRISPR-Based Cancer Therapy

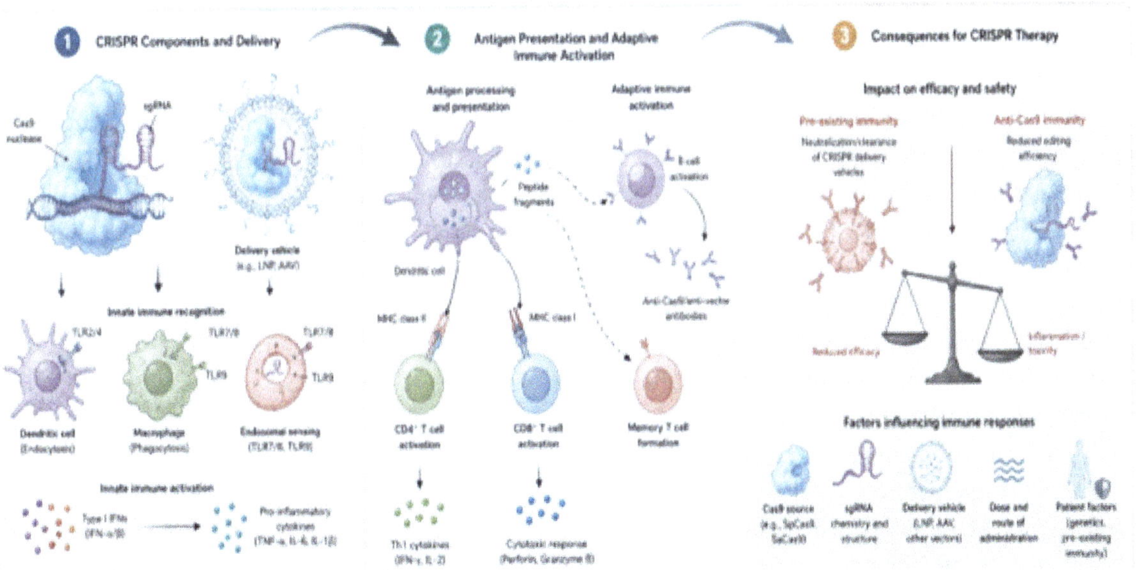

3.4 Long-Term Safety Monitoring

One of the important aspects of safe clinical use of cancer therapeutics based on CRISPR is long-term monitoring for safety (Anwar et al., 2020). Despite significant therapeutic potential, the long-term biological consequences of genome-editing technologies remain incompletely understood, and they have great therapeutic potential. The genetic changes resulting from the treatment may last long and could be associated with late effects like unexpected changes in cell behavior, immune problems, or secondary mutations, genomic instability. Hence, close monitoring is critical to assess durability of treatment, new safety issues and to maintain the therapeutic benefit in the long term. Complete monitoring programs are designed to review patient response following a gene-editing therapy, and include periodic clinical evaluations, molecular testing, and genomic sequencing and measurement of biomarkers (Aziz et al., 2023).

Advanced omics methods like genomics, transcriptomics, proteomics and metabolomics are valuable resources to uncover the changes in biology that can be detected following therapy. Another advantage of the monitoring offered by AI and big data is the ability to integrate and interpret complex, longitudinal data (Besenhard et al., 2023). These can be used for early detection of adverse events and for making informed decisions regarding patient management and therapy optimisation. The long-term safety monitoring is also a key aspect of long-term regulatory concerns and public trust of novel gene editing technologies. As the technologies of omics, nanotechnology and CRISPR join the suite of precision oncology, long-term safety regimes will be important to ensure responsible innovation, patient well-being and successful translation to the clinic (Cheng et al., 2023).

4. NANOTOXICOLOGY AND SAFETY ASSESSMENT

4.1 Toxicological Effects of Nanoparticles

Nanoparticles have become an essential component of precision oncology because they enable targeted drug delivery, molecular imaging, and site-specific therapeutic interventions while reducing systemic toxicity (Song et al., 2011). But their special physicochemical characteristics, such as small size, large surface, high reactivity and ability to penetrate biological barriers, might also cause unwanted toxicological effects. The nanoparticles can undergo interactions with cell membranes, proteins, nucleic acids and organelles upon entry into the body, resulting in oxidative stress, inflammation, DNA damage, mitochondrial dysfunction and cellular toxicity. The effects can vary with the composition, size, shape, surface charge, concentration, and duration of exposure of the particles. A knowledge of these toxicological mechanisms is crucial for the design of safer nanomaterials, as well as for the reduction of adverse effects on cancer patients (Ding et al., 2022).

One of the main prerequisites for clinical translation of nanomedicine-based oncology products is the comprehensive toxicological assessment (Zhou and Simonian, 2024). *In vitro* and *in vivo* exposure studies are used to determine the cytotoxicity, genotoxicity, immunotoxicity and long-term biological response of nanoparticles. New analytical methods and computational models are becoming more sophisticated and are increasingly being used to predict the behaviour and toxicity profiles of nanoparticles prior to clinical use. There is also a great need for regulatory agencies to conduct thorough safety assessments of nano-enabled therapeutics to ensure that these products do not present any safety concerns. There are ongoing studies seeking to develop biodegradable, biocompatible and low-toxicity nanomaterials to enhance the safety profile of nanotechnology based cancer therapies and to increase their use in precision medicine (Badea and Feeney, 2025). Figure 2 illustrates the major toxicological effects of nanoparticles, including oxidative stress, inflammation, cytotoxicity, genotoxicity, neurotoxicity, immunotoxicity, and organ toxicity, highlighting their adverse impacts on cells, tissues, and organ systems.

Figure 2. Major Toxicological Effects of Nanoparticles

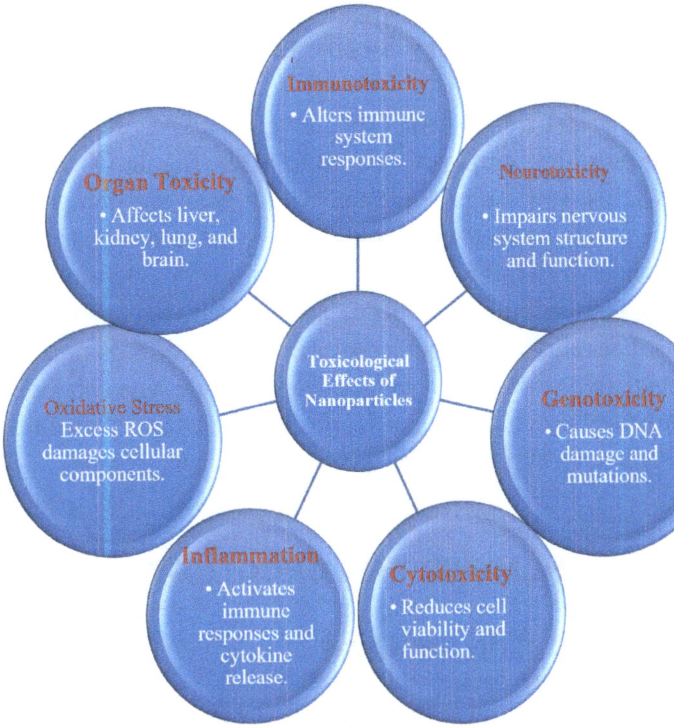

4.2 Biodistribution and Biocompatibility

Biodistribution is the process of movement, accumulation and clearance of nanoparticles in biological systems after administration; biocompatibility is the ability of nanoparticles to interact with living tissues without harm (Ward et al., 2018). For cancer, the most important benefit of effective biodistribution is that therapeutic nanoparticles need to be delivered to the tumor locations in adequate amounts while sparing accumulation in healthy organs. The circulation time, penetration into tissues, uptake by cells and elimination from the body will depend very much on the particle size, surface chemistry, shape and functionalization. Physiological filtration and immune surveillance mechanisms often lead to the deposition of nanoparticles in organs like the liver, spleen, kidneys and lungs (Hassan et al., 2025).

Biodistribution is, therefore, an important consideration for maximizing therapeutic efficacy and minimizing off-target toxicity. Storage stability is also crucial for safe clinical use of nano-delivery systems, along with biocompatibility (Prabhudesai et al., 2025). Ideally, biocompatible nanoparticles should not lead to any significant immune response, inflammation, thrombosis or damage to tissue, while retaining their therapeutic properties.Surface modifications like polymer coatings, PEGylation (conjugation of polyethylene glycol), biomimetic designs are widely used to enhance the compatibility with biological systems and increase circulation time. The use of advanced imaging technologies and pharmacokinetic studies gives insight in the *in vivo* behavior of nanoparticles and allows to optimize the design of the formulation. The challenge of optimizing biodistribution and biocompatibility is an ongoing priority in the development of nanomedicine for precision oncology, as these factors are crucial for improving the safety and efficacy of the interventions using nanotechnology (Yang et al., 2025).

4.3 Risk Evaluation of Nano-Delivery Systems

Biosafety evaluation of nano-delivery systems is an important aspect of risk assessment in the field of nanomedicine and precision oncology (Arriaga-Canon et al., 2022). The aim of the nano-delivery platforms is to increase the targeted delivery of drugs, nucleic acids, proteins, and components of a gene-editing system, but the complexity of interactions with biological systems can introduce potential safety risks. Risk assessment of nano-delivery systems focuses on evaluating nanoparticle interactions with biological systems, including toxicity, biodistribution, biodegradability, pharmacokinetics, immunogenicity, and long-term biosafety, rather than reiterating their therapeutic advantages. Comprehensive analyses can aid in the identification of possible hazards from nanoparticle exposure, and can facilitate the preparation of safer therapeutic formulations. Nanomaterials have properties that are often different from traditional pharmaceuticals and medical devices and call for specific assessment techniques (Wang et al., 2022).

5. ETHICAL ISSUES IN GENOME EDITING ONCOLOGY

5.1 Human Genome Modification Concerns

One of the biggest ethical concerns regarding the use of CRISPR and other genome editing technologies in oncology is human genome modification (Casati, 2018). These technologies provide unprecedented opportunities for the correction of disease-causing mutations, the enhancement of immune responses against tumour cells and the creation of very personalized cancer treatments. But deliberate changes to the human genome are not without their risks, potential side effects and implications for society. The off-target mutations and genomic instability may have long-term effects on individual patients, and, if the genetic changes are passed on to future generations, they can also affect future generations. Further, there is some concern over non-therapeutic uses of genome-editing technologies, which pose ethical questions regarding the limits of scientific manipulation of human biology. With the quick growth of genome editing technologies, there has also been debate about social justice, equity and access to new medical technologies (Sadek et al., 2023).

As highly advanced treatments for gene manipulation become available to only those who can afford them, health disparities may rise dramatically (Sakolish et al., 2018). The ethical situation becomes further complicated with genetic discrimination and privacy issues as well as the issue of ownership of genetic information. Ethical oversight and supporting scientific progress will depend on the cooperation among researchers, health care providers, bioethicists and policymakers. Effective governance, public participation and ongoing ethical review will help to ensure that human genome modification technologies are used responsibly and principally for purposes of patient health and wellbeing (Shamjetsabam et al., 2024).

5.2 Germline Editing and Ethical Debates

Germ-line editing is genetic modification of reproductive cells, embryos or early developmental stages that can be passed on to future generations (Ding et al., 2022). While germ-line editing may be one of the most controversial uses of CRISPR technology, it could be used to cure hereditary diseases and genetic risk factors for specific cancers. Germ-line editing involves more than just the treated person; it involves their offspring, with lasting ethical, social and biological implications. Some critics say there is still a lack of understanding about the consequences of these inherited genetic changes, and that modifying a person's genes in this way can have irreversible effects (Kumar et al., 2013).

While concerns about the potential risks to human safety dominate the scientific discourse on the ethics of germ-line editing, the topic also raises questions on human dignity, reproductive autonomy, social inequality, and the prospect of "Designer Babies" (Alsulami, 2021). Social pressure to change desirable traits, like intelligence, physical appearance, or athletic ability, is a concern about genetic enhancement for non-medical purposes. This may further perpetuate discrimination and instigate further social inequalities. International organizations and regulatory bodies have thus urged cautious approaches and, in many instances, restrictions, to clinical germ-line editing. Scientists, ethicists, policy makers and the public all need to be engaged in ongoing discussions about acceptable limits for genome modifications to the germ line and their future applications, and how they are consistent with ethical principles and societal values (Song et al., 2011).

5.3 Ethical Use of AI in Cancer Care

AI is increasingly transforming the way cancer is diagnosed, treatment is planned, prognosis is predicted, and decisions are made in the clinic using the analysis of vast amounts of biomedical and patient data (Sharma et al., 2013). Additionally, patterns that are hard for practitioners to see may be unveiled through the use of AI algorithms, resulting in more precise diagnoses and treatment plans. However, as AI continues to play a more significant role in oncology, a range of ethical challenges arises, including questions of transparency, accountability, fairness, and patient autonomy. AI algorithms are used in clinical decision making and patient decision making, and are complex and difficult to understand how they have come up with a specific recommendation, because they are a "Black Box". This lack of interpretation can make this difficult to build trust and to responsibly make clinical decisions (Green and Sambrook, 2018).

One ethical concern is the potential for bias in the data used to train AI algorithms that might result in unequal treatment and outcomes in healthcare settings (Olesen, 2018). The data sets may not be representative of the diversity of patients, such that the resulting predictions or recommendations for some patients may be false. Also, patient confidentiality, responsibility for algorithms, human judgment and machine recommendations, and a balance between the two are crucial. Regulatory measures, ethical principles, and continuous monitoring are required to ensure that AI is used responsibly and ethically in the field of cancer care. Ethical governance of AI in oncology should primarily emphasize algorithmic transparency, fairness, explainability, clinical accountability, and human oversight (Veryaskina et al., 2022).

6. DATA PRIVACY AND CYBERSECURITY IN PRECISION ONCOLOGY

6.1 Protection of Genomic Data

Genomic information can include extremely sensitive and personally identifiable data regarding an individual's disease susceptibility, genetic makeup, and response to treatment; safeguarding this information is an essential part of precision oncology (Yadav et al., 2023). The advent of omics research, the CRISPR technologies, and genomics has led to the generation of vast quantities of genetic data that must be securely stored, processed and shared. Unauthorized accessing and misuse of genomic information has the potential to cause breaches in privacy, genetic discrimination and loss of patient trust. Given this, robust cybersecurity measures, encryption, access control, and compliance guidelines play a vital role in safeguarding genomic data. Furthermore, responsible governance policies and clear management of genomic data within the programs promote ethical use and safeguard patient privacy and the integrity of precision medicine initiatives (Satam et al., 2023).

6.2 AI and Big Data Security Challenges

AI and big data analytics have been a game changer in the healthcare sector, particularly in oncology, where they have improved the prediction, diagnosis, treatment optimization, and clinical decision-making processes (Ward et al., 2018). But the storage and handling of very large datasets are significant security issues in health care. Malicious actors and unauthorized users are interested in AI systems due to their wealth of patient information, such as genomic, clinical, imaging, and behavioral data. The risks of security can be data breaches, ransom ware attacks, manipulation of algorithms, threats of adversarial machine learning, and unauthorized data extraction (Capone et al., 2024).

Additionally, the complexity of AI models makes them difficult to understand and identify vulnerabilities and guarantee the integrity of the data via their analysis (Kuppan et al., 2024). Addressing these challenges requires comprehensive cybersecurity strategies that include encrypted data storage, multi-factor authentication, continuous system monitoring, secure cloud infrastructures, AI-specific risk assessment, and regulatory compliance to protect sensitive genomic and clinical information while maintaining healthcare system resilience (Meaburn and Schulz, 2012).

6.3 Block-chain and Secure Healthcare Systems

Block-chain technology's promise of providing enhanced security, transparency, and trust for healthcare data management systems has proven to be a potential solution (Satam et al., 2023). The incorporation of block-chain in precision oncology can provide a decentralized and tamper-proof way to store genomic, clinical, and treatment-related data, which is essential for its analysis and decision-making. Block-chain technology can be applied in precision oncology to establish a decentralized and secure system for handling, sharing, and verifying the authenticity of valuable genomic, clinical, and treatment information utilized for analysis and choice-making. With a centralized database system, the failure or unauthorized change or modification at one point can cause the entire system to fail (Lynch et al., 2024).

For block-chain systems, the data is replicated on multiple nodes and therefore more difficult to corrupt or lose. Smart contracts and cryptographic techniques further bolster other dimensions of data integrity, access control, and auditability, thus improving patients' control over their health information

(Steel et al., 2025). Moreover, block-chain can facilitate secure sharing of patient information across different stakeholders, including healthcare providers, researchers, and regulatory bodies, and it can also assist with regulatory compliance. Overall, blockchain technology enhances secure data sharing, improves data integrity, supports regulatory compliance, and strengthens patient control over sensitive genomic information, making it a promising platform for precision oncology data management (Ewim et al., 2023).

7. REGULATORY FRAMEWORKS FOR OMICS-NANO-CRISPR TECHNOLOGIES

7.1 International Regulatory Guidelines

International regulatory frameworks provide standardized guidelines to ensure the safe, ethical, and effective clinical implementation of genome editing, nanomedicine, artificial intelligence, and multi-omics technologies in precision oncology. The U.S. Food and Drug Administration (FDA) regulates genome-editing therapies, nanomedicines, and AI-enabled medical devices through evidence-based clinical evaluation and post-marketing surveillance. The European Medicines Agency (EMA) oversees Advanced Therapy Medicinal Products (ATMPs), including gene-editing technologies and nanomedicines. The International Council for Harmonisation (ICH) establishes globally accepted standards for pharmaceutical quality, safety, efficacy, and Good Clinical Practice (GCP). The World Health Organization (WHO) provides ethical governance recommendations for human genome editing, while the Organisation for Economic Co-operation and Development (OECD) develops internationally accepted principles for nanomaterial safety assessment and responsible AI implementation. Collectively, these organizations facilitate regulatory harmonization and promote the safe clinical translation of advanced precision oncology technologies (Alsulami, 2021).

7.2 Quality Control and Standardization

Quality control and standardization are key pillars of the development and clinical application of advanced technologies in oncology (Fatumo et al., 2020). Highly reproducible, reliable and accurate procedures are necessary for omics analyses, formulation of nanoparticles, gene editing systems and AI based diagnostic platforms. Quality control involves standardized validation procedures that ensure the reproducibility, reliability, safety, and analytical accuracy of omics technologies, genome-editing systems, nanomedicines, and AI-based diagnostic platforms before clinical implementation. These measures can detect errors, contamination and variation, as well as performance problems that could potentially affect patient safety or treatment outcomes. One of the main principles of Precision Oncology is the need for high standards of quality, particularly when therapeutic decisions are made on the basis of very sensitive molecular and genetic data (Olesen, 2018).

Standardization ensures consistency in research institutions, health care centers and industrial manufacturing by providing common protocols, terms, performance criteria, and validation processes (Casati, 2018). Lack of uniformity might result in data interpretation, clinical outcomes and regulatory assessment that are inconsistent. Collaboration and comparability of results across regions is important, so it is crucial to develop harmonized standards for genomics, nanotechnology, and CRISPR applications internationally. The automation and digital quality management systems and artificial intelligence are also enhancing quality assurance processes. Quality control and standardization systems improve the scientific credibility, help to regulate and facilitate the widespread adoption of precision oncology technologies (Viergever and Hendriks, 2016).

7.3 Challenges in Regulatory Harmonization

Regulatory harmonisation means bringing regulatory requirements, the basis for evaluation, and compliance requirements into line between countries and jurisdictions (Ezeanolue et al., 2019). The process of regulatory harmonization is especially difficult for the field of omics-nano-CRISPR oncology due to the rapidly evolving technology and the wide variation in legal, ethical and healthcare systems across the globe. Countries may have divergent strategies with respect to gene editing, privacy concerns with respect to genomic data, safety testing of nanomaterials, governance of artificial intelligence and clinical trial policies. These differences can make it difficult to collaborate with other countries, slow product approval processes, raise development expenses and make it harder to roll out innovative cancer treatments internationally. Other hurdles include the multi-disciplinary character of precision oncology technologies that frequently include both biological and digital and engineering elements, all integrated into the same therapeutic platform (Ward et al., 2018).

8. ARTIFICIAL INTELLIGENCE AND EXPLAINABILITY IN ONCOLOGY

8.1 Explainable AI in Cancer Diagnosis

Explainable Artificial Intelligence (XAI) is an essential component of the cancer diagnosis process that aims to make all AI decisions explainable, understandable and transparent for human doctors (Wang et al., 2022). Many of the traditional AI models, particularly deep learning algorithms, are 'black boxes' that make very accurate predictions without providing any explanation to the conclusions that they have reached. When doctors' recommendations have direct impact on patient care, such as in oncology, they must rely on the rationale for the decision. Explainable AI improves model transparency by identifying the biological features, imaging biomarkers, and clinical variables that contribute to prediction outcomes, thereby increasing clinician trust, supporting regulatory acceptance, and facilitating safer clinical decision-making (Sakolish et al., 2018).

8.2 Machine Learning for Treatment Prediction

In the field of oncology, AI is being applied in many different ways to forecast responses to therapies and to tailor treatment plans (Abinaya and Viswanathan, 2021). These patterns observed in the data collected from genomics, transcriptomics, proteomics, medical imaging, EHRs and clinical trials can be detected by machine learning algorithms and linked to treatment success, resistance, toxicity, disease progression, and more. These predictive capabilities enable doctors to tailor treatment to the individual patient, potentially leading to improved treatment results and reduced unnecessary side effects. In the field of precision oncology, machine learning models are increasingly being used to inform decisions about drug selection, dosing, prognosis, and stratification, which can help optimize and ensure evidence-based cancer treatment (Alsulami, 2021).

Continuous growth in multi-omics datasets and high-performance computing has substantially improved the predictive capability, scalability, and clinical applicability of machine-learning models for precision oncology. Algorithms can adapt to new clinical information, get better in accuracy as they are used more and more, and adjust to the different treatment context (Wang et al., 2022). Machine-learning models in oncology are increasingly applied to predict therapeutic response, estimate survival probability, identify drug resistance mechanisms, and optimize personalized treatment selection using integrated clinical and molecular datasets. Other factors to consider are the quality of the data, the interpretability of the models, algorithmic bias and clinical validation. Rigorous testing and a transparent methodology and regulatory oversight must be implemented to ensure the safety, reliability, and clinical meaningfulness of machine learning based treatment prediction systems to support cancer management (Anwar et al., 2020). Table 1 demonstrates that explainability techniques such as SHAP, LIME, Grad-CAM, and saliency maps improve the interpretability of AI models across various oncology applications, thereby increasing clinician confidence and supporting regulatory acceptance.

Table 1. AI Techniques and Explainable AI (XAI) Methods for Oncology Applications

S. No.	AI Technique	Explainability Method	Oncology Application	References
1	Machine Learning	Feature Importance	Cancer Risk Prediction	(Ashique et al., 2023)
2	Deep Learning	Grad-CAM	Tumor Detection	(Aziz et al., 2023)
3	CNN	Saliency Maps	Medical Imaging	(Besenhard et al., 2023)
4	Random Forest	SHAP	Treatment Response	(Cheng et al., 2023)
5	XGBoost	LIME	Cancer Classification	(Ding et al., 2022)
6	NLP	Attention Visualization	Clinical Notes Analysis	(Zhou and Simonian, 2024)
7	Survival Models	SHAP Values	Survival Prediction	(Badea and Feeney, 2025)

9. CLINICAL TRANSLATION AND PRECISION CANCER THERAPY

9.1 Personalized Treatment Strategies

Personalized treatment strategies integrate molecular biomarkers, genomic alterations, clinical characteristics, and patient-specific therapeutic responses to design individualized cancer management plans that maximize efficacy while minimizing adverse effects (Hassan et al., 2025). A standard therapy might treat a large number of patients even though a personalized one could help identify changes in the molecular level, biomarkers and genetic differences that influence the course of a disease and how a patient will respond to therapy in a given patient. Next-generation sequencing (NGS), artificial intelligence, nanomedicine, and CRISPR-based genome editing collectively enable highly personalized therapeutic strategies by integrating molecular profiling with patient-specific clinical information. Treatment strategies are also personalized, enabling the tracking of disease progression and therapeutic responses in real time while treatment plans are continually modified as per patient needs. This not only results in better clinical outcomes, but also increases the quality of life and plays a crucial role in progressing towards patient-centered cancer care (Prabhudesai et al., 2025).

9.2 Clinical Trials and Safety Evaluation

Clinical trials generate the scientific evidence required by international regulatory agencies to establish the safety, efficacy, and quality of advanced precision oncology technologies before widespread clinical implementation. Clinical trials complement long-term biosafety studies by validating treatment efficacy, optimizing dosing strategies, identifying rare adverse events, and generating the clinical evidence required for regulatory approval and routine implementation in precision oncology (Yang et al., 2025). These evaluations can be used to detect adverse reactions, risks, and therapeutic advantages prior to the technology's wide-spread clinical deployment. Precision oncology uses clinical trials that better match patients to treatment, implement adaptive study designs, and employ sophisticated monitoring systems to achieve better treatment evaluation and patient outcomes. New cancer treatments are critical to patient protection and public confidence, and comprehensive safety assessments including genomic analyses, toxicology testing, and immunologic testing are essential to patient protection and confidence in new cancer therapeutics. The careful clinical testing and regulatory regulation continue to be essential to the responsible implementation of new and innovative biomedical advances into medical practice (Arriaga-Canon et al., 2022).

9.3 Biomarker-Based Therapeutic Design

Therapeutic design using biomarkers is the identification and use of biological markers to inform the development and selection of targeted cancer treatments (Wang et al., 2022). Biomarkers can be genetic mutations, gene expression profiles, protein signatures, metabolic characteristics, and molecular pathways involved in disease progression and therapeutic response among others. Enhanced discovery, and validation, of clinically relevant biomarkers through advances in multi-omics technologies has significantly contributed to the precision stratification of patients and optimization of treatment. Precision oncology involves using biomarkers to determine who is most likely to respond to a certain treatment, be it a medicine, immunotherapy, or gene editing, and spare others from exposure to treatments that have limited efficacy. The development of AI-powered tools, bioinformatics, and high throughput analytical platforms further contribute to biomarker discovery and therapeutic design. This means biomarker-based strategies help to enhance the effectiveness of cancer treatment, minimize toxicity, and provide more tailored treatment strategies (Casati, 2018).

9.4 Translational Oncology Approaches

The field of translational oncology centers on the interaction between basic science and clinical practice, and includes strategies that aim to translate the results of basic science into actionable diagnostic, therapeutic, and prevention strategies for the treatment of cancer (Sadek et al., 2023). It is a multidisciplinary field combining molecular biology, genomics, nanotechnology, artificial intelligence and clinical medicine to speed up the development of novel cancer therapies. Translational research is the process of identifying disease mechanisms, validating therapeutic targets, developing experimental therapies, assessing clinical outcome in well designed studies. The integration of omics technologies, smart nanocarriers and genome editing systems based on CRISPR has made a tremendous impact on the conversion of basic science into clinical applications for health care. Translational oncology also encourages sharing of knowledge and research between scientists, doctors, regulators, and industry stakeholders, leading to the timely and responsible use of emerging technologies to benefit patients and drive the future of precision oncology (Shamjetsabam et al., 2024).

10. FUTURE PERSPECTIVES AND EMERGING INNOVATIONS

10.1 Next-Generation Genome Editing Systems

Next-generation genome-editing technologies have been specifically developed to overcome the biosafety limitations of conventional CRISPR-Cas9 systems by improving editing precision, minimizing unintended genomic alterations, and enabling clinically safer therapeutic applications (Kumar et al., 2013). New platforms like base editing, prime editing, Crisp-associated transposases, epigenome editing systems and RNA targeting technologies are allowing specific genetic modifications without generating large double-stranded DNA breaks (dsDNA). These advances alleviate potential off-target effects, genomic instability, and mutations that have been a worry in conventional gene-editing methods (Song et al., 2011). In oncology, next-generation genome editing systems offer a powerful suite of tools that can be used for correcting cancer-associated mutations, for the regulation of gene expression, for the engineering of immune cells, and as a means of developing personalized treatment strategies that can be matched to patient-specific profiles. The potential of these genome-editing platforms is enormous, when combined with the latest developments in artificial intelligence, omics technologies and nanocarrier-based delivery systems, they will help revolutionize precision oncology, allowing safer and more effective therapeutic interventions to be delivered and the creation of highly targeted and personalized cancer treatments (Sheeja et al., 2021).

Recent advances in genome editing have extended beyond conventional CRISPR-Cas9 to include base editing, prime editing, CRISPR activation (CRISPRa), CRISPR interference (CRISPRi), RNA-targeting Cas13 systems, and programmable epigenome editing. These next-generation platforms enable highly precise genetic modifications while minimizing double-strand DNA breaks and reducing off-target mutations. Several clinically relevant CRISPR-based therapies, including engineered CAR-T cells and lipid nanoparticle-mediated CRISPR delivery systems, are currently being evaluated in early-phase clinical trials for liver, prostate, lung, and hematological malignancies (Casati, 2018).

10.2 Future Intelligent Precision Oncology Ecosystems

Omics technologies, artificial intelligence, nanomedicine, genome editing, digital health platforms and advanced clinical decision-support systems will be brought together in a single healthcare ecosystem in the future through innovative precision oncology approaches (Babu et al., 2021). These interwoven ecosystems will continually receive, process, and interpret enormous amounts of data from the genome, molecules, images, clinical data, and real patient data to deliver highly personalized cancer prevention, diagnosis, treatment, and monitoring solutions. AI will play a crucial role in predicting disease trajectories, helping to make treatment decisions, identifying new markers, and assisting with real-time medical decision-making (Maqsood et al., 2023). Concurrently, innovative nanocarrier systems as well as next generation genome editing tools will enable targeted delivery and precise intervention within molecular systems with greater safety and efficiency. Successful clinical implementation will depend on coordinated multidisciplinary partnerships that integrate scientific innovation, ethical governance, regulatory oversight, industrial development, and patient-centered healthcare delivery. These smart precision oncology systems can revolutionize cancer therapy, with the capacity to augment treatment achievements, cut healthcare bills, speed up scientific advancement, and enable a more predictive, preventative, personalised and participatory approach to medicine (Sharma et al., 2013).

10.3 Knowledge Gaps and Future Research Priorities

Despite rapid technological progress, several knowledge gaps remain before integrated precision oncology platforms can be routinely implemented in clinical practice. Large-scale longitudinal studies are required to evaluate long-term genomic stability, nanoparticle biodistribution, immune responses, and clinical safety. Future investigations should also focus on integrating explainable AI, federated learning, adaptive clinical trial designs, and standardized regulatory frameworks to improve reproducibility, transparency, and patient safety (Ward et al., 2018).

11. CONCLUSION

Future progress in precision oncology will require multidisciplinary efforts to improve genome-editing accuracy, develop safer nanocarriers, implement explainable and privacy-preserving AI, standardize multi-omics integration, and harmonize international regulatory policies to ensure equitable and responsible clinical translation. The convergence of omics technologies, nanomedicine, CRISPR genome editing, and artificial intelligence represents a transformative paradigm in precision oncology by enabling highly personalized cancer diagnosis, targeted therapy, and real-time clinical decision-making. The potential for off-target gene manipulations, genomic instability, toxicity of nanoparticles, long-term monitoring of safety and efficacy, patient privacy, informed consent, data security, and ethical oversight of genome editing must all be thoroughly examined and overseen on an ongoing basis. Moreover, there are concerns about transparency, accountability, algorithmic bias, and equitable access to healthcare innovations as AI and big data analytics are incorporated into healthcare practices. The key for balancing technological innovation with patient protection and the well-being of society lies in the adoption of effective regulatory frameworks, the support of international cooperation, the implementation of standardized quality control measures and harmonized safety guidelines. Research and application will continue to advance, and scientists, clinicians, the ethical community and policy makers, industry, and regulatory agencies must collaborate to address emerging challenges and maximize the potential of these game-changers. Although remarkable progress has been achieved in integrating omics technologies, nanomedicine, CRISPR genome editing, and artificial intelligence, successful clinical translation requires overcoming substantial scientific, ethical, regulatory, and technological challenges. Future research should emphasize safer genome-editing platforms, biodegradable nanocarriers, explainable and privacy-preserving AI models, standardized multi-omics integration, and globally harmonized regulatory policies. Multidisciplinary collaboration among clinicians, researchers, bioengineers, ethicists, regulatory agencies, and policymakers will be essential to establish robust biosafety standards, ensure equitable access to advanced therapies, and accelerate the responsible implementation of precision oncology technologies for improved patient outcomes.

REFERENCES

Abinaya, R. V., & Viswanathan, P. (2021). Biotechnology-based therapeutics. In *Translational Biotechnology: A Journey from Laboratory to Clinic* (pp. 27–52). Elsevier., DOI: 10.1016/B978-0-12-821972-0.00019-8

Alsulami, M. N. (2021). Eradication of malaria: Present situations and new strategies. *Journal of Pharmaceutical Research International*, *33*(47A), 17–37. DOI: 10.9734/jpri/2021/v33i47A32986

Anwar, A., Mohammed, B. S., Wahab, M. A., & Liew, M. S. (2020). Enhanced properties of cementitious composite tailored with graphene oxide nanomaterial: A review. *Developments in the Built Environment*, *1*, 100002. DOI: 10.1016/j.dibe.2019.100002

Arriaga-Canon, C., Contreras-Espinosa, L., Rebollar-Vega, R., Montiel-Manríquez, R., Cedro-Tanda, A., García-Gordillo, J. A., Álvarez Gómez, R. M., Jiménez-Trejo, F., Castro-Hernández, C., & Herrera, L. A. (2022). Transcriptomics and RNA-based therapeutics as potential approaches to manage SARS-CoV-2 infection. *International Journal of Molecular Sciences*, *23*(19), 11058. DOI: 10.3390/ijms231911058 PMID: 36232363

Ashique, S., Garg, A., Mishra, N., Raina, N., Ming, L. C., & Tuli, H. S. (2023). Nano-mediated strategy for targeting and treatment of non-small cell lung cancer (NSCLC). *Naunyn-Schmiedeberg's Archives of Pharmacology*, *396*(11), 2769–2792. DOI: 10.1007/s00210-023-02522-5 PMID: 37219615

Aziz, A., Rehman, U., Sheikh, A., Abourehab, M. A. S., & Kesharwani, P. (2023). Lipid-based nanocarrier-mediated CRISPR/Cas9 delivery for cancer therapy. *Journal of Biomaterials Science. Polymer Edition*, *34*(3), 398–418. DOI: 10.1080/09205063.2022.2121592 PMID: 36083788

Babu, K. N., Sheeja, T. E., Minoo, D., Rajesh, M. K., Samsudeen, K., Suraby, E. J., & Kumar, I. P. V. (2021). Random amplified polymorphic DNA (RAPD) and derived techniques. In Besse, P. (Ed.), *Molecular Plant Taxonomy: Methods and Protocols* (pp. 219–247). Springer., DOI: 10.1007/978-1-0716-0997-2_13

Badea, A. R., & Feeney, O. (2025). Genome editing dilemma: Navigating dual-use potential and charting the path forward. *Journal of Bioethical Inquiry*, *22*(1), 101–110. DOI: 10.1007/s11673-024-10358-8 PMID: 39046699

Besenhard, M. O., Pal, S., Gkogkos, G., & Gavriilidis, A. (2023). Non-fouling flow reactors for nanomaterial synthesis. *Reaction Chemistry & Engineering*, *8*(5), 955–977. DOI: 10.1039/D2RE00412G

Capone, P., Chiarella, P., & Sisto, R. (2024). Advanced technologies in genomic toxicology: Current trends and future directions. *Current Opinion in Toxicology*, *37*, 100444. DOI: 10.1016/j.cotox.2023.100444

Casati, S. (2018). Integrated approaches to testing and assessment. *Basic & Clinical Pharmacology & Toxicology*, *123*(S5, Suppl. 5), 51–55. DOI: 10.1111/bcpt.13018 PMID: 29604238

Cheng, X., Xie, Q., & Sun, Y. (2023). Advances in nanomaterial-based targeted drug delivery systems. *Frontiers in Bioengineering and Biotechnology*, *11*, 1177151. DOI: 10.3389/fbioe.2023.1177151 PMID: 37122851

Ding, H., Zhang, J., Zhang, F., Xu, Y., Liang, W., & Yu, Y. (2022). Nanotechnological approaches for diagnosis and treatment of ovarian cancer: A review of recent trends. *Drug Delivery*, *29*(1), 3218–3232. DOI: 10.1080/10717544.2022.2132032 PMID: 36259505

Ewim, D. R. E., Orikpete, O. F., Scott, T. O., Onyebuchi, C. N., Onukogu, A. O., Uzougbo, C. G., & Onunka, C. (2023). Survey of wastewater issues due to oil spills and pollution in the Niger Delta area of Nigeria: A secondary data analysis. *Bulletin of the National Research Center*, *47*(1), 116. DOI: 10.1186/s42269-023-01090-1

Ezeanolue, E. E., Iheanacho, T., Patel, D. V., Patel, S., Sam-Agudu, N., Obiefune, M., Dakum, P., Okonkwo, P., Olutola, A., Khamofu, H., Oyeledun, B., Aliyu, S., Aina, M., Eyo, A., Oko, J., Akinmurele, T., Oleribe, O., Gebi, U., Aliyu, M. H., & Siberry, G. (2019). Challenges and strategies for improving training of mid-level research personnel in Nigeria. *Annals of Global Health*, *85*(1), 87. DOI: 10.5334/aogh.2405 PMID: 31225955

Fatumo, S., Ebenezer, T. E., Ekenna, C., Isewon, I., Ahmad, U., Adetunji, C., Oladipo, E. K., Adebiyi, M., Adebiyi, E., & Nashiru, O. (2020). The Nigerian Bioinformatics and Genomics Network (NBGN): A collaborative platform to advance bioinformatics and genomics in Nigeria. *Global Health, Epidemiology and Genomics*, *5*, e3. DOI: 10.1017/gheg.2020.3 PMID: 32742665

Green, M. R., & Sambrook, J. (2018). Analysis and normalization of real-time polymerase chain reaction (PCR) experimental data. *Cold Spring Harbor Protocols, 2018*(6), pdb.top095000. DOI: 10.1101/pdb.top095000

Hassan, Y. M., Mohamed, A. S., Hassan, Y. M., & El-Sayed, W. M. (2025). Recent developments and future directions in point-of-care next-generation CRISPR-based rapid diagnosis. *Clinical and Experimental Medicine*, *25*(1), 33. DOI: 10.1007/s10238-024-01540-8 PMID: 39789283

Kumar, A., Misra, P., & Dube, A. (2013). Amplified fragment length polymorphism: An adept technique for genome mapping, genetic differentiation, and intraspecific variation in protozoan parasites. *Parasitology Research*, *112*(2), 457–466. DOI: 10.1007/s00436-012-3238-6 PMID: 23254590

Kuppan, N., Padman, M., Mahadeva, M., Srinivasan, S., & Devarajan, R. (2024). A comprehensive review of sustainable bioremediation techniques: Eco-friendly solutions for waste and pollution management. *Waste Management Bulletin*, *2*(3), 154–171. DOI: 10.1016/j.wmb.2024.07.005

Low, L. A., & Tagle, D. A. (2017). Tissue chips: Innovative tools for drug development and disease modeling. *Lab on a Chip*, *17*(18), 3026–3036. DOI: 10.1039/C7LC00462A PMID: 28795174

Lynch, C., Sakamuru, S., Ooka, M., Huang, R., Klumpp-Thomas, C., Shinn, P., Gerhold, D., Rossoshek, A., Michael, S., Casey, W., Santillo, M. F., Fitzpatrick, S., Thomas, R. S., Simeonov, A., & Xia, M. (2024). High-throughput screening to advance *in vitro* toxicology: Accomplishments, challenges, and future directions. *Annual Review of Pharmacology and Toxicology*, *64*(1), 191–209. DOI: 10.1146/annurev-pharmtox-112122-104310 PMID: 37506331

Maqsood, Q., Sumrin, A., Waseem, R., Hussain, M., Imtiaz, M., & Hussain, N. (2023). Bioengineered microbial strains for detoxification of toxic environmental pollutants. *Environmental Research*, *227*, 115665. DOI: 10.1016/j.envres.2023.115665 PMID: 36907340

Meaburn, E., & Schulz, R. (2012). Next generation sequencing in epigenetics: Insights and challenges. *Seminars in Cell & Developmental Biology*, *23*(2), 192–199. DOI: 10.1016/j.semcdb.2011.10.010 PMID: 22027613

Olesen, O. F. (2018). European & Developing Countries Clinical Trials Partnership (EDCTP). *Human Vaccines & Immunotherapeutics*, *14*(1), 14–16. DOI: 10.1080/21645515.2017.1356649 PMID: 29136388

Prabhudesai, M., Wu, M., Zadeh, A., Fragkiadaki, K., & Pathak, D. (2025). Diffusion beats autoregressive in data-constrained settings. *arXiv*. /arXiv.2507.15857DOI: 10.52202/085713-0354

Reardon, A. J. F., Farmahin, R., Williams, A., Meier, M. J., Addicks, G. C., Yauk, C. L., Matteo, G., Atlas, E., Harrill, J., Everett, L. J., Shah, I., Judson, R., Ramaiahgari, S., Ferguson, S. S., & Barton-Maclaren, T. S. (2023). From vision toward best practices: Evaluating *in vitro* transcriptomic points of departure for application in risk assessment using a uniform workflow. *Frontiers in Toxicology*, *5*, 1194895. DOI: 10.3389/ftox.2023.1194895 PMID: 37288009

Sadek, H. E., Ebadah, I. M., & Mahmoud, Y. A. (2023). Importance of biotechnology in controlling insect pests. *Journal of Modern Agriculture and Biotechnology*, *2*(1), 2023005. DOI: 10.53964/jmab.2023005

Sakolish, C., Weber, E. J., Kelly, E. J., Himmelfarb, J., Mouneimne, R., Grimm, F. A., House, J. S., Wade, T., Han, A., Chiu, W. A., & Rusyn, I. (2018). Technology transfer of the microphysiological systems: A case study of the human proximal tubule tissue chip. *Scientific Reports*, *8*(1), 14882. DOI: 10.1038/s41598-018-33099-2 PMID: 30291268

Satam, H., Joshi, K., Mangrolia, U., Waghoo, S., Zaidi, G., Rawool, S., Thakare, R. P., Banday, S., Mishra, A. K., Das, G., & Malonia, S. K. (2023). Next generation sequencing technology: Current trends and advancements. *Biology (Basel)*, *12*(7), 997. DOI: 10.3390/biology12070997 PMID: 37508427

Shamjetsabam, N. D., Rana, R., Malik, P., & Ganguly, N. K. (2024). CRISPR/Cas9: An overview of recent developments and applications in cancer research. *International Journal of Surgery*, *110*(10), 6198–6213. DOI: 10.1097/JS9.0000000000001081 PMID: 38377059

Sharma, A. K., Sharma, M., Sharma, A. K., Sharma, M., & Sharma, M. (2023). Mapping the impact of environmental pollutants on human health and environment: A systematic review and meta-analysis. *Journal of Geochemical Exploration*, *255*, 107325. DOI: 10.1016/j.gexplo.2023.107325

Sharma, G., Verma, H. N., & Sharma, R. (2013). RAPD analysis to study metagenome diversity in soil microbial community of arid zone plants. *Proceedings of the National Academy of Sciences. India. Section B, Biological Sciences*, *83*(2), 135–139. DOI: 10.1007/s40011-012-0041-1

Sheeja, T. E., Kumar, I. P. V., Giridhari, A., Minoo, D., Rajesh, M. K., & Babu, K. N. (2021). Amplified fragment length polymorphism: Applications and recent developments. In Besse, P. (Ed.), *Molecular Plant Taxonomy: Methods and Protocols* (pp. 187–218). Springer., DOI: 10.1007/978-1-0716-0997-2_12

Song, N., Nwafili, S. A., & Gao, T. X. (2011). Genetic diversity and population structure of *Chrysichthysnigrodigitatus* from Niger Delta based on AFLP analysis. *Biochemical Systematics and Ecology*, *39*(4), 320–327. DOI: 10.1016/j.bse.2011.04.002

Steel, A., Gallego-Perez, D. F., Ijaz, N., Gall, A., Bangpan, M., Dos Santos Boeira, L., Schveitzer, M. C., Chutaputti, A., Mahlanza-Langer, L., Pillai, G. K. G., Posayanonda, T., Sungchol, K., Shankar, D., & Kuchenmüller, T. (2025). Integration of traditional, complementary, and integrative medicine in the institutionalization of evidence-informed decision-making: The World Health Organization meeting report. *Journal of Integrative and Complementary Medicine*, *31*(4), 388–394. DOI: 10.1089/jicm.2024.0837 PMID: 39819104

Thomas, R. S., Bahadori, T., Buckley, T. J., Cowden, J., Deisenroth, C., Dionisio, K. L., Frithsen, J. B., Grulke, C. M., Gwinn, M. R., Harrill, J. A., Higuchi, M., Houck, K. A., Hughes, M. F., Hunter, E. S.III, Isaacs, K. K., Judson, R. S., Knudsen, T. B., Lambert, J. C., Linnenbrink, M., & Williams, A. J. (2019). The next generation blueprint of computational toxicology at the U.S. Environmental Protection Agency. *Toxicological Sciences : An Official Journal of the Society of Toxicology*, *169*(2), 317–332. DOI: 10.1093/toxsci/kfz058 PMID: 30835285

Veryaskina, Y. A., Titov, S. E., & Zhimulev, I. F. (2022). Reference genes for qPCR-based miRNA expression profiling in 14 human tissues. *Medical Principles and Practice : International Journal of the Kuwait University, Health Science Centre*, *31*(4), 322–332. DOI: 10.1159/000524283 PMID: 35354155

Viergever, R. F., & Hendriks, T. C. (2016). The 10 largest public and philanthropic funders of health research in the world: What they fund and how they distribute their funds. *Health Research Policy and Systems*, *14*(1), 12. DOI: 10.1186/s12961-015-0074-z PMID: 26892771

Wang, B., Chen, J. Z., Luo, X. Q., Wan, G. H., Tang, Y. L., & Wang, Q. P. (2022). The application of genome-wide CRISPR-Cas9 screens to dissect the molecular mechanisms of toxins. *Computational and Structural Biotechnology Journal*, *20*, 5076–5084. DOI: 10.1016/j.csbj.2022.09.012 PMID: 36187925

Ward, C. L., Shaw, D., Sprumont, D., Sankoh, O., Tanner, M., & Elger, B. (2018). Good collaborative practice: Reforming capacity building governance of international health research partnerships. *Globalization and Health*, *14*(1), 1. DOI: 10.1186/s12992-017-0319-4 PMID: 29310698

Yadav, D., Patil-Takbhate, B., Khandagale, A., Bhawalkar, J., Tripathy, S., & Khopkar-Kale, P. (2023). Next-generation sequencing transforming clinical practice and precision medicine. *Clinica Chimica Acta; International Journal of Clinical Chemistry*, *551*, 117568. DOI: 10.1016/j.cca.2023.117568 PMID: 37839516

Yang, C., Zhou, C., Wipf, D., & Li, Z. (2025). On powerful ways to generate: Autoregression, diffusion, and beyond. *arXiv*. /arXiv.2510.06190DOI: 10.48550

Zhou, L., & Simonian, A. L. (2024). CRISPR/Cas technology: The unique synthetic biology genome-editing tool shifting the paradigm in viral diagnostics, defense, and therapeutics. *Annual Review of Biomedical Engineering*, *26*(1), 247–272. DOI: 10.1146/annurev-bioeng-081723-013033 PMID: 38346278

Compilation of References

Abdollahi, H., & Rahmim, A. (2022). DTs for personalized healthcare: Application to radiopharmaceutical therapies. *Frontiers in Biomedical Technologies*, *10*(1), 1–5.

Abinaya, R. V., & Viswanathan, P. (2021). Biotechnology-based therapeutics. In *Translational Biotechnology: A Journey from Laboratory to Clinic* (pp. 27–52). Elsevier., DOI: 10.1016/B978-0-12-821972-0.00019-8

Abo Qoura, L., Churov, A. V., Maltseva, O. N., Eruslanova, K. A., & Zeng, G. (2026). Senescent cells in systemic aging: SASP heterogeneity, immune escape, and endocrine modulation. *Biogerontology*, *27*(3), 103. DOI: 10.1007/s10522-026-10449-6 PMID: 42113304

Achary, P. G. R. (2020, September 1). Applications of Quantitative Structure-Activity Relationships (QSAR) based Virtual Screening in Drug Design: A Review. *Mini Reviews in Medicinal Chemistry*, *20*(14), 1375–1388. DOI: 10.2174/1389557520666200429102334 PMID: 32348219

Achi, F., Attar, A. M., & Ait, L. A. (2024). Electrochemical nanobiosensors for the detection of cancer biomarkers in real samples: Trends and challenges. *Trends in Analytical Chemistry*, *170*, 117423. DOI: 10.1016/j.trac.2023.117423

Achir, A., Debbarh, I., Zoubir, N., Battas, I., Medromi, H., & Moutaouakkil, F. (2025). Advances in leukemia detection and classification: A systematic review of AI and image processing techniques. *F1000 Research*, *13*, 1536. DOI: 10.12688/f1000research.159318.2 PMID: 41200664

Adeniji, A. A., Dulal, S., & Martin, M. G. (2021, June). Personalized Medicine in Oncology in the Developing World: Barriers and Concepts to Improve *Status Quo. World Journal of Oncology*, *12*(2–3), 50–60. DOI: 10.14740/wjon1345 PMID: 34046099

Adeniyi, M., Ayoola, V. B., Samuel, T. E., & Awosan, W. (2024). Artificial intelligence-driven wearable electronics and smart nanodevices for continuous cancer monitoring and enhanced diagnostic accuracy. [IJSRMT]. *Int J Sci Res Modern Technol*, *3*(11), 3–18. DOI: 10.38124/ijsrmt.v3i11.106

Afrose, N., Chakraborty, R., Bhowmick, P., & Bhowmick, M. (2025). Advancing drug delivery systems through biomedical engineering: Innovations and future directions. In Innovations in Biomedical Engineering [Internet]. (pp. 1–32). Elsevier., Available from https://linkinghub.elsevier.com/retrieve/pii/B9780443301469000010, Retrieved October 28, 2025, from. DOI: 10.1016/B978-0-443-30146-9.00001-0

Afrose, N., Chakraborty, R., & Rajendran, K. (2025, August). Plasma-polymerized chitosan/agarose/chitosan-agarose composite coatings for antibacterial and antibiofilm wound dressings: A review. [Internet]. *International Journal of Biological Macromolecules, 320 (Pt 1)*, 145873. Retrieved October 27, 2025, from https://linkinghub.elsevier.com/retrieve/pii/S0141813025064281. DOI: 10.1016/j.ijbiomac.2025.145873 PMID: 40645242

Aftab, M., Mehmood, F., Zhang, C., Nadeem, A., Dong, Z., Jiang, Y., & Liu, K. (2025). AI in oncology: Transforming cancer detection through machine learning and deep learning applications. *arXiv preprint arXiv:2501.15489*.

Afzal, O., Altamimi, A. S. A., Nadeem, M. S., Alzarea, S. I., Almalki, W. H., Tariq, A., Mubeen, B., Murtaza, B. N., Iftikhar, S., Riaz, N., & Kazmi, I. (2022). Nanoparticles in drug delivery: From history to therapeutic applications. *Nanomaterials (Basel, Switzerland)*, *12*(24), 4494. DOI: 10.3390/nano12244494

Agarwal, N., & Nupur, P. K. (2024). Paul, and S. K. Mishra, "Artificial Intelligence and Machine Learning for Analysis of Multi-omics. In Mani, I., & Singh, V. (Eds.), *Multi-Omics Analysis of the Human Microbiome: From Technology to Clinical Applications* (pp. 339–354). Springer Nature Singapore., DOI: 10.1007/978-981-97-1844-3_16

Aggarwal, N., & Weinberg, O. K. (2021). Update on acute leukemias of ambiguous lineage. *Clinics in Laboratory Medicine*, *41*(3), 453–466. DOI: 10.1016/j.cll.2021.03.016 PMID: 34304775

Aghamiri, S. S., & Amin, R. (2025). The Potential Use of Digital Twin Technology for Advancing CAR-T Cell Therapy. *Current Issues in Molecular Biology*, *47*(5), 321. DOI: 10.3390/cimb47050321 PMID: 40699720

Ahluwalia, P., Ballur, K., Leeman, T., Vashisht, A., Singh, H., Omar, N., Mondal, A. K., Vaibhav, K., Baban, B., & Kolhe, R. (2024, January 23). Incorporating Novel Technologies in Precision Oncology for Colorectal Cancer: Advancing Personalized Medicine. *Cancers (Basel)*, *16*(3), 480. DOI: 10.3390/cancers16030480 PMID: 38339232

Ahmad, S., Bashar, A., Khanna, K., Bano, N., & Raza, K. (2025). Deep learning in variant detection and annotation. In Deep Learning in Genetics and Genomics [Internet]. (pp. 383–396). Elsevier., Available from https://linkinghub.elsevier.com/retrieve/pii/B9780443275746000102, Retrieved October 27, 2025, from. DOI: 10.1016/B978-0-443-27574-6.00010-2

Ahmad, Z. (2018). Flux balance analysis identifies distinct NADPH production strategies across NCI 60 cancer cell lines. *bioRxiv*, 332262.

Ahmed, Z. (2020, December). Practicing precision medicine with intelligently integrative clinical and multi-omics data analysis. [Internet]. *Human Genomics*, *14*(1), 35. Retrieved November 2, 2025, from https://humgenomics.biomedcentral.com/articles/10.1186/s40246-020-00287-z. DOI: 10.1186/s40246-020-00287-z PMID: 33008459

Ahmed, Z. (2022, April 15). Multi-omics strategies for personalized and predictive medicine: Past, current, and future translational opportunities. [Internet]. *Emerging Topics in Life Sciences*, *6*(2), 215–225. Retrieved November 2, 2025, from https://portlandpress.com/emergtoplifesci/article/6/2/215/230971/Multi-omics-strategies-for-personalized-and. DOI: 10.1042/ETLS20210244 PMID: 35234253

Ahn, J. S., Shin, S., Yang, S. A., Park, E. K., Kim, K. H., Cho, S. I., Ock, C.-Y., & Kim, S. (2023). Artificial Intelligence in Breast Cancer Diagnosis and Personalized Medicine. [Internet]. *Journal of Breast Cancer, 26*(5), 405. Retrieved November 2, 2025, from https://ejbc.kr/DOIx.php?id=10.4048/jbc.2023.26.e45. DOI: 10.4048/jbc.2023.26.e45 PMID: 37926067

Ai, K., Liu, B., Chen, X., Huang, C., Yang, L., Zhang, W., & Lai, P. (2024). Optimizing CAR-T cell therapy for solid tumors: Current challenges and potential strategies. *Journal of Hematology & Oncology, 17*(1), 105. DOI: 10.1186/s13045-024-01625-7 PMID: 39501358

Ajmal, C. S., Yerram, S., Abishek, V., Nizam, V. P. M., Aglave, G., Patnam, J. D., Raghuvanshi, R. S., & Srivastava, S. (2025). Innovative Approaches in Regulatory Affairs: Leveraging Artificial Intelligence and Machine Learning for Efficient Compliance and Decision-Making. *The AAPS Journal, 27*(1), 22. DOI: 10.1208/s12248-024-01006-5 PMID: 39776314

Akbani, R., Akdemir, K. C., Aksoy, B. A., Albert, M., Ally, A., Amin, S. B., Arachchi, H., Arora, A., Auman, J. T., Ayala, B., Baboud, J., Balasundaram, M., Balu, S., Barnabas, N., Bartlett, J., Bartlett, P., Bastian, B. C., Baylin, S. B., Behera, M., & Zou, L. (2015, June). Genomic Classification of Cutaneous Melanoma. *Cell, 161*(7), 1681–1696. DOI: 10.1016/j.cell.2015.05.044 PMID: 26091043

Akter, S. (2024). AI-driven precision medicine: Transforming personalized cancer treatment. *JAPMI, 2*(1), 10–21. DOI: 10.60087/vol2iisue1.p21

Al Saber, M., Biswas, P., Dey, D., Kaium, M. A., Islam, M. A., Tripty, M. I. A., Rahman, M. D. H., Rahaman, T. I., Biswas, M. Y., Paul, P., Rahman, M. A., Hasan, M. N., & Kim, B. (2021, December 23). A Comprehensive Review of Recent Advancements in Cancer Immunotherapy and Generation of CAR T Cell by CRISPR-Cas9. *Processes (Basel, Switzerland), 10*(1), 16. DOI: 10.3390/pr10010016

Alamri, A. M., Assiri, A. A., Khan, B., & Khan, N. U. (2025). Next-generation oncology: Integrative therapeutic frontiers at the crossroads of precision genomics, immuno-engineering, and tumor micro-environment modulation. *Medical Oncology, 42*(11), 482. DOI: 10.1007/s12032-025-03042-3 PMID: 40974488

Albalawi, A. F., Yassen, M. H., Almuraydhi, K. M., Althobaiti, A. D., Alzahrani, H. H., & Alqahtani, K. M. (2024). Ethical obligations and patient consent in the integration of artificial intelligence in clinical decision-making. *JOHS, 04*(12), 957–963. DOI: 10.52533/JOHS.2024.41241

Al-Daffaie, F. M., Al-Mudhafar, S. F., Alhomsi, A., Tarazi, H., Almehdi, A. M., El-Huneidi, W., Abu-Gharbieh, E., Bustanji, Y., Alqudah, M. A., Abuhelwa, A. Y., Guella, A., Alzoubi, K. H., & Semreen, M. H. (2024, May 7). Metabolomics and proteomics in prostate cancer research: Overview, analytical techniques, data analysis, and recent clinical applications. *International Journal of Molecular Sciences, 25*(10), 5071. DOI: 10.3390/ijms25105071 PMID: 38791108

Aldosari, B. N., Alfagih, I. M., & Almurshedi, A. S. (2021). Lipid nanoparticles as delivery systems for RNA-based vaccines. *Pharmaceutics, 13*(2), 206. DOI: 10.3390/pharmaceutics13020206 PMID: 33540942

Alexander, T. B., & Orgel, E. (2021). Mixed phenotype acute leukemia: Current approaches to diagnosis and treatment. *Current Oncology Reports, 23*(2), 22. DOI: 10.1007/s11912-020-01010-w PMID: 33544265

Alexandrov, L. B., Kim, J., Haradhvala, N. J., Huang, M. N., Tian Ng, A. W., Wu, Y., & Stratton, M. R. (2020). The repertoire of mutational signatures in human cancer. *Nature*, *578*(7793), 94–101. DOI: 10.1038/s41586-020-1943-3

Alfano, C. M., Suarez-Almazor, M., Rodriguez, M. A., Palos, G. R., & Gilmore, K. R. (2021). Innovating the personalization of stratified survivorship care pathways: Using a cancer data ecosystem to improve care access, outcomes, efficiency, and costs. *Journal of the National Cancer Institute. Monographs*, *2021*(57), 3–9. DOI: 10.1093/jncimonographs/lgaa011 PMID: 34478510

Algarvio, R. C., Conceição, J., Rodrigues, P. P., Ribeiro, I., & Ferreira-da-Silva, R. (2025). Artificial intelligence in pharmacovigilance: A narrative review and practical experience with an expert-defined Bayesian network tool. *International Journal of Clinical Pharmacy*, *47*(4), 932–944. DOI: 10.1007/s11096-025-01975-3 PMID: 40736603

Ali, K. A., Choudhuri, S., Chakraborty, R., Pradhan, A., Kotal, A., & Shaharyar, M. A. A. I. (2025). Technology-Powered Precision Medicine and Smart Healthcare. In AI and Precision Medicine in Infectious Disease Management [Internet]. 1st ed., pp. 51–70). CRC Press., Available from https://www.taylorfrancis.com/books/9781003615699/chapters/10.1201/9781003615699-3, Retrieved October 29, 2025, from. DOI: 10.1201/9781003615699-3

Ali, K. A., Mohin, S., Mondal, P., Goswami, S., Ghosh, S., & Choudhuri, S. Influence of artificial intelligence in modern pharmaceutical formulation and drug development. Futur J Pharm Sci [Internet]. 2024 Mar 29 [cited 2025 Feb 20];10(1):53. Available from: https://fjps.springeropen.com/articles/10.1186/s43094-024-00625-1

Ali, A. M., & Mohammed, M. A. (2024). A comprehensive review of artificial intelligence approaches in omics data processing: Evaluating progress and challenges. *Int. J. Math. Stat. Comput. Sci.*, *2*, 114–167.

Aliferis, C., & Simon, G. (2024). Overfitting, underfitting and general model overconfidence and under-performance pitfalls and best practices in machine learning and AI. *Artificial intelligence and machine learning in health care and medical sciences: Best practices and pitfalls*, 477-524.

Ali, H. (2023). Artificial intelligence in multi-omics data integration: Advancing precision medicine, biomarker discovery and genomic-driven disease interventions. *Int J Sci Res Arch*, *8*(1), 1012–1030. DOI: 10.30574/ijsra.2023.8.1.0189

Ali, K. A., Choudhuri, S., Bishal, A., Das, T., Nandy, S., & Hossain, C. M. (2025). Approaches of AI and Advanced Computational Biology for Combating HIV Infections. In *AI and Precision Medicine in Infectious Disease Management* (1st ed., pp. 149–166). CRC Press. DOI: 10.1201/9781003615699-8

Ali, K. A., Mohin, S., Mondal, P., Goswami, S., Ghosh, S., & Choudhuri, S. (2024, March 29). Influence of artificial intelligence in modern pharmaceutical formulation and drug development*Future Journal of Pharmaceutical Sciences*, *10*(1), 53. DOI: 10.1186/s43094-024-00625-1

Alinejad, T., Modarressi, S., Sadri, Z., Hao, Z., & Chen, C. S. (2023). Diagnostic applications and therapeutic option of Cascade CRISPR/Cas in the modulation of miRNA in diverse cancers: Promises and obstacles. *Journal of Cancer Research and Clinical Oncology*, *149*(12), 9557–9575. DOI: 10.1007/s00432-023-04747-6 PMID: 37222810

Allam, S., Nasr, K., Khalid, F., Shah, Z., Suheb, M. Z. K., Mulla, S., & Chaulagain, C. P. (2023). Liquid biopsies and minimal residual disease in myeloid malignancies. *Frontiers in Oncology*, *13*, 1164017. DOI: 10.3389/fonc.2023.1164017 PMID: 37213280

Allemailem, K. S., Alsahli, M. A., Almatroudi, A., Alrumaihi, F., Alkhaleefah, F. K., Rahmani, A. H., & Khan, A. A. (2022, December). Current updates of CRISPR/Cas9-mediated genome editing and targeting within tumor cells: An innovative strategy of cancer management. *Cancer Communications*, *42*(12), 1257–1287. DOI: 10.1002/cac2.12366 PMID: 36209487

Alshorman, J., Mehran, M. J., Miyanda Tembo, K., Bolideei, M., & Wang, Y. (2026). Applications of genome editing technologies in the treatment of human diseases. *Gene Therapy*, •••, 1–29. DOI: 10.1038/s41434-026-00627-z PMID: 42298089

Alsulami, M. N. (2021). Eradication of malaria: Present situations and new strategies. *Journal of Pharmaceutical Research International*, *33*(47A), 17–37. DOI: 10.9734/jpri/2021/v33i47A32986

Aman, R., Mahas, A., & Mahfouz, M. (2020). Nucleic acid detection using CRISPR/Cas biosensing technologies. *ACS Synthetic Biology*, *9*(6), 1226–1233. DOI: 10.1021/acssynbio.9b00507

Amiri, S., & Mahjoub, M. A. (2019). HMDHBN: Hidden Markov inducing a dynamic hierarchical Bayesian network for tumor growth prediction. In *International Conference on Computer Analysis of Images and Patterns* (pp. 3-14). Cham: Springer International Publishing. DOI: 10.1007/978-3-030-29888-3_1

Anderson, J. W., & Visweswaran, S. (2025). Algorithmic individual fairness and healthcare: A scoping review. *JAMIA Open*, *8*(1), ooae149. DOI: 10.1093/jamiaopen/ooae149 PMID: 39737346

Andoh, V., Ocansey, D. K. W., Naveed, H., Wang, N., Chen, L., Chen, K., & Mao, F. (2024). The advancing role of nanocomposites in cancer diagnosis and treatment. *International Journal of Nanomedicine*, *19*, 6099–6126. DOI: 10.2147/IJN.S471360 PMID: 38911500

André, T., Shiu, K.-K., Kim, T. W., Jensen, B. V., Jensen, L. H., Punt, C., Smith, D., Garcia-Carbonero, R., Benavides, M., Gibbs, P., de la Fouchardiere, C., Rivera, F., Elez, E., Bendell, J., Le, D. T., Yoshino, T., Van Cutsem, E., Yang, P., Farooqui, M. Z. H., & Diaz, L. A.Jr. (2020). Pembrolizumab in microsatellite-instability-high advanced colorectal cancer. *The New England Journal of Medicine*, *383*(23), 2207–2218. DOI: 10.1056/NEJMoa2017699 PMID: 33264544

Andrianandrianina Johanesa, T. V., Equeter, L., & Mahmoudi, S. A. (2024). Survey on AI Applications for Product Quality Control and Predictive Maintenance in Industry 4.0. *Electronics (Basel)*, *13*(5), 976. DOI: 10.3390/electronics13050976

AnishRuban. (1880). S., Raj, F. J., & Thangaraj, P. (2025). Phytochemical intervention in BCRP-driven cancer drug resistance: A comprehensive review. Biochimica et Biophysica Acta (BBA)-. *Revisiones en Cáncer*, (3), 189349.

Anitha, K., Posinasetty, B., Kumari, K. N., Chenchula, S., Padmavathi, R., Prakash, S., & Radhika, C. (2024). Liquid biopsy for precision diagnostics and therapeutics. *Clinica Chimica Acta; International Journal of Clinical Chemistry*, *554*, 117746. DOI: 10.1016/j.cca.2023.117746 PMID: 38151071

Anwar, A., Mohammed, B. S., Wahab, M. A., & Liew, M. S. (2020). Enhanced properties of cementitious composite tailored with graphene oxide nanomaterial: A review. *Developments in the Built Environment, 1*, 100002. DOI: 10.1016/j.dibe.2019.100002

Anzalone, A. V., Randolph, P. B., Davis, J. R., Sousa, A. A., Koblan, L. W., Levy, J. M., Chen, P. J., Wilson, C., Newby, G. A., Raguram, A., & Liu, D. R. (2019, December). Search-and-replace genome editing without double-strand breaks or donor DNA. *Nature, 576*(7785), 149–157. DOI: 10.1038/s41586-019-1711-4 PMID: 31634902

Arden, N. S., Fisher, A. C., Tyner, K., Yu, L. X., Lee, S. L., & Kopcha, M. (2021). Industry 4.0 for pharmaceutical manufacturing: Preparing for the smart factories of the future. *International Journal of Pharmaceutics, 602*, 120554. DOI: 10.1016/j.ijpharm.2021.120554 PMID: 33794326

Argelaguet, R., Velten, B., Arnol, D., Dietrich, S., Zenz, T., Marioni, J. C., Buettner, F., Huber, W., & Stegle, O. (2018). Multi-Omics Factor Analysis—A framework for unsupervised integration of multi-omics data sets. *Molecular Systems Biology, 14*(6), e8124. DOI: 10.15252/msb.20178124 PMID: 29925568

Arigbabu, A. T., Olaniyi, O. O., Adigwe, C. S., Adebiyi, O. O., & Ajayi, S. A. (2024). Data governance in AI-enabled healthcare systems: A case of the project nightingale. *Asian Journal of Research in Computer Science, 17*(5), 85–107. DOI: 10.9734/ajrcos/2024/v17i5441

Ariyanti, S., & Suryanegara, M. (2024). Current research themes and future research needs on making ai's energy consumption efficient: a review. In *2024 4th International Conference on Electronic and Electrical Engineering and Intelligent System (ICE3IS)* (pp. 99-104). IEEE. DOI: 10.1109/ICE3IS62977.2024.10775966

Arora, G., Joshi, J., Mandal, R. S., Shrivastava, N., Virmani, R., & Sethi, T. (2021). Artificial Intelligence in Surveillance, Diagnosis, Drug Discovery and Vaccine Development against COVID-19. *Pathogens (Basel, Switzerland), 10*(8), 1048. DOI: 10.3390/pathogens10081048 PMID: 34451513

Arriaga-Canon, C., Contreras-Espinosa, L., Rebollar-Vega, R., Montiel-Manríquez, R., Cedro-Tanda, A., García-Gordillo, J. A., Álvarez Gómez, R. M., Jiménez-Trejo, F., Castro-Hernández, C., & Herrera, L. A. (2022). Transcriptomics and RNA-based therapeutics as potential approaches to manage SARS-CoV-2 infection. *International Journal of Molecular Sciences, 23*(19), 11058. DOI: 10.3390/ijms231911058 PMID: 36232363

Arthur, J. C., Perez-Chanona, E., Mühlbauer, M., Tomkovich, S., Uronis, J. M., Fan, T.-J., Campbell, B. J., Abujamel, T., Dogan, B., Rogers, A. B., Rhodes, J. M., Stintzi, A., Simpson, K. W., Hansen, J. J., Keku, T. O., Fodor, A. A., & Jobin, C. (2012). Intestinal inflammation targets cancer-inducing activity of the microbiota. *Science, 338*(6103), 120–123. DOI: 10.1126/science.1224820 PMID: 22903521

Artificial Intelligence in Drug Manufacturing. (n.d.). US FDA. Retrieved November 3, 2025, from https://www.fda.gov/media/165743/download

Asadi, M., Zafari, V., Sadeghi-Mohammadi, S., Shanehbandi, D., Mert, U., Soleimani, Z., & Zarredar, H. (2024). The role of tumor microenvironment and self-organization in cancer progression: Key insights for therapeutic development. *BioImpacts, 15*, 30713. DOI: 10.34172/bi.30713 PMID: 40256216

Asan, O., Bayrak, A. E., & Choudhury, A. (2020). Artificial intelligence and human trust in healthcare: Focus on clinicians. *Journal of Medical Internet Research, 22*(6), e15154. DOI: 10.2196/15154 PMID: 32558657

Ashique, S., Garg, A., Mishra, N., Raina, N., Ming, L. C., & Tuli, H. S. (2023). Nano-mediated strategy for targeting and treatment of non-small cell lung cancer (NSCLC). *Naunyn-Schmiedeberg's Archives of Pharmacology, 396*(11), 2769–2792. DOI: 10.1007/s00210-023-02522-5 PMID: 37219615

Ashley, E. A. (2016, September). Towards precision medicine. *Nature Reviews. Genetics, 17*(9), 507–522. DOI: 10.1038/nrg.2016.86 PMID: 27528417

Ashrafizadeh, M., Zarrabi, A., Bigham, A., Taheriazam, A., Saghari, Y., Mirzaei, S., Hashemi, M., Hushmandi, K., Karimi-Maleh, H., Nazarzadeh Zare, E., Sharifi, E., Ertas, Y. N., Rabiee, N., Sethi, G., & Shen, M. (2023, November). (Nano) platforms in breast cancer therapy: Drug/gene delivery, advanced nanocarriers and immunotherapy. *Medicinal Research Reviews, 43*(6), 2115–2176. DOI: 10.1002/med.21971 PMID: 37165896

Asrina, A. (2024). Precision medicine approaches in oncology: Current trends and future directions. *Advances in Healthcare Research, 2*(1), 36–48. DOI: 10.60079/ahr.v2i1.370

Athanasopoulou, K., Michalopoulou, V. I., Scorilas, A., & Adamopoulos, P. G. Integrating Artificial Intelligence in Next-Generation Sequencing: Advances, Challenges, and Future Directions. CIMB [Internet]. 2025 June 19 [cited 2025 Oct 28];47(6):470. Available from: https://www.mdpi.com/1467-3045/47/6/470

Athanasopoulou, K., Daneva, G. N., Adamopoulos, P. G., & Scorilas, A. (2022). Artificial intelligence: The milestone in modern biomedical research. *BioMedInformatics, 2*(4), 727–744. DOI: 10.3390/biomedinformatics2040049

Atul, R. B. (2025). Transforming healthcare delivery: AI-powered clinical decision support systems. *INTERNATIONAL JOURNAL, 11*(1), 339–347.

Au, R., Kolachalama, V. B., & Paschalidis, I. C. (2022). Redefining and validating digital biomarkers as fluid, dynamic multi-dimensional digital signal patterns. *Frontiers in Digital Health, 3*, 751629. DOI: 10.3389/fdgth.2021.751629 PMID: 35146485

Avanzato, V. A., Matson, M. J., Seifert, S. N., Pryce, R., Williamson, B. N., Anzick, S. L., Barbian, K., Judson, S. D., Fischer, E. R., Martens, C., Bowden, T. A., de Wit, E., Riedo, F. X., & Munster, V. J. (2020, December). Case Study: Prolonged Infectious SARS-CoV-2 Shedding from an Asymptomatic Immunocompromised Individual with Cancer. *Cell, 183*(7), 1901–1912.e9. DOI: 10.1016/j.cell.2020.10.049 PMID: 33248470

Awasthi, N. P., Singh, P., & Agrawal, N. (2019). Mixed phenotype acute leukemia, B/myeloid with t (9; 22): Diagnostic and therapeutic challenges. *Indian Journal of Hematology & Blood Transfusion : An Official Journal of Indian Society of Hematology and Blood Transfusion, 35*(1), 179–181. DOI: 10.1007/s12288-018-1020-3 PMID: 30828170

Aziz, A., Rehman, U., Sheikh, A., Abourehab, M. A. S., & Kesharwani, P. (2023). Lipid-based nanocarrier-mediated CRISPR/Cas9 delivery for cancer therapy. *Journal of Biomaterials Science. Polymer Edition, 34*(3), 398–418. DOI: 10.1080/09205063.2022.2121592 PMID: 36083788

Babu, K. N., Sheeja, T. E., Minoo, D., Rajesh, M. K., Samsudeen, K., Suraby, E. J., & Kumar, I. P. V. (2021). Random amplified polymorphic DNA (RAPD) and derived techniques. In Besse, P. (Ed.), *Molecular Plant Taxonomy: Methods and Protocols* (pp. 219–247). Springer., DOI: 10.1007/978-1-0716-0997-2_13

Badano, A. (2021). In silico imaging clinical trials: Cheaper, faster, better, safer, and more scalable. *Trials, 22*(1), 64. DOI: 10.1186/s13063-020-05002-w PMID: 33468186

Badea, A. R., & Feeney, O. (2025). Genome editing dilemma: Navigating dual-use potential and charting the path forward. *Journal of Bioethical Inquiry, 22*(1), 101–110. DOI: 10.1007/s11673-024-10358-8 PMID: 39046699

Bae, H., Ji, H., Konstantinov, K., Sluyter, R., Ariga, K., Kim, Y. H., & Kim, J. H. (2025, October). Artificial intelligence-driven nanoarchitectonics for smart targeted drug delivery. *Advanced Materials, 37*(42), e10239. DOI: 10.1002/adma.202510239 PMID: 40772340

Bahssas, D. M., AlBar, A. M., & Hoque, M. (2015). Enterprise Resource Planning (ERP) Systems: Design, Trends and Deployment. *The International Technology Management Review, 5*(2), 72. DOI: 10.2991/itmr.2015.5.2.2

Bai, S., Li, W., Yin, N., Cao, Y., Zhang, S., Wang, Z., Qu, D., Wang, B., Li, Y., Liu, X., Jiang, Y., & Wang, Y. (2026). Bi-Target Nanomodulator for Glioma-Associated Myeloid Cells Polarization by Inducing Pyroptosis and Regulating Tryptophan Metabolism to Reshape Immunosuppressive Microenvironment of IDH-Mutant Gliomas. *Advanced Materials, 23021*(22), e23021. Advance online publication. DOI: 10.1002/adma.202523021 PMID: 41863418

Bakhshi, A., Bakhshi, M., Hosseine, M., Hasannezhad, H., Rahdar, A., Fosso-Kankeu, E., & Pandey, S. (2026). Integrating Artificial Intelligence and Quantum Computing with Omics Data for Cancer Therapy and Diagnosis. *Intelligent Nanocarriers: AI-Based Tools in Cancer Diagnosis and Therapy*, 413-524.

Balasubramaniam, N., Kauppinen, M., Rannisto, A., Hiekkanen, K., & Kujala, S. (2023). Transparency and explainability of AI systems: From ethical guidelines to requirements. *Information and Software Technology, 159*, 107197. DOI: 10.1016/j.infsof.2023.107197

Balasubramanian, A., & Sundrud, M. S. (2023). ATP-dependent transporters: Emerging players at the crossroads of immunity and metabolism. *Frontiers in Immunology, 14*, 1286696. DOI: 10.3389/fimmu.2023.1286696 PMID: 38022644

Bamodu, O. A., & Chung, C. C. (2024, August). Cancer Care Disparities: Overcoming Barriers to Cancer Control in Low- and Middle-Income Countries. *JCO Global Oncology*, (10), e2300439. DOI: 10.1200/GO.23.00439 PMID: 39173080

Bamodu, O. A., Chung, C. C., & Pisanic, T. R.II. (2023). Harnessing liquid biopsies: Exosomes and ctDNA as minimally invasive biomarkers for precision cancer medicine. *The Journal of Liquid Biopsy, 2*, 100126. DOI: 10.1016/j.jlb.2023.100126

Banda, A., Impomeni, O., Singh, A., Baloch, A. R., Hu, W., & Jaijyan, D. K. (2024). Precision in Action: The Role of Clustered Regularly Interspaced Short Palindromic Repeats/Cas in Gene Therapies. *Vaccines, 12*(6), 636. Advance online publication. DOI: 10.3390/vaccines12060636 PMID: 38932365

Bandla, C. (2024). Distributed Database Architectures for Federated Medical Training. *International Journal of Advanced Research in Science* [IJARSCT]. *Tongxin Jishu, 4*(2), 633–640.

Banerjee, A., Maity, S., & Mastrangelo, C. H. (2021). Nanostructures for biosensing, with a brief overview on cancer detection, IoT, and the role of machine learning in smart biosensors. *Sensors (Basel), 21*(4), 1253. DOI: 10.3390/s21041253

Banerjee, S., Mariya, A., Vangapally, S., & Paidimuddala, B. (2026). Nanobody-drug conjugates as versatile tools for improving therapeutic potential. *Biochemical Engineering Journal, 228*, 110072. DOI: 10.1016/j.bej.2026.110072

Bao, D., Liu, Z., Geng, Y., Li, L., Xu, H., Zhang, Y., Hu, L., Zhao, X., Zhao, Y., & Luo, D. (2022, December). Baseline MRI-based radiomics model assisted predicting disease progression in nasopharyngeal carcinoma patients with complete response after treatment. *Cancer Imaging; the Official Publication of the International Cancer Imaging Society, 22*(1), 10. DOI: 10.1186/s40644-022-00448-4 PMID: 35090572

Bao, M., Chen, Q., Xu, Z., Jensen, E. C., Liu, C., Waitkus, J. T., Yuan, X., He, Q., Qin, P., & Du, K. (2021). Challenges and opportunities for clustered regularly interspaced short palindromic repeats based molecular biosensing. *ACS Sensors, 6*(7), 2497–2522. DOI: 10.1021/acssensors.1c00530

Barani, M., Sabir, F., Rahdar, A., Arshad, R., & Kyzas, G. Z. (2020). Nanotreatment and nanodiagnosis of prostate cancer: Recent updates. *Nanomaterials (Basel, Switzerland), 10*(9), 1696. DOI: 10.3390/nano10091696 PMID: 32872181

Baran, S. W., Bratcher, N., Dennis, J., Gaburro, S., Karlsson, E. M., Maguire, S., & LaFollette, M. R. (2022). Emerging role of translational digital biomarkers within home cage monitoring technologies in preclinical drug discovery and development. *Frontiers in Behavioral Neuroscience, 15*, 758274. DOI: 10.3389/fnbeh.2021.758274 PMID: 35242017

Barbadilla-Martínez, L., Klaassen, N., Franceschini-Santos, V. H., Breda, J., Hernandez-Quiles, M., van Lieshout, T., & van Steensel, B. (2024). The regulatory grammar of human promoters uncovered by MPRA-trained deep learning. bioRxiv, 2024-07.

Barbieri, M. A., Battini, V., Carnovale, C., Cocco, M., Papoutsi, D. G., Heckmann, N. S., Dong, G., Rossi, A., Peker, S., Van Manen, R. P., Thapar, S., & Sessa, M. (2025). Artificial intelligence in pharmacovigilance signal management: A review of tools, implementations, research, and regulatory landscape. *Expert Opinion on Drug Safety*, •••, 1–16. DOI: 10.1080/14740338.2025.2545926 PMID: 40778768

Barenholz, Y. (2012). Doxil®—The first FDA-approved nano-drug: Lessons learned. *Journal of Controlled Release : Official Journal of the Controlled Release Society, 160*(2), 117–134. DOI: 10.1016/j.jconrel.2012.03.020 PMID: 22484195

Barta, J. A., Powell, C. A., & Wisnivesky, J. P. (2019, January 22). Global Epidemiology of Lung Cancer. *Annals of Global Health, 85*(1), 8. DOI: 10.5334/aogh.2419 PMID: 30741509

Bartolomucci, A., Nobrega, M., Ferrier, T., Dickinson, K., Kaorey, N., Nadeau, A., Castillo, A., & Burnier, J. V. (2025, March 24). Circulating tumor DNA to monitor treatment response in solid tumors and advance precision oncology. *NPJ Precision Oncology*, *9*(1), 84. DOI: 10.1038/s41698-025-00876-y PMID: 40122951

Basu, A. (2024). Human digital twins and machine learning applications in precision medicine and surgery: Current state and future directions. *Artificial Intelligence in Medicine and Surgery-An Exploration of Current Trends, Potential Opportunities, and Evolving Threats-Volume 2.*

Batool, Z., Kamal, M. A., & Shen, B. (2024). Advancements in triple-negative breast cancer sub-typing, diagnosis and treatment with assistance of artificial intelligence : A focused review. *Journal of Cancer Research and Clinical Oncology*, *150*(8), 383. DOI: 10.1007/s00432-024-05903-2 PMID: 39103624

Battineni, G., Sagaro, G. G., Chinatalapudi, N., & Amenta, F. (2020, March 31). Applications of Machine Learning Predictive Models in the Chronic Disease Diagnosis. *Journal of Personalized Medicine*, *10*(2), 21. DOI: 10.3390/jpm10020021 PMID: 32244292

Baydoun, A., Jia, A. Y., Zaorsky, N. G., Kashani, R., Rao, S., Shoag, J. E., Vince, R. A.Jr, Bittencourt, L. K., Zuhour, R., Price, A. T., Arsenault, T. H., & Spratt, D. E. (2024, March). Artificial intelligence applications in prostate cancer. *Prostate Cancer and Prostatic Diseases*, *27*(1), 37–45. DOI: 10.1038/s41391-023-00684-0 PMID: 37296271

Becchi, T., Beltrame, L., Mannarino, L., Calura, E., Marchini, S., & Romualdi, C. (2023). A pan-cancer landscape of pathogenic somatic copy number variations. *Journal of Biomedical Informatics*, *147*, 104529. DOI: 10.1016/j.jbi.2023.104529

Becker, M., Farina, K. A., & Mascarenhas, J. (2024). Acute myeloid leukemia: Current understanding and management. *JAAPA : Official Journal of the American Academy of Physician Assistants*, *37*(1), 34–39. DOI: 10.1097/01.JAA.0000995680.52352.b5 PMID: 38128137

Benard, B., Gentles, A. J., Köhnke, T., Majeti, R., & Thomas, D. (2019). Data mining for mutation-specific targets in acute myeloid leukemia. *Leukemia*, *33*(4), 826–843. DOI: 10.1038/s41375-019-0387-y PMID: 30728456

Bennati, C., Paglialunga, L., Ricciuti, B., Metro, G., Marcomigni, L., Gili, A., & Crinò, L. (2016, June). Targeting EGFR and ALK in NSCLC: Current evidence and future perspective. *Lung Cancer Management*, *5*(2), 79–90. DOI: 10.2217/lmt-2016-0005 PMID: 30643552

Berger, M. F., & Van Allen, E. M. (2016). Delivering on the promise of precision cancer medicine. *Genome Medicine*, *8*(1), 110. DOI: 10.1186/s13073-016-0373-1 PMID: 27782852

Bernardi, S., Vallati, M., & Gatta, R. (2024). Artificial Intelligence-based management of adult chronic myeloid leukemia: Where are we and where are we going? *Cancers (Basel)*, *16*(5), 848. DOI: 10.3390/cancers16050848 PMID: 38473210

Berry, D. A. (2012). Adaptive clinical trials in oncology. *Nature Reviews. Clinical Oncology*, *9*(4), 199–207. DOI: 10.1038/nrclinonc.2011.165 PMID: 22064459

Bersani, M. M., Braghin, C., Cortellessa, V., Gargantini, A., Grassi, V., Presti, F. L., & Scandurra, P. (2022). Towards trust-preserving continuous co-evolution of digital twins. In *2022 IEEE 19th International Conference on Software Architecture Companion (ICSA-C)* (pp. 96-99). IEEE. DOI: 10.1109/ICSA-C54293.2022.00024

Bertolaccini, L., Casiraghi, M., Uslenghi, C., Maiorca, S., & Spaggiari, L. (2024, October). Recent advances in lung cancer research: Unravelling the future of treatment. [Internet]. *Updates in Surgery*, *76*(6), 2129–2140. Retrieved November 2, 2025, from https://link.springer.com/10.1007/s13304-024-01841-3. DOI: 10.1007/s13304-024-01841-3 PMID: 38581618

Besenhard, M. O., Pal, S., Gkogkos, G., & Gavriilidis, A. (2023). Non-fouling flow reactors for nanomaterial synthesis. *Reaction Chemistry & Engineering*, *8*(5), 955–977. DOI: 10.1039/D2RE00412G

Bhardwaj, H., Archana, N., Noumani, A., Himanshu, J. K., Chakravorty, S., & Solanki, P. R. (2024). Recent advancement in the detection of potential cancer biomarkers using the nanomaterial integrated electrochemical sensing technique: A detailed review. *Materials Advances*, *5*(2), 475–503. DOI: 10.1039/D3MA00621B

Bhatia, S., Kumar, A., & Tandon, S. (2024). Uncovering the challenges from algorithmic bias affecting the marginalized patient groups in healthcare. *Anuj and Tandon, Stuti, Uncovering the Challenges From Algorithmic Bias Affecting the Marginalized Patient Groups in Healthcare (May 30, 2024).*

Bhushan, A., Gonsalves, A., & Menon, J. U. (2021). Current State of Breast Cancer Diagnosis, Treatment, and Theranostics. *Pharmaceutics*, *13*(5), 723. Advance online publication. DOI: 10.3390/pharmaceutics13050723 PMID: 34069059

Bian, W., Huang, Y., Shan, B., & Chang, Y. (2024). Capsule Defect Detection and Classification Using Enhanced MobileNet Model. *American Journal of Biochemistry and Biotechnology*, *20*(4), 356–364. DOI: 10.3844/ajbbsp.2024.356.364

Bifarin, O. O., & Fernández, F. M. (2024). Automated machine learning and explainable AI (AutoML-XAI) for metabolomics: Improving cancer diagnostics. *Journal of the American Society for Mass Spectrometry*, *35*(6), 1089–1100. DOI: 10.1021/jasms.3c00403

Biggar, K., Ridgeway, N., Chopra, A., Lukinovic, V., Feldman, M., Charih, F., & Green, J. (2024). Machine learning-based exploration of enzyme-substrate networks: SET8-mediated methyllysine and its changing impact within cancer proteomes. https://doi.org/DOI: 10.21203/rs.3.rs-3771179/v1

Binnie, A., Fernandes, E., Almeida-Lousada, H., de Mello, R. A., & Castelo-Branco, P. (2021). CRISPR-based strategies in infectious disease diagnosis and therapy. *Infection*, *49*(3), 377–385. DOI: 10.1007/s15010-020-01554-w PMID: 33393066

Bishal, A., Paul, P., Bera, S., Patra, D., Bandyopadhyay, B., Debnath, B., & Ali, K. A. (2025, August 5). Targeted Drug Delivery for Breast Cancer using Functionalized Liposomes: Preparation Methods, Challenges, and Clinical Translation. *AAPS PharmSciTech*, *26*(7), 207. DOI: 10.1208/s12249-025-03206-4 PMID: 40764411

Biswas, N., & Chakrabarti, S. (2020, October 14). Artificial Intelligence (AI)-Based Systems Biology Approaches in Multi-Omics Data Analysis of Cancer. [Internet]. *Frontiers in Oncology, 10*, 588221. Retrieved November 2, 2025, from https://www.frontiersin.org/article/10.3389/fonc.2020.588221/full. DOI: 10.3389/fonc.2020.588221 PMID: 33154949

Blanco, E., Shen, H., & Ferrari, M. (2015). Principles of nanoparticle design for overcoming biological barriers to drug delivery. *Nature Biotechnology, 33*(9), 941–951. DOI: 10.1038/nbt.3330 PMID: 26348965

Bobo, D., Robinson, K. J., Islam, J., Thurecht, K. J., & Corrie, S. R. (2016, October). Nanoparticle-Based Medicines: A Review of FDA-Approved Materials and Clinical Trials to Date. *Pharmaceutical Research, 33*(10), 2373–2387. DOI: 10.1007/s11095-016-1958-5 PMID: 27299311

Bock, C., Datlinger, P., Chardon, F., Coelho, M. A., Dong, M. B., Lawson, K. A., Lu, T., Maroc, L., Norman, T. M., Song, B., Stanley, G., Chen, S., Garnett, M., Li, W., Moffat, J., Qi, L. S., Shapiro, R. S., Shendure, J., Weissman, J. S., & Zhuang, X. (2022, February 10). High-content CRISPR screening. *Nature Reviews. Methods Primers, 2*(1), 8. DOI: 10.1038/s43586-021-00093-4 PMID: 37214176

Bode, A. M., & Dong, Z. (2018). Recent advances in precision oncology research. *NPJ Precision Oncology, 2*(1), 11. DOI: 10.1038/s41698-018-0055-0 PMID: 30202789

Boehnke, N., Straehla, J. P., Safford, H. C., Kocak, M., Rees, M. G., Ronan, M., Rosenberg, D., Adelmann, C. H., Chivukula, R. R., Nabar, N., Ebrahimkhani, M. R., & Hammond, P. T. (2022). Massively parallel pooled screening reveals genomic determinants of nanoparticle delivery. *Science, 377*(6604), eabm5551. Advance online publication. DOI: 10.1126/science.abm5551 PMID: 35862544

Bogers, J. F., Berghuis, N. F., Busker, R. W., van Booma, A., Paauw, A., & van Leeuwen, H. C. (2021). Bright fluorescent nucleic acid detection with CRISPR-Cas12a and poly(thymine) templated copper nanoparticles. *Biology Methods and Protocols, 6*(1), 20–26. DOI: 10.1093/biomethods/bpaa020

Bolocan, V.-O., Mitoi, A., Nicu-Canareica, O., Băean, M.-L., Medar, C., & Popa, G.-A. (2025). Artificial Intelligence in Prostate MRI: Current Evidence and Clinical Translation Challenges—A Narrative Review. *Journal of Imaging, 11*(10), 335. Advance online publication. DOI: 10.3390/jimaging11100335 PMID: 41150011

Bondy-Denomy, J. (2018). Protein inhibitors of CRISPR-Cas9. *ACS Chemical Biology, 13*(2), 417–423. DOI: 10.1021/acschembio.7b00831

Borgheti-Cardoso, L. N., Viegas, J. S. R., Silvestrini, A. V. P., Caron, A. L., Praça, F. G., Kravicz, M., & Bentley, M. V. L. B. (2020, January). Nanotechnology approaches in the current therapy of skin cancer. *Advanced Drug Delivery Reviews, 153*, 109–136. DOI: 10.1016/j.addr.2020.02.005 PMID: 32113956

Botsis, T., & Kreimeyer, K. (2023). Improving drug safety with adverse event detection using natural language processing. *Expert Opinion on Drug Safety, 22*(8), 659–668. DOI: 10.1080/14740338.2023.2228197 PMID: 37339273

Brash, D. E. (2015, January). '<scp>UV</scp> Signature Mutations'. *Photochemistry and Photobiology, 91*(1), 15–26. DOI: 10.1111/php.12377 PMID: 25354245

Bray, F., Laversanne, M., Sung, H., Ferlay, J., Siegel, R. L., Soerjomataram, I., & Jemal, A. (2024, May). Global cancer statistics 2022: GLOBOCAN estimates of incidence and mortality worldwide for 36 cancers in 185 countries. *CA: a Cancer Journal for Clinicians*, *74*(3), 229–263. DOI: 10.3322/caac.21834 PMID: 38572751

Brennan, C. A., & Garrett, W. S. (2016). Gut microbiota, inflammation, and colorectal cancer. *Annual Review of Microbiology*, *70*(1), 395–411. DOI: 10.1146/annurev-micro-102215-095513 PMID: 27607555

Brezgin, S., Kostyusheva, A., Kostyushev, D., & Chulanov, V. (2019). Dead Cas Systems: Types, Principles, and Applications. *International Journal of Molecular Sciences*, *20*(23), 6041. Advance online publication. DOI: 10.3390/ijms20236041 PMID: 31801211

Broday, E. E. (2022). The evolution of quality: From inspection to quality 4.0. *International Journal of Quality and Service Sciences*, *14*(3), 368–382. DOI: 10.1108/IJQSS-09-2021-0121

Brosch-Lenz, J., Uribe, C., Rahmim, A., & Saboury, B. (2023). Theranostic digital twins: An indispensable prerequisite for personalized cancer care. *Journal of Nuclear Medicine*, *64*(3), 501. DOI: 10.2967/jnumed.122.264929 PMID: 36863782

Budczies, J., Kazdal, D., Menzel, M., Beck, S., Kluck, K., Altbürger, C., Schwab, C., Allgäuer, M., Ahadova, A., Kloor, M., Schirmacher, P., Peters, S., Krämer, A., Christopoulos, P., & Stenzinger, A. (2024, October). Tumour mutational burden: Clinical utility, challenges and emerging improvements. [Internet]. *Nature Reviews. Clinical Oncology*, *21*(10), 725–742. Retrieved November 1, 2025, from https://www.nature.com/articles/s41571-024-00932-9. DOI: 10.1038/s41571-024-00932-9 PMID: 39192001

Bui, T. A., Mei, H., Sang, R., Ortega, D. G., & Deng, W. (2024, August). Advancements and challenges in developing in vivo CAR T cell therapies for cancer treatment. *EBioMedicine*, *106*, 105266. DOI: 10.1016/j.ebiom.2024.105266 PMID: 39094262

Bulashevska, A., Nacsa, Z., Lang, F., Braun, M., Machyna, M., Diken, M., & König, R. (2024). Artificial intelligence and neoantigens: Paving the path for precision cancer immunotherapy. *Frontiers in Immunology*, *15*, 1394003. DOI: 10.3389/fimmu.2024.1394003

Buschmann, M. D., Carrasco, M. J., Alishetty, S., Paige, M., Alameh, M.-G., & Weissman, D. (2021). Nanomaterial Delivery Systems for mRNA Vaccines. *Vaccines*, *9*(1), 65. DOI: 10.3390/vaccines9010065 PMID: 33478109

Byalahunasishetter, B. I., Chowdareddy, S., & Raghavan, R. (2026). Cancer Risk Prediction Based on Gut Microbiota Analysis. *Smart Microbiomes for Personalized Medicine*, 307–342. DOI: 10.4018/979-8-3373-8172-5.ch009

Cadet, J., & Douki, T. (2018). Formation of UV-induced DNA damage contributing to skin cancer development. *Photochemical & Photobiological Sciences*, *17*(12), 1816–1841.

Caf, M., Esmaeilnejad-Ahranjani, P., Kolosnjaj-Tabi, J., Sabotič, J., Berlec, A., Zaveršek, N., Pajk, S., Zahirović, A., Golzio, M., Milosevic, I., & Kralj, S. (2026). Magnetic Field-Driven Strategies for Biofilm Disruption: From Iron Oxide Nanoparticles to Adaptive Swarms of Magnetic Microrobots. *ACS Nano*, *20*(1), 34–58. DOI: 10.1021/acsnano.5c14390 PMID: 41477707

Cai, Y., & Wang, S. (2024). Deeply integrating latent consistent representations in high-noise multi-omics data for cancer subtyping. *Briefings in Bioinformatics, 25*(2), bbae061. DOI: 10.1093/bib/bbae061

Cai, Z., Poulos, R. C., Liu, J., & Zhong, Q. (2022, January 22). Machine learning for multi-omics data integration in cancer. *iScience, 25*(2), 103798. DOI: 10.1016/j.isci.2022.103798 PMID: 35169688

Camps, J., Jimenez-Franco, A., García-Pablo, R., Joven, J., & Arenas, M. (2025). Artificial intelligence-driven integration of multi-omics and radiomics: A new hope for precision cancer diagnosis and prognosis. Biochimica et biophysica acta (BBA)-Molecular basis of disease, 1871(6), 167841.

Cancer Genome Atlas Network. (2012). Comprehensive molecular characterization of human colon and rectal cancer. *Nature, 487*(7407), 330–337. DOI: 10.1038/nature11252 PMID: 22810696

Cao, J., Peek, N., Renehan, A., & Ananiadou, S. (2023). Gaussian distributed prototypical network for few-shot genomic variant detection. In *The 22nd Workshop on Biomedical Natural Language Processing and BioNLP Shared Tasks* (pp. 26-36). DOI: 10.18653/v1/2023.bionlp-1.2

Cao, Y. H., Ding, J., Tang, Q. H., Zhang, J., Huang, Z. Y., Tang, X. M., & Fu, D. (2022). Deciphering cell–cell interactions and communication in the tumor microenvironment and unraveling intratumoral genetic heterogeneity via single-cell genomic sequencing. *Bioengineered, 13*(7-12), 14974–14986. DOI: 10.1080/21655979.2023.2185434

Capone, P., Chiarella, P., & Sisto, R. (2024). Advanced technologies in genomic toxicology: Current trends and future directions. *Current Opinion in Toxicology, 37*, 100444. DOI: 10.1016/j.cotox.2023.100444

Capponi, S., & Daniels, K. G. (2023). Harnessing the power of artificial intelligence to advance cell therapy. *Immunological Reviews, 320*(1), 147–165. DOI: 10.1111/imr.13236 PMID: 37415280

Capurro, D., Coghlan, S., & Pires, D. E. (2022). Preventing digital overdiagnosis. *Journal of the American Medical Association, 327*(6), 525–526. DOI: 10.1001/jama.2021.22969 PMID: 35061004

Cararo Lopes, E., Sawant, A., Moore, D., Ke, H., Shi, F., Laddha, S., Chen, Y., Sharma, A., Naumann, J., Guo, J. Y., Gomez, M., Ibrahim, M., Smith, T. L., Riedlinger, G. M., Lattime, E. C., Trooskin, S., Ganesan, S., Su, X., Pasqualini, R., & White, E. (2023, June). Integrated metabolic and genetic analysis reveals distinct features of human differentiated thyroid cancer. *Clinical and Translational Medicine, 13*(6), e1298. DOI: 10.1002/ctm2.1298 PMID: 37317665

Carini, C., & Seyhan, A. A. (2024). Tribulations and future opportunities for artificial intelligence in precision medicine. *Journal of Translational Medicine, 22*(1), 411. DOI: 10.1186/s12967-024-05067-0 PMID: 38702711

Casati, S. (2018). Integrated approaches to testing and assessment. *Basic & Clinical Pharmacology & Toxicology, 123*(S5, Suppl. 5), 51–55. DOI: 10.1111/bcpt.13018 PMID: 29604238

Castro, G. A., Almeida, J. M., Machado-Neto, J. A., & Almeida, T. A. (2024). A decision support system to recommend appropriate therapy protocol for AML patients. *Frontiers in Artificial Intelligence, 7*, 1343447. DOI: 10.3389/frai.2024.1343447 PMID: 38510471

Cellina, M., Cè, M., Alì, M., Irmici, G., Ibba, S., Caloro, E., & Papa, S. (2023). Digital twins: The new frontier for personalized medicine? *Applied Sciences (Basel, Switzerland)*, *13*(13), 7940. DOI: 10.3390/app13137940

Chakraborty, R., Afrose, N., Bhowmick, P., & Bhowmick, M. (2025). Biomedical sensors in wearable health technologies. In Innovations in Biomedical Engineering [Internet]. (pp. 159–184). Elsevier., Available from https://linkinghub.elsevier.com/retrieve/pii/B9780443301469000058, Retrieved October 28, 2025, from. DOI: 10.1016/B978-0-443-30146-9.00005-8

Chakraborty, R., Jana, B., Koner, S., Kanta Maiti, S., Afrose, N., Bhowmick, P., ... Exploring the transcriptomic landscape of rheumatoid arthritis using next-generation RNA sequencing. ITPS [Internet]. 2024 Nov 22 [cited 2025 Oct 27];8(1):14. Available from: https://accscience.com/journal/ITPS/8/1/10.36922/itps.4449

Chakraborty, S., Sharma, G., Karmakar, S., & Banerjee, S. (2024). Multi-OMICS approaches in cancer biology: New era in cancer therapy. *Biochim Biophys Acta BBA - Mol Basis Dis.*, *1870*(5), 167120.

Chakraborty, D., Gutierrez-Chakraborty, E., Rodriguez-Aguayo, C., Başağaoğlu, H., Lopez-Berestein, G., & Amero, P. (2024). Discovering genetic biomarkers for targeted cancer therapeutics with eXplainable Artificial Intelligence. *Cancer Communications (London, England)*, *44*(5), 584–588. DOI: 10.1002/cac2.12530

Chakraborty, R., Kannan, S. P., Afrose, N., & Narayanasamy, D. (2025, September). Tumor microenvironment expressed enzymes (Matrix metalloproteinases, cathepsins, urokinase-type plasminogen activator) triggered polymersomes for liquid biopsy and cancer diagnostics: A review. [Internet]. *International Journal of Biological Macromolecules*, *321(2)*, 146375. Retrieved October 27, 2025, from https://linkinghub.elsevier.com/retrieve/pii/S0141813025069326. DOI: 10.1016/j.ijbiomac.2025.146375 PMID: 40738429

Chakravarty, K., Antontsev, V., Bundey, Y., & Varshney, J. (2021, June). Driving success in personalized medicine through AI-enabled computational modeling. [Internet]. *Drug Discovery Today*, *26*(6), 1459–1465. Retrieved November 2, 2025, from https://linkinghub.elsevier.com/retrieve/pii/S1359644621000738. DOI: 10.1016/j.drudis.2021.02.007 PMID: 33609781

Chambers, D. A., Feero, W. G., & Khoury, M. J. (2016, May 10). Convergence of Implementation Science, Precision Medicine, and the Learning Health Care System: A New Model for Biomedical Research. *Journal of the American Medical Association*, *315*(18), 1941. DOI: 10.1001/jama.2016.3867 PMID: 27163980

Chan, G. J., & Ding, C. C. (2026, March). Advances in artificial intelligence in prostate cancer pathology. *Seminars in Diagnostic Pathology*, *43*(2), 150995. DOI: 10.1016/j.semdp.2026.150995 PMID: 41807227

Chang, C. Y., Park, H., Malone, D. C., Wang, C. Y., Wilson, D. L., Yeh, Y. M., Van Boemmel-Wegmann, S., & Lo-Ciganic, W.-H. (2020, March 25). Immune Checkpoint Inhibitors and Immune-Related Adverse Events in Patients With Advanced Melanoma: A Systematic Review and Network Meta-analysis. *JAMA Network Open*, *3*(3), e201611. DOI: 10.1001/jamanetworkopen.2020.1611 PMID: 32211869

Chang, L., Liu, J., Zhu, J., Guo, S., Wang, Y., Zhou, Z., & Wei, X. (2025). Advancing precision medicine: The transformative role of artificial intelligence in immunogenomics, radiomics, and pathomics for biomarker discovery and immunotherapy optimization. *Cancer Biology & Medicine*, *22*(1), 33–47. DOI: 10.20892/j.issn.2095-3941.2024.0376

Chang, T. G., Park, S., Schäffer, A. A., Jiang, P., & Ruppin, E. (2025, March). Hallmarks of artificial intelligence contributions to precision oncology. *Nature Cancer*, *6*(3), 417–431. DOI: 10.1038/s43018-025-00917-2 PMID: 40055572

Chaparro-Cárdenas, S. L., Ramirez-Bautista, J.-A., Terven, J., Córdova-Esparza, D.-M., Romero-Gonzalez, J.-A., Ramírez-Pedraza, A., & Chavez-Urbiola, E. A. (2025). A Technological Review of Digital Twins and Artificial Intelligence for Personalized and Predictive Healthcare. *Health Care*, *13*(14), 1763. DOI: 10.3390/healthcare13141763 PMID: 40724790

Chapman, P. B., Hauschild, A., Robert, C., Haanen, J. B., Ascierto, P., Larkin, J., Dummer, R., Garbe, C., Testori, A., Maio, M., Hogg, D., Lorigan, P., Lebbe, C., Jouary, T., Schadendorf, D., Ribas, A., O'Day, S. J., Sosman, J. A., Kirkwood, J. M., & McArthur, G. A. (2011, June 30). Improved Survival with Vemurafenib in Melanoma with BRAF V600E Mutation. *The New England Journal of Medicine*, *364*(26), 2507–2516. DOI: 10.1056/NEJMoa1103782 PMID: 21639808

Chaturvedi, R., & Verma, S. (2023). Opportunities and Challenges of AI-Driven Customer Service. In Sheth, J. N., Jain, V., Mogaji, E., & Ambika, A. (Eds.), *Artificial Intelligence in Customer Service* (pp. 33–71). Springer International Publishing., DOI: 10.1007/978-3-031-33898-4_3

Checcucci, E., Oing, C., Amparore, D., Porpiglia, F., & Rescigno, P. (2025). Digital twins in urological oncology: Precise treatment planning via complex modelling. *European Urology Oncology*, *8*(1), 1–4. DOI: 10.1016/j.euo.2024.10.005 PMID: 39443222

Chehelgerdi, M., Chehelgerdi, M., Khorramian-Ghahfarokhi, M., Shafieizadeh, M., Mahmoudi, E., Eskandari, F., Rashidi, M., Arshi, A., & Mokhtari-Farsani, A. (2024). Comprehensive review of CRISPR-based gene editing: Mechanisms, challenges, and applications in cancer therapy. *Molecular Cancer*, *23*(1), 9. DOI: 10.1186/s12943-023-01925-5 PMID: 38195537

Chen, B., Jing, Z., Kshirsagar, A., Gkanatsios, N., & Chui, H. (2021). *U.S. Patent Application No. 16/944,995.*

Chen, R. J., Chen, T. Y., Lipkova, J., Wang, J. J., Williamson, D. F., Lu, M. Y., & Mahmood, F. (2021). Algorithm fairness in AI for medicine and healthcare. *arXiv preprint arXiv:2110.00603.*

Chen, F., Alphonse, M., & Liu, Q. (2020). Strategies for nonviral nanoparticle-based delivery of CRISPR/Cas9 therapeutics. *Wiley Interdisciplinary Reviews. Nanomedicine and Nanobiotechnology*, *12*(3), e1609. DOI: 10.1002/wnan.1609 PMID: 31797562

Cheng, C., Wang, Y., Zhao, J., Wu, D., Li, H., & Zhao, H. (2025). Deep learning and radiomics in triple-negative breast cancer: Predicting long-term prognosis and clinical outcomes. *Journal of Multidisciplinary Healthcare*, *18*, 319–327. DOI: 10.2147/JMDH.S509004

Cheng, M., Xiong, E., Tian, T., Zhu, D., Ju, H. Q., & Zhou, X. (2021). A CRISPR-driven colorimetric code platform for highly accurate telomerase activity assay. *Biosensors & Bioelectronics*, *172*, 112749. DOI: 10.1016/j.bios.2020.112749

Cheng, X., Xie, Q., & Sun, Y. (2023). Advances in nanomaterial-based targeted drug delivery systems. *Frontiers in Bioengineering and Biotechnology*, *11*, 1177151. DOI: 10.3389/fbioe.2023.1177151 PMID: 37122851

Chen, H., Fang, X., Shao, J., Zhang, Q., Xu, L., Chen, J., & Guo, G. (2024). Pan-Cancer Single-Nucleus Total RNA Sequencing Using snHH-Seq. *Advanced Science (Weinheim, Baden-Württemberg, Germany)*, *11*(5), 2304755. DOI: 10.1002/advs.202304755

Chen, S., Gao, L., Tan, M., Zhang, K., & Lv, F. (2025, June 26). Performance Evaluation of Four Deep Learning-Based CAD Systems and Manual Reading for Pulmonary Nodules Detection, Volume Measurement, and Lung-RADS Classification Under Varying Radiation Doses and Reconstruction Methods. *Diagnostics (Basel)*, *15*(13), 1623. DOI: 10.3390/diagnostics15131623 PMID: 40647622

Chen, S., Zhou, Z., Li, Y., Du, Y., & Chen, G. (2023, October 27). Application of single-cell sequencing to the research of tumor microenvironment. *Frontiers in Immunology*, *14*, 1285540. DOI: 10.3389/fimmu.2023.1285540 PMID: 37965341

Chen, X.-Y., Wu, Z.-X., Wang, J.-Q., Teng, Q.-X., Tang, H., Liu, Q., Chen, Z.-S., & Chen, W. (2024). Multidrug resistance transporters P-gp and BCRP limit the efficacy of ATR inhibitor ceralasertib in cancer cells. *Frontiers in Pharmacology*, *15*, 1400699. DOI: 10.3389/fphar.2024.1400699 PMID: 38756373

Chen, Y. J., Abila, B., & Mostafa Kamel, Y. (2023, January 21). CAR-T: What Is Next? *Cancers (Basel)*, *15*(3), 663. DOI: 10.3390/cancers15030663 PMID: 36765623

Chen, Y., Jarecka, D., Abraham, S. A., Gau, R., Ng, E., Low, D. M., & Ghosh, S. S. (2025). Standardizing survey data collection to enhance reproducibility: Development and comparative evaluation of the ReproSchema ecosystem. *Journal of Medical Internet Research*, *27*, e63343. DOI: 10.2196/63343 PMID: 40644691

Chen, Y.-M., Hsiao, T.-H., Lin, C.-H., & Fann, Y. C. (2025). Unlocking precision medicine: Clinical applications of integrating health records, genetics, and immunology through artificial intelligence. *Journal of Biomedical Science*, *32*(1), 16. DOI: 10.1186/s12929-024-01110-w PMID: 39915780

Chen, Y., Ma, B., Zuo, Y., Chen, G., Hao, Q., Zhao, C., Chen, Y., & Zhang, X. (2023). Versatile sweat bioanalysis on demand with hydrogel-programmed wearables. *Biosensors & Bioelectronics*, *235*, 115412. DOI: 10.1016/j.bios.2023.115412

Chen, Z., Lin, L., Wu, C., Li, C., Xu, R., & Sun, Y. (2021, November). Artificial intelligence for assisting cancer diagnosis and treatment in the era of precision medicine. [Internet]. *Cancer Communications*, *41*(11), 1100–1115. Retrieved November 2, 2025, from https://onlinelibrary.wiley.com/doi/10.1002/cac2.12215. DOI: 10.1002/cac2.12215 PMID: 34613667

Chen, Z., Zhu, L., Yang, Z., & Matsubara, T. (2022). Automated cancer subtyping via vector quantization mutual information maximization. In *Joint European Conference on Machine Learning and Knowledge Discovery in Databases* (pp. 88-103). Cham: Springer International Publishing.

Cheo, H. M., Ong, C. Y. G., & Ting, Y. (2025, June 24). A Systematic Review of AI Performance in Lung Cancer Detection on CT Thorax. *Healthcare (Basel), 13*(13), 1510. DOI: 10.3390/healthcare13131510 PMID: 40648536

Chhabra, R. (2024, October 23). Molecular and modular intricacies of precision oncology. *Frontiers in Immunology, 15*, 1476494. DOI: 10.3389/fimmu.2024.1476494 PMID: 39507541

Chi, A., He, X., Hou, L., Nguyen, N. P., Zhu, G., Cameron, R. B., & Lee, J. M. (2021, June 11). Classification of Non-Small Cell Lung Cancer's Tumor Immune Micro-Environment and Strategies to Augment Its Response to Immune Checkpoint Blockade. [Internet]. *Cancers (Basel), 13*(12), 2924. Retrieved November 2, 2025, from https://www.mdpi.com/2072-6694/13/12/2924. DOI: 10.3390/cancers13122924 PMID: 34208113

Chiang, C. L.-L., Rovelli, R., Sarivalasis, A., & Kandalaft, L. E. (2021). Integrating cancer vaccines in the standard-of-care of ovarian cancer: Translating preclinical models to human. *Cancers (Basel), 13*(18), 4553. DOI: 10.3390/cancers13184553 PMID: 34572778

Chiappetta, M., Lococo, F., Leuzzi, G., Sperduti, I., Bria, E., Petracca Ciavarella, L., Mucilli, F., Filosso, P. L., Ratto, G., Spaggiari, L., Facciolo, F., & Margaritora, S. (2021, March). Survival Analysis in Single N2 Station Lung Adenocarcinoma: The Prognostic Role of Involved Lymph Nodes and Adjuvant Therapy. *Cancers (Basel), 13*(6), 1326. DOI: 10.3390/cancers13061326 PMID: 33809513

Chitoran, E., Rotaru, V., Stefan, D. C., Gullo, G., & Simion, L. (2025, March 27). Blocking Tumoral Angiogenesis VEGF/VEGFR Pathway: Bevacizumab—20 Years of Therapeutic Success and Controversy. *Cancers (Basel), 17*(7), 1126. DOI: 10.3390/cancers17071126 PMID: 40227654

Cho, H. J., Wang, Z., Cong, Y., Bekiranov, S., Zhang, A., & Zang, C. (2024). DARDN: A Deep-learning approach for CTCF binding sequence classification and oncogenic regulatory feature discovery. *Genes, 15*(2), 144. DOI: 10.3390/genes15020144

Chong, P. L., Vaigeshwari, V., Mohammed Reyasudin, B. K., Noor Hidayah, B. R. A., Tatchanaamoorti, P., Yeow, J. A., & Kong, F. Y. (2025, December 31). Integrating artificial intelligence in healthcare: Applications, challenges, and future directions. *Future Science OA, 11*(1), 2527505. DOI: 10.1080/20565623.2025.2527505 PMID: 40616302

Chopra, H., Annu, , Shin, D. K., Munjal, K., Priyanka, , Dhama, K., & Emran, T. B. (2023a). Revolutionizing clinical trials: The role of AI in accelerating medical breakthroughs. *International Journal of Surgery, 109*(12), 4211–4220. DOI: 10.1097/JS9.0000000000000705 PMID: 38259001

Chouaïd, C., Gendarme, S., & Auliac, J. B. (2023, July). Artificial intelligence to finally enable precision medicine for the management of resected non-small-cell lung cancer. [Internet]. *Annals of Oncology : Official Journal of the European Society for Medical Oncology, 34*(7), 565–566. Retrieved November 2, 2025, from https://linkinghub.elsevier.com/retrieve/pii/S0923753423006750. DOI: 10.1016/j.annonc.2023.05.001 PMID: 37182802

Chougrad, H., Zouaki, H., & Alheyane, O. (2018). Deep Convolutional Neural Networks for breast cancer screening. *Computer Methods and Programs in Biomedicine, 157*, 19–30. DOI: 10.1016/j.cmpb.2018.01.011 PMID: 29477427

Chowbina, S., Janes, K. A., Peirce, S. M., & Papin, J. A. (2011). Mathematical and computational models in cancer. In *Targeted Therapies: Mechanisms of Resistance* (pp. 113–126). Humana Press. DOI: 10.1007/978-1-60761-478-4_6

Chugh, V., Basu, A., Kaushik, A., Bhansali, S., & Basu, A. K. (2024). Employing nano-enabled artificial intelligence (AI)-based smart technologies for prediction, screening, and detection of cancer. *Nanoscale*, *16*(11), 5458–5486. DOI: 10.1039/D3NR05648A PMID: 38391246

Chukwunweike, J. N., Ayodele, P. A., & Atata, B. B. (2024, August). HARNESSING MACHINE LEARNING FOR CYBERSECURITY: HOW CONVOLUTIONAL NEURAL NETWORKS ARE REVOLUTIONIZING THREAT DETECTION AND DATA PRIVACY. *Int J Res Publ Rev [Internet]*, *5*(8), 3608–3617. Retrieved November 2, 2025, from https://ijrpr.com/uploads/V5ISSUE8/IJRPR32530.pdf. DOI: 10.55248/gengpi.5.0824.2402

Chung, C., Yeung, V. T. Y., & Wong, K. C. W. (2023, September). Prognostic and predictive biomarkers with therapeutic targets in breast cancer: A 2022 update on current developments, evidence, and recommendations. *Journal of Oncology Pharmacy Practice : Official Publication of the International Society of Oncology Pharmacy Practitioners*, *29*(6), 1343–1360. DOI: 10.1177/10781552221119797 PMID: 35971313

Cohen, A. B., & Schwamm, L. H. (2024). Digital health for oncological care. *The Cancer Journal (Sudbury, Mass.)*, *30*(1), 34–39. DOI: 10.1097/PPO.0000000000000693 PMID: 38265925

Cohen, J. A., Bomben, R., Pozzo, F., Tissino, E., Härzschel, A., Hartmann, T. N., & Gattei, V. (2020). An updated perspective on current prognostic and predictive biomarkers in chronic lymphocytic leukemia in the context of chemoimmunotherapy and novel targeted therapy. *Cancers (Basel)*, *12*(4), 894. DOI: 10.3390/cancers12040894 PMID: 32272636

Ćojbašić, Ž., Ćojbašić, I., Marković, N., & Trajanović, M. (2018). Machine Learning Leukemia Clinical Outcome Prediction. *pathways*, *1*, 2.

Collins, F. S., & Varmus, H. (2015, February). A New Initiative on Precision Medicine. *The New England Journal of Medicine*, *372*(9), 793–795. DOI: 10.1056/NEJMp1500523 PMID: 25635347

Corces, M. R., Granja, J. M., Shams, S., Louie, B. H., Seoane, J. A., Zhou, W., & Chang, H. Y. (2018). The chromatin accessibility landscape of primary human cancers. *Science*, *362*(6413), eaav1898.

Correia, R. P., Bento, L. C., de Sousa, F. A., Barroso, R. D. S., Campregher, P. V., & Bacal, N. S. (2021). How I investigate minimal residual disease in acute lymphoblastic leukemia. *International Journal of Laboratory Hematology*, *43*(3), 354–363. DOI: 10.1111/ijlh.13463 PMID: 33423385

Costa, P. A. C., da Silva, W. N., Moura Prazeres, P. H. D., Lima, A. S., Oliveira, V. P., Silva, J. B., Santos, R. R., Birbrair, A., Azevedo, V., & Guimarães, P. P. G. (2024). siRNA lipid nanoparticles for CXCL12 silencing modulate brain immune response during Zika infection. *Biomedicine & Pharmacotherapy = Biomédecine & Pharmacothérapie*, *170*, 115981. DOI: 10.1016/j.biopha.2023.115981 PMID: 38091634

Coupland, S. E., Sonntag, S. R., Heimann, H., & Grisanti, S. (2024). The concept of liquid biopsy in the treatment of malignant eye tumors. *Die Ophthalmologie*, *121*(12), 946–953. DOI: 10.1007/s00347-024-02132-3

Cousins, H. C., Nayar, G., & Altman, R. B. (2024). Computational approaches to drug repurposing: Methods, challenges, and opportunities. *Annual Review of Biomedical Data Science*, *7*(1), 7. DOI: 10.1146/annurev-biodatasci-110123-025333 PMID: 38598857

Crombie, J. L., & Brown, J. R. (2019). Advances in drug-based therapies in chronic lymphocytic leukemia and future prospects. *Advances in Cell and Gene Therapy*, *2*(3), e51. DOI: 10.1002/acg2.51

Crowgey, E. L., Mahajan, N., Wong, W. H., Gopalakrishnapillai, A., Barwe, S. P., Kolb, E. A., & Druley, T. E. (2020). Error-corrected sequencing strategies enable comprehensive detection of leukemic mutations relevant for diagnosis and minimal residual disease monitoring. *BMC Medical Genomics*, *13*(1), 32. DOI: 10.1186/s12920-020-0671-8 PMID: 32131829

da Silva, R. G. L. (2024, May 21). The advancement of artificial intelligence in biomedical research and health innovation: Challenges and opportunities in emerging economies. *Globalization and Health*, *20*(1), 44. DOI: 10.1186/s12992-024-01049-5 PMID: 38773458

Dai, L., Shen, G., Wang, Y., Yang, P., Wang, H., & Liu, Z. (2021). PSMA-targeted melanin-like nanoparticles as a multifunctional nanoplatform for prostate cancer theranostics. *Journal of Materials Chemistry. B*, *9*(4), 1151–1161. DOI: 10.1039/D0TB02576C PMID: 33434248

Dai, X., & Shen, L. (2022). Advances and trends in omics technology development. *Frontiers in Medicine*, *9*, 911861. DOI: 10.3389/fmed.2022.911861 PMID: 35860739

Dall'Olio, F. G., Marabelle, A., Caramella, C., Garcia, C., Aldea, M., Chaput, N., Robert, C., & Besse, B. (2022, February). Tumour burden and efficacy of immune-checkpoint inhibitors. [Internet]. *Nature Reviews. Clinical Oncology*, *19*(2), 75–90. Retrieved November 2, 2025, from https://www.nature.com/articles/s41571-021-00564-3. DOI: 10.1038/s41571-021-00564-3 PMID: 34642484

Damane, B. P., Mkhize-Kwitshana, Z. L., Kgokolo, M. C., Luvhengo, T., & Dlamini, Z. (2023). Applying artificial intelligence prediction tools for advancing precision oncology in immunotherapy: future perspectives in personalized care. In *Artificial intelligence and precision oncology: bridging cancer research and clinical decision support* (pp. 239–258). Springer Nature Switzerland. DOI: 10.1007/978-3-031-21506-3_12

Damaševičius, R. (2024). Explainable artificial intelligence methods for breast cancer recognition. *Innov Discov*, *1*(3), 25. DOI: 10.53964/id.2024025

Danaei, M., Dehghankhold, M., Ataei, S., Hasanzadeh Davarani, F., Javanmard, R., Dokhani, A., Khorasani, S., & Mozafari, M. R. (2018, May). Impact of Particle Size and Polydispersity Index on the Clinical Applications of Lipidic Nanocarrier Systems. *Pharmaceutics*, *10*(2), 57. DOI: 10.3390/pharmaceutics10020057 PMID: 29783687

Dang, A. (2023). Real-World Evidence: A Primer. *Pharmaceutical Medicine*, *37*(1), 25–36. DOI: 10.1007/s40290-022-00456-6 PMID: 36604368

Dar, K. B., Bhat, A. H., Amin, S., Anjum, S., Reshi, B. A., Zargar, M. A., & Ganie, S. A. (2019). Exploring proteomic drug targets, therapeutic strategies and protein-protein interactions in cancer: Mechanistic view. *Current Cancer Drug Targets*, *19*(6), 430–448. DOI: 10.2174/1568009618666180803104631

Dasgupta, S., Gayen, S., Chakraborty, T., Afrose, N., Pal, R., Mahata, S., Nasare, V., & Roy, S. (2024, March 27). Potential role of immune cell therapy in gynecological cancer and future promises: A comprehensive review. [Internet]. *Medical Oncology*, *41*(5), 98. Retrieved November 2, 2025, from https://link.springer.com/10.1007/s12032-024-02337-1. DOI: 10.1007/s12032-024-02337-1 PMID: 38536512

David, I., Shao, G., Gomes, C., Tilbury, D., & Zarkout, B. (2024). Interoperability of digital twins: Challenges, success factors, and future research directions. In *International Symposium on Leveraging Applications of Formal Methods* (pp. 27-46). Cham: Springer Nature Switzerland.

Davies, H., Bignell, G. R., Cox, C., Stephens, P., Edkins, S., Clegg, S., Teague, J., Woffendin, H., Garnett, M. J., Bottomley, W., Davis, N., Dicks, E., Ewing, R., Floyd, Y., Gray, K., Hall, S., Hawes, R., Hughes, J., Kosmidou, V., & Futreal, P. A. (2002, June). Mutations of the BRAF gene in human cancer. *Nature*, *417*(6892), 949–954. DOI: 10.1038/nature00766 PMID: 12068308

Davodabadi, F., Sajjadi, S. F., Sarhadi, M., Mirghasemi, S., Nadali Hezaveh, M., Khosravi, S., Kamali Andani, M., Cordani, M., Basiri, M., & Ghavami, S. (2023). Cancer chemotherapy resistance: Mechanisms and recent breakthrough in targeted drug delivery. *European Journal of Pharmacology*, *958*, 176013. DOI: 10.1016/j.ejphar.2023.176013 PMID: 37633322

Davri, A., Birbas, E., Kanavos, T., Ntritsos, G., Giannakeas, N., Tzallas, A. T., & Batistatou, A. (2023, August 5). Deep Learning for Lung Cancer Diagnosis, Prognosis and Prediction Using Histological and Cytological Images: A Systematic Review. *Cancers (Basel)*, *15*(15), 3981. DOI: 10.3390/cancers15153981 PMID: 37568797

de Jong, D., Das, J. P., Ma, H., Pailey Valiplackal, J., Prendergast, C., Roa, T., Braumuller, B., Deng, A., Dercle, L., Yeh, R., Salvatore, M. M., & Capaccione, K. M. (2023, May 21). Novel Targets, Novel Treatments: The Changing Landscape of Non-Small Cell Lung Cancer. [Internet]. *Cancers (Basel)*, *15*(10), 2855. Retrieved November 2, 2025, from https://www.mdpi.com/2072-6694/15/10/2855. DOI: 10.3390/cancers15102855 PMID: 37345192

De Santis, M., Mottet, N., Cornford, P., & Gillessen, S. (2020). Precision oncology for metastatic prostate cancer: Translation into practice. *European Urology*, *78*(6), 771–774. DOI: 10.1016/j.eururo.2020.08.021 PMID: 32863055

De Vries, N. L., Mahfouz, A., Koning, F., & De Miranda, N. F. (2020). Unraveling the complexity of the cancer microenvironment with multidimensional genomic and cytometric technologies. *Frontiers in Oncology*, *10*, 1254. DOI: 10.3389/fonc.2020.01254

Deeba, K., Chinnpa Prabu Shankar, K., Gnanavel, S., & Elsisi, M. (2025). Artificial Intelligence, Computer Vision, and Robotics for Industry 5.0. In Tyagi, A. K. (Ed.), *Next Generation Data Science and Blockchain Technology for Industry 5.0* (1st ed., pp. 295–324). Wiley., DOI: 10.1002/9781394245215.ch12

Dee, W., Alaaeldin Ibrahim, R., & Marouli, E. (2024). Histopathological domain adaptation with generative adversarial networks: Bridging the domain gap between thyroid cancer histopathology datasets. *PLoS One*, *19*(12), e0310417. DOI: 10.1371/journal.pone.0310417

Dejea, C. M., Fathi, P., Craig, J. M., Boleij, A., Taddese, R., Geis, A. L., Wu, X., DeStefano Shields, C. E., Hechenbleikner, E. M., Huso, D. L., Anders, R. A., Giardiello, F. M., Wick, E. C., Wang, H., Wu, S., Pardoll, D. M., Housseau, F., & Sears, C. L. (2018). Patients with familial adenomatous polyposis harbor colonic biofilms containing tumorigenic bacteria. *Science*, *359*(6375), 592–597. DOI: 10.1126/science.aah3648 PMID: 29420293

Dekker, E., Tanis, P. J., Vleugels, J. L. A., Kasi, P. M., & Wallace, M. B. (2019). Colorectal cancer. *Lancet*, *394*(10207), 1467–1480. DOI: 10.1016/S0140-6736(19)32319-0 PMID: 31631858

Derek, V., & Collings, P. (2025). *Natural Language Processing (NLP) in Healthcare AI: Enhancing Clinical Insight Extraction from Unstructured Patient Data. Preprints*. DOI: 10.22541/au.174845034.41890518/v1

Desai, K. S., Tiyyala, V. M., Tiyyala, P., Yeola, A., Gallegos-Rangel, A., Montiel-Torres, A., Allen, M. R., Dredze, M., Vandrey, R. G., Thrul, J., Leas, E. C., Hogarth, M., Smith, D. M., & Ayers, J. W. (2025). Waldo: Automated discovery of adverse events from unstructured self reports. *PLOS Digital Health*, *4*(9), e0001011. DOI: 10.1371/journal.pdig.0001011 PMID: 41026671

Desani, N. R., & Reddy, R. R. K. (2020). Edge AI for real-time health monitoring using streaming data. *international Journal of Science and Research, 10*(1), 1669-1674.

Deshmukh, R., Singh, V., Harwansh, R. K., Agrawal, R., Garg, A., Singh, S., Elossaily, G. M., Ansari, M. N., Ali, N., & Prajapati, B. G. (2024, February 20). Emerging trends of nanomedicines in the management of prostate cancer: Perspectives and potential applications. *Pharmaceutics*, *16*(3), 297. DOI: 10.3390/pharmaceutics16030297 PMID: 38543191

Deshmukh, V. G., Sapkal, S. B., Gadekar, S. S., & Deshmukh, V. (2025, September). EGFR inhibitors across generations: Progress, challenges, and future directions. *Journal of Molecular Structure*, *1339*, 142326. DOI: 10.1016/j.molstruc.2025.142326

Dessale, M., Mengistu, G., & Mengist, H. M. (2022). Nanotechnology: A promising approach for cancer diagnosis, therapeutics and theragnosis. *International Journal of Nanomedicine*, *17*, 3735–3749. DOI: 10.2147/IJN.S378074

Dezhakam, E., Vayghan, R. F., Dehghani, S., Kafili-Hajlari, T., Naseri, A., Dadashpour, M., & Hasanzadeh, M. (2024). Highly efficient electrochemical biosensing platform in breast cancer detection based on MOF-COF@Au core-shell like nanostructure. *Scientific Reports*, *14*(1), 1–13. DOI: 10.1038/s41598-024-78836-y

Dhade, P., & Shirke, P. (2024). Federated learning for healthcare: A comprehensive review. *Engineering Proceedings*, *59*(1), 230.

Dhandayuthapani, V. B. (2024). *Advancing Healthcare: AI Integration, Interoperability and Sustainability Challenges*. DOI: 10.25929/J3XXXY63

Dhara, C., Sah, H., Gantayat, S., Rajput, M., & Mishra, S. (2026). CRISPR/Cas9 in perspective: Evaluating efficacy, delivery methods, and ethical challenges in genome editing. *Molecular Biology Reports*, *53*(1), 719. DOI: 10.1007/s11033-026-11815-w PMID: 42090053

Dhatchinamoorthy, K., Colbert, J. D., & Rock, K. L. (2021). Cancer Immune Evasion Through Loss of MHC Class I Antigen Presentation. *Frontiers in Immunology*, *12*, 636568. DOI: 10.3389/fimmu.2021.636568 PMID: 33767702

Dhillon, A., Singh, A., & Bhalla, V. K. (2023). A Systematic Review on Biomarker Identification for Cancer Diagnosis and Prognosis in Multi-omics: From Computational Needs to Machine Learning and Deep Learning. *Archives of Computational Methods in Engineering*, *30*(2), 917–949. DOI: 10.1007/s11831-022-09821-9

Di Carlo, E., & Sorrentino, C. (2024). State of the art CRISPR-based strategies for cancer diagnostics and treatment. *Biomarker Research*, *12*(1), 156. DOI: 10.1186/s40364-024-00701-x PMID: 39696697

Díaz de la Guardia-Bolívar, E., Barrios-Rodríguez, R., Zwir, I., Jiménez-Moleón, J. J., & del Val, C. (2022, July). Identification of novel prostate cancer genes in patients stratified by Gleason classification: Role of antitumoral genes. *International Journal of Cancer*, *151*(2), 255–264. DOI: 10.1002/ijc.33988 PMID: 35234293

Díaz-García, D., Díaz-Sánchez, M., Álvarez-Conde, J., & Gómez-Ruiz, S. (2024). Emergence of quantum dots as innovative tools for early diagnosis and advanced treatment of breast cancer. *ChemMedChem*, *19*(16), e202400172. DOI: 10.1002/cmdc.202400172

Diaz, L. A.Jr, Williams, R. T., Wu, J., Kinde, I., Hecht, J. R., Berlin, J., Allen, B., Bozic, I., Reiter, J. G., Nowak, M. A., Kinzler, K. W., Oliner, K. S., & Vogelstein, B. (2012). The molecular evolution of acquired resistance to targeted EGFR blockade in colorectal cancers. *Nature*, *486*(7404), 537–540. DOI: 10.1038/nature11219 PMID: 22722843

Diaz, M. J., Natarelli, N., Aflatooni, S., Aleman, S. J., Neelam, S., Tran, J. T., Taneja, K., Lucke-Wold, B., & Forouzandeh, M. (2023, July). Nanoparticle-Based Treatment Approaches for Skin Cancer: A Systematic Review. *Current Oncology (Toronto, Ont.)*, *30*(8), 7112–7131. DOI: 10.3390/curroncol30080516 PMID: 37622997

Dienstmann, R., Vermeulen, L., Guinney, J., Kopetz, S., Tejpar, S., & Tabernero, J. (2017). Consensus molecular subtypes and the evolution of precision medicine in colorectal cancer. *Nature Reviews. Cancer*, *17*(2), 79–92. DOI: 10.1038/nrc.2016.126

Dimitrov, G., & Popov, E. (2026). In Vaschetto, L. M., & Srivastava, S. (Eds.), *Precision Medicine in Prostate Cancer BT - Molecular Precision Medicine in Cancer* (pp. 105–135). Springer Nature Switzerland., DOI: 10.1007/978-3-032-24071-2_4

Dimitsaki, S., Natsiavas, P., & Jaulent, M.-C. (2024). Applying AI to Structured Real-World Data for Pharmacovigilance Purposes: Scoping Review. *Journal of Medical Internet Research*, *26*, e57824. DOI: 10.2196/57824 PMID: 39753222

Ding, H., Zhang, J., Zhang, F., Xu, Y., Liang, W., & Yu, Y. (2022). Nanotechnological approaches for diagnosis and treatment of ovarian cancer: A review of recent trends. *Drug Delivery*, *29*(1), 3218–3232. DOI: 10.1080/10717544.2022.2132032 PMID: 36259505

Ding, L., Agrawal, P., Singh, S. K., Chhonker, Y. S., Sun, J., & Murry, D. J. (2024). Polymer-Based Drug Delivery Systems for Cancer Therapeutics. *Polymers, 16*(6), 843. DOI: 10.3390/polym16060843 PMID: 38543448

Dinh-Le, C., Chuang, R., Chokshi, S., & Mann, D. (2019, September 11). Wearable Health Technology and Electronic Health Record Integration: Scoping Review and Future Directions. *JMIR mHealth and uHealth, 7*(9), e12861. DOI: 10.2196/12861 PMID: 31512582

Divate, M., Tyagi, A., Richard, D. J., Prasad, P. A., Gowda, H., & Nagaraj, S. H. (2021). Deep learning-based pan-cancer classification model reveals cancer-specific gene expression signatures. bioRxiv, 2021-03.

Dixit, R. R. (2025). Applying reinforcement learning to optimize cancer treatment protocols in machine learning frameworks. In *Artificial Intelligence Revolutionizing Cancer Care* (pp. 20–35). CRC Press.

Dixit, S., Kumar, A., Srinivasan, K., Vincent, P. D. R., & Ramu Krishnan, N. (2024). Advancing genome editing with artificial intelligence: Opportunities, challenges, and future directions. *Frontiers in Bioengineering and Biotechnology, 11*, 1335901. DOI: 10.3389/fbioe.2023.1335901 PMID: 38260726

Dkhar, D. S., Kumari, R., Patel, V., Srivastava, A., Prasad, R., Srivastava, R., & Chandra, P. (2024). Versatile approaches of quantum dots in biosensing and imaging. *Wiley Interdisciplinary Reviews. Nanomedicine and Nanobiotechnology, 16*(5), e1998. DOI: 10.1002/wnan.1998

Doench, J. G. (2018, February). Am I ready for CRISPR? A user's guide to genetic screens. *Nature Reviews. Genetics, 19*(2), 67–80. DOI: 10.1038/nrg.2017.97 PMID: 29199283

Döhner, H., Wei, A. H., Appelbaum, F. R., Craddock, C., DiNardo, C. D., Dombret, H., & Löwenberg, B. (2022). Diagnosis and management of AML in adults: 2022 recommendations from an international expert panel on behalf of the ELN. *Blood, 140*(12), 1345–1377. DOI: 10.1182/blood.2022016867 PMID: 35797463

Dong, M., Wang, L., Hu, N., Rao, Y., Wang, Z., & Zhang, Y. (2025, August 29). Integration of multi-omics approaches in exploring intra-tumoral heterogeneity. *Cancer Cell International, 25*(1), 317. DOI: 10.1186/s12935-025-03944-2 PMID: 40883767

Dorado, C. Q., Ramírez, A. S., Pérez, M. P., Centeno, M. S., Mici, L. P., Güemez, C. M., Albers Acosta, E., Luis, G. C., Costal, M., Diez, P. T., Laorden, N. R., Díaz, R. J., Velasco Balanza, C., & Manso, L. S. J. (2025, November 24). Redefining Prostate Cancer Precision: Radiogenomics, Theragnostics, and AI-Driven Biomarkers. *Cancers (Basel), 17*(23), 3747. DOI: 10.3390/cancers17233747 PMID: 41374949

Doudican, N. A., Kumar, A., Singh, N. K., Nair, P. R., Lala, D. A., Basu, K., Talawdekar, A. A., Sultana, Z., Tiwari, K., Tyagi, A., Abbasi, T., Vali, S., Vij, R., Fiala, M., King, J., Perle, M. A., & Mazumder, A. (2015). Personalization of cancer treatment using predictive simulation. *Journal of Translational Medicine, 13*(1), 43. DOI: 10.1186/s12967-015-0399-y PMID: 25638213

Drost, J., van Jaarsveld, R. H., Ponsioen, B., Zimberlin, C., van Boxtel, R., Buijs, A., Sachs, N., Overmeer, R. M., Offerhaus, G. J., Begthel, H., Korving, J., van de Wetering, M., Schwank, G., Logtenberg, M., Cuppen, E., Snippert, H. J. G., Medema, J. P., Kops, G. J. P. L., & Clevers, H. (2015). Sequential cancer mutations in cultured human intestinal stem cells. *Nature, 521*(7550), 43–47. DOI: 10.1038/nature14415 PMID: 25924068

Duan, L., Zhang, S., Yan, Q., Chen, Y., Su, Y., & Hu, X. (2026). Comparative efficacy of e-health interventions for quality of life in patients with cancer: A network meta-analysis of randomized controlled trials. *Worldviews on Evidence-Based Nursing*, *23*(1), e70103. Advance online publication. DOI: 10.1111/wvn.70103

Duarah, R., Torné-Morató, H., Zhang, G., Amin, Y., Pabast, M., Sharma, N., He, S., Das, M. R., & Pompa, P. P. (2026). Next-Generation Wearable Optical Sensors for Personalized Health and Point-of-Care Diagnostics—A Systematic Review. *Advanced Healthcare Materials*, *15*(10), e04419. DOI: 10.1002/adhm.202504419 PMID: 41486722

Duarte, J. G., Duarte, M. G., Piedade, A. P., & Mascarenhas-Melo, F. (2025). Rethinking Pharmaceutical Industry with Quality by Design: Application in Research, Development, Manufacturing, and Quality Assurance. *The AAPS Journal*, *27*(4), 96. DOI: 10.1208/s12248-025-01079-w PMID: 40389714

Du, B., Tian, X., Chen, Z., Ge, Y., Chen, C., Gao, H., Li, S., & Zhang, Y. (2025). Ultrasensitive optoelectronic biosensor arrays based on twisted bilayer graphene superlattice. *National Science Review*, *12*(10), nwaf357. Advance online publication. DOI: 10.1093/nsr/nwaf357

Dubale, A. T., Tareke, A. A., Butta, F. W., Shibabaw, A. A., Eniyew, E. B., Ahmed, M. H., Kassie, S. Y., Demsash, A. W., Chereka, A. A., Dube, G. N., Walle, A. D., & Kitil, G. W. (2024). Healthcare professionals' willingness to utilize a mobile health application for adverse drug reaction reporting in a limited resource setting: An input for digital health, 2023. *European Journal of Obstetrics & Gynecology and Reproductive Biology: X*, *23*, 100324. DOI: 10.1016/j.eurox.2024.100324 PMID: 39050924

Dubey, S., & Sikarwar, S. S. (2025). Applications of AI in Cancer Detection — A Review of the Specific Ways in which AI Is Being Used to Detect and Diagnose Various Types of Cancer. In R. Singh, A. Gehlot, N. Rathour, & S. V. Akram (Eds.), AI in Disease Detection [Internet]. 1st ed., pp. 147–166). Wiley., Available from https://onlinelibrary.wiley.com/doi/10.1002/9781394278695.ch7, Retrieved November 2, 2025, from. DOI: 10.1002/9781394278695.ch7

Dubey, A. K., Gupta, V. K., Kujawska, M., Orive, G., Kim, N.-Y., Li, C.-Z., Mishra, Y. M., & Kaushik, A. (2022). Exploring nano-enabled CRISPR-Cas-powered strategies for efficient diagnostics and treatment of infectious diseases. *Journal of Nanostructure in Chemistry*, *12*(5), 833–864. DOI: 10.1007/s40097-022-00472-7

Dudhwala, Y. D., Mehta, R. K., Kapoor, D. U., Shah, D. P., Shah, K. V., & Saini, V. (2026). Ligand-functionalized nanocarriers for targeted breast cancer therapy: Advances in receptor-mediated drug delivery. *Heal. Nanotechnol.*, *2*(1), 7. DOI: 10.1186/s44301-026-00029-5

Dupras, C., & Bunnik, E. M. (2021). Toward a framework for assessing privacy risks in multi-omic research and databases. *The American Journal of Bioethics : AJOB*, *21*(12), 46–64. DOI: 10.1080/15265161.2020.1863516 PMID: 33433298

Durga, D., Gadhiya, D., Suhani, F., Singh, N. P., Lakkimsetti, M., Syed, S., & Desai, H. D. (2024). Epidemiological Trends of Acute Lymphoid Leukemia in Individuals Younger Than 20 Years across 204 Countries and Territories from 1990-2021: A Benchmarking Global Analysis. *Blood*, *144*(Supplement 1), 1428. DOI: 10.1182/blood-2024-212326

Dutta, A., Deb, T., & Pathak, S. (2021). Automated Data Harmonization (ADH) using Artificial Intelligence (AI). *OPSEARCH*, *58*(2), 257–275. DOI: 10.1007/s12597-020-00467-4

Duttke, S. H. C., Lacadie, S. A., Ibrahim, M. M., Glass, C. K., Corcoran, D. L., Benner, C., Heinz, S., Kadonaga, J. T., & Ohler, U. (2015, November). Perspectives on Unidirectional versus Divergent Transcription. *Molecular Cell*, *60*(3), 348–349. DOI: 10.1016/j.molcel.2015.10.014 PMID: 26545075

Dziopa, K., Lekadir, K., van der Harst, P., & Asselbergs, F. W. (2025). Digital twins: Reimagining the future of cardiovascular risk prediction and personalised care. *Hellenic Journal of Cardiology : HJC = Hellēnikē Kardiologikē Epitheōrēsē*, *81*, 4–8. DOI: 10.1016/j.hjc.2024.06.001 PMID: 38852883

Edelman, L. B., Eddy, J. A., & Price, N. D. (2010). In silico models of cancer. *Wiley Interdisciplinary Reviews. Systems Biology and Medicine*, *2*(4), 438–459. DOI: 10.1002/wsbm.75 PMID: 20836040

Eftekhari, A., Kryschi, C., Pamies, D., Gulec, S., Ahmadian, E., Janas, D., Davaran, S., & Khalilov, R. (2023). Natural and synthetic nanovectors for cancer therapy. *Nanotheranostics*, *7*(3), 236–257. DOI: 10.7150/ntno.77564 PMID: 37064613

Eichler, H., Bloechl-Daum, B., Broich, K., Kyrle, P. A., Oderkirk, J., Rasi, G., Santos Ivo, R., Schuurman, A., Senderovitz, T., Slawomirski, L., Wenzl, M., & Paris, V. (2019). Data Rich, Information Poor: Can We Use Electronic Health Records to Create a Learning Healthcare System for Pharmaceuticals? *Clinical Pharmacology and Therapeutics*, *105*(4), 912–922. DOI: 10.1002/cpt.1226 PMID: 30178490

Ekundayo, F. (2024). Reinforcement learning in treatment pathway optimization: A case study in oncology. *International Journal of Science and Research Archive*, *13*(02), 2187–2205. DOI: 10.30574/ijsra.2024.13.2.2450

El Alaoui, Y., Elomri, A., Qaraqe, M., Padmanabhan, R., Yasin Taha, R., El Omri, H., & Aboumarzouk, O. (2022). A review of artificial intelligence applications in hematology management: Current practices and future prospects. *Journal of Medical Internet Research*, *24*(7), e36490. DOI: 10.2196/36490 PMID: 35819826

el Ata, N. A. (2024). Perspective chapter: making the shift to personalized preventive medicine with human digital twins. In *Personalized Medicine-New Perspectives*. IntechOpen. DOI: 10.5772/intechopen.1003639

El Mahdaoui, C., Dehbi, H., & Cherkaoui, S. (2024). A Review of Emerging Cytogenetic and Molecular Approaches in Identifying Abnormalities in Acute Lymphoblastic Leukemia.

El Sabrouty, R., & Elouadi, A. (2024, August 2). Harnessing AI for Precision Oncology: Transformative Advances in Non-Small Cell Lung Cancer Treatment. [Internet]. *Statistics, Optimization & Information Computing*, *13*(1), 347–367. Retrieved November 2, 2025, from http://www.iapress.org/index.php/soic/article/view/2078. DOI: 10.19139/soic-2310-5070-2078

El-Ashmawy, N. E., Al-Ashmawy, G. M., Hamada, O. B., & Khedr, N. F. (2025). The role of ABCG2 in health and disease: Linking cancer therapy resistance and other disorders. *Life Sciences*, *360*, 123245. DOI: 10.1016/j.lfs.2024.123245 PMID: 39561874

Elhaddad, M., & Hamam, S. (2024, April 6). AI-Driven Clinical Decision Support Systems: An Ongoing Pursuit of Potential. [Internet]. *Cureus, 16*(4), e57728. Retrieved November 2, 2025, from https://www.cureus.com/articles/244463-ai-driven-clinical-decision-support-systems-an-ongoing-pursuit-of-potential. DOI: 10.7759/cureus.57728 PMID: 38711724

Elsayed, B., Elhadary, M., Elshoeibi, R. M., Elshoeibi, A. M., Badr, A., Metwally, O., & Yassin, M. (2023). Deep learning enhances acute lymphoblastic leukemia diagnosis and classification using bone marrow images. *Frontiers in Oncology, 13*, 1330977. DOI: 10.3389/fonc.2023.1330977 PMID: 38125946

El-Tanani, M., Nsairat, H., Matalka, I. I., Lee, Y. F., Rizzo, M., Aljabali, A. A., & Tambuwala, M. M. (2024). The impact of the BCR-ABL oncogene in the pathology and treatment of chronic myeloid leukemia. *Pathology, Research and Practice, 254*, 155161. DOI: 10.1016/j.prp.2024.155161 PMID: 38280275

Eminaga, O., Abbas, M., Kunder, C., Tolkach, Y., Han, R., Brooks, J. D., Nolley, R., Semjonow, A., Boegemann, M., West, R., Long, J., Fan, R. E., & Bettendorf, O. (2024, March 4). Critical evaluation of artificial intelligence as a digital twin of pathologists for prostate cancer pathology. *Scientific Reports, 14*(1), 5284. DOI: 10.1038/s41598-024-55228-w PMID: 38438436

Emrani, J., Ahmed, M., Newman, R. H., Thomas, M. D., Mowa, C. N., & Teleha, J. (2020, July). Introducing Chemistry Students to Emerging Technologies in Gene Editing, Their Applications, and Ethical Considerations. *Journal of Chemical Education, 97*(7), 1931–1943. DOI: 10.1021/acs.jchemed.9b01154

Emran, T. B., Shahriar, A., Mahmud, A. R., Rahman, T., Abir, M. H., Siddiquee, M. F.-R., Ahmed, H., Rahman, N., Nainu, F., Wahyudin, E., & Sarker, M. R. (2022). Multidrug Resistance in Cancer: Understanding Molecular Mechanisms, Immunoprevention and Therapeutic Approaches. *Frontiers in Oncology, 12*, 891652. DOI: 10.3389/fonc.2022.891652 PMID: 35814435

Eskandari, F., Aali, M., Hadisadegh, S. N., & Azadeh, M. (2024). Advances in breast cancer research using CRISPR/Cas9 system. *Nano Select, 5*(10), 2400015. DOI: 10.1002/nano.202400015

Esteva, A., Kuprel, B., Novoa, R. A., Ko, J., Swetter, S. M., Blau, H. M., & Thrun, S. (2017, February). Dermatologist-level classification of skin cancer with deep neural networks. *Nature, 542*(7639), 115–118. DOI: 10.1038/nature21056 PMID: 28117445

Everall, A., Tapinos, A., Hawari, A., Cornish, A., Sud, A., Chubb, D., & Wedge, D. C. (2023). Comprehensive repertoire of the chromosomal alteration and mutational signatures across 16 cancer types from 10,983 cancer patients. MedRxiv, 2023-06. DOI: 10.1101/2023.06.07.23290970

Ewim, D. R. E., Orikpete, O. F., Scott, T. O., Onyebuchi, C. N., Onukogu, A. O., Uzougbo, C. G., & Onunka, C. (2023). Survey of wastewater issues due to oil spills and pollution in the Niger Delta area of Nigeria: A secondary data analysis. *Bulletin of the National Research Center, 47*(1), 116. DOI: 10.1186/s42269-023-01090-1

Ezeanolue, E. E., Iheanacho, T., Patel, D. V., Patel, S., Sam-Agudu, N., Obiefune, M., Dakum, P., Okonkwo, P., Olutola, A., Khamofu, H., Oyeledun, B., Aliyu, S., Aina, M., Eyo, A., Oko, J., Akinmurele, T., Oleribe, O., Gebi, U., Aliyu, M. H., & Siberry, G. (2019). Challenges and strategies for improving training of mid-level research personnel in Nigeria. *Annals of Global Health, 85*(1), 87. DOI: 10.5334/aogh.2405 PMID: 31225955

Faa, G., Fraschini, M., & Barberini, L. (2024). Reproducibility and explainability in digital pathology: The need to make black-box artificial intelligence systems more transparent. *Journal of Public Health Research, 13*(4), 22799036241284898. DOI: 10.1177/22799036241284898

Faiz, M., Mounika, B. G., Akbar, M., & Srivastava, S. (2024). Deep and machine learning for acute lymphoblastic leukemia diagnosis: A comprehensive review. *ADCAIJ: Advances in Distributed Computing and Artificial Intelligence Journal, 13*, e31420–e31420. DOI: 10.14201/adcaij.31420

Fajrial, A. K., He, Q. Q., Wirusanti, N. I., Slansky, J. E., & Ding, X. (2020). A review of emerging physical transfection methods for CRISPR/Cas9-mediated gene editing. *Theranostics, 10*(12), 5532–5549. DOI: 10.7150/thno.43465 PMID: 32373229

Fang, T., Cao, X., Ibnat, M., & Chen, G. (2022). Stimuli-responsive nanoformulations for CRISPR-Cas9 genome editing. *Journal of Nanobiotechnology, 20*(1), 354. DOI: 10.1186/s12951-022-01570-y PMID: 35918694

Fang, Y., Lin, S., Yang, F., Situ, J., Lin, S., & Luo, Y. (2020). Aptamer-Conjugated Multifunctional Polymeric Nanoparticles as Cancer-Targeted, MRI-Ultrasensitive Drug Delivery Systems for Treatment of Castration-Resistant Prostate Cancer. *BioMed Research International, 2020*(1), 9186583. DOI: 10.1155/2020/9186583 PMID: 32420382

Farahani, B., & Monsefi, A. K. (2023a). Smart and collaborative industrial IoT: A federated learning and data space approach. *Digital Communications and Networks, 9*(2), 436–447. DOI: 10.1016/j.dcan.2023.01.022

Fardale, S., Goyal, A., Palav, M. R., Palav, S. M., Kalaiarasan, C., & Namdeo, A. K. (2024). Real-Time Market Analysis With AI-Enhanced Signal Processing. In Pandey, D., Muniandi, B., Pandey, B. K., & George, A. S. (Eds.), (pp. 45–60). Advances in Computational Intelligence and Robotics. IGI Global., DOI: 10.4018/979-8-3693-7367-5.ch004

Fareed, M. A., Zhang, B., Memon, K., Hao, Y., Liu, Z., & Zhao, G. (2026). Recent advances in electrospun nanofiber-based strategies for diabetic wound healing application. *Materials Today. Chemistry, 51*, 103259. DOI: 10.1016/j.mtchem.2025.103259

Farghama, K., Khan, M. S., Wajid, M., & Shaffique, S. (2026). AI Driven Computational Nanomedicine for Personalized Cancer Therapy. *BioNanoScience, 16*(7), 447. DOI: 10.1007/s12668-026-02681-8

Farooq, M., & Shahid, M. U. (2023). Explainable AI and machine learning algorithms to predict treatment failures for patients with cancer. *Journal of Clinical Oncology : Official Journal of the American Society of Clinical Oncology, 41*(16), e13577–e13577. DOI: 10.1200/JCO.2023.41.16_suppl.e13577

Farzin, M. A., Abdoos, H., & Saber, R. (2022). AuNP-based biosensors for the diagnosis of pathogenic human coronaviruses: COVID-19 pandemic developments. *Analytical and Bioanalytical Chemistry, 414*(24), 7069–7084. DOI: 10.1007/s00216-022-04193-2

Fatemeh, S., Mahboobeh, Z., Khadijeh, A., Amirhossein, M. K., & Pegah, M. (2024). An in-silico study to determine susceptibility to cancer by evaluating the coding and non-coding non-synonymous single nucleotide variants in the SOCS3 gene. *Journal of Biomolecular Structure & Dynamics, 42*(16), 8281–8292. DOI: 10.1080/07391102.2023.2256408

Fatima Qizilbash, F., Sartaj, A., Qamar, Z., Kumar, S., Imran, M., Mohammed, Y., Ali, J., Baboota, S., & Ali, A. (2023, September 14). Nanotechnology revolutionises breast cancer treatment: Harnessing lipid-based nanocarriers to combat cancer cells. *Journal of Drug Targeting*, *31*(8), 794–816. DOI: 10.1080/1061186X.2023.2243403 PMID: 37525966

Fatima, S. S., Sultana, S., Perwaiz, N., Lashari, T. A., & Jabeen, F. (2024). Experimental Analysis of Deep Learning Techniques for Leukemia Cancer Classification using Micro Gene Array Dataset. In *2024 International Conference on Frontiers of Information Technology (FIT)* (pp. 1-6). IEEE. DOI: 10.1109/FIT63703.2024.10838426

Fatumo, S., Ebenezer, T. E., Ekenna, C., Isewon, I., Ahmad, U., Adetunji, C., Oladipo, E. K., Adebiyi, M., Adebiyi, E., & Nashiru, O. (2020). The Nigerian Bioinformatics and Genomics Network (NBGN): A collaborative platform to advance bioinformatics and genomics in Nigeria. *Global Health, Epidemiology and Genomics*, *5*, e3. DOI: 10.1017/gheg.2020.3 PMID: 32742665

Fearon, E. R., & Vogelstein, B. (1990). A genetic model for colorectal tumorigenesis. *Cell*, *61*(5), 759–767. DOI: 10.1016/0092-8674(90)90186-I PMID: 2188735

Fedorov, A., Longabaugh, W. J., Pot, D., Clunie, D. A., Pieper, S. D., Gibbs, D. L., & Kikinis, R. (2023). National cancer institute imaging data commons: Toward transparency, reproducibility, and scalability in imaging artificial intelligence. *Radiographics : A Review Publication of the Radiological Society of North America, Inc.*, *43*(12), e230180. DOI: 10.1148/rg.230180 PMID: 37999984

Feng, L., Yu, P., He, N., Zhang, Q., Tang, Y., Geng, J., Lu, Q., Song, C., & Chen, F. (2026). Label-Free MicroRNA Diagnostics: From CRISPR Nucleases to Nanomaterial-Enhanced Transducers. *ACS Synthetic Biology*, *15*(3), 915–954. DOI: 10.1021/acssynbio.5c00868 PMID: 41709870

Feretzakis, G., Juliebø-Jones, P., Tsaturyan, A., Sener, T. E., Verykios, V. S., Karapiperis, D., Bellos, T., Katsimperis, S., Angelopoulos, P., Varkarakis, I., Skolarikos, A., Somani, B., & Tzelves, L. (2024, February 16). Emerging trends in AI and radiomics for bladder, kidney, and prostate cancer: A critical review. *Cancers (Basel)*, *16*(4), 810. DOI: 10.3390/cancers16040810 PMID: 38398201

Ferlay, J., Colombet, M., Soerjomataram, I., Parkin, D. M., Piñeros, M., Znaor, A., & Bray, F. (2021). Cancer statistics for the year 2020: An overview. *International Journal of Cancer*, *149*(4), 778–789. DOI: 10.1002/ijc.33588

Fink, P. K., & Herren, L. T. (1996). Modelling disease processes for drug development: bridging the gap between quantitative and heuristic models. In *Proceedings of the 28th conference on Winter simulation* (pp. 1183-1190). DOI: 10.1145/256562.256930

Finn, J. D., Smith, A. R., Patel, M. C., Shaw, L., Youniss, M. R., van Heteren, J., Dirstine, T., Ciullo, C., Lescarbeau, R., Seitzer, J., Shah, R. R., Shah, A., Ling, D., Growe, J., Pink, M., Rohde, E., Wood, K. M., Salomon, W. E., Harrington, W. F., & Morrissey, D. V. (2018, February). A Single Administration of CRISPR/Cas9 Lipid Nanoparticles Achieves Robust and Persistent In Vivo Genome Editing. *Cell Reports*, *22*(9), 2227–2235. DOI: 10.1016/j.celrep.2018.02.014 PMID: 29490262

Fiorentino, V., Giordano, W., Pizzimenti, C., Zuccalà, V., Ieni, A., Molinario, C., Cannavò, S., Campennì, A., Tralongo, P., Martini, M., Giuffrè, G., Larocca, L. M., Fadda, G., & Rossi, E. D. (2026). Molecular profiling of thyroid nodules on cytologic samples: Findings from an Italian multi-institutional cohort. *Cancer Cytopathology*, *134*(1), e70065. DOI: 10.1002/cncy.70065 PMID: 41370117

Flores, J. E., Claborne, D. M., Weller, Z. D., Webb-Robertson, B. J. M., Waters, K. M., & Bramer, L. M. (2023). Missing data in multi-omics integration: Recent advances through artificial intelligence. *Frontiers in Artificial Intelligence*, *6*, 1098308. DOI: 10.3389/frai.2023.1098308

Foran, D. J., Chen, W., Chu, H., Sadimin, E., Loh, D., Riedlinger, G., & DiPaola, R. S. (2017). Roadmap to a comprehensive clinical data warehouse for precision medicine applications in oncology. *Cancer Informatics*, *16*, 1176935117694349. DOI: 10.1177/1176935117694349 PMID: 28469389

Forrest, S. J., Geoerger, B., & Janeway, K. A. (2018, February). Precision medicine in pediatric oncology. *Current Opinion in Pediatrics*, *30*(1), 17–24. DOI: 10.1097/MOP.0000000000000570 PMID: 29189430

Frampton, J. E. (2021, September). Isatuximab: A Review of Its Use in Multiple Myeloma. *Targeted Oncology*, *16*(5), 675–686. DOI: 10.1007/s11523-021-00827-0 PMID: 34351561

Freedman, B. S., Bulte, J. W. M., Conklin, B. R., Judge, L. M., Dwinell, M. R., Geurts, A. M., Sitton, M. J., Mahajan, V., Kiani, S., Gersbach, C. A., Ebrahimkhani, M. R., Kelly, J. J., Ronald, J. A., Morizane, R., Gupta, N., Shakeri-Zadeh, A., Vo, N., Saha, K., Saxena, S., & Tsai, S. Q. (2026). Monitoring biological effects of somatic cell genome editing. *Nature Reviews. Genetics*, *27*(4), 1–20. DOI: 10.1038/s41576-025-00916-0 PMID: 41530266

Freitas, L. F., Ferreira, A. H., Thipe, V. C., Varca, G. H., Lima, C. S., Batista, J. G., Riello, F. N., Nogueira, K., Cruz, C. P., Mendes, G. O., Rodrigues, A. S., Sousa, T. S., Alves, V. M., & Lugão, A. B. (2021, September 30). The state of the art of theranostic nanomaterials for lung, breast, and prostate cancers. *Nanomaterials (Basel, Switzerland)*, *11*(10), 2579. DOI: 10.3390/nano11102579 PMID: 34685018

Fujiwara, Y., Kato, S., & Kurzrock, R. (2024, April). Evolution of Precision Oncology, Personalized Medicine, and Molecular Tumor Boards. *Surgical Oncology Clinics of North America*, *33*(2), 197–216. DOI: 10.1016/j.soc.2023.12.004 PMID: 38401905

Fukushima, A., Iessa, N., Balakrishnan, M. R., & Pal, S. N. (2022). Smartphone-based mobile applications for adverse drug reactions reporting: Global status and country experience. *BMC Medical Informatics and Decision Making*, *22*(1), 118. DOI: 10.1186/s12911-022-01832-7 PMID: 35501745

Fu, L., Jia, G., Liu, Z., Pang, X., & Cui, Y. (2025). The applications and advances of artificial intelligence in drug regulation: A global perspective. *Acta Pharmaceutica Sinica. B*, *15*(1), 1–14. DOI: 10.1016/j.apsb.2024.11.006 PMID: 40041903

Fu, Q., Lai, J., Zhong, T., & Ran, L. (2023). Design and Implementation of Clinical LIS360 Laboratory Management System Based on AI Technology. *International Journal of Computational Intelligence Systems*, *16*(1), 33. DOI: 10.1007/s44196-023-00207-8

Fu, X., Shi, Y., Peng, F., Zhou, M., Yin, Y., Tan, Y., Chen, M., Yin, X., Ke, G., & Zhang, X. B. (2021). Exploring the trans-cleavage activity of CRISPR/Cas12a on gold nanoparticles for stable and sensitive biosensing. *Analytical Chemistry*, *93*(11), 4967–4974. DOI: 10.1021/acs.analchem.1c00027

Gallagher, K., Strobl, M. A., Park, D. S., Spoendlin, F. C., Gatenby, R. A., Maini, P. K., & Anderson, A. R. (2024). Mathematical model-driven deep learning enables personalized adaptive therapy. *Cancer Research*, *84*(11), 1929–1941. DOI: 10.1158/0008-5472.CAN-23-2040 PMID: 38569183

Gambacorta, N., Gasperi, V., Guzzo, T., Di Leva, F. S., Ciriaco, F., Sánchez, C., Tullio, V., Rozzi, D., Marinelli, L., Topai, A., Nicolotti, O., & Maccarrone, M. (2023). Exploring the 1,3-benzoxazine chemotype for cannabinoid receptor 2 as a promising anti-cancer therapeutic. *European Journal of Medicinal Chemistry*, *259*, 115647. DOI: 10.1016/j.ejmech.2023.115647 PMID: 37478557

Gambardella, V., Tarazona, N., Cejalvo, J. M., Lombardi, P., Huerta, M., Roselló, S., Fleitas, T., Roda, D., & Cervantes, A. (2020, April 19). Personalized Medicine: Recent Progress in Cancer Therapy. *Cancers (Basel)*, *12*(4), 1009. DOI: 10.3390/cancers12041009 PMID: 32325878

Ganatra, H., Tan, J. K., Simmons, A., Bigogno, C. M., Khurana, V., Ghose, A., & Ayodele, O. (2024). Applying whole-genome and whole-exome sequencing in breast cancer: A review of the landscape. *Breast Cancer (Tokyo, Japan)*, *31*(6), 999–1009. DOI: 10.1007/s12282-024-01628-9

Gandhewar, N., Pimpalkar, A., Jadhav, A., Shelke, N., & Jain, R. (2025). Leveraging Deep Learning for Genomics Analysis: Advances and Applications. In S. Choudhary, S. Kumar, S. Gowroju, M. Gulhane, & R. Sri Lakshmi (Eds.), Genomics at the Nexus of AI, Computer Vision, and Machine Learning [Internet]. 1st ed., pp. 191–225). Wiley., Available from https://onlinelibrary.wiley.com/doi/10.1002/9781394268832.ch9, Retrieved October 27, 2025, from. DOI: 10.1002/9781394268832.ch9

Gangwal, A., & Lavecchia, A. (2025). Artificial intelligence in preclinical research: Enhancing digital twins and organ-on-chip to reduce animal testing. *Drug Discovery Today*, *30*(5), 104360. DOI: 10.1016/j.drudis.2025.104360 PMID: 40252989

Gao, J., Lan, J., Liao, H., Yang, F., Qiu, P., Jin, F., & Zhu, Y. (2023). Promising preclinical patient-derived organoid (PDO) and xenograft (PDX) models in upper gastrointestinal cancers: Progress and challenges. *BMC Cancer*, *23*(1), 1205. DOI: 10.1186/s12885-023-11434-9 PMID: 38062430

Garcia-Fernandez, A., Vivo-Llorca, G., Sancho, M., García-Jareño, A. B., Ramirez-Jimenez, L., Barber-Cano, E., Murguía, J., Orzáez, M., Sancenón, F., & Martinez-Manez, R. (2022). Nanodevices for the efficient codelivery of CRISPR-Cas9 editing machinery and an entrapped cargo: A proposal for dual anti-inflammatory therapy. *Pharmaceutics*, *14*(7), 1495. DOI: 10.3390/pharmaceutics14071495 PMID: 35890389

Garcia, G. G., Schmidt, C. J., & Hajal, C. (2024). The tumor microenvironment in therapy resistance. *Frontiers in Lab on a Chip Technologies*, *3*, 1420233. DOI: 10.3389/frlct.2024.1420233

Garcia, P. L., Miller, A. L., & Yoon, K. J. (2020, May 22). Patient-Derived Xenograft Models of Pancreatic Cancer: Overview and Comparison with Other Types of Models. *Cancers (Basel)*, *12*(5), 1327. DOI: 10.3390/cancers12051327 PMID: 32456018

Gardner, S., Das, S., & Taylor, K. (2020). AI enabled precision medicine: patient stratification, drug repurposing and combination therapies. In *Artificial intelligence in oncology drug discovery and development*. IntechOpen. DOI: 10.5772/intechopen.92594

Garg, P., Pareek, S., Kulkarni, P., Horne, D., Salgia, R., & Singhal, S. S. (2024). Next-Generation Immunotherapy: Advancing Clinical Applications in Cancer Treatment. *Journal of Clinical Medicine*, *13*(21), 6537. DOI: 10.3390/jcm13216537 PMID: 39518676

Garg, S., Williams, N. L., Ip, A., & Dicker, A. P. (2018, December). Clinical Integration of Digital Solutions in Health Care: An Overview of the Current Landscape of Digital Technologies in Cancer Care. *JCO Clinical Cancer Informatics*, (2), 1–9. DOI: 10.1200/CCI.17.00159 PMID: 30652580

Garland, J. (2017). Unravelling the complexity of signalling networks in cancer: A review of the increasing role for computational modelling. *Critical Reviews in Oncology/Hematology*, *117*, 73–113. DOI: 10.1016/j.critrevonc.2017.06.004

Garon, E. B., Rizvi, N. A., Hui, R., Leighl, N., Balmanoukian, A. S., Eder, J. P., Patnaik, A., Aggarwal, C., Gubens, M., Horn, L., Carcereny, E., Ahn, M.-J., Felip, E., Lee, J.-S., Hellmann, M. D., Hamid, O., Goldman, J. W., Soria, J.-C., Dolled-Filhart, M., & Gandhi, L. (2015, May 21). Pembrolizumab for the Treatment of Non–Small-Cell Lung Cancer. *The New England Journal of Medicine*, *372*(21), 2018–2028. DOI: 10.1056/NEJMoa1501824 PMID: 25891174

Garrone, O., & La Porta, C. A. M. (2024). Artificial Intelligence for Precision Oncology of Triple-Negative Breast Cancer: Learning from Melanoma. *Cancers (Basel)*, *16*(4), 692. Advance online publication. DOI: 10.3390/cancers16040692 PMID: 38398083

Gatto, F., Miess, H., Schulze, A., & Nielsen, J. (2015). Flux balance analysis predicts essential genes in clear cell renal cell carcinoma metabolism. *Scientific Reports*, *5*(1), 10738. DOI: 10.1038/srep10738

Gautam, L. (2024). Natural language processing-based structured data extraction from unstructured clinical notes. *Notes (Philadelphia, Pa.)*, *6*, 9.

Gawankar, S., Nair, S., Pawar, V., Vhatkar, A., & Chavan, P. (2024). Patient privacy and data security in the era of AI-driven healthcare. In *2024 8th International Conference on Computing, Communication, Control and Automation (ICCUBEA)* (pp. 1-6). IEEE. DOI: 10.1109/ICCUBEA61740.2024.10775004

Georgescu, I., Dricu, A., Artene, S. A., Vrăjitoru, N. R., Barcan, E., Tache, D. E., & Popescu, S. O. (2024). Digital-focused approaches in cancer patients' management in the post-covid era: Challenges and solutions. *Applied Sciences (Basel, Switzerland)*, *14*(18), 8097. DOI: 10.3390/app14188097

Ghaffar Nia, N., Kaplanoglu, E., & Nasab, A. (2023). Evaluation of artificial intelligence techniques in disease diagnosis and prediction. *Discover Artificial Intelligence*, *3*(1), 5. DOI: 10.1007/s44163-023-00049-5 PMID: 40478140

Ghanem, M., Ghaith, A. K., & Bydon, M. (2024). Artificial intelligence and personalized medicine: Transforming patient care. In *The New Era of Precision Medicine* (pp. 131–142). Elsevier., DOI: 10.1016/B978-0-443-13963-5.00012-1

Gharti, L., Kaur, S. D., Vihal, S., Ramakrishna, K., Thakur, D., Kapoor, D., Sharma, L., & Kaurav, H. (2026). Recent advances in analytical techniques for cancer diagnostics and therapeutics: Combining state-of-the-art technologies for precision oncology. *Critical Reviews in Analytical Chemistry*, •••, 1–26. DOI: 10.1080/10408347.2026.2656439 PMID: 41983744

Ghasemi, A., Hashtarkhani, S., Schwartz, D. L., & Shaban-Nejad, A. (2024). Explainable artificial intelligence in breast cancer detection and risk prediction: A systematic scoping review. *Cancer Innovation*, *3*(5), e136. DOI: 10.1002/cai2.136

Ghoreyshi, N., Heidari, R., Farhadi, A., Chamanara, M., Farahani, N., Vahidi, M., & Behroozi, J. (2025, April 20). Next-generation sequencing in cancer diagnosis and treatment: Clinical applications and future directions. *Discover. Oncology*, *16*(1), 578. DOI: 10.1007/s12672-025-01816-9 PMID: 40253661

Giansanti, D., & Morelli, S. (2025, May 20). Exploring the Potential of Digital Twins in Cancer Treatment: A Narrative Review of Reviews. *Journal of Clinical Medicine*, *14*(10), 3574. DOI: 10.3390/jcm14103574 PMID: 40429568

Gibson, E. G., & Deweese, J. E. (2022). Structural and Biochemical Basis of Etoposide-Resistant Mutations in Topoisomerase IIα. *Symmetry*, *14*(7), 1309. DOI: 10.3390/sym14071309

Giercuszkiewicz-Haśnik, K., Morak-Młodawska, B., & Jeleń, M. (2026). Molecularly Targeted Therapies in Oncology: Mechanisms, Resistance, and Combination Strategies. *Molecules (Basel, Switzerland)*, *31*(7), 1195. DOI: 10.3390/molecules31071195 PMID: 41976235

Gilbert, L. A., Larson, M. H., Morsut, L., Liu, Z., Brar, G. A., Torres, S. E., Stern-Ginossar, N., Brandman, O., Whitehead, E. H., Doudna, J. A., Lim, W. A., Weissman, J. S., & Qi, L. S. (2013, July). CRISPR-Mediated Modular RNA-Guided Regulation of Transcription in Eukaryotes. *Cell*, *154*(2), 442–451. DOI: 10.1016/j.cell.2013.06.044 PMID: 23849981

Gillmore, J. D., Gane, E., Taubel, J., Kao, J., Fontana, M., Maitland, M. L., Seitzer, J., O'Connell, D., Walsh, K. R., Wood, K., Phillips, J., Xu, Y., Amaral, A., Boyd, A. P., Cehelsky, J. E., McKee, M. D., Schiermeier, A., Harari, O., Murphy, A., & Lebwohl, D. (2021, August). CRISPR-Cas9 In Vivo Gene Editing for Transthyretin Amyloidosis. *The New England Journal of Medicine*, *385*(6), 493–502. DOI: 10.1056/NEJMoa2107454 PMID: 34215024

Goel, A. K., Campbell, W. S., & Moldwin, R. (2021). Structured data capture for oncology. *JCO Clinical Cancer Informatics*, *5*(5), 194–201. DOI: 10.1200/CCI.20.00103 PMID: 33591796

Goel, I., Bhaskar, Y., Kumar, N., Singh, S., Amanullah, M., Dhar, R., & Karmakar, S. (2025). Role of AI in empowering and redefining the oncology care landscape: Perspective from a developing nation. *Frontiers in Digital Health*, *7*, 1550407. DOI: 10.3389/fdgth.2025.1550407 PMID: 40103737

Gogoshin, G., & Rodin, A. S. (2023). Graph neural networks in cancer and oncology research: Emerging and future trends. *Cancers (Basel)*, *15*(24), 5858. DOI: 10.3390/cancers15245858 PMID: 38136405

Golder, S., Xu, D., O'Connor, K., Wang, Y., Batra, M., & Hernandez, G. G. (2025). Leveraging Natural Language Processing and Machine Learning Methods for Adverse Drug Event Detection in Electronic Health/Medical Records: A Scoping Review. *Drug Safety*, *48*(4), 321–337. DOI: 10.1007/s40264-024-01505-6 PMID: 39786481

Gomari, D. P., Schweickart, A., Cerchietti, L., Paietta, E., Fernandez, H., Al-Amin, H., & Krumsiek, J. (2022). Variational autoencoders learn transferrable representations of metabolomics data. *Communications Biology*, *5*(1), 645. DOI: 10.1038/s42003-022-03579-3

Gómez-Luque, M. Á., Rodríguez-Marcos, P., Campanario-Pérez, R., Medina-González, A., Cárdenas-Fernández, A., Orlandi-Oliveira, W., & Ruibal-Moldes, M. (2026, February). Towards precision oncology: A new predictive machine learning model for early progression to castration resistant prostate cancer. *The Prostate*, *86*(3), 320–328. DOI: 10.1002/pros.70088 PMID: 41186168

Gonesh, C., Saha, M. B., Khan, M. J., Saha, H., & Avinash, D. K. (2023). Integrative analysis of multi-omics data with deep learning: Challenges and opportunities in bioinformatics. *J. Propuls. Technol*, *44*, 1384–1389.

Gong, S., Zhang, S., Lu, F., Pan, W., Li, N., & Tang, B. (2021). CRISPR/Cas-Based In Vitro Diagnostic Platforms for Cancer Biomarker Detection. *Analytical Chemistry*, *93*(35), 11899–11909. DOI: 10.1021/acs.analchem.1c02533

Göppner, D., & Leverkus, M. (2011). Basal Cell Carcinoma: From the Molecular Understanding of the Pathogenesis to Targeted Therapy of Progressive Disease. *Journal of Skin Cancer*, *2011*, 1–8. DOI: 10.1155/2011/650258 PMID: 21253551

Goswami, S., Pauken, K. E., Wang, L., & Sharma, P. (2024). Next-generation combination approaches for immune checkpoint therapy. *Nature Immunology*, *25*(12), 2186–2199. DOI: 10.1038/s41590-024-02015-4 PMID: 39587347

Gote, V., Nookala, A. R., Bolla, P. K., & Pal, D. (2021). Drug Resistance in Metastatic Breast Cancer: Tumor Targeted Nanomedicine to the Rescue. *International Journal of Molecular Sciences*, *22*(9), 4673. Advance online publication. DOI: 10.3390/ijms22094673 PMID: 33925129

Grady, W. M., & Carethers, J. M. (2008). Genomic and epigenetic instability in colorectal cancer pathogenesis. *Gastroenterology*, *135*(4), 1079–1099. DOI: 10.1053/j.gastro.2008.07.076 PMID: 18773902

Greco, A., Safi, D., Swami, U., Ginader, T., Milhem, M., & Zakharia, Y. (2019, December 5). Efficacy and Adverse Events in Metastatic Melanoma Patients Treated with Combination BRAF Plus MEK Inhibitors Versus BRAF Inhibitors: A Systematic Review. *Cancers (Basel)*, *11*(12), 1950. DOI: 10.3390/cancers11121950 PMID: 31817473

Green, M. R., & Sambrook, J. (2018). Analysis and normalization of real-time polymerase chain reaction (PCR) experimental data. *Cold Spring Harbor Protocols*, *2018*(6), pdb.top095000. DOI: 10.1101/pdb.top095000

Green, S., Prainsack, B., & Sabatello, M. (2024, January). The Roots of (in)Equity in Precision Medicine: Gaps in the Discourse. *Personalized Medicine*, *21*(1), 5–9. DOI: 10.2217/pme-2023-0097 PMID: 38088178

Griffin, A. C., Wang, K. H., Leung, T. I., & Facelli, J. C. (2024). Fairness and Inclusion in Biomedical Artificial Intelligence Research and Clinical Use: Technical and Social Perspectives. *Journal of Biomedical Informatics*, *157*, 104693. DOI: 10.1016/j.jbi.2024.104693 PMID: 39019301

Guilbert, B., Rieu, T., Delamarre, D., Laurent, F., Serre, F., Leroy, S., & Cuggia, M. (2024). ONCO-FAIR project: Improving data interoperability in oncology chemotherapy treatments for data reuse. *Studies in Health Technology and Informatics*, *316*, 1373–1377. DOI: 10.3233/SHTI240667 PMID: 39176636

Guimarães, P. P. G., Figueroa-Espada, C. G., Riley, R. S., Lammers, T., Dreaden, E. C., Sempkowski, M., Bhatia, S. N., Hammond, P. T., Langer, R., Jaklenec, A., & Blankschtein, D. (2023). In vivo bone marrow microenvironment siRNA delivery using lipid–polymer nanoparticles for multiple myeloma therapy. *Proceedings of the National Academy of Sciences of the United States of America*, *120*(25), e2215711120. DOI: 10.1073/pnas.2215711120 PMID: 37310997

Guinney, J., Dienstmann, R., Wang, X., de Reyniès, A., Schlicker, A., Soneson, C., Marisa, L., Roepman, P., Nyamundanda, G., Angelino, P., Bot, B. M., Morris, J. S., Simon, I. M., Gerster, S., Fessler, E., De Sousa, E., Melo, F., Missiaglia, E., Ramay, H., & Tejpar, S. (2015). The consensus molecular subtypes of colorectal cancer. *Nature Medicine*, *21*(11), 1350–1356. DOI: 10.1038/nm.3967 PMID: 26457759

Gu, J., & Chen, J. Y. (2024). MLPA: A multi-scale digital twin framework for personalized cancer simulation and treatment optimization. In *Proceedings of the 15th ACM International Conference on Bioinformatics, Computational Biology and Health Informatics* (pp. 1-1). DOI: 10.1145/3698587.3701425

Gulati, S., Yadav, R., Kumari, V., Nair, S., Gupta, C., & Aishwari, M. (2025). Nanosensors in healthcare: Transforming real-time monitoring and disease management with cutting-edge nanotechnology. *RSC Pharmaceutics*, *2*(5), 1003–1018. Advance online publication. DOI: 10.1039/D5PM00125K

Gunasekaran, A., & Yusuf, Y. Y. (2002). Agile manufacturing: A taxonomy of strategic and technological imperatives. *International Journal of Production Research*, *40*(6), 1357–1385. DOI: 10.1080/00207540110118370

Guo, H., Xu, X., Zhang, J., Du, Y., Yang, X., He, Z., Zhao, L., Liang, T., & Guo, L. (2024, August 9). The pivotal role of preclinical animal models in anti-cancer drug discovery and personalized cancer therapy strategies. *Pharmaceuticals (Basel, Switzerland)*, *17*(8), 1048. DOI: 10.3390/ph17081048 PMID: 39204153

Gupta, D., Boora, A., Thakur, A., & Gupta, T. K. (2023). Green and Sustainable Synthesis of Nanomaterials: Recent Advancements and Limitations. *Environmental Research*, *231*, 116316. DOI: 10.1016/j.envres.2023.116316

Gupta, N., Gupta, G. D., & Singh, D. (2022, October). Localized topical drug delivery systems for skin cancer: Current approaches and future prospects. *Frontiers in Nanotechnology*, *4*, 1006628. Advance online publication. DOI: 10.3389/fnano.2022.1006628

Gupta, P., Neupane, Y. R., Parvez, S., & Kohli, K. (2021, December). Recent Advances in Targeted Nanotherapeutic Approaches for Breast Cancer Management. *Nanomedicine (London)*, *16*(29), 2605–2631. DOI: 10.2217/nnm-2021-0281 PMID: 34854336

Gupta, R., Srivastava, D., Sahu, M., Tiwari, S., Ambasta, R. K., & Kumar, P. (2021). Artificial intelligence to deep learning: Machine intelligence approach for drug discovery. *Molecular Diversity*, *25*(3), 1315–1360. DOI: 10.1007/s11030-021-10217-3 PMID: 33844136

Gupta, S. (2025). Ethical, privacy, and security issues in smart healthcare. In *Driving Global Health and Sustainable Development Goals With Smart Technology* (pp. 83–98). IGI Global Scientific Publishing.

Gupta, U., Pranav, A., Modi, R. K., & Dubey, A. (2025). Ethical implications of AI in CRISPR: responsible genome editing. In *Bioinformatics and Beyond* (pp. 256–275). CRC Press. DOI: 10.1201/9781003508403-13

Gupta, V., Kumar, D., Gupta, S., Tanwar, R., Jaiswal, N. K., Islam, M. M., Singh, S., Choudhary, N., Gowri, S., Webster, T., & Faiyazuddin, M. (2026). Nanotechnology-driven cancer therapies for precision oncology: Advances and clinical outlook. *International Journal of Nanomedicine*, *21*, 568254. DOI: 10.2147/IJN.S568254 PMID: 41869396

Gupta, Y., Srivastava, V., & Singh, R. K. (2025). AI-enhanced patient-centric clinical trial design. In. AIP Conference Proceedings: Vol. 3254. *No. 1* (p. 020020). AIP Publishing LLC.

Hachem, S., Yehya, A., El Masri, J., Mavingire, N., Johnson, J. R., Dwead, A. M., Kattour, N., Bouchi, Y., Kobeissy, F., Rais-Bahrami, S., Mechref, Y., Abou-Kheir, W., & Woods-Burnham, L. (2024, September 25). Contemporary update on clinical and experimental prostate cancer biomarkers: A multi-omics-focused approach to detection and risk stratification. *Biology (Basel)*, *13*(10), 762. DOI: 10.3390/biology13100762 PMID: 39452071

Haenssle, H. A., Fink, C., Schneiderbauer, R., Toberer, F., Buhl, T., Blum, A., Kalloo, A., Hassen, A. B. H., Thomas, L., Enk, A., Uhlmann, L., Alt, C., Arenbergerova, M., Bakos, R., Baltzer, A., Bertlich, I., Blum, A., Bokor-Billmann, T., Bowling, J., & Zalaudek, I. (2018, August). Man against machine: Diagnostic performance of a deep learning convolutional neural network for dermoscopic melanoma recognition in comparison to 58 dermatologists. *Annals of Oncology : Official Journal of the European Society for Medical Oncology*, *29*(8), 1836–1842. DOI: 10.1093/annonc/mdy166 PMID: 29846502

Hafstað, V., Häkkinen, J., Larsson, M., Staaf, J., Vallon-Christersson, J., & Persson, H. (2023). Improved detection of clinically relevant fusion transcripts in cancer by machine learning classification. *BMC Genomics*, *24*(1), 783. DOI: 10.1186/s12864-023-09889-y

Hai, L., Jiang, Z., Zhang, H., & Sun, Y. (2024, December 23). From multi-omics to predictive biomarker: AI in tumor microenvironment. *Frontiers in Immunology*, *15*, 1514977. DOI: 10.3389/fimmu.2024.1514977 PMID: 39763649

Hamato, A., Lee, D., Ota, R., Hemmati, H., Muller, M., Wang, Y., Pandian, L. S., Kim, S., Yamaya, T., Cherry, S. R., Glodo, J., & Kwon, S. I. (2026). Ultra-dense lutetium oxide ceramic scintillators for positron emission tomography. *Physics in Medicine and Biology*, *71*(4), 045014. DOI: 10.1088/1361-6560/ae456f PMID: 41678960

Hamdan, A., & Ewing, A. (2022). Unravelling the tumour genome: The evolutionary and clinical impacts of structural variants in tumourigenesis. *The Journal of Pathology*, *257*(4), 479–493. DOI: 10.1002/path.5901

Hamdy, N. M., Basalious, E. B., El-Sisi, M. G., Nasr, M., Kabel, A. M., Nossier, E. S., & Abadi, A. H. (2024, November). Advancements in current one-size-fits-all therapies compared to future treatment innovations for better improved chemotherapeutic outcomes: A step-toward personalized medicine. *Current Medical Research and Opinion*, *40*(11), 1943–1961. DOI: 10.1080/03007995.2024.2416985 PMID: 39412377

Hammad, M., Panaousis, M., Ali, H., & Khan, W. A. (2026). Industry 4.0 in Pharmaceuticals: A New Era of Smart Manufacturing. In M. Hammad, M. Panaousis, H. Ali, & W. A. Khan, *Smart Manufacturing Blueprint: Navigating Industry 4.0 Across Diverse Sectors* (pp. 165–211). Springer Nature Switzerland. DOI: 10.1007/978-3-032-00214-3_6

Hanahan, D. (2022). Hallmarks of cancer: New dimensions. *Cancer Discovery*, *12*(1), 31–46. DOI: 10.1158/2159-8290.CD-21-1059 PMID: 35022204

Haney, K., Tandon, P., Divi, R., Ossandon, M. R., Baker, H., & Pearlman, P. C. (2017). The role of affordable, point-of-care technologies for cancer care in low-and middle-income countries: A review and commentary. *IEEE Journal of Translational Engineering in Health and Medicine*, *5*, 1–14. DOI: 10.1109/JTEHM.2017.2761764 PMID: 29204328

Han, H. A., Pang, J. K. S., & Soh, B.-S. (2020). Mitigating off-target effects in CRISPR/Cas9-mediated in vivo gene editing. *Journal of Molecular Medicine : Official Organ of the "Gesellschaft Deutscher Naturforscher und Ärzte"*, *98*(5), 615–632. DOI: 10.1007/s00109-020-01893-z PMID: 32198625

Han, J., Xiong, Z., Liu, Y., Li, J., Li, C., Zhang, W., Gao, J., Shuai, X., Wang, Z., Wang, G., Shu, X., Tao, K., & Cai, M. (2025). Circadian rhythm in gastrointestinal cancer: Clinical applications and future perspectives. *Annals of Medicine*, *57*(1), 2529571. DOI: 10.1080/07853890.2025.2529571 PMID: 40616824

Han, X., Xu, K., Taratula, O., & Farsad, K. (2019, January 17). Applications of nanoparticles in biomedical imaging. *Nanoscale*, *11*(3), 799–819. DOI: 10.1039/C8NR07769J PMID: 30603750

Haque, R., Al Sakib, A., Hossain, M. F., Islam, F., Ibne Aziz, F., Ahmed, M. R., & Hasan, M. J. (2024). Advancing early leukemia diagnostics: A comprehensive study incorporating image processing and transfer learning. *BioMedInformatics*, *4*(2), 966–991. DOI: 10.3390/biomedinformatics4020054

Harding, J. J., Fan, J., Oh, D. Y., Choi, H. J., Kim, J. W., Chang, H. M., Bao, L., Sun, H.-C., Macarulla, T., Xie, F., Metges, J.-P., Ying, J., Bridgewater, J., Lee, M.-A., Tejani, M. A., Chen, E. Y., Kim, D. U., Wasan, H., Ducreux, M., & Zhao, H. (2023, July). Zanidatamab for HER2-amplified, unresectable, locally advanced or metastatic biliary tract cancer (HERIZON-BTC-01): A multicentre, single-arm, phase 2b study. *The Lancet. Oncology*, *24*(7), 772–782. DOI: 10.1016/S1470-2045(23)00242-5 PMID: 37276871

Harrer, S., Shah, P., Antony, B., & Hu, J. (2019). Artificial Intelligence for Clinical Trial Design. *Trends in Pharmacological Sciences*, *40*(8), 577–591. DOI: 10.1016/j.tips.2019.05.005 PMID: 31326235

Hart, T., Chandrashekhar, M., Aregger, M., Steinhart, Z., Brown, K. R., MacLeod, G., Mis, M., Zimmermann, M., Fradet-Turcotte, A., Sun, S., Mero, P., Dirks, P., Sidhu, S., Roth, F. P., Rissland, O. S., Durocher, D., Angers, S., & Moffat, J. (2015). High-resolution CRISPR screens reveal fitness genes and genotype-specific cancer liabilities. *Cell*, *163*(6), 1515–1526. DOI: 10.1016/j.cell.2015.11.015 PMID: 26627737

Hasan, J., Hasan, K., Al Noman, A., Hasan, S., Sultana, S., Arafat, M. A., & Ahmed, M. R. (2024). Transforming leukemia classification: A comprehensive study on deep learning models for enhanced diagnostic accuracy. In *2024 IEEE International Conference on Power, Electrical, Electronics and Industrial Applications (PEEIACON)* (pp. 1-6). IEEE. DOI: 10.1109/PEEIACON63629.2024.10800693

Hashemi, M., Khosravinejad, A., Alimohammadi, M., Jamshidian, F., Shirazi, R. M., Ghannad Zadeh, R., Jamali, B., Esfahani, K. S., Shirazi, A., Zali, N., & Farahani, N. 2026. CRISPR/Cas9 in the Treatment of Prostate Cancer. In *INNOVATIVE THERAPEUTIC APPROACHES BY CRISPR/CAS9 IN CANCER THERAPY* (pp. 226-253).

Hashim, S., Ali, M., Nandakumar, K., & Yaqub, M. (2022). SubOmiEmbed: Self-supervised representation learning of multi-omics data for cancer type classification. In *2022 10th International Conference on Bioinformatics and Computational Biology (ICBCB)* (pp. 66-72). IEEE.

Hasin, Y., Seldin, M., & Lusis, A. (2017). Multi-omics approaches to disease. *Genome Biology, 18*(1), 83. DOI: 10.1186/s13059-017-1215-1 PMID: 28476144

Hassane, M. (2024). Artificial intelligence-driven precision medicine in cancer treatment. *Science, 1*, 100041.

Hassan, R. Y. A. (2022). Advances in electrochemical nano-biosensors for biomedical and environmental applications: From current work to future perspectives. *Sensors (Basel), 22*(19), 7539. DOI: 10.3390/s22197539

Hassan, S., Bahar, R., Johan, M. F., Mohamed Hashim, E. K., Abdullah, W. Z., Esa, E., Abdul Hamid, F. S., & Zulkafli, Z. (2023, January 19). Next-Generation Sequencing (NGS) and Third-Generation Sequencing (TGS) for the Diagnosis of Thalassemia. [Internet]. *Diagnostics (Basel), 13*(3), 373. Retrieved November 2, 2025, from https://www.mdpi.com/2075-4418/13/3/373. DOI: 10.3390/diagnostics13030373 PMID: 36766477

Hassan, Y. M., Mohamed, A. S., Hassan, Y. M., & El-Sayed, W. M. (2025). Recent developments and future directions in point-of-care next-generation CRISPR-based rapid diagnosis. *Clinical and Experimental Medicine, 25*(1), 33. DOI: 10.1007/s10238-024-01540-8 PMID: 39789283

Hays, P. (2021). Great strides in precision medicine: personalized oncology, immunotherapies, and molecular diagnostics. In *Advancing Healthcare Through Personalized Medicine* (pp. 141–417). Springer International Publishing. DOI: 10.1007/978-3-030-80100-7_5

Hayward, N. K., Wilmott, J. S., Waddell, N., Johansson, P. A., Field, M. A., Nones, K., Patch, A.-M., Kakavand, H., Alexandrov, L. B., Burke, H., Jakrot, V., Kazakoff, S., Holmes, O., Leonard, C., Sabarinathan, R., Mularoni, L., Wood, S., Xu, Q., Waddell, N., & Mann, G. J. (2017, May). Whole-genome landscapes of major melanoma subtypes. *Nature, 545*(7653), 175–180. DOI: 10.1038/nature22071 PMID: 28467829

Heesen, P., Schelling, G., Birbaumer, M., Jäger, R., Bode, B., Studer, G., & Fuchs, B. (2024). Real-World-Time Data and RCT Synergy: Advancing Personalized Medicine and Sarcoma Care through Digital Innovation. *Cancers (Basel), 16*(14), 2516. DOI: 10.3390/cancers16142516 PMID: 39061155

Hehr, M., Sadafi, A., Matek, C., Lienemann, P., Pohlkamp, C., Haferlach, T., & Marr, C. (2023). Explainable AI identifies diagnostic cells of genetic AML subtypes. *PLOS Digital Health, 2*(3), e0000187. DOI: 10.1371/journal.pdig.0000187 PMID: 36921004

He, L., Tang, J., Andersson, E. I., Timonen, S., Koschmieder, S., Wennerberg, K., & Aittokallio, T. (2018). Patient-customized drug combination prediction and testing for T-cell prolymphocytic leukemia patients. *Cancer Research*, *78*(9), 2407–2418. DOI: 10.1158/0008-5472.CAN-17-3644 PMID: 29483097

He, M., Cao, Y., Chi, C., Yang, M., Zhang, Y., Chen, W., Zhang, H., Li, W., Wang, J., & Liu, Z. (2023). Unleashing novel horizons in advanced prostate cancer treatment: Investigating the potential of prostate specific membrane antigen-targeted nanomedicine-based combination therapy. *Frontiers in Immunology*, *14*, 1265751. DOI: 10.3389/fimmu.2023.1265751 PMID: 37795091

Hemdan, M., Abuelhaded, K., Shaker, A. A. S., Ashour, M. M., Abdelaziz, M. M., Dahab, M. I., Nassar, Y. A., Sarguos, A. M. M., Zakaria, P. S., Fahmy, H. A., & El-Sherbiny, I. M. (2025). Recent advances in nano-enhanced biosensors: Innovations in design, applications in healthcare, environmental monitoring, and food safety, and emerging research challenges. *Sensing and Bio-Sensing Research*, *48*, 100783. DOI: 10.1016/j.sbsr.2025.100783

Hennawi, M., Pakasticali, N., Tashkandi, H., & Hussaini, M. (2022). Genomic landscape of mixed-phenotype acute leukemia. *International Journal of Molecular Sciences*, *23*(19), 11259. DOI: 10.3390/ijms231911259 PMID: 36232559

Heo, Y. J., Hwa, C., Lee, G. H., Park, J. M., & An, J. Y. (2021, July 31). Integrative Multi-Omics Approaches in Cancer Research: From Biological Networks to Clinical Subtypes. *Molecules and Cells*, *44*(7), 433–443. DOI: 10.14348/molcells.2021.0042 PMID: 34238766

Hermain, S., Harshitha, K. S., & Naveen, N. R. (2026). Beyond chemotherapy: The rise of nucleic acid nanoformulations in personalized lung cancer therapy. *Cancer Pathogenesis and Therapy*, *4*, E16–E62. DOI: 10.1016/j.cpt.2026.04.002 PMID: 42597591

Hernández-Sánchez, A., & Bullinger, L. (2023). Recent advances in precision medicine for acute myeloid leukemia. *Current Opinion in Oncology*, *35*(6), 581–588. DOI: 10.1097/CCO.0000000000000965 PMID: 37621173

Heuser, M., Ofran, Y., Boissel, N., Mauri, S. B., Craddock, C., & Janssen, J.ESMO Guidelines Committee. (2020). Acute myeloid leukaemia in adult patients: ESMO Clinical Practice Guidelines for diagnosis, treatment and follow-up. *Annals of Oncology : Official Journal of the European Society for Medical Oncology*, *31*(6), 697–712. DOI: 10.1016/j.annonc.2020.02.018 PMID: 32171751

He, Y., Lu, W., Zhou, X., Mu, J., & Shen, W. (2024). Unraveling Alzheimer's disease: Insights from single-cell sequencing and spatial transcriptomic. *Frontiers in Neurology*, *15*, 1515981. DOI: 10.3389/fneur.2024.1515981 PMID: 39741706

He, Y., Tu, X., Xue, Y., Chen, Y., Ye, B., Li, X., Li, D., Zhong, Z., & Zhong, Q. (2025). CRISPR screening redefines therapeutic target identification and drug discovery with precision and scalability. *Journal of Pharmaceutical Analysis*, *16*(2), 101357. DOI: 10.1016/j.jpha.2025.101357 PMID: 41732321

Hidalgo, I. Fenández-de_Vega, F., Ceberio, J., Garnica, O., Velasco, J. M., Cortés, J. C., & Díaz, J. (2023). Sustainable artificial intelligence systems: an energy efficiency approach. Authorea Preprints.

Hii, A. R. K., Qi, X., & Wu, Z. (2024, February). Advanced strategies for CRISPR/Cas9 delivery and applications in gene editing, therapy, and cancer detection using nanoparticles and nanocarriers. *Journal of Materials Chemistry. B*, *12*(6), 1467–1489. DOI: 10.1039/D3TB01850D PMID: 38288550

Hirani, A., & Zhao, X. B. (2021). *U.S. Patent No. 10,916,252*. Washington, DC: U.S. Patent and Trademark Office.

Hodel, K. V. S., Fiuza, B. S. D., Conceição, R. S., Aleluia, A. C. M., Pitanga, T. N., Fonseca, L. M. D. S., Valente, C. O., Minafra-Rezende, C. S., & Machado, B. A. S. (2024). Pharmacovigilance in Vaccines: Importance, Main Aspects, Perspectives, and Challenges—A Narrative Review. *Pharmaceuticals (Basel, Switzerland)*, *17*(6), 807. DOI: 10.3390/ph17060807 PMID: 38931474

Hodis, E., Watson, I. R., Kryukov, G. V., Arold, S. T., Imielinski, M., Theurillat, J.-P., Nickerson, E., Auclair, D., Li, L., Place, C., DiCara, D., Ramos, A. H., Lawrence, M. S., Cibulskis, K., Sivachenko, A., Voet, D., Saksena, G., Stransky, N., Onofrio, R. C., & Chin, L. (2012, July). A Landscape of Driver Mutations in Melanoma. *Cell*, *150*(2), 251–263. DOI: 10.1016/j.cell.2012.06.024 PMID: 22817889

Hoffmann, J., Eminovic, S., Wilhelm, C., Krause, S. W., Neubauer, A., Thrun, M. C., & Brendel, C. (2023). Prediction of clinical outcomes with explainable artificial intelligence in patients with chronic lymphocytic leukemia. *Current Oncology (Toronto, Ont.)*, *30*(2), 1903–1915. DOI: 10.3390/curroncol30020148 PMID: 36826109

Ho, H. Y., Chung, K. S., Kan, C. M., & Wong, S. C. (2024). Liquid biopsy in the clinical management of cancers. *International Journal of Molecular Sciences*, *25*(16), 8594. DOI: 10.3390/ijms25168594

Honkala, A., Malhotra, S. V., Kummar, S., & Junttila, M. R. (2022, February). Harnessing the predictive power of preclinical models for oncology drug development. *Nature Reviews. Drug Discovery*, *21*(2), 99–114. DOI: 10.1038/s41573-021-00301-6 PMID: 34702990

Hood, L., & Friend, S. H. (2011, March). Predictive, personalized, preventive, participatory (P4) cancer medicine. *Nature Reviews. Clinical Oncology*, *8*(3), 184–187. DOI: 10.1038/nrclinonc.2010.227 PMID: 21364692

Horgan, C. C., Bergholt, M. S., Nagelkerke, A., Thin, M. Z., Pence, I. J., Kauscher, U., & Stevens, M. M. (2020). Integrated photodynamic Raman theranostics for cancer diagnosis, treatment, and post-treatment molecular monitoring. *Theranostics*, *11*(4), 2006–2019. DOI: 10.7150/thno.53031

Hossain, M. R., Mahabub, S., Al Masum, A., & Jahan, I. (2024). Natural language processing (nlp) in analyzing electronic health records for better decision making. *Journal of Computer Science and Technology Studies*, *6*(5), 216–228. DOI: 10.32996/jcsts.2024.6.5.18

Houghton, C., Casey, D., Shaw, D., & Murphy, K. (2013). Rigour in qualitative case-study research. *Nurse Researcher*, *20*(4), 12–17. DOI: 10.7748/nr2013.03.20.4.12.e326 PMID: 23520707

Hou, J., He, Z., Liu, T., Chen, D., Wang, B., Wen, Q., & Zheng, X. (2022). Evolution of molecular targeted cancer therapy: Mechanisms of drug resistance and novel opportunities identified by CRISPR-Cas9 screening. *Frontiers in Oncology*, *12*, 755053. DOI: 10.3389/fonc.2022.755053 PMID: 35372044

Hou, X., Zaks, T., Langer, R., & Dong, Y. (2021). Lipid nanoparticles for mRNA delivery. *Nature Reviews. Materials, 6*(12), 1078–1094. DOI: 10.1038/s41578-021-00358-0 PMID: 34394960

Ho, W., Gao, M., Li, F., Li, X., Zhang, X. Q., & Xu, X. (2021). Next-Generation Vaccines: nanoparticle-Mediated DNA and mRNA Delivery. *Advanced Healthcare Materials, 10*(8), 2001812. DOI: 10.1002/adhm.202001812 PMID: 33458958

Hsu, C. C., & Wu, Y. (2022). Recent advances in nanotechnology-enabled biosensors for detection of exosomes as new cancer liquid biopsy. *Experimental Biology and Medicine (Maywood, N.J.), 247*(23), 2152–2172. DOI: 10.1177/15353702221110813

Huanbutta, K., Burapapadh, K., Kraisit, P., Sriamornsak, P., Ganokratanaa, T., Suwanpitak, K., & Sangnim, T. (2024a). Artificial intelligence-driven pharmaceutical industry: A paradigm shift in drug discovery, formulation development, manufacturing, quality control, and post-market surveillance. *European Journal of Pharmaceutical Sciences : Official Journal of the European Federation for Pharmaceutical Sciences, 203*, 106938. DOI: 10.1016/j.ejps.2024.106938 PMID: 39419129

Huang, D., Ma, N., Li, X., Gou, Y., Duan, Y., Liu, B., & Zhang, X. (2023). Advances in single-cell RNA sequencing and its applications in cancer research. *Journal of Hematology & Oncology, 16*(1), 98. DOI: 10.1186/s13045-023-01494-6

Huang, D., Qian, J., Shi, Z., Zhao, J., Fang, M., & Xu, Z. (2020). CRISPR-Cas12a-assisted multicolor biosensor for semiquantitative point-of-use testing of the Nopaline synthase terminator in genetically modified crops by unaided eyes. *ACS Synthetic Biology, 9*(11), 3114–3123. DOI: 10.1021/acssynbio.0c00365

Huang, J., Chan, S. C., Ngai, C. H., Lok, V., Zhang, L., Lucero-Prisno, D. E.III, & Wong, M. C. (2022). Disease burden, risk factors, and trends of leukaemia: A global analysis. *Frontiers in Oncology, 12*, 904292. DOI: 10.3389/fonc.2022.904292 PMID: 35936709

Huang, P., & Zhang, J. (2025). Global leukemia burden and trends: A comprehensive analysis of temporal and spatial variations from 1990-2021 using GBD (Global Burden of Disease) data. *BMC Public Health, 25*(1), 262. DOI: 10.1186/s12889-025-21428-w PMID: 39838416

Huang, S., Chaudhary, K., & Garmire, L. X. (2017, June). More Is Better: Recent Progress in Multi-Omics Data Integration Methods. *Frontiers in Genetics, 8*, 84. Advance online publication. DOI: 10.3389/fgene.2017.00084 PMID: 28670325

Huang, X., Zhu, Y., & Kianfar, E. (2021). Nano biosensors: Properties, applications and electrochemical techniques. *Journal of Materials Research and Technology, 12*, 1649–1672. DOI: 10.1016/j.jmrt.2021.03.048

Huang, Y. L., Chen, D. R., & Liu, Y. K. (2004). Breast cancer diagnosis using image retrieval for different ultrasonic systems. In *2004 International Conference on Image Processing, 2004. ICIP'04.* (Vol. 5, pp. 2957-2960). IEEE.

Huang, Z., Zhan, X., Xiang, S., Johnson, T. S., Helm, B., Yu, C. Y., Zhang, J., Salama, P., Rizkalla, M., Han, Z., & Huang, K. (2019, March 8). SALMON: Survival analysis learning with multi-omics neural networks on breast cancer. *Frontiers in Genetics, 10*, 166. DOI: 10.3389/fgene.2019.00166 PMID: 30906311

Hugo, W., Zaretsky, J. M., Sun, L., Song, C., Moreno, B. H., Hu-Lieskovan, S., Berent-Maoz, B., Pang, J., Chmielowski, B., Cherry, G., Seja, E., Lomeli, S., Kong, X., Kelley, M. C., Sosman, J. A., Johnson, D. B., Ribas, A., & Lo, R. S. (2016, March). Genomic and Transcriptomic Features of Response to Anti-PD-1 Therapy in Metastatic Melanoma. *Cell, 165*(1), 35–44. DOI: 10.1016/j.cell.2016.02.065 PMID: 26997480

Huhulea, E. N., Huang, L., Eng, S., Sumawi, B., Huang, A., Aifuwa, E., Hirani, R., Tiwari, R. K., & Etienne, M. (2025, April 13). Artificial Intelligence Advancements in Oncology: A Review of Current Trends and Future Directions. *Biomedicines, 13*(4), 951. DOI: 10.3390/biomedicines13040951 PMID: 40299653

Hull, R., Chauke-Malinga, N., Gaudji, G. R., Blenman, K. R., & Dlamini, Z. (2023). The role of digital twinning, the next generation of EMR/EHR in healthcare in a society 5.0: collecting patient data from birth to the grave. In *Society 5.0 and Next Generation Healthcare: Patient-Focused and Technology-Assisted Precision Therapies* (pp. 179–200). Springer Nature Switzerland. DOI: 10.1007/978-3-031-36461-7_8

Human Genome Editing. (2017). National Academies Press., DOI: 10.17226/24623

Hunter, B., Hindocha, S., & Lee, R. W. (2022). The role of artificial intelligence in early cancer diagnosis. *Cancers (Basel), 14*(6), 1524. DOI: 10.3390/cancers14061524 PMID: 35326674

Ichihara, G., Katsumata, Y., Sugiura, Y., Matsuoka, Y., Maeda, R., Endo, J., Anzai, A., Shirakawa, K., Moriyama, H., Kitakata, H., & Fukuda, K. (2023). MRP1-Dependent Extracellular Release of Glutathione Induces Cardiomyocyte Ferroptosis After Ischemia-Reperfusion. *Circulation Research, 133*(10), 861–876. DOI: 10.1161/CIRCRESAHA.123.323517 PMID: 37818671

Idumah, G., Li, L., Yehia, L., Mahdi, H., & Ni, Y. (2025). Metabolic signatures of immune checkpoint inhibitor response in gynecologic cancers: insights from flux balance analysis. bioRxiv, 2025-01.

Ikhtiar, F., Jamal, A., & Bokhari, S. M. S. M. (2025, September 16). Deciphering and targeting oncogenic pathways through integrated approaches and amino acid metabolism in hematologic malignancies. *Discover. Oncology, 16*(1), 1687. DOI: 10.1007/s12672-025-03535-7 PMID: 40956459

Iminova, R., Berganayeva, G., Zhurtbayeva, A., Abdurakhmanova, L., Almabekova, A., Shepilov, D., Vassilina, G., Nurmahanova, A., Kairalapova, G., & Dyusebaeva, M. (2026). From PEGylation to Next-Generation Polymers: Overcoming Biological Barriers-A Review. *Molecules (Basel, Switzerland), 31*(4), 675–675. DOI: 10.3390/molecules31040675 PMID: 41752452

Ioakeim-Skoufa, I., Atkins, K., & Hernández-Rodríguez, M. Á. (2025). Optimizing real-world evidence studies for regulatory decision-making and impact assessment in pharmacovigilance. *British Journal of Clinical Pharmacology, 91*(4), 1092–1095. DOI: 10.1111/bcp.16393 PMID: 39821103

Irizarry, T., DeVito Dabbs, A., & Curran, C. R. (2015, June 23). Patient Portals and Patient Engagement: A State of the Science Review. *Journal of Medical Internet Research, 17*(6), e148. DOI: 10.2196/jmir.4255 PMID: 26104044

Isaic, A., Motofelea, N., Hoinoiu, T., Motofelea, A. C., Leancu, I. C., Stan, E., Gheorghe, S. R., Dutu, A. G., & Crintea, A. (2025, September 23). Next-Generation Sequencing: A Review of Its Transformative Impact on Cancer Diagnosis, Treatment, and Resistance Management. *Diagnostics (Basel), 15*(19), 2425. DOI: 10.3390/diagnostics15192425 PMID: 41095644

Isella, C., Terrasi, A., Bellomo, S. E., Petti, C., Galatola, G., Muratore, A., Mellano, A., Senetta, R., Cassenti, A., Sonetto, C., Inghirami, G., Trusolino, L., Fekete, Z., De Ridder, M., Cassoni, P., Storme, G., Bertotti, A., & Medico, E. (2015). Stromal contribution to the colorectal cancer transcriptome. *Nature Genetics, 47*(4), 312–319. DOI: 10.1038/ng.3224 PMID: 25706627

Islam, S., & Hasan, M. N. (2024). Personalized graph feature-based multi-omics data integration for cancer subtype identification. *arXiv preprint arXiv:2408.08832.*

Iyer, K. A., Koynova-Tenchov, R., Sasso, J. M., Thite, T., Deng, Y., & Zhou, Q. A. (2026). The Future of Epigenetics: Emerging Technologies and Clinical Applications. *ACS Pharmacology & Translational Science, 9*(3), 506–544. DOI: 10.1021/acsptsci.5c00729 PMID: 41852637

Iyer, K. A., Tenchov, R., Lotti Diaz, L. M., Jain, P., Thite, T., Deng, Y., & Zhou, Q. A. (2025, November 24). CRISPR technology: Transforming the future of medicine and diagnostics. *Biochemistry, 64*(24), 4628–4660. DOI: 10.1021/acs.biochem.5c00480 PMID: 41284414

Iyer, P., Jasdanwala, S. S., Wang, Y., Bhatia, K., & Bhatt, S. (2024). Decoding acute myeloid leukemia: A clinician's guide to functional profiling. *Diagnostics (Basel), 14*(22), 2560. DOI: 10.3390/diagnostics14222560 PMID: 39594226

J, S. P., Moger, S. S., Byalahunasishetter, B. I., & Raghavan, R. (2026). The One Health Framework and Antimicrobial Resistance: Role of Herbal Medicine. *Herbal Medicine Approaches to Antimicrobial Resistance and One Health*, 1–36. DOI: 10.4018/979-8-3373-7810-7.ch001

J. Kang, X. Ke, Y. Zhao, and Y. Zhou, "Nanotechnology-Assisted Molecular Profiling: Emerging Advances in Circulating Tumor DNA Detection," *Int. J. Nanomedicine*, vol. 21, no. null, p. 585118, Dec. 2026, .DOI: 10.2147/IJN.S585118

Jagtap, S., Garcia-Garcia, G., & Rahimifard, S. (2021). Optimisation of the resource efficiency of food manufacturing via the Internet of Things. *Computers in Industry, 127*, 103397. DOI: 10.1016/j.compind.2021.103397

Jain, D. (2024). Artificial Intelligence in Quality Control Systems: A Cross-Industry Analysis of Applications, Benefits, and Implementation Frameworks. *International Journal of Scientific Research in Computer Science. Engineering and Information Technology, 10*(6), 1321–1333. DOI: 10.32628/CSEIT241061162

Jalasri, M., Panchal, S. M., Mahalingam, K., Venkatasubramanian, R., Hemalatha, R., & Boopathi, S. (2025). AI-powered smart energy management for optimizing energy efficiency in high-performance computing systems. In *Future of Digital Technology and AI in Social Sectors* (pp. 329–366). IGI Global.

Jamalinia, M., & Weiskirchen, R. (2025, April). Advances in personalized medicine: Translating genomic insights into targeted therapies for cancer treatment. *Annals of Translational Medicine, 13*(2), 18–18. DOI: 10.21037/atm-25-34 PMID: 40438512

Jangama, R., Suresh, C., Korrapati, C., & Raghavan, R. (2026). Hybrid Coating Systems: Bioactive Molecules and Nanoparticles for Cancer Therapy. *Biomedical Applications of Laser-Processed Coatings*, 99–134. DOI: 10.4018/979-8-3373-4104-0.ch003

Jani, C. T., Singh, H., Abdallah, N., Mouchati, C., Arora, S., Kareff, S., Salciccioli, J., Thomson, C. C., & Velcheti, V. (2023, September). Trends in Lung Cancer Incidence and Mortality (1990-2019) in the United States: A Comprehensive Analysis of Gender and State-Level Disparities. *JCO Global Oncology*, (9), e2300255. DOI: 10.1200/GO.23.00255 PMID: 38127772

Jarallah, S. J., Almughem, F. A., Alhumaid, N. K., Fayez, N. A., Alradwan, I., Alsulami, K. A., Tawfik, E. A., & Alshehri, A. A. (2025). Artificial intelligence revolution in drug discovery: A paradigm shift in pharmaceutical innovation. *International Journal of Pharmaceutics, 680*, 125789. DOI: 10.1016/j.ijpharm.2025.125789 PMID: 40451590

Jayachandran, A., Ramakrishna, S., Shukla, R., Prithviraj, P., Kannourakis, G., & Poddar, A. (2026). *Synthetic Biology in Precision Medicine and Nanotechnology*. Handb. Synth. Biol. DOI: 10.1007/978-981-95-3957-4_37

Jayaraman, P., Desman, J., Sabounchi, M., Nadkarni, G. N., & Sakhuja, A. (2024). A primer on reinforcement learning in medicine for clinicians. *NPJ Digital Medicine*, *7*(1), 337. DOI: 10.1038/s41746-024-01316-0 PMID: 39592855

Jazieh, A. R., El Rouby, N., Guinigundo, A., Huelsman, K. M., Curran, E., Khan, R., Grund, J., Calvo, A. R., Claes, J. J., Overton, S. C., Hellard, S., Vasiliadis, L., Liu, M., & Blaxall, B. C. (2024, August 1). A Systematic Approach to Optimize the Implementation of Precision Oncology in Clinical Practice: A Meeting Proceeding. *Journal of Immunotherapy and Precision Oncology*, *7*(3), 210–216. DOI: 10.36401/JIPO-23-41 PMID: 39219991

Jeelani, O. F., Njie, M., & Korzhuk, V. (2024). Methods and algorithms of ensuring data privacy in ai-based healthcare systems and technologies. *Digital Health and Informatics Innovations for Sustainable Health Care Systems*, 1260-1261.

Jena, R., Samal, H. B., Sharma, J., Suresh, P., Mishra, A. P., & Nigam, M. (2025). Biotechnology in drug discovery and development for cancer. In *Biotechnology and cancer therapeutics* (pp. 447–478). Springer. DOI: 10.1007/978-981-96-4959-4_14

Jenna, S., Boucher, B., Dudragne, L., & Diallo, A. B. (2022). Augmented intelligence to define drug targets associated with triple negative breast cancer (TNBC) metabolic reprogramming. *Cancer Research*, *82*(12, Supplement), 6375–6375. DOI: 10.1158/1538-7445.AM2022-6375

Jensen, J., & Deng, J. (2023). Digital twins for radiation oncology. In *Companion Proceedings of the ACM Web Conference 2023* (pp. 989-993). DOI: 10.1145/3543873.3587688

Jerby-Arnon, L., Shah, P., Cuoco, M. S., Rodman, C., Su, M.-J., Melms, J. C., Leeson, R., Kanodia, A., Mei, S., Lin, J.-R., Wang, S., Rabasha, B., Liu, D., Zhang, G., Margolais, C., Ashenberg, O., Ott, P. A., Buchbinder, E. I., Haq, R., & Regev, A. (2018, November). A Cancer Cell Program Promotes T Cell Exclusion and Resistance to Checkpoint Blockade. *Cell*, *175*(4), 984–997.e24. DOI: 10.1016/j.cell.2018.09.006 PMID: 30388455

Jiang, S., Lan, Y., Wang, W., & Guo, Z. (2024). Pyramid: A heterogeneous data integration algorithm based on hierarchical graph. In *ICASSP 2024-2024 IEEE International Conference on Acoustics, Speech and Signal Processing (ICASSP)* (pp. 6220-6224). IEEE. DOI: 10.1109/ICASSP48485.2024.10447879

Jiang, Y. (2007). A multiscale, cell-based framework for modelling cancer development. In *International Conference on Computational Science* (pp. 770-777). Berlin, Heidelberg: Springer Berlin Heidelberg.

Jiang, C., Meng, L., Yang, B., & Luo, X. (2020). Application of CRISPR/Cas9 gene editing technique in the study of cancer treatment. *Clinical Genetics*, *97*(1), 73–88. DOI: 10.1111/cge.13589 PMID: 31231788

Jiang, H. (2024). Latest research progress of liquid biopsy in tumor-a narrative review. *Cancer Management and Research*, *16*, 1031–1042. DOI: 10.2147/CMAR.S479338 PMID: 39165347

Jiang, S., Yang, C., Huang, Z., Guo, Z., Lu, F., Nian, X., Chen, Z., Luo, P., Jiang, J., Gao, X., Li, M., & Liu, F. (2026, April). Multicentre Evaluation of an AI-Assisted Urine Test for Clinically Significant Prostate Cancer in Men Undergoing Initial Biopsy. *Journal of Extracellular Vesicles*, *15*(4), e70233. DOI: 10.1002/jev2.70233 PMID: 41877634

Jiang, X., Hu, Z., Wang, S., & Zhang, Y. (2023). Deep Learning for Medical Image-Based Cancer Diagnosis. *Cancers (Basel)*, *15*(14), 3608. Advance online publication. DOI: 10.3390/cancers15143608 PMID: 37509272

Jiang, Y., Wang, C., & Zhou, S. (2023). Artificial intelligence-based risk stratification, accurate diagnosis and treatment prediction in gynecologic oncology. [). Academic Press.]. *Seminars in Cancer Biology*, *96*, 82–99. DOI: 10.1016/j.semcancer.2023.09.005

Jiang, Y., Wang, C., Zu, C., Rong, X., Yu, Q., & Jiang, J. (2024, October 2). Synergistic Potential of Nanomedicine in Prostate Cancer Immunotherapy: Breakthroughs and Prospects. *International Journal of Nanomedicine*, *19*, 9459–9486. DOI: 10.2147/IJN.S466396 PMID: 39371481

Jillson, L. K., Yette, G. A., Laajala, T. D., Tilley, W. D., Costello, J. C., & Cramer, S. D. (2021). Androgen Receptor Signaling in Prostate Cancer Genomic Subtypes. *Cancers (Basel)*, *13*(13), 3272. Advance online publication. DOI: 10.3390/cancers13133272 PMID: 34208794

Jin, M., Fang, J., Peng, J., Wang, X., Xing, P., Jia, K., Hu, J., Wang, D., Ding, Y., Wang, X., Li, W., & Chen, Z. (2024, November 29). PD-1/PD-L1 immune checkpoint blockade in breast cancer: Research insights and sensitization strategies. *Molecular Cancer*, *23*(1), 266. DOI: 10.1186/s12943-024-02176-8 PMID: 39614285

Ji, Y., Li, R., Tang, G., Xiao, Y., Ye, R., Shi, J., Geng, C., Ran, R., Zhu, C., Wang, W., Chen, C., & Yang, Q. (2025, July 14). Integrative proteomics and metabolomics analyses reveals the regulation of autophagy and ferroptosis by RAB10 through Slc37a2/mTOR pathway in breast cancer. *Cellular and Molecular Life Sciences : CMLS*, *82*(1), 275. DOI: 10.1007/s00018-025-05788-5 PMID: 40658144

John, S. A., Chattree, A., Ramteke, P. W., Shanthy, P., Nguyen, T. A., & Rajendran, S. (2022). Nanosensors for plant health monitoring. *Nanosensors for Smart Agriculture*, *1*, 449–461. DOI: 10.1016/B978-0-12-824554-5.00012-4

Johnson, K. B., Wei, W., Weeraratne, D., Frisse, M. E., Misulis, K., Rhee, K., Zhao, J., & Snowdon, J. L. (2021, January). Precision Medicine, AI, and the Future of Personalized Health Care. *Clinical and Translational Science*, *14*(1), 86–93. DOI: 10.1111/cts.12884 PMID: 32961010

Johnston, M., Dissanayake-Perera, S., Collins, J. J., Stevens, M. M., & Dincer, C. (2025). Convergence of nanotechnology and CRISPR-based diagnostics. *Nature Nanotechnology*, *20*(10), 1365–1373. DOI: 10.1038/s41565-025-02018-8 PMID: 41039096

Jones, M. F., & Smith, C. C. (2024). Mechanisms of resistance to targeted therapies in AML. *Annual Review of Cancer Biology*, *8*(1), 81–96. DOI: 10.1146/annurev-cancerbio-062822-025055

Jones, S., Thompson, K., Porter, B., Shepherd, M., Sapkaroski, D., Grimshaw, A., & Hargrave, C. (2024, June). Automation and artificial intelligence in radiation therapy treatment planning. [Internet]. *Journal of Medical Radiation Sciences*, *71*(2), 290–298. Retrieved November 2, 2025, from https://onlinelibrary.wiley.com/doi/10.1002/jmrs.729. DOI: 10.1002/jmrs.729 PMID: 37794690

Jörgensen, A. M., Wibel, R., & Bernkop-Schnürch, A. (2023). Biodegradable Cationic and Ionizable Cationic Lipids: A Roadmap for Safer Pharmaceutical Excipients. *Small*, *19*(17), 2206968. DOI: 10.1002/smll.202206968 PMID: 36610004

Joshi, S. K., Tognon, C. E., Druker, B. J., & Rodland, K. D. (2022). Oncoproteomic profiling of AML: Moving beyond genomics. *Expert Review of Proteomics*, *19*(7-12), 283–287. DOI: 10.1080/14789450.2023.2176757 PMID: 36734985

Juang, Y. R., Ang, L., & Seow, W. J. (2025, March 4). Predictive performance of risk prediction models for lung cancer incidence in Western and Asian countries: A systematic review and meta-analysis. *Scientific Reports*, *15*(1), 4259. DOI: 10.1038/s41598-024-83875-6 PMID: 40038330

Jupin, J., & Shi, J. Y. (2015). A proposition for resilient graph-based record linkage using parallel processing on distributed networks. In *2015 Resilience Week (RWS)* (pp. 1-3). IEEE. DOI: 10.1109/RWEEK.2015.7287442

Kääriäinen, T. P., & Kortesniemi, Y. (2024). Building trustworthy twin-based systems with self-sovereign identities. *IEEE Access : Practical Innovations, Open Solutions*, *12*, 182101–182123. DOI: 10.1109/ACCESS.2024.3507924

Kabir, M. R., & Ray, S. (2024). DT-IoMT: A digital twin reference model for secure internet of medical things. In *2024 IEEE Computer Society Annual Symposium on VLSI (ISVLSI)* (pp. 433-438). IEEE. DOI: 10.1109/ISVLSI61997.2024.00084

Kalaiselvi, C., & Arunasenbagam, G. (2017). An Overview of Statistical Pattern Recognition—A Review. *Journal of Environmental Nanotechnology*, *6*(1), 75–78. DOI: 10.13074/jent.2017.03.171235

Kamińska, K., Nalejska, E., Kubiak, M., Wojtysiak, J., Żołna, Ł., Kowalewski, J., & Lewandowska, M. A. (2019, February). Prognostic and Predictive Epigenetic Biomarkers in Oncology. *Molecular Diagnosis & Therapy*, *23*(1), 83–95. DOI: 10.1007/s40291-018-0371-7 PMID: 30523565

Kaminski, M. M., Abudayyeh, O. O., Gootenberg, J. S., Zhang, F., & Collins, J. J. (2021). CRISPR-based diagnostics. *Nature Biomedical Engineering*, *5*(7), 643–656. DOI: 10.1038/s41551-021-00760-7

Kanaujia, K. A., Yadav, V. K., Yadav, S. S., Talha, M., Saraf, S. A., & Kumar, S. (2026). 4D Printing in Healthcare: Innovations, Challenges, and Future Directions. *ACS Applied Bio Materials*, *9*(2), 493–528. DOI: 10.1021/acsabm.5c01165 PMID: 41474416

Kann, B. H., Hosny, A., & Aerts, H. J. W. L. (2021, July). Artificial intelligence for clinical oncology. [Internet]. *Cancer Cell*, *39*(7), 916–927. Retrieved November 2, 2025, from https://linkinghub.elsevier.com/retrieve/pii/S1535610821002105. DOI: 10.1016/j.ccell.2021.04.002 PMID: 33930310

Kapsoritakis, N., Tsitoura, F., Stathaki, M., Bourogianni, O., Georgoulias, P., Barmparis, G. D., Bertsias, A., Tsironis, G. P., & Koukouraki, S. (2025, October 17). Clinical Impact of FDG PET/CT in Pulmonary Nodule Characterization: Current Perspectives on Dual-Time-Point Imaging and Semi-Quantitative Imaging Metrics. *Cancers (Basel)*, *17*(20), 3353. DOI: 10.3390/cancers17203353 PMID: 41154408

Karczewski, K. J., & Snyder, M. P. (2018, May). Integrative omics for health and disease. *Nature Reviews. Genetics*, *19*(5), 299–310. DOI: 10.1038/nrg.2018.4 PMID: 29479082

Karthik, A., Patthi, S., Maheswari, B. U., Devi, J. S., Pradhan, A. K., Balaji, G., Nallasivan, G., & Rajaram, A. (2025, February). Advancing idiopathic pulmonary fibrosis prognosis through integrated CNN-LSTM predictive modeling and uncertainty quantification. [Internet]. *Biomedical Signal Processing and Control*, *100*, 106811. Retrieved November 1, 2025, from https://linkinghub.elsevier.com/retrieve/pii/S1746809424008693. DOI: 10.1016/j.bspc.2024.106811

Kather, J. N., Heij, L. R., Grabsch, H. I., Loeffler, C., Echle, A., Muti, H. S., Krause, J., Niehues, J. M., Sommer, K. A. J., Bankhead, P., Kooreman, L. F. S., Schulte, J. J., Cipriani, N. A., Buelow, R. D., Boor, P., Ortiz-Brüchle, N., Hanby, A. M., Speirs, V., Kochanny, S., & Pearson, A. T. (2020). Pan-cancer image-based detection of clinically actionable genetic alterations. *Nature Cancer*, *1*(8), 800–810. DOI: 10.1038/s43018-020-0087-6 PMID: 33763651

Kather, J. N., Pearson, A. T., Halama, N., Jäger, D., Krause, J., Loosen, S. H., Marx, A., Boor, P., Tacke, F., Neumann, U. P., Grabsch, H. I., Yoshikawa, T., Brenner, H., Chang-Claude, J., Hoffmeister, M., Trautwein, C., & Luedde, T. (2019). Deep learning can predict microsatellite instability directly from histology in gastrointestinal cancer. *Nature Medicine*, *25*(7), 1054–1056. DOI: 10.1038/s41591-019-0462-y PMID: 31160815

Kato, H., Iwadate, D., Yamamoto, K., Fujiwara, H., Ijichi, H., Fujishiro, M., & Tateishi, K. (2022). Hi-C analysis in patient-derived organoids reveals the impact of three-dimensional chromatin structures on pancreatic cancer transcriptional subtypes. *Cancer Research*, *82*(22), 47–48. DOI: 10.1158/1538-7445.PANCA22-A069

Kaur, P., Singh, A., & Chana, I. (2021). Computational Techniques and Tools for Omics Data Analysis: State-of-the-Art, Challenges, and Future Directions. *Archives of Computational Methods in Engineering*, *28*(7), 4595–4631. DOI: 10.1007/s11831-021-09547-0 PMID: 33519179

Kaur, R., Gupta, S., Kulshrestha, S., Khandelwal, V., Pandey, S., Kumar, A., Sharma, G., Kumar, U., Parashar, D., & Das, K. (2024). Metabolomics-driven biomarker discovery for breast cancer prognosis and diagnosis. *Cells*, *14*(1), 5. DOI: 10.3390/cells14010005 PMID: 39791706

Kausar, A., Kausar, A., & Das, S. (2024). AI in bioinformatics and computational biology. In *Future of AI in Biomedicine and Biotechnology* (pp. 278–290). IGI Global Scientific Publishing. DOI: 10.4018/979-8-3693-3629-8.ch014

Kavran, A. J., Stuart, S. A., Hayashi, K. R., Basken, J. M., Brandhuber, B. J., & Ahn, N. G. (2022, March 22). Intermittent treatment of $BRAF^{V600E}$ melanoma cells delays resistance by adaptive resensitization to drug rechallenge. *Proceedings of the National Academy of Sciences of the United States of America, 119*(12), e2113535119. DOI: 10.1073/pnas.2113535119 PMID: 35290123

Kaźmierczak-Siedlecka, K., Kucharski, R., Stachowska, E., Pelikant-Małecka, I., Marano, L., Makarewicz, W., Kalinowska, M., Dambrauskas, Ž., & Kalinowski, L. (2026). Intra-tumoural microenvironment and bugs-based drug design: Foreseeable future in oncology and immuno-oncology. *Frontiers in Pharmacology, 17*, 1732712. Advance online publication. DOI: 10.3389/fphar.2026.1732712 PMID: 41693784

Kelkar, A. H., Hantel, A., Koranteng, E., Cutler, C. S., Hammer, M. J., & Abel, G. A. (2024). Digital health to patient-facing artificial intelligence: Ethical implications and threats to dignity for patients with cancer. *JCO Oncology Practice, 20*(3), 314–317. DOI: 10.1200/OP.23.00412

Kemkar, S., Tao, M., Ghosh, A., Stamatakos, G., Graf, N., Poorey, K., & Radhakrishnan, R. (2024). Towards verifiable cancer digital twins: Tissue level modelling protocol for precision medicine. *Frontiers in Physiology, 15*, 1473125. DOI: 10.3389/fphys.2024.1473125 PMID: 39507514

Kenmogne, V. L., Nweke, E. E., Takundwa, M. M., Fru, P. N., & Thimiri Govinda Raj, D. B. (2022). *Application of drug repurposing-based precision medicine platform for leukaemia patient treatment. Cell Biology and Translational Medicine* (Vol. 19). Perspectives in Diverse Human Diseases and Their Therapeutic Options.

Ke, X., & Cardoso, W. V. (2026). Transitional cell states at the crossroads of development, disease, and repair-regeneration. *Physiological Reviews, 106*(3), 2059–2078. DOI: 10.1152/physrev.00041.2025 PMID: 42012134

Keyl, J., Keyl, P., Montavon, G., Hosch, R., Brehmer, A., Mochmann, L., & Kleesiek, J. (2025). Decoding pan-cancer treatment outcomes using multimodal real-world data and explainable artificial intelligence. *Nature Cancer, 6*(2), 307–322. DOI: 10.1038/s43018-024-00891-1

Khalid, R., & Riasat, S. (2023). Molecular pathogenesis and treatment strategies of Chronic Myeloid Leukemia (CML). *Sudan Journal of Medical Sciences, 18*(4), 525–538. DOI: 10.18502/sjms.v18i4.14741

Khan, N., Marvel, F. A., Wang, J., & Martin, S. S. (2017, August). Digital Health Technologies to Promote Lifestyle Change and Adherence. *Current Treatment Options in Cardiovascular Medicine, 19*(8), 60. DOI: 10.1007/s11936-017-0560-4 PMID: 28647844

Khan, S. N., Danishuddin, M. W. A., Khan, M. W. A., Guarnera, L., & Akhtar, S. M. F. (2025). Khan, L. Guarnera, and S. M. F. Akhtar, "Multi-modal AI in precision medicine: Integrating genomics, imaging, and EHR data for clinical insights,". *Frontiers in Artificial Intelligence, 8*, 1743921. DOI: 10.3389/frai.2025.1743921 PMID: 41573270

Khanzode, P. C., Mohod, A. P., & Raut, P. P. (2024). A real-time patient health monitoring system. In *Predictive Data Modelling for Biomedical Data and Imaging* (pp. 189–204). River Publishers. DOI: 10.1201/9781003516859-9

Khazaei, M., Hosseini, M. S., Haghighi, A. M., & Misaghi, M. (2023). Nanosensors and their applications in early diagnosis of cancer. *Sensing and Bio-Sensing Research, 41*, 100569. DOI: 10.1016/j.sbsr.2023.100569

Khemani, D., Malave, D. S., Shinde, S., Shukla, M., Shikalgar, R., & Talwar, H. (2025). AI-driven pharmacovigilance: Enhancing adverse drug reaction detection with deep learning and NLP. *MethodsX, 15*, 103460. DOI: 10.1016/j.mex.2025.103460 PMID: 40678458

Khinvasara, T., Ness, S., & Shankar, A. (2024). Leveraging AI for Enhanced Quality Assurance in Medical Device Manufacturing. *Asian Journal of Research in Computer Science, 17*(6), 13–35. DOI: 10.9734/ajrcos/2024/v17i6454

Khosravi, A., & Roayaei, M. (2024). MOAC-RT: Personalized multi-objective radiotherapy using actor-critic deep reinforcement learning. PREPRINT (Version 1) available at Research Square [https://doi.org/10.21203/rs.3.rs-5637868/v1] DOI: 10.21203/rs.3.rs-5637868/v1

Khosravi, B., Rouzrokh, P., & Erickson, B. J. (2022). Getting more out of large databases and EHRs with natural language processing and artificial intelligence: The future is here. *The Journal of Bone and Joint Surgery. American Volume, 104*(Suppl 3), 51–55. DOI: 10.2106/JBJS.22.00567 PMID: 36260045

Khoury, J. D., Solary, E., Abla, O., Akkari, Y., Alaggio, R., Apperley, J. F., & Hochhaus, A. (2022). The 5th edition of the World Health Organization classification of haematolymphoid tumours: myeloid and histiocytic/dendritic neoplasms. *leukemia, 36*(7), 1703-1719.

Khoury, R., Raffoul, C., Khater, C., & Hanna, C. (2025, July 7). Precision Medicine in Hematologic Malignancies: Evolving Concepts and Clinical Applications. *Biomedicines, 13*(7), 1654. DOI: 10.3390/biomedicines13071654 PMID: 40722725

Kim, E., Duarte, S. E., Yu, E., Hong, I., Lee, G., & Chae, Y. K. (2024). Evaluation of Artificial Intelligence (AI)-based in silico tools for variant classification in clinically actionable NSCLC variants. medRxiv, 2024-04.

Kim, S., Jung, T., Sohn, D. K., & Chang, Y. J. (2023). Enhancing cancer care services with the multi-domain metaverse cancer care digital platform. In *International XR Conference* (pp. 254-264). Cham: Springer Nature Switzerland.

Kim, E. S., Herbst, R. S., Wistuba, I. I., Lee, J. J., Blumenschein, G. R.Jr, Tsao, A., Stewart, D. J., Hicks, M. E., Erasmus, J.Jr, Gupta, S., Alden, C. M., Liu, S., Tang, X., Khuri, F. R., Tran, H. T., Johnson, B. E., Heymach, J. V., Mao, L., Fossella, F., & Hong, W. K. (2011, June 1). The BATTLE Trial: Personalizing Therapy for Lung Cancer. *Cancer Discovery, 1*(1), 44–53. DOI: 10.1158/2159-8274.CD-10-0010 PMID: 22586319

Kim, J., Harper, A., McCormack, V., Sung, H., Houssami, N., Morgan, E., Mutebi, M., Garvey, G., Soerjomataram, I., & Fidler-Benaoudia, M. M. (2025, April). Global patterns and trends in breast cancer incidence and mortality across 185 countries. *Nature Medicine*, *31*(4), 1154–1162. DOI: 10.1038/s41591-025-03502-3 PMID: 39994475

Kimura, S., & Mullighan, C. G. (2024). Molecular basis of acute lymphoblastic leukemia. *Molecular Hematology*, 67-81.

Kim, Y. S., Hsieh, A. C., & Lam, H. M. (2022). Bladder cancer patient-derived organoids and avatars for personalized cancer discovery. *European Urology Focus*, *8*(3), 657–659. DOI: 10.1016/j.euf.2022.07.006 PMID: 35915037

Klar, R., & Angelakis, V. (2023). Standardized and interoperable digital twins. In *2023 IEEE Conference on Standards for Communications and Networking (CSCN)* (pp. 382-382). IEEE. DOI: 10.1109/CSCN60443.2023.10453189

Knaggs, K. L. M., Sun, Y., Walz, B. A., Pang, J., & Khan, O. F. (2025). The role of excipients in lipid nanoparticle metabolism: Implications for enhanced therapeutic effect. *Therapeutic Delivery*, *16*(7), 687–700. DOI: 10.1080/20415990.2025.2506977 PMID: 40438965

Knezevic, I., Liu, M. A., Peden, K., Zhou, T., & Kang, H.-N. (2021). Development of mRNA vaccines: Scientific and regulatory issues. *Vaccines*, *9*(2), 81. DOI: 10.3390/vaccines9020081 PMID: 33498787

Kobayashi, K., & Alam, S. B. (2024). Explainable, interpretable, and trustworthy AI for an intelligent digital twin: A case study on remaining useful life. *Engineering Applications of Artificial Intelligence*, *129*, 107620. DOI: 10.1016/j.engappai.2023.107620

Kobayashi, S., Boggon, T. J., Dayaram, T., Jänne, P. A., Kocher, O., Meyerson, M., Johnson, B. E., Eck, M. J., Tenen, D. G., & Halmos, B. (2005, February 24). *EGFR* Mutation and Resistance of Non–Small-Cell Lung Cancer to Gefitinib. *The New England Journal of Medicine*, *352*(8), 786–792. DOI: 10.1056/NEJMoa044238 PMID: 15728811

Kodumuru, R., Sarkar, S., Parepally, V., & Chandarana, J. (2025a). Artificial Intelligence and Internet of Things Integration in Pharmaceutical Manufacturing: A Smart Synergy. *Pharmaceutics*, *17*(3), 290. DOI: 10.3390/pharmaceutics17030290 PMID: 40142954

Kommiya Mothilal, R., Mahajan, D., Tan, C., & Sharma, A. (2021). Towards unifying feature attribution and counterfactual explanations: Different means to the same end. In *Proceedings of the 2021 AAAI/ACM Conference on AI, Ethics, and Society* (pp. 652-663). DOI: 10.1145/3461702.3462597

Komor, A. C., Kim, Y. B., Packer, M. S., Zuris, J. A., & Liu, D. R. (2016, May). Programmable editing of a target base in genomic DNA without double-stranded DNA cleavage. *Nature*, *533*(7603), 420–424. DOI: 10.1038/nature17946 PMID: 27096365

Kondaveeti, H. K., Simhadri, C. G., Mangapathi, S., & Vatsavayi, V. K. (2024). Federated learning for privacy preservation in healthcare. In *Federated Learning and Privacy-Preserving in Healthcare AI* (pp. 121–136). IGI Global Scientific Publishing. DOI: 10.4018/979-8-3693-1874-4.ch009

Kondra, S., Xu, W., & Raghavan, V. V. (2024). Unlocking the power of clinical notes: natural language processing in healthcare. *Acta Scientific MEDICAL SCIENCES (ISSN: 2582-0931), 8*(6).

Konermann, S., Brigham, M. D., Trevino, A. E., Joung, J., Abudayyeh, O. O., Barcena, C., Hsu, P. D., Habib, N., Gootenberg, J. S., Nishimasu, H., Nureki, O., & Zhang, F. (2015, January). Genome-scale transcriptional activation by an engineered CRISPR-Cas9 complex. *Nature, 517*(7536), 583–588. DOI: 10.1038/nature14136 PMID: 25494202

Kong, H. (2025). Advances in personalized cancer vaccine development: AI applications from neoantigen discovery to mRNA formulation. *Biochem, 5*(2), 5. DOI: 10.3390/biochem5020005

Koo, K. M., Mainwaring, P. N., Tomlins, S. A., & Trau, M. (2019). Merging new-age biomarkers and nanodiagnostics for precision prostate cancer management. *Nature Reviews. Urology, 16*(5), 302–317. DOI: 10.1038/s41585-019-0178-2 PMID: 30962568

Kopetz, S., Grothey, A., Yaeger, R., Van Cutsem, E., Desai, J., Yoshino, T., Wasan, H., Ciardiello, F., Loupakis, F., Hong, Y. S., Steeghs, N., Guren, T. K., Arkenau, H.-T., Garcia-Alfonso, P., Pfeiffer, P., Orlov, S., Lonardi, S., Elez, E., Kim, T.-W., & Tabernero, J. (2019). Encorafenib, binimetinib, and cetuximab in BRAF V600E-mutated colorectal cancer. *The New England Journal of Medicine, 381*(17), 1632–1643. DOI: 10.1056/NEJMoa1908075 PMID: 31566309

Korrapati, C., Chowdareddy, S., Suresh, C., & Raghavan, R. (2026). Cancer Cell Membrane-Coated Nanoparticles: Therapeutic Advancements and Hybrid Systems. *Applications of Nanomaterials for Oncology*, 131–160. DOI: 10.4018/979-8-3373-3765-4.ch005

Kostic, A. D., Gevers, D., Pedamallu, C. S., Michaud, M., Duke, F., Earl, A. M., Ojesina, A. I., Jung, J., Bass, A. J., Tabernero, J., Baselga, J., Liu, C., Shivdasani, R. A., Ogino, S., Birren, B. W., Huttenhower, C., Garrett, W. S., & Meyerson, M. (2012). Genomic analysis identifies association of *Fusobacterium* with colorectal carcinoma. *Genome Research, 22*(2), 292–298. DOI: 10.1101/gr.126573.111 PMID: 22009990

Kruse, A., Abdel-Azim, N., Kim, H. N., Ruan, Y., Phan, V., Ogana, H., & Kim, Y. M. (2020). Minimal residual disease detection in acute lymphoblastic leukemia. *International Journal of Molecular Sciences, 21*(3), 1054. DOI: 10.3390/ijms21031054 PMID: 32033444

Krzywanski, J., Sosnowski, M., Grabowska, K., Zylka, A., Lasek, L., & Kijo-Kleczkowska, A. (2024). Advanced Computational Methods for Modeling, Prediction and Optimization—A Review. *Materials (Basel), 17*(14), 3521. DOI: 10.3390/ma17143521 PMID: 39063813

Kubasch, A. S., Grieb, N., Oeser, A., Haferlach, C., & Platzbecker, U. (2022). RETRACTED ARTICLE: Use of artificial intelligence in the management of acute leukemia. *Die Onkologie, 28*(8), 731–732. DOI: 10.1007/s00761-022-01153-4

Kulkarni, P., Garin, S., Kanhere, A. U., Siegel, E., Yi, P. H., & Parekh, V. S. (2023). High-throughput AI inference for medical image classification and segmentation using intelligent streaming. *Clinical Orthopaedics and Related Research*. arXiv2305.15617v2

Kumar, A. AI Driven Precision Oncology: Predictive Biomarker Discovery and Personalized Treatment Optimization Using Genomic Data. 2025 Mar 17 [cited 2025 May 7]; Available from: https://zenodo .org/doi/10.5281/zenodo.15037946

Kumar, A. (2003). Neural network based detection of local textile defects. *Pattern Recognition, 36*(7), 1645–1659. DOI: 10.1016/S0031-3203(03)00005-0

Kumar, A. (2025). *AI Driven Precision Oncology: Predictive Biomarker Discovery and Personalized Treatment Optimization Using Genomic Data*. DOI: 10.5281/ZENODO.15037946

Kumar, A. R., Devan, A. R., Nair, B., Vinod, B. S., & Nath, L. R. (2021, December). Harnessing the immune system against cancer: Current immunotherapy approaches and therapeutic targets. *Molecular Biology Reports, 48*(12), 8075–8095. DOI: 10.1007/s11033-021-06752-9 PMID: 34671902

Kumar, A., Lunawat, A. K., Kumar, A., Sharma, T., Islam, M. M., Kahlon, M. S., Mukherjee, D., Narang, R. K., & Raikwar, S. (2024, March 6). Recent trends in nanocarrier-based drug delivery system for prostate cancer. *AAPS PharmSciTech, 25*(3), 55. DOI: 10.1208/s12249-024-02765-2 PMID: 38448649

Kumar, A., Misra, P., & Dube, A. (2013). Amplified fragment length polymorphism: An adept technique for genome mapping, genetic differentiation, and intraspecific variation in protozoan parasites. *Parasitology Research, 112*(2), 457–466. DOI: 10.1007/s00436-012-3238-6 PMID: 23254590

Kumar, D., Kaur, A., Shruti, , & Kaur, D. (2025). Artificial Intelligence in Pharmacovigilance. *Current Computer Science, 04*, e29503779340108. DOI: 10.2174/0129503779340108241226103521

Kumar, H., Raj, U., Gupta, S., & Varadwaj, P. K. (2016). In-silico identification of inhibitors against mutated BCR-ABL protein of chronic myeloid leukemia: A virtual screening and molecular dynamics simulation study. *Journal of Biomolecular Structure & Dynamics, 34*(10), 2171–2183. DOI: 10.1080/07391102.2015.1110046 PMID: 26479578

Kumari, R., Sah, D. K., Gupta, S., Cengiz, K., & Ivković, N. (2023). Advancing medical recommendations with federated learning on decentralized data: A roadmap for implementation. *IEEE Transactions on Consumer Electronics, 70*(1), 2666–2674. DOI: 10.1109/TCE.2023.3334159

Kumar, S., Rathaur, S., Maity, D., Varshney, P., & Gayen, J. R. (2026). Nanotechnology based targeted drug delivery systems for breast cancer: Challenges, recent advancement and future perspectives. *Nanomedicine (London), 21*(7), 1027–1048. DOI: 10.1080/17435889.2026.2637591 PMID: 41773625

Kumar, V., Randhawa, P., Bilodeau, R., Mercola, D., McClelland, M., Agrawal, A., Nguyen, J., Castro, P., Ittmann, M. M., & Rahmatpanah, F. (2022, October 8). Spatial profiling of the prostate cancer tumor microenvironment reveals multiple differences in gene expression and correlation with recurrence risk. *Cancers (Basel), 14*(19), 4923. DOI: 10.3390/cancers14194923 PMID: 36230846

Kuntamung, K., Jakmunee, J., & Ounnunkad, K. (2021). A label-free multiplex electrochemical biosensor for the detection of three breast cancer biomarker proteins employing dye/metal ion-loaded and antibody-conjugated polyethyleneimine-gold nanoparticles. *Journal of Materials Chemistry. B, 9*(33), 6576–6585. DOI: 10.1039/D1TB00940K

Kuppan, N., Padman, M., Mahadeva, M., Srinivasan, S., & Devarajan, R. (2024). A comprehensive review of sustainable bioremediation techniques: Eco-friendly solutions for waste and pollution management. *Waste Management Bulletin, 2*(3), 154–171. DOI: 10.1016/j.wmb.2024.07.005

Kurniawan, M. H., Handiyani, H., Nuraini, T., Hariyati, R. T. S., & Sutrisno, S. (2024). A systematic review of artificial intelligence-powered (AI-powered) chatbot intervention for managing chronic illness. *Annals of Medicine*, *56*(1), 2302980. DOI: 10.1080/07853890.2024.2302980 PMID: 38466897

Kurul, F., Turkmen, H., Cetin, A. E., & Topkaya, S. N. (2025). Nanomedicine: How nanomaterials are transforming drug delivery, bio-imaging, and diagnosis. *Next Nanotechnology*, *7*, 100129. DOI: 10.1016/j.nxnano.2024.100129

Kwong, A., Ho, C. Y. S., Shin, V. Y., Au, C. H., Chan, T. L., & Ma, E. S. K. (2022, December). How does re-classification of variants of unknown significance (VUS) impact the management of patients at risk for hereditary breast cancer? [Internet]. *BMC Medical Genomics*, *15*(1), 122. Retrieved November 2, 2025, from https://bmcmedgenomics.biomedcentral.com/articles/10.1186/s12920-022-01270-4. DOI: 10.1186/s12920-022-01270-4 PMID: 35641994

Kwon, W. A., & Joung, J. Y. (2025, April 27). Precision Targeting in Metastatic Prostate Cancer: Molecular Insights to Therapeutic Frontiers. *Biomolecules*, *15*(5), 625. DOI: 10.3390/biom15050625 PMID: 40427518

Kyriakidis, N. C., Echeverría, C. E., Bautista, J., Rivera-Orellana, S., Ramos-Medina, M. J., Salazar-Santoliva, C., Izquierdo-Condoy, J. S., Ortiz-Prado, E., Guerrero, S., & López-Cortés, A. (2025, August 28). Reprogramming cancer immunity with next-generation combination therapies. *Frontiers in Cell and Developmental Biology*, *13*, 1652047. DOI: 10.3389/fcell.2025.1652047 PMID: 40950404

La'ah, A. S., & Chiou, S. H. (2024, March 1). Cutting-Edge Therapies for Lung Cancer. [Internet]. *Cells*, *13*(5), 436. Retrieved November 2, 2025, from https://www.mdpi.com/2073-4409/13/5/436. DOI: 10.3390/cells13050436 PMID: 38474400

Lachowiez, C. A., DiNardo, C. D., & Loghavi, S. (2023). Molecularly targeted therapy in acute myeloid leukemia: Current treatment landscape and mechanisms of response and resistance. *Cancers (Basel)*, *15*(5), 1617. DOI: 10.3390/cancers15051617 PMID: 36900407

Laetsch, T. W., DuBois, S. G., Bender, J. G., Macy, M. E., & Moreno, L. (2021, March). Opportunities and Challenges in Drug Development for Pediatric Cancers. *Cancer Discovery*, *11*(3), 545–559. DOI: 10.1158/2159-8290.CD-20-0779 PMID: 33277309

Lai, J., Luo, Z., Chen, L., & Wu, Z. (2024). Advances in nanotechnology-based targeted-contrast agents for computed tomography and magnetic resonance. *Science Progress*, *107*(1), 00368504241228076. DOI: 10.1177/00368504241228076 PMID: 38332327

Lajmi, N., Alves-Vasconcelos, S., Tsiachristas, A., Haworth, A., Woods, K., Crichton, C., Noble, T., Salih, H., Várnai, K. A., Branford-White, H., Orrell, L., Osman, A., Bradley, K. M., Bonney, L., McGowan, D. R., Davies, J., Prime, M. S., & Hassan, A. B. (2024). Challenges and solutions to system-wide use of precision oncology as the standard of care paradigm. *Cambridge Prisms. Precision Medicine*, *2*, e4. DOI: 10.1017/pcm.2024.1 PMID: 38699518

Lakshmaiah, A., & Raghavan, R. (2026). Deep Learning-Enhanced Hyperspectral Imaging: Applications in Oncology Cancer Diagnosis and Tumor Margin Identification. *Biomedical Applications in Deep Learning-Enhanced Hyperspectral Imaging*, 335–368. DOI: 10.4018/979-8-3373-7183-2.ch011

Lamichhane, P., & Agrawal, A. (2023, April). Precision medicine and implications in medical education. *Annals of Medicine and Surgery (London), 85*(4), 1342–1345. DOI: 10.1097/MS9.0000000000000298 PMID: 37113952

Lammert, J., Pfarr, N., Kuligin, L., Mathes, S., Dreyer, T., Modersohn, L., Metzger, P., Ferber, D., Kather, J. N., Truhn, D., Adams, L. C., Bressem, K. K., Lange, S., Schwamborn, K., Boeker, M., Kiechle, M., Schatz, U. A., Bronger, H., & Tschochohei, M. (2025, July 9). Large language models-enabled digital twins for precision medicine in rare gynecological tumors. *NPJ Digital Medicine, 8*(1), 420. DOI: 10.1038/s41746-025-01810-z PMID: 40634659

Lange, M., Bergen, V., Klein, M., Setty, M., Reuter, B., Bakhti, M., & Theis, F. J. (2022). CellRank for directed single-cell fate mapping. *Nature Methods, 19*(2), 159–170. DOI: 10.1038/s41592-021-01346-6

Langenberg, K. P. S., Looze, E. J., & Molenaar, J. J. (2021, August 27). The Landscape of Pediatric Precision Oncology: Program Design, Actionable Alterations, and Clinical Trial Development. *Cancers (Basel), 13*(17), 4324. DOI: 10.3390/cancers13174324 PMID: 34503139

Lan, J., Chen, W., Yin, G., Haque, A. A., Xie, K., Yu, Q., & Yuan, X. (2023). LDSSNV: A linkage disequilibrium-based method for the detection of somatic single-nucleotide variants. *IEEE/ACM Transactions on Computational Biology and Bioinformatics, 20*(5), 3020–3032. DOI: 10.1109/TCBB.2023.3291134

Larkin, J., Chiarion-Sileni, V., Gonzalez, R., Grob, J. J., Rutkowski, P., Lao, C. D., Cowey, C. L., Schadendorf, D., Wagstaff, J., Dummer, R., Ferrucci, P. F., Smylie, M., Hogg, D., Hill, A., Márquez-Rodas, I., Haanen, J., Guidoboni, M., Maio, M., Schöffski, P., & Wolchok, J. D. (2019, October 17). Five-Year Survival with Combined Nivolumab and Ipilimumab in Advanced Melanoma. *The New England Journal of Medicine, 381*(16), 1535–1546. DOI: 10.1056/NEJMoa1910836 PMID: 31562797

Larocca, C. A., LeBoeuf, N. R., Silk, A. W., & Kaufman, H. L. (2020, December). An Update on the Role of Talimogene Laherparepvec (T-VEC) in the Treatment of Melanoma: Best Practices and Future Directions. *American Journal of Clinical Dermatology, 21*(6), 821–832. DOI: 10.1007/s40257-020-00554-8 PMID: 32767272

Le, D. T., Durham, J. N., Smith, K. N., Wang, H., Bartlett, B. R., Aulakh, L. K., Lu, S., Kemberling, H., Wilt, C., Luber, B. S., Wong, F., Azad, N. S., Rucki, A. A., Laheru, D., Donehower, R., Zaheer, A., Fisher, G. A., Crocenzi, T. S., Lee, J. J., & Diaz, L. A.Jr. (2017). Mismatch repair deficiency predicts response of solid tumors to PD-1 blockade. *Science, 357*(6349), 409–413. DOI: 10.1126/science.aan6733 PMID: 28596308

Le, D. T., Uram, J. N., Wang, H., Bartlett, B. R., Kemberling, H., Eyring, A. D., Skora, A. D., Luber, B. S., Azad, N. S., Laheru, D., Biedrzycki, B., Donehower, R. C., Zaheer, A., Fisher, G. A., Crocenzi, T. S., Lee, J. J., Duffy, S. M., Goldberg, R. M., de la Chapelle, A., & Diaz, L. A.Jr. (2015). PD-1 blockade in tumors with mismatch-repair deficiency. *The New England Journal of Medicine, 372*(26), 2509–2520. DOI: 10.1056/NEJMoa1500596 PMID: 26028255

Lee, H. K., Oh, Y., Hong, J., Lee, S. H., & Hur, J. K. (2021, February). Development of CRISPR technology for precise single-base genome editing: A brief review. *BMB Reports, 54*(2), 98–105. DOI: 10.5483/BMBRep.2021.54.2.217 PMID: 33298245

Lee, H., Menezes, V., Zeng, S., Kim, C., Baseman, C. M., Kim, J. H., Padmanabhan, S., Premdas, P., Djeddar, N., Bryksin, A., Pandey, N., Anastasiadis, P., Kim, A. J., MacDonald, T. J., Bettegowda, C., Woodworth, G. F., Herrmann, F. J., & Arvanitis, C. (2026). Data-Driven Feedback Identifies Focused Ultrasound Exposure Regimens for Improved Nanotheranostic Targeting of the Brain. *Advanced Science (Weinheim, Baden-Württemberg, Germany)*, *13*(13), e17834. DOI: 10.1002/advs.202517834 PMID: 41498583

Lee, H.-O., Hong, Y., Etlioglu, H. E., Cho, Y. B., Pomella, V., Van den Bosch, B., Vanhecke, J., Verbandt, S., Hong, H., Min, J.-W., Kim, N., Eum, H. H., Qian, J., Boeckx, B., Lambrechts, D., Tsantoulis, P., De Hertogh, G., Chung, W., Lee, T., & Park, W.-Y. (2020). Lineage-dependent gene expression programs influence the immune landscape of colorectal cancer. *Nature Genetics*, *52*(6), 594–603. DOI: 10.1038/s41588-020-0636-z PMID: 32451460

Lee, J. H., Long, G. V., Boyd, S., Lo, S., Menzies, A. M., Tembe, V., Guminski, A., Jakrot, V., Scolyer, R. A., Mann, G. J., Kefford, R. F., Carlino, M. S., & Rizos, H. (2017, May). Circulating tumour DNA predicts response to anti-PD1 antibodies in metastatic melanoma. *Annals of Oncology : Official Journal of the European Society for Medical Oncology*, *28*(5), 1130–1136. DOI: 10.1093/annonc/mdx026 PMID: 28327969

Lee, M. (2023). Deep learning techniques with genomic data in cancer prognosis: A comprehensive review of the 2021–2023 literature. *Biology (Basel)*, *12*(7), 893. DOI: 10.3390/biology12070893 PMID: 37508326

Lee, S. M., Lee, G., Lee, S., Song, H., Yoon, S. S., Yun, H., & Kim, H. U. (2022). A computational model of cellular metabolism with mutation data predicts metabolites associated with somatic mutations in cancers. *Cancer Research*, *82*(12, Supplement), 2733–2733. DOI: 10.1158/1538-7445.AM2022-2733

Lee, S., Park, R., & Jung, H. S. (2025). AI-Enhanced Surface-Enhanced Raman Scattering for Accurate and Sensitive Biomedical Sensing. *Advanced Intelligent Discovery*, *202500030*. Advance online publication. DOI: 10.1002/aidi.202500030

Lee, W. C., Blanchette, C., Pokras, S., Shaikh, J., & Miller, J. (2025). The evolution and future of integrated evidence planning. *Expert Review of Pharmacoeconomics & Outcomes Research*, *25*(6), 855–862. DOI: 10.1080/14737167.2025.2497876 PMID: 40266581

Li, Q., Ren, Z., Cao, K., Li, M. M., Wang, K., & Zhou, Y. (2022). CancerVar: An artificial intelligence–empowered platform for clinical interpretation of somatic mutations in cancer. *Science advances, 8*(18), eabj1624.

Liang, A., Kong, Y., Chen, Z., Qiu, Y., Wu, Y., Zhu, X., & Li, Z. (2024). Advancements and applications of single-cell multi-omics techniques in cancer research: Unveiling heterogeneity and paving the way for precision therapeutics. *Biochemistry and Biophysics Reports*, *37*, 101589. DOI: 10.1016/j.bbrep.2023.101589

Liao, C., & Wang, X. (2023). TCGAplot: An R package for integrative pan-cancer analysis and visualization of TCGA multi-omics data. *BMC Bioinformatics*, *24*(1), 483. DOI: 10.1186/s12859-023-05615-3

Liao, Y., Thompson, C., Peterson, S., Mandrola, J., & Beg, M. S. (2019, May). The Future of Wearable Technologies and Remote Monitoring in Health Care. *American Society of Clinical Oncology Educational Book*, (39), 115–121. DOI: 10.1200/EDBK_238919 PMID: 31099626

Li, B., & Nabavi, S. (2024). A multimodal graph neural network framework for cancer molecular subtype classification. *BMC Bioinformatics, 25*(1), 27. DOI: 10.1186/s12859-023-05622-4

Li, C. H., Chan, M. H., Chang, Y. C., & Hsiao, M. (2023). Gold nanoparticles as a biosensor for cancer biomarker determination. *Molecules (Basel, Switzerland), 28*(1), 364. DOI: 10.3390/molecules28010364

Li, C., Lei, S., Ding, L., Xu, Y., Wu, X., Wang, H., Zhang, Z., Gao, T., Zhang, Y., & Li, L. (2023, July 5). Global burden and trends of lung cancer incidence and mortality. *Chinese Medical Journal, 136*(13), 1583–1590. DOI: 10.1097/CM9.0000000000002529 PMID: 37027426

Li, C.-Z., & Hu, T. Y. (2022). Nanotechnology Powered CRISPR-Cas Systems for Point of Care Diagnosis and Therapeutic. *AAAS Research, 9810237*, 2022/9810237. Advance online publication. DOI: 10.34133/2022/9810237

Li, E. W., Tran, N. Y. K., McCulloch, D., Krigstein, M., Catalano, A., Othman, J., & Iland, H. (2024a). FLT3-TKD measurable residual disease detection using droplet digital PCR and clinical applications in acute myeloid leukemia. *International Journal of Molecular Sciences, 25*(11), 5771. DOI: 10.3390/ijms25115771 PMID: 38891959

Li, G., Xu, J., Tian, X., Xiao, J., Long, J., Chen, Y., Shen, W., & Zhao, S. (2026). Intrapatient variation in PD-L1 expression and tumor mutational burden and implications for outcomes to immune checkpoint inhibitor therapy in patients with non-small-cell lung cancer. *Journal for Immunotherapy of Cancer, 14*(1), e013328. Advance online publication. DOI: 10.1136/jitc-2025-013328 PMID: 41571296

Li, H., Yang, Y., Hong, W., Huang, M., Wu, M., & Zhao, X. (2020). Applications of genome editing technology in the targeted therapy of human diseases: Mechanisms, advances and prospects. *Signal Transduction and Targeted Therapy, 5*(1), 1. DOI: 10.1038/s41392-019-0089-y PMID: 32296011

Li, L., Cho, K. H., Yu, X., & Cheng, S. (2025, May). Systematic multi-omics investigation of androgen receptor driven gene expression and epigenetics changes in prostate cancer. *Computers in Biology and Medicine, 189*, 110000. DOI: 10.1016/j.compbiomed.2025.110000 PMID: 40056843

Li, L., Guan, Y., Xiong, H., Deng, T., Ji, Q., Xu, Z., Kang, Y., & Pang, J. (2020, September). Fundamentals and applications of nanoparticles for ultrasound-based imaging and therapy. *Nano Select, 1*(3), 263–284. DOI: 10.1002/nano.202000035

Limeta, A., Gatto, F., Herrgård, M. J., Ji, B., & Nielsen, J. (2023). Leveraging high-resolution omics data for predicting responses and adverse events to immune checkpoint inhibitors. *Computational and Structural Biotechnology Journal, 21*, 3912–3919. DOI: 10.1016/j.csbj.2023.07.032

Lim, S. Y., Lin, Y., Lee, J. H., Pedersen, B., Stewart, A., Scolyer, R. A., & Rizos, H. (2024). Single-cell RNA sequencing reveals melanoma cell state-dependent heterogeneity of response to MAPK inhibitors. *EBioMedicine*, •••, 107.

Li, N., Xiao, L., Li, S., Zhang, L., & Chen, S. (2026). β-Galactosidase-Activated Near-Infrared Fluorescence and 1H/19F MRI Trimodal Probe for Real-Time In Vivo Imaging of Ovarian Cancer. *Analytical Chemistry*, *98*(7), 5261–5270. DOI: 10.1021/acs.analchem.5c05864 PMID: 41672934

Lin, L., Cui, Z., Wang, S., Chen, Y., & Zong, Y. (2024). AI-driven protein engineering for DNA sequence modification. *Journal of Theory and Practice of Engineering Science*, *4*(03), 183–190. DOI: 10.53469/jtpes.2024.04(03).17

Lino, C. A., Harper, J. C., Carney, J. P., & Timlin, J. A. (2018, January). Delivering CRISPR: A review of the challenges and approaches. *Drug Delivery*, *25*(1), 1234–1257. DOI: 10.1080/10717544.2018.1474964 PMID: 29801422

Li, R. (2025). Abstract IA05: Toward multi-modal foundation AI for precision oncology. *Clinical Cancer Research*, *31*(2, Supplement), IA05–IA05. DOI: 10.1158/1557-3265.TARGETEDTHERAP-IA05

Liu, C., Zhang, Y., Liang, Y., Zhang, T., & Wang, G. (2024a). DrugReSC: Targeting disease-critical cell subpopulations with single-cell transcriptomic data for drug repurposing in cancer. *Briefings in Bioinformatics*, *25*(6), bbae490. DOI: 10.1093/bib/bbae490 PMID: 39350337

Liu, D., & Imai, N. (2025, September 24). Decision-Making Biomarkers Guiding Therapeutic Strategies in Hepatocellular Carcinoma: From Prediction to Personalized Care. [Internet]. *Cancers (Basel)*, *17*(19), 3105. Retrieved November 2, 2025, from https://www.mdpi.com/2072-6694/17/19/3105. DOI: 10.3390/cancers17193105 PMID: 41097634

Liu, F., Beck, S., Yang, L., Luo, H., & Zhang, K. (2026). Advancing AI for multi-omics and clinical data integration in basic and translational cancer research. *Nature Reviews. Cancer*, *26*(7), 1–16. DOI: 10.1038/s41568-026-00922-2 PMID: 42014628

Liu, H., Zhu, X., Wei, Y., Song, C., & Wang, Y. (2023). Recent advances in targeted gene silencing and cancer therapy by nanoparticle-based delivery systems. *Biomedicine & Pharmacotherapy = Biomédecine & Pharmacothérapie*, *157*, 114065. DOI: 10.1016/j.biopha.2022.114065 PMID: 36481408

Liu, J. Q., Zhang, C., Zhang, X., Zhao, F., Li, J., Yang, Y., Chen, S., Wang, H., Zhang, Y., & Liu, W. (2022). Intratumoral delivery of IL-12 and IL-27 mRNA using lipid nanoparticles for cancer immunotherapy. *Journal of Controlled Release : Official Journal of the Controlled Release Society*, *345*, 306–313. DOI: 10.1016/j.jconrel.2022.03.021 PMID: 35301053

Liu, J., Yang, L., Liu, D., Wu, Q., Yu, Y., Huang, X., Li, J., & Liu, S. (2025, September). The role of multi-omics in biomarker discovery, diagnosis, prognosis, and therapeutic monitoring of tissue repair and regeneration processes. *Journal of Orthopaedic Translation*, *54*, 131–151. DOI: 10.1016/j.jot.2025.07.006 PMID: 40822515

Liu, W., Puri, A., Fu, D., Chen, L., Wang, C., Kellis, M., & Yang, J. (2024b). Dissecting the tumor microenvironment in response to immune checkpoint inhibitors via single-cell and spatial transcriptomics. *Clinical & Experimental Metastasis*, *41*(4), 313–332. DOI: 10.1007/s10585-023-10246-2 PMID: 38064127

Liu, X., Peng, T., Xu, M., Lin, S., Hu, B., Chu, T., & Wu, P. (2024a). Spatial multi-omics: Deciphering technological landscape of integration of multi-omics and its applications. *Journal of Hematology & Oncology*, *17*(1), 72. DOI: 10.1186/s13045-024-01596-9 PMID: 39182134

Liu, X., Yu, L., Xiao, A., Sun, W., Wang, H., Wang, X., Feng, S., & Zhang, Y. (2025). Analytical methods in studying cell force sensing: Principles, current technologies and perspectives. *Regenerative Biomaterials, 12*, rbaf007. Advance online publication. DOI: 10.1093/rb/rbaf007

Liu, Y., Liang, J., Zhang, Y., & Guo, Q. (2024b). Drug resistance and tumor immune microenvironment: An overview of current understandings. *International Journal of Oncology, 65*(4), 96. DOI: 10.3892/ijo.2024.5684 PMID: 39219258

Liu, Y., Yan, Z., Liu, C., Yang, R., Zheng, Q., Jian, J., & Liu, X. (2025). Integrated RNA sequencing analysis and machine learning identifies a metabolism-related prognostic signature in clear cell renal cell carcinoma. *Scientific Reports, 15*(1), 1691. DOI: 10.1038/s41598-025-85618-7

Liu, Z., Zhang, S., Wu, Z., Mu, E., Wei, H., Liu, Y., Li, Z., & Chen, Y. (2023). High-performance integrated chip-level thermoelectric device for power generation and microflow detection. *Nano Energy, 114*, 108611. DOI: 10.1016/j.nanoen.2023.108611

Liv, B. (2024). *Unlocking the future of biotechnology: The transformative potential of CRISPR and emerging gene-editing technologies*. Next Generation Journal for The Young Researchers.

Li, X., Krivtsov, V., Pan, C., Nassehi, A., Gao, R. X., & Ivanov, D. (2025, February). End-to-end supply chain resilience management using deep learning, survival analysis, and explainable artificial intelligence. [Internet]. *International Journal of Production Research, 63*(3), 1174–1202. Retrieved November 1, 2025, from https://www.tandfonline.com/doi/full/10.1080/00207543.2024.2367685. DOI: 10.1080/00207543.2024.2367685

Li, X., Loscalzo, J., Mahmud, A. F., Aly, D. M., Rzhetsky, A., Zitnik, M., & Benson, M. (2025). Digital twins as global learning health and disease models for preventive and personalized medicine. *Genome Medicine, 17*(1), 11. DOI: 10.1186/s13073-025-01435-7 PMID: 39920778

Li, X., Yang, M., Yang, L., Dang, X., Li, X., & Li, G. (2025). Sequencing of high-frequency mutated genes in breast cancer (BRCA) and associated-functions analysis. *International Journal of Clinical and Experimental Pathology, 18*(2), 46–62. DOI: 10.62347/YODE5431 PMID: 40083350

Li, Y., Shi, W., & Wasserman, W. W. (2018). Genome-wide prediction of cis-regulatory regions using supervised deep learning methods. *BMC Bioinformatics, 19*(1), 1–14. DOI: 10.1186/s12859-018-2187-1

Li, Z., Du, L., Du, B., Ullah, Z., Zhang, Y., Tu, Y., Zhou, Y., & Guo, B. (2025). Inorganic and hybrid nanomaterials for NIR-II fluorescence imaging-guided therapy of Glioblastoma and perspectives. *Theranostics, 15*(12), 5616–5665. DOI: 10.7150/thno.112204 PMID: 40365286

Li, Z., Li, Z., Luo, Y., Chen, W., Fang, Y., Xiong, Y., & Zhu, J. (2024b). Application and new findings of scRNA-seq and ST-seq in prostate cancer. *Cell Regeneration (London, England), 13*(1), 23. DOI: 10.1186/s13619-024-00206-w PMID: 39470950

Lloyd, M. R., Spring, L. M., Bardia, A., & Wander, S. A. (2022, March 1). Mechanisms of Resistance to CDK4/6 Blockade in Advanced Hormone Receptor–positive, HER2-negative Breast Cancer and Emerging Therapeutic Opportunities. *Clinical Cancer Research, 28*(5), 821–830. DOI: 10.1158/1078-0432.CCR-21-2947 PMID: 34725098

Logullo, P., MacCarthy, A., Dhiman, P., Kirtley, S., Ma, J., Bullock, G., & Collins, G. S. (2023). Artificial intelligence in lung cancer diagnostic imaging: A review of the reporting and conduct of research published 2018-2019. *BJR Open*, *5*(1), 20220033. DOI: 10.1259/bjro.20220033 PMID: 37389003

Lokare, R. R., Wadmare, J., Patil, S., Wadmare, G., & Patil, D. (2024). Transparent precision: Explainable AI empowered breast cancer recommendations for personalized treatment. *IAES Int J Artif Intell*, *13*(3), 2694–2694. DOI: 10.11591/ijai.v13.i3.pp2694-2702

Lokhorst, H. M., Plesner, T., Laubach, J. P., Nahi, H., Gimsing, P., Hansson, M., Minnema, M. C., Lassen, U., Krejcik, J., Palumbo, A., van de Donk, N. W. C. J., Ahmadi, T., Khan, I., Uhlar, C. M., Wang, J., Sasser, A. K., Losic, N., Lisby, S., Basse, L., & Richardson, P. G. (2015, September 24). Targeting CD38 with Daratumumab Monotherapy in Multiple Myeloma. *The New England Journal of Medicine*, *373*(13), 1207–1219. DOI: 10.1056/NEJMoa1506348 PMID: 26308596

Lombardo, D., Kiselev, M. A., & Caccamo, M. T. (2019). Smart Nanoparticles for Drug Delivery Application: Development of Versatile Nanocarrier Platforms in Biotechnology and Nanomedicine. *Journal of Nanomaterials*, *2019*(1), 3702518. DOI: 10.1155/2019/3702518

Lopez-Perez, L., Merino, B., Rujas, M., Maccaro, A., Guillén, S., Pecchia, L., & Fico, G. (2024). Regulatory frameworks and validation strategies for advancing artificial intelligence in healthcare. In *European Medical and Biological Engineering Conference* (pp. 260-265). Cham: Springer Nature Switzerland. DOI: 10.1007/978-3-031-61628-0_28

Low, L. A., & Tagle, D. A. (2017). Tissue chips: Innovative tools for drug development and disease modeling. *Lab on a Chip*, *17*(18), 3026–3036. DOI: 10.1039/C7LC00462A PMID: 28795174

Lu, C., Shiradkar, R., & Liu, Z. (2021, October 31). Integrating pathomics with radiomics and genomics for cancer prognosis: A brief review. *Chinese Journal of Cancer Research = Chung-Kuo Yen Cheng Yen Chiu*, *33*(5), 563–573. DOI: 10.21147/j.issn.1000-9604.2021.05.03 PMID: 34815630

Luke, J. J., Flaherty, K. T., Ribas, A., & Long, G. V. (2017, August). Targeted agents and immunotherapies: Optimizing outcomes in melanoma. *Nature Reviews. Clinical Oncology*, *14*(8), 463–482. DOI: 10.1038/nrclinonc.2017.43 PMID: 28374786

Luo, B., Teng, F., Tang, G., Cen, W., Liu, X., Chen, J., & Zhang, J. (2025). StereoMM: A graph fusion model for integrating spatial transcriptomic data and pathological images. *Briefings in Bioinformatics*, *26*(3), bbaf210. DOI: 10.1093/bib/bbaf210 PMID: 40407386

Lu, W., Wang, Y., & Cao, X. (2023). Editorial: Advanced nanobiosensors for non-invasive disease diagnosis. *Frontiers in Bioengineering and Biotechnology*, *11*, 1208679. DOI: 10.3389/fbioe.2023.1208679

Lu, Y., Peng, R., Dong, L., Xia, K., Wu, R., Xu, S., & Wang, J. (2023). Multiomics dynamic learning enables personalized diagnosis and prognosis for pancancer and cancer subtypes. *Briefings in Bioinformatics*, *24*(6), bbad378. DOI: 10.1093/bib/bbad378

Lv, T., Zhang, Y., Liu, J., Kang, Q., & Liu, L. (2024). Multi-omics integration for both single-cell and spatially resolved data based on dual-path graph attention auto-encoder. *Briefings in Bioinformatics*, *25*(5), bbae450. DOI: 10.1093/bib/bbae450 PMID: 39293805

Lynch, C., Sakamuru, S., Ooka, M., Huang, R., Klumpp-Thomas, C., Shinn, P., Gerhold, D., Rossoshek, A., Michael, S., Casey, W., Santillo, M. F., Fitzpatrick, S., Thomas, R. S., Simeonov, A., & Xia, M. (2024). High-throughput screening to advance *in vitro* toxicology: Accomplishments, challenges, and future directions. *Annual Review of Pharmacology and Toxicology*, *64*(1), 191–209. DOI: 10.1146/annurev-pharmtox-112122-104310 PMID: 37506331

M, M. K., P, D. A., J, S. P., Moger, S. S., & Raghavan, R. (2026). Spatial Multi-Omics in Cancers: Integrating Molecular Landscapes for Precision Oncology. *Spatially Resolved Single-Cell and Multiple Omics in Human Diseases*, 217–248. DOI: 10.4018/979-8-3373-4849-0.ch007

M. Maqbool, F. Bekele, and G. Fekadu, "Treatment Strategies Against Triple-Negative Breast Cancer: An Updated Review," *Breast Cancer Targets Ther.*, vol. 14, no. null, pp. 15–24, Dec. 2023, .DOI: 10.2147/BCTT.S348060

Ma, Y., & Guan, J. (2022). MOCSC: a multi-omics data based framework for cancer subtype classification. In *2022 IEEE international conference on bioinformatics and biomedicine (BIBM)* (pp. 2853-2859). IEEE. DOI: 10.1109/BIBM55620.2022.9995564

Mağara, Ş. S., Dietrich, N., Ünal, A. B., & Akgün, M. (2024). Accelerating privacy-preserving medical record linkage: A three-party MPC approach. *arXiv preprint arXiv:2410.21605.*

Maharjan, R., Kim, N. A., Kim, K. H., & Jeong, S. H. (2025). Transformative roles of digital twins from drug discovery to continuous manufacturing: Pharmaceutical and biopharmaceutical perspectives. *International Journal of Pharmaceutics: X*, *10*, 100409. DOI: 10.1016/j.ijpx.2025.100409 PMID: 41089758

Mahato, T. K. (2025). Pharmaceutical Innovation and Integrative Research: Bridging Drug Discovery, Biotechnology, and Life Sciences for Next-Generation Therapies. *Biopress J. Comput. Life Sci.*, *1*(09), 14–17.

Maher, K., & Chang, A. C. (2024). Regulatory frameworks for artificial intelligence in cardiovascular medicine and surgery. In *Intelligence-Based Cardiology and Cardiac Surgery* (pp. 421–426). Academic Press. DOI: 10.1016/B978-0-323-90534-3.00026-3

Mahmood, A. A. R., Jha, A. M., & Manivannan, K. (2024). Precision medicine: personalizing the fight against cancer. *International Journal of Trends in OncoScience*, 10-18.

Majeed, H., Irfan, M., Irfan, M., & Iftikhar, T. (2026). Intelligent multi omics and industry 6.0 and 7.0 enabled technologies for deciphering the tumor immune microenvironment and advancing cancer immunotherapy. *Toxicology Research*, *15*(1), tfag002. Advance online publication. DOI: 10.1093/toxres/tfag002 PMID: 41623584

Ma, L., Gao, W., Hu, X., Zhou, D., Wang, C., Yu, J., & Tang, K. (2024). An improved cancer diagnosis algorithm for protein mass spectrometry based on PCA and a one-dimensional neural network combining ResNet and SENet. *The Analyst*, *149*(23), 5675–5683. DOI: 10.1039/D4AN00784K

Malard, F., & Mohty, M. (2020). Acute lymphoblastic leukaemia. *Lancet*, *395*(10230), 1146–1162. DOI: 10.1016/S0140-6736(19)33018-1 PMID: 32247396

Maleki, E. H., Bahrami, A. R., & Matin, M. M. (2024, January). Cancer cell cycle heterogeneity as a critical determinant of therapeutic resistance. *Genes & Diseases*, *11*(1), 189–204. DOI: 10.1016/j.gendis.2022.11.025 PMID: 37588236

Manguso, R. T., Pope, H. W., Zimmer, M. D., Brown, F. D., Yates, K. B., Miller, B. C., Collins, N. B., Bi, K., LaFleur, M. W., Juneja, V. R., Weiss, S. A., Lo, J., Fisher, D. E., Miao, D., Van Allen, E., Root, D. E., Sharpe, A. H., Doench, J. G., & Haining, W. N. (2017, July). In vivo CRISPR screening identifies Ptpn2 as a cancer immunotherapy target. *Nature*, *547*(7664), 413–418. DOI: 10.1038/nature23270 PMID: 28723893

Manias, G., Kouremenou, E., Alzúaz, A. A., Kranas, P., Melillo, F., & Kyriazis, D. (2023). An optimized pipeline for the processing of healthcare data towards the creation of holistic health records. In *2023 International Conference on Applied Mathematics & Computer Science (ICAMCS)* (pp. 50-56). IEEE. DOI: 10.1109/ICAMCS59110.2023.00015

Maqsood, Q., Sumrin, A., Waseem, R., Hussain, M., Imtiaz, M., & Hussain, N. (2023). Bioengineered microbial strains for detoxification of toxic environmental pollutants. *Environmental Research*, *227*, 115665. DOI: 10.1016/j.envres.2023.115665 PMID: 36907340

Mariam, Z., Niazi, S. K., & Magoola, M. (2024). Unlocking the future of drug development: Generative AI, digital twins, and beyond. *BioMedInformatics*, *4*(2), 1441–1456. DOI: 10.3390/biomedinformatics4020079

Marisa, L., de Reyniès, A., Duval, A., Selves, J., Gaub, M. P., Vescovo, L., Etienne-Grimaldi, M.-C., Schiappa, R., Guenot, D., Ayadi, M., Kirzin, S., Chazal, M., Fléjou, J.-F., Benchimol, D., Berger, A., Lagarde, A., Pencreach, E., Piard, F., Elias, D., & Boige, V. (2013). Gene expression classification of colon cancer into molecular subtypes: Characterization, validation, and prognostic value. *PLoS Medicine*, *10*(5), e1001453. DOI: 10.1371/journal.pmed.1001453 PMID: 23700391

Marshall, M., Ivanovich, J., Schmitt, M., Helvie, A., Langsford, L., Casterline, J., & Ferguson, M. (2023, October 30). Pediatric precision oncology: "better three hours too soon than a minute too late.". *Frontiers in Oncology*, *13*, 1279953. DOI: 10.3389/fonc.2023.1279953 PMID: 38023209

Martínez-García, A., Cangioli, G., Chronaki, C., Löbe, M., Beyan, O., Juehne, A., & Parra-Calderón, C. L. (2022). FAIRness for FHIR: towards making health datasets FAIR using HL7 FHIR. In *MEDINFO 2021: One World, One Health–Global Partnership for Digital Innovation* (pp. 22-26). IOS Press. DOI: 10.3233/SHTI220024

Mary, L. W., & Raj, S. A. A. (2024). Smart wearable devices for prediction using machine learning techniques: depth exploration. In *2024 7th International Conference on Circuit Power and Computing Technologies (ICCPCT)* (Vol. 1, pp. 1522-1526). IEEE. DOI: 10.1109/ICCPCT61902.2024.10672922

Marzec, J., Ross-Adams, H., Pirrò, S., Wang, J., Zhu, Y., Mao, X., Gadaleta, E., Ahmad, A. S., North, B. V., Kammerer-Jacquet, S. F., Stankiewicz, E., Kudahetti, S. C., Beltran, L., Ren, G., Berney, D. M., Lu, Y.-J., & Chelala, C. (2021, January 19). The transcriptomic landscape of prostate cancer development and progression: An integrative analysis. *Cancers (Basel)*, *13*(2), 345. DOI: 10.3390/cancers13020345 PMID: 33477882

Massimino, M., Martorana, F., Stella, S., Vitale, S. R., Tomarchio, C., Manzella, L., & Vigneri, P. (2023). Single-cell analysis in the omics era: Technologies and applications in cancer. *Genes*, *14*(7), 1330. DOI: 10.3390/genes14071330

Masucci, M., Karlsson, C., Blomqvist, L., & Ernberg, I. (2024, May 24). Bridging the Divide: A Review on the Implementation of Personalized Cancer Medicine. *Journal of Personalized Medicine*, *14*(6), 561. DOI: 10.3390/jpm14060561 PMID: 38929782

Masuda, K., Sota, Y., & Matsuda, H. (2024). Gene fusion detection in long-read transcriptome datasets from multiple cancer cell lines. *Frontiers in Bioscience (Landmark Edition)*, *29*(12), 413. DOI: 10.31083/j.fbl2912413

Matano, M., Date, S., Shimokawa, M., Takano, A., Fujii, M., Ohta, Y., Watanabe, T., Kanai, T., & Sato, T. (2015). Modeling colorectal cancer using CRISPR-Cas9-mediated engineering of human intestinal organoids. *Nature Medicine*, *21*(3), 256–262. DOI: 10.1038/nm.3802 PMID: 25706875

Matsuno, Y., Kusumoto-Matsuo, R., Manaka, Y., Asai, H., & Yoshioka, K. I. (2022). Echoed induction of nucleotide variants and chromosomal structural variants in cancer cells. *Scientific Reports*, *12*(1), 20964. DOI: 10.1038/s41598-022-25479-6

Matthies, D. J., Gabrecht, M., & Hellbrück, H. (2023). Cyber-physical & human systems (cphs)–a review and outlook. *Proceedings of Mensch und Computer*, *2023*, 364–369. DOI: 10.1145/3603555.3608527

Maude, S. L., Frey, N., Shaw, P. A., Aplenc, R., Barrett, D. M., Bunin, N. J., Chew, A., Gonzalez, V. E., Zheng, Z., Lacey, S. F., Mahnke, Y. D., Melenhorst, J. J., Rheingold, S. R., Shen, A., Teachey, D. T., Levine, B. L., June, C. H., Porter, D. L., & Grupp, S. A. (2014, October 16). Chimeric Antigen Receptor T Cells for Sustained Remissions in Leukemia. *The New England Journal of Medicine*, *371*(16), 1507–1517. DOI: 10.1056/NEJMoa1407222 PMID: 25317870

McCoach, C. E., & Bivona, T. G. (2019). Engineering multidimensional evolutionary forces to combat cancer. *Cancer Discovery*, *9*(5), 587–604. DOI: 10.1158/2159-8290.CD-18-1196 PMID: 30992280

McGinnis, L. F. (2019). *Formalizing ISA-95 level 3 control with smart manufacturing system models* (NIST GCR 19-022; p. NIST GCR 19-022). National Institute of Standards and Technology. DOI: 10.6028/NIST.GCR.19-022

McMullen, M., Karakasis, K., Madariaga, A., & Oza, A. M. (2020, June 17). Overcoming Platinum and PARP-Inhibitor Resistance in Ovarian Cancer. *Cancers (Basel)*, *12*(6), 1607. DOI: 10.3390/cancers12061607 PMID: 32560564

Meaburn, E., & Schulz, R. (2012). Next generation sequencing in epigenetics: Insights and challenges. *Seminars in Cell & Developmental Biology*, *23*(2), 192–199. DOI: 10.1016/j.semcdb.2011.10.010 PMID: 22027613

Mecklenbrauck, R., & Heuser, M. (2023). Resistance to targeted therapies in acute myeloid leukemia. *Clinical & Experimental Metastasis*, *40*(1), 33–44. DOI: 10.1007/s10585-022-10189-0 PMID: 36318439

Medina, M. A., Oza, G., Sharma, A., Arriaga, L. G., Hernandez Hernandez, J. M., Rotello, V. M., & Ramirez, J. T. (2020). Triple-negative breast cancer: A review of conventional and advanced therapeutic strategies. *International Journal of Environmental Research and Public Health, 17*(6), 2078. DOI: 10.3390/ijerph17062078 PMID: 32245065

Meng, L., Teng, Z., Yang, S., Wang, X., Li, J., Liu, W., Zhang, Y., Wang, X., Chen, Q., & Zhang, Z. (2023). Biomimetic nanoparticles for DC vaccination: A versatile approach to boost cancer immunotherapy. *Nanoscale, 15*(14), 6432–6455. DOI: 10.1039/D2NR07071E PMID: 36916703

Meng, L., Wu, H., Wu, J., Ding, P., He, J., Sang, M., & Liu, L. (2024, January 4). Mechanisms of immune checkpoint inhibitors: Insights into the regulation of circular RNAS involved in cancer hallmarks. *Cell Death & Disease, 15*(1), 3. DOI: 10.1038/s41419-023-06389-5 PMID: 38177102

Menon, A. V., Song, B., Chao, L., Sriram, D., Chansky, P., Bakshi, I., Ulianova, J., & Li, W. (2025, April 11). Unraveling the future of genomics: CRISPR, single-cell omics, and the applications in cancer and immunology. *Frontiers in Genome Editing, 7*, 1565387. DOI: 10.3389/fgeed.2025.1565387 PMID: 40292231

Menyhárt, O., & Győrffy, B. (2021). Multi-omics approaches in cancer research with applications in tumor subtyping, prognosis, and diagnosis. *Computational and Structural Biotechnology Journal, 19*, 949–960. DOI: 10.1016/j.csbj.2021.01.009 PMID: 33613862

Mertins, P., Mani, D. R., Ruggles, K. V., Gillette, M. A., Clauser, K. R., Wang, P., Wang, X., Qiao, J. W., Cao, S., Petralia, F., Kawaler, E., Mundt, F., Krug, K., Tu, Z., Lei, J. T., Gatza, M. L., Wilkerson, M., Perou, C. M., Yellapantula, V., & Carr, S. A. (2016, June). Proteogenomics connects somatic mutations to signalling in breast cancer. *Nature, 534*(7605), 55–62. DOI: 10.1038/nature18003 PMID: 27251275

Miązek, A., & Świątowiec-Szczepańska, J. (2020). The application of content analysis as a research method in management sciences. In Michałkiewicz, A., & Mierzejewska, W. (Eds.), *Contemporary organisation and management. Challenges and trends* (pp. 159–172). Wydawnictwo Uniwersytetu Łódzkiego., DOI: 10.18778/8220-333-2.10

Migliozzi, S., Oh, Y. T., Hasanain, M., Garofano, L., D'angelo, F., Najac, R. D., & Iavarone, A. (2023). Integrative multi-omics networks identify PKCδ and DNA-PK as master kinases of glioblastoma subtypes and guide targeted cancer therapy. *Nature Cancer, 4*(2), 181–202. DOI: 10.1038/s43018-022-00510-x

Mikhael, P. G., Wohlwend, J., Yala, A., Karstens, L., Xiang, J., Takigami, A. K., Bourgouin, P. P., Chan, P. Y., Mrah, S., Amayri, W., Juan, Y.-H., Yang, C.-T., Wan, Y.-L., Lin, G., Sequist, L. V., Fintelmann, F. J., & Barzilay, R. (2023, April 20). Sybil: A Validated Deep Learning Model to Predict Future Lung Cancer Risk From a Single Low-Dose Chest Computed Tomography. *Journal of Clinical Oncology : Official Journal of the American Society of Clinical Oncology, 41*(12), 2191–2200. DOI: 10.1200/JCO.22.01345 PMID: 36634294

Militaru, F. C., Militaru, V., Crisan, N., Bocsan, I. C., Udrea, A. A., Catana, A., Kutasi, E., & Militaru, M. S. (2023, October 1). Molecular basis and therapeutic targets in prostate cancer: A comprehensive review. *Biomolecules and Biomedicine, 23*(5), 760–771. DOI: 10.17305/bb.2023.8782 PMID: 37021836

Mina, R., Oliva, S., & Boccadoro, M. (2020). Minimal residual disease in multiple myeloma: State of the art and future perspectives. *Journal of Clinical Medicine*, *9*(7), 2142. DOI: 10.3390/jcm9072142 PMID: 32645952

Ming, C., Viassolo, V., Probst-Hensch, N., Chappuis, P. O., Dinov, I. D., & Katapodi, M. C. (2019). Machine learning techniques for personalized breast cancer risk prediction: Comparison with the BCRAT and BOADICEA models. *Breast Cancer Research : BCR*, *21*(1), 75. DOI: 10.1186/s13058-019-1158-4 PMID: 31221197

Mishra, A. D., Shrivastava, A. K., & Bhushan, M. (2025). Disease diagnosis and control by reinforcement learning techniques: A systematic literature review. *Discover Computing*, *28*(1), 129. DOI: 10.1007/s10791-025-09539-9

Mitchell, M. J., Billingsley, M. M., Haley, R. M., Wechsler, M. E., Peppas, N. A., & Langer, R. (2021, February). Engineering precision nanoparticles for drug delivery. *Nature Reviews. Drug Discovery*, *20*(2), 101–124. DOI: 10.1038/s41573-020-0090-8 PMID: 33277608

Mitra, U., & Rehman, S. U. (2025). Leveraging AI and machine learning for next-generation clinical decision support systems (CDSS). In *AI-Driven Innovation in healthcare data analytics* (pp. 83-112). IGI Global Scientific Publishing.

Mitsogiannis, I., Tzelves, L., Dellis, A., Issa, H., Papatsoris, A., & Moussa, M. (2022). Prostate cancer immunotherapy. *Expert Opinion on Biological Therapy*, *22*(5), 577–590. DOI: 10.1080/14712598.2022.2027904 PMID: 35037527

Miziak, P., Baran, M., Błaszczak, E., Przybyszewska-Podstawka, A., Kałafut, J., Smok-Kalwat, J., Dmoszyńska-Graniczka, M., Kiełbus, M., & Stepulak, A. (2023, September 23). Estrogen receptor signaling in breast cancer. *Cancers (Basel)*, *15*(19), 4689. DOI: 10.3390/cancers15194689 PMID: 37835383

Modi, N. D., Sorich, M. J., Rowland, A., Logan, J. M., McKinnon, R. A., Kichenadasse, G., Wiese, M. D., & Hopkins, A. M. (2020). A literature review of treatment-specific clinical prediction models in patients with breast cancer. *Critical Reviews in Oncology/Hematology*, *148*, 102908. DOI: 10.1016/j.critrevonc.2020.102908 PMID: 32109714

Moffitt, L. R., Karimnia, N., Wilson, A. L., Stephens, A. N., Ho, G. Y., & Bilandzic, M. (2024, December 18). Challenges in Implementing Comprehensive Precision Medicine Screening for Ovarian Cancer. *Current Oncology (Toronto, Ont.)*, *31*(12), 8023–8038. DOI: 10.3390/curroncol31120592 PMID: 39727715

Mohamed Isa, E. D., Ahmad, H., Abdul Rahman, M. B., & Gill, M. R. (2021). Progress in Mesoporous Silica Nanoparticles as Drug Delivery Agents for Cancer Treatment. *Pharmaceutics*, *13*(2), 152. DOI: 10.3390/pharmaceutics13020152 PMID: 33498885

Mohammady, M., Hadidi, M., Iman Ghetmiri, S., & Yousefi, G. (2021). Design of ultra-fine carvedilol nanococrystals: Development of a safe and stable injectable formulation. *European Journal of Pharmaceutics and Biopharmaceutics : Official Journal of Arbeitsgemeinschaft für Pharmazeutische Verfahrenstechnik e.V*, *168*, 139–151. DOI: 10.1016/j.ejpb.2021.08.015 PMID: 34481906

Mohammed, M. A., Lakhan, A., Abdulkareem, K. H., Abd Ghani, M. K., Marhoon, H. A., Kadry, S., & Zapirain, B. G. (2024). Industrial Internet of Water Things architecture for data standarization based on blockchain and digital twin technology. *Journal of Advanced Research*, *66*, 1–14. DOI: 10.1016/j.jare.2023.10.005 PMID: 37839503

Mohan, B., Modi, K., Singh, G., Paul, A., Garazade, I. M., Pombeiro, A. J. L., & Kumar, S. (2025). Understanding the electrochemical MOF sensors in detecting cancer with special emphasis on breast carcinoma biomarkers. *Topics in Current Chemistry (Cham)*, *383*(1), 1–48. DOI: 10.1007/s41061-025-00493-0

Mohr, A. E., Ortega-Santos, C. P., Whisner, C. M., Klein-Seetharaman, J., & Jasbi, P. (2024). Navigating challenges and opportunities in multi-omics integration for personalized healthcare. *Biomedicines*, *12*(7), 1496. DOI: 10.3390/biomedicines12071496

Molefi, T., Marima, R., Demetriou, D., Basera, A., & Dlamini, Z. (2023). Employing AI-Powered decision support systems in recommending the most effective therapeutic approaches for Individual Cancer patients: Maximising Therapeutic Efficacy. In *Artificial Intelligence and Precision Oncology: bridging Cancer Research and clinical decision support* (pp. 259–275). Springer Nature Switzerland. DOI: 10.1007/978-3-031-21506-3_13

Molla, G., & Bitew, M. (2024, November 30). Revolutionizing Personalized Medicine: Synergy with Multi-Omics Data Generation, Main Hurdles, and Future Perspectives. *Biomedicines*, *12*(12), 2750. DOI: 10.3390/biomedicines12122750 PMID: 39767657

Molla, G., & Bitew, M. (2025). The Future of Cancer Diagnosis and Treatment: Unlocking the Power of Biomarkers and Personalized Molecular-Targeted Therapies. *Journal of Molecular Pathology*, *6*(3), 20. Advance online publication. DOI: 10.3390/jmp6030020

Mollanoori, H., Rahmati, Y., Hassani, B., Mehr, M. H., & Teimourian, S. (2021). Promising therapeutic approaches using CRISPR/Cas9 genome editing technology in the treatment of Duchenne muscular dystrophy. *Genes & Diseases*, *8*(2), 146–156. DOI: 10.1016/j.gendis.2019.12.007 PMID: 33997161

Mondal, D., Pareek, V., & Barthwal, M. (2023, January). Personalized medicine in radiation oncology and radiation sensitivity index: Pathbreaking genomic way to define the role of radiation in cancer management. *Journal of Cancer Research and Therapeutics*, *19*(Suppl 2), S508–S512. DOI: 10.4103/jcrt.jcrt_508_23 PMID: 38384012

Moon, C. I., & Zhang, B. (2024). Optimizing neoadjuvant treatment response prediction for triple-negative breast cancer using clinical trial data and deep auxiliary learning. bioRxiv, 2024-11.

Moreira, D. C., Hashmi, S. K., Andujar, A., Berg, F., Conger, K., Fox Irwin, L., & Rodríguez-Galindo, C. (2025). The St. Jude Global Academy: A multifaceted education program to expand pediatric oncology workforce capacity. *Pediatric Blood & Cancer*, *72*(3), e31499. DOI: 10.1002/pbc.31499 PMID: 39757452

Moreno-Sánchez, R., Saavedra, E., Gallardo-Pérez, J. C., Rumjanek, F. D., & Rodríguez-Enríquez, S. (2016). Understanding the cancer cell phenotype beyond the limitations of current omics analyses. *The FEBS Journal*, *283*(1), 54–73. DOI: 10.1111/febs.13535

Morsi, , M.H., Elawfi, , B., Alsaad, , S.A., Nazar, . A., Mostafa, , H.A., & Awwad, , S.A. (2025, August). Unveiling the Disparities in the Field of Precision Medicine: A Perspective. *Health Science Reports, 8*(8), e71102. DOI: 10.1002/hsr2.71102 PMID: 40735262

Mosleh-Shirazi, S., Abbasi, M., Moaddeli, M., Vaez, A., Shafiee, M., Kasaee, S. R., Amani, A. M., & Hatam, S. (2022, August 21). reza Moaddeli M, Vaez A, Shafiee M, Kasaee SR, Amani AM, Hatam S. Nanotechnology advances in the detection and treatment of cancer: An overview. *Nanotheranostics, 6*(4), 400–423. DOI: 10.7150/ntno.74613 PMID: 36051855

Moztarzadeh, O., Jamshidi, M., Sargolzaei, S., Jamshidi, A., Baghalipour, N., Malekzadeh Moghani, M., & Hauer, L. (2023). Metaverse and healthcare: Machine learning-enabled digital twins of cancer. *Bioengineering (Basel, Switzerland), 10*(4), 455. DOI: 10.3390/bioengineering10040455 PMID: 37106642

Mullighan, C. G. (2012). The molecular genetic makeup of acute lymphoblastic leukemia. *Hematology 2010, the American Society of Hematology Education Program Book, 2012*(1), 389-396.

Munquad, S., & Das, A. B. (2023). DeepAutoGlioma: A deep learning autoencoder-based multi-omics data integration and classification tools for glioma subtyping. *BioData Mining, 16*(1), 32. DOI: 10.1186/s13040-023-00349-7

Murcia-Mejía, M., Canela-Capdevila, M., García-Pablo, R., Jiménez-Franco, A., Jiménez-Aguilar, J. M., Badía, J., & Joven, J. (2024). Combining metabolomics and machine learning to identify diagnostic and prognostic biomarkers in patients with non-small cell lung cancer pre-and post-radiation therapy. *Biomolecules, 14*(8), 898. DOI: 10.3390/biom14080898

Murphy, B., Nasir-Moin, M., von Oiste, G., Chen, V., Riina, H. A., Kondziolka, D., & Oermann, E. K. (2021). Patient level simulation and reinforcement learning to discover novel strategies for treating ovarian cancer. *arXiv preprint arXiv:2110.11872.*

Murugesan, R., & Raman, S. (2022). Recent Trends in Carbon Nanotubes Based Prostate Cancer Therapy: A Biomedical Hybrid for Diagnosis and Treatment. *Current Drug Delivery, 19*(2), 229–237. DOI: 10.2174/18755704MTE08NDki3 PMID: 33655834

Muthalakshmi, M., Jeyapal, K., & Vinoth, M. PS, D., Murugan, N. S., & Sheela, K. S. (2024). Federated learning for secure and privacy-preserving medical image analysis in decentralized healthcare systems. In *2024 5th international conference on electronics and sustainable communication systems (ICESC)* (pp. 1442-1447). IEEE.

Mu, X., Zhang, M., Wei, A., Yang, F., Yin, J., Wang, X., Li, Y., Liu, Y., Chen, H., & Zhang, H. (2021). Doxorubicin and PD-L1 siRNA co-delivery with stem cell membrane-coated polydopamine nanoparticles for the targeted chemoimmunotherapy of PCa bone metastases. *Nanoscale, 13*(19), 8998–9008. DOI: 10.1039/D0NR08024A PMID: 33973580

N. Siva, S. Gupta, A. Gupta, J. N. Shukla, B. Malik, and N. Shukla, "Genome-editing approaches and applications: a brief review on CRISPR technology and its role in cancer," *3 Biotech*, vol. 11, no. 3, p. 146, 2021, .DOI: 10.1007/s13205-021-02680-4

Naddeo, S., Verde, L., Forastiere, M., De Pietro, G., & Sannino, G. (2017). A real-time m-health monitoring system: An integrated solution combining the use of several wearable sensors and mobile devices. In *Special Session on Smart Medical Devices-From Lab to Clinical Practice* (Vol. 6, pp. 545-552). SciTePress.

Nafissi, N., Heiranizadeh, N., Shirinzadeh-Dastgiri, A., Vakili-Ojarood, M., Naseri, A., Danaei, M., Saberi, A., Aghasipour, M., Shiri, A., Yeganegi, M., & Rahmani, A. (2024). The application of artificial intelligence in breast cancer. *Eurasian Journal of Medicine and Oncology*, *8*(3), 235. DOI: 10.14744/ejmo.2024.45903

Nagar, A., Gobburu, J., & Chakravarty, A. (2025a). Artificial intelligence in pharmacovigilance: Advancing drug safety monitoring and regulatory integration. *Therapeutic Advances in Drug Safety*, *16*, 20420986251361435. DOI: 10.1177/20420986251361435 PMID: 40756610

Nahata, L., Khare, V., Madnoorkar, N., Khare, S., & Jain, R. (2026). Multidisciplinary Cooperation: Combining Viewpoints to Provide the Best Possible Patient Care. In *Individualized Treatment Strategies in the Radiotherapy Era* (pp. 136–150). CRC Press.

Naimi, A., Mohammed, R. N., Raji, A., Chupradit, S., Yumashev, A. V., Suksatan, W., & Razeghian, E. (2022). Tumor immunotherapies by immune checkpoint inhibitors (ICIs); the pros and cons. *Cell Communication and Signaling : CCS*, *20*(1), 44. DOI: 10.1186/s12964-022-00854-y PMID: 35392976

Naresh, V., & Lee, N. (2021). A review on biosensors and recent development of nanostructured materials-enabled biosensors. *Sensors (Basel)*, *21*(4), 1109. DOI: 10.3390/s21041109

Narushima, Y., Kozuka-Hata, H., Tsumoto, K., Inoue, J. I., & Oyama, M. (2016). Quantitative phosphoproteomics-based molecular network description for high-resolution kinase-substrate interactome analysis. *Bioinformatics (Oxford, England)*, *32*(14), 2083–2088. DOI: 10.1093/bioinformatics/btw164

Nasiri, H., & Abbasian, K. (2025). Surface plasmon resonance detection of anti-cancer drug flutamide by graphitic carbon nitride/chitosan nanocomposite. *Scientific Reports*, *15*(1), 1–9. DOI: 10.1038/s41598-025-86665-w

Naskar, S., Sharma, S., Kuotsu, K., Halder, S., Pal, G., Saha, S., Mondal, S., Biswas, U. K., Jana, M., & Bhattacharjee, S. (2025, May 28). The biomedical applications of artificial intelligence: An overview of decades of research. *Journal of Drug Targeting*, *33*(5), 717–748. DOI: 10.1080/1061186X.2024.2448711 PMID: 39744873

Natarajan, A., Thangarajan, R., & Kesavan, S. (2019). Repurposing drugs by in silico methods to target BCR kinase domain in chronic myeloid leukemia. *Asian Pacific Journal of Cancer Prevention : APJCP*, *20*(11), 3399–3406. DOI: 10.31557/APJCP.2019.20.11.3399 PMID: 31759365

Navarro, C. L. A., Damen, J. A., Takada, T., Nijman, S. W., Dhiman, P., Ma, J., & Hooft, L. (2021). Risk of bias in studies on prediction models developed using supervised machine learning techniques: systematic review. *bmj, 375*.

Nazar Saeed, T., Al-Hussainy, A. F., Sanghvi, G., Ballal, S., Singh, A., Sudhakar, T., Mishra, S., Kubaev, A., Taher, S. G., Alwan, M., Jawad, M., & Mushtaq, H. (2026). Tunable silk fibroin platforms for controlled drug delivery: A biocompatible approach. *Journal of Drug Delivery Science and Technology*, *115*, 107649. DOI: 10.1016/j.jddst.2025.107649

Nazaretyan, L., Rentzsch, P., & Kircher, M. (2024). varCADD: large sets of standing genetic variation enable genome-wide pathogenicity prediction. bioRxiv, 2024-09.

Nazer, L. H., Zatarah, R., Waldrip, S., Ke, J. X. C., Moukheiber, M., Khanna, A. K., & Mathur, P. (2023). Bias in artificial intelligence algorithms and recommendations for mitigation. *PLOS Digital Health*, *2*(6), e0000278. DOI: 10.1371/journal.pdig.0000278 PMID: 37347721

Neagu, A.-N., Whitham, D., Bruno, P., Morrissiey, H., Darie, C. A., & Darie, C. C. (2023). Omics-Based Investigations of Breast Cancer. *Molecules (Basel, Switzerland)*, *28*(12), 4768. Advance online publication. DOI: 10.3390/molecules28124768 PMID: 37375323

Nevedomskaya, E., & Haendler, B. (2022, June 3). From Omics to Multi-Omics Approaches for In-Depth Analysis of the Molecular Mechanisms of Prostate Cancer. *International Journal of Molecular Sciences*, *23*(11), 6281. DOI: 10.3390/ijms23116281 PMID: 35682963

Neves, A. R., Mineiro, R., Quintela, T., & Costa, D. (2026). The Potential of Chronotherapy and Nanotherapy-Based Strategies for Glioblastoma Treatment. *Pharmaceutics*, *18*(2), 235. DOI: 10.3390/pharmaceutics18020235 PMID: 41754976

Ngo, V. M., Sood, G., Donohue, F., Kearney, P., Buckley, C., & Roantree, M. (2024). Using hl7-fhir as an integration platform for chronic disease services management and planning in the irish healthcare sector. In *The Joint Conference of ISEH ICEPH & ISEG on Environment and Health.* (pp. 324) DOI: 10.20944/preprints202408.0681.v1

Nguyen, H., Do, C., Tran, B., & Trinh, H. (2025, June 18). Rare sequential EGFR and ALK mutations in non-small cell lung cancer: A case report and literature review. *Oncology Letters*, *30*(2), 1–6. DOI: 10.3892/ol.2025.15144 PMID: 40606307

Niarakis, A., Laubenbacher, R., An, G., Ilan, Y., Fisher, J., Flobak, Å., & Glazier, J. A. (2024). Immune digital twins for complex human pathologies: Applications, limitations, and challenges. *NPJ Systems Biology and Applications*, *10*(1), 141. DOI: 10.1038/s41540-024-00450-5 PMID: 39616158

Niazi, S. K. (2025). Regulatory Perspectives for AI/ML Implementation in Pharmaceutical GMP Environments. *Pharmaceuticals (Basel, Switzerland)*, *18*(6), 901. DOI: 10.3390/ph18060901 PMID: 40573297

Nicolo, C., Sips, F., Vaghi, C., Baretta, A., Carbone, V., Emili, L., & Bursi, R. (2023). Accelerating digitalization in healthcare with the insilicotrials cloud-based platform: Four use cases. *Annals of Biomedical Engineering*, *51*(1), 125–136. DOI: 10.1007/s10439-022-03052-6 PMID: 36074307

Nicora, G., Zucca, S., Limongelli, I., Bellazzi, R., & Magni, P. (2022). A machine learning approach based on ACMG/AMP guidelines for genomic variant classification and prioritization. *Scientific Reports*, *12*(1), 2517. DOI: 10.1038/s41598-022-06547-3

Nie, C., Shi, Z., Song, Z., Wang, F., & Min, J. (2026). Artificial intelligence and multiomics for longevity across species. *Element, 1*(1), n/a, n/a. DOI: 10.20517/element.2026.01

Nielsen, R. L., Monfeuga, T., Kitchen, R. R., Egerod, L., Leal, L. G., Schreyer, A. T. H., Gade, F. S., Sun, C., Helenius, M., Simonsen, L., Willert, M., Tahrani, A. A., McVey, Z., & Gupta, R. (2024, April 1). Data-driven identification of predictive risk biomarkers for subgroups of osteoarthritis using interpretable machine learning. [Internet]. *Nature Communications, 15*(1), 2817. Retrieved November 1, 2025, from https://www.nature.com/articles/s41467-024-46663-4. DOI: 10.1038/s41467-024-46663-4 PMID: 38561399

Nikolov, S., Dimitrov, A., & Vera, J. (2019). Hierarchical levels of biological systems and their integration as a principal cause for tumour occurrence. *Nonlinear Dynamics Psychology and Life Sciences, 23*(3), 315. PMID: 31173701

Nikshya, J. E., Karthikeyan, M. S., Prasad, S., Krishnan, R. S., Balamurugan, S., & Raj, J. R. F. (2024). A Machine Learning Framework for Integrating Multi-Omics Data for Early Leukemia Detection. In *2024 8th International Conference on I-SMAC (IoT in Social, Mobile, Analytics and Cloud)(I-SMAC)* (pp. 1348-1356). IEEE. DOI: 10.1109/I-SMAC61858.2024.10714596

Nwebonyi, N., & McKay, F. (2024). Exploring bias risks in artificial intelligence and targeted medicines manufacturing. *BMC Medical Ethics, 25*(1), 113. DOI: 10.1186/s12910-024-01112-1 PMID: 39415204

O'Doherty, K. C., Shabani, M., Dove, E. S., Bentzen, H. B., Borry, P., Burgess, M. M., Chalmers, D., De Vries, J., Eckstein, L., Fullerton, S. M., Juengst, E., Kato, K., Kaye, J., Knoppers, B. M., Koenig, B. A., Manson, S. M., McGrail, K. M., McGuire, A. L., Meslin, E. M., & Burke, W. (2021, January). Toward better governance of human genomic data. *Nature Genetics, 53*(1), 2–8. DOI: 10.1038/s41588-020-00742-6 PMID: 33414545

O'Leary, B., Skinner, H., Schoenfeld, J. D., Licitra, L., Le Tourneau, C., Esdar, C., Schroeder, A., Salmio, S., & Psyrri, A. (2024). Evasion of apoptosis and treatment resistance in squamous cell carcinoma of the head and neck. *Cancer Treatment Reviews, 129*, 102773. DOI: 10.1016/j.ctrv.2024.102773 PMID: 38878677

Ogut, E. (2025, September 16). Artificial Intelligence in Clinical Medicine: Challenges Across Diagnostic Imaging, Clinical Decision Support, Surgery, Pathology, and Drug Discovery. [Internet]. *Clinics and Practice, 15*(9), 169. Retrieved November 2, 2025, from https://www.mdpi.com/2039-7283/15/9/169. DOI: 10.3390/clinpract15090169 PMID: 41002784

Oh, T. R. (2024). Integrating predictive modelling and causal inference for advancing medical science. *Childhood Kidney Diseases, 28*(3), 93–98. DOI: 10.3339/ckd.24.018

Okenwa, C. D., Adeyemi, A. F., Orelaja, A., & Nasimbwa, R. (2024). Predictive Analytics In Financial Regulation: Advancing Compliance Models For Crime Prevention. *Iosr Journal Of Economics And Finance., 15*(4), 1–7.

Olatunde Omisola, J., Oluwasegun Shiyanbola, J., & Omotunde Osho, G. (2024). A Predictive Quality Assurance Model Using Lean Six Sigma: Integrating FMEA, SPC, and Root Cause Analysis for Zero-Defect Production Systems. *International Journal of Advanced Multidisciplinary Research and Studies, 4*(6), 1481–1497. DOI: 10.62225/2583049X.2024.4.6.4051

Olawade, D. B., Clement David-Olawade, A., Adereni, T., Egbon, E., Teke, J., & Boussios, S. (2025). Integrating ai into cancer immunotherapy-a narrative review of current applications and future directions. *Diseases (Basel, Switzerland), 13*(1), 24. DOI: 10.3390/diseases13010024

Olesen, O. F. (2018). European & Developing Countries Clinical Trials Partnership (EDCTP). *Human Vaccines & Immunotherapeutics, 14*(1), 14–16. DOI: 10.1080/21645515.2017.1356649 PMID: 29136388

Olumuyiwa, B. I., Han, T. A., & Shamszaman, Z. U. (2024). Enhancing cancer diagnosis with explainable & trustworthy deep learning models. *arXiv preprint arXiv:2412.17527.*

Ong, T. C., Kahn, M. G., Kwan, B. M., Yamashita, T., Brandt, E., Hosokawa, P., & Schilling, L. M. (2017). Dynamic-ETL: A hybrid approach for health data extraction, transformation and loading. *BMC Medical Informatics and Decision Making, 17*(1), 134. DOI: 10.1186/s12911-017-0532-3 PMID: 28903729

Oni, O. (2019). *Multi-Platform Genomic Data Fusion with Integrative Deep Learning* (Doctoral dissertation).

Opderbeck, D. W. (2019). Artificial intelligence in pharmaceuticals, biologics, and medical devices: Present and future regulatory models. *Fordham Law Review, 88*, 553.

Orsini, A., Diquigiovanni, C., & Bonora, E. (2023). Omics Technologies Improving Breast Cancer Research and Diagnostics. *International Journal of Molecular Sciences, 24*(16), 12690. Advance online publication. DOI: 10.3390/ijms241612690 PMID: 37628869

Orstad, S., Fløtten, Ø., Madebo, T., Gulbrandsen, P., Strand, R., Lindemark, F., Fluge, S., Tilseth, R. H., & Schaufel, M. A. (2023, September). "The challenge is the complexity" – A qualitative study about decision-making in advanced lung cancer treatment. [Internet]. *Lung Cancer : Journal of the International Association for the Study of Lung Cancer, 183*, 107312. Retrieved November 2, 2025, from https://linkinghub.elsevier.com/retrieve/pii/S0169500223008504. DOI: 10.1016/j.lungcan.2023.107312 PMID: 37481888

Ortiz, M. M. O., & Andrechek, E. R. (2023). Molecular Characterization and Landscape of Breast cancer Models from a multi-omics Perspective. *Journal of Mammary Gland Biology and Neoplasia, 28*(1), 12. DOI: 10.1007/s10911-023-09540-2 PMID: 37269418

Oualikene-Gonin, W., Jaulent, M.-C., Thierry, J.-P., Oliveira-Martins, S., Belgodère, L., Maison, P., & Ankri, J.The Scientific Advisory Board of ANSM. (2024). Artificial intelligence integration in the drug lifecycle and in regulatory science: Policy implications, challenges and opportunities. *Frontiers in Pharmacology, 15*, 1437167. DOI: 10.3389/fphar.2024.1437167 PMID: 39156111

Overman, M. J., McDermott, R., Leach, J. L., Lonardi, S., Lenz, H.-J., Morse, M. A., Desai, J., Hill, A., Axelson, M., Moss, R. A., Goldberg, M. V., Cao, Z. A., Ledeine, J.-M., Maglinte, G. A., Kopetz, S., & André, T. (2017). Nivolumab in patients with metastatic DNA mismatch repair-deficient or microsatellite instability-high colorectal cancer (CheckMate 142): An open-label, multicentre, phase 2 study. *The Lancet. Oncology*, *18*(9), 1182–1191. DOI: 10.1016/S1470-2045(17)30422-9 PMID: 28734759

Padmavathi, U., Harshitha, R. S., & Jayashre, K. (2024). Integrating cyber-physical systems for enhanced efficiency in healthcare solutions. In *Technologies for Sustainable Healthcare Development* (pp. 300–324). IGI Global.

Pai, S., Bontempi, D., Hadzic, I., Prudente, V., Sokač, M., Chaunzwa, T. L., Bernatz, S., Hosny, A., Mak, R. H., Birkbak, N. J., & Aerts, H. J. W. L. (2024, March 15). Foundation model for cancer imaging biomarkers. *Nature Machine Intelligence*, *6*(3), 354–367. DOI: 10.1038/s42256-024-00807-9 PMID: 38523679

Palacín-Aliana, I., García-Romero, N., Asensi-Puig, A., Carrión-Navarro, J., González-Rumayor, V., & Ayuso-Sacido, Á. (2021, July 28). Clinical Utility of Liquid Biopsy-Based Actionable Mutations Detected via ddPCR. *Biomedicines*, *9*(8), 906. DOI: 10.3390/biomedicines9080906 PMID: 34440110

Palaz, F., Kalkan, A. K., Can, O., Demir, A. N., Tozluyurt, A., Ozcan, A., & Ozsoz, M. (2021, May 26). CRISPR-Cas13 system as a promising and versatile tool for cancer diagnosis, therapy, and research. *ACS Synthetic Biology*, *10*(6), 1245–1267. DOI: 10.1021/acssynbio.1c00107 PMID: 34037380

Pallavi, V. N., Chowdareddy, S., Byalahunasishetter, B. I., & Raghavan, R. (2026). CRISPR-Cas9 Technology for Precision Cancer Medicine: Target Discovery, Disease Modeling, and Therapeutics. *CRISPR-Cas9 Technology for Precision Medicine*, 35–72. DOI: 10.4018/979-8-3373-4857-5.ch002

Panda, V. K., Mishra, B., Mahapatra, S., Swain, B., Malhotra, D., Saha, S., Khanra, S., Mishra, P., Majhi, S., Kumari, K., Nath, A. N., Saha, S., Jena, S., & Kundu, G. C. (2025, January 13). Molecular insights on signaling cascades in breast cancer: A comprehensive review. *Cancers (Basel)*, *17*(2), 234. DOI: 10.3390/cancers17020234 PMID: 39858015

Pande, S. (2018). Causality or Relatedness assessment in adverse drug reaction and its relevance in dermatology. *Indian Journal of Dermatology*, *63*(1), 18. DOI: 10.4103/ijd.IJD_579_17 PMID: 29527021

Pandey, H., Amod, A., Jaggi, K., Garg, R., Jain, A., & Tantia, V. (2024). Digital twin ecosystem for oncology clinical operations. *arXiv preprint arXiv:2409.17650*.

Pandey, A., Dey, S., & Gayen, D. (2026). Advances in CRISPR-Cas technology, mechanisms, applications, and ethical challenges. *Discover Plants*, *3*(1), 243. DOI: 10.1007/s44372-026-00705-x

Pandey, G. A., Kashikar, P. M., Nathani, K., Mangrulkar, S., Sawarkar, S. P., & Omri, A. (2026). Biomimetic Scaffolds and Extracellular Matrix-Based Strategies for Myofiber Regeneration in Volumetric Muscle Loss. *Drug Design, Development and Therapy*, *20*(0), 544862. DOI: 10.2147/DDDT.S544862 PMID: 41924553

Pandey, K., Lee, E., Park, N., Hur, J., Cho, Y. B., Katuwal, N. B., Kim, S. K., Lee, S. A., Kim, I., An, H. J., Hwang, S., & Moon, Y. W. (2021, January 25). Deregulated immune pathway associated with palbociclib resistance in preclinical breast cancer models: Integrative genomics and transcriptomics. *Genes*, *12*(2), 159. DOI: 10.3390/genes12020159 PMID: 33504001

Pandey, S., & Yadav, P. (2025, January). Liquid biopsy in cancer management: Integrating diagnostics and clinical applications. *Practical Laboratory Medicine*, *43*, e00446. DOI: 10.1016/j.plabm.2024.e00446 PMID: 39839814

Pandit, S., Sarkar, S. K., Barik, S., & Sahu, S. (2024). AI in Skill Development, Critical Thinking, Digital Literacy, and AI-Driven Job Preparation. In Mobo, F. D. (Ed.), (pp. 23–76). Advances in Educational Technologies and Instructional Design. IGI Global., DOI: 10.4018/979-8-3693-8191-5.ch002

Pang, W., Chen, M., & Qin, Y. (2024, May 9). Prediction of anticancer drug sensitivity using an interpretable model guided by deep learning. *BMC Bioinformatics*, *25*(1), 182. DOI: 10.1186/s12859-024-05669-x PMID: 38724920

Pan, J., Li, B., Wang, Y., Han, Y., Liu, G., & Sun, S.-K. (2026). Physical stimuli-responsive CRISPR-Cas9 systems for spatiotemporally precise control of genome engineering. *Theranostics*, *16*(6), 2984–3010. DOI: 10.7150/thno.122238 PMID: 41510162

Pan, L., Liu, G., Lin, F., Zhong, S., Xia, H., Sun, X., & Liang, H. (2017). Machine learning applications for prediction of relapse in childhood acute lymphoblastic leukemia. *Scientific Reports*, *7*(1), 7402. DOI: 10.1038/s41598-017-07408-0 PMID: 28784991

Pan, R., Li, Y., Chen, Y., Li, G., Zhang, Y., Wang, X., & Li, J. (2022). Ultrasensitive CRISPR/Cas12a-driven SERS biosensor for on-site nucleic acid detection and its application to milk authenticity testing. *Journal of Agricultural and Food Chemistry*, *70*(14), 4484–4491. DOI: 10.1021/acs.jafc.1c08262

Pant, S., Choudhari, J. K., Pathak, R. K., Yadav, A. K., & Singh, D. B. (2026). Biological big data analysis in the post-genomic era: Applications and limitations. *Network Modeling and Analysis in Health Informatics and Bioinformatics*, *15*(1), 62. DOI: 10.1007/s13721-026-00749-9

Pan, Y., Sheng, W., Shi, L., Jing, D., Jiang, W., Chen, J. C., Wang, H., & Qiu, J. (2023, June 1). Whole slide imaging-based deep learning to predict the treatment response of patients with non-small cell lung cancer. *Quantitative Imaging in Medicine and Surgery*, *13*(6), 3547–3555. DOI: 10.21037/qims-22-1098 PMID: 37284119

Papachristou, K., Katsakiori, P. F., Papadimitroulas, P., Strigari, L., & Kagadis, G. C. (2024). Digital twins' advancements and applications in healthcare, towards precision medicine. *Journal of Personalized Medicine*, *14*(11), 1101. DOI: 10.3390/jpm14111101 PMID: 39590593

Papatsirou, M., Artemaki, P. I., Karousi, P., Scorilas, A., & Kontos, C. K. (2021). Circular RNAs: Emerging regulators of the major signaling pathways involved in cancer progression. *Cancers (Basel)*, *13*(11), 2744. DOI: 10.3390/cancers13112744

Parhi, S., Afrose, N., Rajendran, K., & Narayanasamy, D. (2025, March 4). Central composite design (CCD) based formulation, optimization, *in-vitro* and *ex-vivo* characterization of 5-fluorouracil-loaded emulgel for enhanced dermal penetration and psoriasis management. [Internet]. *Drug Development and Industrial Pharmacy, 51*(3), 244–261. Retrieved November 2, 2025, from https://www.tandfonline.com/doi/full/10.1080/03639045.2025.2464782. DOI: 10.1080/03639045.2025.2464782 PMID: 39932381

Park, J., Brattoli, B., Shi, J., Senior, M., Qaiser, T., Rawson, J., Hwang, W., Shin, S., Lee, T., Ock, C.-Y., Pereira, S., Bannister, H., Riboni-Verri, E., Ali, S., & Moore, L. (2025, April 21). Abstract 2463: AI-based EGFR-mutation prediction from haematoxylin and eosin (H&E) images in non-small cell lung cancer (NSCLC): A global multi-cohort validation study. *Cancer Research, 85*(8, Supplement_1), 2463–2463. DOI: 10.1158/1538-7445.AM2025-2463

Park, J., Park, E., Lee, D., Kang, S., Lee, T., Lee, S., & Lee, U. (2023). (2023). Human digital twins for pervasive healthcare: A scoping review. *Interactive Journal of Medical Research.* Advance online publication. DOI: 10.2196/46012

Park, J., Shim, J. K., Yoon, S. J., Kim, S. H., Chang, J. H., & Kang, S. G. (2019). Transcriptome profiling-based identification of prognostic subtypes and multi-omics signatures of glioblastoma. *Scientific Reports, 9*(1), 10555. DOI: 10.1038/s41598-019-47066-y

Park, M., Kang, B. H., & Jeong, K. H. (2018). Paper-based biochip assays and recent developments: A review. *Biochip Journal, 12*(1), 1–10. DOI: 10.1007/s13206-017-2101-3

Park, W. Y., Jeon, K., Schmidt, T. S., Kondylakis, H., Alkasab, T., Dewey, B. E., & Nagy, P. (2024). Development of medical imaging data standardization for imaging-based observational research: OMOP common data model extension. *Journal of Imaging Informatics in Medicine, 37*(2), 899–908. DOI: 10.1007/s10278-024-00982-6 PMID: 38315345

Parra-Medina, R., Guerron-Gomez, G., Mendivelso-González, D., Gil-Gómez, J. H., Alzate, J. P., Gomez-Suarez, M., Polo, J. F., Sprockel, J. J., & Mosquera-Zamudio, A. (2025, May 30). Deep learning in histopathology images for prediction of oncogenic driver molecular alterations in lung cancer: A systematic review and meta-analysis. *Translational Lung Cancer Research, 14*(5), 1756–1769. DOI: 10.21037/tlcr-2024-1196 PMID: 40535093

Parvin, N., Joo, S. W., Jung, J. H., & Mandal, T. K. (2025). Multimodal AI in Biomedicine: Pioneering the Future of Biomaterials, Diagnostics, and Personalized Healthcare. *Nanomaterials (Basel, Switzerland), 15*(12), 895. Advance online publication. DOI: 10.3390/nano15120895 PMID: 40559258

Pasas-Farmer, S., & Jain, R. (2025). From discovery to delivery: Governance of AI in the pharmaceutical industry. *Green Analytical Chemistry, 13*, 100268. DOI: 10.1016/j.greeac.2025.100268

Pasupuleti, V., Thuraka, B., Kodete, C. S., & Malisetty, S. (2024). Enhancing Supply Chain Agility and Sustainability through Machine Learning: Optimization Techniques for Logistics and Inventory Management. *Logistics (Basel), 8*(3), 73. DOI: 10.3390/logistics8030073

Patel, S. J., Sanjana, N. E., Kishton, R. J., Eidizadeh, A., Vodnala, S. K., Cam, M., Gartner, J. J., Jia, L., Steinberg, S. M., Yamamoto, T. N., Merchant, A. S., Mehta, G. U., Chichura, A., Shalem, O., Tran, E., Eil, R., Sukumar, M., Guijarro, E. P., Day, C.-P., & Restifo, N. P. (2017, August). Identification of essential genes for cancer immunotherapy. *Nature*, *548*(7669), 537–542. DOI: 10.1038/nature23477 PMID: 28783722

Patel, S., Goldsack, J. C., Cordovano, G., Downing, A., Fields, K. K., Geoghegan, C., & Robinson, E. J. (2023). Advancing digital health innovation in oncology: Priorities for high-value digital transformation in cancer care. *Journal of Medical Internet Research*, *25*, e43404. DOI: 10.2196/43404 PMID: 36598811

Patra, S., Shand, H., & Ghorai, S. (2026). Nanobiosensor-Based Strategies for Early Cancer Detection and Therapeutic Monitoring. *Biomedical Materials & Devices*, •••. Advance online publication. DOI: 10.1007/s44174-026-00688-z

Pavlova, N. N., & Thompson, C. B. (2016, January). The Emerging Hallmarks of Cancer Metabolism. *Cell Metabolism*, *23*(1), 27–47. DOI: 10.1016/j.cmet.2015.12.006 PMID: 26771115

Pecoraro, M., Messina, E., Novelli, S., Laschena, L., Blasilli, G., Tronci, E., & Panebianco, V. (2026, June 19). Digital twin technologies in prostate cancer as a frontier for precision medicine. *European Radiology Experimental*, *10*(1), 93. DOI: 10.1186/s41747-026-00753-8 PMID: 42319650

Peer, D., Karp, J. M., Hong, S., Farokhzad, O. C., Margalit, R., & Langer, R. (2007). Nanocarriers as an emerging platform for cancer therapy. *Nature Nanotechnology*, *2*(12), 751–760. DOI: 10.1038/nnano.2007.387 PMID: 18654426

Pellegrini, C., Maturo, M., Di Nardo, L., Ciciarelli, V., Gutiérrez García-Rodrigo, C., & Fargnoli, M. (2017, November). Understanding the Molecular Genetics of Basal Cell Carcinoma. *International Journal of Molecular Sciences*, *18*(11), 2485. DOI: 10.3390/ijms18112485 PMID: 29165358

Penault-Llorca, F., & Socinski, M. A. (2025, March 10). Emerging molecular testing paradigms in non-small cell lung cancer management—Current perspectives and recommendations. *The Oncologist*, *30*(3), oyae357. DOI: 10.1093/oncolo/oyae357 PMID: 40126879

Pérez-Callejo, D., González-Rincón, J., Sánchez, A., Provencio, M., & Sánchez-Beato, M. (2015, September). Action and resistance of monoclonal CD20 antibodies therapy in B-cell Non-Hodgkin Lymphomas. *Cancer Treatment Reviews*, *41*(8), 680–689. DOI: 10.1016/j.ctrv.2015.05.007 PMID: 26045227

Pérez-Romero, K., Rodríguez, R. M., Amedei, A., Barceló-Coblijn, G., & Lopez, D. H. (2020). Immune landscape in tumor microenvironment: Implications for biomarker development and immunotherapy. *International Journal of Molecular Sciences*, *21*(15), 5521. DOI: 10.3390/ijms21155521 PMID: 32752264

Peroz, M., Mananet, H., Roussot, N., Kaderbhai, C. G., Derangère, V., Truntzer, C., & Ghiringhelli, F. (2024). Clinical interest in exome-based analysis of somatic mutational signatures for non-small cell lung cancer. *Cancers (Basel)*, *16*(17), 3115. DOI: 10.3390/cancers16173115

Peters, A. A., Decasper, A., Munz, J., Klaus, J., Loebelenz, L. I., Hoffner, M. K. M., Hourscht, C., Heverhagen, J. T., Christe, A., & Ebner, L. (2021, May). Performance of an AI based CAD system in solid lung nodule detection on chest phantom radiographs compared to radiology residents and fellow radiologists. *Journal of Thoracic Disease*, *13*(5), 2728–2737. DOI: 10.21037/jtd-20-3522 PMID: 34164165

Petursson, B., & Petsalaki, E. (2022). Data-driven extraction of human kinase-substrate relationships from omics datasets.

Pfeifer, G. P., You, Y.-H., & Besaratinia, A. (2005, April). Mutations induced by ultraviolet light. *Mutation Research. Fundamental and Molecular Mechanisms of Mutagenesis*, *571*(1–2), 19–31. DOI: 10.1016/j.mrfmmm.2004.06.057 PMID: 15748635

Phiri, V. J., Battas, I., Semmar, A., Medromi, H., & Moutaouakkil, F. (2025). Towards enterprise-wide pharma 4.0 adoption. *Scientific African*, *28*, e02771. DOI: 10.1016/j.sciaf.2025.e02771

Piergentili, R., Del Rio, A., Signore, F., Umani Ronchi, F., Marinelli, E., & Zaami, S. (2021). CRISPR-Cas and Its Wide-Ranging Applications: From Human Genome Editing to Environmental Implications, Technical Limitations, Hazards and Bioethical Issues. *Cells*, *10*(5), 969. Advance online publication. DOI: 10.3390/cells10050969 PMID: 33919194

Pierobon, M., Wulfkuhle, J., Liotta, L. A., & Petricoin, E. F.III. (2012). Development and clinical implementation of reverse phase protein microarrays for protein network activation mapping: personalized cancer therapy. In *Systems Biology in Cancer Research and Drug Discovery* (pp. 309–323). Springer Netherlands. DOI: 10.1007/978-94-007-4819-4_13

Pietrantoni, L., San Román-Niaves, M., & De Angelis, M. (2024). Human Trust and Digital Twins in a Human Factors and Ergonomic Framework. In *2024 IEEE 29th International Conference on Emerging Technologies and Factory Automation (ETFA)* (pp. 1-9). IEEE. DOI: 10.1109/ETFA61755.2024.10711109

Pillai, P. M., & Carroll, W. L. (2022). Acute lymphoblastic leukemia. In *Lanzkowsky's Manual of Pediatric Hematology and Oncology* (pp. 413–438). Academic Press. DOI: 10.1016/B978-0-12-821671-2.00004-0

Pindi, V. (2019). A Ai-assisted clinical decision support systems: Enhancing diagnostic accuracy and treatment recommendations. *International Journal of Innovations in Engineering Research and Technology*, *6*(10), 1–10. DOI: 10.26662/ijiert.v6i10.pp1-10

Piroumian, V. (2021). Digital twins: Universal interoperability for the digital age. *Computer*, *54*(1), 61–69. DOI: 10.1109/MC.2020.3032148

Poirion, O. B., Chaudhary, K., Huang, S., & Garmire, L. X. (2020). Multi-omics-based pan-cancer prognosis prediction using an ensemble of deep-learning and machine-learning models. *medRxiv*, 19010082.

Pomerantz, M. M., Qiu, X., Zhu, Y., Takeda, D. Y., Pan, W., Baca, S. C., Gusev, A., Korthauer, K. D., Severson, T. M., Ha, G., Viswanathan, S. R., Seo, J.-H., Nguyen, H. M., Zhang, B., Pasaniuc, B., Giambartolomei, C., Alaiwi, S. A., Bell, C. A., O'Connor, E. P., & Freedman, M. L. (2020, August). Prostate cancer reactivates developmental epigenomic programs during metastatic progression. *Nature Genetics*, *52*(8), 790–799. DOI: 10.1038/s41588-020-0664-8 PMID: 32690948

Ponti, G., De Angelis, C., Ponti, R., Pongetti, L., Losi, L., Sticchi, A., Tomasi, A., & Ozben, T. (2023). Hereditary breast and ovarian cancer: From genes to molecular targeted therapies. *Critical Reviews in Clinical Laboratory Sciences*, *60*(8), 640–650. DOI: 10.1080/10408363.2023.2234488 PMID: 37455374

Pont, M., Marqués, M., Sorolla, M. A., Parisi, E., Urdanibia, I., Morales, S., Salud, A., & Sorolla, A. (2023). Applications of CRISPR technology to breast cancer and triple negative breast cancer research. *Cancers (Basel), 15*(17), 4364. DOI: 10.3390/cancers15174364 PMID: 37686639

Popel, A. S. (2020). Immunoactivating the tumor microenvironment enhances immunotherapy as predicted by integrative computational model. *Proceedings of the National Academy of Sciences of the United States of America, 117*(9), 4447–4449. DOI: 10.1073/pnas.2001050117 PMID: 32102915

Prabhakar, P. K., Paliya, N., Malik, V., & Sharma, K. (2024). Liquid biopsies in oncology: the future of non-invasive cancer detection. *International Research Journal of Medicine and Surgery, 1*(3),15-19.

Prabhudesai, M., Wu, M., Zadeh, A., Fragkiadaki, K., & Pathak, D. (2025). Diffusion beats autoregressive in data-constrained settings. *arXiv*. /arXiv.2507.15857DOI: 10.52202/085713-0354

Prakash, G., & Harish, S. (2024). Health monitoring system by using IoT. In *2024 International Conference on Power, Energy, Control and Transmission Systems (ICPECTS)* (pp. 1-9). IEEE.

Praveen, J., Kumar Cm, K., & Channappa, A. H. (2023). Transforming Pharmacovigilance Using Gen AI: Innovations in Aggregate Reporting, Signal Detection, and Safety Surveillance. *Journal of Multidisciplinary Research*, •••, 9–16. DOI: 10.37022/tjmdr.v3i3.484

Presciuttini, A., Cantini, A., & Portioli-Staudacher, A. (2024). From explanations to actions: leveraging shap, lime, and counterfactual analysis for operational excellence in maintenance decisions. In *2024 4th International Conference on Electrical, Computer, Communications and Mechatronics Engineering (ICECCME)* (pp. 1-6). IEEE. DOI: 10.1109/ICECCME62383.2024.10796106

Procureur, A., Simonaggio, A., Bibault, J. E., Oudard, S., & Vano, Y. A. (2021, February 8). Enhance the Immune Checkpoint Inhibitors Efficacy with Radiotherapy Induced Immunogenic Cell Death: A Comprehensive Review and Latest Developments. *Cancers (Basel), 13*(4), 678. DOI: 10.3390/cancers13040678 PMID: 33567530

Pulumati, A., Pulumati, A., Dwarakanath, B. S., Verma, A., & Papineni, R. V. L. (2023). Technological advancements in cancer diagnostics: Improvements and limitations. *Cancer Reports, 6*(2), e1764. DOI: 10.1002/cnr2.1764

Qi, L., Li, Z., Liu, J., & Chen, X. (2024). Omics-enhanced nanomedicine for cancer therapy. *Advanced Materials, 36*(50), 2409102. DOI: 10.1002/adma.202409102 PMID: 39473316

Qoronfleh, M. W., & Al-Dewik, N. (2025, September 12). Cancer Biomarkers: Reflection on Recent Progress, Emerging Innovations, and the Clinical Horizon. *Cancers (Basel), 17*(18), 2981. DOI: 10.3390/cancers17182981 PMID: 41008828

Qureshi, M. A., Khan, M. Y., Imran, A., Maqsood, Q., Hussain, N., & Ali, S. W. (2024). Revolutionizing Breast Cancer Care: Cutting-Edge Breakthroughs and Future Frontiers in Precision Medicine. In Rezaei, N. (Ed.), *Breast Cancer Treatment: An Interdisciplinary Approach* (pp. 115–141). Springer Nature Switzerland., DOI: 10.1007/16833_2024_254

Qu, Y., Gao, L., Xiang, Y., Shen, S., & Yu, S. (2022). FedTwin: Blockchain-enabled adaptive asynchronous federated learning for digital twin networks. *IEEE Network*, *36*(6), 183–190. DOI: 10.1109/MNET.105.2100620

Rabaan, A. A., Bakhrebah, M. A., AlSaihati, H., Alhumaid, S., Alsubki, R. A., Turkistani, S. A., Al-Abdulhadi, S., Aldawood, Y., Alsaleh, A. A., Alhashem, Y. N., Almatouq, J. A., Alqatari, A. A., Alahmed, H. E., Sharbini, D. A., Alahmadi, A. F., Alsalman, F., Alsayyah, A., & Mutair, A. A. (2022, November 14). Artificial intelligence for clinical diagnosis and treatment of prostate cancer. *Cancers (Basel)*, *14*(22), 5595. DOI: 10.3390/cancers14225595 PMID: 36428686

Rabiee, N. (2025). Programmable Biointerfaces and Adaptive Functionality in Next-Generation Green Nanomaterials. *ACS Synthetic Biology*, *14*(10), 3862–3879. DOI: 10.1021/acssynbio.5c00620

Rabilloud, N., Allaume, P., Acosta, O., De Crevoisier, R., Bourgade, R., Loussouarn, D., Rioux-Leclercq, N., Khene, Z. E., Mathieu, R., Bensalah, K., Pecot, T., & Kammerer-Jacquet, S.-F. (2023, August 14). Deep learning methodologies applied to digital pathology in prostate cancer: A systematic review. *Diagnostics (Basel)*, *13*(16), 2676. DOI: 10.3390/diagnostics13162676 PMID: 37627935

Rafii, S., Mukherji, D., Komaranchath, A. S., Khalil, C., Iqbal, F., Abdelwahab, S. I., Abyad, A., Abuhelwa, A. Y., Gandikota, L., & Al-Shamsi, H. O. (2025, September 3). Advancing CAR T-Cell Therapy in Solid Tumors: Current Landscape and Future Directions. *Cancers (Basel)*, *17*(17), 2898. DOI: 10.3390/cancers17172898 PMID: 40940995

Rafiq, Z., Bashir, T., Lu, W., Puebla-Osorio, N., & Integrating, A. I. (2025, October 10). Machine Learning, and Animal Models for Precision Oncology: Bridging Preclinical and Clinical Gaps. *ACS Pharmacology & Translational Science*, *8*(10), 3677–3683. DOI: 10.1021/acsptsci.5c00543 PMID: 41098564

Raheja, I. K., Kumar, P., Rajashekarappa, K. K., & Mahadevan, G. D. (2026). Nanobiosensors for Early Detection of Cancer: A Recent Update. *Biomedical Materials & Devices*, *4*(1), 78–96. DOI: 10.1007/s44174-024-00263-4

Rahman, S., Arun, A. S., Kim, I. Y., Oh, W. K., Kim, J. W., & Kim, W. J. (2026). Navigating the winding road toward precision prostate cancer care. *Gene*, *984*, 149966. DOI: 10.1016/j.gene.2025.149966 PMID: 41419150

Rajesh, M. V., & Elumalai, K. (2025). The transformative power of artificial intelligence in pharmaceutical manufacturing: Enhancing efficiency, product quality, and safety. *Journal of Holistic Integrative Pharmacy*, *6*(2), 125–135. DOI: 10.1016/j.jhip.2025.03.007

Rajput, S., Khandia, R., Gaur, T., Anant, A., Kamal, M. A., Gurjar, P., Al-Hussain, S. A., & Zaki, M. E. A. (2026). Advancements in nanobiosensors for early cancer detection, challenges, treatment, and future prospects: A comprehensive review. *Discover Nano*, *21*(1), 83. DOI: 10.1186/s11671-026-04478-w PMID: 41874885

Rakholia, R., Suárez-Cetrulo, A. L., Singh, M., & Carbajo, R. S. (2025). Integrating AI and IoT for Predictive Maintenance in Industry 4.0 Manufacturing Environments: A Practical Approach. *Information (Basel)*, *16*(9), 737. DOI: 10.3390/info16090737

Ramachandran, V., Pradhan, A., Kumar, A., Sarvepalli, B., Rao, S., Oswal, K., Kommu, R., Sharma, M., Pathak, S., Kunnambath, R., Kuriakose, M. A., Rengaswamy, S., Alajlani, M., & Arvanitis, T. N. (2022, September 1). A Distributed Cancer Care Model with a Technology-Driven Hub-and-Spoke and further Spoke Hierarchy: Findings from a Pilot Implementation Programme in Kerala, India. *Asian Pacific Journal of Cancer Prevention : APJCP*, *23*(9), 3133–3139. DOI: 10.31557/APJCP.2022.23.9.3133 PMID: 36172676

Ramakrishnan, R., Washington, A., Suveena, S., Rani, J. R., & Oommen, O. V. (2025). From DNA to Big Data: NGS technologies and their applications. *Artificial Intelligence (AI) in Cell and Genetic Engineering*, 459-482.

Ramakrishnan, A., Wangensteen, G., Kim, S., Nestler, E. J., & Shen, L. (2024). DeepRegFinder: Deep learning-based regulatory elements finder. *Bioinformatics Advances*, *4*(1), vbae007. DOI: 10.1093/bioadv/vbae007

Rambow, F., Rogiers, A., Marin-Bejar, O., Aibar, S., Femel, J., Dewaele, M., Karras, P., Brown, D., Chang, Y. H., Debiec-Rychter, M., Adriaens, C., Radaelli, E., Wolter, P., Bechter, O., Dummer, R., Levesque, M., Piris, A., Frederick, D. T., Boland, G., & Marine, J.-C. (2018, August). Toward Minimal Residual Disease-Directed Therapy in Melanoma. *Cell*, *174*(4), 843–855.e19. DOI: 10.1016/j.cell.2018.06.025 PMID: 30017245

Ramesh, R. N., Palled, S. B., Shivaraju, A., & Raghavan, R. (2026). Reproductive Toxicities and Transgenerational Cancer Risk From Climate-Related Occupational Hazards. *One Health Perspectives on Climate Change and Emerging Occupational Health Risks*, 191–230. DOI: 10.4018/979-8-3373-8470-2.ch007

Ramesh, M., Janani, R., Deepa, C., & Rajeshkumar, L. (2023). Nanotechnology-enabled biosensors: A review of fundamentals, design principles, materials, and applications. *Biosensors (Basel)*, *13*(1), 40. DOI: 10.3390/bios13010040

Ram, M., Afrash, M. R., Moulaei, K., Parvin, M., Esmaeeli, E., Karbasi, Z., & Sabahi, A. (2024). Application of artificial intelligence in chronic myeloid leukemia (CML) disease prediction and management: A scoping review. *BMC Cancer*, *24*(1), 1026. DOI: 10.1186/s12885-024-12764-y PMID: 39164653

Ramos, R., Moura, C. S., Costa, M., Lamas, N. J., Correia, R., Garcez, D., Pereira, J. M., Lindahl, T., Sousa, C., & Vale, N. (2025, February 21). Lung Cancer Therapy: The Role of Personalized Medicine. *Cancers (Basel)*, *17*(5), 725. DOI: 10.3390/cancers17050725 PMID: 40075573

Ran, D., Li, J., Zhao, M., Du, L., Zhang, Y., & Zhu, J. “Artificial intelligence integrates multi-omics data for precision stratification and drug resistance prediction in breast cancer,” *Front. Oncol.*, vol. Volume 15-2025, 2025, DOI: 10.3389/fonc.2025.1612474

Ranga, P. (2026). CRISPR and genome editing: precision tools for personalized therapies. In Ahmad, S. R. (Ed.), *The Gut Microbiome in Precision Medicine* (pp. 863–873). Academic Press., DOI: 10.1016/B978-0-443-40483-2.00047-9

Rankin, A. W., Duncan, B. B., Allen, C., Silbert, S. K., & Shah, N. N. (2025, March). Evolving strategies for addressing CAR T-cell toxicities. [Internet]. *Cancer Metastasis Reviews*, *44*(1), 17. Retrieved November 2, 2025, from https://link.springer.com/10.1007/s10555-024-10227-1. DOI: 10.1007/s10555-024-10227-1 PMID: 39674824

Rao, S., Mehta, S., Kulkarni, S., Dalvi, H., Katre, N., & Narvekar, M. (2022). *A study of LIME and SHAP model explainers for autonomous disease predictions. In 2022 ieee bombay section signature conference (ibssc)*. IEEE.

Rashid, Z., Folarin, A. A., Ranjan, Y., Conde, P., Sankesara, H., Zhang, Y., & Dobson, R. J. (2023). disease insight through digital biomarkers developed by remotely collected wearables and smartphone data. *arXiv preprint arXiv:2308.02043*.

Raut, R., Lokare, S., Gavkare, S., & Narute, M. (2024). Leukemia subtype classification via gene expression using machine learning algorithms. In *2024 IEEE International Conference on Information Technology, Electronics and Intelligent Communication Systems (ICITEICS)* (pp. 1-7). IEEE. DOI: 10.1109/ICITEICS61368.2024.10625370

Rawla, P. (2019, April). Epidemiology of Prostate Cancer. *World Journal of Oncology*, *10*(2), 63–89. DOI: 10.14740/wjon1191 PMID: 31068988

Raza, A. (2023). *Secure and privacy-preserving federated learning with explainable artificial intelligence for smart healthcare system*. University of Kent.

Reardon, A. J. F., Farmahin, R., Williams, A., Meier, M. J., Addicks, G. C., Yauk, C. L., Matteo, G., Atlas, E., Harrill, J., Everett, L. J., Shah, I., Judson, R., Ramaiahgari, S., Ferguson, S. S., & Barton-Maclaren, T. S. (2023). From vision toward best practices: Evaluating *in vitro* transcriptomic points of departure for application in risk assessment using a uniform workflow. *Frontiers in Toxicology*, *5*, 1194895. DOI: 10.3389/ftox.2023.1194895 PMID: 37288009

Rebersek, A. P. M. (2020). Consensus molecular subtypes (CMS) in metastatic colorectal cancer-personalized medicine decision. *Radiology and Oncology*, *54*(3), 272–277. DOI: 10.2478/raon-2020-0031

Reinert, T., Henriksen, T. V., Christensen, E., Sharma, S., Salari, R., Sethi, H., Knudsen, M., Nordentoft, I., Wu, H.-T., Tin, A. S., Heilskov Rasmussen, M., Vang, S., Shchegrova, S., Frydendahl Boll Johansen, A., Srinivasan, R., Assaf, Z., Balcioglu, M., Olson, A., Dashner, S., & Lindbjerg Andersen, C. (2019). Analysis of plasma cell-free DNA by ultradeep sequencing in patients with stages I to III colorectal cancer. *JAMA Oncology*, *5*(8), 1124–1131. DOI: 10.1001/jamaoncol.2019.0528 PMID: 31070691

Ren, S., Li, J., Dorado, J., Sierra, A., Gonzalez-Diaz, H., Duardo, A., & Shen, B. (2023, December 18). From molecular mechanisms of prostate cancer to translational applications: Based on multi-omics fusion analysis and intelligent medicine. *Health Information Science and Systems*, *12*(1), 6. DOI: 10.1007/s13755-023-00264-5 PMID: 38125666

Repici, A., Badalamenti, M., Maselli, R., Correale, L., Radaelli, F., Rondonotti, E., Ferrara, E., Spadaccini, M., Alkandari, A., Fugazza, A., Anderloni, A., Galtieri, P. A., Pellegatta, G., Carrara, S., Di Leo, M., Craviotto, V., Lamonaca, L., Lorenzetti, R., Andrealli, A., & Wallace, M. B. (2020). Efficacy of real-time computer-aided detection of colorectal neoplasia in a randomized trial. *Gastroenterology, 159*(2), 512–520.e7. DOI: 10.1053/j.gastro.2020.04.062 PMID: 32371116

Rigatti, M., Chapman, B., Chai, P. R., Smelson, D., Babu, K., & Carreiro, S. (2023). Digital biomarker applications across the spectrum of opioid use disorder. *Cogent Mental Health, 2*(1), 2240375. DOI: 10.1080/28324765.2023.2240375 PMID: 37546179

Rituraj, , Pal, R. S., Wahlang, J., Pal, Y., Chaitanya, M. V. N. L., & Saxena, S. (2025, June 9). Rituraj, Pal RS, Wahlang J, Pal Y, Chaitanya M, Saxena S. Precision oncology: Transforming cancer care through personalized medicine. *Medical Oncology, 42*(7), 246. DOI: 10.1007/s12032-025-02817-y PMID: 40488843

Rives, T. A., Collard, J., Li, N., Yan, D., Dietrich, C. S., Miller, R. W., Ueland, F. R., Pickarski, J., & Kolesar, J. M. (2024, September). Clinical Utility of Molecular Tumor Board Review for Identification of Possible Germline Pathogenic Variants on Tumor Next-Generation Sequencing Reports. *JCO Precision Oncology*, (8), e2400301. DOI: 10.1200/PO.24.00301 PMID: 39259913

Rodriguez, J. R., Fernandez, S., Swartz, N., Alonge, A., Bhullar, F., Betros, T., & Jacobs, R. J. (2024). A chronological overview of using deep learning for leukemia detection: A scoping review. *Cureus, 16*(5). Advance online publication. DOI: 10.7759/cureus.61379 PMID: 38846204

Rodriguez, N. G., Mercado, N. C., Panjiyar, K., & Kunwor, R. (2025, October 8). Artificial intelligence versus radiologists in predicting lung cancer treatment response: A systematic review and meta-analysis. *Frontiers in Oncology, 15*, 1634694. DOI: 10.3389/fonc.2025.1634694 PMID: 41132734

Roelofsen, L. M., Kaptein, P., & Thommen, D. S. (2022). Multimodal predictors for precision immunotherapy. *Immuno-oncology Technology, 14*, 100071. DOI: 10.1016/j.iotech.2022.100071 PMID: 35755892

Roesler, D., Czypionka, T., Eisenberg, S., Kraus, M., Reiss, M., & Zech, C. (2024). Impact of in-silico technologies in the development of high-risk medical devices on healthcare. *European Journal of Public Health, 34*(Supplement_3), ckae144–1210. DOI: 10.1093/eurpub/ckae144.1210

Rohart, F., Gautier, B., Singh, A., & Lê Cao, K.-A. (2017). mixOmics: An R package for 'omics feature selection and multiple data integration. *PLoS Computational Biology, 13*(11), e1005752. DOI: 10.1371/journal.pcbi.1005752 PMID: 29099853

Rohilla, J., Mishra, R., Varshney, S., Mazumder, R., Mazumder, A., Chauhan, A., & Tyagi, P. K. (2024). An insight into signalling pathways in cancer: Hedgehog, PI3K, and notch pathways and therapeutic perspectives. *Letters in Drug Design & Discovery, 21*(18), 4148–4160. DOI: 10.2174/0115701808353919241126110520

Roisman, L. C., Kian, W., Anoze, A., Fuchs, V., Spector, M., Steiner, R., Kassel, L., Rechnitzer, G., Fried, I., Peled, N., & Bogot, N. R. (2023, November 21). Radiological artificial intelligence - predicting personalized immunotherapy outcomes in lung cancer. *NPJ Precision Oncology, 7*(1), 125. DOI: 10.1038/s41698-023-00473-x PMID: 37990050

Ronemus, M., Bradford, D., Laster, Z., & Li, S. (2025, February 27). Exploring genome-transcriptome correlations in cancer. *Biochemical Society Transactions*, *53*(01), 113–120. DOI: 10.1042/BST20240108 PMID: 39910794

Rossi, C., Cicalini, I., Cufaro, M. C., Consalvo, A., Upadhyaya, P., Sala, G., Antonucci, I., Del Boccio, P., Stuppia, L., & De Laurenzi, V. (2022, April 14). Breast cancer in the era of integrating "Omics" approaches. *Oncogenesis*, *11*(1), 17. DOI: 10.1038/s41389-022-00393-8 PMID: 35422484

Rossini, R., Oshaghi, M., Nekrasov, M., Bellanger, A., Domaschenz, R., Dijkwel, Y., & Paulsen, J. (2024). Loss of multi-level 3D genome organization during breast cancer progression. bioRxiv, 2023-11.

Rout, T., Mohapatra, A., Kar, M., & Muduly, D. K. (2024). Essential cancer protein identification using graph-based random walk with restart. *Computer Methods in Biomechanics and Biomedical Engineering*, •••, 1–14.

Rozony, F. Z., Aktar, M. N. A., Ashrafuzzaman, M., & Islam, A. (2024). A systematic review of big data integration challenges and solutions for heterogeneous data sources. *Academic Journal on Business Administration. Innovation & Sustainability*, *4*(04), 1–18.

Runnels, J., Bloom, J. R., Hsieh, K., Chen, Y., Wang, Z., Li, X., Zhang, M., Liu, W., Yang, C., & Wang, X. (2023). Combining Radiotherapy and Immunotherapy in Head and Neck Cancer. *Biomedicines*, *11*(8), 2097. DOI: 10.3390/biomedicines11082097 PMID: 37626594

Russo, V., Lallo, E., Munnia, A., Spedicato, M., Messerini, L., D'Aurizio, R., & Peluso, M. (2022). Artificial intelligence predictive models of response to cytotoxic chemotherapy alone or combined to targeted therapy for metastatic colorectal cancer patients: A systematic review and meta-analysis. *Cancers (Basel)*, *14*(16), 4012. DOI: 10.3390/cancers14164012

Saadeh, A., Al-Bitar, F., Al Ta'ani, O., Abdulelah, A. A., Usmani, A. A., Jibrin, D., & Scott, K. (2024). The burden of leukemia in the united states from 1990 to 2021. *Blood*, *144*(Supplement 1), 2421. DOI: 10.1182/blood-2024-208967

Saadh, M. J., Khidr, W. A., Alfarttoosi, K. H., Bishoyi, A. K., Ganesan, S., Shankhyan, A., Gayathri, S., Rizaev, J., Taher, W. M., Alwan, M., Jawad, M. J., & Al-Nuaimi, A. M. A. (2025, February 23). Metal nanoparticles as a promising therapeutic approach for prostate cancer diagnosis and therapy: A comprehensive review. *Medical Oncology*, *42*(4), 83. DOI: 10.1007/s12032-025-02633-4 PMID: 39987535

Saariluoma, P., Cañas, J., & Karvonen, A. (2020). Human digital twins and cognitive mimetic. In *International conference on human interaction and emerging technologies* (pp. 97-102). Cham: Springer International Publishing.

Saber Sichani, A., Ranjbar, M., Baneshi, M., Torabi Zadeh, F., & Fallahi, J. (2023). A Review on Advanced CRISPR-Based Genome-Editing Tools: Base Editing and Prime Editing. *Molecular Biotechnology*, *65*(6), 849–860. DOI: 10.1007/s12033-022-00639-1 PMID: 36547823

Sabha, S. U. (2025). Artificial Intelligence and Omics in Health and Diseases. In Macha, M. A., Assad, A., & Bhat, M. R. (Eds.), *Artificial Intelligence in Human Health and Diseases* (pp. 75–92). Springer Nature Singapore., DOI: 10.1007/978-981-96-8176-1_5

Sabit, H., Arneth, B., Pawlik, T. M., Abdel-Ghany, S., Ghazy, A., Abdelazeem, R. M., Alqosaibi, A., Al-Dhuayan, I. S., Almulhim, J., Alrabiah, N. A., & Hashash, A. (2025). Leveraging single-cell multiomics to decode tumor microenvironment diversity and therapeutic resistance. *Pharmaceuticals (Basel, Switzerland), 18*(1), 75. DOI: 10.3390/ph18010075 PMID: 39861138

Sadée, C., Testa, S., Barba, T., Hartmann, K., Schuessler, M., Thieme, A., Church, G. M., Okoye, I., Hernandez-Boussard, T., Hood, L., Shmulevich, I., Kuhl, E., & Gevaert, O. (2025, July). Medical digital twins: Enabling precision medicine and medical artificial intelligence. *The Lancet. Digital Health, 7*(7), 100864. DOI: 10.1016/j.landig.2025.02.004 PMID: 40518342

Sadeghi, M., Sadeghi, S., Naghib, S. M., & Garshasbi, H. R. (2023). A comprehensive review on electrochemical nano biosensors for precise detection of blood-based oncomarkers in breast cancer. *Biosensors (Basel), 13*(4), 481. DOI: 10.3390/bios13040481

Sadek, H. E., Ebadah, I. M., & Mahmoud, Y. A. (2023). Importance of biotechnology in controlling insect pests. *Journal of Modern Agriculture and Biotechnology, 2*(1), 2023005. DOI: 10.53964/jmab.2023005

Sadik, H., Pritchard, D., Keeling, D. M., Policht, F., Riccelli, P., Stone, G., Finkel, K., Schreier, J., & Munksted, S. (2022, October). Impact of Clinical Practice Gaps on the Implementation of Personalized Medicine in Advanced Non–Small-Cell Lung Cancer. *JCO Precision Oncology*, (6), e2200246. DOI: 10.1200/PO.22.00246 PMID: 36315914

Sadiya Inamdar, T. K. (2024). *Harnessing AI And Machine Learning in Pharmaceutical Quality Assurance*. DOI: 10.5281/ZENODO.14186540

Saghir, S. A. M. (2023). A new insight updates in diagnosis and management of acute lymphoblastic leukemia, cytogenetics, immunophenotyping, and proteomic profile. *Electronic Journal of General Medicine, 20*(5), em519. DOI: 10.29333/ejgm/13386

Sahu, A., Blätke, M. A., Szymański, J. J., & Töpfer, N. (2021). Advances in flux balance analysis by integrating machine learning and mechanism-based models. *Computational and Structural Biotechnology Journal, 19*, 4626–4640. DOI: 10.1016/j.csbj.2021.08.004

Saini, J. P. S., Thakur, A., & Yadav, D. (2025). AI-driven innovations in pharmaceuticals: Optimizing drug discovery and industry operations. *RSC Pharmaceutics, 2*(3), 437–454. DOI: 10.1039/D4PM00323C

Sakolish, C., Weber, E. J., Kelly, E. J., Himmelfarb, J., Mouneimne, R., Grimm, F. A., House, J. S., Wade, T., Han, A., Chiu, W. A., & Rusyn, I. (2018). Technology transfer of the microphysiological systems: A case study of the human proximal tubule tissue chip. *Scientific Reports, 8*(1), 14882. DOI: 10.1038/s41598-018-33099-2 PMID: 30291268

Salciccia, S., Capriotti, A. L., Lagana, A., Fais, S., Logozzi, M., De Berardinis, E., Busetto, G. M., Di Pierro, G. B., Ricciuti, G. P., Del Giudice, F., Sciarra, A., Carroll, P. R., Cooperberg, M. R., Sciarra, B., & Maggi, M. (2021, April 22). Biomarkers in prostate cancer diagnosis: From current knowledge to the role of metabolomics and exosomes. *International Journal of Molecular Sciences, 22*(9), 4367. DOI: 10.3390/ijms22094367 PMID: 33922033

Sampene, A. K., & Nyirenda, F. (2024). Evaluating the effect of artificial intelligence on pharmaceutical product and drug discovery in China. *Future Journal of Pharmaceutical Sciences*, *10*(1), 58. DOI: 10.1186/s43094-024-00632-2

Sandner, P., Gross, J., & Richter, R. (2020). Convergence of Blockchain, IoT, and AI. *Frontiers in Blockchain*, *3*, 522600. DOI: 10.3389/fbloc.2020.522600

Sandoval, S., Gonçalves, G., Barrio, J. P., Kharlamova, M. V., & Tobías-Rossell, G. (2026). A Comprehensive Review on Filled Carbon Nanotubes: Synthesis, Properties and Applications. *Chemical Reviews*, *126*(4), 2283–2390. DOI: 10.1021/acs.chemrev.5c00219 PMID: 41636552

Sanson, K. R., Hanna, R. E., Hegde, M., Donovan, K. F., Strand, C., Sullender, M. E., Vaimberg, E. W., Goodale, A., Root, D. E., Piccioni, F., & Doench, J. G. (2018, December). Optimized libraries for CRISPR-Cas9 genetic screens with multiple modalities. *Nature Communications*, *9*(1), 5416. DOI: 10.1038/s41467-018-07901-8 PMID: 30575746

Saranya, G., & Sivaraman, K. (2024). Applying reinforcement learning to optimize treatment plans for chronic disease management. In *2024 International Conference on Advances in Computing, Communication and Applied Informatics (ACCAI)* (pp. 1-6). IEEE. DOI: 10.1109/ACCAI61061.2024.10601732

Sarhangi, N., Hajjari, S., Heydari, S. F., Ganjizadeh, M., Rouhollah, F., & Hasanzad, M. (2022). Breast cancer in the era of precision medicine. *Molecular Biology Reports*, *49*(10), 10023–10037. DOI: 10.1007/s11033-022-07571-2 PMID: 35733061

Sarkar, B., & Paul, R. K. (2025). AI-Driven Manufacturing Processes. In B. Sarkar & R. K. Paul, *AI for Advanced Manufacturing and Industrial Applications* (pp. 19–59). Springer Nature Switzerland. DOI: 10.1007/978-3-031-86091-1_2

Sartori, F., Codicè, F., Caranzano, I., Rollo, C., Birolo, G., Fariselli, P., & Pancotti, C. (2025, May 28). A Comprehensive Review of Deep Learning Applications with Multi-Omics Data in Cancer Research. *Genes*, *16*(6), 648. DOI: 10.3390/genes16060648 PMID: 40565540

Sarvepalli, S., & Vadarevu, S. (2025). Role of artificial intelligence in cancer drug discovery and development. *Cancer Letters*, *627*, 217821. DOI: 10.1016/j.canlet.2025.217821 PMID: 40414522

Satam, H., Joshi, K., Mangrolia, U., Waghoo, S., Zaidi, G., Rawool, S., Thakare, R. P., Banday, S., Mishra, A. K., Das, G., & Malonia, S. K. (2023, July 13). Next-Generation Sequencing Technology: Current Trends and Advancements. *Biology (Basel)*, *12*(7), 997. DOI: 10.3390/biology12070997 PMID: 37508427

Scarpini, S., Morigi, F., Betti, L., Dondi, A., Biagi, C., & Lanari, M. (2021). Development of a vaccine against human cytomegalovirus: Advances, barriers, and implications for the clinical practice. *Vaccines*, *9*(6), 551. DOI: 10.3390/vaccines9060551 PMID: 34070277

Schiavone, M. L., Scarpitta, R., Ravera, F., Bleve, S., Reduzzi, C., Di Cocco, F., Dameri, M., Zoppoli, G., Naccarato, A. G., Nuzzo, P. V., Cristofanilli, M., Fanelli, G. N., & Scatena, C. (2025, September). Liquid biopsy in breast cancer: Redefining precision medicine. *The Journal of Liquid Biopsy*, *9*, 100312. DOI: 10.1016/j.jlb.2025.100312 PMID: 40740670

Schleidgen, S., Dederer, H. G., Sgodda, S., Cravcisin, S., Lüneburg, L., Cantz, T., & Heinemann, T. (2020, September 11). Human germline editing in the era of CRISPR-Cas: Risk and uncertainty, inter-generational responsibility, therapeutic legitimacy. *BMC Medical Ethics*, *21*(1), 87. DOI: 10.1186/s12910-020-00487-1 PMID: 32912206

Schlicker, A., Beran, G., Chresta, C. M., McWalter, G., Pritchard, A., Weston, S., Runswick, S., Davenport, S., Heathcote, K., Alferez Castro, D., Orphanides, G., French, T., & Wessels, L. F. A. (2012). Subtypes of primary colorectal tumors correlate with response to targeted treatment in colorectal cell lines. *BMC Medical Genomics*, *5*(1), 66. DOI: 10.1186/1755-8794-5-66 PMID: 23272949

Schreiber, A. R., Andress, M., & Diamond, J. R. (2021, December 2). Tackling metastatic triple-negative breast cancer with sacituzumab govitecan. *Expert Review of Anticancer Therapy*, *21*(12), 1303–1311. DOI: 10.1080/14737140.2021.1993065 PMID: 34651524

Schreuder, A., Scholten, E. T., Van Ginneken, B., & Jacobs, C. (2021, May). Artificial intelligence for detection and characterization of pulmonary nodules in lung cancer CT screening: Ready for practice? *Translational Lung Cancer Research*, *10*(5), 2378–2388. DOI: 10.21037/tlcr-2020-lcs-06 PMID: 34164285

Schubach, M., Maass, T., Nazaretyan, L., Röner, S., & Kircher, M. (2024). CADD v1. 7: Using protein language models, regulatory CNNs and other nucleotide-level scores to improve genome-wide variant predictions. *Nucleic Acids Research*, *52*(D1), D1143–D1154. DOI: 10.1093/nar/gkad989

Schulz, W. L., Durant, T. J., & Krumholz, H. M. (2019). Validation and regulation of clinical artificial intelligence.

Schürch, M., Boos, L., Heinzelmann-Schwarz, V., Gut, G., Krauthammer, M., Wicki, A., & the Tumor Profiler Consortium. (2024). Towards AI-based precision oncology: a machine learning framework for personalized counterfactual treatment suggestions based on multi-omics data. *arXiv preprint arXiv:2402.12190.*

Sebastian, A. M., & Peter, D. (2022). Artificial intelligence in cancer research: Trends, challenges and future directions. *Life (Basel, Switzerland)*, *12*(12), 1991. DOI: 10.3390/life12121991 PMID: 36556356

Sebastiani, F., Yanez Arteta, M., Lerche, M., Porcar, L., Lang, C., Bragg, R., Elmore, C. S., Krishnamurthy, R., Cárdenas, M., Nylander, T., & Dabkowska, A. P. (2021). Apolipoprotein E Binding Drives Structural and Compositional Rearrangement of mRNA-Containing Lipid Nanoparticles. *ACS Nano*, *15*(4), 6709–6722. DOI: 10.1021/acsnano.0c10064 PMID: 33754708

Sedano, R., Solitano, V., Vuyyuru, S. K., Yuan, Y., Hanžel, J., Ma, C., & Jairath, V. (2025). Artificial intelligence to revolutionize IBD clinical trials: A comprehensive review. *Therapeutic Advances in Gastroenterology*, *18*, 17562848251321915. DOI: 10.1177/17562848251321915 PMID: 39996136

Seijas, A., Cora, D., Novo, M., Al-Soufi, W., Sánchez, L., & Arana, Á. J. (2025). CRISPR/Cas9 Delivery Systems to Enhance Gene Editing Efficiency. *International Journal of Molecular Sciences*, *26*(9), 4420. Advance online publication. DOI: 10.3390/ijms26094420 PMID: 40362657

Sekhoacha, M., Riet, K., Motloung, P., Gumenku, L., Adegoke, A., & Mashele, S. (2022, September 5). Prostate Cancer Review: Genetics, Diagnosis, Treatment Options, and Alternative Approaches. *Molecules (Basel, Switzerland)*, *27*(17), 5730. DOI: 10.3390/molecules27175730 PMID: 36080493

Sembiring, M., & Novagusda, F. (2024). Enhancing Data Security Resilience in AI-Driven Digital Transformation: Exploring Industry Challenges and Solutions Through ALCOA+ Principles. *Acta Informatica Medica*, *32*(1), 65. DOI: 10.5455/aim.2024.32.65-70 PMID: 38585596

Sendak, M., Balu, S., & Hernandez, A. F. (2023). Proactive algorithm monitoring to ensure health equity. *JAMA Network Open*, *6*(12), e2345022–e2345022. DOI: 10.1001/jamanetworkopen.2023.45022 PMID: 38100115

Sepich-Poore, G. D., Zitvogel, L., Straussman, R., Hasty, J., Wargo, J. A., & Knight, R. (2021). The microbiome and human cancer. *Science*, *371*(6536), eabc4552. Advance online publication. DOI: 10.1126/science.abc4552 PMID: 33766858

Shaaban, S., & Ji, Y. (2024). Challenges of Clinical Pharmacogenomics Implementation in the Era of Precision Medicine. *Medical Research Archives*, *12*(11). DOI: 10.18103/mra.v12i11.5955

Shademan, B., Masjedi, S., Karamad, V., Isazadeh, A., Sogutlu, F., Rad, M. H., & Nourazarian, A. (2022, October). CRISPR technology in cancer diagnosis and treatment: Opportunities and challenges. *Biochemical Genetics*, *60*(5), 1446–1470. DOI: 10.1007/s10528-022-10193-9 PMID: 35092559

Shahid, M. S., & Imran, A. (2025). Breast cancer detection using deep learning techniques: Challenges and future directions. *Multimedia Tools and Applications*, *84*(6), 3257–3304.

Shah, M., Patel, M., Shah, M., Patel, M., & Prajapati, M. (2024, October). Computational transformation in drug discovery: A comprehensive study on molecular docking and quantitative structure activity relationship (QSAR). *Intell Pharm.*, *2*(5), 589–595. DOI: 10.1016/j.ipha.2024.03.001

Shahriyari, L., Abdel-Rahman, M., Asadpoure, A., Cebulla, C., Chang, Y. H., Hao, W., & Zervantonakis, I. (2022). *My Virtual Cancer* (No. DOE-UMASSAMHERST-21630). Univ. of Massachusetts, Amherst, MA (United States).

Shain, A. H., & Bastian, B. C. (2016, June). From melanocytes to melanomas. *Nature Reviews. Cancer*, *16*(6), 345–358. DOI: 10.1038/nrc.2016.37 PMID: 27125352

Shalem, O., Sanjana, N. E., Hartenian, E., Shi, X., Scott, D. A., Mikkelson, T., Heckl, D., Ebert, B. L., Root, D. E., Doench, J. G., & Zhang, F. (2014). Genome-scale CRISPR-Cas9 knockout screening in human cells. *Science*, *343*(6166), 84–87. DOI: 10.1126/science.1247005 PMID: 24336571

Shamim, M. A., Shamim, M. A., Arora, P., & Dwivedi, P. (2024). Artificial intelligence and big data for pharmacovigilance and patient safety. *Journal of Medicine, Surgery, and Public Health*, *3*, 100139. DOI: 10.1016/j.glmedi.2024.100139

Shamjetsabam, N. D., Rana, R., Malik, P., & Ganguly, N. K. (2024). CRISPR/Cas9: An overview of recent developments and applications in cancer research. *International Journal of Surgery*, *110*(10), 6198–6213. DOI: 10.1097/JS9.0000000000001081 PMID: 38377059

Shao, Y., Lv, X., Ying, S., & Guo, Q. (2024). Artificial intelligence-driven precision medicine: Multi-omics and spatial multi-omics approaches in Diffuse Large B-Cell Lymphoma (DLBCL). *Frontiers in Bioscience (Landmark Edition)*, *29*(12), 404. DOI: 10.31083/j.fbl2912404

Shariaty, F., Pavlov, V. A., Fedyashina, S. V., & Serebrennikov, N. A. (2024). Integrating deep learning and explainable AI for non-invasive prediction of EGFR and KRAS mutations in NSCLC: a novel radiogenomic approach. In *2024 V International Conference on Neural Networks and Neurotechnologies (NeuroNT)* (pp. 32-35). IEEE. DOI: 10.1109/NeuroNT62606.2024.10585441

Sharma, A., Verma, A., Malviya, R., & Yadav, S. (2022). Applications of Deep Learning in Radiation Therapy. In R. Malviya, G. Ghinea, R. K. Dhanaraj, B. Balusamy, & S. Sundram (Eds.), Deep Learning for Targeted Treatments [Internet]. 1st ed., pp. 247–288). Wiley., Available from https://onlinelibrary.wiley.com/doi/10.1002/9781119857983.ch9, Retrieved November 2, 2025, from. DOI: 10.1002/9781119857983.ch9

Sharma, A. K., Sharma, M., Sharma, A. K., Sharma, M., & Sharma, M. (2023). Mapping the impact of environmental pollutants on human health and environment: A systematic review and meta-analysis. *Journal of Geochemical Exploration*, *255*, 107325. DOI: 10.1016/j.gexplo.2023.107325

Sharma, A., Lövgren, S. K., Eriksson, K. L., Wang, Y., Robertson, S., Hartman, J., & Rantalainen, M. (2024, August 14). Validation of an AI-based solution for breast cancer risk stratification using routine digital histopathology images. [Internet]. *Breast Cancer Research : BCR*, *26*(1), 123. Retrieved November 1, 2025, from https://breast-cancer-research.biomedcentral.com/articles/10.1186/s13058-024-01879-6. DOI: 10.1186/s13058-024-01879-6 PMID: 39143539

Sharma, G., Verma, H. N., & Sharma, R. (2013). RAPD analysis to study metagenome diversity in soil microbial community of arid zone plants. *Proceedings of the National Academy of Sciences. India. Section B, Biological Sciences*, *83*(2), 135–139. DOI: 10.1007/s40011-012-0041-1

Sharma, P., & Singh, S. (2026). AI-nanotech synergies: Recent advances in sustainable bio-manufacturing. *World Journal of Microbiology & Biotechnology*, *42*(7), 332. DOI: 10.1007/s11274-026-05067-w PMID: 42250160

Sharma, S., Parveen, R., & Chatterji, B. P. (2021). Toxicology of Nanoparticles in Drug Delivery. *Current Pathobiology Reports*, *9*(4), 133–144. DOI: 10.1007/s40139-021-00227-z PMID: 34840918

Sharma, S., Sharma, K., & Grover, S. (2024). Real-Time Data Analysis with Smart Sensors. In Gulati, S. (Ed.), *Application of Artificial Intelligence in Wastewater Treatment* (pp. 127–153). Springer Nature Switzerland., DOI: 10.1007/978-3-031-69433-2_5

Shehadi, I. A., Rashdan, H. R., & Abdelmonsef, A. H. (2020). Homology modeling and virtual screening studies of antigen MLAA-42 protein: Identification of novel drug candidates against leukemia-An in silico approach. *Computational and Mathematical Methods in Medicine*, *2020*(1), 8196147. DOI: 10.1155/2020/8196147 PMID: 32256683

Shen, M. D., Chen, S. B., & Ding, X. D. (2024). The effectiveness of digital twins in promoting precision health across the entire population: A systematic review. *NPJ Digital Medicine*, *7*(1), 145. DOI: 10.1038/s41746-024-01146-0 PMID: 38831093

Shen, S., Qi, W., Liu, X., Zeng, J., Li, S., Zhu, X., Dong, C., Wang, B., Shi, Y., Yao, J., Wang, B., Jing, L., Cao, S., & Liang, G. (2025, March 19). From virtual to reality: Innovative practices of digital twins in tumor therapy. *Journal of Translational Medicine*, *23*(1), 348. DOI: 10.1186/s12967-025-06371-z PMID: 40108714

Sherlock, L., Martin, B. R., Behsangar, S., & Mok, K. H. (2023). Application of novel AI-based algorithms to biobank data: Uncovering of new features and linear relationships. *Frontiers in Medicine*, *10*, 1162808. DOI: 10.3389/fmed.2023.1162808 PMID: 37521348

Shi, J., Kantoff, P. W., Wooster, R., & Farokhzad, O. C. (2017). Cancer nanomedicine: Progress, challenges and opportunities. *Nature Reviews. Cancer*, *17*(1), 20–37. DOI: 10.1038/nrc.2016.108 PMID: 27834398

Shi, J., Wang, E., Milazzo, J. P., Wang, Z., Kinney, J. B., & Vakoc, C. R. (2015). Discovery of cancer drug targets by CRISPR-Cas9 screening of protein domains. *Nature Biotechnology*, *33*(6), 661–667. DOI: 10.1038/nbt.3235 PMID: 25961408

Shi, P., Ren, C., & Tong, H. (2026). Research progress of stem cells in the treatment of atherosclerosis. *Frontiers in Cell and Developmental Biology*, *13*, 1722416. DOI: 10.3389/fcell.2025.1722416 PMID: 41552013

Shirzad, M., Shaban, M., Mohammadzadeh, V., Rahdar, A., Fathi-karkan, S., Hoseini, Z. S., Najafi, M., Kharaba, Z., & Aboudzadeh, M. A. (2025, September). Artificial intelligence-assisted design of nanomedicines for breast cancer diagnosis and therapy: Advances, challenges, and future directions. *BioNanoScience*, *15*(3), 354. DOI: 10.1007/s12668-025-01980-w

Shiva, S., & Kalluri, H. K. (2024). Lung Cancer Detection Using Fusion-Based Deep Learning Techniques. In *2024 IEEE Students Conference on Engineering and Systems (SCES)* (pp. 1-5). IEEE. DOI: 10.1109/SCES61914.2024.10652562

Shokoya, M. M., Szabó, E., Nagy, Z. K., Sebe, I., & Zelkó, R. (2026). Therapeutic nanofibrous scaffolds in precision drug delivery: Formulation design, challenges and regulatory frameworks. *Expert Opinion on Drug Delivery*, *23*(5), 823–839. DOI: 10.1080/17425247.2026.2619643 PMID: 41549842

Sholikah, T. A. (2017). Fusion gene bcr-abl: from etiopathogenesis to the management of Chronic Myeloid Leukemia. *JKKI: Jurnal Kedokteran dan Kesehatan Indonesia*, 29-37.

Short, N. J., Rytting, M. E., & Cortes, J. E. (2018). Acute myeloid leukaemia. *Lancet*, *392*(10147), 593–606. DOI: 10.1016/S0140-6736(18)31041-9 PMID: 30078459

Shreve, J. T., Khanani, S. A., & Haddad, T. C. (2022). Artificial intelligence in oncology: Current capabilities, future opportunities, and ethical considerations. In *American Society of Clinical Oncology Educational book. American Society of Clinical Oncology. Annual Meeting* (Vol. 42, pp. 1-10).

Silvestris, N., Ciliberto, G., De Paoli, P., Apolone, G., Lavitrano, M. L., Pierotti, M. A., & Stanta, G. (2017). Liquid dynamic medicine and N-of-1 clinical trials: A change of perspective in oncology research. *Journal of Experimental & Clinical Cancer Research : CR*, *36*(1), 128. DOI: 10.1186/s13046-017-0598-x

Sim, J. A., Huang, X., Horan, M. R., Stewart, C. M., Robison, L. L., Hudson, M. M., & Huang, I. C. (2023). Natural language processing with machine learning methods to analyze unstructured patient-reported outcomes derived from electronic health records: A systematic review. *Artificial Intelligence in Medicine*, *146*, 102701.

Singhal, S. (2024). Data privacy, compliance, and security including AI ML: healthcare. In *Practical Applications of Data Processing, Algorithms, and Modelling* (pp. 111-126). IGI Global Scientific Publishing. DOI: 10.4018/979-8-3693-2909-2.ch009

Singha, M., Pu, L., Stanfield, B. A., Uche, I. K., Rider, P. J., Kousoulas, K. G., & Brylinski, M. (2022). Artificial intelligence to guide precision anticancer therapy with multitargeted kinase inhibitors. *BMC Cancer*, *22*(1), 1211. DOI: 10.1186/s12885-022-10293-0

Singh, D., Dhiman, V. K., Pandey, M., Dhiman, V. K., Sharma, A., Pandey, H., Verma, S. K., & Pandey, R. (2024). Personalized medicine: An alternative for cancer treatment. *Cancer Treatment and Research Communications*, *42*, 100860. DOI: 10.1016/j.ctarc.2024.100860 PMID: 39827574

Singh, R., Johnsten, T., Raghavan, V., & Xie, Y. (2009). An efficient algorithm for discovering positive and negative patterns. *2009 IEEE International Conference on Granular Computing*, 507–512. DOI: 10.1109/GRC.2009.5255068

Sinisi, S., Alimguzhin, V., Mancini, T., Tronci, E., Mari, F., & Leeners, B. (2020). Optimal personalised treatment computation through in silico clinical trials on patient digital twins. *Fundamenta Informaticae*, *174*(3-4), 283–310. DOI: 10.3233/FI-2020-1943

Sîrbu, C. L., & Mercioni, M. A. (2024). Fostering trust in AI-driven healthcare: a brief review of ethical and practical considerations. In *2024 International Symposium on Electronics and Telecommunications (ISETC)* (pp. 1-4). IEEE. DOI: 10.1109/ISETC63109.2024.10797264

Smallwood, A. J., Schapira, M. M., Fedders, M., & Neuner, J. M. (2017, February). A pilot randomized controlled trial of a decision aid with tailored fracture risk tool delivered via a patient portal. *Osteoporosis International : A Journal Established as Result of Cooperation between the European Foundation for Osteoporosis and the National Osteoporosis Foundation of the USA*, *28*(2), 567–576. DOI: 10.1007/s00198-016-3767-4 PMID: 27647529

Smith, R., Geary, S. M., & Salem, A. K. (2020). Silicon nanowires and their impact on cancer detection and monitoring. *ACS Applied Nano Materials*, *3*(9), 8522–8536. DOI: 10.1021/acsanm.0c01572

Smokovski, I., Steinle, N., Behnke, A., Bhaskar, S. M., Grech, G., Richter, K., & Golubnitschaja, O. (2024). Digital biomarkers: 3PM approach revolutionizing chronic disease management-EPMA 2024 position. *The EPMA Journal*, *15*(2), 149–162. DOI: 10.1007/s13167-024-00364-6 PMID: 38841615

Smolarz, B., Łukasiewicz, H., Samulak, D., Piekarska, E., Kołaciński, R., & Romanowicz, H. (2025, February 26). Lung Cancer—Epidemiology, Pathogenesis, Treatment and Molecular Aspect (Review of Literature). *International Journal of Molecular Sciences*, *26*(5), 2049. DOI: 10.3390/ijms26052049 PMID: 40076671

Smolarz, B., Nowak, A. Z., & Romanowicz, H. (2022). Breast Cancer—Epidemiology, Classification, Pathogenesis and Treatment (Review of Literature). *Cancers (Basel), 14*(10), 2569. Advance online publication. DOI: 10.3390/cancers14102569 PMID: 35626173

Snow, A., Chen, D., & Lang, J. E. (2019, November). The current status of the clinical utility of liquid biopsies in cancer. *Expert Review of Molecular Diagnostics, 19*(11), 1031–1041. DOI: 10.1080/14737159.2019.1664290 PMID: 31482746

Sohrabei, S., Moghaddasi, H., Hosseini, A., & Ehsanzadeh, S. J. (2024). Investigating the effects of artificial intelligence on the personalization of breast cancer management: A systematic study. *BMC Cancer, 24*(1), 852. DOI: 10.1186/s12885-024-12575-1 PMID: 39026174

Soleymanjahi, S., Huebner, J., Elmansy, L., Rajashekar, N., Lüdtke, N., Paracha, R., Thompson, R., Grimshaw, A. A., Foroutan, F., Sultan, S., & Shung, D. L. (2024). Artificial intelligence-assisted colonoscopy for polyp detection: A systematic review and meta-analysis. *Annals of Internal Medicine, 177*(12), 1652–1663. DOI: 10.7326/ANNALS-24-00981 PMID: 39531400

Sonalisha, S., Kirti, A., Asima, S. P., Parashar, R., Pal, S., Barik, D., Naser, S. S., Souto, E. B., & Verma, S. K. (2026). Translational paradigm of Advanced Nanoscale Strategies for Triple Negative Breast Cancer (TNBC): Mechanistic Insights, Metastatic Pathways, and emerging theragnosis. *Materials Advances, 7*(6), 3072–3100. DOI: 10.1039/D5MA00866B

Song, A. H., Jaume, G., Williamson, D. F. K., Lu, M. Y., Vaidya, A., Miller, T. R., & Mahmood, F. (2023, October 2). Artificial intelligence for digital and computational pathology. [Internet]. *Nature Reviews Bioengineering, 1*(12), 930–949. Retrieved November 2, 2025, from https://www.nature.com/articles/s44222-023-00096-8. DOI: 10.1038/s44222-023-00096-8

Song, M., Liu, C., Chen, S., & Zhang, W. (2021, February). Nanocarrier-Based Drug Delivery for Melanoma Therapeutics. *International Journal of Molecular Sciences, 22*(4), 1873. DOI: 10.3390/ijms22041873 PMID: 33668591

Song, N., Nwafili, S. A., & Gao, T. X. (2011). Genetic diversity and population structure of *Chrysichthysnigrodigitatus* from Niger Delta based on AFLP analysis. *Biochemical Systematics and Ecology, 39*(4), 320–327. DOI: 10.1016/j.bse.2011.04.002

Song, Y., Liu, T., Hao, Q., Fang, Q., Gong, X., Li, Y., & Mi, Y. (2025). Comprehensive omics-based classification system in adult patients with B-cell acute lymphoblastic leukemia. *Molecular Oncology, 19*(12), 3578–3595. DOI: 10.1002/1878-0261.70053 PMID: 40388565

Song, Y., Wang, Z., Zhang, G., Hou, J., Liu, K., Wei, S., & Mi, Y. (2023). Integrative multi-omic analysis for prognosis stratification in acute myeloid leukemia. *Blood, 142*(Supplement 1), 5984. DOI: 10.1182/blood-2023-173211

Soori, M., Arezoo, B., & Dastres, R. (2023). Internet of things for smart factories in industry 4.0, a review. *Internet of Things and Cyber-Physical Systems, 3*, 192–204. DOI: 10.1016/j.iotcps.2023.04.006

Soos, B. (2012). *Multi-scale Models of Tumor Growth and Invasion* (Doctoral dissertation, University of Waterloo).

Sopruchi, A. D., & Rashid, A. (2024). The Integration of AI-Driven Decision Support Systems in Healthcare: Enhancements, Challenges, and Future Directions. *IDOSR Journal of Computer and Applied Sciences*, *9*(2), 17–25. DOI: 10.59298/JCAS/2024/92.1725

Sorel, N., Cayssials, É., Brizard, F., & Chomel, J. C. (2017). Treatment and molecular monitoring update in chronic myeloid leukemia management. *Annales de Biologie Clinique*, *75*(2), 129–145. DOI: 10.1684/abc.2017.1233 PMID: 28377326

Sousa, R., Abelha, V., Peixoto, H., & Machado, J. (2024, December 17). Unlocking Healthcare Data Potential: A Comprehensive Integration Approach with GraphQL, openEHR, Redis, and Pervasive Business Intelligence. [Internet]. *Technologies*, *12*(12), 265. Retrieved November 2, 2025, from https://www.mdpi.com/2227-7080/12/12/265. DOI: 10.3390/technologies12120265

Soussi, T., Taschner, P. E., & Samuels, Y. (2017). Synonymous somatic variants in human cancer are not infamous: A plea for full disclosure in databases and publications. *Human Mutation*, *38*(4), 339–342. DOI: 10.1002/humu.23163

Srivastav, A. K., Mishra, M. K., Lillard, J. W.Jr, & Singh, R. (2025). Transforming Pharmacogenomics and CRISPR Gene Editing with the Power of Artificial Intelligence for Precision Medicine. *Pharmaceutics*, *17*(5), 555. Advance online publication. DOI: 10.3390/pharmaceutics17050555 PMID: 40430848

Stadtmauer, E. A., Fraietta, J. A., Davis, M. M., Cohen, A. D., Weber, K. L., Lancaster, E., Mangan, P. A., Kulikovskaya, I., Gupta, M., Chen, F., Tian, L., Gonzalez, V. E., Xu, J., Jung, I.-Y., Melenhorst, J. J., Plesa, G., Shea, J., Matlawski, T., Cervini, A., & June, C. H. (2020). CRISPR-engineered T cells in patients with refractory cancer. *Science*, *367*(6481), eaba7365. Advance online publication. DOI: 10.1126/science.aba7365 PMID: 32029687

Stahlberg, E. A., Abdel-Rahman, M., Aguilar, B., Asadpoure, A., Beckman, R. A., Borkon, L. L., & Zervantonakis, I. (2022). Exploring approaches for predictive cancer patient digital twins: Opportunities for collaboration and innovation. *Frontiers in Digital Health*, *4*, 1007784. DOI: 10.3389/fdgth.2022.1007784 PMID: 36274654

Steel, A., Gallego-Perez, D. F., Ijaz, N., Gall, A., Bangpan, M., Dos Santos Boeira, L., Schveitzer, M. C., Chutaputti, A., Mahlanza-Langer, L., Pillai, G. K. G., Posayanonda, T., Sungchol, K., Shankar, D., & Kuchenmüller, T. (2025). Integration of traditional, complementary, and integrative medicine in the institutionalization of evidence-informed decision-making: The World Health Organization meeting report. *Journal of Integrative and Complementary Medicine*, *31*(4), 388–394. DOI: 10.1089/jicm.2024.0837 PMID: 39819104

Steele, C. D., Abbasi, A., Islam, S. A., Bowes, A. L., Khandekar, A., Haase, K., & Pillay, N. (2022). Signatures of copy number alterations in human cancer. *Nature*, *606*(7916), 984–991. DOI: 10.1038/s41586-022-04738-6

tefănigă, S. A., Cordo , A. A., Ivascu, T., Feier, C. V. I., Muntean, C., Stupinean, C. V., Călinici, T., Alua , M., & Bolboacă, S. D. (2024, November 13). Advancing Precision Oncology with Digital and Virtual Twins: A Scoping Review. *Cancers (Basel)*, *16*(22), 3817. DOI: 10.3390/cancers16223817 PMID: 39594772

Stilgenbauer, S. (2007). Progress and prognosis in chronic lymphocytic leukaemia. In *Hematology Meeting Reports,1*(5), 3-11.

Stowers, C. E., Wu, C., Kumar, S., Lima, E. A., Zu, X., Yam, C., & Yankeelov, T. E. (2023). Developing MRI-based digital-twins via mathematical modeling and deep learning to predict the response of triple-negative breast cancer to neoadjuvant therapy. *Cancer Research, 83*(7, Supplement), 845–845. DOI: 10.1158/1538-7445.AM2023-845

Stratigos, A. J., Garbe, C., Dessinioti, C., Lebbe, C., Bataille, V., Bastholt, L., Dreno, B., Fargnoli, M. C., Forsea, A. M., Frenard, C., Harwood, C. A., Hauschild, A., Hoeller, C., Kandolf-Sekulovic, L., Kaufmann, R., Kelleners-Smeets, N. W. J., Malvehy, J., del Marmol, V., Middleton, M. R., & Grob, J.-J. (2020, March). European interdisciplinary guideline on invasive squamous cell carcinoma of the skin: Part 1. epidemiology, diagnostics and prevention. *European Journal of Cancer : Official Journal for European Organization for Research and Treatment of Cancer (EORTC) [and] European Association for Cancer Research (EACR), 128*, 60–82. DOI: 10.1016/j.ejca.2020.01.007 PMID: 32113941

Strielkowski, W., Vlasov, A., Selivanov, K., Muraviev, K., & Shakhnov, V. (2023). Prospects and Challenges of the Machine Learning and Data-Driven Methods for the Predictive Analysis of Power Systems: A Review. *Energies, 16*(10), 4025. DOI: 10.3390/en16104025

Subasi, A., & Subasi, M. E. (2024). Digital twins in healthcare and biomedicine. In *Artificial Intelligence, Big Data, Blockchain and 5G for the Digital Transformation of the Healthcare Industry* (pp. 365-401). Academic Press. DOI: 10.1016/B978-0-443-21598-8.00011-7

Subha, S., Shanmugathai, M., Prasanth, A., Varagi, S. S., & Dhanashree, V. (2024). Digital Transformation in the Pharmaceutical and Biotech Industry: Challenges and Research Directions. In Dhanaraj, R. K., Balusamy, B., Samuel, P., Bashir, A. K., & Kadry, S. (Eds.), *Digital Twins in Industrial Production and Smart Manufacturing* (1st ed., pp. 297–324). Wiley., DOI: 10.1002/9781394195336.ch13

Sumini, M., Teodori, F., & Isolan, L. (2024). Digital twins in dosimetry and radiotherapy, a survey and some applications. *Radiation Physics and Chemistry, 218*, 111649. DOI: 10.1016/j.radphyschem.2024.111649

Sumiyoshi, T., Chi, K. N., & Wyatt, A. W. (2021, June). Clinical implications of genomic alterations in metastatic prostate cancer. *Prostate Cancer and Prostatic Diseases, 24*(2), 310–322. DOI: 10.1038/s41391-020-00308-x PMID: 33452452

Sung, H., Ferlay, J., Siegel, R. L., Laversanne, M., Soerjomataram, I., Jemal, A., & Bray, F. (2021, May). Global Cancer Statistics 2020: GLOBOCAN Estimates of Incidence and Mortality Worldwide for 36 Cancers in 185 Countries. *CA: a Cancer Journal for Clinicians, 71*(3), 209–249. DOI: 10.3322/caac.21660 PMID: 33538338

Sun, S., & Chen, J. (2024). Recent advances in hydrogel-based biosensors for cancer detection. *ACS Applied Materials & Interfaces, 16*(36), 46988–47002. DOI: 10.1021/acsami.4c02317

Sun, T., He, X., Song, X., Shu, L., & Li, Z. (2022, July 14). The Digital Twin in Medicine: A Key to the Future of Healthcare? *Frontiers in Medicine, 9*, 907066. DOI: 10.3389/fmed.2022.907066 PMID: 35911407

Sun, X., Liu, K., Lu, S., He, W., & Du, Z. (2022). Targeted Therapy and Immunotherapy for Heterogeneous Breast Cancer. *Cancers (Basel), 14*(21), 5456. Advance online publication. DOI: 10.3390/cancers14215456 PMID: 36358874

Sun, X., Zhang, M., Xie, L., Ren, Q., Chang, J., & Jiang, W. (2023). Material-enhanced biosensors for cancer biomarkers detection. *Microchemical Journal : Devoted to the Application of Microtechniques in All Branches of Science, 194*, 109298. DOI: 10.1016/j.microc.2023.109298

Sun, X., Zhao, P., Lin, J., Chen, K., & Shen, J. (2023). Recent advances in access to overcome cancer drug resistance by nanocarrier drug delivery system. *Cancer Drug Resistance (Alhambra, Calif.), 6*(2), 390–415. DOI: 10.20517/cdr.2023.16 PMID: 37457134

Sun, Y., Qiao, G., Gao, B., & Li, Y. (2024). DAPM-CDR: A domain adaptation prompting model for drug response prediction. *Future Generation Computer Systems, 160*, 316–324. DOI: 10.1016/j.future.2024.06.009

Sun, Y., Zhu, S., Ma, K., Liu, W., Yue, Y., Hu, G., Lu, H., & Chen, W. (2019, November 21). Identification of 12 cancer types through genome deep learning. *Scientific Reports, 9*(1), 17256. DOI: 10.1038/s41598-019-53989-3 PMID: 31754222

Suryawanshi, M. V., Bagban, I., & Patne, A. Y. (2025). Integrating Nanotechnology and Artificial Intelligence for Early Detection and Prognostication of Glioblastoma: A Translational Perspective. *Targets (Basel), 3*(4), 31. Advance online publication. DOI: 10.3390/targets3040031

Sutiene, K., Schwendner, P., Sipos, C., Lorenzo, L., Mirchev, M., Lameski, P., Kabasinskas, A., Tidjani, C., Ozturkkal, B., & Cerneviciene, J. (2024). Enhancing portfolio management using artificial intelligence: Literature review. *Frontiers in Artificial Intelligence, 7*, 1371502. DOI: 10.3389/frai.2024.1371502 PMID: 38650961

Suzuki, E. (2002). UNDIRECTED DISCOVERY OF INTERESTING EXCEPTION RULES. *International Journal of Pattern Recognition and Artificial Intelligence, 16*(08), 1065–1086. DOI: 10.1142/S0218001402002155

Swain, B., & Maddi, S. (2024). Patient-derived Tumor Organoids: Generation and Applications in Disease Modeling and Personalized Therapy. *Nature Cell and Science, 2*(3), 185–197. DOI: 10.61474/ncs.2024.00008

Swathi, Y., & Challa, M. (2023). A comparative analysis of explainable AI techniques for enhanced model interpretability. In *2023 3rd international conference on pervasive computing and social networking (ICPCSN)* (pp. 229-234). IEEE.

Syed, S. A., Nisar, H., Jaffari, R., Hum, Y. C., Jen, L. Y., & Qaisar, M. (2024). Enhancing leukemia detection: an automated approach using deep learning and ensemble techniques. http://hdl.handle.net/20.500.14131/1460

T A, A., Vallazhath, A., Maniyal, V., & Menon, D. (. (2026). Advances in therapeutic strategies for atherosclerosis: From pharmacologics to stents and stent coatings. *Journal of Materials Chemistry. B, 14*(8), 2407–2437. DOI: 10.1039/D5TB02293B PMID: 41668608

Tabibian, M., Moghaddam, F. S., Motevaseli, E., & Ghafouri-Fard, S. (2024, October 17). Targeting mRNA-coding genes in prostate cancer using CRISPR/Cas9 technology with a special focus on androgen receptor signaling. *Cell Communication and Signaling : CCS*, *22*(1), 504. DOI: 10.1186/s12964-024-01833-1 PMID: 39420406

Tafirenyika, S. (2023). *AI in healthcare: Predictive modeling, explainability, and clinical impact.* World J Adv Res Rev.

Tai, J., Fan, S., Ding, S., & Ren, L. (2022). Gold nanoparticles based optical biosensors for cancer biomarker proteins: A review of the current practices. *Frontiers in Bioengineering and Biotechnology*, *10*, 877193. DOI: 10.3389/fbioe.2022.877193

Talasila, P., Gomes, C., Mikkelsen, P. H., Arboleda, S. G., Kamburjan, E., & Larsen, P. G. (2023). Digital twin as a service (DTaaS): a platform for digital twin developers and users. In *2023 IEEE Smart World Congress (SWC)* (pp. 1-8). IEEE. DOI: 10.1109/SWC57546.2023.10448890

Tambi, R., Zehra, B., Vijayakumar, A., Satsangi, D., Uddin, M., & Berdiev, B. K. (2024). Artificial intelligence and omics in malignant gliomas. *Physiological Genomics*, *56*(12), 876–895. DOI: 10.1152/physiolgenomics.00011.2024

Tang, C., Yi, W., Occhipinti, E., Dai, Y., Gao, S., & Occhipinti, L. G. (2024). A roadmap for the development of human body digital twins. *Nature Reviews Electrical Engineering*, *1*(3), 199–207. DOI: 10.1038/s44287-024-00025-w

Tang, F. H., Fong, Y. W., Yung, S. H., Wong, C. K., Tu, C. L., Chan, M. T., & Radiomics-Clinical, A. I. (2023, July 25). Model with Probability Weighted Strategy for Prognosis Prediction in Non-Small Cell Lung Cancer. *Biomedicines*, *11*(8), 2093. DOI: 10.3390/biomedicines11082093 PMID: 37626590

Tao, Y., Rajaraman, A., Cui, X., Cui, Z., Eaton, J., Kim, H., & Schwartz, R. (2019). Improving personalized prediction of cancer prognoses with clonal evolution models. *bioRxiv*, 761510.

Tarighati, E., Keivan, H., & Mahani, H. (2022). A review of prognostic and predictive biomarkers in breast cancer. *Clin Exp Med*. https://link.springer.com/10.1007/s10238-021-00781-1

Taruttis, A., & Ntziachristos, V. (2012). Translational optical imaging. *AJR. American Journal of Roentgenology*, *199*(2), 263–271. DOI: 10.2214/AJR.11.8431 PMID: 22826386

Tătaru, O. S., Vartolomei, M. D., Rassweiler, J. J., Virgil, O., Lucarelli, G., Porpiglia, F., Amparore, D., Manfredi, M., Carrieri, G., Falagario, U., Terracciano, D., de Cobelli, O., Busetto, G. M., Giudice, F. D., & Ferro, M. (2021, February 20). Artificial intelligence and machine learning in prostate cancer patient management—Current trends and future perspectives. *Diagnostics (Basel)*, *11*(2), 354. DOI: 10.3390/diagnostics11020354 PMID: 33672608

Tavakoli, H., Mohammadi, S., Li, X., Fu, G., & Li, X. (2022). Microfluidic platforms integrated with nano-sensors for point-of-care bioanalysis. *Trends in Analytical Chemistry*, *157*, 116806. DOI: 10.1016/j.trac.2022.116806

Tayefi, M., Ngo, P., Chomutare, T., Dalianis, H., Salvi, E., Budrionis, A., ... Challenges and opportunities beyond structured data in analysis of electronic health records. WIREs Computational Stats [Internet]. 2021 Nov [cited 2025 Nov 2];13(6):e1549. Available from: https://wires.onlinelibrary.wiley.com/doi/10.1002/wics.1549

Teixeira, M., Silva, F., Ferreira, R. M., Pereira, T., Figueiredo, C., & Oliveira, H. P. (2024, May 30). A review of machine learning methods for cancer characterization from microbiome data. *NPJ Precision Oncology*, *8*(1), 123. DOI: 10.1038/s41698-024-00617-7 PMID: 38816569

Tenchov, R., Sasso, J. M., & Zhou, Q. A. (2023). PEGylated Lipid Nanoparticle Formulations: Immunological Safety and Efficiency Perspective. *Bioconjugate Chemistry*, *34*(6), 941–960. DOI: 10.1021/acs.bioconjchem.3c00174 PMID: 37162501

Terranova, N., & Venkatakrishnan, K. (2024). Machine learning in modeling disease trajectory and treatment outcomes: An emerging enabler for model-informed precision medicine. *Clinical Pharmacology and Therapeutics*, *115*(4), 720–726.

Terzi, T. A. (2024). Using artificial intelligence for personalized cancer treatment. *Next Front. Life Sci. AI*, *8*(1), 133–136. DOI: 10.62802/mpfsy492

Testa, U., Castelli, G., & Pelosi, E. (2019, July 30). Cellular and Molecular Mechanisms Underlying Prostate Cancer Development: Therapeutic Implications. *Medicines (Basel, Switzerland)*, *6*(3), 82. DOI: 10.3390/medicines6030082 PMID: 31366128

Thirupathi, L. SrihaReddy, E., & Deepika, J. U. (2024). Challenges and innovations in the creation of digital twins in healthcare. In *Exploring the Advancements and Future Directions of Digital Twins in Healthcare 6.0* (pp. 69-85). IGI Global.

Thomas, A. M., Manghi, P., Asnicar, F., Pasolli, E., Armanini, F., Zolfo, M., Beghini, F., Manara, S., Karcher, N., Pozzi, C., Gandini, S., Serrano, D., Tarallo, S., Francavilla, A., Gallo, G., Trompetto, M., Ferrero, G., Mizutani, S., Shiroma, H., & Segata, N. (2019). Metagenomic analysis of colorectal cancer datasets identifies cross-cohort microbial diagnostic signatures and a link with choline degradation. *Nature Medicine*, *25*(4), 667–678. DOI: 10.1038/s41591-019-0405-7 PMID: 30936548

Thomas, R. S., Bahadori, T., Buckley, T. J., Cowden, J., Deisenroth, C., Dionisio, K. L., Frithsen, J. B., Grulke, C. M., Gwinn, M. R., Harrill, J. A., Higuchi, M., Houck, K. A., Hughes, M. F., Hunter, E. S.III, Isaacs, K. K., Judson, R. S., Knudsen, T. B., Lambert, J. C., Linnenbrink, M., & Williams, A. J. (2019). The next generation blueprint of computational toxicology at the U.S. Environmental Protection Agency. *Toxicological Sciences : An Official Journal of the Society of Toxicology*, *169*(2), 317–332. DOI: 10.1093/toxsci/kfz058 PMID: 30835285

Tian, X., & Liu, Z. (2024). Single nucleotide variants in lung cancer. *Chinese Medical Journal Pulmonary and Critical Care Medicine*, *2*(2), 88–94. DOI: 10.1016/j.pccm.2024.04.004

Tian, Y., Liu, Z., Wang, J., Huang, W., Li, S., Zhang, N., Wang, H., Xu, Y., Cao, Y., & Liu, Q. (2023). Nanomedicine for Combination Urologic Cancer Immunotherapy. *Pharmaceutics*, *15*(2), 546. DOI: 10.3390/pharmaceutics15020546 PMID: 36839868

Tie, J., Cohen, J. D., Lahouel, K., Lo, S. N., Wang, Y., Kosmider, S., Wong, R., Shapiro, J., Lee, M., Harris, S., Khattak, A., Burge, M., Harris, M., Lynam, J., Nott, L., Day, F., Hayes, T., McLachlan, S.-A., Lee, B., & Gibbs, P. (2022). Circulating tumor DNA analysis guiding adjuvant therapy in stage II colon cancer. *The New England Journal of Medicine*, *386*(24), 2261–2272. DOI: 10.1056/NEJMoa2200075 PMID: 35657320

Timilsina, S. S., & Li, X. J. (2024). A paper-in-polymer-pond (PiPP) hybrid microfluidic microplate for multiplexed ultrasensitive detection of cancer biomarkers. *Lab on a Chip*, *24*(21), 4962–4973. DOI: 10.1039/D4LC00485J

Tirosh, I., Izar, B., Prakadan, S. M., Wadsworth, M. H., Treacy, D., Trombetta, J. J., & Garraway, L. A. (2016). Dissecting the multicellular ecosystem of metastatic melanoma by single-cell RNA-seq. *Science*, *352*(6282), 189–196.

Tiwari, A. (2023). Simulating molecular markers in acute myeloid leukemia using quantum computing. In *2023 IEEE MIT Undergraduate Research Technology Conference (URTC)* (pp. 1-4). IEEE. DOI: 10.1109/URTC60662.2023.10534988

Tiwari, A., Mishra, S., & Kuo, T. R. (2025, June 2). Current AI technologies in cancer diagnostics and treatment. [Internet]. *Molecular Cancer*, *24*(1), 159. Retrieved November 2, 2025, from https://molecular-cancer.biomedcentral.com/articles/10.1186/s12943-025-02369-9. DOI: 10.1186/s12943-025-02369-9 PMID: 40457408

Tomljenovic-Hanic, S., & Khalid, A. (2026). Silk Fibroin for Biomedical Applications with Emphasis on Bioimaging, Biosensing and Regenerative Systems: A Review. *Molecules (Basel, Switzerland)*, *31*(7), 1142. DOI: 10.3390/molecules31071142 PMID: 41976184

Topalian, S. L., Hodi, F. S., Brahmer, J. R., Gettinger, S. N., Smith, D. C., McDermott, D. F., Powderly, J. D., Sosman, J. A., Atkins, M. B., Leming, P. D., Spigel, D. R., Antonia, S. J., Drilon, A., Wolchok, J. D., Carvajal, R. D., McHenry, M. B., Hosein, F., Harbison, C. T., Grosso, J. F., & Sznol, M. (2019, October 1). Five-Year Survival and Correlates Among Patients With Advanced Melanoma, Renal Cell Carcinoma, or Non–Small Cell Lung Cancer Treated With Nivolumab. *JAMA Oncology*, *5*(10), 1411. DOI: 10.1001/jamaoncol.2019.2187 PMID: 31343665

Topol, E. J. (2019, January). High-performance medicine: The convergence of human and artificial intelligence. *Nature Medicine*, *25*(1), 44–56. DOI: 10.1038/s41591-018-0300-7 PMID: 30617339

Torchilin, V. P. (2014, November). Multifunctional, stimuli-sensitive nanoparticulate systems for drug delivery. *Nature Reviews. Drug Discovery*, *13*(11), 813–827. DOI: 10.1038/nrd4333 PMID: 25287120

Torres Quintas, S., Canha-Borges, A., Oliveira, M. J., Sarmento, B., & Castro, F. (2024). Special Issue: Nanotherapeutics in Women's Health Emerging Nanotechnologies for Triple-Negative Breast Cancer Treatment. *Small*, *20*(41), 2300666. DOI: 10.1002/smll.202300666 PMID: 36978237

Tosca, E. M., De Carlo, A., Ronchi, D., & Magni, P. (2024). Model-informed reinforcement learning for enabling precision dosing via adaptive dosing. *Clinical Pharmacology and Therapeutics*, *116*(3), 619–636. DOI: 10.1002/cpt.3356 PMID: 38989560

Tran, P., Lee, S.-E., Kim, D.-H., Pyo, Y.-C., & Park, J.-S. (2020). Recent advances of nanotechnology for the delivery of anticancer drugs for breast cancer treatment. *Journal of Pharmaceutical Investigation*, *50*(3), 261–270. DOI: 10.1007/s40005-019-00459-7

Trauer, J., Schweigert-Recksiek, S., Schenk, T., Baudisch, T., Mörtl, M., & Zimmermann, M. (2022). A digital twin trust framework for industrial application. *Proceedings of the Design Society*, *2*, 293–302. DOI: 10.1017/pds.2022.31

Trilla-Fuertes, L., Gámez-Pozo, A., Díaz-Almirón, M., Prado-Vázquez, G., Zapater-Moros, A., López-Vacas, R., & Fresno Vara, J. A. (2019). Computational metabolism modeling predicts risk of distant relapse-free survival in breast cancer patients. *Future Oncology*, *15*(30), 3483–3490. DOI: 10.2217/fon-2018-0698

Tripura, U., Nagrale, N. V., Singh, O. G., Dey, A., & Venkatesh, J. (2025). Ethical, Social, and Legal Issues Related to Precision Medicine in India: Challenges and Solutions. *Ann Afr Med*. https://journals.lww.com/10.4103/aam.aam_13_25

Truhn, D., Arasteh, S. T., Saldanha, O. L., Müller-Franzes, G., Khader, F., Quirke, P., & Kather, J. N. (2024). Encrypted federated learning for secure decentralized collaboration in cancer image analysis. *Medical Image Analysis*, *92*, 103059. DOI: 10.1016/j.media.2023.103059 PMID: 38104402

Tsai, S. Q., & Joung, J. K. (2016, May). Defining and improving the genome-wide specificities of CRISPR–Cas9 nucleases. *Nature Reviews. Genetics*, *17*(5), 300–312. DOI: 10.1038/nrg.2016.28 PMID: 27087594

Tschandl, P., Rosendahl, C., & Kittler, H. (2018, August). The HAM10000 dataset, a large collection of multi-source dermatoscopic images of common pigmented skin lesions. *Scientific Data*, *5*(1), 180161. DOI: 10.1038/sdata.2018.161 PMID: 30106392

Tsimberidou, A. M., Fountzilas, E., Nikanjam, M., & Kurzrock, R. (2020, June). Review of precision cancer medicine: Evolution of the treatment paradigm. *Cancer Treatment Reviews*, *86*, 102019. DOI: 10.1016/j.ctrv.2020.102019 PMID: 32251926

Tsou, P., & Wu, C. J. (2020). Seeing glycolysis on PDAC: Applying deep learning convolutional neural network model. *Cancer Research*, *80*(16, Supplement), 2107–2107. DOI: 10.1158/1538-7445.AM2020-2107

Tuan, D. A. (2024). Bridging the gap between black box AI and clinical practice: Advancing explainable AI for trust, ethics, and personalized healthcare diagnostics. DOI: 10.20944/preprints202409.1974.v2

Tupsakhare, P. (2023). Improving clinical decision support in health care through ais. progress in medical sciences. PMS-E118. *Prog Med Sci*, *7*(3), 1–4.

Turabi, K., Klute, K., & Radhakrishnan, P. (2024). Decoding the dynamics of circulating tumor DNA in liquid biopsies. *Cancers (Basel)*, *16*(13), 2432. DOI: 10.3390/cancers16132432

Turab, M., & Jamil, S. (2023). A comprehensive survey of digital twins in healthcare in the era of metaverse. *BioMedInformatics*, *3*(3), 563–584. DOI: 10.3390/biomedinformatics3030039

Turk, A., Čeh, E., Calin, G. A., & Kunej, T. (2024). Multiple omics levels of chronic lymphocytic leukemia. *Cell Death Discovery, 10*(1), 293. DOI: 10.1038/s41420-024-02068-2 PMID: 38906881

Tyner, J. W., Tognon, C. E., Bottomly, D., Wilmot, B., Kurtz, S. E., Savage, S. L., & Druker, B. J. (2018). Functional genomic landscape of acute myeloid leukaemia. *Nature, 562*(7728), 526–531. DOI: 10.1038/s41586-018-0623-z PMID: 30333627

Uddin, R., Darwish, N. H., & Mousa, S. A. (2021). Acute myeloid leukemia mutations and future mechanistic target to overcome resistance. *Current Treatment Options in Oncology, 22*(9), 76. DOI: 10.1007/s11864-021-00880-x PMID: 34213682

Upadhyaya, P. (2025). The Role of AI in Identifying and Preventing Fraudulent Activities. In Gupta, S. K., Rosak-Szyrocka, J., Rena, R., Shieh, C.-S., & Erkol Bayram, G. (Eds.), *The Impact of Artificial Intelligence on Finance: Transforming Financial Technologies* (Vol. 53, pp. 367–381). Springer Nature Switzerland., DOI: 10.1007/978-3-031-92916-8_18

Urban, G., Tripathi, P., Alkayali, T., Mittal, M., Jalali, F., Karnes, W., & Baldi, P. (2018). Deep learning localizes and identifies polyps in real time with 96% accuracy in screening colonoscopy. *Gastroenterology, 155*(4), 1069–1078.e8. DOI: 10.1053/j.gastro.2018.06.037 PMID: 29928897

Vakili, S., Beheshti, I., Barzegar Behrooz, A., Łos, M. J., Vitorino, R., & Ghavami, S. (2025). Transforming Prostate Cancer Care: Innovations in Diagnosis, Treatment, and Future Directions. *International Journal of Molecular Sciences, 26*(11), 5386. Advance online publication. DOI: 10.3390/ijms26115386 PMID: 40508192

Valla, V., Tzelepi, K., Charitou, P., Lewis, A., Polatidis, B., Koukoura, A., Karapatsia, A., Antonopoulou, K., Prodromidou, K., Papadaki, E., & Vassiliadis, E. (2023). Use of Real-World Evidence for International Regulatory Decision Making in Medical Devices. *International Journal of Digital Health, 3*(1), 1. DOI: 10.29337/ijdh.50

Vallée, A. (2025, August 5). Digital Twins for Personalized Medicine Require Epidemiological Data and Mathematical Modeling: Viewpoint. *Journal of Medical Internet Research, 27*, e72411. DOI: 10.2196/72411 PMID: 40762974

van Beljouw, S. P. B., Sanders, J., Rodríguez-Molina, A., & Brouns, S. J. J. (2023). RNA-targeting CRISPR–Cas systems. *Nature Reviews. Microbiology, 21*(1), 21–34. DOI: 10.1038/s41579-022-00793-y PMID: 36171275

van Belzen, I. A., van Tuil, M., Badloe, S., Janse, A., Verwiel, E. T., Santoso, M., & Hehir-Kwa, J. Y. (2024). Complex structural variation is prevalent and highly pathogenic in pediatric solid tumors. *Cell Genomics, 4*(11), 100675. DOI: 10.1016/j.xgen.2024.100675

van Dongen, J. E., Berendsen, J. T. W., Steenbergen, R. D. M., Wolthuis, R. M. F., Eijkel, J. C. T., & Segerink, L. I. (2020). Point-of-care CRISPR/Cas nucleic acid detection: Recent advances, challenges and opportunities. *Biosensors & Bioelectronics, 166*, 112445. DOI: 10.1016/j.bios.2020.112445

van Wijk, Y., Halilaj, I., van Limbergen, E., Walsh, S., Lutgens, L., Lambin, P., & Vanneste, B. G. (2019). Decision support systems in prostate cancer treatment: An overview. *BioMed Research International, 2019*(1), 4961768. DOI: 10.1155/2019/4961768 PMID: 31281840

Varghese, J., Holz, C., Neuhaus, P., Bernardi, M., Boehm, A., Ganser, A., & Dugas, M. (2016). Key data elements in myeloid leukemia. In *Exploring Complexity in Health: An Interdisciplinary Systems Approach* (pp. 282–286). IOS Press.

Varpe, A., Bhujbal, R., & Kumar, R. (2024). Enhancing predictive power in personalized breast cancer treatment through the integration of multi-omics data and clinical information with deep learning models. In *2024 MIT Art, Design and Technology School of Computing International Conference (MITADTSoCiCon)* (pp. 1-7). IEEE. DOI: 10.1109/MITADTSoCiCon60330.2024.10575890

Vasaikar, S., Huang, C., Wang, X., Petyuk, V. A., Savage, S. R., Wen, B., Dou, Y., Zhang, Y., Shi, Z., Arshad, O. A., Gritsenko, M. A., Zimmerman, L. J., McDermott, J. E., Clauss, T. R., Moore, R. J., Zhao, R., Monroe, M. E., Wang, Y.-T., Chambers, M. C., & Zhang, B. (2019). Proteogenomic analysis of human colon cancer reveals new therapeutic opportunities. *Cell, 177*(4), 1035–1049.e19. DOI: 10.1016/j.cell.2019.03.030 PMID: 31031003

Vashani, D., Paul, H., Nguyen, S., Topkiran, U. C., Valimukhametova, A. R., Dorsky, A., Sottile, O., Durmaz, L., Gonzalez-Rodriguez, R., & Naumov, A. V. (2026). Precursor-dependent optical and structural properties of eleven NIR-emissive graphene quantum dots for bioimaging applications. *2D Materials, 13*(2), 025017. DOI: 10.1088/2053-1583/ae4e41

Vatta, S. (2025). *Role of Artificial Intelligence and Machine Learning in Advancing Nanomedicine for Breast Cancer.* Next-Gen Nanomedicine Breast Cancer From Bench to Bedside Beyond. DOI: 10.70593/978-93-7185-537-2_11

Vaughan, O. (2024). Building brain–computer interfaces with graphene. *Nature Electronics, 7*(12), 1061–1062. DOI: 10.1038/s41928-024-01308-8

Večurkovská, I., Roškovičová, V., & Kaťuchová, J. (2025, September 5). Challenges in the Clinical Application of Machine Learning for Pancreatic Cancer. *Bratislavské Lekárske Listy, 126*(10), 2437–2450. DOI: 10.1007/s44411-025-00312-4

Vegesna, V. V. (2024). Medicine: a frame work for tailored cancer treatment. *International journal of innovative research in advanced engineering, 6*(11), 747-752.

Vegesna, V. V. (2024). Medicine: A framework for tailored cancer treatment. *International journal of innovative research in advanced engineering, 6*(11), 747-752.

Velpula, T., & Buddolla, V. (2025, June). Enhancing detection and monitoring of circulating tumor cells: Integrative approaches in liquid biopsy advances. *The Journal of Liquid Biopsy, 8*, 100297. DOI: 10.1016/j.jlb.2025.100297 PMID: 40453103

Venkatesh, K. P., Brito, G., & Kamel Boulos, M. N. (2024). Health digital twins in life science and health care innovation. *Annual Review of Pharmacology and Toxicology, 64*(1), 159–170. DOI: 10.1146/annurev-pharmtox-022123-022046 PMID: 37562495

Ventelou, L., Ginzac, A., Ferreira, M.-C., Canetti, L., Philippe, S., Pinard, C., Durando, X., & Bernadach, M. (2026). Circulating tumour DNA in head and neck squamous-cell carcinomas: A literature review. *Cancer Treatment Reviews, 142*, 103070. Advance online publication. DOI: 10.1016/j.ctrv.2025.103070 PMID: 41435530

Venunadh, A., & Joy, S. P. (2024). A comprehensive study on technological advancements in treatment and prognosis of acute myeloid leukemia. In *2024 International Conference on Intelligent and Innovative Technologies in Computing, Electrical and Electronics (IITCEE)* (pp. 1-7). IEEE. DOI: 10.1109/IITCEE59897.2024.10467781

Verma, V., & Kumar, D. (2025). Artificial intelligence and machine learning in drug discovery: From lead discovery to clinical validation (2020–2025). *Letters in Drug Design & Discovery*, *22*(12), 100341. DOI: 10.1016/j.lddd.2026.100341

Veryaskina, Y. A., Titov, S. E., & Zhimulev, I. F. (2022). Reference genes for qPCR-based miRNA expression profiling in 14 human tissues. *Medical Principles and Practice : International Journal of the Kuwait University, Health Science Centre*, *31*(4), 322–332. DOI: 10.1159/000524283 PMID: 35354155

Vial, A., Stirling, D., Field, M., Ros, M., Ritz, C., Carolan, M., & Miller, A. A. (2018). The role of deep learning and radiomic feature extraction in cancer-specific predictive modelling: A review. *Translational Cancer Research*, *7*(3), 803–816. DOI: 10.21037/tcr.2018.05.02

Vidal, E., Sayols, S., Moran, S., Guillaumet-Adkins, A., Schroeder, M. P., Royo, R., & Esteller, M. (2017). A DNA methylation map of human cancer at single base-pair resolution. *Oncogene*, *36*(40), 5648–5657. DOI: 10.1038/onc.2017.176

Viduedo, P. H. Z., Candeia, V. C. U., Martins, V. B. L., Destefani, A. C., & Destefani, V. C. (2024). Harnessing the power of AI and machine learning for next-generation sequencing data analysis: A comprehensive review of applications, challenges, and future directions in precision oncology. *Revista Ibero-Americana de Humanidades. Ciência & Educação (Bauru)*, *10*(8), 2898–2904.

Viergever, R. F., & Hendriks, T. C. (2016). The 10 largest public and philanthropic funders of health research in the world: What they fund and how they distribute their funds. *Health Research Policy and Systems*, *14*(1), 12. DOI: 10.1186/s12961-015-0074-z PMID: 26892771

Vijayakumar, S. (2023). Cancer targeted contrast studies and photothermal therapy using engineered gold nanoparticles. *Applied Physics. A, Materials Science & Processing*, *129*(6), 1–9. DOI: 10.1007/s00339-023-06736-5

Vikal, A., Maurya, R., Bhowmik, S., Patel, P., Singh, R., Das Gupta, G., & Das Kurmi, B. (2025, April). Precision Genome Editing: The Synergy of CRISPR and Nanotechnology in Cancer Treatment. *Current Cancer Therapy Reviews*, *21*(3), 265–277. DOI: 10.2174/0115733947290474240315062214

Villiger, L., Joung, J., Koblan, L., Weissman, J., Abudayyeh, O. O., & Gootenberg, J. S. (2024). CRISPR technologies for genome, epigenome and transcriptome editing. *Nature Reviews. Molecular Cell Biology*, *25*(6), 464–487. DOI: 10.1038/s41580-023-00697-6 PMID: 38308006

Vlachogiannis, G., Hedayat, S., Vatsiou, A., Jamin, Y., Fernández-Mateos, J., Khan, K., Lampis, A., Eason, K., Huntingford, I., Burke, R., Rata, M., Koh, D.-M., Tunariu, N., Collins, D., Hulkki-Wilson, S., Ragulan, C., Spiteri, I., Moorcraft, S. Y., Chau, I., & Valeri, N. (2018). Patient-derived organoids model treatment response of metastatic gastrointestinal cancers. *Science*, *359*(6378), 920–926. DOI: 10.1126/science.aao2774 PMID: 29472484

Vobugari, N., Raja, V., Sethi, U., Gandhi, K., Raja, K., & Surani, S. R. (2022). Advancements in Oncology with Artificial Intelligence—A Review Article. *Cancers (Basel), 14*(5), 1349. Advance online publication. DOI: 10.3390/cancers14051349 PMID: 35267657

Vogelstein, B., Fearon, E. R., Hamilton, S. R., Kern, S. E., Preisinger, A. C., Leppert, M., Nakamura, Y., White, R., Smits, A. M., & Bos, J. L. (1988). Genetic alterations during colorectal-tumor development. *The New England Journal of Medicine, 319*(9), 525–532. DOI: 10.1056/NEJM198809013190901 PMID: 2841597

Vora, L. K., Gholap, A. D., Jetha, K., Thakur, R. R. S., Solanki, H. K., & Chavda, V. P. (2023a). Artificial Intelligence in Pharmaceutical Technology and Drug Delivery Design. *Pharmaceutics, 15*(7), 1916. DOI: 10.3390/pharmaceutics15071916 PMID: 37514102

Vosoughi, P., Naghib, S. M., & Takdehghan, G. (2025, December). Machine learning in cancer prognostic and diagnostic biomarkers: A promising approach for early cancer detection. *Sensors and Actuators Reports, 10*, 100385. DOI: 10.1016/j.snr.2025.100385

Vu, S., Ebah, K., Zaibi, F., Qaiss, A., Siaj, M., & Izquierdo, R. (2026). Wet-Spun Graphene-Enhanced PVDF Fibers for Flexible Nanocomposites. *Materials (Basel), 19*(7), 1376. DOI: 10.3390/ma19071376 PMID: 41976663

Vyskočil, J., Douda, P., Novák, P., & Wally, B. (2023). A Digital Twin-Based Distributed Manufacturing Execution System for Industry 4.0 with AI-Powered On-The-Fly Replanning Capabilities. *Sustainability (Basel), 15*(7), 6251. DOI: 10.3390/su15076251

W.-A. Kwon, H. Kim, and Y. S. Song, "Prostate Cancer at the Microbial Crossroads: Illuminating a New Frontier in Precision Medicine," *World J. Mens. Health*, vol. 43, 2025.

Wadan, A.-H. S., Ali, M. H., & Ellakwa, D. E.-S. (2026). Sustainable nanomaterials for precision dental medicine: Green synthesis, therapeutic applications, and future directions. *Journal of Nanobiotechnology, 24*(1), 190. Advance online publication. DOI: 10.1186/s12951-025-04008-3 PMID: 41652613

Wang, B., Chen, J. Z., Luo, X. Q., Wan, G. H., Tang, Y. L., & Wang, Q. P. (2022). The application of genome-wide CRISPR-Cas9 screens to dissect the molecular mechanisms of toxins. *Computational and Structural Biotechnology Journal, 20*, 5076–5084. DOI: 10.1016/j.csbj.2022.09.012 PMID: 36187925

Wang, B., Hu, S., Teng, Y., Chen, J., Wang, H., Xu, Y., Wang, K., Li, Y., & Zhang, J. (2024). Current advance of nanotechnology in diagnosis and treatment for malignant tumors. *Signal Transduction and Targeted Therapy, 9*(1), 1–65. DOI: 10.1038/s41392-024-01889-y

Wang, H., Huang, J., Fang, X., Liu, M., Fan, X., & Li, Y. (2025, April 3). Advances in next-generation sequencing (NGS) applications in drug discovery and development. [Internet]. *Expert Opinion on Drug Discovery, 20*(4), 537–550. Retrieved November 2, 2025, from https://www.tandfonline.com/doi/full/10.1080/17460441.2025.2481262. DOI: 10.1080/17460441.2025.2481262 PMID: 40099494

Wang, K., Wang, Y., Li, Y., Fan, X., Xiao, S., & Hu, L. (2024). A review of the technology standards for enabling digital twin [version 2; peer review: 2 approved]. *Digital Twin, 1*(1), 4. DOI: 10.12688/digitaltwin.17549.2

Wang, M., Jin, L., Hang-Mei Leung, P., Wang-Ngai Chow, F., Zhao, X., Chen, H., Pan, W., Liu, H., & Li, S. (2024). Advancements in magnetic nanoparticle-based biosensors for point-of-care testing. *Frontiers in Bioengineering and Biotechnology*, *25*(12), 1393789. DOI: 10.3389/fbioe.2024.1393789

Wang, M., Liu, H., Huang, J., Cai, T., Xu, Z. P., & Zhang, L. (2025). Advancing cancer gene therapy: The emerging role of nanoparticle delivery systems. *Journal of Nanobiotechnology*, *23*(1), 362. DOI: 10.1186/s12951-025-03433-8 PMID: 40394591

Wang, Q., Atluri, K., Tiwari, A. K., & Babu, R. J. (2023). Exploring the Application of Micellar Drug Delivery Systems in Cancer Nanomedicine. *Pharmaceuticals (Basel, Switzerland)*, *16*(3), 433. DOI: 10.3390/ph16030433 PMID: 36986532

Wang, Q., Liu, Y., Li, C., Xu, B., Xu, S., & Liu, B. (2025). Machine learning-enhanced nanoparticle design for precision cancer drug delivery. *Advancement of Science*, *12*(30), e03138.

Wang, S. W., Gao, C., Zheng, Y. M., Yi, L., Lu, J. C., Huang, X. Y., Cai, J.-B., Zhang, P.-F., Cui, Y.-H., & Ke, A. W. (2022). Current applications and future perspective of CRISPR/Cas9 gene editing in cancer. *Molecular Cancer*, *21*(1), 57. DOI: 10.1186/s12943-022-01518-8 PMID: 35189910

Wang, S., Zhu, H., Zhou, H., Cheng, J., & Yang, H. (2020). MSpectraAI: A powerful platform for deciphering proteome profiling of multi-tumor mass spectrometry data by using deep neural networks. *BMC Bioinformatics*, *21*(1), 1–15. DOI: 10.1186/s12859-020-03783-0

Wang, T. W., Wang, C. K., Hong, J. S., Chao, H. S., Chen, Y. M., & Wu, Y. T. (2025, February 12). Deep Learning in Thoracic Oncology: Meta-Analytical Insights into Lung Nodule Early-Detection Technologies. *Cancers (Basel)*, *17*(4), 621. DOI: 10.3390/cancers17040621 PMID: 40002216

Wang, X., Yan, J., Ye, Z., Zhang, Z., Wang, S., Hao, S., & Wei, G. (2022). Reorganization of 3D chromatin architecture in doxorubicin-resistant breast cancer cells. *Frontiers in Cell and Developmental Biology*, *10*, 974750. DOI: 10.3389/fcell.2022.974750

Wang, Y., Carter, B. Z., Li, Z., & Huang, X. (2022, January 7). Application of machine learning methods in clinical trials for precision medicine. *JAMIA Open*, *5*(1), ooab107. DOI: 10.1093/jamiaopen/ooab107 PMID: 35178503

Wang, Y., Zhang, W., Liu, X., Tian, L., Li, W., He, P., Huang, S., He, F., & Pan, X. (2025, January). Artificial intelligence in precision medicine for lung cancer: A bibliometric analysis. [Internet]. *Digital Health*, *11*, 20552076241300229. Retrieved November 2, 2025, from https://journals.sagepub.com/doi/10.1177/20552076241300229. DOI: 10.1177/20552076241300229 PMID: 39758259

Wang, Z. Z., Li, X. H., Wen, X. L., Wang, N., Guo, Y., Zhu, X., Fu, S. H., Xiong, F. F., Bai, J., Gao, X. L., & Wang, H. J. (2023, March 21). Integration of multi-omics data reveals a novel hybrid breast cancer subtype and its biomarkers. *Frontiers in Oncology*, *13*, 1130092. DOI: 10.3389/fonc.2023.1130092 PMID: 37064087

Wang, Z., & Maini, P. K. (2017). Editorial special section on multiscale cancer modelling. *IEEE Transactions on Bio-Medical Engineering*, *64*(3), 501–503. DOI: 10.1109/TBME.2017.2655439 PMID: 28824199

Wang, Z., & Wang, Y. (2019). Extracting a biologically latent space of lung cancer epigenetics with variational autoencoders. *BMC Bioinformatics, 20*(S18, Suppl 18), 568. DOI: 10.1186/s12859-019-3130-9

Wang, Z., Xing, H., Liu, A., Zhang, J., Chen, L., Zhu, W., Li, X., Wang, Y., Zhang, X., & Liu, Q. (2023). Multifunctional nano-system for multi-mode targeted imaging and enhanced photothermal therapy of metastatic prostate cancer. *Acta Biomaterialia, 166*, 581–592. DOI: 10.1016/j.actbio.2023.05.014 PMID: 37172637

Wang, Z., Zhou, L., & Wu, X. (2024). Influencing factors and solution strategies of chimeric antigen receptor T-cell therapy (CAR–T) cell immunotherapy. *Oncology Research, 32*(9), 1479–1516. DOI: 10.32604/or.2024.048564 PMID: 39220130

Wan, J. C. M., Massie, C., Garcia-Corbacho, J., Mouliere, F., Brenton, J. D., Caldas, C., Pacey, S., Baird, R., & Rosenfeld, N. (2017, April). Liquid biopsies come of age: Towards implementation of circulating tumour DNA. *Nature Reviews. Cancer, 17*(4), 223–238. DOI: 10.1038/nrc.2017.7 PMID: 28233803

Waqas, A., Naveed, J., Shahnawaz, W., Asghar, S., Bui, M. M., & Rasool, G. (2024). Digital pathology and multimodal learning on oncology data. *BJR|. Artificial Intelligence, 1*(1), ubae014.

Ward, C. L., Shaw, D., Sprumont, D., Sankoh, O., Tanner, M., & Elger, B. (2018). Good collaborative practice: Reforming capacity building governance of international health research partnerships. *Globalization and Health, 14*(1), 1. DOI: 10.1186/s12992-017-0319-4 PMID: 29310698

Warner, J., Prada Jardim, A., & Albera, C. (2025). Artificial Intelligence: Applications in Pharmacovigilance Signal Management. *Pharmaceutical Medicine, 39*(3), 183–198. DOI: 10.1007/s40290-025-00561-2 PMID: 40257538

Wasilewski, T., Kamysz, W., & Gębicki, J. (2024). AI-assisted detection of biomarkers by sensors and biosensors for early diagnosis and monitoring. *Biosensors (Basel), 14*(7), 356. DOI: 10.3390/bios14070356 PMID: 39056632

Wei, F., Kouro, T., Nakamura, Y., Ueda, H., Iiizumi, S., Hasegawa, K., & Sasada, T. (2024). Enhancing Mass spectrometry-based tumor immunopeptide identification: Machine learning filter leveraging HLA binding affinity, aliphatic index and retention time deviation. *Computational and Structural Biotechnology Journal, 23*, 859–869. DOI: 10.1016/j.csbj.2024.01.023

Wei, L., Niraula, D., Gates, E. D., Fu, J., Luo, Y., Nyflot, M. J., & Cui, S. (2023a). Artificial intelligence (AI) and machine learning (ML) in precision oncology: A review on enhancing discoverability through multiomics integration. *The British Journal of Radiology, 96*(1150), 20230211. DOI: 10.1259/bjr.20230211

Wei, R., & Mahmood, A. (2020). Recent advances in variational autoencoders with representation learning for biomedical informatics: A survey. *IEEE Access : Practical Innovations, Open Solutions, 9*, 4939–4956. DOI: 10.1109/ACCESS.2020.3048309

Wei, Y., Li, L., Zhao, X., Yang, H., Sa, J., Cao, H., & Cui, Y. (2023b). Cancer subtyping with heterogeneous multi-omics data via hierarchical multi-kernel learning. *Briefings in Bioinformatics, 24*(1), bbac488. DOI: 10.1093/bib/bbac488

Welti, M., Dimitriou, F., Gutzmer, R., & Dummer, R. (2022, November 8). Triple Combination of Immune Checkpoint Inhibitors and BRAF/MEK Inhibitors in BRAFV600 Melanoma: Current Status and Future Perspectives. *Cancers (Basel), 14*(22), 5489. DOI: 10.3390/cancers14225489 PMID: 36428582

Wess, M., Andersen, M. K., Midtbust, E., Guillem, J. C. C., Viset, T., Størkersen, Ø., & Tessem, M. B. (2024). Spatial integration of multi-omics data using the novel multi-omics imaging integration toolset. bioRxiv, 2024-06.

Whiteman, D. C., Green, A. C., & Olsen, C. M. (2016, June). The Growing Burden of Invasive Melanoma: Projections of Incidence Rates and Numbers of New Cases in Six Susceptible Populations through 2031. *The Journal of Investigative Dermatology, 136*(6), 1161–1171. DOI: 10.1016/j.jid.2016.01.035 PMID: 26902923

Wiegrebe, S., Kopper, P., Sonabend, R., Bischl, B., & Bender, A. (2024, February 19). Deep learning for survival analysis: A review. [Internet]. *Artificial Intelligence Review, 57*(3), 65. Retrieved November 1, 2025, from https://link.springer.com/10.1007/s10462-023-10681-3. DOI: 10.1007/s10462-023-10681-3

Wilhelm, S., Tavares, A. J., Dai, Q., Ohta, S., Audet, J., Dvorak, H. F., & Chan, W. C. W. (2016). Analysis of nanoparticle delivery to tumours. *Nature Reviews. Materials, 1*(5), 16014. DOI: 10.1038/natrevmats.2016.14

Williams, J., Xu, B., Putnam, D., Thrasher, A., Li, C., Yang, J., & Chen, X. (2021). MethylationToActivity: A deep-learning framework that reveals promoter activity landscapes from DNA methylomes in individual tumors. *Genome Biology, 22*(1), 1–27. DOI: 10.1186/s13059-020-02220-y

Williamson, S. M., & Prybutok, V. (2024). Balancing privacy and progress: A review of privacy challenges, systemic oversight, and patient perceptions in AI-driven healthcare. *Applied Sciences (Basel, Switzerland), 14*(2), 675. DOI: 10.3390/app14020675

Wills, Q. F., & Mead, A. J. (2015). Application of single-cell genomics in cancer: Promise and challenges. *Human Molecular Genetics, 24*(R1), R74–R84. DOI: 10.1093/hmg/ddv235

Wirbel, J., Pyl, P. T., Kartal, E., Zych, K., Kashani, A., Milanese, A., Fleck, J. S., Voigt, A. Y., Palleja, A., Ponnudurai, R., Sunagawa, S., Coelho, L. P., Schrotz-King, P., Vogtmann, E., Habermann, N., Niméus, E., Thomas, A. M., Manghi, P., Gandini, S., & Zeller, G. (2019). Meta-analysis of fecal metagenomes reveals global microbial signatures that are specific for colorectal cancer. *Nature Medicine, 25*(4), 679–689. DOI: 10.1038/s41591-019-0406-6 PMID: 30936547

Wolf, F. A., Hamey, F. K., Plass, M., Solana, J., Dahlin, J. S., Göttgens, B., & Theis, F. J. (2019). PAGA: Graph abstraction reconciles clustering with trajectory inference through a topology preserving map of single cells. *Genome Biology, 20*(1), 1–9. DOI: 10.1186/s13059-019-1663-x

Wong, K. H., Yang, D., Chen, S., He, C., & Chen, M. (2022). Development of nanoscale drug delivery systems of dihydroartemisinin for cancer therapy: A review. *Asian Journal of Pharmaceutical Sciences, 17*(4), 475–490. DOI: 10.1016/j.ajps.2022.04.005 PMID: 36105316

Wu, A., Liu, X., Fruhstorfer, C., & Jiang, X. (2024). Clinical insights into structure, regulation, and targeting of ABL kinases in human leukemia. *International Journal of Molecular Sciences, 25*(6), 3307. DOI: 10.3390/ijms25063307 PMID: 38542279

Wu, C. J., Acun, B., Raghavendra, R., & Hazelwood, K. (2024a). Beyond efficiency: Scaling AI sustainably. *IEEE Micro*, *44*(5), 37–46. DOI: 10.1109/MM.2024.3409275

Wu, C., Lima, E., Stowers, C., Xu, Z., Yam, C., Son, J. B., & Yankeelov, T. (2024b). Abstract PO5-03-06: Optimizing therapeutic regimens via digital twins to improve triple negative breast cancer response to neoadjuvant therapy. *Cancer Research*, *84*(9, Supplement), PO5–PO03. DOI: 10.1158/1538-7445.SABCS23-PO5-03-06

Wu, C., Lorenzo, G., Hormuth, D. A., Lima, E. A., Slavkova, K. P., DiCarlo, J. C., & Yankeelov, T. E. (2022). Integrating mechanism-based modelling with biomedical imaging to build practical digital twins for clinical oncology. *Biophysics Reviews*, *3*(2), 021304. DOI: 10.1063/5.0086789 PMID: 35602761

Wu, J., Wang, W., Zhang, J., Zhou, B., Zhao, W., Su, Z., & Chen, S. (2019). DeepHLApan: A deep learning approach for neoantigen prediction considering both HLA-peptide binding and immunogenicity. *Frontiers in Immunology*, *10*, 2559. DOI: 10.3389/fimmu.2019.02559

Wu, X., Yang, X., Dai, Y., Zhao, Z., Zhu, J., Guo, H., & Yang, R. (2024). Single-cell sequencing to multi-omics: Technologies and applications. *Biomarker Research*, *12*(1), 110. DOI: 10.1186/s40364-024-00643-4 PMID: 39334490

Wu, Y., & Xie, L. (2025). AI-driven multi-omics integration for multi-scale predictive modeling of genotype-environment-phenotype relationships. *Computational and Structural Biotechnology Journal*, *27*, 265–277. DOI: 10.1016/j.csbj.2024.12.030

Wu, Z., Wang, Y., Zhang, Y., Chen, W., Wang, Y., & Ji, Q. (2020). Strategies for developing CRISPR-based gene editing methods in bacteria. *Small Methods*, *4*(2), 1900560. DOI: 10.1002/smtd.201900560

Xia, K., Saidy, C., Kirkpatrick, M., Anumbe, N., Sheth, A., & Harik, R. (2021). Towards Semantic Integration of Machine Vision Systems to Aid Manufacturing Event Understanding. *Sensors (Basel)*, *21*(13), 4276. DOI: 10.3390/s21134276 PMID: 34206522

Xia, W., Tao, Z., Zhu, B., Zhang, W., Liu, C., Chen, S., & Song, M. (2021, August 24). Targeted delivery of drugs and genes using polymer nanocarriers for cancer therapy. *International Journal of Molecular Sciences*, *22*(17), 9118. DOI: 10.3390/ijms22179118 PMID: 34502028

Xie, L., Li, J., Wang, L., & Dai, Y. (2023). Engineering metal-phenolic networks for enhancing cancer therapy by tumor microenvironment modulation. *Wiley Interdisciplinary Reviews. Nanomedicine and Nanobiotechnology*, *15*(3), e1864. DOI: 10.1002/wnan.1864 PMID: 36333962

Xie, Y., Ruan, F., Li, Y., Luo, M., Zhang, C., Chen, Z., & Tang, C. (2024). Spatial chromatin accessibility sequencing resolves high-order spatial interactions of epigenomic markers. *eLife*, *12*, RP87868. DOI: 10.7554/eLife.87868

Xin, L., Huang, C., Li, H., Huang, S., Feng, Y., Kong, Z., & Tang, H. (2024). Artificial intelligence for central dogma-centric multi-omics: challenges and breakthroughs. *arXiv preprint arXiv:2412.12668.*

Xiu, X., Wu, Y., Li, J., Lin, D., Sun, X., Su, X., Weng, Z., Zuo, X., Yang, X., Fan, C., Wang, Y., Zhang, D. Y., & Song, P. (2026). Anchored random reverse primer sequencing for quantitative detection of novel gene fusions. *Nature Biomedical Engineering*, •••, 1–16. DOI: 10.1038/s41551-025-01564-9 PMID: 41482592

Xu, J., Jiang, T., Guo, J., Pan, T., & Li, Y. (2023). Landscape and perturbation of enhancer-driven core transcription regulatory circuits in cancer. *Molecular Therapy. Nucleic Acids*, *32*, 872–874. DOI: 10.1016/j.omtn.2023.05.011

Xu, J., Xiao, Y., Wang, W. H., Ning, Y., Shenkman, E. A., Bian, J., & Wang, F. (2022). Algorithmic fairness in computational medicine. *EBioMedicine*, *84*, 104250. DOI: 10.1016/j.ebiom.2022.104250 PMID: 36084616

Xu, X., Li, J., Zhu, Z., Zhao, L., Wang, H., Song, C., Chen, Y., Zhao, Q., Yang, J., & Pei, Y. (2024, February 25). A Comprehensive Review on Synergy of Multi-Modal Data and AI Technologies in Medical Diagnosis. [Internet]. *Bioengineering (Basel, Switzerland)*, *11*(3), 219. Retrieved November 2, 2025, from https://www.mdpi.com/2306-5354/11/3/219. DOI: 10.3390/bioengineering11030219 PMID: 38534493

Xu, Y., Cao, L., Chen, Y., Zhang, Z., Liu, W., Li, H., & Xu, W. (2024). Integrating machine learning in metabolomics: A path to enhanced diagnostics and data interpretation. *Small Methods*, *8*(12), 2400305. DOI: 10.1002/smtd.202400305

Xuyan, H., Meng, S., Chengxing, S., Haoxuan, L., & Jianlin, Z. (2025, October). Visual-language reasoning large language models for primary care: Advancing clinical decision support through multimodal AI. [Internet]. *The Visual Computer*, *41*(13), 11327–11348. Retrieved November 2, 2025, from https://link.springer.com/10.1007/s00371-025-04109-y. DOI: 10.1007/s00371-025-04109-y

Y. Cheng, J. Kong, X. Liu, and S. Li, "Recent Advances and Emerging Directions in Machine Learning-Based Breast Cancer Drug Discovery: A Comprehensive Review," *Breast Cancer Targets Ther.*, vol. 18, no. null, p. 586786, Dec. 2026, .DOI: 10.2147/BCTT.S586786

Yadav, D., Patil-Takbhate, B., Khandagale, A., Bhawalkar, J., Tripathy, S., & Khopkar-Kale, P. (2023). Next-generation sequencing transforming clinical practice and precision medicine. *Clinica Chimica Acta; International Journal of Clinical Chemistry*, *551*, 117568. DOI: 10.1016/j.cca.2023.117568 PMID: 37839516

Yaeger, R., Weiss, J., Pelster, M. S., Spira, A. I., Barve, M., Ou, S.-H. I., Leal, T. A., Bekaii-Saab, T. S., Paweletz, C. P., Heavey, G. A., Christensen, J. G., Velastegui, K., Kheoh, T., Der-Torossian, H., Klempner, S. J., & Hong, D. S. (2023). Adagrasib with or without cetuximab in colorectal cancer. *The New England Journal of Medicine*, *388*(1), 44–54. DOI: 10.1056/NEJMoa2212419 PMID: 36546659

Yamagishi, Y., Baba, Y., Suzuki, J., Okada, Y., Kanao, K., & Oyama, M. (2025). Few-shot learning for prostate cancer detection on MRI: Comparative analysis with radiologists' performance. *Journal of Imaging Informatics in Medicine*, *39*(2), 1–13. DOI: 10.1007/s10278-025-01581-9

Yamamoto, H., & Hirasawa, A. (2022). Homologous Recombination Deficiencies and Hereditary Tumors. *International Journal of Molecular Sciences*, *23*(1), 348. Advance online publication. DOI: 10.3390/ijms23010348 PMID: 35008774

Yang, C., Zhou, C., Wipf, D., & Li, Z. (2025). On powerful ways to generate: Autoregression, diffusion, and beyond. *arXiv*. /arXiv.2510.06190DOI: 10.48550

Yang, C. H., Moi, S. H., Chuang, L. Y., & Lin, Y. D. (2023a). An information fusion system-driven deep neural networks with application to cancer mortality risk estimate. *IEEE Transactions on Neural Networks and Learning Systems*.

Yang, F., He, Q., Dai, X., Zhang, X., & Song, D. (2023). The potential role of nanomedicine in the treatment of breast cancer to overcome the obstacles of current therapies. *Frontiers in Pharmacology*, *14*, 1143102. DOI: 10.3389/fphar.2023.1143102 PMID: 36909177

Yang, H. Y., Ma, D., Liu, Y. R., Hu, X., Zhang, J., Wang, Z. H., Di, G.-H., Hu, X.-C., & Shao, Z.-M. (2017, April 25). Impact of hormone receptor status and distant recurrence-free interval on survival benefits from trastuzumab in HER2-positive metastatic breast cancer. *Scientific Reports*, *7*(1), 1134. DOI: 10.1038/s41598-017-00663-1 PMID: 28442763

Yang, R., Tsigelny, I. F., Kesari, S., & Kouznetsova, V. L. (2024a). Colorectal cancer detection via metabolites and machine learning. *Current Issues in Molecular Biology*, *46*(5), 4133–4146. DOI: 10.3390/cimb46050254

Yang, X., Vladmirovich, R. I., Georgievna, P. M., Sergeevna, A. Y., He, M., Zeng, Z., & Sergeevich, K. T. (2024b). Personalized chemotherapy selection for patients with triple-negative breast cancer using deep learning. *Frontiers in Medicine*, *11*, 1418800. DOI: 10.3389/fmed.2024.1418800

Yang, X., & Zhang, B. (2023). A review on CRISPR/Cas: A versatile tool for cancer screening, diagnosis, and clinic treatment. *Functional & Integrative Genomics*, *23*(2), 182. DOI: 10.1007/s10142-023-01117-w PMID: 37231285

Yang, Z., Chen, Z., Matsubara, Y., & Sakurai, Y. (2023b). MoCLIM: Towards accurate cancer subtyping via multi-omics contrastive learning with omics-inference modeling. In *Proceedings of the 32nd ACM International Conference on Information and Knowledge Management* (pp. 2895-2905).

Yankeelov, T. E., An, G., Saut, O., Luebeck, E. G., Popel, A. S., Ribba, B., & Genin, G. M. (2016). Multi-scale modelling in clinical oncology: Opportunities and barriers to success. *Annals of Biomedical Engineering*, *44*(9), 2626–2641. DOI: 10.1007/s10439-016-1691-6 PMID: 27384942

Yan, X., Yeh, C., & Zou, L. (2020). Clinical Applications of Circulating Tumor DNA, Circulating Tumor Cells, and Exosomes as Liquid Biopsy-Based Tumor Biomarkers. *Journal of Applied Bioanalysis*, *6*(3), 107–130. DOI: 10.17145/jab.20.013

Yao, L., Shen, H., Laird, P. W., Farnham, P. J., & Berman, B. P. (2015). Inferring regulatory element landscapes and transcription factor networks from cancer methylomes. *Genome Biology*, *16*(1), 1–21. DOI: 10.1186/s13059-015-0668-3

Yapar, O. (2024). Federated learning for national healthcare systems: Balancing privacy and innovation. *World Journal of Advanced Engineering Technology and Sciences*, *13*(1), 10–30574.

Yasamineh, S., Nikben, N., Hamed Ahmed, M., Abdul Kareem, R., Kadhim Al Aridhy, A., & Hosseini Hooshiar, M. (2024). Increasing the sensitivity and accuracy of detecting exosomes as biomarkers for cancer monitoring using optical nanobiosensors. *Cancer Cell International*, *24*(1), 1–44. DOI: 10.1186/s12935-024-03379-1

Yavuz, Y., Kartal, I., Cesur, S., Kanli, Z., Kaya, E., Tinaz, G., & Gunduz, O. (2026). Electrospun PVA/CS/HA/BA Nanofiber Scaffolds with Enhanced Mechanical Stability and Antifungal Activity for Bone Tissue Engineering. *Materials (Basel)*, *19*(2), 412. DOI: 10.3390/ma19020412 PMID: 41598123

Yazzourh, S., Savy, N., Saint-Pierre, P., & Kosorok, M. R. (2025). Medical Knowledge Integration Into Reinforcement Learning Algorithms for Dynamic Treatment Regimes. *International Statistical Review = Revue Internationale de Statistique*.

Yetgin, A. (2025, April 7). Revolutionizing multi-omics analysis with artificial intelligence and data processing. *Quantitative Biology*, *13*(3), e70002. DOI: 10.1002/qub2.70002 PMID: 41675959

Yi, M., Zheng, X., Niu, M., Zhu, S., Ge, H., & Wu, K. (2022, January 21). Combination strategies with PD-1/PD-L1 blockade: Current advances and future directions. *Molecular Cancer*, *21*(1), 28. DOI: 10.1186/s12943-021-01489-2 PMID: 35062949

Yin, B., Wang, J., Zhang, Y., Liu, C., Wang, Y., & Li, Q. (2022). A CRISPR-Cas12a integrated SERS nanoplatform with chimeric DNA/RNA hairpin guide for ultrasensitive nucleic acid detection. *Theranostics*, *12*(13), 5914–5930. DOI: 10.7150/thno.75816

Yin, G., Liu, L., Yu, T., Yu, L., Feng, M., Zhou, C., Wang, X., Teng, G., Ma, Z., Zhou, W., Ye, C., Zhang, J., Ji, C., Zhao, L., Zhou, P., Guo, Y., Meng, X., Fu, Q., Zhang, Q., & Yu, Z. (2024). Genomic and transcriptomic analysis of breast cancer identifies novel signatures associated with response to neoadjuvant chemotherapy. *Genome Medicine*, *16*(1), 11. DOI: 10.1186/s13073-024-01286-8 PMID: 38217005

Ying, J., Yang, L., Yin, J. C., Xia, G., Xing, M., Chen, X., Pang, J., Wu, Y., Bao, H., Wu, X., Shao, Y., Zhu, L., & Cheng, X. (2021, September). Additive effects of variants of unknown significance in replication repair-associated DNA polymerase genes on mutational burden and prognosis across diverse cancers. *Journal for Immunotherapy of Cancer*, *9*(9), e002336. DOI: 10.1136/jitc-2021-002336 PMID: 34479923

Ying, S., Guan, Z., Ofoegbu, P. C., Clubb, P., Rico, C., He, F., & Hong, J. (2022). Green Synthesis of Nanoparticles: Current Developments and Limitations. *Environmental Technology & Innovation*, *26*, 102336. DOI: 10.1016/j.eti.2022.102336

Yokota, T., & Kanakura, Y. (2016). Genetic abnormalities associated with acute lymphoblastic leukemia. *Cancer Science*, *107*(6), 721–725. DOI: 10.1111/cas.12927 PMID: 26991355

Yoo, H., Kim, Y., Kim, J., Cho, H., & Kim, K. (2024). Overcoming Cancer Drug Resistance with Nanoparticle Strategies for Key Protein Inhibition. *Molecules (Basel, Switzerland)*, *29*(17), 3994. DOI: 10.3390/molecules29173994 PMID: 39274842

Young, A. L., Davis, H. C., & Challen, G. A. (2023). Droplet digital PCR for oncogenic KMT2A fusion detection. *The Journal of Molecular Diagnostics : JMD*, *25*(12), 898–906. DOI: 10.1016/j.jmoldx.2023.09.006 PMID: 37813299

Young, A. L., Oxtoby, N. P., Garbarino, S., Fox, N. C., Barkhof, F., Schott, J. M., & Alexander, D. C. (2024, February). Data-driven modelling of neurodegenerative disease progression: Thinking outside the black box. [Internet]. *Nature Reviews. Neuroscience, 25*(2), 111–130. Retrieved November 1, 2025, from https://www.nature.com/articles/s41583-023-00779-6. DOI: 10.1038/s41583-023-00779-6 PMID: 38191721

Yousefi, S., Amrollahi, F., Amgad, M., Dong, C., Lewis, J. E., Song, C., Gutman, D. A., Halani, S. H., Velazquez Vega, J. E., Brat, D. J., & Cooper, L. A. D. (2017, September 15). Predicting clinical outcomes from large scale cancer genomic profiles with deep survival models. *Scientific Reports, 7*(1), 11707. DOI: 10.1038/s41598-017-11817-6 PMID: 28916782

Yuan, C., Tian, T., Sun, J., Hu, M., Wang, X., Xiong, E., Cheng, M., Bao, Y., Lin, W., Jiang, J., Yang, C., Chen, Q., Zhang, H., Wang, H., Wang, X., Deng, X., Liao, X., Liu, Y., Wang, Z., & Zhou, X. (2020). Universal and naked-eye gene detection platform based on the clustered regularly interspaced short palindromic repeats/Cas12a/13a system. *Analytical Chemistry, 92*(5), 4029–4037. DOI: 10.1021/acs.analchem.9b05597

Yue, L., Niu, P., & Wang, Y. (2019). Guidelines for defining user requirement specifications (URS) of manufacturing execution system (MES) based on ISA-95 standard. *Journal of Physics: Conference Series, 1168*, 032065. DOI: 10.1088/1742-6596/1168/3/032065

Yu, J., Feng, Q., Wong, S. H., Zhang, D., Liang, Q. Y., Qin, Y., Tang, L., Zhao, H., Stenvang, J., Li, Y., Wang, X., Xu, X., Chen, N., Wu, W. K. K., Al-Aama, J., Nielsen, H. J., Kiilerich, P., Jensen, B. A. H., Yau, T. O., & Wang, J. (2017). Metagenomic analysis of faecal microbiome as a tool towards targeted non-invasive biomarkers for colorectal cancer. *Gut, 66*(1), 70–78. DOI: 10.1136/gutjnl-2015-309800 PMID: 26408641

Yun, Z., Zhang, Y., Wang, J., Xue, C., Wu, J., Han, X., & Hou, L. (2024). Global, regional and national burden of leukemia, 1990-2021: Analysis of data from the global burden of disease study 2021. *Blood, 144*(Supplement 1), 7907. DOI: 10.1182/blood-2024-205478

Yu, Y., Cai, G., Lin, R., Wang, Z., Chen, Y., Tan, Y., He, Z., Sun, Z., Ouyang, W., Yao, H., & Zhang, K. (2024, December). Multimodal data fusion AI model uncovers tumor microenvironment immunotyping heterogeneity and enhanced risk stratification of breast cancer. [Internet]. *MedComm, 5*(12), e70023. Retrieved November 1, 2025, from https://onlinelibrary.wiley.com/doi/10.1002/mco2.70023. DOI: 10.1002/mco2.70023 PMID: 39669975

Zaccariotto, G. D., Bistaffa, M. J., Zapata, A. M., Rodero, C., Coelho, F., Quitiba, J. V., Lima, L., Sterman, R., Cardoso, V. M., & Zucolotto, V. (2025, April 9). Cancer nanovaccines: Mechanisms, design principles, and clinical translation. *ACS Nano, 19*(17), 16204–16223. DOI: 10.1021/acsnano.4c15765 PMID: 40202241

Zack, M., Stupichev, D. N., Moore, A. J., Slobodchikov, I. D., Sokolov, D. G., Trifonov, I. F., & Gobbs, A. (2025, September 1). Artificial intelligence and multi-omics in pharmacogenomics: A new era of precision medicine. *Mayo Clinic Proceedings. Digital Health, 3*(3), 100246. DOI: 10.1016/j.mcpdig.2025.100246 PMID: 40881104

Zadeh Shirazi, A., Tofighi, M., Gharavi, A., & Gomez, G. A. (2024). The application of artificial intelligence to cancer research: A comprehensive guide. *Technology in Cancer Research & Treatment, 23*, 15330338241250324. DOI: 10.1177/15330338241250324 PMID: 38775067

Zang, Z., Chen, J., Dong, Y., Chen, L., Yang, M., Mu, M., Zhou, L., Zhang, W., Zou, G., & Liu, C. (2026). CRISPR-Cas12a biosensing technology advances and applications in precision diagnostics and cancer research. *Talanta, 301*, 129298. DOI: 10.1016/j.talanta.2025.129298 PMID: 41500123

Zaripov, E. A., Khraibah, A., Kasyanchyk, P., Radchanka, A., Hüttmann, N., & Berezovski, M. V. (2025, May). CE–MS Metabolomic and LC–MS Proteomic Analyses of Breast Cancer Exosomes Reveal Alterations in Purine and Carnitine Metabolism. *Journal of Proteome Research, 24*(5), 2505–2516. DOI: 10.1021/acs.jproteome.4c00795 PMID: 40036676

Zeng, J., Han, Z., Peng, X., Xiao, J., Wang, P., & Wang, Y. (2023). Polymerized feature-based domain adaptation for cervical cancer dose map prediction. In *2023 IEEE 20th International Symposium on Biomedical Imaging (ISBI)* (pp. 1-4). IEEE. DOI: 10.1109/ISBI53787.2023.10230491

Zeng, W. F., Mann, M., & Strauss, M. T. (2024). Artificial intelligence in mass spectrometry-based proteomics. In *Artificial Intelligence in Clinical Practice* (pp. 389–394). Academic Press. DOI: 10.1016/B978-0-443-15688-5.00010-3

Zeng, W., & Zhang, P. (2022). Resistance and recurrence of malignancies after CAR-T cell therapy. *Experimental Cell Research, 410*(2), 112971. DOI: 10.1016/j.yexcr.2021.112971 PMID: 34906583

Zhang, B. (2021). CRISPR/Cas gene therapy. *Journal of Cellular Physiology, 236*(4), 2459–2481. DOI: 10.1002/jcp.30064 PMID: 32959897

Zhang, B., Wang, J., Wang, X., Zhu, J., Liu, Q., Shi, Z., Chambers, M. C., Zimmerman, L. J., Shaddox, K. F., Kim, S., Davies, S. R., Wang, S., Wang, P., Kinsinger, C. R., Rivers, R. C., Rodriguez, H., Townsend, R. R., Ellis, M. J. C., Carr, S. A., & Liebler, D. C. (2014). Proteogenomic characterization of human colon and rectal cancer. *Nature, 513*(7518), 382–387. DOI: 10.1038/nature13438 PMID: 25043054

Zhang, C., Li, N., Zhang, P., Jiang, Z., Cheng, Y., Li, H., & Pang, Z. (2024). Advancing precision and personalized breast cancer treatment through multi-omics technologies. *American Journal of Cancer Research, 14*(12), 5614–5627. DOI: 10.62347/MWNZ5609 PMID: 39803662

Zhang, D. Y., Ye, F., Gao, L., Liu, X., Zhao, X., Che, Y., & Lee, P. (2009). Proteomics, pathway array and signaling network-based medicine in cancer. *Cell Division, 4*(1), 20. DOI: 10.1186/1747-1028-4-20

Zhang, H., Qin, C., An, C., Zheng, X., Wen, S., Chen, W., Liu, X., Lv, Z., Yang, P., Xu, W., Gao, W., & Wu, Y. (2021, October 1). Application of the CRISPR/Cas9-based gene editing technique in basic research, diagnosis, and therapy of cancer. *Molecular Cancer, 20*(1), 126. DOI: 10.1186/s12943-021-01431-6 PMID: 34598686

Zhang, N., Yang, X., Piao, M., Xun, Z., Wang, Y., Ning, C., Zhang, X., Zhang, L., Wang, Y., Wang, S., Chao, J., Lu, Z., Yang, X., Wang, H., & Zhao, H. (2024, February 14). Biomarkers and prognostic factors of PD-1/PD-L1 inhibitor-based therapy in patients with advanced hepatocellular carcinoma. [Internet]. *Biomarker Research, 12*(1), 26. Retrieved November 1, 2025, from https://biomarkerres.biomedcentral.com/articles/10.1186/s40364-023-00535-z. DOI: 10.1186/s40364-023-00535-z PMID: 38355603

Zhang, S., & Kipps, T. J. (2014). The pathogenesis of chronic lymphocytic leukemia. *Annual Review of Pathology*, *9*(1), 103–118. DOI: 10.1146/annurev-pathol-020712-163955 PMID: 23987584

Zhang, T., Zhang, L., Payne, P. R., & Li, F. (2020). Synergistic drug combination prediction by integrating multiomics data in deep learning models. In *Translational bioinformatics for therapeutic development* (pp. 223–238). Springer US.

Zhang, W. S., Pan, J., Li, F., Zhu, M., Xu, M., Zhu, H., Yu, Y., & Li, G. (2021). Reverse transcription recombinase polymerase amplification coupled with CRISPR-Cas12a for facile and highly sensitive colorimetric SARS-CoV-2 detection. *Analytical Chemistry*, *93*(8), 4126–4133. DOI: 10.1021/acs.analchem.1c00013

Zhang, W., Guo, Y., & Jin, Q. (2023, September 27). Radiomics and Its Feature Selection: A Review. *Symmetry*, *15*(10), 1834. DOI: 10.3390/sym15101834

Zhang, X. A., Yates, A., Vasilevsky, N., Gourdine, J. P., Callahan, T. J., Carmody, L. C., & Robinson, P. N. (2019). Semantic integration of clinical laboratory tests from electronic health records for deep phenotyping and biomarker discovery. *NPJ Digital Medicine*, *2*(1), 32. DOI: 10.1038/s41746-019-0110-4 PMID: 31119199

Zhang, X., Zhang, J., Sun, K., Yang, X., Dai, C., & Guo, Y. (2019, November). Integrated multi-omics analysis using variational autoencoders: application to pan-cancer classification. In *2019 IEEE International Conference on Bioinformatics and Biomedicine (BIBM)* (pp. 765-769). IEEE. DOI: 10.1109/BIBM47256.2019.8983228

Zhang, Y., Wang, D., Peng, M., Tang, L., Ouyang, J., Xiong, F., & Xiong, W. (2021). Single-cell RNA sequencing in cancer research. *Journal of Experimental & Clinical Cancer Research : CR*, *40*(1), 1–17. DOI: 10.1186/s13046-021-01874-1

Zhang, Y., Zhang, X., Lane, A. N., Fan, T. W., & Liu, J. (2018, August). TFmeta: a machine learning approach to uncover transcription factors governing metabolic reprogramming. In *Proceedings of the 2018 ACM International Conference on Bioinformatics, Computational Biology, and Health Informatics* (pp. 351-359). DOI: 10.1145/3233547.3233580

Zhang, Z., Huang, J., Zhang, Z., Shen, H., Tang, X., Wu, D., & Chen, S. (2024). Application of omics in the diagnosis, prognosis, and treatment of acute myeloid leukemia. *Biomarker Research*, *12*(1), 60. DOI: 10.1186/s40364-024-00600-1 PMID: 38858750

Zhang, Z., Tian, W., Gu, J., Li, J., Liu, M., Soleymani, L., & Li, Y. (2026). A general strategy to enhance aptamer affinity by suppressing dissociation through symmetric assembly. *Nucleic Acids Research*, *54*(6), gkag268. Advance online publication. DOI: 10.1093/nar/gkag268 PMID: 41873762

Zhang, Z., Xiao, Y., Liu, J., Xiao, F., Zeng, J., Zhu, H., Tu, W., & Guo, H. (2025, June 19). Machine learning-based MRI radiomics predict IL18 expression and overall survival of low-grade glioma patients. *NPJ Precision Oncology*, *9*(1), 196. DOI: 10.1038/s41698-025-00966-x PMID: 40533462

Zhao, B., Tian, Y., Yu, M., Han, J., Chen, Z., Zhang, J., Liu, J., Geng, L., & Lei, Y. (2026). The Protective Effect of N, Ce-doped Carbon Dots Against H2O2-Induced Oxidative Damage in MOVAS Cells. *Applied Biochemistry and Biotechnology*, *198*(3), 1855–1868. DOI: 10.1007/s12010-025-05540-6 PMID: 41504846

Zhao, Q., Yu, Y., Pu, C., Zheng, S., Chen, L., Zeng, F., Liu, L., & Li, D. (2026). Research progress on the spatiotemporal dynamics of therapy-induced senescence in remodeling the tumor microenvironment. *Frontiers in Immunology*, *17*, 1727142. DOI: 10.3389/fimmu.2026.1727142 PMID: 41694369

Zhao, S., Zhao, H., Yang, W., & Zhang, L. (2025, August). The next generation of immunotherapies for lung cancers. *Nature Reviews. Clinical Oncology*, *22*(8), 592–616. DOI: 10.1038/s41571-025-01035-9 PMID: 40528044

Zhao, W., Hu, J., Liu, J., Li, X., Sun, S., Luan, X., & Zhang, Y. (2022). Si nanowire Bio-FET for electrical and label-free detection of cancer cell derived exosomes. *Microsystems & Nanoengineering*, *8*(1), 1–12. DOI: 10.1038/s41378-022-00387-x

Zhao, X., Liu, D., Li, G., Xu, W., & Wu, G. (2026). Nanomedicines Reshape the Tumor Microenvironment: Multidimensional Strategies from Modulating "Barriers" to Metabolic Intervention. *International Journal of Nanomedicine*, *21*(0), 1–24. DOI: 10.2147/IJN.S570411 PMID: 41858588

Zheng, X., Cheng, M., Fu, B., Fan, X., Wang, Q., Yu, X., & Wei, H. (2015). Targeting LUNX inhibits non–small cell lung cancer growth and metastasis. *Cancer Research*, *75*(6), 1080–1090. DOI: 10.1158/0008-5472.CAN-14-1831 PMID: 25600649

Zhong, H., Zhou, Z., Wang, H., Wang, R., Shen, K., Huang, R., & Wang, Z. (2024). The biological roles and clinical applications of the PI3K/AKT pathway in targeted therapy resistance in HER2-positive breast cancer: A comprehensive review. *International Journal of Molecular Sciences*, *25*(24), 13376. DOI: 10.3390/ijms252413376 PMID: 39769140

Zhong, L., Li, Y., Xiong, L., Wang, W., Wu, M., Yuan, T., Yang, W., Tian, C., Miao, Z., Wang, T., & Yang, S. (2021, May 31). Small molecules in targeted cancer therapy: Advances, challenges, and future perspectives. *Signal Transduction and Targeted Therapy*, *6*(1), 201. DOI: 10.1038/s41392-021-00572-w PMID: 34054126

Zhong, Y., Yu, L., Zhao, J., Fang, Y., Yang, Y., Wu, Z., Wang, J., & Hu, W. (2021, June 24). Clinical Implementation of Automated Treatment Planning for Rectum Intensity-Modulated Radiotherapy Using Voxel-Based Dose Prediction and Post-Optimization Strategies. [Internet]. *Frontiers in Oncology*, *18(1)*, 697995. Retrieved November 2, 2025, from https://www.frontiersin.org/articles/10.3389/fonc.2021.697995/full. DOI: 10.3389/fonc.2021.697995 PMID: 34249757

Zhou, E., Shen, Q., & Hou, Y. (2024). Integrating artificial intelligence into the modernization of traditional Chinese medicine industry: A review. *Frontiers in Pharmacology*, *15*, 1181183. DOI: 10.3389/fphar.2024.1181183 PMID: 38464717

Zhou, H., Ye, P., Xiong, W., Duan, X., Jing, S., He, Y., Zeng, Z., Wei, Y., & Ye, Q. (2024). Genome-scale CRISPR-Cas9 screening in stem cells: Theories, applications and challenges. *Stem Cell Research & Therapy*, *15*(1), 218. DOI: 10.1186/s13287-024-03831-z PMID: 39026343

Zhou, H., Zhang, H., Zhang, R., Yuan, X., & Chang, H. (2025). AI-driven photonic noses: From conventional sensors to cloud-to-edge intelligent microsystems. *Microsystems & Nanoengineering*, *11*(1), 209. DOI: 10.1038/s41378-025-01058-3 PMID: 41198617

Zhou, J., Abasolo, S. A., Yang, Y., Guo, F., Xing, M., & Zhu, D. (2026). Biomimetic Woven Bone Organoids with Multifocal Centers for Retaining Stemness and Prolonging Osteogenic Activity. *Advanced Functional Materials, 36*(36), e27855. Advance online publication. DOI: 10.1002/adfm.202527855

Zhou, J., He, J., Li, G., & Liu, Y. (2020). Identifying Capsule Defect Based on an Improved Convolutional Neural Network. *Shock and Vibration, 2020*, 1–9. DOI: 10.1155/2020/8887723

Zhou, J., Huang, C., & Gao, X. (2024a). Patient privacy in AI-driven omics methods. *Trends in Genetics : TIG, 40*(5), 383–386. DOI: 10.1016/j.tig.2024.03.004

Zhou, L., & Simonian, A. L. (2024). CRISPR/Cas technology: The unique synthetic biology genome-editing tool shifting the paradigm in viral diagnostics, defense, and therapeutics. *Annual Review of Biomedical Engineering, 26*(1), 247–272. DOI: 10.1146/annurev-bioeng-081723-013033 PMID: 38346278

Zhou, Y., Tao, L., Qiu, J., Xu, J., Yang, X., Zhang, Y., Tian, X., Guan, X., Cen, X., & Zhao, Y. (2024, May 20). Tumor biomarkers for diagnosis, prognosis and targeted therapy. *Signal Transduction and Targeted Therapy, 9*(1), 132. DOI: 10.1038/s41392-024-01823-2 PMID: 38763973

Zhou, Z., Wang, J., Wang, J., Yang, S., Wang, R., Zhang, G., & Lu, Q. (2024b). Deciphering the tumor immune microenvironment from a multidimensional omics perspective: Insight into next-generation CAR-T cell immunotherapy and beyond. *Molecular Cancer, 23*(1), 131. DOI: 10.1186/s12943-024-02047-2

Zhuang, J., Zhao, Y., Yang, M., Li, J., Chen, Y., & Liu, H. (2022). SERS-based CRISPR/Cas assay on microfluidic paper analytical devices for supersensitive detection of pathogenic bacteria in foods. *Biosensors & Bioelectronics, 207*, 114167. DOI: 10.1016/j.bios.2022.114167

Zhuang, L., Lian, Y., & Zhu, T. (2025). Multifunctional gold nanoparticles: Bridging detection, diagnosis, and targeted therapy in cancer. *Molecular Cancer, 24*(1), 228. DOI: 10.1186/s12943-025-02431-6 PMID: 40993729

Zhuang, X., Pei, H. Z., Li, T., Huang, J., Guo, Y., Zhao, Y., & Chen, Y. (2022). The molecular mechanisms of resistance to IDH inhibitors in acute myeloid leukemia. *Frontiers in Oncology, 12*, 931462. DOI: 10.3389/fonc.2022.931462 PMID: 35814406

Zhu, C. X., & Qin, J. J. (2023). Single-cell RNA sequencing and spatial transcriptomic technologies and applications in exploring gastric cancer: A review. *Oncology Advances, 1*(1), 6–16. DOI: 10.14218/OnA.2023.00039

Zhu, G., He, J., Shi, N., Cai, Z., Zhang, J., Wang, S. C. Y., Li, J., Zhang, M., & Wang, Y. (2026). Prognostic, chemotherapy and immunotherapy roles of GPR37/GPR37L1 in pan-cancer. *Medicine, 105*(10), e47813. DOI: 10.1097/MD.0000000000047813 PMID: 41790623

Zhu, J., Ji, L., Chen, Y., Li, H., Huang, M., Dai, Z., & Chu, X. (2023). Organoids and organs-on-chips: Insights into predicting the efficacy of systemic treatment in colorectal cancer. *Cell Death Discovery, 9*(1), 72. DOI: 10.1038/s41420-023-01354-9 PMID: 36813783

Zhu, J., Zhang, K., Chen, Y., Ge, X., Wu, J., Xu, P., & Yao, J. (2024). Progress of single-cell RNA sequencing combined with spatial transcriptomics in tumour microenvironment and treatment of pancreatic cancer. *Journal of Translational Medicine, 22*(1), 563. DOI: 10.1186/s12967-024-05307-3 PMID: 38867230

Zhu, M., Sali, R., Baba, F., Khasawneh, H., Ryndin, M., Leveillee, R. J., Hurwitz, M. D., Lui, K., Dixon, C., & Zhang, D. Y. (2024, August 25). Artificial intelligence in pathologic diagnosis, prognosis and prediction of prostate cancer. *American Journal of Clinical and Experimental Urology*, *12*(4), 200–215. DOI: 10.62347/JSAE9732 PMID: 39308594

Zhu, Z., Wang, S.-H., & Zhang, Y.-D. (2023, March). A Survey of Convolutional Neural Network in Breast Cancer. *Computer Modeling in Engineering & Sciences*, *136*(3), 2127–2172. DOI: 10.32604/cmes.2023.025484 PMID: 37152661

Zuo, S. Q., Li, C., Liu, Y. L., Tan, Y. H., Wan, X., Xu, T., Li, Q., Wang, L., Wu, Y. L., Deng, F. M., & Tang, B. (2021, October). Cordycepin inhibits cell senescence by ameliorating lysosomal dysfunction and inducing autophagy through the AMPK and mTOR–p70S6K pathway. *FEBS Open Bio*, *11*(10), 2705–2714. DOI: 10.1002/2211-5463.13263 PMID: 34448542

About the Contributors

Jen-Tsung Chen is a professor of cell biology at the National University of Kaohsiung in Taiwan. He teaches cell biology, genomics, proteomics, plant physiology, and plant biotechnology. Dr. Chen's research interests include bioactive compounds, chromatography techniques, plant molecular biology, plant biotechnology, bioinformatics, and systems pharmacology. He is an active editor of academic books and international journals to advance the exploration of multidisciplinary knowledge involving plant physiology, plant biotechnology, nanotechnology, materials science, ethnopharmacology, and systems biology. He serves as an associate editor, editorial board member, and guest editor in reputed journals. Dr. Chen published books in collaboration with international publishers, and he is handling book projects on diverse topics such as AI technologies, drug discovery, drug development, herbal medicine, medicinal biotechnology, nanotechnology, bioengineering, plant functional genomics, plant speed breeding, epigenetics, functional RNAs, and CRISPR-based genome editing. Dr. Chen is a productive author in academic publications and was recognized as one of the World's Top 2% Scientists 2023 and 2024 by Elsevier and Stanford University. Dr. Chen received the Springer Nature Editor of Distinction Award in 2025.

***

Naureen Afrose is an Assistant Professor of Pharmaceutics at the Dr. B.C. Roy College of Pharmacy and AHS, Durgapur, West Bengal, India. Ms. Afrose is driven by a passion to revolutionize drug delivery systems. Her research delves into a variety of areas, including traditional Solid formulations, but her true focus lies in pushing boundaries with Novel drug delivery formulations, Nanoformulations, and utilizing Transdermal and Film-based methods for more efficient drug delivery. She's keenly interested in the potential of AI-related systems and their role in shaping the future of drug delivery, actively exploring how these advancements can further optimize drug effectiveness and patient experience.

Lokeshwari B is a dedicated and accomplished researcher in Microbiology, Biotechnology, and Nanotechnology, with a strong academic foundation in life sciences. She holds a Bachelor's degree in Biotechnology and a Master's degree in Microbiology. She has developed an expertise in Agricultural Microbiology, Microbial Pigments, Nanobiotechnology, and Environmental Biotechnology. Her research primarily focuses on microbial pigments and their pharmacological properties, exploring their potential applications in medicine, industry, and environmental sustainability. A testament to her innovative contributions is her UK-patented invention, "Heavy Metal Identification Gloves Using Nanotechnology," which enables real-time detection of toxic heavy metals, demonstrating her commitment to developing practical, nanotechnology-driven solutions for environmental and industrial challenges. Ms. Lokeshwari has made significant contributions through peer-reviewed research papers, review articles, book chapters, and monographs published in internationally recognized journals.

Rideb Chakraborty currently works as an Assistant Professor of Pharmaceutics at Division of Pharmaceutics, Dr. B.C. Roy College of Pharmacy and Allied Health Sciences, Dr. Meghnad Saha Sarani, Bidhannagar, Durgapur, Paschim Bardhaman, West Bengal. He attained his postgraduate degree with a focus on Pharmaceutics from the Department of Pharmaceutical Technology at Jadavpur University, Kolkata. His research spans various domains including Oral Solid Dosage (OSD) formulation, Nanoformulation, Transdermal drug delivery systems, and Novel Drug Delivery Systems (NDDS). Additionally, he holds a strong interest in leveraging novel artificial intelligence for drug discovery and development.

Tanushree Das is an Assistant Professor in the Department of Pharmaceutics at Eminent College of Pharmaceutical Technology. She is deeply committed to advancing pharmaceutical sciences through both teaching and research. With a strong passion for academic excellence, she actively engages in mentoring students and fostering a learning environment that encourages critical thinking and innovation. Her research interests focus on contemporary challenges in pharmaceutics, contributing to the development of effective and accessible drug delivery systems. Tanushree Das continues to strive for excellence in both education and scientific inquiry.

Sumit Durgapal is Professor of Pharmaceutics at Uttaranchal Institute of Pharmaceutical Sciences, Uttaranchal University, Dehradun, India, with over seventeen years of teaching and research experience across institutions including Kumaun University and Teerthanker Mahaveer University, along with a research stint at Pusan National University, South Korea. He holds a Ph.D. from IFTM University, Moradabad, and works primarily on microemulsions, nanoemulsions, gastroretentive and chronomodulated drug delivery systems, and plant-based novel drug delivery formulations. He has authored over 60 research articles in national and international journals and holds four filed patents. He currently supervises five research scholars pursuing Ph.D. and M.Pharm degrees.

Sourav Dutta is a pharmaceutical professional with a Master of Pharmacy (M.Pharm) currently working as a Raw Material Analyst in the R&D department. Experienced in the analysis and evaluation of pharmaceutical raw materials according to pharmacopeial standards such as IP and BP. Skilled in performing analytical tests including assay, identification tests, TLC, titrimetric analysis, and preparation of Certificates of Analysis (COA). Possesses strong knowledge of pharmaceutical quality control practices, analytical techniques, and laboratory documentation.

Sachin Kuila is currently pursuing his M.Pharm in Pharmaceutics from Maulana Abul Kalam Azad University of Technology. He completed his B.Pharm from Bengal School of Technology in 2024 with a CGPA of 8.68. He completed his Secondary and Higher Secondary education from Midnapore Town School (H.S.) under the West Bengal Board of Secondary Education and the West Bengal Council of Higher Secondary Education. He is from Jhargram, West Bengal, India. His academic interests include pharmaceutical formulation and drug delivery systems, and his hobbies are cricket, football, writing, and photography.

Rahul Kumar is an Assistant Professor at the Faculty of Pharmacy, Vidya University, Meerut. He holds an M.Pharm in Pharmaceutical Chemistry from Galgotias University and a B.Pharm from Swami Vivekanand Subharti University, Meerut. His research interests include Computer-Aided Drug Design (CADD), pharmaceutical chemistry, and molecular modeling. He has authored research articles, book chapters with international publishers (Springer Nature, Elsevier, CRC Press, IGI Global), and is an inventor on a design patent for a MUC-1 based SPR biosensor for Triple-Negative Breast Cancer diagnosis. He has actively participated in national and international conferences, contributing to discussions on drug discovery, computational approaches, and pharmaceutical innovation. Rahul has been recognized with several awards for research and innovation, including top positions at Anveshan 2025, Health-A-Thon 2.0, Pharmathon 2024, and Idea Hackathon 2024. His ongoing work emphasizes integrating computational and experimental strategies to advance drug discovery and therapeutic development.

Rabab Mahmoud is an assistant professor at A' Sharqiyah University, College of Health and Applied Sciences, MDLS department. She holds a PhD in hematology and immunohematology and brings extensive experience from academic research, higher education, and clinical laboratory practice. Dr. Rabab is an active researcher who has published numerous scholarly papers and contributed to several academic books, with research interests spanning diagnostic hematology, immunohematology, transfusion science, and quality management in clinical laboratories. She serves as a keynote speaker at national and international conferences and actively contributes to multiple academic and institutional committees. She is certified as a specialist in hematology (ASCP-SH), demonstrating her commitment to professional excellence.

Manoj Mani is a research scholar with a specialized background in human genetics and a profound interest in animal research. They have an exceptional research acumen, authoring over 27 scientific publications and earning several awards for my outstanding presentation skills at academic and scientific forums. Their research interests also extend to phytotherapeutics and molecular biology, where they explore the molecular underpinnings of genetic and reproductive processes.

Saranraj P holds a Ph.D. in Microbiology from Annamalai University, Chidambaram, India, and is NET (ASRB–ICAR) qualified in Microbiology. He is currently serving as Assistant Professor and Head, Department of Microbiology, at Sacred Heart College (Autonomous), Tirupattur, Tamil Nadu, India. In addition, he is the Proprietor and Managing Director of JPS Scientific Publications, contributing actively to academic publishing and research dissemination. Dr. Saranraj has over 15 years of research experience and 10 years of teaching experience in the fields of Environmental Microbiology, Agricultural Microbiology, Industrial Microbiology, Bioremediation, Microbial Biotechnology, and Sustainable Ecosystem Management. His research work has focused on Microbial degradation of pollutants, Effluent treatment, Soil health improvement, Plant–microbe interactions, and Applied environmental sustainability. He has an extensive publication record, with more than 240 international research articles, over 90 books (international and national), and 70+ book chapters published with reputed academic publishers.

Jyoti Pandit is an Assistant Professor in the School of Pharmaceutical Sciences at Jigyasa University, Dehradun. She completed her Bachelor of Pharmacy and Master of Pharmacy in Pharmaceutical Chemistry from Uttaranchal University. She has experience in teaching, laboratory training, hospital training, and industrial training. Her interest lies in pharmaceutical chemistry, drug development, and research. She has published research and review articles and actively participates in national and international conferences. Jyoti believes in continuous learning and helping students build strong academic knowledge. She is dedicated to teaching, research, and contributing to the growth of pharmaceutical education with honesty and commitment

Veda Paragonnavar is a research scholar at Sri Sathya Sai University for Human Excellence Kalaburagi, Karnataka 585313, India

Paramita Paul is an Assistant Professor at the Department of Pharmaceutical Technology, University of North Bengal, Darjeeling, India. She pursued B.Pharm., M.Pharm., and Ph.D. degrees from Jadavpur University, Kolkata, India. She has been awarded DST Women Scientist-A, Govt. of India, and worked in the twining program of Department of Biotechnology, Government of India. She has more than 9 years of teaching experience in PCI-approved Pharmacy Institutions. She has published more than 55 peer- reviewed research and review papers each with high impact. She has authored 6 books and 20 book chapters in edited books published by renowned national and international publishers.

Arvind Raghav is an Assistant Professor in the Department of Pharmacy at Shri Ram Murti Smarak College of Engineering and Technology, Bareilly with over 4 years of academic and research experience. He has published 30+ research and review papers, authored 3+ books, contributed 8+ book chapters, and holds 7+ Indian and UK design patents. His expertise includes pharmaceutical formulation development, QbD-based optimization using DOE, systematic review and meta-analysis, and network pharmacology.

Ramya Raghavan is an Assistant Professor in the Department of Life Sciences at the Sri Sathya Sai University for Human Excellence in Kamalapur, Karnataka India. Her academic work focuses on cellular and molecular physiology, developmental programming, and the intersection of environmental health risks with social determinants.

Ashok Kumar Sah is an Assistant Professor in the Department of Medical Laboratory Science, College of Applied Health Sciences, A' Sharqiyah University, Oman. A certified Medical Laboratory Scientist (ASCPi), he earned his PhD in Medical Laboratory Science from Glocal University, India, following his Bachelor's and Master's degrees from HNB Garhwal University. His expertise spans Molecular Pathology, Clinical Biochemistry, Microbiology, Immunology, Clinical Pathology, Transplant Technology, and Cancer Biomarkers. Dr. Sah has published 55 peer-reviewed papers, five books, and three book chapters, with contributions to leading journals including The Lancet, NPJ Clean Water, Biomarker Research, and Molecular and Cellular Biology. With over 15 years of academic and research experience across India, Nepal, and Oman, he actively presents at national and international conferences. Beyond academia, he enjoys reading, traveling, and engaging with diverse cultures and is committed to global public health and scientific advancement.

Index

A

B

C

D

E

G

H

I

L

M

N

O

P

Q

R

S

T

www.ingramcontent.com/pod-product-compliance
Lightning Source LLC
LaVergne TN
LVHW062030101126
852251LV00005B/66